Biological Diversity of Mexico:
Origins and Distribution

Biological Diversity of Mexico: Origins and Distribution

Edited by

T.P. Ramamoorthy

Robert Bye

Antonio Lot

Instituto de Biología,
Universidad Nacional Autónoma de México,
Mexico City, Mexico.

John Fa

Centro de Ecología
Universidad Nacional Autónoma de México,
Mexico City, Mexico.

New York   Oxford
Oxford University Press
1993

Oxford University Press

Oxford   New York   Toronto
Delhi   Bombay   Calcutta   Madras   Karachi
Kuala Lumpur   Singapore   Hong Kong   Tokyo
Nairobi   Dar es Salaam   Cape Town
Melbourne   Auckland

and associated companies in
Berlin and Ibadan

Published by Oxford University Press, Inc.,
200 Madison Avenue, New York, New York 10016

Oxford is a registered trademark of Oxford University Press

Library of Congress Cataloging-in-Publication Data
Biological Diversity of Mexico: Origins and Distribution / edited by
T.P. Ramamoorthy, Robert Bye, Antonio Lot and John Fa
p. cm.
"Most of the chapters in this book were originally presented at the
Symposium on the Biological Diversity of Mexico organized and
partly funded by the Instituto de Biología (IBUNAM), Universidad
Nacional Autónoma de México (UNAM, the National Autonomous
University of Mexico) in 1988."
Includes bibliographical references and index.
ISBN 0-19-506674-X
1. Biological Diversity—Mexico—Congresses. 2. Biogeography—
Mexico—Congresses. I. Ramamoorthy, T. P. II. Symposium on the
Biological Diversity of Mexico (1988: Oaxtepec, Morelos, Mexico)
QH107.B55   1992   574.972—dc20   91-3684

1 2 3 4 5 6 7 8 9

Printed in the United States of America
on acid-free paper

We dedicate this volume to scholars, past and present,
who have made the study of biodiversity a global science.
We hope that this contribution will help future students
in their endeavour to provide a worldwide conservation agenda.

Acknowledgments

Most of the chapters in this book were originally presented at the Symposium on the Biological Diversity of Mexico organized and partly funded by the Instituto de Biología (IBUNAM), Universidad Nacional Autónoma de México (UNAM, The National Autonomous University of Mexico) in 1988. Cosponsors of the Symposium included the Agency for International Development of the United States (Mexico City Embassy), the British Council (Mexico City), Conservation International (Washington, D.C.), the Dirección General de Asuntos del Personal Académico (DGAPA) of UNAM, and World Wildlife Fund (Washington, D.C.). We express our grateful thanks to these institutions for their help. Among the many who assisted with the organization of the symposium and who are recognized in the symposium program and abstracts, we would like to mention the following whose contributions were critical: Marguerite Elliott (presently with the Texas Union, the University of Texas at Austin); Marú García P., Pedro Ramírez G., and Mario Sousa P. (Department of Botany, IBUNAM); Edelmira Linares and Virginia Romo (Jardín Botánico del IBUNAM); Verónica López P. and Susana Hernández L. (Dirección del IBUNAM); and Jesús Wong L. (Coordinación de la Investigación Científica de la UNAM). Financial coordination for the Symposium was provided by Pedro Enrique M. of IBUNAM. Elvia Esparza A. (IBUNAM) planned and created the symposium poster. Rodolfo Dirzo (Centro de Ecología, UNAM) was an enthusiastic participant during the early stages.

T.P. Ramamoorthy expresses his special thanks to Hilda Flores O. for her invaluable support during his sabbatical leave and appreciates the assistance and kindnesses of Gloria Andrade, Fernando Chiang C., Patricia Davila A., Claudio Delgadillo M., Pedro Enrique M., Eloísa and Ismael Ferrusquía, Héctor M. Hernández, Helga Ochoterena B., Gilda Ortíz C., Alejandro Novelo, and Alfredo Wong during his frequent visits to Mexico in 1990 and 1991. During these trips he enjoyed the wonderful hospitality of the Durbecq-Marino family (Alain, Cecilia, and Elise) and the Byes (Bob and Edelmira); he was a welcome guest at their homes in Tlalpan and Tacubaya.

We extend our thanks to Alfonso Delgado S. and Mario Sousa S. (Department of Botany, IBUNAM). The former provided computer facilities during the initial stages of the preparation of this volume, and the latter moral support during the earlier stages of the work. Many colleagues provided "courier services," and two institutions (Field Museum at Chicago and Missouri Botanical Garden) assisted with correspondence during the early stages.

José Sarukhán Kermez (formerly Coordinador de la Investigación Científica de la UNAM, and currently Rector) and Peter H. Raven (Director, Missouri Botanical Garden) lent critical support by providing contacts.

The editorial work was carried out at the Plant Resources Center (PRC) of the University of Texas at Austin during T.P. Ramamoorthy's sabbatical leave there as a visiting scholar; it is with great appreciation that he thanks Billie L. Turner and the staff of the Center for providing facilities and encouragement. His stay in Texas was partly supported by a grant from DGAPA of UNAM. A volume of this nature would have been impossible without the help of the biological community. Many reviewers gave us their time, and shared with us their knowledge of the various groups discussed in the book, and we are particularly grateful to William Anderson, John Beaman, Bruce Benz, Keith Brown, William Buck, Gerrit Davidse, Robert Dickerson, William Durkee, Ismael Ferrusquía, Alwyn Gentry, Alan Graham, Raymond Harley, James Henrickson, David Hillis, Colin Hugh, Garrison Wilkes, Gwelym Lewis, Fred McDowell, Rogers McVaugh, Robert Miller, Robbin Moran, Gary Nabham, John Neff, Guy Nesom, Roger Polhill, Carlos Rincón-M., David Riskind, Velva Rudd, Rafael de Sá, Beryl Simpson, Arthur Shapiro, Alan Smith, Carol Todzia, Billie Turner, Victor Toledo, Rolla Tryon, and Dieter Wasshaussen for reviews of chapters. Michael Barnett (Department of Computer Sciences, UT) arranged for and aided in transferring parts of text to diskette via an optical reader. We are most grateful to Felipe Villegas M. (IBUNAM) who skillfully provided technical assistance with graphic art work. We express our appreciation to Monique Williamson and Mahinda Martínez who translated several chapters from Spanish into English.

The editors express their delight at the ease and comfort with which they were able to work constructively with Kirk Jensen, Stanley George, and Ann Hegeman, all of Oxford University Press, New York. We thank William Curtis for valuable counsel during the early stages of the preparation of the volume. Nancy Schatken's copy-editing is gratefully acknowledged. T.P. Ramamoorthy thanks Mrs. Kiyo Elliott, whose fine collection of teas he exhausted while proofing the copy-edits.

Benjamin Moss prepared the camera-ready copy. We appreciate the diligence he brought to the task as well as his patience when it came to incorporating the numerous changes, often requiring re-formatting. The indices (subject and taxonomic) were prepared by T.P. Ramamoorthy, Benjamin Moss, and Marguerite Elliott.

We are especially indebted to Marguerite Elliott, whose contributions to this volume are too varied to list. They include both translation of Spanish papers and generous, critical editorial assistance. The successful completion of the volume owes much to her constant, unfailing support and encouragement throughout its preparation.

Contents

I
Historical Background

    Ismael Ferrusquía-Villafranca
    *Baja California Peninsula Morphotectonic Province (M.P.) • Northwestern
    Plains and Sierras M.P. • Sierra Madre Occidental M.P. • Chihuahuan-
    Coahuilan Plateaus and Ranges M.P. • Sierra Madre Oriental M.P. • Gulf
    Coast Plain M.P. • Central Plateau M.P. • Trans-Mexican Volcanic Belt
    M.P. • Sierra Madre del Sur M.P. • Sierra Madre de Chiapas M.P. •
    Yucatan Platform M.P. • Review of Morphotectonic Provinces and
    Conclusions*

    Alan Graham
    *Comparison of Diversity in Tertiary Floras from Mexico and Panama—
    Physiographic Diversity, Climatic Changes, Source Areas*

    Jerzy Rzedowski
    *Floristic Richness • Endemism • Evolution of Plant Lineages in Mexico •
    Geographical Affinities and Origins of the Flora*

II
Selected Faunistic Groups of Mexico

III
Selected Floristic Groups of Mexico

IV
Phytogeography of
Selected Vegetation Types in Mexico

V
Plant Diversity and Humans

VI
Review of Habitats

List of Figures

List of Tables

Contributors

Virgilio Arenas Instituto de Limnología y Ciencias del Mar, Universidad Nacional Autónoma de México, 04510 México D.F., México

Ricardo Ayala   Estación de Biología Chamela, Instituto de Biología, Universidad Nacional Autónoma de México, Apartado Postal 21, 48980 San Patricio, Jalisco, México

Stephen H. Bullock Estación de Biología Chamela, Instituto de Biología, Universidad Nacional Autónoma de México, Apartado Postal 21, 48980 San Patricio, Jalisco, México

Robert Bye Jardín Botánico del Instituto de Biología, Universidad Nacional Autónoma de México, Apartado Postal 70-614, 04510 México D.F., México

Ismael Cabral C. Departamento de Botánica, Universidad Autónoma Agraria Antonio Narro, Saltillo 25315, Coahuila, México

Thomas Daniel Department of Botany, California Academy of Sciences, Golden Gate Park, San Francisco, California 94118, U.S.A.

Claudio Delgadillo M. Departamento de Botánica, Instituto de Biología, Universidad Nacional Autónoma de México, Apartado Postal 70-233, 04510 México D.F., México

Alfonso Delgado S. Departamento de Botánica, Instituto de Biología, Universidad Nacional Autónoma de México, Apartado Postal 70-233, 04510 México D.F., México

Marguerite Elliott Texas Union, The University of Texas, Austin, Texas 78713, U.S.A.

Patricia Escalante P. Museo de Zoología, Facultad de Ciencias, Universidad Nacional Autónoma de México, Apartado Postal 70-399, 04510 México D.F., México. Present address: Dept. of Ornithology, American Museum of Natural History, Central Park West at 79th street, New York, NY 10024, U.S.A.

Héctor Espinosa-Pérez Departamento de Zoología, Universidad Nacional Autónoma de México, Apartado Postal 70-157, 04510 México D.F., México

John E. Fa Centro de Ecología, Universidad Nacional Autónoma de México, Apartado Postal 70-275, 04510 México D.F., México. Present address: Medambios, P.O. Box 438, Gibraltar

Ismael Ferrusquía-Villafranca Instituto de Geología, Universidad Nacional Autónoma de México, 04510 México D.F., México

Patricia Fuentes-Mata Departamento de Zoología, Universidad Nacional Autónoma de México, Apartado Postal 70-157, 04510 México D.F., México

Oscar Flores-Villela Centro de Ecología, Universidad Nacional Autónoma de México, Apartado Postal 70-275, 04510 México D.F., México. Present address: Museo de Zoología, Facultad de Ciencias, Universidad Nacional Autónoma de México, Apartado Postal 70-399, 04510 México D.F., México

Alan Graham Department of Biology, Kent State University, Kent, Ohio 44262, U.S.A.

Terry L. Griswold U.S.D.A., Bee Biology and Systematics Laboratory, Agricultural Research Service, Utah State University, Logan, Utah 84322-5310, U.S.A.

Efraím Hernández X. Centro de Botánica, Colegio de Postgraduados, 56320 Chapingo, México, México. Deceased February 21st, 1991.

David Hunt The Herbarium, Royal Botanical Gardens, Kew, Richmond, Surrey, TW9 3AB, England

Ma. de Jesús Ordoñez Centro de Ecología, Universidad Nacional Autónoma de México, Apartado Postal 70-275, 04510 México D.F., México

Antonio Lot Departamento de Botánica, Instituto de Biología, Universidad Nacional Autónoma de México, Apartado Postal 70-233, 04510 México D.F., México

Jorge LLorente-Bousquets Museo de Zoología, Facultad de Ciencias, Universidad Nacional Autónoma de México, Apartado Postal 70-399, 04510 México D.F., México

J. Andrew McDonald Department of Botany, The University of Texas, Austin, Texas 78713, U.S.A.

Armando Luis-Martínez Museo de Zoología, Facultad de Ciencias, Universidad Nacional Autónoma de México, Apartado Postal 70-399, 04510 México D.F., México

Luis Miguel Morales Instituto de Geografía, Universidad Nacional Autónoma de México, Apartado Postal 20-850, 04510 México D.F., México

Adolfo G. Navarro-Sigüenza Museo de Zoología, Facultad de Ciencias, Universidad Nacional Autónoma de México, Apartado Postal 70-399, 04510 México D.F., México

Guy L. Nesom Department of Botany, The University of Texas, Austin, Texas 78713, U.S.A.

Kevin C. Nixon L.H. Bailey Hortorium, Cornell University, Ithaca, New York 14853. U.S.A.

Alejandro Novelo Departamento de Botánica, Instituto de Biología, Universidad Nacional Autónoma de México, Apartado Postal 70-233, 04510 México D.F., México

A. Townsend Peterson   Division of Birds, Field Museum of Natural History, Chicago, Illinois 60605–2496, U.S.A.

T.P. Ramamoorthy   Departamento de Botánica, Instituto de Biología, Universidad Nacional Autónoma de México, Apartado Postal 70-233, 04510 México D.F., México

Pedro Ramírez-García   Departamento de Botánica, Instituto de Biología, Universidad Nacional Autónoma de México, Apartado Postal 70-233, 04510 México D.F., México

Ramón Riba   Departamento de Biología, Div. C.B.S., Universidad Autónoma Metropolitana, Unidad Iztapalapa, Av. Michoacán y La Purisima, Iztapalapa, 09340 D.F., México

Jerzy Rzedowski Instituto de Ecología, Centro Regional del Bajío, Apartado Postal 386, 61600 Pátzuaro, Michoacán, México

Mario Sousa S.   Departamento de Botánica, Instituto de Biología, Universidad Nacional Autónoma de México, Apartado Postal 70-233, 04510 México D.F., México

Brian T. Styles   Oxford Forestry Institute, Department of Plant Sciences, University of Oxford, Oxford OX1 3RB, England

Ma. Teresa Gaspar-Dillanes   Departamento de Zoología, Universidad Nacional Autónoma de México, Apartado Postal 70-157, 04510 México D.F., México

Victor M. Toledo   Centro de Ecología, Universidad Nacional Autónoma de México, Apartado Postal 70–275, 04510 México D.F., México

Billie Turner   Department of Botany, The University of Texas, Austin, Texas 78713, U.S.A.

Jesús Valdés R.   Departamento de Botánica, Universidad Autónoma Agraria Antonio Narro, 25315 Saltillo, Coahuila, México

Noé Vargas B.   Museo de Zoología & ENEP-Iztacala, Universidad Nacional Autónoma de México, México D.F., México

Tom Wendt   Centro de Botánica, Colegio de Postgraduados, 56320 Chapingo, México, México. Present Address: Department of Botany, Lousiana State University, Baton Rouge, Lousiana 70803–1705, U.S.A.

Acronyms of Mexican Institutions

CICYT	Centro de Investigaciones Científicas de Yucatán
COTECOCA	Comisión Técnico Consultiva de Coeficientes de Agastadero, SARH
CP	Colegio de Posgraduados, Chapingo
DGAPA	Direción General de Asuntos Personal Académico, UNAM
DIC-UAT	Direción de Investigación Científica, Universidad Autónoma de Tamaulipas
ENCB	Escuela Nacional de Ciencias Biológicas, IPN
IBUNAM	Instituto de Biología, UNAM
IMRNAR	Instituto Mexicano de Recursos Natural Renovables
IMSS	Instituto Mexicano de Seguro Social
INAH	Instituto Nacional de Antropología e Historia
INEGI	Instituto Nacional de Estadística, Geografía e Informática
INI	Instituto Nacional de Indigenista
INIF	Instituto Nacional de Investigaciones Forestales
INIFAP	Instituto Nacional de Investigación Forestal y Agropecuaria
INIREB	Instituto Nacional de Investigación de Recursos Bióticos (1975–1988; incorporated into Instituto de Ecología, April 1989)
IPN	Instituto Politécnico Nacional
SAHOP	Secretaría de Asistancia Humana y Obras Públicas
SARH	Secretaría de Agricultura y Recursos y Hidráulicos
SEDUE	Secretaría de Desarrollo Urbano y Ecología
SEP	Secretaría de Educación Pública
SPP	Secretaría de Programación y Presupuesto
UAM	Universidad Autónoma Metropolitana
UNAM	Universidad Nacional Autónoma de México

Introduction

The diversity of life has occupied the interest of human beings ever since they first appeared on Earth. Early peoples, through their interaction with the elements of surrounding biological diversity, developed systems for the recognition, exploitation, and management of their natural resources. These systems and their accompanying quest for knowledge have often been the predecessors of major economic and cultural developments. For instance, Renaissance Europe benefited enormously from the discovery of the considerable biological diversity encountered during the exploration of the New World. Indeed, many voyagers returned with extraordinary tales of not only the material wealth but also the remarkable plants and animals encountered. These tales—fancied and real—expanded the European concept of diversity and had profound effects on the course of world history, not only in the history of nations and their conquests but also in the development of human intellect. Such was the result of great voyages of discovery (Columbus, da Gama, Darwin, Humboldt, and others).

At another level, the study of biological diversity guided the birth of modern systematics. It also eventually led to the demise of typological thinking and essentialism, and helped develop the concepts of population biology and biological species. An appreciation of biodiversity has also taught us the uniqueness of every species and its irreplaceability, which is the basis of modern conservation thinking. The study of diversity will no doubt continue to play an important role in the further development of concepts and ideas regarding its function in the general scheme of life on Earth of which humans are a part.

Recent interests and reports on the biological diversity of the world highlight two facts: that knowledge of the biodiversity of our environment is incomplete and that mass extinctions of taxa, particularly in the tropics where biodiversity is greatest, are proceeding at a fast rate (Wilson, 1988). The resulting concerns address the global need for conservation; but it is especially needed in the tropics where, lacking a reliable taxonomic data base, there is little knowledge of what is being lost. Among most biologists there is a consensus that loss of forests, both tropical and temperate, may

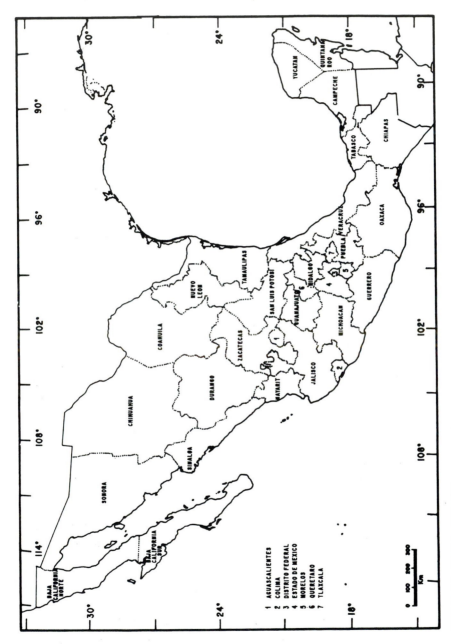

Figure I.1. States of Mexico.

lead to catastrophic changes in the complex ecosystems of the world, with unforeseen consequences for life on Earth. Information is emerging on the role of species in their communities and the ecological and economic consequences of their loss. The aforesaid concerns are matched only by the immensity of the challenge of documenting biodiversity.

The appreciation of the differences in biodiversity in various parts of the world has led to the recognition of certain nations as mega-diversity countries (Mittermeier, 1988). Mexico (Fig. I.1) is included among these countries; it follows Brazil and Colombia in order of richness and ranks ahead of such nations as Zaire, Madagascar, and Indonesia. Mexico exemplifies in many ways the challenges and opportunities in tropical biology. No complete floristic or faunistic treatments exist for the entire country. Large areas of Mexico that are species-rich (the tropical south) or centers of endemism (the morphotectonic provinces of the Trans-Mexican Volcanic Belt and the Sierra Madre del Sur, as defined by Ferrusquía-V. in this volume) remain only partially explored. The taxonomy of numerous biotic groups in Mexico is obscure. Scientific literature on the Mexican biota is fairly abundant, but these publications are generally not available in Mexico. Locally produced data in the form of dissertations remain for the most part unpublished and hence unavailable to both Mexican and foreign scholars. The need to put together this information on the one hand and to undertake intensive exploration toward completion of the biological inventory on the other is great. The primary purpose of the present contribution, which is largely a product of a symposium on Mexican biodiversity held in 1988, is to assemble data pertaining to different but representative groups of Mexican biota and make it available to Mexicans and the world community alike.

The book is organized in six parts. The first section consists of background material such as geological and historical factors, including an essay on the diversity of Mexico's flowering plant flora and its origins. The second and third parts contain chapters on selected major faunistic and floristic groups, respectively. Several contributions in these sections discuss diversity (species richness), endemism, and distribution, among other themes, including an essay on the ecological diversity in scrub jays. The fourth part deals with phytogeographical patterns in contrasting ecosystems: the tropical rain forests and the alpine floras of Mexico. The fifth section presents two ethnobiological essays: one on the influence of man on plant species diversification and the other on aspects of the domestication of plants in Mexico. The last section provides a brief scenario of Mexican biodiversity and a review of terrestrial habitats in Mexico.

VIEW OF MEXICAN BIODIVERSITY AND ITS SIGNIFICANCE

Although Mexico, with 1,972,544 km^2, is the fourteenth largest country in the world, it ranks third in biological diversity (Mittermeier, 1988). The

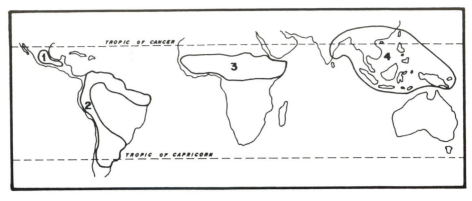

Figure I.2. "Gene belt" and centers and noncenters of agricultural origins. 1. Meso-american center. 2. South American noncenter. 3. African noncenter. 4. Southeast Asian and South Pacific noncenter (centers and noncenters of agricultural origins, after Harlan, 1975).

wealth of species, floristic and faunistic, is the result of a varied biogeo-graphical history over a range of climates, encompassing a Nearctic realm to the north and a Neotropical one in the south. Mexico harbors approxi-mately 30,000 species of plants, of which over 21,600 in approximately 2,500 genera are flowering plants (Rzedowski, this volume). Over 300 genera and 50–60% of the species are endemic to the country (Ramamoorthy & Lorence, 1987). Coniferous trees dominate vast tracts of the territory with about 15 genera and over 150 species. There are 49 species of pines, which is more than 50% of the world total (Styles, this volume). There are 900 to 1,000 fern species (Riba, this volume) and 2,300 species of bryophytes (Delgadillo, this volume). In addition, the species of fungi and algae in Mexico is known to be exceptionally rich.

In terms of animal life, although Mexico is two and a half times smaller than Brazil, it supports 449 mammalian species of which 142 are endemic, in comparison to 394 (65 endemic) for Brazil (McNeely et al., 1990). Over 1,000 bird species are known to inhabit the Mexican national territory (Escalante et al., this volume). Fifty-three percent of the reptiles (of 693 species) and 45% of the amphibians (of 285 species) are endemic to Mexico (Flores-V., this volume). Among the fishes there exist over 2,000 species in Mexico (Espinosa-P. et al., this volume). Insect species number in the hundreds of thousands, of which about 25,000 are Lepidopterans (LLorente-B. & Luis-M., this volume) suggesting that there are more than one species of Lepidopteran for every flowering plant species in the country. The bees alone comprise 154 genera and 1,580 species (Ayala et al., this volume).

Mexico's biodiversity is one of world importance. Many of the crop plants used by humans have their origins in Mexico. A large portion of a critically important "gene belt" that circles the world between the Tropics of Cancer and Capricorn lies in Mexico. This map coincides with the distribution of centers and noncenters of agricultural origins proposed by Harlan (1975). The superimposition of these regions (Fig. I.2) highlights

Mexico's uniqueness in that it is the only nation where a megadiversity country and a center of agricultural origin coincide within the "gene belt." The two agricultural centers of origin (Neareast and N. China) are outside the "gene belt." The legume genus *Phaseolus*, for example, which is a rich source of plant protein, is represented by 35 species in the country. All four major species and several varieties of it have been domesticated in Mexico. More importantly, several relatives of the domesticated bean species are still to be found in the field, and geneflow between these and their wild progenitors has been documented. As another example, the all-important corn *Zea mays*, the mainstay of American civilizations, spread to other areas after its initial domestication in Mexico.

The biological consequences of species extinctions are only poorly understood just as the understanding of the role of species in the ecosystems is emerging. However, some economic extrapolations of species deaths have been made. Farnsworth and Soejarto (1985), for example, have calculated that the economic cost resulting from the extinction of a single plant species, a product of which may form a part of a prescription drug in the United States, may be around U.S. $203 million for 15 years. For modern agriculture, which relies on high-yielding monocultures susceptible to blights, disease-resistant genes that are found in their wild progenitors are an invaluable resource. For example, *Zea diploperennis*, a relative of cultivated corn, contains genes resistant to at least seven diseases that commonly afflict cultivated maize. It is immune to or tolerant of seven of the nine important viral and mycoplasmal diseases of maize and in many cases represents the only known source of such germplasm for maize (Nault & Findley, 1982). The extinction of *Zea diploperennis*, a species endemic to a small portion of the state of Jalisco and called "madre de maiz" by farmers, would be a great loss. *Phaseolus* is an important source of plant protein in Mexico and other places. Several species are cultivated in Mexico, and seeds of both the progenitors and their relatives are collected for consumption. The role of this genus in the Mexican economy is incalculable, and the financial cost resulting from the loss of, for example, even one population of one of these useful species with genetically desirable qualities (e.g., disease resistance, high yield) would be difficult to estimate.

The biological and social aspects of cultivation and domestication of plants in Mexico is discussed by Hernández X. (this volume) who estimates that over 105 economically important species were manipulated by peoples of Mexico prior to Spanish conquest. Bye (this volume) estimates that 50% of the useful plants have been derived from natural communities, that weeds tend to be mostly indigenous, and that over 9% of the ornamental species grown in the temperate world originated in Mexico.

Although many of these resources may be locally based, their use ranges far beyond the borders of this country. An examination of human diet everywhere bears out this point. Maize, beans, tomatoes, squash, and chili peppers are a few examples of Mexican plants consumed elsewhere. Mexico's genetic resources have been employed by mankind, and ongoing studies show that others have great potential use, e.g., *Amaranthus* spp., *Chenopodium* spp.

The well-being of a nation depends upon sensible exploitation of its biological resources. The work of the genetic engineers is of great promise, but for the foreseeable future humans will have to do with presently extant plants and animals for their nutritional and other needs. Moreover, the raw materials for genetic engineering are largely contained in the natural resources of the biological world. The importance of these resources, nature's germplasm, in modern biotechnology cannot be overemphasized. Dougall (1979) has suggested, for example, that tissue and cell suspension cultures play a crucial role in the production of commercially valuable compounds. The maintenance of high-yielding cell cultures is facilitated through selection of "high-yielding individuals." The decline of high-yielding clones can be offset through introduction of high-yielding materials from nature. The maintenance of high yield *in vitro* cultures requires that the genetic determinants for high production be present in these clones. These facts suggest that the success of the industrial exploitation of these techniques depends on the appreciation of nature as a living laboratory.

CENTERS OF ENDEMISM AND SPECIES RICHNESS IN MEXICO

The distribution of many groups of animals and plants suggests that the centers of endemism or species richness of various taxonomic groups are dispersed. The Trans-Mexican Volcanic Belt and the Sierra Madre del Sur, as defined by Ferrusquía-V. (this volume), are two of the areas that may be considered major centers of endemism in Mexico for many groups (mammals: Fa & Morales, this volume; Asteraceae: Turner and Nesom, this volume; oaks: Nixon, this volume; many plant groups: Rzedowski, this volume). Several of the contributions herein (Daniel; Flores-V.; Hunt; Ramamoorthy & Elliott) highlight the state of Oaxaca, part of the Sierra Madre del Sur, as a major area of endemism, but centers of endemism are, in fact, distributed throughout the country. The fragmented topography, along with climatic extremes, has resulted in many biological islands housing numerous endemics. The alpine areas sheltering many highly restricted species, for example, occur in great isolation from one another (McDonald, this volume). Centers of endemism for butterflies are widely separated from one another (LLorente-B. & Luis-M., this volume). The tropical south is generally species-rich but poor in endemics compared to the other vegetational regions of Mexico (Rzedowski, this volume; Wendt, this volume). Endemism in mammals is high in the Trans-Mexican Volcanic Belt and the Sierra Madre del Sur (Fa & Morales, this volume). Among the birds the highest number of species (240) is found in the semideciduous tropical forests, followed by 225 in tropical rain forests, 211 in tropical deciduous forests, and 206 in desert scrub (Escalante et al., this volume). The scattered mangroves discussed by Lot et al. (this volume) harbor probably unique marine fauna. The pine-oak forests shelter numerous endemic reptiles, but endemism in this group is notably high in the arid zones of northwestern Mexico (Flores-V., this volume). The highest concentration of apifauna is along the borderlands between Mexico and the United States (Ayala et al., this volume). Endemism in the all-important grass family is highest in the Chihuahuan-Coahuilan Plateaus and Ranges Morphotectonic Province in northern Mexico (Valdés & Cabral, this volume). Such scattered centers of endemism and concentrated centers of species richness pose peculiar and challenging problems for conservation biologists and policy makers (see below).

NATURE AND DYNAMICS OF REGIONAL AND LOCAL DIVERSITIES

Nothing in biology, it is said, makes sense in the absence of organic evolution, and it may equally be true that nothing in biogeography makes sense in the absence of plate tectonics. The regional diversity of Mexico, whose affinities are largely to the Nearctic and Neotropical realms, is a product of the dynamics of the lithosphere. The plates that have had a major influence here are the North American, South American, Cocos, Farollan, and Caribbean.

The consensus about time and arrival of the various biotic elements into the Mexican theater differs and will remain necessarily a subject of speculation. Graham (this volume) discusses the historical basis for diversity in Mexico. Laurasian elements in Mexico may be older than the Neotropical elements in the country, which may be quite recent. Several contributors have identified Mexican elements with affinities to groups in Asia and Africa (Rzedowski; Sousa & Delgado; Wendt; all this volume). The possibility that some of the species-rich tropical elements with a Laurasian origin may have arrived in Mexico directly, before reaching South America, is good. Subsequently, the South American descendants of these elements may have reinvaded Mexico during the Pleistocene. Wendt (this volume) suggests several examples of vascular plants.

Mexico, needless to say, has been an active theater of speciation for many taxa that may have originated elsewhere. The scattered distribution of centers of endemism of various taxonomic groups suggests that the complex geological history of Mexico has been a major evolutionary force. Ferrusquía-V. (this volume), who divides Mexico into 11 morphotectonic provinces, further subdivides the Sierra Madre del Sur into as many as 14 subprovinces; the complex geological history of this area is reflected by the extreme richness of endemism here. The epochs of orogeny and mountain building surely contributed to fragmentation of vegetation and ancestral populations. The pattern of variation in many groups suggests evolution by fragmentation as well. *Salvia*, with over 300 species, is a model plant example (Ramamoorthy & Elliott, this volume). Moreover, many faunistic and floristic groups in Mexico are considered to be autochthonous (several authors, this volume).

It is interesting to compare Mexico's regional and local diversities with other equally large areas of the world that fall in the same latitudinal ranges. India is a case in point. The dynamics of diversity of the Indian biota dramatically contrast with that of Mexico. India's present diversity is probably a product of invasion and colonization of taxa from adjoining regions; its original biota presumably suffered massive extinctions either during its transit through various latitudinal/climatic belts or through volcanism (Sahni, 1988). It is not suggested here that biota in Mexico did not experience extinctions; they did, but not on the same scale as occurred in India. The high generic and specific endemisms, e.g., vascular plant genera exceeding 300 and species 55–60% (Ramamoorthy & Lorence, 1987; Rzedowski, this volume), highlight Mexico as a major theater of speciation, which contrasts sharply with the low levels of endemism in the Indian biota. Most endemics in India may be found in the Himalayan range and may have evolved in response to tectonic events associated with collision of the Indian plate with the Asian plate. Mexico's diversity is perhaps more complex than that of India. The two regions contrast each other in the histories of their biota. Although India may have remained largely isolated while in transit, Mexico has served as a corridor and barrier to biota with diverse origins.

MEXICO AS THEATER FOR TRANSSPECIFIC SPECIATION

Generic and suprageneric endemisms in Mexico suggest that Mexico has long been an active center for the evolution and differentiation of supraspecific categories. Generic endemism in flowering plants, as already mentioned above, is high. Turner and Nesom (this volume) recognize over 40 endemic genera in the Asteraceae, many of which are believed to have originated *in situ*. Sousa and Delgado (this volume) recommend recognition of several natural tribes of many genera in the legumes in their Mexican Phytogeographic Legume Area. Among the grasses, several sections in the Eragrostidae seem to have differentiated in the Chihuahuan-Coahuilan Plateaus and Ranges Morphotectonic Province (Valdés & Cabral, this volume). Ramamoorthy and Elliott (this volume) stress the evolution of sections in *Salvia* (Lamiaceae). In Rzedowski's Mega-Mexico 1 (this volume), there are four flowering plant families endemic to the area. The radiation in mammals (Fa & Morales, this volume), in quails and jays among the birds (Escalante et al., this volume), and in several groups of reptiles (Flores-V., this volume) suggests that Mexico has been an active theater of speciation for specific and supraspecific categories. The underlying dynamics in the evolution and differentiation of such categories is subject to different interpretations. Mexican biodiversity affords excellent opportunities for the study of this aspect of organic evolution.

CONTRIBUTION TO CONSERVATION POLICIES FOR MEXICO

The present global concern for conservation is based on understanding the uniqueness of each species, its irreplaceability, its ecological functions, the present economic uses of many extant species, and the great potential various inadequately known species hold as well as those yet to be discovered may bear. As widely observed, the loss of vegetational cover of the Earth may bring on changes with unanticipated consequences for life on the planet. Conservation is a global issue, but the disparities in the standard of living in different parts of the world inject unique sets of problems in diverse areas. Industrial pollution (e.g., toxic waste sites, acid rain) and the resulting degradation of the environment are the primary focus in the highly industrialized societies. In the tropics these same problems are compounded by an expanding population, growing unemployment, and poverty. A new agenda seems to be called for, wherein governments, academicians, and people's representatives together plan environmentally sensible use of our biological resources. Indeed, our natural heritage needs to be seen within a historical context. The musings of Toynbee (1974) on austerity and frugal use of land may be relevant here.

The challenge facing conservation biologists and policy-makers in Mexico and other countries with similar socioeconomic crises is formidable. The task in Mexico is doubly difficult because the components of its biodiversity, which is one of the largest in the world, are mostly restricted and highly scattered throughout the country. Depending on the site and kind of vegetation, one may be dealing with a species-rich zone low in endemics or a region high in endemic species. The species-rich tropical forests of Chiapas are as important as the pine-oak forests in the Trans-Mexican Volcanic Belt or the arid lands of the north, which are rich in endemics. Ayala et al. (this volume) report that "the arid and semiarid Madrean region straddling the Mexico-United States border has the richest Apifauna [bee fauna] in the world."

Conservation of habitats and biodiversity will greatly benefit from the public's awareness of the uniqueness of species as well as its appreciation of natural phenomena. In Mexico, people are unfailingly astounded and inspired by the annual migration of hundreds of thousands of monarch butterflies that begin arriving in November to overwinter at pockets of surviving forest sites near Angangueo, Michoacán. These butterflies leave the harsh winter of northern latitudes to settle on a relatively small number of firs and pines. Their sheer numbers make the branches droop. The ability of new generations of these fragile voyagers, "which are four or five generations removed from their forebears who flew south the previous year" (Brower, 1988) to home in on sometimes the very same trees winter after winter elicits admiration from those familiar with this remarkable phenomenon. The loss of their habitats will surely result in extinction of this biological phenomenon.

Given the socioeconomic conditions, what are the priorities? Should species-rich areas or centers of endemism be off-limits to land exploitation and development with all the inherent political risks that such decisions run? Should we enforce stricter control and better management in the existing national parks? Create additional national parks? Start an intensive program toward completion of the biological inventory of the country so we know what is where? These themes are critical in Mexico, as elsewhere, even though the problem is partly offset by a great under-standing among the Mexican public of the need for conservation. Devel-opment of curricula including a strong emphasis on biodiversity and resource management is especially necessary in the national universities so that the urgently required taxonomic data base with scientific documenta-tion can be produced. If not, the development and application of the much-needed biotic resource information will be delayed by years as the number of trained people to do the job shrinks. Conservation of habitats with known and documented biota is certainly an easier objective than trying to preserve those whose biota have not even been recorded.

The greatest threat to the biodiversity in Mexico comes from loss of habitats (Toledo & Ordoñez; and other authors, this volume). An educa-tional campaign toward habitat preservation (e.g., low intensity grazing, reforestation with indigenous species, prevention of soil erosion, careful management of water resources, and meticulous planning of industrial development and disposal of industrial wastes) will go a long way to help save Mexico's fragile habitats and their ecosystems. Careful supervision and management of the existing national parks should contribute signifi-cantly. These programs will benefit enormously from a reliable taxonomic data base. The chapters in this volume bring together a rich bibliography that can contribute to a harmonious balance between satisfying ever-increasing human needs on the one hand and maintaining the natural biological diversity on the other.

The scattered centers of endemism, species richness, and biological islands (some examples, among many, are Cuatro Cienegas in Coahuila, Barranca de Tolantongo in Hidalgo, Rancho El Cielo in Tamaulipas) vex conservation policy-makers. One of the ways to preserve these habitats may be to plan collaborative research and study, involving educational and research institutions and government agencies. Ideally, such programs include local citizens. The selection of critically important sites involves consultation among local residents, planners, academics, and government agencies. This volume brings together data pertaining to the diverse geological substrates of the country and to many taxonomic groups repre-senting both plants and animals. The authors have discussed and indicated centers of endemism in their respective study groups as well as areas of species richness. This background information will strengthen the deci-sion-making ability of both conservation biologists and policy-makers.

The concern regarding global conservation is more than a quarter of a century old, but the practical solutions continue to elude policy-makers. They, along with the scientific community, have the evolutionary respon-sibility (Frankel & Soulé, 1981) of maintaining the broad ecological theater in order to maintain its intrinsic values and to meet unforeseen economic needs. Species richness results from the intertwining of the ecological and evolutionary processes. Its dynamics require constant inventorying, reanalysis, and reevaluation. In order to realize the shared evolutionary responsibilities, the current unidirectional transformation of information (from the preliminary data base to the ultimate policy-makers) should be reintegrated in such a way that the expertise of several sectors (genetics, systematics, ecology, conservation) work in unison (e.g., networking). The incorporation of information from different levels of biotic organization (gene to species to community) requires the active participation of scientists from socioeconomic, genetic, taxonomic, ecological and conservation disciplines in concert with policy- makers. Such an approach will result in a more practical selection and maintenance of *in situ* conservation (e.g., national parks, natural reserves) and *ex situ* germplasm conservation (e.g., seed banks).

April, 1992

T.P. Ramamoorthy
Robert Bye
Antonio Lot
John Fa

REFERENCES

Brower, L.P. 1988. A place in the sun. Animal Kingdom 91(4):42–51.

Dougall, D.K. 1979. Factors affecting the yields of the secondary products in plant tissue cultures. *In* W.R. Sharp, P.O. Larsen, E.F. Paddock & V. Raghavan (eds.), Plant Cell and Tissue Culture, Principles and Applications. Columbus: OH: Ohio State Univ. Press.

Farnsworth, N.R. & D.D. Soejarto. 1985. Potential consequence of plant extinction in the United States on the current and future availability of prescription drugs. Econ. Bot. 39(3):231–240.

Frankel O.H. & M.E. Soule. 1981. Conservation and Evolution. Cambridge: Cambridge Univ. Press.

Harlan, J.R. 1975. Crops and Man. American Society of Agronomy. Madison, WI: Crop Science Society of America.

McNeely, J.A., K.R. Miller, W.V. Reid, R.A. Mittermeier & T.B. Werner. 1990. Conserving the World's Biological Diversity. Gland, Switzerland and Washington, DC, U.S.A.: IUCN, WRI, CI, WWF-US, The World Bank.

Mittermeier, R.A. 1988. Primate diversity and the tropical forest: case studies from Brazil and Madagascar and the importance of the megadiversity countries. *In* E. Wilson (ed.). Biodiversity. Washington, DC: National Academic Press.

Nault, L.R. & W.R. Findley. 1982. *Zea diploperennis*: a primitive relative offers new traits to improve corn. Desert Plants 3(4):203–205.

Ramamoorthy, T.P. & D.H. Lorence. 1987. Species vicariance in Mexican flora and a new species of *Salvia* from Mexico. Adansonia 2:167–175.

Sahni, A. 1988. Cretaceous-Tertiary boundary events: mass extinctions, iridium enrichment and Deccan volcanism. Current Sci. 57(10):513–519.

Toynbee, A.J. 1974. *In* G.R. Urban (ed.), Toynbee on Toynbee. A Conversation Between A.J. Toynbee and G.R. Urban. New York: Oxford Univ. Press.

Wilson, E.O. (ed.). 1988. Biodiversity. Washington, DC: National Academic Press.

I
HISTORICAL BACKGROUND

Geology of Mexico: A Synopsis

ISMAEL FERRUSQUÍA-VILLAFRANCA

The Mexican territory has been divided into 11 morphotectonic provinces that have physiographic and geologic-tectonic features distinctive enough to individualize them from such other neighboring ones: (1) Baja California Peninsula, (2) Northwestern Plains and Sierras, (3) Sierra Madre Occidental, (4) Chihuahua-Coahuila Plateaus and Ranges, (5) Sierra Madre Oriental, (6) Gulf Coast Plain, (7) Central Plateau, (8) Trans-Mexican Volcanic Belt, (9) Sierra Madre del Sur (10) Sierra Madre de Chiapas, and (11) Yucatan Platform. Mexico spans 18° of latitude and longitude, but its surface area is only one-fourth that of this geodetic quadrangle; the remainder is taken up by the Pacific Ocean, Gulf of Mexico, and Caribbean Sea and has an unique, horn-like shape. It is set in a northwest-southeast fashion, with the greater part toward the northwest. This geographic setting allows the country to have more littoral and greater marine influx than it would have otherwise. Two-thirds of the territory lies 800 m or more above sea level, which is somewhat above the world and continental mean altitudes. The relief, dominated by mountains, is rugged; these mountains in turn are of various kinds, e.g., isolated or associated with long or short chains (sierras or ranges); small to large, low to high, rounded, crest-like, peaked, table-shaped, folded, block-faulted, volcanic, complex. The climate of this territory, which covers arid, steppe-like, temperate, and tropical regions, is varied. The general trend is toward increasing dryness north of the Trans-Mexican Volcanic Belt as well as the Yucatan Platform.

The geological composition is varied and complex; the rock types span the lithic spectrum: (1) *sedimentary* (marine: clastic, chemical, or mixed; shallow to deep; near shore—transitional: beach, dunes, lagoon, marsh, tidal complex, and estuarine, to name a few; and continental: fluvial complex, lake, desert); (2) *volcanic* (pyroclastic to lavic; silicic, intermediate, mafic and ultramafic, erupted through foci or fissures; continental or marine); (3) *intrusive* (shallow, subvolcanic, hypabyssal to deep, i.e., plutonic); and (4) *metamorphic* (thermaly or dynamically generated, local or regional; low to high grade; phyllitic to gneissic in texture; and of silicic to mafic composition).

These rocks could be arranged into stratigraphic units of various kinds. Those chiefly made up of sedimentary rock tend to be sheet-like: one dimension is much smaller than the other two. The marine rocks tend to be larger and more homogeneous than the continental rocks; each may or may not bear fossils. The pyroclastic rocks also tend to be sheets or aprons that reduce their thickness away from the eruption area; the mafic lavic-basaltic rocks in particular are sheets or accumulate in the topographic lows, in view of the low viscosity of the basaltic lavas, whereas those of silicic—particularly rhyolitic—composition form dome-shaped bodies because of the high viscosity of the rhyolitic magmas. The intermediate or andesitic bodies stand in between and may form large stratovolcanic edifices or extensive volcanic fields. The plutonic bodies may vary from a small or local size, forming stocks, to a regional extent (>100 km²), in the form of batholiths. Their outcropping is the result of removal of roof, brought about through various processes. The hypabyssal and subvolcanic bodies are usually much smaller. The regional metamorphic bodies are large. Their genesis implies normally high pressures and moderate to high temperatures possible only at the depth in the crust; hence their outcropping also suggests regional uplift and extensive removal of the cover. Structurally, a wide spectrum of features is present in Mexico and includes mainly folds (open or tight, isolated or arranged in systems: anticlinoria and synclinoria) and faults (of vertical or lateral displacement, local to regional extent, normal to thrust in kind, and isolated or arranged in systems). Their presence bears strongly on the tectonic history of the region; for instance, the lack of these features indicates tectonic stability. Folding and thrust-faulting call for compression (probably occurring in collision or convergent margin tectonics) and faulting, particularly rifting, i.e., parallel faults that produce a down-thrown elongated block, indicates tension (probably occurring in divergent margin tectonics). Finally, in terms of age, the rock bodies extend from the Precambrian–Mid-Proterozoic, some 1,800 Ma to the Recent; but by and large, the Precambrian and Paleozoic make up a much smaller portion of the territory than the Mesozoic and Cenozoic ones.

Few territories the size of Mexico have such a complex make-up that is expressed in its rugged relief and varied climate. This make-up sets a heterogeneous physical-geographical scenario for one of the most diverse biotas of the world. The current seismic and volcanic activities attest to the prevailing existence of some of the geodynamic processes responsible for such a composition. Yet the information concerning Mexico's geology, found in more than 7,000 bibliographic entries, is incomplete. Past works were chiefly of reconnaissance type and are for the most part unpublished reports of government agencies or private companies. Their availability is limited. Such literature covers only a small portion of Mexico and is inadequate for understanding its origin, complex history, and regional evolution. The revolutionary Theory of Plate Tectonics suggests a highly dynamic and mobile lithosphere and necessitates the reevaluation and interpretation of past data—an immense challenge, as description of the Mexican territory remains incomplete.

    The main works that summarize the geologic information of this country include those of Aguilera and Muñoz (1896), Garfias and Chapin (1944), Alvarez (1954), López-Ramos (1979), Morán (1984), and de Cserna (1989). The first and next to the last were written as companion volumes to corresponding geologic maps: Aguilera and Muñoz (1896) and INEGI (1982), respectively. In these works, data have been interpreted according to the then-prevailing theory, and the descriptions are frequently biased by extrapolations, interpretations, or inferences derived from the theories that have no objective factual support or, worse, are treated as proved facts. In

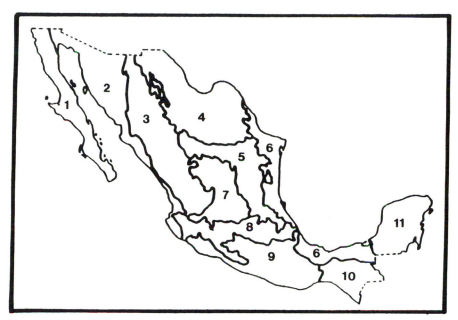

Figure 1.1. Morphotectonic provinces of Mexico. 1. Baja California Peninsula. 2. Northwestern Plains and Sierras. 3. Sierra Madre Occidental. 4. Chihuahua-Coahuila Plateaus and Ranges. 5. Sierra Madre Oriental. 6. Gulf Coast Plain. 7. Central Plateau. 8. Trans-Mexican Volcanic Belt. 9. Sierra Madre del Sur. 10. Sierra Madre de Chiapas. 11. Yucatan Platform.

many cases it is difficult to extricate the data from the inferences. In addition to the geological maps mentioned, those by de Cserna (1961) and Comité de la Carta Geológica (1976) provide summary reports on Mexican geology. The first is tectonic (scale 1:5,000,000), and the second chronostratigraphic (1:2,000,000), although the latter is fast becoming obsolete as the chief reference for regional works.

The preparation of a modern cognitive synthesis of the geology of Mexico for a nonspecialist readership becomes difficult under these circumstances but is a necessary task nonetheless. This goal is the purpose of the present contribution, which emphasizes the data specifically in accordance with the Theory of Plate Tectonics. Hidden and unresolved problems are highlighted, as the result would otherwise be a disorderly collage of data and bits of interpretations or a one-sided account of the history and supposed evolution of the country. This possibility is a real danger, because in geology, perhaps more than in any other discipline, the same body of data may be interpreted not only in different but even in contradictory forms because of the intrinsic incompleteness of the record and its sampling.

Given the fact that the physiography of a region, i.e., its relief, land forms, and other main physical-geographic features, reflects its geology,

Table 1.1. Location and basic features of the morphotectonic provinces of Mexico

Province[a]	Location	Surface area of province (km²)	Altitude ranges (m)[c]	Climate[d]	Chief land form
1	Northwestern Mexico 109°30'–117°00' WL 23°00'–32°30' NL	144,000 ≈ 7.34%[b]	0–2,130 0–1,000	BWh, BShs, Csa	Sierras & plains
2	Northwestern Mexico 107°00'–116°00' WL 23°00'–32°30'NL	236,800 ≈ 12.02%	0–2,200 200–1,000	BWh, BSh	Sierras & plains
3	Western and northwestern Mexico 102°20'–109°40' WL 20°30'–31°20' NL	289,000 ≈ 14.68%	200–3,000 2,000–3,000	Cfb, Aw	Sierras & plateaus
4	Northern Mexico 101°31'–110°31' WL 26°00'–31°45' NL	255,900 ≈ 12.52%	200–2,000 800–1,200	BShw, BWh, BSk	Sierras & plateaus
5	Northeastern and northcentral Mexico Transverse sector 100°00'–105°00' WL 24°30'–26°00' NL Eastern sector 97°30'–101°20' WL 19°40'–26°00' NL	145,500 ≈ 7.54%	200–3,000 1,000–2,000	Transverse sector BWh, BSk Eastern sector Cfa, Cwa, BSh	Sierras

Table 1.1. (cont.)

Province[a]	Location	Surface area of province (km²)	Altitude ranges (m)[c]	Climate[d]	Chief land form
6	Eastern Mexico Northern sector 96°30'–100°20' WL 20°00'–26°00' NL Southern sector 91°15'–96°46' WL 17°10'–19°20' NL	170,600 ≈ 8.66%	0–200	Northern sector Aw', Cw, Cx'w' Southern sector Afw', Amw'	Plains
7	Central Mexico 100°00'–104°00' WL 21°00'–24°00' NL	85,300 ≈ 4.33%	1,000–3,300 2,000–3,000	BSh	Plateaus
8	Central Mexico 96°20'–105°20' WL 17°30'–20°25' NL Main sector 19°00'–21°00' NL	175,700 ≈ 9.17%	1,000–5,000 1,000–2,000	Aw', Cfa, Cwa, BSh, Cw, Cfb, Aw	Peaks & plateaus
9	Southern Mexico 94°45'–104°40' WL 15°40'–19°40' NL	195,700 ≈ 9.93%	0–3,500 1,200–1,800	Aw', Aw, BShw, Cwa, Cfa	Sierras & depressions

Table 1.1. (cont.)

Province[a]	Location	Surface area of province (km²)	Altitude ranges (m)[c]	Climate[d]	Chief land form
10	Southeastern Mexico 90°30'–95°00' WL 14°30'–17°40' NL	105,400 ≈ 5.35%	0–2,500 200–1,000	Aw, Cw, Cf	Sierras, depressions, & plains
11	Eastern Mexico 87°00'–91°00' WL 17°50'–21°30' NL	167,600 ≈ 8.46%	0–200	BShw, Amw	Plains & karst topography

[a] These numbers correspond to those on the map in figure 1.1. 1. Baja California morphotectonic province (mp). 2. Northwestern Plains and Sierras mp. 3. Sierra Madre Occidental mp. 4. Chihuahuan-Cohuilan Plateaus and Ranges mp. 5. Sierra Madre Oriental mp. 6. Gulf Coast Plain mp. 7. Central Plateau mp. 8. Trans-Mexican Volcanic Belt mp. 9. Sierra Madre del Sur mp. 10. Sierra Madre de Chiapas mp. 11. Yucatan Platform mp.

[b] Percentage = ratio of mp to total surface area of Mexico.

[c] First entry = total range, second entry = dominant range. See text for details.

[d] BWh, desert-like, Mat >18°C; BShs Csa, temperate with dry winter; BSh, dry Mat >18°C; Cfb, temperate humid with no dry season; BShw, steppe-like, winter dry season, Mat >18°C; BSk, steppe-like, Mat >18°C; Cfa, temperate, no defined dry season; Cwa, temperate with dry winter; Aw', tropical with dry winter and rainy fall; Cw, temperate with dry winter; Cx'w', temperate with little rain throughout the year; Atw', tropical rainy with no defined dry season; Cf, temperate with no defined dry season. The key to the letter symbology is: A, warm humid and subhumid Climate Group (lack of a well-defined dry season); m, rainy season restricted to the summer; w, dry winter and warm season from April to September; w', less rainy summer with a short dry season. B, warm to cold and very arid to semiarid Climate Group; BS, warm to semicold and arid to semiarid Climate Subgroup; BW, warm to semicold and very arid Climate Subgroup; h, semiwarm with cool winter; k, temperate with a warm summer; s, rainy winter. C, temperate to semicold and humid to semihumid Climate Group; a, warm summer; b, cool and long summer; x', rainy fall. *Source:* García (1988).

structure, and history, Mexico's territory is described herein through morphotectonic provinces (Table 1.1, Fig. 1.1), which have physiographic and geologic-tectonic features that are distinctive enough to distinguish and differentiate them from other such neighboring provinces. They include the Baja California Peninsula, Northwestern Plains and Sierras, Sierra Madre Occidental, Chihuahua-Coahuila Plateaus and Ranges, Sierra Madre Oriental, Gulf Coast Plain, Central Plateau, Trans-Mexican Volcanic Belt, Sierra Madre de Sur, Sierra Madre de Chiapas and Yucatan Peninsula. This approach is not novel (Alvarez, 1954; Guzmán & de Cserna, 1963; de Cserna, 1989), but it is the first time that a physiographic characterization is given along with the geologic description of each province. Thus an objective and systematically arranged body of data for each province is presented and clearly separated from possible interpretation(s). The reader may then draw a fairly objective picture of each province's geology without hindrance of a given interpretation or opinion. The purpose is to best serve the readership.

The pattern adopted to describe the provinces includes three major aspects, i.e., the geographic and physiographic setting, geology, and geologic story. The choice of words and phrasing is intended to connote the separation of description from interpretation. For clarity, use of technical terms has been kept to a minimum, and few references are included in the text. In a synthetic work of this nature, individually citing the originators of particular concepts or contributions each time may needlessly distract the reader from following the trend of ideas being presented, in addition to occupying much space; yet omitting citations would be unfair to the authors, or, worse, might convey the impression of piracy by this writer. As a compromise, in the descriptive text, only the unavoidable references are included; the rest, relevant to the province, are listed at the end of each province. A complete bibliography is provided at the end of the chapter.

Selected References

Aguilera & Muñoz (1896); Clark et al. (1982); Comité de la Carta Geológica. Méx. (1976); Campa & Coney (1983); X & XX Congreso Geológica Internacional (1906, 1956); Cook & Bally (1975); Damon et al. (1984); de Cserna (1960, 1961, 1974, 1989); Eardly (1951); Garfias & Chapin (1944); Guzmán & de Cserna (1963); INEGI (1982); Instituto de Geología (1986, 1987b, 1989); King, P.B. (1969, 1977); López-Ramos (1979); Morán (1984, 1986); Murray (1961); Pindell et al. (1988); Raisz (1959); Ross & Scotese (1988); Ruiz et al. (1988); Santiago et al. (1984); Santillán (1936); Schuchert (1935); Urrutia (1986b); Urrutia et al. (1986).

BAJA CALIFORNIA PENINSULA MORPHOTECTONIC PROVINCE

Geographic and Physiographic Setting

The Baja California Peninsula Morphotectonic Province is located in north-western-most Mexico between long. 109°30'–117°00' W and lat. 23°00'–32°30' N. It is about 1,260 km long and 95 km wide on average (Fig 1.2). Its orientation is N35°W–S35°E, with a surface area of about 144,000 km² (about 7.34% of Mexico). The dominant climate is BWh (desert-like, mean annual temperature [Mat] >18°C), cooling northward to BShs and CSa (temperate with dry winter). The altitude ranges from 0 to 2,130 meters above sea level (masl). The Pacific side of the southern half includes a wide coastal plain (<200 masl). Most of the sierra country lies below 1,000 masl; the northern half is more mountainous and higher, and about one-third of it lies above 1,000 masl.

The peninsula includes a series of mountain ranges collectively known as Peninsular Ranges, that extend lengthwise through the entire peninsula. Its main sierras include La Rumorosa, Juárez, San Pedro Martir, and La Libertad in Northern Baja and La Giganta and San Lázaro in Southern Baja. The Peninsular Ranges are much narrower in Southern Baja, where they are located in the eastern part, leaving a broad coastal plain on the Pacific side. In Northern Baja, the plain is much narrower. In the central portion of the peninsula, this plain is broadest and forms the Vizcaíno Desert. The coastal plain continues offshore with a continental shelf that is, on average, as wide as the peninsula but becomes narrow north of Punta San Fernando.

On the Gulf side and at the southern end, the peninsula has practically no shelf except in La Paz and north of parallel 30°, where it broadens and meets the mainland shelf. Adjacent to this margin of the peninsula are several small islands.

Geology

The geology of the peninsula is complex and different in its two halves (Fig. 1.3).

Northern Baja

Small to medium-sized Paleozoic metamorphic and plutonic rock bodies occur in isolation throughout the northern part of the Peninsular Ranges, roughly aligned in a zone that significantly narrows to about one-third of its width south of parallel 41°30'. The chief portion of the Peninsular Ranges is formed by plutonic and metamorphic rock bodies of Mesozoic age. The plutons are thought to be genetically related, forming a large intrusive body referred to as the Baja California Batholith, whose compo-

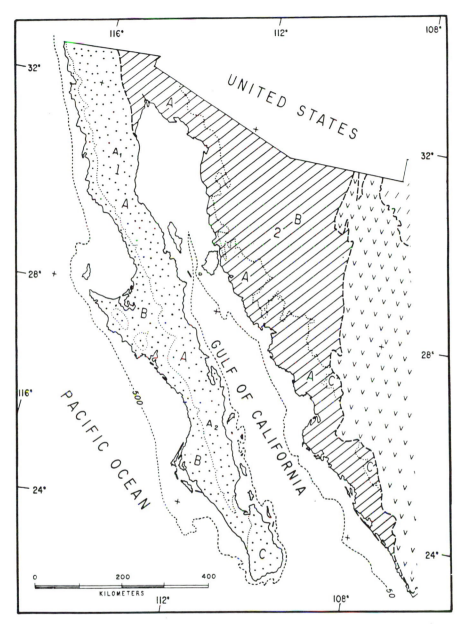

Figure 1.2. Morphotectonic provinces of Mexico. 1. Baja California Peninsula: A, Peninsular Ranges Subprovince; A_1, northern portion; A_2, southern portion; B, Pacific Coastal Subprovince; C, La Paz-Los Cabos Subprovince. 2. Northwestern Plains and Sierras: A, Pacific Coastal Plain Subprovince; B, Sonoran Basin and Range Subprovince; C, Sinaloan Minor Sierras Subprovince. Dotted line offshore indicates the 500 m isobath.

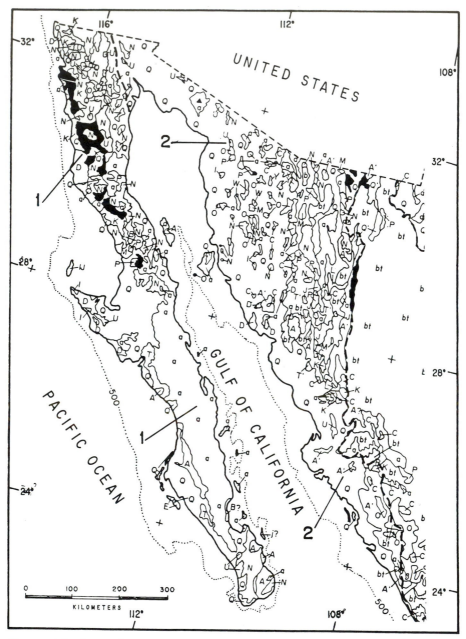

Figure 1.3. Generalized geologic map of Baja California Peninsula (1) and Northwestern Plains and Sierras (2) Morphotectonic Provinces and adjacent areas.

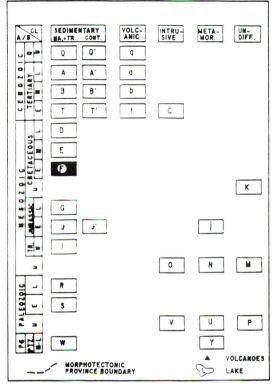

Abbreviations

A/S, age/stage*; CL, chief lithology of rock bodies; CONT, continental; E, early/lower*; L, late/upper*; M, middle/middle*; MA + TR, marine and transitional; METAMOR, metamorphic; P-€, Precambrian; PTZ, Proterozoic; Q, Quaternary; TR, Triassic; UNDIFF, undifferentiated. Dotted line offshore indicates 500 m isobath.

*First term is geochronological, i.e., it refers to time; second term is chronostratigraphic, i.e., it pertains to kind of rock bodies bound by time divisions.

Modified from Comité de la Carta Geológica, 1976 and other sources.

Legend to Figure 1.3.

sition is dominantly granitic to granodioritic but includes small gabbroic bodies as well.

The western portion of the Peninsular Ranges consists of a group of irregularly shaped volcanic, volcaniclastic, and continental and marine sedimentary rock bodies of Medial Cretaceous age (Aptian-Albian), collectively known as the Alisitos Formation, which forms a quasicontinuous belt that nearly spans Northern Baja.

The narrow Pacific coastal plain consists of Late Cretaceous (Campanian-Maestrichtian) sedimentary clastic marine rock bodies, named the Formation Rosario. Its strata are nearly horizontal and unconformably rest on the Alisitos Formation, which in turn unconformably and tectonically rests on the batholithic plutons. Quaternary near-shore and beach deposits around San Quintin and Late Tertiary volcanic rock bodies in the northern tip complete the make-up of the Pacific coastal plain.

The Gulf Coastal Plain is widest north of parallel 31°, and consists of Quaternary near-shore and beach deposits that surround small isolated exposures of Mesozoic age plutons and Late Tertiary volcanic silicic pyroclastic units as well as andesitic lavas; they are probably related to similar bodies irregularly appearing in the Peninsular Ranges south of the

30°30' parallel, interposed among the Mesozoic plutonic and Paleozoic metamorphic rock bodies previously mentioned.

Southern Baja

The greatest land feature is the Sierra de La Giganta made up of volcanic (silicic pyroclastic, andesitic to basaltic lavas) and continental sedimentary (arkosic and tuffaceous sandstones and conglomerates) rock bodies of Late Tertiary (Late Miocene-Early Pliocene) age. Their aggregate thickness reaches locally up to 1,200 m; they are collectively referred to as the "Comondú Formation." The islands of the Gulf of California seem to be genetically related to the geologic processes that generated La Sierra de La Giganta. West of this Sierra, the Pacific Coastal Plain broadens, forming in the north the Vizcaíno Desert and associated lagoons, the Villa Constitución Plain, and the Bahía de Magdalena area. The plain is made up of Late Tertiary and Quaternary fine-grained shallow marine and beach deposits that unconformably cover older Tertiary sedimentary marine formations to the east, and even older rock bodies to the west.

This region has a core of small deep marine sedimentary (chert, volcanic litharenites, and reefal limestone blocks) bodies resting on pillow basalts, both of the Late Triassic age. They are unconformably overlain by metamorphic rock bodies (interpreted as ophiolitic sequences, i.e., metamorphosed oceanic floor basalts and associated sediments) of the Jurassic age, complexly associated to volcaniclastic and sedimentary marine rock bodies of the Late Jurassic–Early Cretaceous age. A similar geologic setting is seen in Santa Margarita Island, where, in addition, ultramafic rock bodies of seemingly Jurassic age are present.

The southern end of the Peninsula (south of La Paz and east of Todos Santos) is geologically very different from the rest. The major land feature is the mountain range defined by the Sierras de Los Novillos, del Triunfo, de la Laguna, and de San Lázaro, chiefly made up of a batholithic pluton of granodioritic to granitic composition and of Mesozoic age, very similar to that of northern Baja. This batholith is separated in the west from the rest of the Peninsula by a north–south trending fault (from La Paz to Todos Los Santos). It is also faulted to the east in the Santiago–Río San José Plain and is limited to the east by the Sierra de la Trinidad, which is formed by a Mesozoic pluton very similar to that previously mentioned. The plain is made up of Medial and Late Tertiary, largely sedimentary, marine rock bodies. Small, isolated metamorphic units, seemingly of Mesozoic age and intrusive plutons, probably of Paleozoic age complete the geology of this region.

Geologic Story

The interpretation and integration of Baja California's geologic and physiographic data into a meaningful evolutionary synthesis is hampered by the complexity of the region as well as the incompleteness and uncertainties of the data base. Current versions of the peninsular geologic evolution (Gastil et al., 1975, 1981; Lozano, 1976; Rangin, 1979; Gastil, 1982; and others) follow the paradigm of Plate Tectonic theory. The account presented below draws heavily from them.

The Paleozoic rock bodies seem to be related to those of the Rocky Mountain Cordillera; they represent marine deposition within a tectonic framework of passive plate margins for this portion of the Paleozoic North American Plate (= Craton). Elsewhere in California two collision events (Antler and Sonoma orogenies) have been documented (Dickinson, 1979).

The Batholithic Pluton and the Alisitos Formation are interpreted to be related to a large Jurassic-Cretaceous volcano-oceanic island arc (Formación Alisitos), developed off-shore but near the southwest margin of the North American Plate (Craton). This island arc was part of an East Pacific Oceanic Plate referred to as the Farallon Plate.

The Farallon and the North American Plates had opposing relative motions that developed an active convergent margin, resulting in the subduction of the Farallon Plate under the North American one. The collision caused strong deformation, metamorphism and partial melting of the subducting plate, regional uplift, volcanic activity, and the emplacement of the batholiths. Apparently the fragmentation of the Farallon Plate by transform faults produced collision episodes that were not simultaneous throughout the peninsular length, and varied from Medial Cretaceous to Late Cretaceous and Early Tertiary in different segments of the "Peninsula." The extensive volcanic activity that generated the Sierra de la Giganta and associated volcanic land forms, seems to represent another effect of the Farallon Plate subduction. Through this collision and its associated geotectonic processes the former oceanic island arc became accreted to the North American Plate.

The Cenozoic regional uplift was uneven so that the eastern portion of the present-day Peninsula was lower and became flooded by a shallow epicontinental sea ("Ancestral Gulf").

During the Oligocene (about 30 Ma) the East-Pacific Rise, which formed the eastern boundary of the Farrollan plate, met the southwestern margin of the North American Plate at about the present position of Sinaloa or Nayarit. This terminated the subduction of the Farallon plate and began the separation and displacement of a portion of southwestern North America, now known as the Baja California Peninsula. The gap became the Gulf of California which developed its oceanic crust during the last 4 Ma. The displacement took place via a series of step-sided dextral faults whose present-day representatives define the San Andrés Fault System, and express the relative motion of the North American and Pacific Plates.

Selected References

Abbot & Gastil (1979); Atwater (1970); Beal (1948); Chase et al. (1970); D'Anglejan (1963); Darton (1921); Demant (1975); Dickinson (1979); Fiala et al. (1982); Filmer & Kirschvink (1989); Finch & Abbot (1977); Finch et al. (1979); Frizzell (1984); Gastil et al. (1975, 1976, 1981); Hagstrum, et al. (1985); Hausback (1988); Heim (1922); Hisazumi (1922); Jones et al. (1976); King (1968); Larson (1972); Lozano (1976); Mina (1957); Morris & Busby-Spera (1988); Ojeda et al. (1965); Ortlieb & Roldán (1981); Pantoja & Carillo (1966); Rangin (1979); Santillán & Barrera (1930); Silver et al. (1963); Urrutia (1986a); Van Andel et al. (1964); Wilson & Rocha (1957); Wilson & Vetia (1949).

NORTHWESTERN PLAINS AND SIERRAS MORPHOTECTONIC PROVINCE

Geographic and Physiographic Setting

The Northwestern Plains and Sierras Morphotectonic Province occupies the northwestern corner of Mexico's mainland, and lies between lat. 23°00'–32°30' N and long. 107°00'–116°00' W (Fig. 1.2). It forms an obtuse triangle whose hypotenuse is the 1,200 km long Pacific margin including the states of Sonora and Sinaloa. The eastern limit is largely north-south trending, 520 km long; and the northern limit is a portion of the Mexico–United States border, which is 640 km long. The area is approximately 236,800 km², or about 12.02% of Mexico. The altitudinal range is 0 to 2,200 masl, but the 200 to 1,000 m zone is dominant. The climate is desert-like (BWh group, Mat >18°C) in the northwestern part and dry (Bsh group, Mat >18°C) in the rest. This province has several major river systems (Colorado, Magdalena-Concepción, Sonora-San Miguel, Yaqui, Mayo, Del Fuerte, Culiacán, and San Lorenzo), but there are no present-day lakes. Instead, there are several dams. Physiographically, this vast area includes three subprovinces: Pacific Coastal Plain, Sonoran Basin and Range, and Sinaloan Minor Sierras.

Pacific Coastal Plain Subprovince

The Pacific Coastal Plain Subprovince is outlined by the 200 m contour line and is best developed in northern Sinaloa–southern Sonora. It is narrower in northern Sonora, particularly so across Tiburón Island. In Sinaloa the coastline is diverse, including delta complexes, estuaries, lagoons, and point and island bars related to the major river systems that discharge here. By contrast, the northern Sonora coast is simpler.

The Sonoran Basin and Range Subprovince

The Sonoran Basin and Range Subprovince is the southern extension of the homonymous physiographic province in the southwestern United States. It has a typical pattern of north-northwest/south-southeast (NNW–SSE) elongated block mountains separated by similarly trending basins. The size and elevation of the blocks increase eastward, contrasting with the size (especially the width) of the basins, which diminish to the east. Among the chief Sierras are de Sonoyta, de San Juan, de la Escondida, de Cananea, de Santa Rosalía, de Bamori, de Ancachi, de Batue, de Nacozari, de Baviacori, and de Tecoripa.

Sinaloan Minor Sierras Subprovince

The Sinaloan Minor Sierras Subprovince is an artificial subdivision that includes the mountainous country located between the Pacific Coastal Plain and the typical Sierra Madre Occidental. Its geology is different and shows no discernible basin and range pattern. It is best represented in the drainage basins of the Sinaloan river systems. Altitude exceeds 2,000 masl only in the easternmost portions (Sierra de Surotato, Altos de Topia, or Espinazo del Diablo). The altitudinal zones of 200–1,000 masl are about the same size.

Geology

Pacific Coastal Plain

The Pacific Coastal Plain (Fig. 1.3) is made up of Quaternary sedimentary clastic (continental-marine) transitional "rock" bodies developed by the evolution of complex delta systems prograding to the Pacific and distributed by long shore currents, waves, and tides that have produced a complex array of bars, lagoons, and estuaries. In Sinaloa, between the 24° and 26° parallels, the eastern portion of the plain is made up of Late Tertiary Continental Sedimentary clastic deposits. The plain ends around the 24° parallel where the Sierra Madre Occidental reaches the coastline.

Sonoran Basin and Range Subprovince

The dominant geologic feature is the presence of numerous high angle, NNW–SSE trending faults that divide this province into narrow uplifted blocks (the local Sierras) and down-thrown blocks (the basins), where Late Tertiary deposits have accumulated. Because of their particular geologic constitution, actual uplift, and erosional stage, the block-faulted sierras show individually different segments of the geologic column. Hence a broad age-discriminated account of the geology is suitable for a description of this subprovince.

Precambrian. Precambrian formations include small isolated metamorphic rock bodies, chiefly of amphibolites and greenschists intruded by granites and diorites. These bodies are concentrated in the northeastern and northwestern Sonora and show different isotopic ages: those around Caborca are the oldest in Mexico, 1,700 to 1,800 Ma (Anderson & Silver, 1979), and those of the northwest 1,600 to 1,700 Ma. Their boundary is attributed to a large, left lateral fault system developed during the Jurassic, named by these authors the Mohave-Sonora Megashear. The sedimentary portion is younger (Late Precambrian) and consists of quartzitic and dolomitic rock bodies located south of Caborca. Included are stromatolites (fossil algal mats) which are the oldest fossils in Mexico.

Paleozoic. Numerous small and medium-sized sedimentary marine rock bodies of fine-grained litharenites and micritic limestone deposits in near-shore and platform environments broadly define a triradiate pattern centered in Hermosillo. One arm is directed to Caborca, another to Cananea, and the third to Naco, where it bends southward. The ages of these bodies span the whole Paleozoic, although those of the Mid-Paleozoic are uncommon. The Paleozoic rock bodies show folding and minor faulting that apparently took place at the end of the Permian.

Mesozoic. Rock bodies of the Mesozoic age are widely, if irregularly, scattered throughout the subprovince. Those occurring in the Caborca, Antimonio, and Santa Rosa areas (northwestern Sonora) are Late Triassic–Early Jurassic and are sedimentary marine (near-shore and highly fossiliferous) and continental. Those around San Marcial (southeast of Hermosillo) are made up of siltstones and fine-grained sandstones with coal interbeds, interpreted to represent shallow marine and continental sedimentation. In other areas the rock bodies are volcaniclastic and continental sedimentary in origin. The Late Triassic–Early Jurassic rock bodies rest unconformably on Paleozoic units and show little structural deformation. The numerous Jurassic units are scattered along the wide northwest–southeast trending zone that excludes the northwestern sector of this province and the Pacific Coastal Plain. The rock bodies are largely volcanic and volcaniclastic of andesitic composition. Some are interbedded with clastic and carbonate marine strata, and others are metamorphosed. Isotopic ages for these bodies span 150–180 Ma.

Cretaceous. Cretaceous rock bodies are numerous and show marked geographic, geologic, and age control. Those of the Early Cretaceous, located in central and western Sonora, consist of andesites (lavic and pyroclastic) partly interbedded with limestone. Those located in eastern Sonora are largely sedimentary marine. The Late Cretaceous rock bodies are mainly granitic to monzonitic plutons (some of batholithic size) or volcanic sequences, dominantly consisting of andesitic lavas, whose size and number increase westward. Their emplacement structurally deformed preexisting rock units.

Cenozoic. Early Tertiary rock bodies geologically similar to those of Late Cretaceous occur in northwestern, central, and southern Sonora. Those of northwestern Sonora are genetically related to copper deposits. The Mid-Tertiary rock bodies (Oligocene–Miocene) are genetically related to the Sierra Madre Occidental. They consist of volcanic (silicic ignimbrites) rocks that form plateaus and mesas, and they unconformably cover preexisting bodies. The Late Tertiary-Quaternary rock units (of medium size) are located in northeastern, southern, and northwestern Sonora. Their composition is basaltic (alkaline), and they form lava-capped plateaus and volcanic mountains such as the Pinacate Volcano. It appears that, coevally with the emplacement of these volcanic bodies, block-faulting tectonically overprinted all preexisting bodies, giving this province its characteristic basin and range structure. Continental sediments, collectively referred to the Baucarit Formation, developed in the basins thus formed.

Sinaloan Minor Sierras

The complex geology of this area (Fig. 1.3) shows some similarities with that of the previous subprovinces as well as with that of Baja California. The main area roughly corresponds to the major river basins of the state and contacts the Sierra Madre Occidental in the east. The rock bodies are described below, by age.

Precambrian and Paleozoic. The Precambrian Sonobari Metamorphic Complex in the Río Fuerte area is the oldest rock body in Mexico. A few relatively large Paleozoic sedimentary marine (clastic and calcareous) rock bodies occur in the Río Sinaloa and other areas but are not in contact with the Sonobari Metamorphic Complex. In one locality, the presence of fusulinids suggests the late Paleozoic age. Some of the Paleozoic bodies are metamorphosed.

Mesozoic. Rock bodies of the Mesozoic age are more numerous and lithologically diverse than the older ones. The most abundant are granitic and monzonitic plutons, of the (?Late) Cretaceous age, that occur north of the 24° parallel. The largest is a batholith located in the Ríos Culiacán and San Lorenzo drainage basins area in northernmost Sinaloa and in the vicinity of Mazatlan.

The volcanic (rhyolitic, andesitic to basaltic lavas, and pyroclasts) and sedimentary (clastic) marine bodies are less numerous: They form three small clusters in Mazatlán, the Río Sinaloa eastern area, and the Río Fuerte area. Some of them are metamorphosed. The last two clusters are contiguous to the plutons and show structural deformation.

The Cretaceous limestone bodies occur irregularly in this subprovince. Some are affected by the plutonic intrusions, and others seem to be associated with the volcanic rock bodies.

Cenozoic. Early Tertiary plutons and volcanic rock bodies similar to those of Cretaceous age occur in the Ríos El Fuerte, Piaxtla, and Presidio areas.

Mid-Cenozoic volcanic (rhyolitic ignimbrites) bodies occur spatially close to the Sierra Madre Occidental, to which they are genetically related. Late Tertiary volcanic bodies (andesitic to basaltic flows) usually occur surrounding the Cretaceous and Early Tertiary Plutons.

Geologic Story

The two age groups of Precambrian metamorphic bodies are interpreted (Anderson & Silver, 1979) to be parts of two northeast–southwest trending orogenic-magmatic belts eventually accreted to the North American Craton (oldest part of the North American Plate). Their anomalous geographic juxtaposition is explained by major left lateral faulting that occurred during the Jurassic; the resulting fault system has been named the Mohave-Sonora Megashear.

The Paleozoic rock bodies are interpreted to be evidence of an epicontinental sea that flooded a Precambrian basement and that perhaps is related to the sea bordering the southern and western parts of the North American Plate. The scarcity (or lack?) of Mid-Paleozoic rock bodies may represent shallowing or even emergence of much of this area. The configuration of the sea and of the southern margin of this plate for the whole era is largely unknown. Tectonic activity during the Late Permian is suggested by folding, faulting, and uplifting of the Paleozoic (and Precambrian) rock bodies.

The paleogeography for the Mesozoic remains poorly known. In northwestern Sonora (Antimonio area), the Late Triassic–Early Jurassic marine rock bodies represent a paleobay, whereas those of southeastern Hermosillo (San Marcial area) represent shallow marine and evaporitic conditions. The coal seams suggest paludal (coastal marsh?) sedimentation. The clastic and volcanic rock bodies testify to the existence of land above sea level.

The Jurassic volcanic and volcaniclastic bodies are interpreted to represent the results of convergent margin tectonics (plate collision) involving the subduction and partial melting of a hypothetical Paleopacific Plate (Farallon or PreFarallon?) under the North American Plate. The Mohave-Sonora Megashear was perhaps related to this collision. The collision zone would have to be found west of Sonora. If this hypothesis is correct, the Jurassic rock bodies are remnants of a volcanic island arc. The net result of the process must have caused regional uplift.

The Cretaceous rock bodies (largely plutonic and volcanic in composition with minor sedimentary marine formations) are interpreted as resulting from the same tectonic process, i.e., subduction of an eastward-moving hypothetical Paleopacific (Farallon?) Plate under the North American one. This process produced regional uplift at the end of the Medial Cretaceous (suggested by an unconformity underlying the Late Cretaceous rock bodies in the Cabullona Area), emplacement of the Late Cretaceous-Early

Tertiary plutons and volcanic bodies, and the overthrusting of Paleozoic and Early Tertiary bodies over Late Cretaceous ones (seen in the Cabullona and Naco areas) (Rangin, F., 1978). As a result of the regional uplift, this province wholly emerged from the sea during the Early Tertiary. By the Late Miocene, the subduction episode apparently ended and another tectonic regime started, as the spreading East Pacific Rise system began to deform the continent. The tensional stresses related to this process are thought to be responsible for the opening of the Gulf of California and development of the extensive block-faulting that generated the basin-and-range geomorphic pattern so characteristic of the province, as well as the effusion of the Late Tertiary volcanic bodies.

Selected References

Alencáster (1961); Anderson & Silver (1979, 1984); Anderson et al. (1978); Barrera (1931); Bonneau (1971); Brunner (1975, 1976); Clark et al. (1980); Cooper et al. (1952, 1954, 1965); Damon et al. (1962); De Cserna & Kent (1961); Drewes (1978); Fiala et al. (1982); Flores (1927); Fries (1962); Hisazumi (1929); Holguín (1978); Imlay (1939b); Keller (1973); King (1939); Mina (1950); Nieto et al. (1985); Ortlieb & Roldan (1981); Pastor (1930); Pineda et al. (1969); Rangin, C. (1984); Rangin, F. (1978); Rodríguez & Córdoba (1978); Roldán (1984); Roldán & Solano (1978); Salas (1970); Santillán (1929a); Stewart (1988); Stewart et al. (1984); Taliferro (1933); Urrutia (1986a); Valentine (1936); Wilson & Rocha (1946).

SIERRA MADRE OCCIDENTAL MORPHOTECTONIC PROVINCE

Geographic and Physiographic Setting

The Sierra Madre Occidental Morphotectonic Province is located between lat. 20°30'–31°20' N and long. 102°20'–109°40' W. It includes parts of Sonora, Chihuahua, Durango, Sinaloa, Zacatecas, Nayarit, and Jalisco States. It covers an area of 289,000 km² (14.68% of Mexico) making it the largest province of Mexico and one of the world's largest volcanic fields (Fig. 1.2). The altitudinal range is great, from 200 masl to slightly over 3,000 masl. The altitudinal band of 2,000–3,000 masl covers about 65% of the area. The province roughly defines an elongated rectangle (6.8 times longer than it is wide), about 1,300 km long and 190 km wide on the average, set in a NW–SE trend (N60°W–S60°E) that narrows northward from the 28° parallel.

The 10° latitudinal spread has little effect on the climate, the altitude being more significant. The dominant climate in the central region is temperate humid: Cfb without a dry season in the higher parts, and Cwa with a dry winter and a warm summer in the lower parts. Only in the southwestern portion, close to the coast, is the climate tropical (Aw with dry winter).

Physiographically, the province consists of a series of closely spaced volcanic sierras and plateaus that locally coalesce to form even larger ranges. Examples from the numerous individual sierras are the Sierra Tarahumara (southern Chihuahua); Cumbres del Gato; Altos de Tarahumara; and Sierra de Tepehuanes, de Durango, de Sombrerete, and de Fresnillo.

The individual sierras and plateaus are separated by river systems; those draining to the Pacific were cited previously. The Mezquital and Atengo Rivers (in the southern portion) should be added. The main east-draining rivers are tributaries of the Conchos and Nazas. The northern segment of this province is drained by the Casas Grandes and Santa María Rivers.

Geology

Few systematic detailed studies (e.g., McDowell & Clabaugh, 1979) exist for this vast area. Therefore only generalizations from the best known areas are offered (Fig. 1.3). Apparently the volcanic bodies that dominate this province can be stratigraphically divided into lower and upper volcanic complexes. The lower volcanic complex, dominantly andesitic in composition, includes lavas and pyroclastic sheets, though some are silicic ignimbrites. Structurally, this complex is slightly "folded" and intensely (block) faulted. Isotopic ages vary from 100 to 45 Ma. The outcrops of this unit are not numerous and have been identified only on the Pacific side of the Sierra Madre Occidental. The basement of the lower complex is largely unknown.

The upper volcanic complex unconformably lies on the lower complex. It consists of extensive silicic ignimbrite sequences up to 1,000 m thick and spans the entire province. Structurally, this complex is nearly horizontal and shows minor faulting. The emission of ignimbrites occurred through numerous calderas (estimated between 200 and 400), some up to 40 km in diameter. Isotopic ages of this complex are 54 to 34 Ma. Locally, Late Tertiary and Quaternary sedimentary clastic bodies, resting discordantly upon the volcanic complexes, occupy structural or topographic depressions.

Geologic Story

Both volcanic complexes are apparently the results of the tectonic process previously outlined, i.e., subduction of a hypothetical Paleopacific (Farallon) Plate under the North American Plate. Apparently, the subduction ceased 30 Ma (the age of the latest ignimbrites). The 45 to 34 Ma gap of volcanic activity is not well understood or well documented, nor is the change from a compressive tectonic regime obviously related to the plate collision. It is expressed by the andesitic composition of the lower volcanic complex to a seemingly distensive or tensional regime, related to the caldera formation and ignimbrite emplacement that generated the upper volcanic complex.

Selected References

Bobier & Robin (1983); Cameron et al. (1986, 1987); Cárdenas, F. (1983); Clabaugh (1972); Clark, et al. (1980); Clark & Goodell (1983); Cortés et al. (1964); Demant & Robin (1975); Henry et al. (1983); Henry & Price (1986); Keller et al. (1982); King (1939); Mauger (1983a, 1983b, 1983c); Mauger & Dayvault, (1983); McDowell & Clabaugh (1979, 1984); McDowel & Keizer (1977); Nieto et al. (1985); Rodríguez & Córdoba (1978); Santillán, (1929a); Swanson et al. (1978); Swanson & McDowell (1984); Urrutia, (1986a); Velasco (1956); Whal (1976).

CHIHUAHUAN-COAHUILAN PLATEAUS AND RANGES MORPHOTECTONIC PROVINCE

Geographic and Physiographic Setting

The Chihuahuan-Coahuilan Plateaus and Ranges Morphotectonic Province is bounded on the north by the 850 km long United States–Mexico boundary, to the west by the Sierra Madre Occidental, to the south by the Sierra Madre Oriental's Transverse Sector (actually the Mayrán–La Laguna lowlands), and to the east by the Gulf Coastal Plain (arbitrarily delimited by the 200 masl contour line) (Fig. 1.4). It includes the northeastern two-thirds of Chihuahua, most of Coahuila, the northwestern corner of Nuevo León, and northeastern Durango. Lying between lat. 26°00'–31°45' N and long. 101°31'–110°31' W, its area measures about 255,900 km² (about 12.52% of Mexico). The altitudinal range is 200 masl to slightly over 2,000 masl. The province is higher (>1,200 m) in the western half, due south of Big Bend,Texas (except along the Río Conchos Valley), than in the eastern half, where only Coahuila, west of long. 102°, lies above 1,200 m. The remainder gradually slopes down to the Coastal Plain. The climate is BShw (i.e., steppe-like, winter dry season with Mat >18°C) in the northeastern third, BWh (desert, Mat >18°C) in the central third, and BSk (steppe-like, Mat <18°C) in the eastern third.

The physiographic characterization of the province is disputed. It has been regarded as a northwestern extension of the Sierra Madre Oriental, a northern extension of the Central Plateau, a southern extension of the Rocky Mountain Cordillera, or a southern extension of the United States Basin and Range Province plus the Great Plains in the northeastern part. Because it does not correspond in land morphology and geology to the

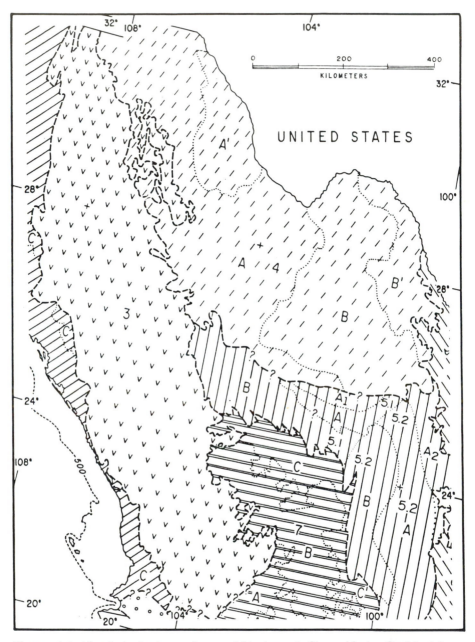

Figure 1.4. Morphotectonic provinces of Mexico. 3. Sierra Madre Occidental.
4. Chihuahua-Coahuila Plateaus and Ranges: A, Chihuahuan Subprovince;
B, Coahuilan Subprovince (A' and B' predominantly plains). Dotted line offshore
indicates the 500 m isobath.

provinces listed, it has been set apart here as an independent province. It may best be described as forming two subprovinces: Chihuahuan and Coahuilan.

Chihuahuan Subprovince

The Chihuahuan Subprovince includes low NW–SE trending block-fold ranges and block mesas separated by intermontane flat-lying basins and plateaus. The latter form lacustrine lowlands in the northwest. The ranges cluster in the north-central part of the province and are chiefly made up of Early Cretaceous sedimentary marine rock bodies. The remaining ranges and mesas are formed by Late Tertiary volcanic bodies. The largest forms the Llanos del Chilicote–Llanos de los Caballos Mesteños Mesa, south of Big Bend. The ranges and mesas appear to be predominantly horsts. The Río Conchos is the main river traversing this subprovince.

Coahuilan Subprovince

The Coahuilan Subprovince includes low dominantly NW–SE trending folded ranges and mesas separated by narrow intermontane flat-lying basins. The narrow north–south trending Mapimí Bolsón and adjacent lowlands separate the Chihuahuan from the Coahuilan subprovince. The narrow east–west trending La Laguna–Mayrán Desert area separates the Coahuilan Ranges and Mesas from the Sierra Madre Oriental's Transverse Sector (see below). The ranges and mesas, which are made up of Early Cretaceous sedimentary marine bodies, cluster in western Coahuila and usually stand above 1,200 masl. The largest feature is the Burro Mesa (south of the Río Bravo's northward bend). Others are the Australia and Cebollas Mesas (in southern Coahuila).

Altitudinally, lower ranges made up of Late Cretaceous sedimentary marine rock bodies are concentrated in northeastern and southeastern Coahuila (and adjacent Nuevo León). The remaining Nuevo León ranges are formed by Early Cretaceous rock bodies. The rivers, which occur mostly in the eastern part, are tributaries of the Bravo. The main rivers include the Sabinas-Salado, San Antonio, and Huizache-Sabinas Hidalgo.

Geology

Chihuahuan Subprovince

Precambrian and Paleozoic. Small Precambrian rock bodies have been described in Chihuahua (Villa Aldama and Carrizolillo areas (McDowell, pers. comm.). Rock bodies of Paleozoic age occur in only two areas: Sierra de la Boca Grande (in the northwest) and Villa Aldama-Placer de Guadalupe (in the west). They are some 320 km apart and lie hundreds of kilometers away from comparable bodies in the region (eastern Sonora, southwestern United States, and northern Mexico). Those of Placer de Guadalupe are the best known. They include an Ordovician-Pennsylvanian sequence (largely composed of limestone bodies deposited in shallow to moderately deep marine conditions and unconformably overlaid by a Permian sequence) made up of fossiliferous (reefal?) limestone, quartzitic sandstone, polymictic conglomerate, and rhyolite bodies, laid down in a marine environment. Both sequences are deformed in tight to recumbent, north–south trending folds that are partly overthrust eastward. In the Sierra de la Boca Grande, the sequence includes Mississippian to Permian sedimentary marine bodies deformed into east–west trending folds.

Mesozoic. Bodies of Jurassic age occur chiefly in Placer de Guadalupe. They form a sequence unconformably resting on the Paleozoic and are made up of micrite limestone (partly metamorphosed to marble), fine-grained red arkose, and a (basal) calclithitic conglomerate body laid down in a continental-marine transitional environment. Late Jurassic (marine) bodies occur sparsely in northern Chihuahua.

Cretaceous. Early Cretaceous sedimentary marine rock bodies crop out extensively in north-central Chihuahua. The older bodies are clastic and carbonate in composition and occur in the "western" (Villa Aldama) and "eastern" (Cuchillo Parado–San Antón) areas, whereas purer limestones of younger age occur in the intervening sierras. This pattern has been interpreted as a marine transgression over an area limited by uplands in the northwest-southeast. This area has been designated the Chihuahua Trough. These bodies are folded into NNW–SSE trending structures and are block-faulted in places.

The Late Cretaceous bodies, which are chiefly confined to the Ojinaga–Big Bend area, consist of clastic marine and continental rocks that bear dinosaur remains in some localities.

Cenozoic. The Cenozoic period is poorly known. The Cenozoic sequence unconformably rests on the Cretaceous one. The rock bodies include a volcanic set (of dominantly silicic composition and of Oligocene-Miocene age) that forms northwest–southeast trending sierras and mesas, frequently bound by high angle faults, and a clastic set (laid down by fluviolacustrine processes) that occupies the intermontane basins and lowlands. Near Ojinaga it contains the only Early Oligocene vertebrate fauna of Mexico. A few isolated small silicic plutons and mafic lavas complete the Cenozoic sequence, which is affected by block-faulting, forming sierras and mesas: horsts and intermontane basins and lowlands (grabens). Geochemically, the volcanic rocks are slightly more alkaline than those of the upper complex of the Sierra Madre Occidental. The age of the sedimentary unit may vary from the Early Oligocene to the Quaternary.

Coahuilan Subprovince

Paleozoic. Rock bodies of the Paleozoic-Permian age (Fig. 1.5) are composed of micritic limestone, fine- to medium-grained (and locally conglomeratic) arkosic and phyllarenitic sandstone, and andesitic lavic flows and sills; they crop out within 40 km northwest of Delicias (southwest Coahuila) form a thick sequence deposited in moderate to deep marine environments. The bodies are folded into a north-south trending anticline. The basement upon which this sequence was deposited is unknown as is its relation to other Late Paleozoic sequences in northern Mexico.

Mesozoic. Rock bodies of Jurassic age are also scarce and isolated, occurring mostly in central-eastern Coahuila. The largest outcrops are in the Sierra de La Gloria and Sierra Azul. They are made up of gypsum, fine arkosic to phyllarenitic sandstone and siltstone, micritic limestone, and coal-bearing siltstone and shale. All are interpreted as Late Jurassic near-shore, shallow marine deposits. The basement for these bodies is unknown. The same formal stratigraphic nomenclature developed in the Sierra Madre Oriental has been uncritically applied to these rock bodies, although no physical continuity or lithologic identity exists between these areas.

Cretaceous rock bodies are the most abundant and consist largely of limestones, siltstones, and shales deposited in a marine environment that was variable with respect to depth, proximity to the coast, terrigenous contamination, and reefal development. The sequence defines a complex depositional-geographical and historical pattern, controlled by a basin and adjacent lowlands of varying relief anddynamics. As a result, the Early Cretaceous bodies of most of Coahuila consist of admixed lime, terrigenous sandstones, and siltstones plus commonly dolomitized limestones. They indicate shallow marine near-shore deposition in a narrow epicontinental sea bounded by upland to the north and south. By contrast, in southeastern Coahuila the presence of extensive bioclastic limestone bodies followed to the south and east by clean, micritic to spathic limestone bodies (both of Early Cretaceous age) indicate large reef-complex developments that gave way to lime deposition in somewhat deeper, offshore conditions.

Medial Cretaceous (Albian-Cenomanian) bodies are of dominantly clean micritic and spathitic limestone composition in central Coahuila, and of dolomitic in northern (Burro Mesa) and southern (Australia and Caballo uplift) Coahuila with reefal limestone bodies interposed. The sedimento-logical interpretation indicates offshore deposition in central and near-shore shallow marine in northern and southern Coahuila. The stratigraphically lower units are limy and become marly to silty and shaley upward, including thick sequences of coal-bearing strata chiefly in the Sabinas–Nueva Rosita area. Sedimentologically, they indicate near-shore marine deposition, becoming even continental and locally marshy, where dinosaur remains have been found.

The Cretaceous bodies are folded into northwest-southeast anticlinoria in central and eastern Coahuila that become open, arc-like structures in the

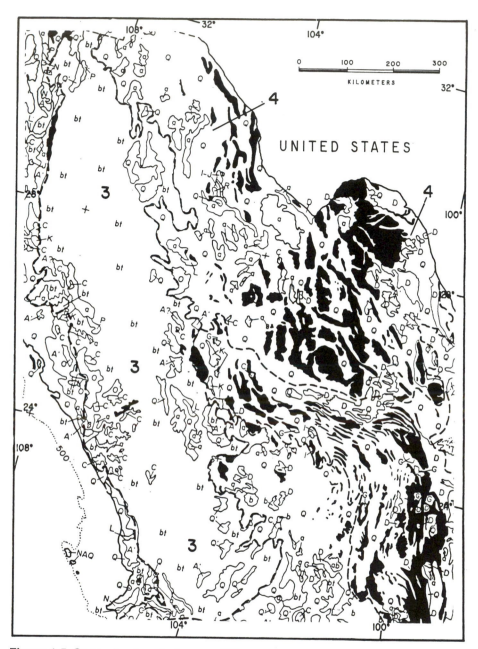

Figure 1.5. Generalized geologic map of Sierra Madre Occidental (3) and Chihua-hua-Coahuila Plateaus and Ranges Morphotectonic province (4) and adjacent areas. Modified from Comité de la Carta Geológica, 1976 and other sources.

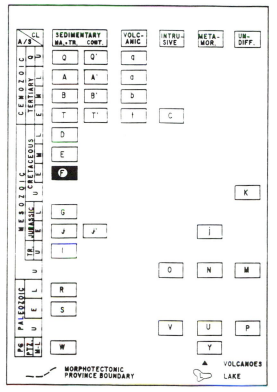

Abbreviations

A/S, age/stage*; CL, chief lithology of rock bodies; CONT, continental; E, early/lower*; L, late/upper*; M, middle/middle*; MA + TR, marine and transitional; METAMOR, metamorphic; P€, Precambrian; PTZ, Proterozoic; Q, Quaternary; TR, Triassic; UNDIFF, undifferentiated. Dotted line offshore indicates 500 m isobath.

*First term is geochronological, i.e., it refers to time; second term is chronostratigraphic, i.e., it pertains to kind of rock bodies bound by time divisions.

Modified from Comité de la Carta Geológica, 1976 and other sources.

Legend to Figure 1.5.

southern and northern parts. They frequently are bound by high-angle normal or reverse faults. The deformation of less competent Late Cretaceous bodies is more complex.

The formal nomenclature is largely the same as that of the Sierra Madre Oriental, although no physical continuity or lithologic identity justifies this usage.

Cenozoic. Rock bodies of the Cenozoic age form a poorly known dominantly clastic sequence laid down by fluviolacustrine processes during the Tertiary and Quaternary; they unconformably rest on the Cretaceous sequence. Tertiary silicic pyroclastic units of Mid(?) to Late Tertiary and Quaternary ages occur in the northeastern (Santa Elena-Víbora and Santo Domingo) and eastern (Sabinas) parts of the State. A few small felsic plutons and some basalts complete the roster of Cenozoic bodies.

Geologic Story

The Paleozoic rock bodies are too few and far between to tell an integrated coherent story. Furthermore, there is no objective evidence to link them.

The Early Paleozoic sequence at Placer de Guadalupe indicates marine deposition in an epicontinental sea whose floor belonged to an unidentified plate. Tectonic activity affected the area, and epicontinental marine deposition took place during the Permian. Late in this period compressive deformation occurred, and the area eventually became upland.

The Paleozoic sequences in northwestern Chihuahua and southwestern Coahuila are also marine but differ among themselves and from that of Placer de Guadalupe. The basements to which their individual sea floors belonged remain unidentified. In both places the Paleozoic sequence is unconformably overlaid by Cretaceous bodies.

The Jurassic sequences are few and far between (a distance of 400 km separates their largest and best known outcrops: Placer de Guadalupe, Chihuahua, and Sierra Azul, Coahuila). Therefore it is not warranted to assume physical continuity and similar plate allocation to their basements. Sedimentologically, their settings are similar: an upland gradually invaded by a shallow epicontinental sea. In Chihuahua, however, there is no record of Late Jurassic marine sedimentation, and in Coahuila it is restricted to the central-eastern portion. Yet there are several hypothetical paleogeographic and paleosedimentological reconstructions that make uncritical assumptions about the extent of the epicontinental sea(s), configurations of the coastline, sea floor topography, and plate tectonic allocation of their basement.

The Early and Medial Cretaceous record indicates marine deposition by an epicontinental sea of heterogeneous floor topography having shallow portions, probably associated with emergent uplands ("Coahuila Island" in the Las Delicias area, "Península de Tamaulipas" southeast of Big Bend, and northeastern Nuevo León), where shallow marine carbonate and terrigenous deposition ("platform deposits") has taken place and reef complexes have developed. Elsewhere, e.g., southeastern Coahuila, offshore carbonate deposition occurred. By the Late Cretaceous most of this province had undergone regional uplift, becoming upland except in the east and south where stagnant transitional deposition occurred, finally becoming continental.

The tectonic activity heralded by this regional uplift seems to have climaxed during the Early Tertiary (Laramide Orogeny), when the extensive folding that structurally characterizes the Cretaceous bodies (and others as well) apparently occurred. The precise age of the folding remains unknown because the intermontane basins thus formed have yet to be studied. The driving deformation mechanism in terms of modern plate tectonic theory also remains to be understood. The style of deformation suggests a compressive tectonic regime. The Mid-Tertiary volcanics in turn suggest a tensional regime, as does the block faulting so apparent in this province. Again, detailed plate-tectonic driving mechanisms remain to be defined. Regardless of their origin (folding, faulting, or both) fluviolacustrine sediments did accumulate in the intermontane basins.

Selected References

Amin (1987); Böse (1910, 1921, 1923); Bridges (1965); Bridges & De Ford (1961); Burckhardt (1930); Burrows (1910); Clark et al. (1980); Córdoba (1969, 1970); Córdoba et al. (1969); Daughtery (1963); De Cserna (1956); Denison et al. (1970); Díaz & Navarro (1964); Drewes (1978); Flawn (1961); Flawn & Maxwell (1958); Franco (1978); Garza (1973); González, J. (1952); González, R. (1976); Gries & Haenggi (1970); Gunderson et al. (1986); Handschy & Dyer (1987); Henry et al. (1983); Henry & Price (1986); Humphrey (1956); Imlay (1936, 1938a, b, 1940, 1944a,b); Kellum (1936, 1944); Kellum et al. (1936); Kelly (1936); King (1934); King & Adkins (1946); Ledezma (1967); Longoria (1984); Mauger et al. (1983a,b); Mayer (1967); McBride et al. (1974); Murray (1961); Murray et al. (1962); Navarro & Tovar (1970); Nimz et al. (1986); Padilla (1985, 1986); Ramírez & Acevedo (1957); Robeck et al. (1956); Rogers et al. (1957, 1961); Schmidt (1930); Seewald & Sundeen (1970); Singewald (1936); Smith (1970); Wall et al. (1961); Weber (1972); Young (1983); Zwanzinger (1978).

SIERRA MADRE ORIENTAL MORPHOTECTONIC PROVINCE

Geographic and Physiographic Setting

This Sierra Madre Oriental Morphotectonic Province includes the mountain ranges and associated lowlands of north-central Mexico and its western prolongation off the Monterrey–Saltillo area (Fig. 1.6). Thus defined, this province has a north-south (eastern) sector and an east-west (frequently designated the transverse) sector. The Eastern Sector lies between lat. 19°40'–26°00' N and long. 97°30'–101°20' W. Its area covers about. 77,000 km²; it is 250 km long and on average 140 km wide. This sector includes parts of the following states: northern Querétaro and Hidalgo, eastern San Luis Potosí, western Tamaulipas, southern Nuevo León, and the southeastern corner of Coahuila. The Transverse Sector lies between lat. 24°30'–26°00' N and long. 100°00'–105°00' W. Its area covers about 68,000 km²; it is 455 km long, and on average 120 km wide. It includes parts of the following states: southern Coahuila, northern Zacatecas, and Durango. The aggregate area of both sectors is about 145,500 km² (about 7.54% of Mexico).

The altitudinal range is wide (200 to >3,000 masl) and uneven. The lowest altitudinal range (200–1,000 m) corresponds to the eastern cuestas that join the Gulf Coastal Plain to the higher Sierras; the cuestas are narrowest (20 km wide on average) between lat. 23° and 24° N. To the south they show a two- to three-fold increase in width, and to the north the increase is even greater, where they form wide mesas (Mesa de Solis) and low-altitude sierras. The greater part of the province has an altitude of 1,000–2,000 masl: barely 20% of it stands at 2,000–3,000 m (or higher). This

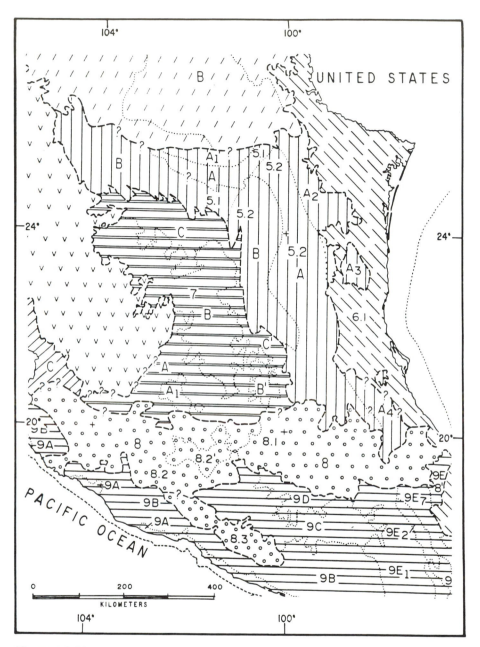

Figure 1.6. Morphotectonic provinces of Mexico. 5. Sierra Madre Oriental: 5.1, Transverse Sector; 5.2, Eastern Sector: A, Close Ridges Subprovince, B, Spaced Ridges Subprovince. 7, the Central Plateau: A, Southern Valleys and Sierras (A₁, Sierras de Arandas & de Tepatitlán); B, Central Sierras and Mesas (B₂, Sierras de San Luis Potosí); C, Northern Lowlands. 8, Trans-Mexican Volcanic Belt (see **Figure 1.10**): 8.1, Eastern Portion; 8.2, Western Portion; 8.2', Tarascan Plateau; 8.3, Southern Extension. 9, Sierra Madre del Sur (see **Figure 1.10**). Dotted line offshore indicates the 500 m isobath.

last portion forms a nearly continuous band in the northeastern part of the province and isolated Sierras elsewhere.

The climate in the Transverse Sector is predominantly BWh (desertic, Mat >18°C), and becomes BSk (arid or steppe-like, cooler) in the southern and western parts. The Eastern Sector is more varied and has a narrow north-south zone of climate Cfa (temperate, without a defined dry season) followed to the west by another narrow zone of Cwa climate (temperate with a winter dry season). The western half of the province has a BSh climate (dry or steppe-like with a dry winter season and Mat <18°C).

Physiographically, the Sierra Madre Oriental Province consists of folded ridges and intermontane, elongated valleys, and plateaus; their position and extent may best be described in terms of two subprovinces: Closely Spaced Ridges and Widely Spaced Ridges, which are present in both geographic sectors.

Subprovince of Closely Spaced Ridges

The Closely Spaced Ridges Subprovince includes the high sierras found in the northeastern part (e.g., Sierra de Playa Madero and Sierra de Las Mitras in southeastern Coahuila), the low ridges of eastern San Luis Potosí (northern Río Moctezuma drainage basin area), the eastern slope (Cuestas) to the high sierras, the low meseta-like folded mountains of Central Tamaulipas, and the medium high, folded and thrust ranges of northern Hidalgo and northeastern Querétaro.

The Subprovince of Widely Spaced Ridges

The Widely Spaced Ridges Subprovince occupies the remainder of the province. The ridges usually form high sierras (2,000–3,000 masl), e.g., Sierras of Mazapil and de la Candelaria in Zacatecas or de Catorce, de Guadalcazar, or de los Alaquines in San Luis, and Cuchillos de la Zarca, Peñoles, and San Pedro del Gallo in Durango. The Sierras are separated by broad, high-altitude flatlands that may correspond to valleys when traversed by rivers; or they may form endorheic basins, with or without present-day lakes, such as La Laguna de Mayrán Area, near Torreón. The eastern and southern portions of this subprovince grade into the Central Plateau Morphotectonic Province (see below) without noticeable breaks, making their boundary somewhat arbitrary.

River Systems

To complete the description of land features, it should be added that the rivers traversing this province have already been named in the description

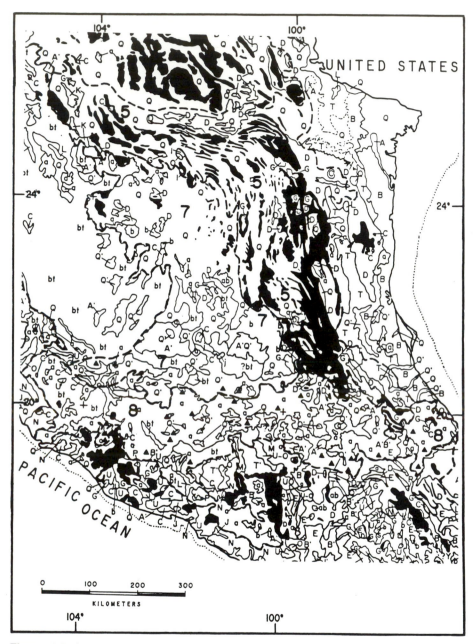

Figure 1.7. Generalized geologic map of Sierra Madre Oriental (5), Central Plateau (7), and Trans-Mexican Volcanic Belt (8) Morphotectonic Provinces and adjacent areas. Modified from Comité de la Carta Geológica, 1976 and other sources.

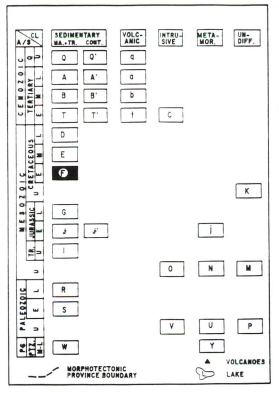

A/S, age/stage*; CL, chief lithology of rock bodies; CONT, continental; E, early/lower*; L, late/upper*; M, middle/middle*; MA + TR, marine and transitional; METAMOR, metamorphic; P℄, Precambrian; PTZ, Proterozoic; Q, Quaternary; TR, Triassic; UNDIFF, undifferentiated. Dotted line offshore indicates 500 m isobath.

*First term is geochronological, i.e., it refers to time; second term is chronostratigraphic, i.e., it pertains to kind of rock bodies bound by time divisions.

Modified from Comité de la Carta Geológica, 1976 and other sources.

Legend to Figure 1.7.

of the neighboring provinces. The Eastern Sector rivers are more developed than those of the Transverse Sector.

Geology

The scarcity of published studies that comprehensively describe this province is surprising, as it is geologically one of the better known parts of Mexico. Even the geologic delimitation of this province is a matter of debate. Up to the mid-1960s it was customary to include the folded ranges of southeastern Puebla and adjacent Veracruz and Oaxaca as part of the Sierra Madre Oriental (De Cserna, 1961; Murray, 1961), thus extending it south of the Trans-Mexican Volcanic Belt. This idea has now largely been abandoned. Several authors include as well the limestone mountain ranges and mesas of Coahuila and north-central Chihuahua within the Sierra Madre Oriental, conceptually linking it to the Rocky Mountain Cordillera and even regarding it as its southern continuation. That view is not supported here because the structure and geology of the Coahuila-Chihuahua Ranges and Mesas differ from those of the Sierra Madre

Oriental. Finally, the western and southern portions of this province grade into the Altiplanicie Mexicana, making its delimitation largely arbitrary. In summary, the Sierra Madre Oriental Morphotectonic Province, as defined here, includes the series of nearly continuous folded mountain ranges and associated land forms that extend north from the Trans-Mexican Volcanic Belt.

Most of the province is regionally made up of extensive Jurassic and Cretaceous sedimentary marine rock bodies that are complexly folded (and faulted) into anticlinoria and synclinoria, and by Cenozoic sedimentary clastic continental rock bodies that occupy the intermontane valleys, basins, and plateaus (Fig. 1.7). Older rock bodies occur only locally.

Precambrian and Paleozoic

The chief isolated exposures of the Precambrian and Paleozoic ages are those present in the northeastern corner of Zacatecas, the area west of Ciudad Victoria, Tamaulipas and the area of Calnalli, in northeastern Hidalgo. The last two are best known. The Ciudad Victoria pre-Mesozoic sequence includes the Late Proterozoic Novilla Gneiss, unconformably overlaid by the Cambrian La Presa Quartzite, in turn overlaid by the Granjeno Schist of apparently Late Paleozoic age and of disputed allochthonous or autochthonous origin. The post-Cambrian Paleozoic sequence consists of sedimentary marine, dominantly clastic (siltstones and phyllarenetic sandstones), and subordinately calcareous and silicic (novaculitic) slightly metamorphosed rock bodies, assigned to seven formations and whose ages span the whole Paleozoic Era. They have been interpreted as having been deposited in a relatively deep marine environment (the uppermost unit particularly so: the Guacamaya Formation whose floor, the Novilla Gneiss, was evidently part of the continental crust).

In the Concepción del Oro area (northeast Zacatecas), the metamorphic rock bodies (Caopas and Rodeo Schists) of supposed Precambrian or Paleozoic age are apparently Early Triassic in actuality.

The Calnali pre-Mesozoic sequence includes early Paleozoic metamorphic bodies (formed by gneiss, schist, and metaconglomerates). They are overlaid by sedimentary clastic marine units (siltstones and phyllarenitic sandstones) of Mississippian age and a 2,000 m thick body of marine phyllarenitic sandstones apparently deposited in a relatively deep marine environment.

Mesozoic–Triassic

Rock bodies supposedly of the Mesozoic–Triassic age are dominantly continental clastic units consisting of reddish, finely to coarsely grained phyllarenites that are largely unfossiliferous and stratigraphically inter-posed between the Paleozoic and Jurassic (fossiliferous). Thus largely on the basis of negative evidence, they have been assigned to the Triassic. These bodies occur in association with the Paleozoic in northwestern Zacatecas, west of Ciudad Victoria, and in the Calnali areas. The best known body is the Huizachal Formation from the area west of Ciudad Victoria, which is up to 2,000 m thick and composed of red, dominantly fine-grained phyllarenites interbedded by similarly composed conglomer-ates. Its lower and upper contacts are angular unconformities. Other supposed Triassic bodies show some resemblance to this formation and were either identified as such or correlated to it.

A rich continental vertebrate assemblage, the Huizachal local fauna (Hopson & Clark, 1985), has been described from the type area of the homonymous formation and has proved to be Middle Jurassic in age. Hence this finding leaves the Triassic age claimed for these bodies without objective support. This subject requires further study.

Jurassic

Rock bodies of the Jurassic age are more numerous and widespread than the older ones. They occur in the southeastern part of the province where they form a nearly continuous band extending from Tamazunchale (south-ernmost San Luis Potosí) to Teziutlán-Zacatlán (northern Puebla). Other exposures are in southern Nuevo León, starting south of Aramberri, proceeding northward to Galeana, and from there northwest and west, forming elongated discontinuous ranges in the southern portion of the Transverse Sector. Finally, other isolated units occur in the province mainly in and around Concepción del Oro (northern Zacatecas) west of Matehuala (northern San Luis Potosí) and east of Tolimán (Central Querétaro).

In the best known sequences (of the southeastern band and southern Nuevo León) the Medial to Late Jurassic sequence unconformably overlies the Huizachal Formation or its southward correlative. In the latter, the rock bodies (Joya, Zuluaga, Olvido, and La Casita Formations) consist of red, fine-grained litharenitic siltstones overlaid by dolomitic limestones, gyp-sum, and other evaporitic strata (Minas Viejas Gypsum) as well as marly and argillaceous sandstones of Medial to Late Jurassic age that represent continental and shallow marine depositions, which partly occur in an evaporitic basin, not unlike that of present-day Guerrero Negro in Baja California.

The lowest unit in the southwestern band is the Huayacocotla Forma-tion. It consists of fine- to coarse-grained phyllarenetic sandstones and conglomerates bearing Early Jurassic plants and ammonites. It is unconformably overlaid by a thin clastic unit, which is in turn unconformably overlaid by a thick sedimentary marine rock body largely composed of gray micritic limestones and calcareous siltstones, named the Tamán Formation of Medial to Late Jurassic age. Another Late Jurassic limestone body conformably covers this unit.

Cretaceous

The bulk of the province is formed by rock bodies of the Cretaceous age. They include an Early and a Late Cretaceous set. The Early Cretaceous set forms a wide, nearly continuous zone that occupies most of the Eastern Sector and becomes separated into numerous elongated and discontinuous east–west trending sierras in the Transverse Sector and in north–south to northwest–southwest trending sierras in the western part to the Eastern Sector. These discontinuous sierras seem to form a broad zone of major arcuate structures, starting with a northwest-southeast trend between lat. 21° and 23° N, shifting to a north–south (to northeast–southwest) trend between lat. 23°00'and 24°30' N, then making a pronounced flexure to the west in the area south of Monterrey (between lat. 24°30' N and long. 100° W). This peculiar arcuate structural pattern is one of the characteristic geologic features of the Sierra Madre Oriental Province as defined here.

The general stratigraphy is composed of different sequences throughout the length of the province. In the southern portion (Huayacocotla area, central-eastern Veracruz) the Early Cretaceous (Neocomian to Albian) includes the following Formations: Tamaulipas Inferior (cherty limestone), Otates (micritic limestone), Tamaulipas Superior (micritic limestone), and Cuesta del Cura (cherty limestone). This sequence is interpreted as representing marine deposition in moderately deep, offshore conditions.

In the central area of the Eastern Sector (the Ciudad Valles area in eastern San Luis Potosí) the Early Cretaceous sequence includes the following Formations: Guaxcoma-Tamaulipas Inferior (cherty limestone) and El Abra (reefal, highly fossiliferous limestones), interpreted as marine deposition occurring in a shallow tropical sea (evidenced by the reefal limestones).

Finally, in the eastern portion of the Transverse Sector the Early Cretaceous sequence includes these Formations: Taraises (limestones and calcareous siltstones), Cupido (reefal and micritic limestones), La Peña (cherty limestone), and the Cuesta del Cura–Tamaulipas Superior facies couplet (micritic-silty cherty limestones). They are interpreted as marine deposition in shallow marine conditions changing upward (and in time) to deeper (moderate depth) depositions.

The rock bodies of the Late Cretaceous (Cenomanian to Maestrichtian) occur chiefly in four separated areas: (1) Saltillo–General Cepeda in southern Coahuila where they form a wide, nearly continuous zone, located to the north of the arcuate structure previously mentioned; (2) the eastern part of the Eastern Sector, mainly in the Montemorelos–Ciudad Mante–Ciudad Valles area in western Tamaulipas and eastern San Luis

Potosí—here the bodies are nearly continuous; (3) Doctor Arroyo in southern Nuevo León where these bodies are isolated, surrounded by Tertiary continental clastic units; and (4) the south-eastern Sector (south of Ciudad del Maíz, San Luis Potosí to Cadereyta, Querétaro), where the elongated Late Cretaceous outcrop area alternates with that of the Early Cretaceous bodies.

In southern Coahuila, the Late Cretaceous sequence includes these Formations: Aurora-Viesca facies couplet (dolomite, micritic, and reefal limestone), Indidura (chert-bearing limestone), San Felipe (chert-bearing limestone), Parras Shale (calcareous shale), and the Difunta Group (largely clastic, both marine and continental). This sequence is interpreted as representing marine carbonate sedimentation in a shallow to moderately deep condition that shallows upward until becoming largely terrestrial.

The same nomenclature with minor changes (e.g., Méndez Shale instead of Parras Shale) has been applied in the other areas.

The Cenozoic is much less known than the Mesozoic. The rock bodies that occupy the intermontane valleys and basins consist of fine- to coarse-grained calclithitic sandstones and conglomerates (mainly in the lower part) and may become tuffaceous upward. They represent continental sedimentation in dominantly fluviolacustrine conditions. The Cenozoic sequence of rock bodies largely remains to be studied; its age spans the whole Cenozoic, and the interrelations remain unknown.

The structural geology of this province, particularly of the Cretaceous, is regionally known and largely involves folding into tight to open, linearly arranged anticlinoria and synclinoria that generate the arcuate structures described in connection with the Early Cretaceous sequence. The stratigraphic (and structural) succession is locally to subregionally affected by faulting (varied: block-faulting, normal and reverse high angle faulting, and even overthrusting), complicating the geology. Characteristically, the Cretaceous sequences show more structural deformation than does the Jurassic.

The geometry of the arcuate structures calls for a complex stress field to account for the observed structures; clearly, a major northeast force acted to generate the great flexure around Monterrey, and northern and eastern components must have been present to generate the Transverse and Eastern Sectors.

Because the western sierras of the Eastern Sector structurally consist of open to moderately tight folds, widely spaced in comparison with the eastern ones which show tighter folding and even thrusting, it appears that the compressive forces involved in the folding increased their intensity eastward. The age of this episode of structural deformation (or tectonic activity) occurred during the Early Tertiary and has been thought of as being related to the major episode of tectonic activity that generated the Rocky Mountain Cordillera, known as Laramide Orogeny; however, the nature of the supposed relations remains obscure.

The net result of this tectonic process was the regional uplift of the area and the generation of the folded arcuate ranges in both sectors of this

province. No information is available on the rate of uplift. In the intermontane lowlands, continental clastic deposition generated the rock bodies that
form the Cenozoic sequence.

Geologic Story

Despite the extensive geologic data base available for this province, only
some aspects of its geologic history are well understood; others remain
largely speculative. The information on the Precambrian and Paleozoic
rock bodies establishes only that, at least locally, the pre-Mesozoic basement was continental (cratonic?) and was invaded by an epicontinental sea
whose depth varied from shallow to deep. Whether the sequences observed in the three areas were part of the same craton and associated
Paleozoic orogenic belt and furthermore, whether this craton was actually
a southern extension of the (ancestral) North American Plate is not clear.
    Whatever the pre-Mesozoic tectonic evolution, the Middle Jurassic rock
bodies represent continental deposition (via fluviolacustrine sedimentary
processes) in an upland area. Again, no proof exists that the now-
discontinuous bodies were ever physically continuous, although custom-
arily they have been so interpreted. There is no published information on
their paleolatitudinal position. The unconformity separating these bodies
from those of the Middle and Late Jurassic indicates tectonic instability and
activity. The latter units represent shallow marine and continental deposition with the exception of the Late Jurassic in the Huayacocotla area, which
corresponds to fully marine (moderately deep) deposition, i.e., transgression of an epicontinental sea. This transgressive episode continued throughout the Cretaceous. The sea floor had an uneven topography, creating
shallower ("shelves") and relatively deeper areas ("basins") where different kinds of sediment were deposited. The Valles–San Luis Potosí Platform
is the main shallow area, and adjacent to it a reef complex developed.
Essentially the same depositional pattern continued until the Late Cretaceous (Campanian), when a regressive episode began and the epicontinental sea gradually withdrew eastward. This phenomenon seems to be linked
to the regional uplift of Mexico's mainland. The uplifted areas shed
terrigenous sediments eastward that eventually formed thick accumula-
tions in the receding seas (Parras and Méndez Shales, Difunta Group). Also
associated with the regional uplift were folding and faulting of the Cretaceous rock bodies, forming folded ranges during the Early Tertiary. It is
believed that this extensive folding involved truncation and gliding eastward ("decollement") of the Late Jurassic–Cretaceous marine sequence
upon the Middle and earlier Jurassic evaporitic bodies. Later during the
Cenozoic, after the ranges were formed, fluviolacustrine deposits formed
in the intermontane valleys and basins. This process continues today.
    There are, however, some problems. (1) The southern extent, configu-
ration, and paleolatitudinal position of the southern end of the North

American Plate are practically unknown for the Precambrian and Paleozoic. Therefore, it is a gratuitous assumption to interpret the Precambrian and Paleozoic basement of the Sierra Madre Oriental as part of this plate, as they are not precisely of the same composition, age, and structure. (2) The reconstructions of Pangea for the Late Paleozoic–Early Mesozoic (Permo-Triassic), as presently configured, leave no room for Mexico's continental area because this area is taken up by the major North American, South American, African, and Paleo-Pacific Plates and Platelets (minor Plates). (3) Therefore it has become more apparent that the regional major Jurassic-Cenozoic evolutionary tectonic events, such as the opening of the Middle Atlantic, the birth and development of the Gulf of Mexico, and the genesis and development of Middle America and the Antillean–Caribbean Region, can no longer be considered individually but must be regarded in an integrated manner, as all are related. Reports listed below have attempted this integration. (4) This goal cannot be accomplished by modeling the geologic and tectonic evolution of Mexico upon present-day configurations or upon untested assumptions conceptually derived from such configurations. (5) Current hypotheses concerning the origin and development of the Sierra Madre Oriental attempt to comply with the tenets of the plate tectonic theory but still make use of unproved assumptions and present-day configuration-based models. The hypotheses propose the following: (a) that this province's basement belongs to the North American Plate; (b) an epicontinental sea transgressed upon the Mexican territory during the Late Jurassic-Cretaceous, laying down an extensive sedimentary marine sequence supposedly in relation to the opening of the Mid-Atlantic and the birth and development of the Gulf of Mexico; (c) there exists an unevenness of the epicontinental sea floor with shallow flooded "shelves" or platforms and moderately deep to very deep "basins" (both seemingly actively subsiding coevally to sedimentation); (d) the "shelves" and basins are related to inherent paleotopographic features of this part of the North American Craton; and (e) subduction of the Paleo-Pacific Plate under the North American one caused the regional uplift and eventual folding and faulting of the Late Jurassic-Cretaceous sequence, hence generating the folded ranges of the province.

These observations remain to be clarified. Further work is needed to resolve the points raised in item 3, above.

Selected References

Baker (1971); Böse (1906, 1923); Burckhardt, (1930); Carillo (1961, 1965, 1971); Clemons & McLeroy (1966); De Cserna (1956); De Cserna et al. (1977); Erben (1956a); Galloway (1989); Heim (1940); Humphrey (1949, 1956); Imlay (1937, 1938a, 1939a, 1940, 1943a,b,c, 1944a); Imlay et al. (1948); Jones (1938); Limón (1950); McLeroy & Clemons (1965); Murray (1961); Padilla (1985, 1986); Palmer (1927); Rogers et al. (1957, 1961); Schmidt-Effing

(1980); Sutter (1980, 1984, 1987); Tardy (1973); Tardy et al. (1974, 1975, 1976); Thalmann (1935); Young (1983).

GULF COAST PLAIN MORPHOTECTONIC PROVINCE

Geographic and Physiographic Setting

The Gulf Coast Plain Morphotectonic Province includes the lowlands bordering the Gulf of Mexico, exclusive of the Yucatan Platform (Fig. 1.8). It is limited to the west and south by mountainous country belonging to several provinces. The Plain is divided into two sectors by the Teziutlán Massif (for some, the eastern end of the Trans-Mexican Volcanic Belt). The Northern Sector lies in the states of Veracruz and Tamaulipas between lat. 20°00'–26°00' N and long. 96°30'–100°20' W. It covers an area of about 107,000 km² and on average is 100 km wide, becoming wider northward and narrower southward. The Southern Sector lies in the states of Veracruz, Tabasco, and Campeche, with minor portions of Oaxaca and Chiapas included. It covers an area of about 63,000 km²., is about 90 km wide on the average, and is bounded by lat. 17°10'–19°20' N and long. 91°15'–96°46' W. The total area measures about 170,600 km² (about 8.66% of Mexico).

The altitudinal range is 0–200 masl. Locally, the plain surrounds isolated sierras (de Tamaulipas and de Tontima in the Northern Sector and the Sierra de los Tuxtlas in the Southern Sector) that do not belong to the plain, but will be briefly described.

The climate in the Southern Sector is dominantly Afw' (tropical rainy with no defined dry season) and becomes drier westward (Amw'). In the Northern Sector, the southern third has an Aw' climate (tropical with dry winter and rainy fall), the central portion has a Cw climate (temperate with dry winter), and the remaining portion has a Cx'w' climate (temperate, with little rain throughout the year).

Physiographically, the following land features of the plain are discernible: coastline, flats, river systems, and associated sierras. The coastline is largely affected by fluvial prograding sedimentation, related to the development of offshore islands, enclosing lagoons, estuaries, and marshes. The last two are common in the Southern Sector (Tabasco–Campeche). Among the main lagoons are Lagunas Madre, de Tamiahua, and de Términos. The flats are the dominant features and include flood plains, alluvial fans, marine terraces, and beach lands. They are largely sandy to muddy but may also be hard when the sediments are well consolidated.

All major rivers draining into the Gulf of Mexico traverse the plain; from north to south the following rivers are found: Bravo, Soto La Marina, Pánuco, Tuxpan, Cazones, and Tecolutla in the Northern Sector; and Jamapa, Papaloapan, Coatzacoalcos, Tonalá, Grijalva, and Usumacinta in the Southern Sector.

The main sierras within the Plain are the Sierra de Teziutlán and Sierra de Tuxtla, both of Late Cenozoic volcanic origin. The first is the landmark

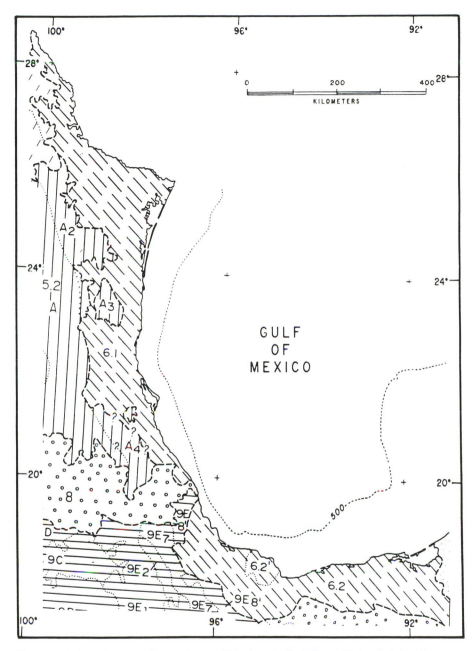

Figure 1.8. Morphotectonic provinces of Mexico. 6. Gulf Coast Plain: 6.1, Northern Section; 6.2, Southern Section; 6.2', Tuxtlas Region. Dotted line offshore indicates the 500 m isobath.

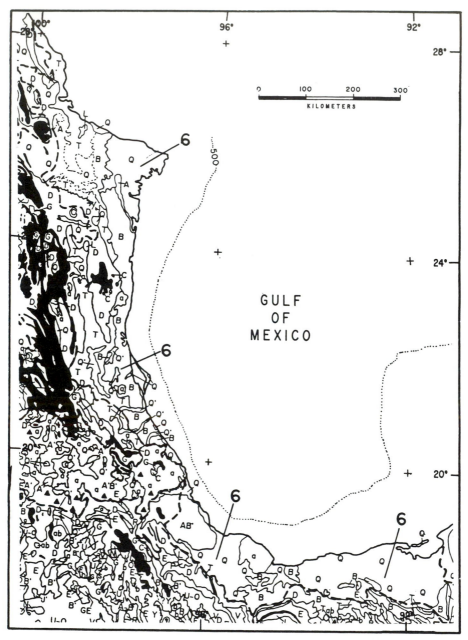

Figure 1.9. Generalized geologic map of Gulf Coast Plain Morphotectonic Province (6). Modified from Comité de la Carta Geológica, 1976 and other sources.

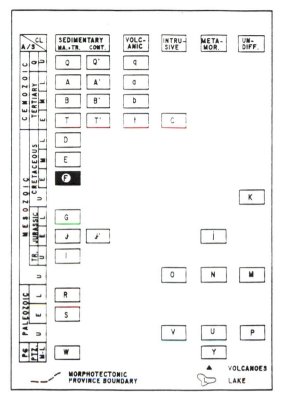

A/S, age/stage*; CL, chief lithology of rock bodies; CONT, continental; E, early/lower*; L, late/upper*; M, middle/middle*; MA + TR, marine and transitional; METAMOR, metamorphic; P€, Precambrian; PTZ, Proterozoic; Q, Quaternary; TR, Triassic; UNDIFF, undifferentiated. Dotted line offshore indicates 500 m isobath.

*First term is geochronological, i.e., it refers to time; second term is chronostratigraphic, i.e., it pertains to kind of rock bodies bound by time divisions.

Modified from Comité de la Carta Geológica, 1976 and other sources.

Legend to Figure 1.9.

that separates this province into its two sectors, and the other marks the northern end of the Isthmus of Tehuantepec. Other minor sierras include San Carlos, Las Rusias, de Tamaulipas, and de Otontepec, all in the Northern Sector. None exceeds 450 masl. The Gulf and Pacific Coastal plains are only 30 km apart in the Isthmian area (separated by the Sierra de Niltepec) as the Gulf Coast penetrates deeply southward here.

The Gulf Coast Plain continues seaward into the gently sloping Gulf Continental Shelf. Northwestward of the Isthmus of Tehuantepec the shelf is narrow (some 40 km wide on average) and runs closely parallel to the coast. East of the isthmus, the shelf widens considerably, meeting that of Yucatan around lat. 20°N.

Geology

This area, because of its oil riches, is one of the best known in Mexico. Numerous rock bodies have been stratigraphically differentiated and named. Their outcrops show a regional pattern of band-like zones of successively younger age gulfward (Fig. 1.9). Lithologically, they consist of marine limestones, sandstones, siltstones, and claystones, which make up tabular bodies that gently dip gulfward. The Cenozoic sequence unconformably rests on the Cretaceous. The distribution of exposures by age is unequal in size and distribution. The oldest units are Paleocene and are exposed only in the Northern Sector, in a 20–30 km wide band. The unit appears best exposed along the Río Pánuco. The Eocene unit is much more developed in the Northern Sector as well, where it is nearly continuously exposed. It reaches deeply inland along the Río Bravo area (lat. 26°–28° N), indicating that the Eocene sea and coastline reached inland as far as Laredo, Texas, thus defining a paleobay (Río Bravo Embayment). The Oligocene "band" is also better developed in the Northern Sector, as it is wider in its southern half, as is the Miocene "band" in the Northern Sector (more so in the northern half). In the Southern Sector the Miocene rock bodies are centered around Minatitlán, Villahermosa, and Tenosique. They do not form a continuous outcrop. The Pliocene "band" is narrowest here, only discernible in the Northern Sector (Río San Fernando area in part).

The Quaternary bodies are typically transitional, being generated in shore and river environments. They are much more extensive in the Southern Sector than in the Northern Sector, where they are present mainly around river Bravo.

Subsurface information shows significant thickenings of the Cenozoic sequences in both sectors that must have corresponded to areas of enhanced subsidence. Such areas include the Burgos, Tampico-Misantla, Veracruz, and Tabasco-Campeche Basins, all of which hold important oil reservoirs. In the southern portion of the Northern Sector there is a thick clastic marine body deposited in deep waters (the Chicontepec Formation of the Early Tertiary age) that seems to represent erosion of adjacent uplands.

Igneous rock bodies are also found in this province. Sierra de Teziutlán and Sierra de Los Tuxtlas, the largest, are made up of andesitic to basaltic alkaline lavas. It is debated whether they belong to the Trans-Mexican Volcanic Belt. The Teziutlán area is geologically complex, includes a Pre-Jurassic(?) pluton that is unconformably covered by Jurassic and Cretaceous sedimentary rock bodies and intruded by an Early Tertiary Pluton. A Paleocene sedimentary marine unit completes the sequence. The Minor Sierras (Tepizintla, Villa Aldama, and Llera areas) are composed of alkaline basalt flows. The Sierra de San Carlos and many small hills southeast of the Sierra de Las Rusias are small plutons (of nepheline syenitic to gabbroic composition) emplaced during the Mid-Tertiary (28–30 Ma).

Geologic Story

The sedimentary rock bodies of this province represent marine deposition upon a continental shelf in a tropical regressive sea (i.e., an epicontinental sea retreating from the continent toward the open ocean) that persisted throughout the Cenozoic. The coastline had paleobays (e.g., Río Bravo, Burgos, Tampico, Misantla, Veracruz, and Tabasco-Campeche). The sea floor had down-warped areas (coinciding with the paleobays) and even truly abyssal fossae (e.g., Chicontepec) where thick sedimentary piles accumulated.

The Cretaceous-Paleocene unconformity indicates tectonic activity at that time, which is expressed by tilting or uplift of the sea floor. The marine regression indicates regional uplift of Mexico's mainland advancing from west to east. The Quaternary features were generated in a (subdued) coastal plain by rapid sea-level changes (related to glaciation) and the interplay of fluvial and marine environments. The Late Tertiary volcanic bodies were emplaced in upland areas where the regression had already occurred (Northern Sector) or they were positive (topographic highs) areas never submerged during the Cenozoic (Teziutlán area). The Mid-Cenozoic plutons were also emplaced in upland areas. In summary, the Gulf Coast Plain is the emerged (emerging?) continental shelf bordering the Gulf of Mexico.

According to plate tectonic theory, the Atlantic continental margin (including that of the Gulf) is interpreted as passive (i.e., occurring on the trailing side of a moving plate yet not totally inactive) and is thought to have behaved as such throughout the Cenozoic. Yet there are geologic features in the Gulf Coast Plain that seem not to fit this interpretation. The active subsidence of the basins and fossae (and indeed their origin) and the volcanic and plutonic activity, as evidenced by the resulting rock bodies, are some such features. Tensional crustal fracturing (related to the proposed Río Grande Rift Valley?), which ultimately related to plate motions, has been offered as an explanation, but there is no agreement on the subject.

Selected References

Antoine et al. (1974); Baker (1971); Barker & Blow (1975); Barnetche & Illing (1956); Belt (1925); Bloomfield & Cepeda (1973); Böse (1905); Bryant et al. (1968, 1969); Buffler et al. (1979); Burckhardt (1930); Cabrera (1953); Calderón (1951); Castillo (1955); Contreras (1958); Coogan et al. (1972); Cruz et al. (1977); Dickerson & Kew (1917); Dumble (1918); Enos (1974); Galloway (1989); Garrison & Martin (1973); González (1976); Hernández, S. (1959); Lesser (1951); Limón (1950); Masson & Alencáster (1951); Mena (1960, 1962); Meneses de Gyves (1980); Mossman & Viniegra (1976); Muir (1936); Murray (1961); Nájera (1952); Padilla (1985, 1986); Ríos (1952, 1959); Robin (1976); Salas & López (1951); Salmerón (1970); Salvador (1987); Schmidt-Effing (1980); Staub (1928); Thorpe (1977); Viniegra (1965, 1971); Viniegra & Castillo (1975); Walpert & Rowett (1972); Wilhem & Ewing (1972); Wilson (1987).

CENTRAL PLATEAU MORPHOTECTONIC PROVINCE

Geographic and Physiographic Setting

The Central Plateau Morphotectonic Province is limited by the Sierra Madre Occidental and Oriental and by the Trans-Mexican Volcanic Belt. It lies between lat. 21°00'–24°00' N and long. 100°00'–104°00' W (Fig 1.6). The province includes most of Guanajuato and western Querétaro, western San Luis Potosí, eastern Zacatecas, eastern Aguascalientes, and eastern Jalisco. Its area conforms to a northwest–southeast oriented parallelogram, with dimensions of 135 km north–south and 225 km east–west, measuring about 85,300 km^2 (i.e., 4.33'% of Mexico). Altitudinally, the province ranges from 1,000 to 3,300 masl, having one small low (1,000–2,000 m) zone in the north and one large high (2,000–3,000 m) in the northwest-southeasternly zone (Sierra de Zacatecas- Guanajuato and associated dams) and one small low zone (1,000–2,000 m) in the south and southwest that includes higher isolated sierras. The climate is BSh (i.e., arid or steppe-like, with a summer dry season and a Mat <18°C). Physiographically, this province can be divided into three subprovinces roughly corresponding to the altitudinal zones: Southeastern Valleys, Central Sierras and Mesas, and Northern Lowlands.

Southern Valleys and Sierras

The Southern Valleys and Sierras zone includes the drainage basin areas of the Río Lerma-Santiago Tributaries (Río Verde, San Juan Turbio, Irapuato, and Apaxco). The valleys are cut in both a plain and rolling land country (the plateau), altitudinally ranging between 1,000 and 2,000 masl, that surrounds the Sierras de Tepatitlán and de Penjamo, which are volcanic mesas.

Central Sierras and Mesas

The Central Sierras and Mesas form a contiguous land feature, obliquely oriented (northwest-southeast), altitudinally ranging between 2,000 and 3,000 masl. The main Sierras include the Zacatecas and the Guanajuato (both include several minor sierras) and the Mesa de La Herradura and Sierra del Bozal in eastern San Luis Potosí. They are complex mountains containing diverse geologic bodies.

Northern Lowlands

The Northern Lowlands include the valley of the river Nieves and the rolling land to the east, which is dotted by low, isolated mountains and hills of diverse geologic composition. The eastern portion is an endorheic basin.

Geology

This province has a complex and diverse make-up that can best be described according to the physiographic subdivision presented above as well as chronologically within each subprovince (Fig. 1.7).

Southern Valleys and Sierras Subprovince

Rock bodies of Pre-Cenozoic age occur only in the eastern part (Río de la Laja Area, Guanajuato). They are small, sedimentary, and apparently of Jurassic(?) age. The Cenozoic rock bodies can be divided into three sets. The first set consists of silicic ignimbrites and tuff-bodies of Mid-Cenozoic age that largely form the Sierras de Tepatitlán, Penjamo and San Miguel de Allende. They are similar to those of the Sierra Madre Occidental upper volcanic complex. The second set is formed by continental clastic bodies of fluviolacustrine origin, probably of Late Tertiary age, occurring in the San Juan de Los Lagos area (northern Jalisco). The last set is formed by Late Tertiary-Quaternary sedimentary rock bodies occupying the lowlands. A large fossil mammal fauna collected in the eastern part (San Miguel de Allende area) establishes its age as Hemphillian to Blancan (latest Miocene, Pliocene, and Early Pleistocene).

The structural geology is even less well known. It appears that the limit between the Trans-Mexican Volcanic Belt and the Central Plateau is a major fault zone (partly documented in the Río Lerma-Santiago area). The narrow valley of Aguascalientes has been interpreted as a graben structure interposed between the Sierra Madre Occidental and the Central Sierras and Mesas subprovinces. It appears that the narrow valley separating the western part of the Trans-Mexican Volcanic Belt and the Sierra Madre Occidental between Guadalajara-Tequila and the Río Santiago is a graben. Finally, the southwestern limit of the Sierra de Guanajuato has also been interpreted as a fault. (Should all these interpretations be correct, the whole subprovince could be a large graben broken up by local horsts, consisting of volcanic sierras.)

Central Sierras and Mesas Subprovince

The Central Sierras and Mesas Subprovince is the largest and most complex subprovince. It includes numerous sierras and mesas, regionally (if informally) known as the Sierras de Zacatecas, San Luis Potosí, and Guanajuato. A voluminous literature on the mining districts and associated areas exists because of the mineral wealth concentrated in the region. A comprehensive integrated synthesis, however, has not been attempted, and major uncertainties remain because the Pre-Cenozoic bodies occur in isolation and do not appear to form a discernible pattern.

Pre-Cenozoic rock bodies are metamorphic, plutonic, and sedimentary marine in origin; their age ranges from Pretriassic(?) to Cretaceous, and they characteristically occur as discontinuous, isolated small bodies.

Pretriassic–Triassic. The only proved Pretriassic rock body occurs northwest of the city of Zacatecas. It is small, made up of sericitic schist, and unconformably underlies small units of quartzitic sandstone, argillocarbonaceous, mollusk-bearing phyllites, and phyllarenitic conglomerates. The mollusks appear to be of early Late Triassic (Carnian) age, but the age of the schistose body remains unknown. Another body said to be Triassic is the Chilitos Formation of the Fresnillo area, interpreted as a metamorphosed submarine basaltic andesitic lava flow ("greenstone") sequence; no evidence is given for the age assignment. The greenstone body in the Zacatecas area has also been assigned to the Triassic without evidence.

Jurassic. Rock bodies assigned to the Jurassic Period include sedimentary marine and low-grade metamorphic units, but only the first is unquestionably assigned. Such units occur as small outcrops near the cities of Zacatecas, Charcas, and San Luis Potosí and in southeastern Zacatecas state, where the same Mesozoic sequence recognized in the Sierra Madre Oriental is thought to be present. A small metamorphic body of ultrabasic (pyroxenitic) composition occurring in easternmost Jalisco (at the longitude of León, Guanajuato) is said to be Late Jurassic, but no evidence has been offered.

Cretaceous. Rock bodies assigned to the Cretaceous Period are sedimentary marine and low-grade metamorphic. The first are more numerous and cluster in southeastern Zacatecas and adjacent San Luis Potosí states as well as in northeastern Zacatecas (Sombrerete-Nieves area). Again, the same Early and Late Cretaceous sequences present in the Sierra Madre Oriental are supposedly recognized in this province, although as for those in the Jurassic sequence, no physical continuity exists between the sequences of the two provinces nor is there lithologic identity among them.

The so-called low-grade metamorphic bodies (called the Esperanza Formation) form the bulk of the Sierra de Guanajuato (between Lagos de Moreno and the city of Guanajuato), covering an area some 900 km², for which detailed information has been published on only about 4% of the area (the Guanajuato Mining District). The sierra includes a sedimentary marine part (dark gray argillaceous and carbonaceous shale interbedded by a dark micritic limestone) and a low- to moderate-grade metamorphic part (sericitic phyllite to serpentine schist, associated quartzites and metabasal "greenstones," and marble). Both parts are interpreted as intergradational, but the shaley part is volumetrically dominant. Elsewhere (road between Valenciana and Santa Rosa, outside the mapped area of the Guanajuato Mining District) it exhibits tight folding and intercalated silicic lava flows. Some (unpublished) work on this sierra has disclosed the presence of a sequence resembling the one described in which phyllarenitic sandstones are dominant in the sedimentary part. The metamorphic portion consists of green schists, greenstones with relict pillow-lava structures, and silica-rich schists, interpreted as metamorphosed andesitic to basaltic lavas and pyroclastic sheets laid down on a sea floor. Identifiable radiolarians found in the Arperos area (some 25 km northwest of Silao) allow dating the sequence here as Cretaceous and suggest that the whole metavolcanic–sedimentary complex of the Guanajuato Sierra may be of this age.

Cenozoic. A well-defined unconformity separates the rock bodies of the Cenozoic age from earlier ones. The Cenozoic rock bodies in this subprovince form a widespread sequence that has been differentiated in only a few isolated areas (the cities of Zacatecas, Fresnillo, San Luis Potosí, and Guanajuato). The general pattern includes a continental coarse clastic (polymictic in composition) unit(s), unconformably overlaid by a volcanic sequence (lavic and pyroclastic of andesitic to rhyolitic composition), usually forming mesas (and occasionally peaks). These formations occur chiefly in northwestern Querétaro, eastern Guanajuato, and southern San Luis Potosí (here isotopic dating indicates that they are Mid-Tertiary in age). Locally, other volcanic flows (lavas of andesitic to basanitic composition) may cover the mesas or the fine-grained (locally coarse-grained) clastic deposits that developed in the adjacent lowlands. This sequence was intruded by small to medium-size plutons (of granitic to dioritic or, even more, mafic composition) that show partial metamorphism and that seem to cut only the earliest clastic units. Another set of plutons, also of small to medium size and andesitic to rhyolitic composition, intruded the volcanic sequence.

The Cenozoic bodies have local names such as Conglomerate Rojo de Zacatecas, de Guanajuato, Comanja Granite, Zacatecas, Andesitic Porphyry, and La Luz-Mota Diorite; but because they have been reliably dated and mapped in only a few places their age, correlation, and other stratigraphic relations remain largely unknown.

In the Sierra de Guanajuato, the Red Conglomerate yielded fossil mammals that date it as Late Eocene. The Comanja Granite, also in this sierra, gave an isotopic age of 53 Ma (Early Eocene), and in San Luis Potosí the lower volcanic sequence gave Oligocene isotopic ages, roughly equivalent to those of the Sierra Madre Occidental upper volcanic complex. Finally, in the San Miguel de Allende area, fossil mammals date the upper fine clastic sequence as latest Miocene to Early Pleistocene.

Structural geology of this subprovince is poorly known. The Pre-Cenozoic metamorphic rock bodies do not seem to show a consistent structural pattern; for instance, the Chilitos Formation is said to have pillows overturned to the northwest, whereas the schists of Zacatecas show foliation directions without a discernible pattern. In the Sierra de Guanajuato the foliation trend is said to be northwest–southeast, but in the Guanajuato District the metamorphic foliation trend is north–south.

The sedimentary marine Jurassic and Cretaceous bodies form anticlines and synclines that may be open, tight, or even overturned. Their axis trends vary from northwest–southeast to north–south or even to nearly east–west in the Sombrerete area. These rock bodies also show secondary faulting along a northwest–southeast trend that in some places becomes almost north–south (as in Moctezuma, San Luis Potosí). The plutons are too scarce to discern a pattern for the whole subprovince, but those of the Sierra de Guanajuato roughly trend northwest–southeast.

The Cenozoic rock bodies are dominantly affected by faulting, which appears to conform to a broadly northwest–southeast trending system, seemingly associated with another system at roughly right angles to it. The faulting is of the block type with high angle normal and reverse faults, so that grabens and horsts were formed. Some of the grabens have been documented e.g., the Villa de Reyes and Carmen de Bledo, S.L.P., or the Sauceda Graben, Guanajuato and appear to have been coeval to the Mid-Tertiary volcanic sequence. Some major faults have been partly documented, e.g., the Bajío Fault, which is the southwest limit of the Sierra de Guanajuato. It appears, then, that the whole subprovince may correspond to a complex system of horsts and grabens oriented at different attitudes, and that the resulting topographic lows have been filled by either clastic debris or post-faulting volcanic products.

The regional tectonic development remains largely speculative. The Chilitos Formation and the ultrabasic body in Guanajuato undoubtedly represent ocean floor (i.e., oceanic crust) and may indicate a spreading tectonic regime; yet its plate affiliation is unknown. The metamorphic rock bodies represent a convergent tectonic regime (plate collision), but the Plates involved remain unknown. The regional folding, faulting, and pluton emplacements seem to be coeval and of Early Tertiary age (Late Eocene?), again in a compressive (convergent?) tectonic regime. The Mid- and Late-Cenozoic volcanic units and the major faulting that produces grabens and horsts corresponds to a tensional (distensive) tectonic regime whose regional significance is not well understood.

Northern Lowland Subprovince

Cretaceous. The small, relatively low, elongated hills that dot the subprovince are largely composed of Cretaceous sedimentary marine rock bodies that have been folded into anticlines and synclines trending northwest–southeast. The units of Late Cretaceous age occur in the central part, and the Early Cretaceous ones occur toward both east and west. These bodies are said to represent the same sequence observed in the Transverse Sector of the Sierra Madre Oriental.

Cenozoic rock bodies are poorly known and largely remain to be differentiated. The volcanic units include a Mid-Cenozoic (?) set occurring in the west, chiefly formed by silicic pyroclastic units, and a Late(?) Cenozoic set, made of andesitic to basaltic lavic flows concentrated in the northwest. The continental clastic bodies also include a coarse-grained set and a seemingly younger (Late Cenozoic?) finer-grained set. The clastic bodies form the lowlands.

Geologic Story

The oldest rock bodies of the province are restricted to Zacatecas and represent marine (?) fine clastic sediments metamorphosed and uplifted before the early Late Triassic. The terrain thus formed was temporarily covered by an epicontinental sea by this time (or shortly after). The isolated occurrence of these bodies renders futile attempts to link them to any plate.

The sedimentary marine Jurassic rock bodies represent a sedimentary environment not unlike that described for comparable bodies in the Sierra Madre Oriental. The basement for this sequence and its plate affiliation remain unknown. Another unsettled problem is the precise time–space relation of the Jurassic bodies among themselves and with those of the Sierra Madre Oriental, their claimed identity notwithstanding.

Also perhaps in the Jurassic, ocean crust was present in two small, separated areas of the province, denoting a spreading (divergent plate margin) tectonic regime that later changed to a convergent one, resulting in their metamorphism. Again, the tectonic Plates involved and their evolution remain unknown.

The sedimentary marine Cretaceous rock bodies suggest a story similar to that of the Sierra Madre Oriental. Like there, it is plagued by the same uncertainties, compounded by unknown precise time–space relations among the rock units and with those of the Sierra Madre Oriental.

The metamorphic Cretaceous(?) sequence indicates (if indeed the sedimentary bodies grade into the metamorphic ones) that marine sedimentation took place on a sea floor of unknown plate affiliation and was associated with marine volcanism occurring in a convergent tectonic regime. Again, the plates involved are unknown.

The convergent conditions remained in the Late Cretaceous and Early Tertiary, producing the partial metamorphism of the sequence and the emplacement and partial metamorphism of the Early Tertiary plutons. Folding of the sedimentary sequence also took place during the Early Tertiary, perhaps in association with this tectonic process. The silicic volcanic activity of the early Mid-Tertiary can also be envisioned in this process as a result of subduction and partial melting of an oceanic plate under a continental one. The relation of this activity to the one that generated the Sierra Madre Occidental is also unknown.

By Late Tertiary (latest Miocene?) extensive block-faulting occurred, generating the horsts and grabens that give this province its typical landscape. This process is evidence of a change in the tectonic regime from a compressive to a distensive (or tensional) one whose regional significance is not well understood. Finally, clastic (fluviolacustrine) sedimentation took place in the grabens, sometimes in connection with andesitic-basaltic volcanism, also interpreted as typical of regions where distensive tectonic processes are occurring.

Selected References

Aranda et al. (1983); Böse (1923); Burckhardt (1930); Cepeda (1967); Clemons & McLeroy (1966); Córdoba (1963); de Cserna (1976); de Cserna & Bello (1963); Echegoyen et al. (1970); Edwards (1955); Enciso (1963, 1968); Fries et al. (1955); Gallagher (1952); García (1968); Geyne et al. (1963); González & White (1947); Hernández L. (1981); Imlay (1938a, 1944a); Instituto de Geol. (1987a); Labarthe et al. (1982); McLeroy & Clemons (1965); Martínez (1972); Ordóñez (1901); Pérez et al. (1961); Ranson et al. (1982); Rogers et al. (1957, 1961, 1963); Schulze (1953); Segerstrom (1961); Serna et al. (1959); Waitz(1926).

TRANS-MEXICAN VOLCANIC BELT MORPHOTECTONIC PROVINCE

Geographic and Physiographic Setting

The Trans-Mexican Volcanic Belt Morphotectonic Province located between lat. 17°30'–20°25' N and long. 96°20'–105°20' W spans the country from coast to coast (Figs 1.6 and 1.10). and includes parts of the following states: southern Jalisco and Nayarit, most of Michoacán, northeastern Colima, western Guerrero, Morelos, Distrito Federal, Mexico, southern Querétaro, southern Guanajuato, southern Hidalgo, Tlaxcala, northern Puebla, and adjacent Veracruz. It is oriented generally east–west between lat. 19°00' and 21°00' N, and is about 930 km long and 120 km wide on average, covering an area of about 175,700 km² (about 9.17% of Mexico). Altitudinally, it ranges from 1,000 to over 5,000 masl, but the dominant altitudinal band is between 1,500 and 2,500 m. The eastern half of the province chiefly includes this band, whereas the western half (except for

---

Figure 1.10. Morphotectonic provinces of Mexico. 8. Trans-Mexican Volcanic Belt: 8.1, Eastern Portion; 8.2, Western Portion; 8.2', Tarascan Plateau; 8.3, Southern Extension. 9. Sierra Madre del Sur: 9A, Pacific Coastal Plain; 9B, Pacific Ranges and Cuestas; 9C, Balsas Depression; 9D, Northern Balsas Upland; 9E, Oaxaca-Puebla Uplands; $9E_1$, Mixteco-Zapotecan Sierras and Highlands; $9E_2$, La Cañada or Tehuacán-Cuicatlán-Quiotepec Depression; $9E_3$, Central Valleys; $9E_4$, Miahuatlán Ranges; $9E_5$, Pacific Cuestas; $9E_6$, Sierra de Ixtlán; $9E_7$, Northeastern Ranges; $9E_8$, Northeastern Cuestas; $9E_9$, Rio Lachixonaxe Uplands and Associated Valleys. Dotted line offshore indicates the 500 m isobath.

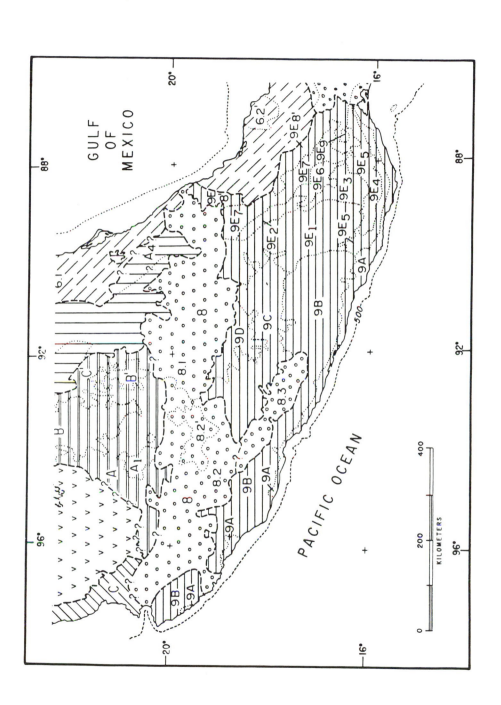

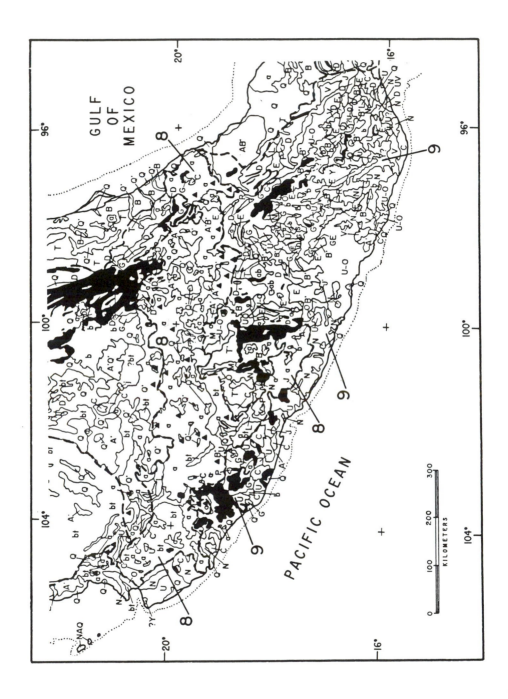

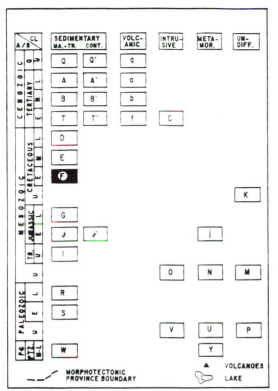

Abbreviations

A/S, age/stage*; CL, chief lithology of rock bodies; CONT, continental; E, early/lower*; L, late/upper*; M, middle/middle*; MA + TR, marine and transitional; METAMOR, metamorphic; PЄ, Precambrian; PTZ, Proterozoic; Q, Quaternary; TR, Triassic; UNDIFF, undifferentiated. Dotted line offshore indicates 500 m isobath.

*First term is geochronological, i.e., it refers to time; second term is chronostratigraphic, i.e., it pertains to kind of rock bodies bound by time divisions.

Modified from Comité de la Carta Geológica, 1976 and other sources.

Legend to Figure 1.11.

the Tarascan Plateau in eastern Michoacán) lies below 1,500 m. Climatically, this province is diverse and includes portions of the largely north–south trending climatic zones of the country. From east to west, the following zones are present: Aw' (tropical rainy with dry winter), Cfa (temperate with a dry season), Cwa (temperate with hot summer), BSh (arid and hot, as the eastern mountains block the humid winds); Cw (temperate with mild dry winter and rainy summer; it occupies most of the province and becomes cooler south and southeasterly), Cfb (temperate, with no defined dry season), and Aw (tropical rainy with dry winter restricted to a northwestern corner).

Physiographically, the dominant land forms have been produced by volcanism. The greater part of the area is a volcanic plateau, much higher on the eastern half than on the western. The plateau is dotted with both volcanic peaks (e.g., Nevado de Colima, Paricutín, Popocatepetl, Nevado

Figure 1.11. Generalized geologic map of Trans-Mexican Volcanic Belt (8) and Sierra Madre del Sur Morphotectonic Provinces (9) and adjacent areas. Modified from Comité de la Carta Geológica, 1976 and other sources.

de Toluca, Pico de Orizaba) and sierras (e.g., Sierra de Cacama, del Tigre, de Mil Cumbres, de Tlalpujahua and de Puebla).

The high portion of the Volcanic Plateau in Michoacán state is found on the Meseta Tarasca. Other portions of the plateau include the Llanos de Apan and the Llanos de San Juan.

The Volcanic Plateau is dissected and drained by numerous rivers, such as the Lerma-Santiago to the northwest, the tributaries of the Balsas-Mexcala to the south, and the Tula-Moctezuma to the northeast. Local depressions have become endorheic basins, and some have developed lakes, such as the Chapala, Patzcuaro, Cuitzeo, and Yuridia (in the western half) and those of the Mexico Basin (in the eastern half).

Geology

The province includes Mid to Late Cenozoic volcanic and Late Cenozoic sedimentary clastic bodies (Figs. 1.7 & 1.11). Their cartographic differentiation is largely lacking. The Mid-Cenozoic volcanic rock bodies chiefly occur in western Jalisco and south-central Michoacán (northern Cutzamala drainage basin). They consist of silicic ignimbrites that form the tilted mesas, and include other pyroclastic products and dacitic lavas. These bodies are of Oligo-Miocene age and closely resemble those of the Sierra Madre Occidental upper volcanic complex.

Late Cenozoic rock bodies are more widespread. The large stratovolcanoes (e.g., Popocatepetl, Nevado de Toluca, and Pico de Orizaba), so prominent in the province, belong to this group. They consist of alternating lavas (andesitic to basaltic) and pyroclastic bodies (andesitic and dacitic to rhyodacitic in composition). Other Late Cenozoic rock bodies include the sierras and plateaus largely formed of basaltic andesites (not emplaced through central vents), which are seemingly associated with east–west trending fractures. The volcanic make-up of this province is completed by adventitious cones (associated with the large stratovolcanoes), cineritic cones (e.g., Paricutín), lava fields, and calderas (of both explosion and of collapse types, some rhyolitic in composition). The active or recently active volcanic apertures (volcanoes) from west to east include San Juan, Songangüey, Ceboruco, de la Primavera Caldera, Nevado de Colima, Colima, Apaxtepec, Paricutín, Tequila, Ixtacihuatl, Popocatepetl, La Malinche, Orizaba Peak, Cofre de Perote, and Teziutlán.

In addition to the strictly volcanic rock bodies, endorheic sedimentary basins were formed in the lowland areas where fluvial and lacustrine deposits accumulated. The age of these sedimentary rock bodies (largely made up of arkosic sandstones, siltstones, and clays) is Late Tertiary to Quaternary, as suggested by their vertebrate fossil contents. Some of the large Mexican urban centers occur in these lowland areas, such as the cities of Tepic, Guadalajara, Morelia, and Mexico.

The structural geology and tectonics of this province are still imperfectly known. Lake Chapala is an east–west elongated depression bound by faults, designated the Chapala Graben. Similar graben structures are interpreted for the Tepic–Río Santiago and Colima areas. It is possible that the northern limit of the province may prove to be related to graben structures. The andesitic bodies of the Mil Cumbres area (eastern Michoacán) and the ignimbritic bodies of the Huetamo–Titzio area (southeastern Michoacán) appear to be folded, forming broad arc-like domes, seemingly of Late Tertiary age. The east–west fractures previously mentioned complete the roster of known structural features for this province.

Geologic Story

From a descriptive standpoint it appears that this east-west trending belt started to develop during the Mid-Tertiary with the emplacement of silicic to andesitic volcanic bodies in its western half, later followed during the Plio-Quaternary by the generation of dominantly andesitic to basaltic rock bodies and associated sedimentary ones that produced the plateaus, sierras, and peaks so prominent throughout the belt and so spectacularly developed in its eastern half. Extensive faulting seemingly associated with the volcanism resulted in the formation of grabens, some of which developed into lakes.

However, the genesis and development of this province is still a subject of debate and speculation because of the scarcity of data on its geology. Even the definition and actual boundaries are disputed.

Traditionally, this belt is said to include the Plio-Quaternary volcanic land features and associated sedimentary ones, plus the Mid-Tertiary volcanic bodies that form a nearly continuous, east–west trending zone between lat. 19° and 21° that spans Mexico's mainland.

A different view (Demant, 1978; Demant & Robin,1975;) excludes the Mid-Tertiary volcanic bodies because they seem to be genetically associated with the Sierra Madre Occidental volcanism rather than with the Plio-Quaternary one. Also it is suggested that the Mid-Tertiary volcanic bodies do not form a continuous belt but, rather, a series of five discrete "foci," each with its own distinct characteristics.

Yet another view holds that the belt (as traditionally conceived) should also include the southern (to the Balsas-Mexcala Basin to the south) and northern (Querétaro, Guanajuato, San Luis Potosí) branches defined by the Plio-Quaternary volcanic bodies occurring there and that are probably related to the same volcanic processes that generated the greater part of the belt. A modified version of the last view is held here. It excludes the northern branch because the volcanic activity has proved to be Mid rather than Late Cenozoic.

Modern hypotheses on the origin of the "Volcanic Belt" relate it to interaction between the Cocos Plate and the North American Plate, resulting in subduction of the former and in genesis of volcanism largely of calcalkaline character (of the andesite suite), which is common in the belt. The process occurs in a compressional stress field. The widespread andesitic-basaltic bodies related to east-west trending fractures, however, call for a tensional stress field, not readily accounted for in the compressional model described above. Also tensional stress fields must have occurred to generate the Chapala Graben and the other structures so interpreted. These facts, plus the oblique alignment of the belt with the supposed collision zone, i.e., the Middle American Trench, pose significant objections to this hypothesis, which has been ingeniously modified to explain these and other facts; yet no satisfactory explanation has been offered. Finally, still in the hypothetical realm, this belt is thought of as occurring in a major plate boundary, or as representing a major former boundary reactivated during the Plio-Quaternary. This reasoning in turn, is used to substantiate speculation on the preMid-Cenozoic tectonics of Mexico.

Selected References

Allan (1986); Besch et al. (1988); Bloomfield, (1975); Carrasco et al. (1988); De Cserna et al. (1988); Demant (1978, 1984); Demant & Robin (1975); Ferriz & Mahood (1986); Fries (1960, 1966); Gasca & Reyes (1977); Gastil & Jensky (1973); Geyne et al. (1963); Hasenaka & Carmichael (1985); Hubp et al. (1985); Ledezma (1970); Luhr et al. (1985); Mahood (1980); Mauvaois (1977); Mooser (1972, 1975); Mooser et al. (1974); Mooser & Ramírez (1987); Negendank (1972); Negendank et al. (1985); Nelson & Livieres (1986); Nelson & Sánchez (1986); Nieto et al. (1985); Nixon (1982); Nixon et al. (1987); Ordoñez (1895); Pasquare et al. (1987); Schlaepfer (1968); Thorpe (1977); Urrutia & del Castillo (1977); Viniegra (1965); Williams (1950); Wright (1981).

SIERRA MADRE DEL SUR MORPHOTECTONIC PROVINCE

Geographic and Physiographic Setting

The Sierra Madre Del Sur Morphotectonic Province includes the territory south of the Trans-Mexican Volcanic Belt between lat. 15°40'–19°40' N and long. 94°45'–104°40' W excluding the Gulf Coastal Plain. It spreads over southwestern Jalisco, Colima, southern Michoacán and the states of Mexico, Morelos, Puebla, Guerrero, and Oaxaca (Fig. 1.10) with an area of about 195,700 km². (about 9.93% of Mexico). The province is 820 km long from east to west, 940 km long on the Pacific, but only 100 km wide north-south on the Tehuantepec Isthmus.

The terrain is rugged, ranging in altitude from 0 to 3,500 masl. The lowlands (<1,200 masl) correspond to the oceanward cuestas and plains and the Balsas Basin. The uplands (>1,200 masl) correspond for the most part to the Oaxaca-Guerrero-Puebla uplands. The climate is varied: Aw' (tropical, rainy autumn and dry winter) in the lowlands, Aw (tropical, drier than the former) in most of the Balsas Basin, BShw (steppe-like, dry winter) in the western part of the Balsas Basin and in the Tehuacán-Cuicatlán Valley, Cwa (temperate, hot summer and dry winter) in most of Oaxaca, and Cfa (temperate, without a defined dry season) in the higher parts of the Pacific Cuestas.

The rugged topography and complex physiography can be divided into the following subprovinces.

Pacific Coastal Plain

The Pacific Coastal Plain, a narrow land strip is limited by the 0- to 200- masl contour lines. West of Acapulco to Cabo Corrientes it is, on average, less than 10 km wide, becoming 20 km wide (or more) between the Papagayo and Verde Rivers and on the Tehuantepec Isthmus. There the plain is broad and forms a large estuarine lagoonal area, partly closed by point bars and fed by several short rivers. The coastline is straight and devoid of lagoons elsewhere. The rivers drain without forming significant delta complexes because near-shore currents promptly disperse the sediments. The continental shelf is likewise narrow, being widest in the isthmus area.

Pacific Ranges and Cuestas

The Pacific Ranges and Cuestas occur in a relatively low (<15% of the area is above 2,000 masl) and narrow (rarely exceeding 80 km in width) belt parallel to the Coastal plain between Cabo Corrientes (Jalisco) and Río Verde (in southwestern Oaxaca). About 65% of the area lying between 200 and 1,000 masl and forming cuestas sloping toward the Pacific is cut by numerous short rivers perpendicular to it. The cuestas give way to plateaus and sierras in the upland such as the Sierra de Coalcoman (Michoacán); de la Cuchilla, de la Tentación, de Iguatlaco, de la Brea, and de Malinaltepec (Guerrero); and de Yucuyagua (Oaxaca). The Sierras de la Tentación and de la Cuchilla, because of their geology, appear to be a southwestern extension of the Trans-Mexican Volcanic Belt and are hence excluded from the Sierra Madre del Sur. This subprovince is located in the states of Michoacán and Guerrero, the southern upland country of the Balsas Depression. Among the main rivers that cut or stem from the Pacific Ranges and Cuestas are the Armería (Colima), Coahuayana (forming the border between Colima and Michoacán), Balsas (forming the border between Michoacán and Guerrero), Papagayo, Ometepec (Guerrero), and Verde (Oaxaca).

Balsas Depression

The Balsas Depression is an east-west trending basin that forms the lowlands (between 200 and 1,000 masl) of the Balsas river drainage basin and includes plateaus lying above 1,000 m in the eastern third. The depression, which is widest on its western third in the Cutzamala and Tacambaro River areas, narrows toward the Balsas and Mexcala River junction where it ends. The central third is known as the Tierra Caliente.

Northern Balsas Upland

The Northern Balsas Upland is the high plateau (>1,000 masl) that forms the northward limit of the Tierra Caliente. It is crowned by sierras, some of which exceed 2,000 m. The main Sierras are de Teloloapan, de Arcelia, de Taxco, and de Sultepec.

Oaxaca-Puebla Uplands

The Oaxaca-Puebla Uplands is the largest and most complex subprovince. It has been informally subdivided into "zones" ("infraprovinces") to simplify the description. Some are more distinct than others.

Mixteco-Zapotecan Sierras and Highlands. This zone is the largest and occupies the western third of the subprovince. Its western area is a high plateau (1,000 masl) that grades toward the Balsas Depression. The remaining country consists of sierras lying at 2,000 masl. It includes the Mixtecas and Zapotecas Sierras and the narrow Sordo River Valley draining southward.

La Cañada or Tehuacán-Cuicatlán-Quiotepec Depression. This narrow northwest–southeast trending rift valley is interposed between the Sierras Mixtecas (to the west) and the Sierra de Zongolica and Juarez (to the east).

Central Valleys. These valleys are set on a plateau (1,000 masl) draining to the south and east. They form a triangular area limited by the Sierras Zapotecas (to the west), Sierra de Miahuatlán (to the south), and Sierra de Ixtlan and de Villalta (to the northeast). Structurally they may form a graben.

The Miahuatlán Ranges. The Miahuatlán comprise a chain of an east–west trending range that includes a narrow plateau (1,000 masl) toward the south which grades northward to a sierra (2,000 masl).

Pacific Cuestas. These cuestas consist of a narrow land strip (200–1,000 masl) descending from the Miahuatlán plateau toward the coastal plain. The Sierra de Ixtlan is a narrow northwest–southeast trending sierra that limits the Central Valleys and forms the southern end of the Cañada.

Northeastern Ranges. These ranges form a narrow northwest–southeast trending mountainous belt lying largely above 2,000 m. The belt spans the whole Oaxaca-Puebla Uplands subprovince and includes several Sierras: Sierra de Zongolica in the north separated southward from the Sierra de Juarez by the Tecomavaca Canyon; and the Sierra de Juarez, which is in turn separated from the north-south trending Sierra Villalta by the Río de los Cajones and connects southeastward to the Sierra de Los Mixes. Structurally, these sierras are complex mountains.

Northeastern Cuestas. These cuestas correspond to the land strip that gradually descends from the Northern Ranges toward the Gulf Coastal Plain. It is broadest in the Cordoba area.

Lachixonaxe River Uplands and Associated Valleys. These areas form a small circular plateau or horst (1,000 masl) separated from the Sierra Mixes by the Tehuantepec River and crowned by a short east–west trending sierra (>2,000 m). This plateau is surrounded by the narrow, low-lying valleys of the Tehuantepec River and its tributaries.

Geology

The published geological literature is meager on this province, geologically the most complex part of Mexico. The 1:50,000 map coverage does not exceed 10% of the area. Ideas regarding its geology are largely speculative. The summary presented below stresses facts and remains, of necessity, incomplete. The geologic description is patterned after the physiographic subdivision used (Fig. 1.11).

Pacific Ranges, Cuestas and Adjacent Coastal Plain

Precambrian and Paleozoic. It is supposed that rock bodies of the Precambrian and Paleozoic ages commonly occur toward the Pacific side. Exposures are isolated and discontinuous in Jalisco, Colima, and Michoacán and are contiguous, forming a belt in Guerrero (and of westernmost Oaxaca), which seems to continue further eastward in southern Oaxaca. The composition of these bodies is roughly known, but their age remains largely hypothetical. Most of them are metamorphic or plutonic. The former, which occurs in Michoacán (schist, gneiss, and greenstone: Paleozoic? or Mesozoic?) and Guerrero (biotite schists, gneiss, and greenstone of Paleozoic? age that form in part the so-called Xolapa Complex), covers a larger area. The overall metamorphic grade is high. These bodies are intruded by granitic, granodioritic, or dioritic plutons, seemingly of different ages: those of the Guerrero-Oaxaca state boundary (Paleozoic?); those of Michoacán (Early Cenozoic?); and many others (Mesozoic?).

Mesozoic. Sedimentary and volcanic bodies of this age are restricted to the western (Colima-Michoacán) and north-central (central Guerrero) portions of these subprovinces. The sedimentary units that form the core of the

Sierra de Coalcoman and de Picila consist of sequences of limestones, marine sandstones, and siltstones, andesitic flows, and red, nonmarine phyllarenites. Partial intertonguing of the sedimentary and volcanic units has been documented. The age of the limestones is largely Early Cretaceous. The north-central exposures seem to be a southern extension of those in the Balsas Depression and are described in connection with that subprovince.

Cenozoic. Units of the Cenozoic age are insignificant volumetrically and occupy the topographic lowlands. Those associated with the Coastal Plain are transitional to shallow near-shore marine. A small marine sequence of Miocene age in the Guerrero-Michoacán border shows a molluscan fauna of Caribbean affinities. The units not associated with the Coastal Plain are continental sedimentary and poorly known. They may be covered by pyroclastic sheets and lava flows of Late Cenozoic age.

Balsas Depression and Northern Uplands

Paleozoic. Units of the Paleozoic age are predominantly metamorphic and form three sets: a western north–south trending body along the Tacambaro River; a northwest-southeast trending range system occupying the Northern Uplands and extending with interruptions from Zitácuaro, Tejupilco, Taxco to Iguala (including the Tierra Caliente Metamorphic Complex); and finally an eastern set, largely coincidental with the Acatlan Metamorphic Complex. The first group, whose bulk is made up of schists and pyllites, is poorly known. The second set includes three medium-sized, discontinuous bodies of mica schist, quartz feldspathic schist, greenstone, and mylonitic-adamelitic gneiss. In Tejupilco the last forms the lowest unit. Their age is variously considered to be Paleozoic, Mesozoic, or Late Paleozoic to Early Mesozoic. The Acatlan Complex crops out widely in the Oaxaca–Puebla Upland Subprovince and is described in connection with it. Finally, a small isolated metamorphic unit occurs west of Arcelia. The only Paleozoic sedimentary unit is the small, recently described Olinalá–Los Arcos Formation (marine) on the eastern part of this subprovince.

Mesozoic. Sedimentary bodies of Mesozoic age form the bulk of these subprovinces. No objective evidence of Triassic age for the oldest units has been presented; hence they are assigned to the Jurassic.

Jurassic. Rock bodies of the Jurassic age occur in eastern Guerrero (Mochitlán, Chilapa, Olinalá, Huamuchtitlán, and Tlapa areas) and southeastern Michoacán (Huetamo area). These units have been classified as the Chapolapa, Rosario, Acahuizotla, and Simon Formations). Stratigraphically, the lower units are composed of red, fine- to coarse-grained phyllarenites sometimes bearing plants, with coal seams; and the upper ones consist of limestones and fine-grained terrigenous clastic marine sediments, indicating shallow marine, near-shore to moderately offshore deposition.

Cretaceous. The Cretaceous System is represented predominantly by sedimentary marine (limestone and fine- to medium-grained terrigenous clastics) rock bodies spanning the whole period. Those of the Early Cretaceous form a north–south trending belt about 120 km long between Tetela del Río (Guerrero) to Corupo (Mexico state) and extending westward from there. Other bodies occur in isolation in Huetamo and western Morelos. Middle Cretaceous bodies occur between Iguala and Chilpancingo and Late Cretaceous ones between Estacion Balsas and Chaucingo. In general these bodies represent offshore, shallow to moderately deep marine sedimentation such as that occurring now upon the continental shelf of the tropical seas. Hence a platform in an epicontinental sea has been inferred and named the "Guerrero-Morelos Platform." The sedimentary regime for the Jurassic-Cretaceous time span indicates transgression during the Late Jurassic and most of the Cretaceous, followed by regression during the Late Cretaceous. The Cretaceous sequence unconformably covers the Jurassic bodies, indicating tectonic activity during the Late Jurassic–Early Cretaceous.

Cenozoic. The rock bodies of the Cenozoic age are strictly continental sedimentary and volcanic in nature. There is a large polymictic conglomerate in the western part, seemingly of Early Tertiary age, that was affected by granodioritic plutons, also of the same age. Elsewhere, the Cenozoic bodies are formed by (1) fluviolacustrine deposits, or (2) rhyolitic pyroclastic sheets, andesitic to basaltic lava flows, or both. Their precise age is not known.

The foliation of the metamorphic units shows two or more trends. They appear to be different in various blocks and not to be related to the structural trends discerned in the Jurassic and Cretaceous bodies. The dominant structures are a series of vaguely sigmoid north–south trending anticlines and synclines (modified by faults) that affect the Cretaceous bodies. The Jurassic units form large homoclinal (tilted) blocks commonly displaced by normal faults. The Cenozoic bodies also show homoclinal structure, and locally the strata may be vertical. Lateral contact between the Cretaceous and Cenozoic bodies is frequently by faults, suggesting that the Cenozoic bodies, at least in the western portion, occupy grabens.

Oaxaca-Puebla Uplands

Precambrian. The Mid-Proterozoic Oaxacan Complex, which is the only proved Precambrian unit of southeastern Mexico, forms the Zapotecan Sierras (west of Oaxaca City) and the Miahuatlán Ranges (in part). The

Complex near Oaxaca includes high-grade metamorphic rocks such as banded gneiss, charnockites, and pegmatitic plutons, and it is intruded by a Paleozoic granitic stock.

Paleozoic. Rock bodies of the Paleozoic age are largely metamorphic and plutonic, but a few small, disperse sedimentary exposures have been recognized. It appears that the Early Paleozoic Acatlán Metamorphic Complex occurs in the western plateau area of the Sierras Mixtecas "Infraprovince" (southwestern Puebla–eastern Guerrero–northwestern Oaxaca). Its main part trends northeast–southwest. It includes migmatites, schists of various kinds, greenstones, amphibolites, serpentinites, eclogites, augen gneiss, and granitoid to gabbroic metamorphosed plutons. It appears that this complex may represent a former collision zone, but the Plates involved remain unknown.

Post-Devonian Paleozoic? or Mesozoic? metamorphic and plutonic bodies occur in the Pacific Coastal Plain, the adjacent Cuestas, and the Sierras de Ixtlan, de Juarez, and de Villalta, as well as in the Mixes portion of the Northeastern Ranges and Cuestas "infraprovinces." They are poorly known, and their age is highly controversial.

The isolated Paleozoic sedimentary units include an Early Ordovician–Permian marine sequence in Nochixtlán (Oaxaca) and Pennsylvanian–Permian Marine Sequences in Mixtepec (Oaxaca), Tuxtepeque (Puebla), and Patlanoaya (Puebla). The Pennsylvanian Matzitzi Formation is the only continental Paleozoic body in the region.

Mesozoic. Again, as in other parts of the province, no documented Triassic bodies have been described. The Jurassic and Cretaceous units unconformably rest on the older ones; because they show an uneven distribution in the "infraprovinces," they are described separately.

Northeastern Ranges and Cuestas

Rock bodies of Jurassic and Cretaceous age in the Zongolica Block are formed by fine-grained marine clastics and limestones; elsewhere they consist of red, fine- to medium-grained continental phyllarenites (the "Todos Los Santos" Formation). These bodies show a homoclinal structure dipping toward the Gulf and are unconformably covered by the largely marine Cretaceous sequence. They represent shallow marine, moderately offshore deposition along a narrow, elongated zone, grossly coincidental with the Oaxacan–Veracruz state boundary (the Cordoba Platform paleofeature). The zone is flanked by areas where somewhat deeper deposition has taken place (the paleofeatures Zongolica and Veracruz Basins). The sedimentary regime indicates regressive conditions from the Late Cretaceous onward with the sea gradually retreating gulfward. The Cretaceous bodies are folded into symmetrical to eastward recumbent anticlines and synclines.

Mixteco-Zapotecan Sierras and Highlands

The Jurassic (plus associated Paleozoic?-Mesozoic? bodies) and Cretaceous rock bodies usually stand out as elongated, northwest-southeast trending block-fold mountains upon a plateau terrain made up of continental sedimentary units, rhyodacitic, pyroclastic mesetas, and andesitic to dacitic lava flowpiles. In some instances normal faults limiting these block mountains have been documented.

Jurassic. Rock bodies of the Jurassic age occur mostly in the western plateau (Huajuapan–Tezoatlán area and a small area near Tlaxiaco). The stratigraphically lower ones are continental and grade to marine upward. Structurally, they are homoclinal blocks, with no discernible pattern.

Cretaceous. Bodies of the Cretaceous age do not occur north of the 19° N parallel. They form three large elongated blocks (Copala–Yosondua in the southwest, Tamazulapan–Tilontongo–Sola de Vega in the center, and the Tepexi–Zapotitlán–Coixtlahuaca–Jaltepetongo in the northeast) separated by intermontane basins. The lithologies include limestones (locally reefal), marls, and fine-grained terrigenous clastics deposited in shallow to moderately deep marine conditions from an epicontinental sea that was transgressive during the Early and Middle Cretaceous and regressive during the Late Cretaceous. Structurally, these blocks are formed by north-south trending, tight to open folds sometimes overturned and block-faulted. At least in the Tamazulapan area it was demonstrated that the block was a horst surrounded by grabens.

The precise stratigraphic and structural relations between these large blocks have not been established. In addition, there are other minor blocks of similar geology, such as the Izúcar–Huehuetlán–Tepeyahualco and the Huamuchtitlán–Tonalá–Arteaga, which are even less well understood. Some medium-sized granitic to dioritic plutons appear to have been emplaced within the blocks during the Late Mesozoic. Also, andesitic lava flows of Cretaceous age have been described in the Tonalá–Arteaga block.

Other Infraprovinces

Elsewhere, mainly in the Central Valleys and the Tehuantepec region, the Cretaceous limestone units form small to medium-sized isolated hills or sierras of the folded and block-faulted types. Their structural trends do not define a recognizable pattern. The precise relations of these bodies remain to be established.

Cenozoic. The geology is described for the whole subprovince. Rock bodies of Cenozoic age are sedimentary continental (fluviolacustrine) and are Early to Late Cenozoic. They include volcanic formations (rhyolitic to rhyodacitic pyroclastic sheets and dacitic to andesitic and doleritic lava flows) of Mid to Late Cenozoic age and small to medium-sized hypabyssal to deep plutons (doleritic or andesitic to granitic in composition) of Early and Mid-Cenozoic age. The sedimentary units dominate the Mixteco–Zapotecan Sierras and Highlands and occupy the intermontane terrain. In the Tehuacán–Cuicatlán–Quiotepec Depression they form the Cañada floor. Elsewhere the volcanic rocks dominate. Structurally, the Cenozoic sedimentary and volcanic bodies vary from flat-lying to steeply homoclinal blocks (sometimes with vertically standing strata) that are affected by block faulting, which commonly forms the grabens. The intensity of the deformation increases southeasterly toward the Tehuantepec Isthmus, where the Mid-Miocene pyroclastic bodies (near Ixtepec) are faulted and steeply inclined northward.

Geologic Story

The Precambrian and Paleozoic geologic evolution of this province is confusing, as the geology is not well known. Their stratigraphic, geochronological, and structural relations remain largely to be established. Therefore basic questions involving the configuration, paleolatitudinal position, and plate affiliation of geologic units for these eons cannot be answered. As a result, it is not known whether the pre-Jurassic basement of present-day Mexican territory south of the Volcanic Belt is (1) a southern extension of the North American Plate; (2) a portion of this plate but accreted to it by small Plates of unknown affiliation; (3) a mosaic of independent Plates of different size accreted in unknown fashion; or (4) any other surmised tectonic scenario.

Sedimentologically, the Jurassic and Cretaceous units of the eastern part indicate Early Jurassic continental deposition in a terrestrial region that was invaded by a shallow to moderately deep epicontinental transgressive sea during the Late to Middle Jurassic to Early to Late Cretaceous. A later regression toward the ocean (the Late Cretaceous "ocean" perhaps included both the Pacific and the Gulf of Mexico) again allowed continental land to emerge during the Early Tertiary. The uncritical assumption usually made about the territorial continuity of both the basement and the Late Jurassic–Cretaceous epicontinental seas is not supported. Better understanding is required of: (1) the discontinuous occurrence of these bodies throughout the province; (2) their unestablished time–space relation; (3) the short extent of the Jurassic units (coupled with their lack of physical continuity); (4) the largely unknown configuration and evolution of the coastlines during this interval; and (5) the nearly unknown nature, extent, and actual relief of the crystalline basement.

In the western part, the association of sedimentary marine bodies and andesitic lava units of Cretaceous age suggests an offshore island arc scenario that was later accreted to the "mainland" by collision tectonic processes. The numerous (Late?) Mesozoic and Early Tertiary plutons of the Pacific Range, Cuestas, and Plains attest to strong magmatic activity, but again their tectonic significance cannot yet be established.

The regressive character of the Late Cretaceous ocean implies regional uplift, perhaps related to subduction of an oceanic (Paleo? Pacific or Farallon? or Cocos? or Protococos?) plate or plateau. The folding and faulting of these units show different trends that perhaps require different stress fields: east–west in the western part of the Pacific Ranges and in the Balsas Depression; north–northeast to south–southwest in the Mixteca–Zapotecan Sierras and Highlands; northwest–southeast in the northeastern Ranges and Cuestas; and nearly north–south in the Tehuantepec Isthmus. The age of the deformation seems to be different, seemingly younger in the isthmus; the tectonic significance of these facts is not well understood.

The compressive stress-field ceased and changed toward a tensional one during the Mid-Cenozoic causing, block faulting, horst and graben formation, accumulation of continental debris in the lowlands, the extrusion of extensive volcanic products, and the emplacement of plutons. Again these geologic processes seem to have occurred somewhat later toward the isthmus.

Selected References

Aguilera (1906); Aguilera & Muñoz (1896); Barrera (1946); Bazán (1984); Böhnel et al. (1989); Bolívar (1963); Burckhardt (1930); Calderón (1956); Campa (1978); Campa et al. (1974, 1977); Campa & Ramírez (1979); Campa & Coney (1983); Cárdenas (1966); Carfantan (1983); Carr et al. (1974); Cortés et al. (1957); de Cserna (1965, 1978); de Cserna et al. (1962, 1975, 1978); Elías (1989); Erben (1956a,b); Ferrusquía (1970, 1976); Ferrusquía et al. (1974, 1978); Fisher (1961); Fries (1960, 1966); Fries et al. (1962); Fries & Rincón (1965a,b); Fries & Valencia (1965); Gonzáles, J. (1961); Göse & Sánchez (1981); Guzmán (1950); Hanus & Vanek (1978); Hisazumi (1932); Klesse (1969); Köhler et al. (1988); McCabe et al. (1988); Meave & Echegoyen (1961); Morán et al. (1988); Mota et al. (1986); Nava et al. (1988); Ontiveros (1973); Ortega (1978, 1981a, 1984); Pantoja (1959, 1970); Pedrazzini et al. (1982); Pérez, I. et al. (1965); Rodríguez (1970); Salas (1949); Santillán (1929b); Urrutia (1981, 1983); Young (1983).

SIERRA MADRE DE CHIAPAS MORPHOTECTONIC PROVINCE

Geographic and Physiographic Setting

The Sierra Madre de Chiapas Morphotectonic Province is located in southeastern Mexico between lat. 14°30'–17°40' N and long. 90°30'–95°00' W. It includes the isthmian area of Oaxaca and Tabasco east of the 95° meridian (excluding the Coastal Plain portion). It has an east–west length of 360 km and a maximum north–south width of 340 km; but because of its rectangular shape and northwest–southeast geographic setting, the average width is only 299 km, narrowing to 120 km in the isthmus (Fig. 1.12).

Its area is about 105,400 km² (about 5.35% of Mexico). The altitude ranges from 0 to 2,500 masl, but 60% of the province lies between 200 and 1,000 masl. Climatically, most of the province is of the Aw type (tropical with dry winter) and becomes cooler with a Cw (temperate, with dry winter) to Cf (temperate without a defined dry season) type in the higher mountain areas of the Sierra Madre and the Northern Plateaus and Fold Ranges Subprovinces.

Physiographically, the following subprovinces are recognized.

Pacific Coastal Plain

The Pacific Coastal Plain is a narrow strip limited by the coastline and the 200 m contour line; it is narrowest between Tonalá and Tigresa, widening at both ends. At the Tehuantepec Isthmus, the coastline is complex, defining two estuarine-lagoonal complexes (Lagunas Superior-Inferior and Mar Muerto) nearly separated from the sea by sand barriers. The adjacent lowlands are marshy. Elsewhere the coast is straight, and rivers drain without developing deltas. The coast appears to be rising. The adjacent continental shelf is moderately wide—about one-third as wide as the subprovince. A few short rivers (including de Los Perros, Chilapa, Niltepec, and Ostuta in the isthmus, and Sesecapa, Vado Ancho, Carton, and Suchiate in Chiapas) traverse the Plain, descending from the Sierra Madre Ranges.

Sierra Madre Ranges

The Sierra Madre Ranges form a narrow belt bordering the plain. They include plateaus and cuestas 200 to 1,000 masl and the higher discontinuous Sierras (1,000–2,000 masl) toward the Pacific side. The adjacent cuestas are narrow on the Pacific side, whereas on the opposite side they become wider and gradually merge with the southern slope of the Central Depression. This subprovince is highest east of the 93° meridian. A series of transverse (i.e., northeast–southwest trending) straight rivers, tributaries of the Río Grande de Chiapa (actually the southern portion of the Río Grijalva), cut this subprovince into "blocks" separated by deep gorges. The main rivers are Encajonado, Cintalapa, San Juan, Santo Domingo, de la Concordia, de Salinas, and Cuilco.

Central Depression

The Central Depression corresponds to the Río Grande de Chiapa basin, which drains northwesternly. Its altitude range is 200–1,000 masl, and it has a sausage-shaped outline, elongated northwest–southeast. In the northwest it becomes wider and contains the flat-lying Mesa de

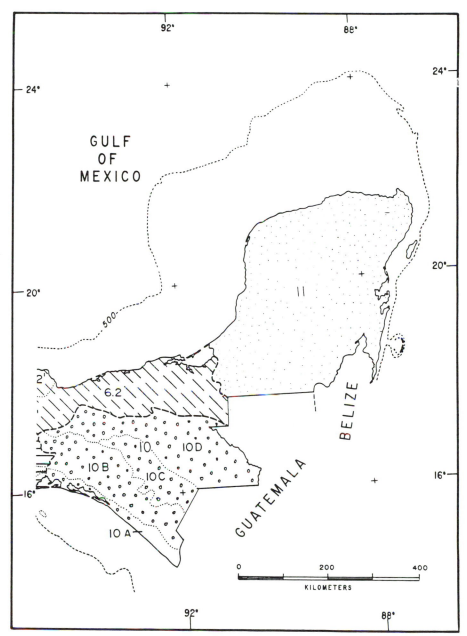

Figure 1.12. Morphotectonic provinces of Mexico. 10. Sierra Madre de Chiapas: 10A, Pacific Coastal Plain; 10B, Sierra Madre Ranges; 10C, Central Depression; 10D, Northern Folded Ranges and Plateaus. 11. Yucatan Platform. Dotted line offshore indicates the 500 m isobath.

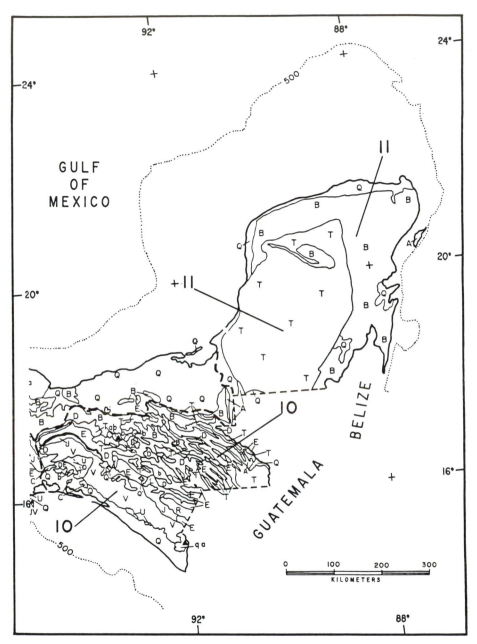

Figure 1.13. Generalized geologic map of Sierra Madre de Chiapas (10) and Yucatan Platform Morphotectonic Provinces (11) and adjacent areas. Modified from Comité de la Carta Geológica, 1976 and other sources.

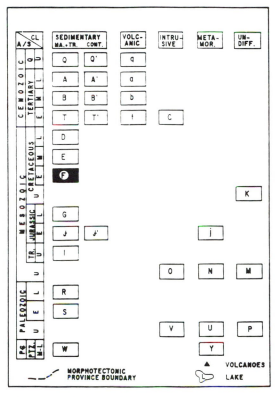

Legend to Figure 1.13.

Ocozocuautla, spectacularly cut by the Río Grijalva's Sumidero Canyon within which the largest dam of Mexico has been built. The northern slope is much narrower and steeper than the southern slope and has no rivers.

Northern Folded Ranges and Plateaus

The subprovince containing the Northern Folded Ranges and Plateaus makes up the northeastern two-thirds of the whole province. The plateaus (Mesetas de San Cristobal and de Comitán, as well as the Ixtapa Depression) are adjacent to the Central Depression. The first plateau is at 2,000 masl and the other at 1,000 masl. The folded ranges are elongated northeast-southeast, forming narrow closely spaced sierras separated by likewise narrow intermontane basins. The eastern ranges and basins occupy the Lacandon Region of Chiapas, which drains southeasternly in a series of parallel, straight northwest–southeast trending tributaries (Ríos Santo Domingo, Jataté, Tzendales, San Pedro, and Lancajá) of the Río Ixcan-Lacantún, which is in turn a tributary of the Río Usumacinta. The western ranges and basins drain in similar fashion to the northwest through tributaries of the Ríos Macuspana (itself a large tributary of the Usumacinta) and Grijalva in Chiapas and of the Ríos Tonalá, Uxpanapa, and Grijalva in Tabasco and Oaxaca. The folded ranges become noticeably attenuated (i.e., lower) and more separated toward the Gulf Coastal Plain, where they disappear altogether.

Geology

Pacific Coastal Plain

Most of the Chiapas (Fig. 1.13) segment consists of largely unconsolidated Quaternary beach and fluvial deposits. About 20 km southeast of La Tigrera, a small granitoid pluton crops out. The isthmian segment (between Juchitan-Arriaga) shows greater complexity because the Quaternary deposits are dotted by small, low exposures of varied age and composition. The oldest occurs just northwest of Chahuites and is made up of phyllite and schist, seemingly of Paleozoic age. Small granitoid plutons of Early Cenozoic age flank the metamorphic body, and even smaller ones crop out by the Superior-Inferior Lagoons. Isolated Cretaceous limestone and Late Cenozoic volcanic bodies complete the make-up of the subprovince.

Sierra Madre Ranges

The most complex subprovince contains the Sierra Madre Ranges (Fig. 1.13). It includes a large, nearly continuous core of crystalline rock bodies (metamorphic and plutonic) of Precambrian? and Paleozoic age, commonly designated the Chiapas Massif, which is partly covered at both (northwestern and southeastern) ends by moderately thick sequences of Early? Jurassic sedimentary continental rock bodies. The northwestern bodies occur toward the Pacific side of the isthmian region, whereas the southeastern bodies occur in the opposite side toward the Central Depression. The latter includes the only known marine sedimentary sequence of this province.

Precambrian? and Paleozoic. The Chiapas Massif is largely composed of granitic, granodioritic, and dioritic plutons that intruded older metamorphic bodies. In the isthmian region, there are three small bodies of micaschists and gneiss, apparently of Paleozoic age. The easternmost of these bodies is in contact with a medium-sized gneissic? body of Precambrian? age. The other metamorphic exposures are in the Pijijiapan-Mapastepec and southern Chicomuselo areas (Chiapas), mostly on the Pacific side of the massif. They consist of schists, amphibolites, and metaquartzites.

The plutons have yielded ages that vary from the Proterozoic to the Cretaceous, but current opinion favors a Late Paleozoic age for the bulk of the massif. Further work is needed to understand its complex geochronology.

The Late Paleozoic sedimentary units occur in the Chicomuselo area (southeastern Chiapas) extending further east into Guatemala, in the Polochic–Montagua Fault Zone. They lie unconformably upon the massif and are composed of fine-grained marine clastics grading upward into limestones. Unconformities separate the various units of this sequence; its lower part is moderately metamorphosed.

Mesozoic. No objective data document the presence of Triassic rock bodies in this province.

Jurassic. The Early? Jurassic Todos Santos Formation and its equivalents are the oldest Mesozoic unit. They consist of unfossiliferous, continental, red phyllarenites that unconformably overlie the massif or the Late Paleozoic sequence just described. They are folded into open structures or in homoclinal blocks, both of which are intensively affected by faults (particularly in the isthmian region).

Cretaceous. Rock bodies of the Cretaceous age are present on both southern and northern flanks of the massif, but only the southern ones are described here, as the others are closely related to those of the Northern Fold Ranges and Plateaus and are treated in connection with them.

The Cretaceous bodies of the isthmian area, which form part of the Sierra de Niltepec, lie unconformably on the massif or the Early? Jurassic bodies. They consist of marine fine-grained clastics and micritic limestones of Middle and Late Cretaceous age and are tightly folded and faulted into arcuate, east–west trending structures that become northwest-southeast trending toward the ends. Some of these bodies are intruded by small Early Cenozoic granitic plutons.

Cenozoic. Mid-Cenozoic silicic pyroclastic bodies occur in the Isthmian (Niltepec area) and southwestern (north of Tapachula) parts of the Massif. Those of the isthmus are deformed into homoclinal block ridges dipping north. They are also faulted and intruded by dioritic to doleritic hypabyssal and subvolcanic bodies. In the south, a medium-sized dioritic pluton of Early Cenozoic age lies in tectonic contact with elongated, east–west trending ophiolitic (antigoritic) bodies associated with the western end of the Polochic-Motagua fault zone. Their age appears to be Late Cretaceous or Early Cenozoic. The associated volcanic bodies consist of flat to moderately inclined Mid-Mesozoic silicic pyroclastic deposits of Late Tertiary to Quaternary age that form the Tacana Volcanic area, apparently the northern end of the Late Cenozoic Central American Volcanic Belt. Late

Cenozoic fluvial deposits located chiefly on the river basins of the Cintalapa to Santo Domingo Rivers complete the geology of this subprovince.

Central Depression

Jurassic. The Early? Jurassic Todos Santos Formation is the only unit of this age that crops out in the subprovince. However, in the isthmian region, close to the northwestern end of the Central Depression, small elongated Late Jurassic bodies occur associated with the massif, unconformably resting on it or on the Todos Santos Formation. They consist of micritic limestone and fine-grained terrigenous sediments indicative of shallow marine, near-shore deposition. Elsewhere, subsurface exploration has disclosed the local presence of fine-grained clastics, indicative of deeper marine sedimentation.

Cretaceous. Middle and Late Cretaceous bodies of largely calcareous composition form the bulk of the Central Depression. The older ones lie on the southeastern slope, and the younger ones on the northeastern slope. Both form competent limestone strata indicative of shallow marine deposition (shelf or platform carbonated and reef-complex developments). They gently warp into open folds trending roughly northwest–southeast, geomorphically expressed as nearly flat-lying plateaus (such as the Ocozocuautla Mesa, near Tuxtla Gutiérrez).

Cenozoic. Bodies of Cenozoic age are restricted to a small area south of Tuxtla Gutiérrez, where Early and Mid-Cenozoic fine-grained clastic marine units are present, apparently lying unconformably on the Cretaceous formation. Late Cenozoic fluvial deposits associated with the rivers complete the composition of this subprovince.

Northern Fold Ranges and Plateaus

Cretaceous. Medial and Late Cretaceous limestone and fine-grained clastic bodies occur extensively in the subprovince comprising the Northern Fold Ranges and Plateaus (Fig. 1.13). They represent shallow marine deposition on a shelf or platform in the Mid-Cretaceous formation. This regime changed during the Late Cretaceous to the dual condition of being shallow and platform-like in most of Chiapas and moderately deep in northwestern Chiapas and adjacent Tabasco and Campeche. The competent thick limestone strata adjacent to the Central Depression are gently folded, forming plateaus: the Comitán Mesa and the even higher San Cristóbal Mesa. Elsewhere they are tightly folded into anticlinoria and synclinoria so characteristic of this subprovince. The age of the deformation must be at least Mid-Cenozoic because nearby Early and Mid-Cenozoic sedimentary units are likewise deformed.

Cenozoic. Early and Mid-Cenozoic sedimentary marine bodies crop out extensively in the northeastern half of the subprovince. They are largely made up of terrigenous clastics deposited in a shallow marine environment; they also include continental to transitional units as well as some micritic and reefal limestone ones. These bodies become younger toward the Gulf Coastal Plain, indicating regional uplift progressing northeasterly, concomitantly associated with a marine regression gulfward. A small graben in the Ixtapa–Soyaló area includes a similar Cenozoic sequence that is surrounded by Cretaceous limestone bodies, much as in the already mentioned basin southeast of Tuxtla Gutiérrez.

Mid and Late Cenozoic volcanic (largely silicic pyroclastic sheets) bodies crop out southeast of Chiapa de Corzo (northern slope of the Central Depression) in San Cristobal and northeastern Chiapas. The recently erupted Chichón is located in this area. Small Early Tertiary granitic plutons occur here. Late Cenozoic fluviolacustrine deposits (mostly around Comitán and Lacantún River) complete the make-up of the subprovince.

In addition to folding, extensive left-lateral faulting is apparent in central Chiapas, mostly in the Central Depression and Plateaus portion of the Folded Ranges Subprovince. The faults have a net northwest–southeast trend. In northern Chiapas, largely northwest–southeast trending thrust-faulting is dominant. The faulting, which varies from coetaneous to posterior folding, probably took place during the Mid to Late Tertiary.

Geologic Story

Precambrian? and Paleozoic

The Precambrian? and Paleozoic history is poorly understood. The crystalline metamorphic and plutonic basement (the Chiapas Massif) is typically continental crust, but its heterogeneity and lack of study do not establish (1) whether it is part of one or several lithospheric plates; (2) its configuration and paleolatitudinal position through time; and (3) even its true extension with any certainty. Under these circumstances, hypotheses about the tectonic evolution of Middle America must be made cautiously, avoiding uncritical assumptions about the plate affiliation, extent, configuration, and age of any recognizable rigid tectonic element of the Chiapas basement.

Whatever its history, paleogeographic configuration, and position, this basement was transgressed during the Late Paleozoic by a shallow epicontinental sea that deposited the Late Paleozoic sedimentary sequence now cropping out in the Chicomuselo area (extending eastward to Guatemala).

The deformation of the units and the unconformities separating them indicate tectonic (compressive?) activity during the Late Paleozoic. If the bulk of the Chiapas Massif plutonic bodies were emplaced in the Paleozoic, it is tempting to think that such emplacement occurred during this tectonically active Late Paleozoic period.

Mesozoic

The unconformity separating the Late Paleozoic sequence from the Early? Jurassic Todos Santos (continental) Formation represents an interval of at least 50 Ma, including regional uplift of the basement and its sedimentary cover above sea level, installing a continental regime of erosion in the topographically high areas and deposition in the low areas. Most of the Todos Santos and subsequent Mesozoic bodies occur on the northeastern side of the present Chiapas Massif, suggesting that this area contains the downslope and topographic low areas (basins) where the terrigenous sediments (derived from ranges located to the southwest) accumulated.

The extent, configuration, and paleogeographic position of this continental region during the Late Paleozoic to Early Jurassic, as well as their relations to the now contiguous Yucatán, Oaxacan, and northern Central America regions, are not known. To assume the same space continuity as it is today for the Early Jurassic (or earlier) is unwarranted according to plate tectonic theory because of the obvious geologic complexity of the region and its location in a highly dynamic part of the Earth.

This ("ancestral continental Chiapas") region was transgressed during the Late Jurassic by an epicontinental sea that generated transitional to shallow marine, near-shore and offshore sedimentary units, including evaporitic deposits such as salt and gypsum.

Again, the coast configuration and its evolution during the Late Jurassic is unknown. The evaporitic deposits are thought to have facilitated deformation of the overlying Cretaceous sequence.

The epicontinental sea remained in most of this region (i.e., "ancestral Chiapas") throughout the Cretaceous. The sea floor behaved as an uninterrupted platform up to the early Late Cretaceous, when a narrow northwest–southeast trending depression was developed in northern Chiapas, fostering deeper marine deposition. This depression seems to have widened during the Early Cenozoic. The sea floor was shallow in some places, allowing the development of reefal complexes. The configuration, extent, and evolution of the Cretaceous sea (and hence of its coastline) is only roughly known. It is thought that the Chiapas Massif experienced partial flooding at one time.

Intermittent tectonic activity occurred during the Cretaceous, as suggested by the emplacement of plutons during the Late? Cretaceous (and Early Cenozoic as well) by way of the unconformity between the Neocomian and Albian bodies in western Chiapas, as well as the terrigenous clastic character of some Late Cretaceous formations. Of critical importance is the

tectonic evolution of this region ("ancestral Chiapas") during the Cretaceous, but it is largely unknown. Its history is linked to that of Middle
America. The only accepted concept is that the whole region could not
possibly have had its present configuration and geographic position during the Mesozoic.

Cenozoic

During the Early and Mid-Cenozoic, shallow marine sedimentation (upon
a sea floor covered by Cretaceous sediments) continued in what is now the
Central Depression (northern slopes) and in the Folded Ranges and Plateaus. Contrasting with largely calcareous deposition during the Cretaceous, the Cenozoic involved large amounts of terrigenous clastics. The
sequence further includes some true continental (i.e., terrestrial) units. The
sedimentary sequence indicates that the shallow epicontinental sea gradually receded (regressed) away from the Chiapas Massif, which furnished
the terrigenous clastics and apparently became regionally uplifted during
the Late Cretaceous or Early Tertiary. Perhaps associated with this tectonic
activity was the emplacement of some granitoid plutons and the ophiolite
complexes.

The Cenozoic marine deposition ceased regionally during the Middle or
Late Miocene as the epicontinental sea left the area. Locally (e.g., in the
Ixtapa area or the area southeast of Chiapa de Corzo), Late Miocene
fluviolacustrine and silicic pyroclastic volcanic units were deposited.
Shortly thereafter intense tectonic deformation occurred so the Cretaceous
and Cenozoic sedimentary layers became tightly folded into N 30° W–S 30°
E trending anticlinoria and synclinoria. In some areas these structures
were further affected by thrust-faulting resulting, in the partial piling up of
the folds. In others (e.g., in Central Chiapas) left-lateral faulting caused
extensive displacements along the strike of the strata. Many river channels
traverse northwest–southeast trending paths marking their linkage to the
structural pattern just described understandable. The Mid-Cenozoic bodies were deformed into strongly tilted (frequently with vertical strata)
homoclinal blocks, rather than folds. These blocks are also affected by
faulting.

Late Cenozoic volcanic activity (in northwestern and southeastern
Chiapas) and fluviolacustrine sedimentation (in the local low areas) complete the geologic processes that occurred in this province.

The present-day Middle American tectonic framework is reasonably
well known; hence it is tempting to speculate how far back in time its
evolution can be followed. This evolution is important for understanding
the geology of southeastern Mexico and the Yucatan Platforms. Yet the
scarcity of basic published geologic information prevents any real break-
through toward this end. The creation of a reliable data base seems at
present a more promising strategy than that of additional tectonic hypotheses (models) with little or no foundation.

Selected References

Bishop (1980); Blair (1988); Böse (1905); Burbach et al. (1984); Burke et al. (1984); Carfantan (1977a,b, 1984); Case et al. (1984); Castro et al. (1975); Chubb (1959); Damon et al. (1981); de Cserna (1969, 1971); Dengo (1973); Dengo & Bohnenberger (1969); Duffield et al. (1984); Frost & Langenheim (1974); Gutiérrez (1956); Hanus & Vanek (1978); Hernández, G. (1973); Imlay (1944a,b); Malfait & Dinkleman (1972); Molnar & Sykes (1969); Mullerried (1957); Murray (1961); Pantoja et al. (1974); Richard (1963); Sánchez (1979); Shor (1974); Steele (1986); Tardy (1980); Waite (1986).

YUCATAN PLATFORM MORPHOTECTONIC PROVINCE

Geographic and Physiographic Setting

The Yucatan Platform Morphotectonic Province lies between lat. 17°50'–21°30' N and long. 87°00'–91°00' W (Fig. 1.12). It includes the States of Quintana Roo, Yucatán, and most of Campeche, as well as Belize and northern Guatemala (Petén Lowlands) and has an area of about 167,600 km² (about 8.46% of Mexico). More than 95% of the province lies below the altitude of 200 m, making it a vast plain. Offshore, the platform extends north and westward to the 92°30' meridian and to the 22°–24° N parallel along an oblique line, roughly defined by the closely spaced 0 to 3,000 m below sea level isobaths. In strong contrast, the eastern margin has no continental platform. The predominant climate is Amw (tropically rainy, with a short winter dry season), and occupies the southwestern two-thirds of the province. The climate BShw (desert with dry winter) occupies only the northwestern corner of the Yucatan Peninsula. The rest has an Aw climate (tropical rainy, with a longer dry winter season).

Only limestone land forms occur because of the calcareous nature of the platform. In the northern sector an area of pitted flats is present that has numerous small sink holes (cenotes) and no surface drainage. Toward the east the coastal zones are marshy and have coral reefs on the seaward side. The rest of the platform is slightly higher and knobbier (i.e., karst topography not fully developed) than the pitted flats from which it is separated by a northwest–southeast trending scarp, the Sierrita de Ticul, in southern Yucatán state.

Geology

The surface geology (Fig. 1.13) of the Yucatan Platform's Mexican portion is relatively simple, with flat-lying limestone, dolomite, and evaporitic bodies spanning the whole Cenozoic. Paleocene units crop out in a narrow band located in southeastern Campeche. The Eocene bodies conformably overlie the Paleocene, and have a greater area, occupying most of Campeche, Yucatán, and eastern Quintana Roo. Surprisingly, the Oligocene bodies are restricted to two narrow, short east-west trending bands in eastern Yucatán (Mérida and Maxcanú areas). The Miocene and Pliocene bodies form a wide zone surrounding the Eocene, cropping out in Yucatán and most of Quintana Roo. Finally, a narrow band of Quaternary limestone units and beach deposits borders the Peninsula, mainly in its northern portion.

This Cenozoic sequence indicates shallow marine calcareous sedimentation on a bank or continental shelf set in a tropical scenario not unlike that of the present-day Bahamas Bank. The nearly horizontal structural attitude of the strata making up the bodies means that they have not been affected by postdepositional deformation, in turn suggesting tectonic stability for this region throughout the Cenozoic. The sequence becomes younger toward the present-day coastline, disclosing the gradual emergence of the Platform during the Cenozoic.

The few wells drilled in the Platform have disclosed the following subsurface information: The Cenozoic sequence is underlaid by Medial and Late Cretaceous limestones, in turn unconformably underlaid by an evaporite and red bed sequence of Mesozoic age that rest on a crystalline metamorphic and plutonic basement. Isotopic ages from the plutons are 410 Ma, thus locally making the basement at least as old as the Silurian.

The pre-Cenozoic geologic record must be sought in Guatemala and Belize. A brief description of this record is presented here.

Paleozoic

The Maya Mountains of southern Belize consist largely of marine clastic metasedimentary rock bodies (consisting of metaquartzites, metaphyllarenites, sericitic phyllites) and of small schist and gneiss bodies. The low-grade metasedimentary bodies bear invertebrate fossils suggestive of a Late Paleozoic (Pennsylvanian?) age. Further south, in Guatemala (eastern Motagua River area), a small body of greenstone (metabasalts) is present that is associated with metasedimentary phyllites resembling those of the Maya Mountains.

Mesozoic

In south-central and central Guatemala a Mesozoic sequence crops out that is roughly similar to that of Chiapas. The oldest unit is the Early? Jurassic Todos Santos Formation, a unit formed by red, fine- to medium-grained phyllarenites of continental origin, structurally forming a homoclinal block tilted northward and affected by faults. This unit lies in the southern portion of the area. It is unconformably overlaid by Cretaceous limestone units structurally deformed into east–west trending folds that become attenuated northward.

Cenozoic

The Cenozoic sequence crops out in the Petén Lowlands (northern Guatemala) and consists mainly of Paleocene and Eocene sedimentary marine rock bodies similar to those of Campeche.

Geologic Story

The Pre-Jurassic geologic record is too sparse to be able to indicate unequivocally the history of the region. Objectively, all that can be said is that a Silurian pluton was emplaced into a necessarily older crystalline basement located somewhere underneath present-day Yucatán. The next datum indicates marine deposition during the Late Paleozoic (Pennsylvanian?) upon an unknown basement now located in southern Belize. This marine sedimentary sequence was later metamorphosed and eventually became upland. The phyllarenitic Todos Santos Formation, apparently of Early Jurassic age, was deposited in a continental sedimentary environment and had a metamorphic upland as a source area. The extent of this unit prior to its transgression by a shallow sea is unknown. The angular unconformity that separates the Todos Santos from the sedimentary marine Cretaceous sequence indicates that tectonic activity occurred sometime between the Early? Jurassic and Cretaceous. From the Cretaceous onward the story becomes clearer: Shallow marine calcareous deposition on a bank or continental shelf, set in a tropical, tectonically passive setting, took place from the Cretaceous to the Cenozoic, with the gradual emergence of the platform occurring during the latter time. Only the southern part of the Yucatan Platform shows geologic complexities: The presence of folded ranges of Cretaceous limestone is interpreted to be related to the geologic processes that generated the Northern Fold Ranges and Plateaus Morphotectonic Subprovince of Chiapas. The neotectonic Polochic–Montagua Fault Zone, actually a large left-lateral strike-slip fault system, is related to the motion of the Caribbean, Cocos, and North American Plates, which seems to have started at least during the Mid-Cenozoic.

   Some problems remain. Major questions surround the limits, configuration, and paleolatitude of the Yucatan Platform throughout its recorded geologic history, as well as the constitution of its basement and to its relations to the adjacent regions (Chiapas, Gulf of Mexico, Caribbean oceanic floor, and northern Central America). Geophysical data are shedding light on the tectonic evolution of this region, but much additional interdisciplinary work remains to be done before a coherent geologic history can be satisfactorily outlined.

Selected References

Alvarez (1954); Antoine et al. (1974); Bateson (1972); Bryant et al. (1968, 1969); Burke et al. (1984); Butterlin & Bonet (1958, 1962); Case et al. (1984);

Dengo (1973); Dengo & Bohnenberger (1969); Flores, G. (1952); Hall & Bateson (1972); Kirkland & Gerhard (1971); Logan et al. (1969); López-Ramos (1975); Malfait & Dinkelman (1972); Molnar & Sykes (1969); Murray (1961); Murray & Weidie (1962); Sapper (1896); Schuchert (1935); Uchipi (1973); Viniegra (1971, 1981); Vinson & Brineman (1963); Wilhem & Ewing (1972).

REVIEW OF MORPHOTECTONIC PROVINCES AND CONCLUSIONS

Baja California Peninsula

Northern Baja includes a Mesozoic crystalline core of plutonic and meta-morphic rocks—the Baja California Batholith—flanked by Cretaceous to Recent sedimentary and Late Cenozoic volcanic bodies. Southern Baja, north of La Paz, includes extensive silicic to mafic volcanic bodies of Mid to Late Cenozoic age that make up the Sierra de la Giganta. The wide Pacific Coast Plain largely consists of Late Cenozoic sedimentary marine units, unconformably covering a complex Triassic basement of the only local outcrop. The peninsular tip is dominated by Mesozoic silicic plutons covered by a thin veneer of Late Cenozoic sedimentary marine units. The precise history and origin of the Peninsula is still unclear, and it may have involved accretion of a Mesozoic island arc associated with the subduction of an oceanic plate—the Farallon—and subsequent detachment from the mainland during the Cenozoic, probably related to the deep-seated pro-cesses that generated the East Pacific Rise.

Northwestern Plains and Sierras

The basin-and-range structure of the Northwestern Plains and Sierras Province has generated block mountains (horsts), chiefly made up of Paleozoic to Cretaceous sedimentary marine and volcanic rock bodies, partly covered by Cenozoic volcanic ones, set on a basement of Mid-Proterozoic crystalline (plutonic and metamorphic) and sedimentary ma-rine units, both of which are the oldest in Mexico. The mountains are separated by elongated basins filled by Mid to Late Cenozoic continental sedimentary units. Its history, related to Baja California (in part), is complex and probably involves accretion of one or more Mesozoic island arcs, and Mid to Late Cenozoic development of the basin-and-range structure (perhaps related to the development of the Gulf of California). Its precise relation to the North America Plate is not clear.

Sierra Madre Occidental

The Sierra Madre Occidental Province, the largest in Mexico, is made up of volcanic rock bodies of Late Cretaceous to Early Tertiary age, overlaid by others of Early to Mid-Tertiary age. Its origin seems to be related to the subduction of the Farallon Plate.

Chihuahuan-Coahuilan Plateaus and Ranges

The Chihuahuan-Coahuilan Plateaus and Ranges Province is largely composed of Jurassic and Cretaceous sedimentary marine rock bodies forming folded and block mountains, separated by topographically lower terrains consisting of poorly known Cenozoic volcanic and continental sedimentary sequences. The Paleozoic sequences, which are few and far apart, are small and consist mostly of sedimentary marine rocks. The Mesozoic history involves marine transgression upon a topographically complex area including highs and lows, followed during the Late Cretaceous and Early Tertiary by regional uplift and regression. Associated with it, folding and faulting occurred, forming mountains; and their erosion products accumulated in the lowlands. The block-faulting and Mid-Cenozoic volcanism suggest a tensional regime, whose regional and tectonic significance is not well understood.

Sierra Madre Oriental

The Sierra Madre Oriental Province consists mostly of Jurassic and Cretaceous sedimentary marine rock bodies forming intricately folded mountain ranges separated by elongated intermontane basins where Cenozoic continental sedimentary sequences are present. The Pre-Jurassic basement, which crops out only in a few places, includes Pre-Cambrian and Paleozoic metamorphic and Paleozoic sedimentary units. The Mesozoic history involves marine transgression of an epicontinental sea upon a land whose basement has not been well studied. During the Late Cretaceous to Early Tertiary, regional uplift was followed by folding and faulting associated with marine regression and continental sedimentation in the emerged lowlands.

Gulf Coast Plain

The Gulf Coast Plain Province is made up of Cenozoic sedimentary marine units, decreasing in age gulfward. They were laid down in the course of a regional regression in a relatively passive continental margin.

Central Plateau

The geologic constitution of the smallest province, the Central Plateau, is complex. The core consists of Jurassic to Cretaceous low- to moderate-grade metamorphic and sedimentary marine rock bodies forming mountains and sierras. Stratigraphically, these structures were unconformably overlaid by Early Cenozoic sedimentary continental units, which were in turn covered by Mid-Cenozoic volcanic ones, locally overlaid by younger lavic bodies. Small to medium-sized Early and Middle Cenozoic plutons complete the make-up. The structural features are not precisely known. A horsts-grabens pattern is suspected with Late Cenozoic sediments accumulating in the lowlands. Pre-Mesozoic metamorphic units are restricted to a small area in Zacatecas. The geologic history is not well understood. The Jurassic and Cretaceous sedimentary marine bodies imply a setting similar to that described for the Sierra Madre Oriental, yet their relation to the metamorphic bodies is not clear. Regional uplift and structural deformation occurred during the Late Cretaceous to Early Tertiary, as did plutonic emplacement. Later the products of the thus-formed uplands accumulated, followed by Mid-Cenozoic magmatic activity, whose regional significance is not clear either.

Trans-Mexican Volcanic Belt

Mid-Cenozoic, silicic bodies occur only in the western sector of the Trans-Mexican Volcanic Belt, and are less extensive than the Late Cenozoic bodies. They are made up of basaltic andesitic to rhyodacitic volcanoes, plateaus, lava fields, calderas, and cineritic cones. The emplacement occurs through both foci and fissures. Apparently, extensive graben development has taken place. Locally, continental sediments have accumulated in the lowlands. The geologic history and plate tectonic significance of this province is not well understood.

Sierra Madre del Sur

The Sierra Madre del Sur is the most complex of all provinces. Demonstrably Pre-Cambrian to Late Proterozoic units are represented by the Oaxaca Metamorphic Complex. Apparently, Paleozoic and questionable Mesozoic metamorphic bodies include complexes present in the Pacific and Oaxaca-Pueblan Upland subprovinces. Paleozoic sedimentary units have small, isolated outcrops. Mesozoic and Early Cenozoic plutons are extensive in the Pacific subprovinces (especially in Michoacán), where andesitic volcanics of Mesozoic age have also been identified. In the Oaxaca uplands, only Late Jurassic ones of smaller size are present.

The Mesozoic and Cenozoic sedimentary rock record is more complete and better known. It starts with the Early Jurassic continental sequence, unconformably overlaid by younger Jurassic and Cretaceous marine formations, separated by unconformities. These sequences are intricately folded and faulted, forming isolated or associated folded block-mountains and ranges, which in turn are congregated into various unconnected sets.

The Cenozoic bodies are largely continental sedimentary and volcanic, and they occupy grabens; the volcanic bodies are somewhat younger than the sedimentary bodies. The geologic history remains obscure. The variety of the crystalline basement rock bodies may indicate complex accretion of an unknown number of platelets. The Mesozoic sedimentary record indicates emerged land later transgressed by an epicontinental sea, followed by structural deformation and regional uplift during the Late Cretaceous to Early Tertiary. The apparently Mesozoic metamorphic bodies, plutons, and volcanics indicate one or more phases of tectonic and magmatic activity throughout the region or alternatively, independent pulses in unconnected areas (from a now single or unified region). During the Cenozoic, continental sediments accumulated in the low areas, particularly during the early part, followed by extensive volcanic activity, whose regional tectonic significance does not appear clear, although it may be related to the complex evolution of the relations of the Plates present there.

Sierra Madre de Chiapas

The Sierra Madre de Chiapas Province is made up of a core of metamorphic and plutonic units—the Chiapas Massif—probably of Late Paleozoic age, unconformably covered in the northeastern part by a Late Paleozoic sedimentary marine sequence (Chicomuselo area). The massif is partly covered by an Early Jurassic continental sequence of red beds, in turn unconformably overlaid by a Cretaceous sedimentary marine one, now intricately folded and faulted. An unconformity between the Late Cretaceous and Early Tertiary marine bodies, as well as the presence of plutons of the same age, attest to coetaneous tectonic activity. Marine sedimentation continued up to the Mid-Miocene, but in a somewhat pulsated manner. The continental sedimentation occurred much later, associated with silicic volcanism and strong regional tectonic activity, that included folding and faulting of both vertical and lateral displacement, probably related to the plate evolution of northern Central America.

Yucatan Platform

Largely consisting of a Cenozoic sedimentary marine sequence that decreases in age northward, the Yucatan Platform is deposited on a stable basement of older Mesozoic sedimentary and Paleozoic crystalline and sedimentary rock bodies.

Conclusions

The picture that emerges of the geologic history of Mexico is rather obscure for the Pre-Cambrian and most of the Paleozoic, becoming clearer for the

Mesozoic. Yet uncertainties remain about the precise configuration of lands and seas (as well as mountain ranges, river systems, basins, and the like) during this vast span of time that attested to the origin and dispersal of modern lineages of vertebrates and angiosperms. The picture does not improve greatly for the Early Cenozoic, yet gradually gets better so that the major current physiographic features such as mountain ranges, plateaus, plains, and large river systems may be reliably delineated. This information contributes to the geographic scenario where one of the most diverse biotas of the world has developed. For the Middle and Late Cenozoic, the delineation significantly improves, as do the resulting physical-geographic models tested against data obtained from the fossils or from current distribution of lineages. These studies, which seem promising for a better understanding of both the geological and biological make-up of the country, call for close collaboration between geologists and biologists.

ACKNOWLEDGMENTS

I am grateful to Fred McDowell (Department of Geosciences, University of Texas at Austin) for his intellectual input, which clarified many of the ideas presented here. I thank T.P. Ramamoorthy and Marguerite Elliott, whose careful review improved the text. Mr. Felipe García Villega's assistance with literature while the chapter was in preparation was invaluable. Luis Burgos Peraita and Felipe Villegas prepared the Figures. Héctor Gómez de Silva, José A. Santos M., and Clara Galindo H. assisted with proof reading.

SELECTED REFERENCES

Abbot, P.L. & Gastil, G. (eds.). 1979. Baja California Geology, Field Guides and Papers: Geol. Soc. America Ann. Mtg. San Diego. Field Trips 10, 12, 13 and 26.

Aguilera, J.G. 1906. Excursión de Tehuacán a Zapotitlán y San Juan Raya. X Congr. Geol. Int. Méx. 7:1–27.

———. & Muñoz, B. 1896. Bosquejo geológico de México. Inst. Geol. Méx. Bol. 4–6:2–267.

Alencáster, G. 1961. Estratigrafía del Triásico Superior de la parte central del Estado de Sonora. UNAM. Inst. Geol. Paleontol. Méx. No. 11, Pt. 1. pp. 1–18.

Allan, J.F. 1986. Geology of the northern Colima and Zacoalco grabens, southwest Mexico: Late Cenozoic rifting in the Mexican Volcanic Belt. Bull. Geol. Soc. Amer. 97(4):473–485.

Alvarez, M. Jr. 1954. Exploración geológica preliminar del Río Hondo, Quintana Roo. Asoc. Mex. Geol. Petrol. Bol. 6:207–213.

Amin, R.D. 1987. Sedimentology and tectonic significance of Wolfcampian– (Lower Permian) conglomerates in the Pedregosa Basin: Southeastern Arizona, southwestern New Mexico and northern Mexico. Bull. Geol. Soc. Amer. 99(1):42–65.

Anderson, T.H., J.H. Eells & L.T. Silver. 1978. Rocas precambricas y paleozoicas de la región de Caborca Sonora, México. UNAM. Inst. Geol. Primer Simposio sobre la Geología y Potencial Minera del Estado de Sonora, Hermosillo, Son. Libro guia. pp. 5–34.

————. & L.T. Silver. 1979. The role of the Mojave–Sonora Megashear in the tectonic evolution of Northern Sonora. Geol. Soc. America, Ann. Mtg. San Diego. Guidebook Field Trip 27. Geology of Northern Sonora. pp. 59–66.

————.1984. An Overview of precambian rocks in Sonora. UNAM. Inst. Geol. Rev. 5(2):131–139.

Antoine, J.W., R.G. Martin, T.G. Pyle & W.R. Bryant. 1974. Continental margins of the Gulf of Mexico. In C.A. Burk & C.L. Drake (eds.), The Geology of Continental Margins. New York: Springer–Verlag. pp. 683–693.

Aranda G., J., H.G. Labarthe & M. Tristán G. 1983. El Volcaniso Cenozoico en San Luis Potosí y su relación con la Provincia Volcánica de la Sierra Madre Occidental. Asoc. Ings. Min. Met. Geol. Méx., 157. Conv. Geol. Nal. Mem. pp. 261–287.

Atwater, T. 1970. Implications of plate tectonics for the Cenozoic tectonic evolution of western North America. Bull. Geol. Soc. Amer. 81:3513– 3536.

Baker, Ch., L. 1971. Geologic Reconnaissance in the Eastern Cordillera of Mexico. Geol. Soc. Amer. Special Paper 131, pp. 1–83.

Barker, R.W. & W.H. Blow. 1975. Biostratigraphy of some Tertiary formations in the Tampico–Misantla Embayment, Mexico. J. Foraminiferal Res. 6(1):39–58

Barnetche, A. & L.V. Illing. 1956. The Tamabra Limestone of the Poza Rica Oil Field, Ver. XX Int. Geol. Congr. Méx. pp. 1–38.

Barrera, T. 1931. Zonas Mineras de los Estados de Jalisco y Nayarit. Inst. Geol. Méx. Bol. 51:1–140.

————. 1946. Guía geologica de Oaxaca. UNAM. Inst. Geol. pp. 1–101.

Bateson, J.H. 1972. New interpretation of geology of Maya Mountains, British Honduras. Amer. Assoc. Petrol. Geol. Bull. 56(5):956–963.

Bazán B., S. 1984. Litoestratigrafía y rasgos estructurales del Complejo Oaxaqueño, Mixteca Alta, Oaxaca. Geomimet. No. 129. pp. 35–63.

Beal, C.H. 1948. Reconnaissance of the geology and oil possibilities of Baja California, Mexico. Geol.Soc. Amer. Mem. 31:1–138.

Belt, B.C. 1925. Stratigraphy of the Tampico District of Mexico. Amer. Assoc. Petrol. Geol. Bull. 9(1):136–144.

Besch, T.H., J.F.W. Negendank, R. Emermann & H.J. Tobschall. 1988. Geochemical constraints of the origin of calcalkaline and alkaline magmas of the eastern Trans–Mexican Volcanic Belt. Geof. Int. 27(4):641–663.

Bishop, W.F. 1980. Petroleum geology of northern Central America. J. Petrol. Geol. 3:3–59.

Blair, T.C. 1988. Mixed siliciclastic-carbonate marine and continental syn-rift sedimentation, Upper Jurassic-lowermost Cretaceous Todos Santos and San Ricardo Formations, western Chiapas, Mexico. J. Sedim. Petrol. 58:623–636.

Bloomfield, K. 1975. A Late Quaternary monogenetic field in Central Mexico. Geol. Rundschau 64(2):476–497.

————. & L. Cepeda D. 1973. Oligocene alkaline igneous activity in N.E. Mexico. Geol. Mag. 110:551–559.

Bobier, C. & C. Robin. 1983. Paleomagnetismc de la Sierra Madre Occidental dans les états de Durango et Sinaloa (Mexique): variations du champ au rotations de blocs au Paléocéne et au Neogéne. Geof. Int. 22(1):57–86.

Böhnel, H., L. Alva V., S. González H., J.H. Urrutia F., J.D. Morán Z. & P. Schaaf. 1989. Paleomagnetic data and the accretion of the Guerrero Terrane, Southern Mexico Continental Margin. UNAM. Inst. Geof. Comun. Tecn. Ser. Invest. No. 88, pp. 1–24.

Bolívar, J. Ma. 1963. Geología del área limitada por El Tomatal, Huitzuco y Mayanalán, Edo. de Guerrero. UNAM. Inst. Geol. Bol. 69:1–35.

Bonneau, M. 1971. Nueva area cretácica fosilifera en Sinaloa. Soc. Geol. Mex. Bol. 32(2):159–167.

Böse, E. 1905. Reseña acerca de la geología de Chiapas y Tabasco. Inst. Geol. Méx. Bol. 20:1–150.

———. 1906. La Fauna de Moluscos del Cononiano de Cárdenas, S.L.P. Inst. Geol. Méx. Bol. 24:1–95.

———. 1910. Monografía geológica y paleontológica del Cerro de Muleros cerca de Cd. Juárez, Edo. de Chihuahua y descripción de la fauna cretácea de la Encantada, Placer de Gpe., Edo. de Chihuahua, Mex. Inst. Geol. Méx. Bol. 25:91.

———. 1921. On the Permian of Coahuila, Northern Mexico. Amer. J. Sci. (5th Ser.) 1:187–194.

———. 1923. Algunas faunas cretácicas de Zacatecas, Dgo. y Gro. Inst. Geol. Méx. Bol. 42:1–219.

Bridges, L.W. 1965. Geología del Area de Plomosas, Chihuahua. UNAM. Inst. Geol. Bol. 74:1–134.

———. & R.K. De Ford. 1961. Pre–carboniferous paleozoic rocks in central Chihuahua, Mexico. Amer. Assoc. Petrol. Geol. Bull. 45(1):98–104.

Brunner, P. 1975. Estudio Estratigráfico del Devónico en el Area de Bisani, Caborca, Sonora. Inst. Mex. Petrol. Rev. 7(1):16–45.

———. 1976. Litología y Bioestratiqrafía Mississípica en el Area de El Bisani, Sonora. Inst. Mex. Petrol. Rev. 8(3):7–41.

Bryant, W.R., J. Antoine, M. Ewing & B. Jones. 1968. Structure of the Mexican continental shelf and slope Gulf of Mexico. Amer. Assoc. Petrol. Geol. Bull. 52:1204–1228.

———, A.A. Meyerhoff, N.K. Brown Jr., M.A. Furrer, T.E. Pyle & J.W. Antoine. 1969. Escarpments, reef trends and diapiric structures in eastern Gulf of Mexico. Amer. Assoc. Petrol. Geol. Bull. 53(12):2506–2542.

Buffler, R.T., F.J. Shaub, J.S. Watkins & J.L. Worzel. 1979. Anatomy of the Mexican Ridges Foldbelt, southwestern Gulf of Mexico. *In* J.S. Watkins, L. Montadert & P.W. Dickerson (eds.), Geological and Geophysical Investigations of Continental Margins. Amer. Assoc. Petrol. Geol. Mem. 29:319–327.

Burbach, G.V., C. Frohlich, W.D. Pennington, & T. Matumoto. 1984. Seismicity and tectonics of the subducted Cocos Plate. J. Geophys. Res. 89(B9):7719–7735.

Burckhardt, C. 1930. Etude synthetique sur de Mesozoique Mexicain. Soc. Paleont. Suisse Mem. 49–50:1–280.

Burke, K., C. Cooper, J.F. Dewey, P. Mann & J.L. Pindell. 1984. Caribbean tectonics and relative plate motions. *In* W.E. Bonini, R.B. Hargrave & R. Shagom (eds.), The Caribbean–South American Plate Boundary and Regional Tectonics. Geol. Soc. Amer. Mem. 162:31–63.

Burrows, R.H. 1910. Geology of northern Mexico. Soc. Geol. Mex. Bol. 7:85–103.

Butterlin, J. & F. Bonet. 1958. Reconocimiento geológico preliminar del territorio de Quintana Roo. Asoc. Mex. Geol. Petrol. Bol. 10(9–10):531–570.

——— & ———. 1962. Las Formaciones Cenozoicas de la parte Mexicana de la Península de Yucatán. UNAM. Inst. Geol. pp. 4–44 .

Cabrera, R. 1953. Estudio biostratigráfico de la porción occidental de la Cuenca Salina del Istmo de Tehuantepec. Asoc. Mex. Geol. Petrol. Bol. 7(5–6):178–212.

Calderón G., A. 1951. Condiciones estratigráficas de las formaciones Miocénicas de la Cuenca Salina del Istmo de Tehuantepec. Asoc. Mex. Geol. Petrol. Bol. 3(7–8):229–258.

———. 1956. Bosquejo geológico de la región de San Juan Raya, Puebla. XX Cong. Geol. Int. Méx. Libro Guía Excursión A–ll. pp. 9–33.

Cameron, K. L., M. Cameron, & B. Barreiro. 1986. Origin of voluminous Mid–Tertiary ignimbrites of the Batopilas Region, Chihuahua: Implications for the formation of continental crust beneath the Sierra Madre Occidental. Geof. Int. 25(1):39–60.

Cameron, M., K. Spaulding & K.L. Cameron. 1987. A synthesis and comparison of the geochemistry of volcanic rocks of Sierra Madre Occidental and Mexican Volcanic Belt. Geof. Int. 26(1):29–84.

Campa, M.F. 1978. La evolución tectónica de Tierra Caliente. Soc. Geol. Mex. Bol. 39(2):52–64.

——, M. Campos, R. Flores & R. Oviedo. 1974. La secuencia Mesozoica volcano–sedimentaria–metamorfizada de Ixtapan de la Sal, Teloloapan, México. Soc. Geol. Mex. Bol. 35:7–28.

——, R. Flores, M. Limón, R. Ramírez B., J. Ramírez E. & M. Vazquez. 1977. La evolución tectónica y la mineralización en la región de Valle de Bravo, Méx. e Iguala, Gro. Asoc. Ing. Min. Met. Geol. Mex. XII Cong. Nal. Mem. pp. 143–170.

——& J. Ramírez. 1979. La evolución geológica y la metalogénesis del noroccidente de Guerrero. Univ. Autón. Guerrero, Serie Técnico–Científico. No. 1. pp. 1–102.

—— & P.J. Coney. 1983. Tectonostratigraphic terranes and mineral resource distributions in Mexico. Can. J. Earth Sci. 20:1040–1051.

Cárdenas V., J. 1966. Contribución al conocimiento geológico de la Mixteca Oaxaqueña. Asoc. Mex. Ings. Min. Met. Geol. Min. Metal. 38:15–107.

Cárdenas D., F. 1983. Volcanic stratigraphy of Sierra Peña Blanca, Chihuahua, Mexico. *In* K.F. Clark & P.C. Goodell (eds.), Geology and Mineral Resources of North Central Chihuahua. El Paso Geol. Soc. Guidebook, Field Conf. pp. 325–334.

Carfantan, J.C. 1977a. La cobijadura de Motozintla–Un paleoarco volcánico en Chiapas. UNAM. Inst. Geol. Rev. 1(1):133–137.

——. 1977b. El prolongamiento del sistema de fallas Polochic Motagua en el SE de Méx: una frontera entre dos provincias. UNAM. Inst. Geol. Rev. 1(2):133–138.

——. 1984. Evolución estructural del sureste de México, paleogeografía e historia tectónica de las zonas internas Mesozoicas. UNAM. Inst. Geol. Rev. 5(2):207–216.

——. 1983. Les ensembles géologiques du Mexique Meridional: evolution géodynamique durante leMesozóique et le Cénozoique. Geof. Int. 22(1):39–56.

Carr, M.J., R.E. Stoiber & C.L. Drake. 1974. The segmented nature of some continental margins. *In* C.A. Burk & C.L. Drake (eds.), The Geology of Continental Margins. New York: Springer–Verlag, pp. 105-114.

Carrasco N., G., M. Milan & S.P. Verma. 1988. Fases volcánicas de la Caldera de Amealco. Geomimet. 153:69–83.

Carrillo B., J. 1961. Geología del Anticlinorio Huizachal–Peregrina al Noroeste de Ciudad Victoria, Tamaulipas. Asoc. Mex. Geol. Petrol. Bol. 13:1–98.

——. 1965. Estudio de una parte del Anticlinorio de Huayacocotla. Asoc. Mex. Geol. Petrol. Bol. 17:73–96.

——. 1971. La Plataforma Valles–San Luis Potosí. Asoc. Mex. Geol. Petrol. Bol. 23(1–6):1–112.

Case, J.E., T.I. Holcombe & R.G. Martin. 1984. Map of geologic provinces in the Caribbean region. *In* W.E. Bonini, R.B. Hargraves & R. Shagam (eds.), The Caribbean–South American Plate Boundary and Regional Tectonics. Geol. Soc. Amer. Mem. 162:1–30.

Castillo T., C. 1955. Bosquejo estratigráfico de la cuenca Salina del Istmo de Tehuantepec. Asoc. Mex. Geol. Petrol. Bol. 7(5–6):175–212.

Castro M., J., J. Shlaepfer C. & E. Martínez. 1975. Estratigráfia y microfacies del Mesozoico de la Sierra Madre del Sur, Chiapas. Asoc. Mex. Geol. Petrol. Bol. 27(1–3):1–103.

Cepeda, L. 1967. Estudio petrológico y mineralogico de la región de El Cubo, Municipio de Guanajuato. Asoc. Mex. Geol. Petrol. Bol. 19(7–I2):39–107.

Chase, C.G., H.W. Menard, R.L. Larson, G.F. Sharman III & S.M. Smith. 1970. History of sea–floor spreading west of Baja California. Bull. Geol. Soc. Amer. 81:491–498.

Chubb, L.J. 1959. Upper Cretaceous of Central Chiapas, Mexico. Amer. Assoc. Geol. Petrol. Bull. 43(4):725–755.

Clabaugh, S.E. 1972. Geological roadlog Durango–Mazatlan. Soc. Geol. Mexicana, II Conv. Nal., Mazatlan, Sin., Libro–Guía Excursión Durango–Mazatlán, pp. 80–96.

Clark, K.F., P.E. Damon, S.R. Schutte & M. Shaffiquillah. 1980. Magmatismo en el norte de México en relación con los yacimientos metaliferos. Geomimet 106:49–71.

———. C.T. Foster & P.E. Damon. 1982. Cenozoic mineral deposits and subduction related magmatic arcs in Mexico. Bull. Geol. Soc. Amer. 93:533–544.

——— & P.C. Goodell (eds.). 1983. Geology and mineral resources of northcentral Chihuahua. El Paso Geol. Soc. Field Conference, Guidebook.

Clemons, R.E. & D.F. McLeroy. 1966. Hoja Torreón 13R–l(l), estados de Coahuila y Durango, con resumen de la geología. UNAM. Inst. Geol. Carta Geol. México, Serie 1:100,000.

Comité de la Carta Geológica. 1976. Carta Geológica de la Republica Mexicana. 4th Ed. Comité de la Carta Geológica de México, Esc. 1:2,000,000.

Coney, P. 1983. Un Modelo Tectónico de México y sus Relaciones con América del Norte, América del Sur y El Caribe. Inst. Mex. Petrol. Rev. 15(1):6–15.

Congreso Geológico Internacional. X Sesión, México, 1906 (33 publications were issued).

———. XX Sesión, México, 1956 (64 publications were issued).

Contreras, H. 1958. Resumen de la geología de la parte media del estado de Tabasco y del norte del estado de Chiapas. Asoc. Mex. Geol. Petrol. Bol. 10(3–4):193–210.

Coogan, A.H., D.G. Bebout & C. Maggio. 1972. Depositional enviroments and geologic history of golden line and Poza Rica trend, an alternative view. Amer. Assoc. Petrol. Geol. Bull. 56(8):1419–1457.

Cook, T.D. & A.W. Bally (eds.); S. Milner, R.T. Buffler, R.E. Farmer & D.K. Clark (compilers); W.C. Hever, J.W. Loofs, J.F. Lavarde & H.G. Villarreal Jr. (cartogr. & draft). 1975. Stratigraphic Atlas of North and Central America. Princeton, N.J., Houston, TX: Shell Oil Co. & PrincetonUniv. Press.

Cooper, G.A., A.R. Arellano, J.H. Johnson, V. Okulitch, A. Stoyanow & C. Lochman. 1952. Cambrian stratigraphy and paleontology near Caborca, Northwest Sonora, Mexico. Smithsonian Misc. Coll. 119:1–184.

———. 1956. Geología y paleontología de la región de Caborca, Norponiente de Sonora. XX Congr. Geol. Int. México. Contrib. UNAM. Inst. Geol. pp. 1–259.

———, O.C. Dunbar, H. Duncan, K.A. Miller & B.J. Knight. 1965. La fauna pérmica del oeste del Antimonio, Sonora. UNAM. Inst. Geol. Bol. 58(3):1–119.

Córdoba M., D.A. 1963. Estudios geológicos en los estados de Durango y San Luis Potosí (Parte 1). UNAM. Inst. Geol. Bol. 71:1–63.

———. 1969. Hoja Cd. Juárez 13R–4(3). Edo. de Chihuahua, con resumen de la geología. UNAM. Inst. Geol, Carta Geol. Méx. Ser. 1:100,000

————. 1970. Mesozoic stratigraphy of northeastern Chihuahua, Mexico. In K. Seewald & D. Sundeen (eds.), The Geologic Framework of the Chihuahua Tectonic Belt. West Texas Geol. Soc. & Univ. Texas–Austin Symposium in honor of Professor Ronald K. DeFord. Midland, TX: West Texas Geol. Soc. pp. 91–96.

————., S.A. Wengerd, & J. Shomaker (eds.). 1969. Guidebook of the Border Region. New Mexico Geol. Soc.

Cortés O., S., L. Torón V., J.J. Martínez B., J. Pérez L., J. Gamboa A., A. Cruz C., & M. Puebla P. 1957. La cuenca carbonífera de la Mixteca. México City Banco de México, S. A. Vols. 1 & 2.

————, R. Elvir A., J. Gamboa A. & F. García C. 1964. Recorrido Geológico–Minero de Culiacán, Sin. a Tepehuanes, Dgo. Consj, Rec. Nat. no Renov. Publ. 14–E. pp. 1–11.

Cruz H., P., R. Verdugo V. & R. Barcenas P. 1977. Origin and distribution of Tertiary Conglomerates, Veracruz Basin Mexico. Amer. Assoc. Geol. Petrol. Bull. 61(2):207–228.

Damon, E.P. 1984. Evolución de los Arcos Magmáticos en México y su relación con la metalogenesis. UNAM. Inst. Geol. Rev. 5(2):223–238.

————, E.V. Livingston, L.R. Mauger, J.B. Gile H. & J. Pantoja A. 1962. Edad del Precambrico anterior y de otras rocas del zócalo de la región de Carborca-Altar, de la parte NW del estado de Sonora, estudios geocronológicos de Rocas Mexicanas. UNAM., Inst. Geol. Bol. 64:11–44.

————, M. Shafiquíllah & K.F. Clark. 1981. Age trends of igneous activity in relation to metallogenesis in southern Cordillera. In W.R. Dickinson et al. (eds.), Relations of Tectonics to Ore Deposits in the Southern Cordillera. Arizona Geol. Soc. Digest. 14:137–154.

D'Anglejan, B.F. 1963. Sobre la presencia de fosforitas marinas frente a Baja California. Soc. Geol. Mex. Bol. 24(2):1–95.

Darton, N.H. 1921. Geologic reconnaissance in Baja California. J. Geol. 29:720–748.

Daugherty, F.W. 1963. Late Cretaceous stratigraphy in Northern Coahuila. Amer. Assoc. Petrol. Geol. Bull. 47(12):2059–2064.

De Cserna, Z. 1956. Tectónica de la Sierra Madre Oriental de México, entre Torreon y Monterrey. XX Internatl. Geol. Congr. México. pp. 1–87.

————. 1960. Orogenesis in time and space in Mexico. Geol. Rundschau 50:595–605.

————. 1961. Tectonic Map of Mexico. Boulder, CO, Geol. Soc. America. Scale 1:5,000,000.

————. 1965. Reconocimiento geológico en la Sierra Madre del Sur de México, entre Chilpancingo y Acapulco, estado de Guerrero. UNAM. Inst. Geol. Bol. 62:1–77.

————. 1969. Tectonic framework of southern Mexico and its bearing on the problem of continental drift. Soc. Geol. Mex. Bol. 30:159–168.

————. 1971. Precambrian sedimentation, tectonics and magmatism in Mexico. Geol. Rundschau 60:1488–1513.

————. 1974. La evolución geológica del panorama fisográfico actual de México. In Z. De Cserna, P. Mosiño, A. & O. Benassini. El Escenario Geográfico, Introducción Ecológica, México, D. F. Inst. Nal. Antropol. Hist. pp. 19–56.

————. 1976. Geology of the Fresnillo, Area Zac. Mexico. Bull. Geol. Soc. Amer. 87:1191–1199.

————. 1978. Notas sobre la geología de la región comprendida entre Iguala, Cd. Altamirano y Temascaltepec, Estados de Guerrero y México. Soc. Geol. Mex. Libreto–Guía Excursión Geol. Tierra Caliente. pp. 1–25.

————. 1989. An outline of the geology of Mexico. In A.W. Bally & A.R. Palmer (eds.), The Geology of North America—an overview. Geol. Soc. America Volume A. 223–264.

————— & B.H. Kent. 1961. Mapa geológico de reconocimiento y secciones estructurales de la región de San Blas y El Fuerte, estados de Sinaloa y Sonora. UNAM. Inst. Geol. Cartas Geol. y Min. No. 4.

—————, E. Schmitter, E.P. Damon, D.E. Livingston & L.J. Kulp. 1962. Edades isotópicas de rocas metamórficas del centro y sur de Guerrero y de una monzonita cuarcífera del norte de Sinaloa. UNAM. Inst. Geol. Bol. 64:71–84.

————— & B. Bello. 1963. Estudios geológicos de los estados de Durango y San Luis Potosí. UNAM. Inst. Geol. Bol. 71:1–63.

—————, C. Fries, C. Rincón, H. Westley, J. Solorio & E. Schimitter. 1975. Edad precámbrica tardía del Esquisto Taxco, Estado de Guerrero. Asoc. Mex. Geol. Petrol. Bol. 26:183–193..

—————, J.L. Graf Jr. & F. Ortega, G. 1977. Aloctono del Paleozóico interior en la región de ciudad Victoria, estado de Tamaulipas. UNAM. Inst. Geol. Rev. l:33–43.

—————, C. Armtrong, R. Y. & J. Solorio. 1978. Rocas metavolcánicas e intrusivos relacionados paleozóicos de la región de Petatlán, estado de Guerrero. UNAM. Inst. Geol. Rev. 2(1):1–7.

—————, M. De la Fuente–Duch, M. Palacios–Nieto, L. Triay, L.M. Mitre–Salazar, R. Mota–Palomino. 1988. Estructura geológica, gravimetria, sismicidad y relaciones neotectónicas regionales de la cuenca de México. UNAM. Inst. Geol. Bol. 104:1–71.

Demant, A. 1975. Caracteres químicos principales del vulcanismo terciario y cuaternario de Baja California Sur: Relaciones con la evolución del margen contienental Pacífico de México. UNAM. Inst. Geol. Rev. 1(1):19–69.

—————. 1978. Características del Eje Neovolcánico Transmexicano y sus problemas de interpretación. UNAM. Inst. Geol. Rev. 2(2):172–187.

—————. 1984. Interpretación Geodinámica del Volcanismo del Eje Neovolcánico Transmexicano. UNAM. Inst. Geol. Rev. 5(2):217–222.

————— & C. Robin. 1975. Las fases del volcanismo en México, una síntesis en relación con la evolución geodinámica desde el Cretácico. UNAM. Inst. Geol. Rev. 75(1):70–83.

Dengo, G. 1973. Estructura Geológica, Historia Tectónica y Morfología de América Central. 2nd ed. Guatemala: Instituto Centroamericano de Investigación y Tecnología Industrial, Centro Regional de Ayuda Técnica, Agencia para el Desarrollo Internacional.

————— & O. Bohnenberger. 1969. Structural development of northern Central America. *In* A.R. McBirney (ed.), Tectonic Relations of Northern Central America and the Western Caribbean, the Bonnacca Expedition. Amer. Assoc. Petrol. Geol. Mem. 11:203–220.

Denison, R.E., W.H. Burke Jr., E.A. Hetherington & J.B. Otju. 1970. Basement rock framework of parts of Texas, southern New Mexico and northern Mexico. *In* K. Seewald & D. Sundeen (eds.), The Geologic Framework of the Chihuahua Tectonic Belt. West Texas Geol. Soc. & Univ. Texas-Austin Symposium in honor of Professor Ronald K. DeFord. Midland, TX: West Texas Geol. Soc. pp. 3–14.

Díaz G., T. & G.A. Navarro. 1964. Litología y correlación estratigráfica del Paleozóico Superior en a Región de Palomas, Chih., Méx. Asoc. Mex. Geol. Petrol. Bol. 16:107–120.

Dickerson, R.E. & W.S. Kew. 1917. The fauna of a medial tertiary formation and the associated horizons of northeastern Mexico. Proc. Cali. Acad. Sci. (4th. Ser). 7:125–156.

Dickinson, W.R. 1979. Plate tectonics and the continental margin of California. *In* W.G. Ernest (ed.), The Geotectonic Development of California (Rubey Vol. I). New York: Prentice Hall. pp. 1–28.

Drewes, H. 1978. The Cordilleran orogenic belt between Nevada and Chihuahaua. Bull. Geol. Soc. Amer. 89:641–657.

Duffield, W.A., R.I. Trilling & R.F. Canul. 1984. Geology of the Chichón volcano, Chiapas. J. Volcanol. Geotherm. Res. 20:117–132.

Dumble, E.T. 1918. Geology of the northern end of the Tampico embayment area. Proc. Cali. Acad. Sci. (4th. Ser). 8:113–156.

Eardly, A.J. 1951. Structural Geology of North America. New York: Harper & Row.

Echegoyen S., J., S. Romero M. & S. Velázquez, S. 1970. Geología y yacimientos minerales de la parte central del Distrito Minero de Guanajuato. Cons. Rec. Nat. No Renov. Bol. 75:1–36.

Edwards, J.D. 1955. Studies of some Early Tertiary red conglomerates of central Mexico. U.S. Geol. Serv. Prof. Paper 264 H. pp. 153–185.

Elías H., M. 1989. Geología Metamórfica del área de San Lucas del Maiz, Estado de México. UNAM. Inst. Geol. Bol. 105:1–78.

Enciso de la Vega, S. 1963. Hoja Nazas 13 R–K(6), estado de Durango, con resumen de la geología. UNAM. Inst. Geol. Carta Geol. Méx. Ser. 1:100.000.

————. 1968. Hoja Cuencame 13 R–1(7), estados de Durango y Chihuahua, con resumen de la geología. UNAM. Inst. Geol. Carta Geol. Méx. Ser. 1:100,000 .

Enos, P. 1974. Reefs, platforms and basins of Middle Cretaceous in northeast Mexico. Asoc. Mex. Geol. Petrol. Bol. 26(4):1–9.

Erben, H.K. 1956a. El Jurásico Medio y el Calloviano de México. XX Congr. Geol. Int. Mexico. 1–140 .

————. 1956b. El Jurásico Inferior de México y sus Amonitas. XX Congr. Geol. Int. México. 1–393.

Ferriz, H. & G.A. Mahood. 1986. Volcanismo Riolitico en el Eje Neovolcanico Mexicano. Geof. Int. 25(1):117–156.

Ferrusquía V., I. 1970. Geología del área Tamazulapan–Teposcolula–Yanhuitlán, Mixteca Alta, Estado de Oaxaca. Soc. Geol. Mex. Exc. México–Oaxaca, Libro–Guía. pp. 97–119.

————. 1976. Estudios geológico–paleontológicos en la región Mixteca. Parte 1. Geología del área Tamazulapan–Teposcolula–Yanhuitlán, Mixteca Alta, Estado de Oaxaca, México. UNAM. Inst. Geol. Bol. 97:1–160.

————, J.A. Wilson, R.E. Denison, F.W. McDowell & J.G. Solorio M. 1974. Tres edades radiométricas oligocénicas y miocénicas de rocas volcánicas de las regiones Mixteca Alta y Valle de Oaxaca, estado de Oaxaca. Asoc. Mex. Geol. Petrol. Bol. 26(4–6):249–262.

————, S.P. Applegate & L. Espinosa. 1978. Rocas volcanosedimentarias mesozoicas y huellas de dinosaurios en la región suroccidental pacifica de México. UNAM. Inst. Geol. Rev. 2(2):150–162.

Fiala, J. Hanus, V., J. Vanek & V. Vankova. 1982. Comparison of radioactivity and geochemistry of rocks from plutonic bodies adjacent to the southern part of the Golfo de California. Geof. Int. 21(1):11–40.

Filmer, P.E. & J.L. Kirschvink. 1989. A paleomagnetic constraint on the Late Cretaceous paleoposition of northwestern Baja California, Mexico. J. Geophys. Res. 94:7332–7342.

Finch, J.W. & P.L. Abbott. 1977. Petrology of a Triassic marine section, Vizcaino Peninsula, Baja California Sur, Mexico. Sediment. Geol. 19:253–273.

————, E.A. Pessagno & P.L. Abbott. 1979. San Hipólito Formación: Triassic marine rocks of the Vizcaino Peninsula, field guides and papers of Baja California. Geol. Soc. Amer. Ann. Mtg. San Diego. pp 117–120.

Fisher, R.L. 1961. Middle American Trench, topography and structure. Bull. Geol. Soc. Amer. 72:703-720.

Flawn, P.T. 1961. Rocas metamórficas en el armazón tectónico de la parte septentrional de México. Asoc. Mex. Geol. Petrol. Bol. 13:105–116.

——— & R.A. Maxwell. l958. Metamorphic rocks in the Sierra del Carmen, Coah. Amer. Assoc. Petrol. Geol. Bull. 42:2245–2249.

Flores, T. 1927. Reconocimientos geológicos en la región central del estado de Sonora. Inst. Geol. Méx. Bol. 49:1–287.

Flores, G. 1952. Geology of northern British Honduras. Amer. Assoc. Petrol. Geol. Bull. 36(2):404–408.

Franco R., M. 1978. Estratigráfia del Albiano–Cenomaniano en la región de Naica, Chihuahua. UNAM. Inst. Geol. Rev. 2(2):132–149.

Fries, C. Jr. 1960. Geología del estado de Morelos y de partes adyacentes de México y Guerrero, región central meridional de México. UNAM. Inst. Geol. Bol. 60:1–236.

———. 1962. Reseña geológica del estado de Sonora, con enfasis en el Paleozóico. Asoc. Mex. Geol. Petrol. Bol. 14:257–273.

———. 1966. Hoja Cuernavaca 14 Q–H(8), estado de Morelos, con resumen de la geología. UNAM. Inst. Geol. Carta Geol. Mex. Ser. 1:100,000.

———, C.W. Hibbard & D.H. Dunkle. 1955. Early Cenozoic vertebrates in the red conglomerates at Guanajuato. Smithsonian Misc. Coll. 123(7):1–25.

———, E. Shmitter, E.P. Damon & D.F. Livingston. 1962. Rocas Precámbricas de edad grenvilliana de la parte central de Oaxaca en el sur de México. UNAM. Inst. Geol. Bol. 64:45–53

——— & C. Rincon, O. 1965a. Nuevas aportaciones geocronológicas y tectónicas empleadas en el Laboratorio de Geocronómetría. UNAM. Inst. Geol. Bol. 73:57–133.

———. 1965b. Contribuciones del Laboratorio de Geocronómetría. UNAM. Inst. Geol. Bol. 75:59–133.

——— & J. Valencia. 1965. Contribuciones del Laboratorio de Geocronómetria (parte 3). UNAM. Inst. Geol. Bol. 73:135–191.

Frizzell V., A., Jr. (ed.). 1984. Geology of the Baja California Peninsula. Los Angeles: Soc. Econom. Palent. Min. Pacific. Sect.

Frost, S.H. & R.L. Langenheim Jr. 1974. Cenozoic Reef Biofacies. De Kalb, IL: Northern Illinois Univ. Press.

Gallagher, D. 1952. Geologic pap of the Canoas Quick Silver district, state of Zacatecas, Mex. U.S. Geol. Surv. Bull. 975:47–85

Galloway, W.E. 1989. Genetic stratigraphic sequences in basin analysis II. Application to Northwest Gulf of Mexico Cenozoic Basin. Amer. Assoc. Petrol. Geol. Bull. 73(2):143–154.

García C., J. 1968. Hoja El Salado 14R–j(ll), estados de Zacatecas y San Luis Potosí, con resumen de la geología. UNAM. Inst. Geol. Carta Geol. Méx. Ser. 1:100.000.

García, E. 1988. Modificaciones al sistema de classificación climática de Köppen (para adaptarlo a las condiciones de la República Mexicana). UNAM. Inst. Geogr.

Garfias, V. & T. Chapin. 1944. Geología de México. México City: Editorial Jus.

Garrison, L. & R.G. Martin Jr. 1973. Geological structures in the Gulf of Mexico Basin. U.S. Geol. Surv. Prof. Paper 773:1–85.

Garza F., R. 1973. Modelo sedimentario del Albiano–Cenomaniano en la porción SE de la Plataforma de Coahuila. Asoc. Mex. Geol. Petrol. Bol. 25 (7–9):311–399.

Gasca D., A. & M. Reyes C. 1977. La cuenca lacustre Plio–Pleistocénica de Tula-Zumpango. INAH. Inf. Tec. No. 2. 1–85.

Gastil, G. & W. Jensky. 1973. Evidence for strike–slip displacement beneath the Trans–Mexican Volcanic Belt. Stanford Univ. Publ. Geol. Sci. 13:171–180.

————, R.P. Phillips & E.C. Allison. 1975. Reconnaissance geology of the state of Baja California Norte. Geol. Soc. Amer. Mem. 14:1–170.

————, D. Krummenacher, J. Doupont, J. Bushee, W. Jensky & D. Barthelmy. 1976. La zona Batolítica del sur de California y el Occidente de México. Soc. Geol. Mex. Bol. 37:80–90.

————, G. Morgan & D. Krummenacher. 1981. The tectonic history of peninsular California. *In* W.G. Ernest (ed.), The Geotectonic Development of California (Rubey vol. 1). New York: Prentice Hall. pp. 285–305.

Geyne, A. R., C. Fries Jr., K. Segerstrom, R.F. Black & I.F. Wilson. 1963. Geología y Yacimientos Minerales del Distrito Pachuca–Real del Monte, Hgo. México. Cons. Rec. Nat. no Renov. Publ. 5–E. pp. 1–215.

González A., J. 1976. Resultados obtenidos en la exploración de la Plataforma de Córdoba y principales campos productores: Soc. Geol. Mex. Bol. 37(2):53–60.

González, R. 1976. Bosquejo geológico de la Zona Norte. Asoc. Mex. Geol. Petrol. Bol. 28(1–2):2–49.

González R., J. 1956. Memoria Geologica Minera del Estado de Chihuahua: XX Congr. Geol. Int. Méx.

————. 1961. Las pegmatitas graníticas de Santa Ana. Telixtlahuaca, Oax. Soc. Geol. Mex. Bol. 24(2):39–50.

———— & D.E. White. 1947. Los yacimientos de Antimonio de San José, Sierra del Catorce, Edo. de San Luis Potosí Comit. Direct. Invest. Rec. Min. Méx. Bol. 14:1–31.

Göse, W.A. & L.A. Sanchez, B. 1981. Paleomagnetic results from southern Mexico. Geof. Int. 20:163–175.

Gries, C. J. & W.T. Haenggi. 1970. Structural evolution of the eastern Chihuahua Tectonic Belt. *In* K. Seewald & D. Sundeen (eds.), The geologic framework of the Chihuahua Tectonic Belt. West Texas Geol. Soc. & Univ. Texas–Austin Symposium in honor of Professor Ronald K. DeFord. Midland, TX: West Texas Geol. Soc. pp. 119–137.

Gunderson, R., K. Cameron & M. Cameron. 1986. Mid–Cenozoic k–calc–Alcalik and Alkalik volcanism in eastern Chihuahua, Mexico: geology and geochemistry of the Benavides–Pozas area. Bull. Geol. Soc. Amer. 97(6):737–753.

Gutiérrez, R. 1956. Bosquejo geológico del estado de Chiapas. XX Congr. Geol. Int. Méx. Excursión C–15 (Geología del Mesozóico y Estratigráfia Pérmica del estado de Chiapas). pp. 7–32.

Guzmán, E.J. 1950. Geología del Noreste de Guerrero. Asoc. Mex. Geol. Petrol. Bol. 1(2):79–156.

————. & de Cserna, Z. 1963. Tectonic history of Mexico. *In* O.E. Chiids & W.B. Beebe (eds.), Backbone of the Americas Tectonic History from Pole to Pole. Amer. Assoc. Petrol. Geol. Mem. 2:113–129.

Hagstrum, J.T., M. McWilliams, D.G. Howell & S. Grommé. 1985. Mesozoic paleomagnetism and northward translation of the Baja California Peninsula. Bull. Geol. Soc. Amer. 96(8):1077–1090.

Hall, I.H.S. & J.H.A. Bateson. 1972. Late Paleozoic Lavas in Maya Mountains, British Honduras and their possible regional significance. Amer. Assoc. Petrol. Geol. Bull. 56:950–956.

Handschy, J.W. & R. Dyer. 1987. Polyphase deformation in Sierra del Cuervo Chihuahua, Mexico: evidence for ancestral Rocky Mountain tectonics in the Ouachita foreland of northern Mexico. Bull. Geol. Soc. Amer. 99(5):618–632.

Hanus, V. & J. Vanek. 1978. Subduction of the Cocos Plate and deep active fracture zones of Mexico. Geof. Int. 17:14–53.

Hasenaka, T. & I.S.E. Carmichel. 1985. The cinder cones of Michoacan-Guanajuato, Central Mexico, their age, volume, distribution and magama discharge rate. J. Volcanol. Geotherm. Res. 25:105–124.

Hausback, B.P. 1988. Miocene paleomagnetism of Baja California Sur; evidence concerning the structural development of western Mexico. Geof. Int. 27(4):463–484.

Heim, A. 1922. Notes on the Tertiary of southern lower California. Geol. Mag. 59:529–547.

———. 1940. The front ranges of the Sierra Madre Oriental Mexico, from C. Victoria to Tamazunchale. Eclog. Geol. Helv. 33:313–362.

Henry, C.D., G.J. Price & F.W. McDowell. 1983. Presence of the Rio Grande Rift in West Texas and Chihuahua. *In* K.F. Clark & P.C. Goodell (eds.), El Paso Geol. Soc. Guidebook Field Conference. pp. 108–119.

——— & G.J. Price. 1986. Early basin and range development in Trans–Pecos Texas and adjacent Chihuahua: magmatism and orientation, timing and style of deformation. J. Geophys. Res. 91(B–6):6213–622

Hernández G., R. 1973. Paleogeografía del Paleozoico de Chiapas. Asoc. Mex. Geol. Petro. Bol. 25:79–113

Hernández L., D. 1981. Estratigrafía de la región central de Aguascalientes, Ags. México. Soc. Geol. Mex. Gaceta Geol. 6(31):17–40.

Hernández, S. 1959. Posibilidades Petrolíferas en la porción Norte de la Cuenca de Macuspana, Tab. Asoc. Mex. Geol. Petrol. Bol. 11(11–12):619– 668.

Hisazumi, H. 1922. El distrito sur de Baja California. Inst. Geol. Méx. 5:41– 82.

———. 1929. Informe geológico preliminar de la parte norte del Estado de Sinaloa. Inst. Geol. Mex. 3:95–109.

———. 1932. Geología de la región Mixteca del estado de Oaxaca. UNAM. Inst. Geol. Unpublished Reporte.

Holguín O., N. 1978. Estudio Estratigráfico del Cretácico Inferior en el Norte de Sinaloa. Inst. Mex. Petrol. Rev. 10(l):1–13.

Hubp J.L., A. Ortiz M., L. Palacios J. & C.B. Vendinelle. 1985. Las zonas más activas en el cinturón volcánico mexicano (entre Michoacán y Tlaxcala). Geof. Int. 24(2):83–96.

Humphrey, W.E. 1949. Geology of the Sierra de los Muertos Area, Mexico (with descriptions of Aptian Cephalopods from the La Peña Formation). Bull. Geol. Soc. Amer. 60:89–176.

———. 1956. Tectonic framework of northeastern Mexico. Gulf Coast Assoc. Geol. Soc. Trans. 6:25–35.

Imlay, R.W. 1936. Evolution of the Coahuila Peninsula, Mexico. Part IV. Geology of the Western part of the Sierra de Parras. Bull. Geol. Soc. Amer. 47 (7):1091–1152.

———. 1937. Stratigraphy and paleontology of the upper Cretaceous beds along the eastern side of the Laguna de Mayran, Coahuila, Mexico. Bull. Geol. Soc. Amer. 48:1785–1872.

———. 1938a. Studies of the Mexican Geosyncline. Bull. Geol. Soc. Amer. 49:1651–1694.

———. 1938b. Ammonites of the Taraises Formation of Northern Mexico. Bull. Geol. Soc. Amer. 49:539–602.

———. 1939a. Upper Jurassic Ammonites from Mexico. Bull. Geol. Soc. Amer. 50:1–78.

———. 1939b. Paleographic studies in northeastern Sonora. Bull. Geol. Soc. Amer. 50:1723–1744.

————. 1940. Neocomian faunas of northern Mexico. Bull. Geol. Soc. Amer. 51:117–190.

————. 1943a. Evidence for upper Jurassic landmass in eastern Mexico. Amer. Assoc. Petrol. Geol. Bull. 27(4):524–549.

————. 1943b. Upper Jurassic Ammonites from the Placer de Guadalupe District, Chihuahua, Mexico. J. Paleontol. 17:527–654.

————. 1943c. Jurassic formations of the Gulf region. Amer. Assoc. Petrol. Geol. Bull 27(11):1407–1533.

————. 1944a. Correlations of the Cretaceous formations of the Greater Antilles, Central America and Mexico. Bull. Geol. Soc. Amer. 55:1005 –1045.

————. 1944b. Cretaceous formations of Central America and Mexico. Amer. Assoc. Petrol. Geol. Bull. 28(8):1077–1195.

————, M. Alvarez Jr. & T. Díaz. 1948. Stratigraphic relations of certain Jurassic formations in eastern Mexico. Amer. Assoc. Petrol. Geol. Bull. 32(9):1750–1761.

Instituto Nacional de Estadística, Geografía Informatica (INEGI). 1982. Carta Geol. Méx. México City: Secretaría de Programación y Presupuesto, Inst. Nal. Estad. Geogr. Inform., Mapa Edit. 4 Hojas Esc. 1:1,000,000.

Instituto de Geología, UNAM. 1986. Geología Regional de México. Programa y resúmenes. UNAM. Inst. Geol. Primer Symposio.

————. 1987a. Simposio sobre la geología de la región de la Sierra de Guanajuato. Programa, Resumenes y Guía de Excursión. UNAM. Inst. Geol.

————. 1987b. Segundo Simposio sobre Geología Regional de México. UNAM. Inst. Geol. Memoria.

————. 1989. Tercer Simposío sobre geología regional de México. Memoria. Idem.

Jones, T.S. 1938. Geology of Sierra de la Peña and Paleontology of the Indidura Formation. Bull. Geol. Soc. Amer. 49:69–150.

Jones, D.L., M.C. Blake Jr. & C. Rangin. 1976. The four Jurassic belts of northern California and their significance on the geology of the southern California borderland. *In* D.G. Howell (ed.), Aspects of the Geological History of California Continental Borderland. Amer. Assoc. Petrol. Geol. Miscell. Publ. 24. pp. 343–362.

Keller, W.T. 1973. Observaciones estratigráficas en Sonora Noroeste de México. Asoc. Mex. Geol. Petrol. Bol. 25(1–3):1–22.

Keller, P.C., N.T. Bockoven & F.W. McDowell. 1982. Tertiary volcanic history of the Sierra del Gallego area, Chihuahua, Mexico. Bull. Geol. Soc. Amer. 93:303–314.

Kellum, L.B. 1936. Evolution of the Coahuila Peninsula, Mexico. Part III. Geology of the mountains west of the La Laguna District. Bull. Geol. Soc. Amer. 47(7):1039–1090.

————. 1944. Geologic history of northern Mexico and its bearing on petroleum exploration. Amer. Assoc. Petrol. Geol. Bull. 28:301–325.

————, R.W. Imlay & W.G. Kane. 1936. Evolution of the Coahuila Peninsula, Mexico, Part I. Relation of Structure, stratigraphy and igneous activity to an early continental margin. Bull. Geol. Soc. Amer. 47(7):969–1008.

Kelly, W.A. 1936. Evolution of the Coahuila Peninsula, Mexico. Part II. Geology of the Mountains bordering the valleys of Acatita and Las Delicias. Bull. Geol. Soc. Amer. 47(7):1009–1038.

King, P.B. 1968. Geologic history of California. Cali. Div. Mines & Geol. Mineral Inf. Serv. 21(3):39–48.

————. 1969. Tectonic Map of North America: United States Geol. Surv. Scale 1:5,000,000.

————. 1977. The Evolution of North America. Princeton, NJ: Princeton Univ. Press.

King, R.E. 1934. The Permian of Southern Coahuila, Mexico. Amer. J. Sci.(5th. Ser.) 27:28–112.

————. 1939. Geological reconnaissance in northern Sierra Madre Occidental of Mexico. Bull. Geol. Soc. Amer. 50:1625–1722.

———— & S.W. Adkins. 1946. Geology of a part of the lower Conchos Valley, Chihuahua, Mexico. Soc. Geol. America Bull. 57:275–294.

Kirkland, D.W. & J.E. Gerhard. 1971. Jurassic salt, central Gulf of Mexico and its temporal relation to circum gulf evaporites. Amer. Assoc. Petr. Geol. Bull. 55:680–686.

Klesse, H. 1971. Geology of the El Ocotito–Ixcuinatoyac region and of La Dicha stratifonm sulphide deposit, state of Guerrero. Mex. Soc. Geol. Mex. Bol. 31(2):107–140.

Köhler, H., P. Schaaf, D. Müller–Schenius, R. Emmermann, J.F.W. Negendank & H.J. Tobschall. 1988. Geochronological and geochemical investigations on plutonic rocks from the complex of Puerto Vallarta, Sierra Madre del Sur. Geof. Int. 27(4):485–518.

Labarthe, H. G., M. Tristán G. & J. Aranda G. 1982. Revisión estratigráfica del Cenozoico de la parte central del Edo. de San Luis Potosí. Univ. Autón. San Luis Potosí, Inst. Geología y Metalurgia. Foll. Técnico 85:1–208.

Larson, R.L. 1972. Bathymetry, magnetic anomalies and plate tectonic history of the mouth of the Gulf of California. Bull. Geol. Soc. Amer. 83:3345–3360.

Ledezma G., O. 1967. Hoja Parras 13R–1(6), estados de Coahuila y Durango, con resumen de Geología. UNAM. Inst. Geol. Carta Geol. Méx. Ser. 1:100,000.

————. 1970. Hoja Calpulalpan 14Q–h(3), estados de México, Tlaxcala, Puebla e Hidalgo, con resumen de la Geología. UNAM. Inst. Geol. Carta Geol. Méx. Ser. 1:100,000.

Lesser J., H. 1951. Geología del área de Vernet y Amate Morales, Tab. Asoc. Mex. Geol. Petrol. Bol. 3(7–8):305–319.

Limón G., L. 1950. Las capas Sorites del Oligoceno Superior de México y sus Foraminiferos. Asoc. Mexicana Geol. Petrol. Bol. 2(10):617–630.

Logan, B.L., J.L. Harding, W.M. Ahs, J.D. Willimas & R.G. Snead. 1969. Carbonate sediments and reefs, Yucatan Shelf, Mexico. Amer. Assoc. Petrol. Geol. Mem. 11:1–198.

Longoria, J.F. 1984. Stratigraphic studies in the Jurassic of northeastern Mexico, evidence of the Sabinas Basin. Soc. Econ. Paleont. & Mineral. Gulf Coast Section, 3rd Ann. Res. Conf., Proc. pp. 171–193.

López-Ramos, E. 1975. Geological summary of the Yucatan Peninsula. *In* The Ocean Basins and Margins 3. London: Plenum Press. pp. 257–282.

————. 1979. Geología de México. Mexico City: Published by author.

Lozano, F. 1976. Evaluación petrolífera de la Península de Baja California, Mexico. Asoc. Mex. Geol. Petrol. Bol. 37(4–6):106–303.

Luhr, J.F., S.A. Nelson, J.F. Allan & I.S.E. Carmichel. 1985. Active rifting in southwestern Mexico, manifestations of an incipient eastward spreading rift jump. Geology 13:54–57.

Mahood, G.A. 1980. Geological evolution of a pleistocene rhyolitic center, Sierra La Primavera, Jalisco, Mexico. J. Volcanol. Geotherm. Res. 8:199–230.

Malfait, B.T. & M.G. Dinkleman. 1972. Circum-Caribbean tectonic and igneous activity and the evolution of the Caribbean Plate. Bull. Geol. Soc. Amer. 83:251–272.

Martínez P., J. 1972. Exploración Geológica del Area Estribos–San Francisco, S.L.P. Asoc. Mex. Geol. Petrol. Bol. 24(7–9):327–398.

Masson, P. & I.G. Alencáster. 1951. Estratigráfia y paleontología del Mioceno de San Andrés Tuxtla, Veracruz, Mexico. Asoc. Mex. Geol. Petrol. Bol. 3(5–6):199–216.

Mauger, R.L. 1983a. The geology and volcanic stratigraphy of the Sierra de Sacramento block near Chihuahua City, Chihuahua, Mexico. *In* K.F. Clark & P.C. Goodell (eds.), Geology and Mineral Resources of North–Central Chihuahua. El Paso Geol. Soc. Field Conf. Guidebook. pp. 113–156.

————. 1983b. Geologic map of the Majalca–Punta de Agua area, central Chihuahua, Mexico. *In* K.F. Clark & P.C. Goodell (eds.), Geology and Mineral Resources of North–Central Chihuahua. El Paso Geol. Soc. Field Conf. Guidebook. pp. 169–174.

————. 1983c. A geologic study of the Provincia El Nido area, northeast flank of Sierra El Nido, Central Chihuahua, Mexico. *In* K.F. Clark & P.C. Goodell (eds.), Geology and Mineral Resources of North–Central Chihuahua. El Paso Geol. Soc. Field Conf. Guidebook. pp. 187–186.

———— & R.D. Dayvault. 1983. The Tertiary volcanic rocks in lower Santa Clara Canyon, Central Chihuahua, Mexico. *In* K.F. Clark & P.C. Goodell (eds.), Geology and Mineral Resources of North–Central Chihuahua. El Paso Geol. Soc. Field Conf. Guidebook. pp. 175–186.

————, F.W. McDowell & J.C. Blount. 1983. Greenville Precambrian rocks of the Los filtros, near Aldama, Chihuahua, Mexico. *In* K.F. Clark & P.C. Goodell (eds.), Geology and MineralResource of Northcentral Chihuahua. El Paso Geol. Soc. Field Conf. Guidebook. pp. 194–201.

Mauvaois, R. 1977. Cabalgamiento miocénico (?) en la parte centro–meridional de Mexico. UNAM. Inst. Geol. Rev. 1(1):48–63.

Mayer P.R., F.A. 1967. Hoja Viesca 13–R–l (5), Estados de Coahuila y Durango, con resumen de la Geología. UNAM. Inst. Geol. Carta Geol. Ser. 1:100,000.

McBride, E. F., A.E. Weidie, J.A. Wolleben & R.C. Laudon. 1974. Stratigraphy and structure of the Parras and La Popa Basins, northeastern Mexico. Bull. Geol. Soc. Amer. 84(10):1603–1622

McCabe Ch., van der Voo, R. & J.H. Urrutia F. 1988. Late Paleozoic or Early Mesozoic magnetizations in remagnetized Paleozoic rocks in Oaxaca, Mexico. Earth Planet Sci. Lett. 91:205–213.

McDowell, F.W. & R.P. Keizer. 1977. Timing of Mid–Tertiary volcanism in the Sierra Madre Occidental, between Durango City and Mazatlan, Mexico. Bull. Geol. Soc. Amer. 88:1479–1487.

———— & S.E. Clabaugh. 1979. Ignimbrites of the Sierra Madre Occidental and their relation to the tectonic history of western Mexico. *In* C.E. Chapin & E.E. Wolfaugh (eds.), Ash–Flow Tuffs. Geol. Soc. Amer. Special Paper 80:113–125.

————. 1984. The igneous history of the Sierra Madre Occidental and its relation to the tectonic evolution of western Mexico. UNAM. Inst. Geol. Rev. 5(2):195–207.

McLeroy, D.F. & R.E. Clemons. 1965. Hoja Pedriceñas 13R–l (4), estados de Coahuila y Durango, con resumen de la geología. UNAM. Inst. Geol. Carta Geol. Mex. Ser. Esc. 1:100,000.

Meave T., E. & S.J. Echegoyen. 1961. Estudio geológico–económico sobre algunos yacimientos de mineral de hierro de los municipios de Pihuamo y Tecalitlan. Inst. Nal. Inv. Rec. Min. Bol. 53:1–88.

Mena R., E. 1960. El Jurásico Marino de la región de Córdoba. Asoc. Mex. Geol. Petrol. Bol. 12(7–8):243–252.

————. 1962. Geología y posibilidades petroliferas del Jurásico Marino en la región de Córdoba, Ver. Asoc. Mex. Geol. Petrol. Bol.19(3–4):77–84.

Meneses de Gyves, J. 1981. Geología de la Sonda de Campeche. Asoc. Mex. Geol. Petrol. Bol. 32:1–26.

Mina U., I.F. 1950. Notas para la geología de Sinaloa. Asoc. Mex. Geol. Petrol. Bol. 2(1):345–370.

———. 1957. Bosquejo geológico del territorio sur de Baja California. Asoc. Mex. Geol. Petrol. Bol. 18(1–2):139–266.

Molnar, P. & L. Sykes. 1969. Tectonics of the Caribbean and Middle American regions from focal mechanisms and seismicity. Bull. Geol. Soc. Amer. 80:1639–1684.

Mooser, F. 1972. The Mexican volcanic belt: structure and tectonics. Geof. Int. 12(2):55–70.

———. 1975. Historia geológica de la cuenca de México. *In* Departamento del Distrito Federal, Memoria de las Obras del Sistema del Drenaje Profundo del Distrito Federal 1:7–38

———, A.E. Nairn & J.F. Negendank. 1974. Paleomagenetic investigations of the Tertiary and Quaternary igneus rocks. VII a Paleomagnetic and petrologic study of volcanics of the Valley of Mexico. Geol. Rundschau 63 (2):451–483.

——— & T. Ramírez. M. 1987. Faja Volcánica Transmexicana, morfo-estructura, tectónica y vulcanotectónica. Soc. Geol. Mex. Bol. 48(2):75–80

Morán Z., J.D. 1984. Geología de la República Mexicana. Mexico City: UNAM., Fac. Ingenieria & Secr. Proqr. Presup., Inst. Nal. Estad. Geogr. Inform.

———. 1986. Breve revisión de la evolución tectónica de México. Geof. Int. 25(1):9–38.

———, J. Urrutia F., H. Bohnel & E. González T. 1988. Paleomagnetismo de rocas Jurásicas del norte de Oaxaca y sus implicaciones tectónicas. Geofísica Intenacional 27(4):485–518.

Morris, W.R. & C.J. Busby–Spera. 1988. Sedimentological evolution of a submarine Canyon in a fore–arc basin, Upper Cretaceous Rosario Formation, San Carlos, Mexico. Amer. Assoc. Petrol. Geol. Bull. 72(6):117–737.

Mossman, R.W. & F. Viniegra O. 1976. Complex fault structures in Veracruz Province of Mexico. Amer. Assoc. Petrol. Geol. Bull. 60(3):379–388.

Mota P., R., J. Adrieux & J. Bonnin. 1986. Bosquejo sismotectonico del Sur de México. Geof. Int. 25(1):207–232.

Muir, J.M. 1936. Geology of the Tampico region, Mexico. Tulsa, Okla: Amer. Assoc. Petrol. Geol. (special publ.).

Müllerried, F.K.G. 1957. La Geología de Chiapas. Publicación del Gobierno del Estado de Chiapas.

Murray, G.E. 1961. Geoloqy of the Atlantic and Gulf Coastal Province of North America. New York: Harper & Brothers.

——— & A.E. Weidie (eds.). 1962. Yucatan Peninsula Guidebook. New Orleans Geol. Society Guidebook.

———, ———, D.R. Boyd, R.H. Forde & P.D. Lewis. 1962. Formational divisions of the Difunta group, Parras Basin, Coahuila and Nuevo Leon states, Mexico. Amer. Assoc. Petrol. Geol. Bull. 46:373– 383.

Nájera C., H. 1952. Estudio de las formaciones del Eoceno en la región de Poza Rica, Veracruz. Asoc. Mex. Geol. Petrol. Bol. 4(3–4):71–115.

Nava, F., F. Nuñez C., D. Córdoba, M. Mena, J. Ansorge, J. González, J. Rodríguez M., E. Banda, S. Mueller, A. Udias, G. García M. & G. Calderón. 1988. Structure of the Middle American Trench in Oaxaca. Tectonophysics 154:241–251.

Navarro G., A. & R. Tovar. 1970. Stratigraphy and tectonics of the State of Chihuahua. *In* K. Seewald & D. Sundeen (eds.), The Geologic Framework of the Chihuahua Tectonic Belt. West Texas Geol. Soc. & Univ. Texas–Austin Symposium in honor of Professor Ronald K. DeFord. Midland, TX: West Texas Geol. Soc. pp. 83–138.

Negendank, J.F.W. 1972. Volcanics of the Valley of Mexico. N. Jb. Miner. Abhand. 116:308–320.

———, R. Emmerdan, R. Krawkzyc, F. Mooser, H. Tobschal & D. Werle. 1985. Geological, geochemical investigations on the eastern Transmexican Volcanic Belt. Geof. Int. 24(4):477–575.

Nelson, S.A. & R.A. Livieres. 1986. Contemporaneous calc–alkaline and alkaline volcanism at Sangangüey Volcano, Nayarit, Mexico. Bull. Geol. Soc. Amer. 97(7):798–808.

——— & R.G. Sánchez. 1986. Transmexican Volcanic Belt Field Guide. Geol. Assoc. Canada Volc. Division/UNAM. Inst. Geol.

Nieto, O.J., L. Delgado & P.E. Damon. 1985. Geochronologic, petrologic and structural data related to large morhologic features between the Sierra Madre Occidental and the Mexican Volcanic Belt. Geof. Int. 24(4):623–665.

Nimz, G.J., K.L. Cameron & S.L. Morris. 1985. Petrology of the crust and upper mantle beneath southeastern Chihuahua, Mexico: a progress report. Geof. Int. 25(1):85–117.

Nixon, G.T. 1982. The relationship between Quaternary volcanism in Central Mexico and the seismicity and structure of subducted ocean lithosphere. Bull. Geol. Soc. Amer. 93:514–523.

———, A. Demant, R.L. Armstron & J.I. Harakal. 1987. K–Ar and geologic data bearing on the age and evolution of the Trans-Mexican Volcanic Belt. Geof. Int. 26(91):109–158.

Ojeda R., J., A. Osorio H. & J. Bravo N. 1965. Geología regional y yacimientos minerales de la porción central del estado de Baja California. Cons. Rec. Nat. no Renov. Proy. Explor. Min. Metal. pp. 1–307.

Ontiveros T., G. 1973. Estudio Estratigráfico de la porción Nor-occidental de la Cuenca de Morelos Guerrero. Asoc. Mex. Geol. Petrol. Bol. 25(4–6):190–234.

Ordoñez, E. 1895. Las Rocas eruptivas del suroeste de la cuenca de México. Inst. Geol. Méx. Bol. 2:1–56.

———. 1900. Las rhyolitas de México (primera parte). Inst. Geol. Méx. Bol. 14:1–78; 15:1–78

Ortega G., F. 1978. Estratigráfia del Complejo Acatlán en la Mixteca Baja, Estados de Puebla y Oaxaca. UNAM. Inst. Geol. Rev. 2(2):112–131.

———. 1981a. Metamorphic belts of southern Mexico and their tectonic significance. Geof. Int. 20(3):177–202.

———. 1984. La evolución Tectónica Premisísipica del Sur de México. UNAM. Inst. Geol. Rev. 5(2):140–157.

Ortlieb, L. & J. Roldan Q. (eds.). 1981. Geology of Northwestern Mexico and Southern Arizona. UNAM. Inst. Geol. (Estn. Reg. Noroeste).

Padilla S., R.J. 1985. Las estructuras de la curvatura de Monterrey, estados de Coahuila, Nuevo León, Zacatecas y San Luis Potosí. UNAM. Inst. Geol. Rev. 6(1):1–20.

———. 1986. Post–Paleozoic tectonics of northwease Mexico and its role in the evolution of the Gulf of Mexico. Geof. Int. 25:157–206.

Palmer, R.H. 1927. Geology of Eastern Hidalgo and adjacent parts of Veracruz, Mexico. Amer. Assoc. Petrol. Geol. Bull. 2(12):1321–1328.

Pantoja A., J. 1959. Estudio geológico de econocimiento de la región de Huetamo, estado de Michoacán. Cons. Rec. Nat. no Renov. Bol. 50:1–36.

———. 1970. Rocas Sedimentarias Paleozoicas de la Región Centro–septentrional de Oaxaca: Libro Guía de la Excursión Mexico–Oaxaca. Soc. Geol. Mex. Exc. Mexico–Oaxaca, Libro–Guía. pp. 67–84.

—————— & J. Carrillo B. 1966. Bosquejo geológico de la región de Santiago, San José del Cabo, Baja California. Asoc. Mex. Geol. Petrol. Bol. 17(1–2):1–11.

——————, C. Fries Jr., C.O. Rincón, T.S. Silver & M.J. Solorio. 1974. Contribución a la geocronología del estado de Chiapas. Asoc. Mex. Geol. Petrol. Bol. 26(6):205–223.

Pasquare, G., L. Ferrare, V. Perazzole, M. Tiberi & F. Turchetti. 1987. Morphological and structural analysis of the central sector of the Transmexican Volcanic Belt. Geof. Int. 26(2):177–194.

Pastor, A. 1930. Informe geológico del Exdistrito de San Ignacio, Sinaloa. UNAM. Inst. Geol. Ant. 5:85–118.

Pedrazzini J, C., N. Holguín & R. Moreno. 1982. Evaluación geológico–geoquimica de la parte noroccidental del Golfo de Tehuantepec. Inst. Mex. Petrol. Rev. 14(4):6–26.

Pérez I., J.M., A. Hokuto & Z. de Cserna. 1965. Reconocimiento geológico del área de Petlalcingo–Santa Cruz, Municipio de Acatlán, Estado de Puebla. UNAM. Inst. Geol. Paleontol. Mex. 21(1):1–22.

Pérez M., J., E. Mapes V. & R. Pesquera V. 1961. Bosquejo geológico del distrito Minero de Zacatecas. Cons. Rec. Nat. no Renov. Bol. 52:1–38.

Pindell, J.L., S.C. Cande, W.C. Pittman III, D.B. Rowley, J.F. Dewey, J. Labrecque & W. Haxby. 1988. A plate kinematic framework for models of Caribbean evolution. Tectonophysics 155:121–138.

Pineda, A., H. Lopez & A. Peña. 1969. Estudio geológico–magnetometrico de los yacimientos ferríferos de Peña Colorada, municipio de Minatitlán, Colima. Cons. Rec. Nat. no Renov. Bol. 77:1–44.

Raisz, E. 1959. Landforms of Mexico. Cambridge, Mass. Map Scale 1:3,000,000.

Ramírez J., C. & C. Acevedo F. 1957. Notas sobre la Geología de Chihuahua. Asoc. Mex. Geol. Petrol. Bol. 9:583–766.

Rangin, C. 1979. Evidence for superimposed subduction and collision processes during Jurassic–Cretaceous time along Baja California continental borderland. *In* P. Abbot & G. Gastil (eds.), Field Guide and Papers on Baja California. Geol. Soc. Amer. Ann. Mtg. San Diego. pp. 37–52.

——————. 1984. Aspectos Geodinámicos de la Región Noroccidental de México. UNAM. Inst. Geol. Rev. 5(2):186–194.

Rangin, F. 1978. Consideraciones Sobre el Paleozoico Sonorense. UNAM. Inst. Geol. I Simposio sobre la Geología y el Potencial Minero del Estado de Sonora, Resúmenes. pp. 35–56.

Ranson, W.A., A. Fernández L., B. W. Simons Jr & S. Enciso De la Vega. 1982. Petrology of the metamorphic rocks of Zacatecas, Zac. Mexico. Soc. Geol. Mex. Bol. 43(1):37–59.

Richard, H. G. 1963. Stratigraphy of Early Mesozoic sediments in southwest Mexico and western Guatemala. Amer. Assoc. Petrol. Geol. Bull. 47:1861–1970.

Ríos M., F. 1952. Estudio geológico de la región de los Tuxtlas, Veracruz. Asoc. Mex. Geol. Petrol. Bol. 4(9–10):325–376.

——————. 1959. Bosquejo geológico de la cuenca de Veracruz y parte de la Cuenca Salina del Istmo de Tehuantepec. Asoc. Mex. Geol. Petrol. Bol. 11 (7–12):389–400.

Robeck, D., V. Pesquera R. & S. Ulloa. 1956. Geología y Depósitos de Carbon en la Región de Sabinas, Coah. XX Congr. Geol. Int. Méx.

Robin, C., 1976. Las series volcánicas de la Sierra Madre Oriental (Basaltos e Ignimbritas): Descripción y caracteres químicos. UNAM. Inst. Geol. Rev. 2(2):13–42.

Rodríguez T., R. 1970. Geología Metamórfica del area de Acatlán, estado de Puebla. Soc. Geol. Mexicana, Exc. Mexico–Oaxaca, Libro–Guía. pp. 51–54.

————, & D.A. Córdoba M. (eds.). 1978. Atlas Geológico y Evaluación Geológica Minera del Estado de Sinaloa. UNAM. Inst. Geol. y Secre. Desarr. Econ. Edo. Sinaloa.

Rogers, L.C., Z. De Cserna, E. Tavera A., S. Ulloa. 1957. Geología general y depositos de fosfatos del Distrito de Concepcion del Oro, Estado de Zacatecas. Inst. Nal. Invest. Rec. Min. Bol. 8:1–123.

————, ————, ————, R. Van Vloten & J. Ojeda R. 1961. Reconocimiento geológico y depositos de fosfatos del norte de Zacatecas y áreas adyacentes en Coahuila, Nuevo León y San Luis Potosí. Cons. Rec. Nat. no Renov. Bol. 56:1–322.

————, R. Van Vloten, J. Rivera O., T. Amezcua E. & Z. De Cserna. 1963. Plutonic rocks of northern Zacatecas and adjacent areas, Mexico. U.S. Geol. Surv. Prof. Paper 475–C, Article 61. pp. 7–11.

Roldán, Q., J. 1984. Evolución tectónica del estado de Sonora. UNAM. Inst. Geol. Rev. 5(2):178–185.

———— & B. Solano. 1978. Contribución a la estratigráfia de las rocas volcanicas del Estado de Sonora. Univ. Auton. Sonora, Dept. Geol. Bol. 1(1):19–26.

Ross, M.I. & C.R. Scotese. 1988. A hierarchical tectonic model of the Gulf of Mexico and Caribbean region. Tectonophysics 155:139–168.

Ruiz, J., J. Patchett P & F. Ortega G. 1988. Proterozoic and Phanerozoic Basement terranes of Mexico from Nd isotopic studies. Bull. Geol. Soc. Amer. 100(2):274–281.

Salas, G. P. 1949. Bosquejo Geológico de La Cuenca Sedimentaria de Oaxaca. Asoc. Mex. Geol. Petrol. Bol. I:79–156.

———— & E. Lopez-Ramos. 1951. Geología y tectónica de la región de Macuspana, Tabasco y Norte de Chiapas. Asoc. Mex. Geol. Petrol. Bol. 3(1–8):3–56.

Salas, G.A. 1970. Areal geology and petrology of the Igeneous rocks of the Santa Ana region N–W of Sonora. Soc. Geol. Mex. Bol. 31(1):11–64.

Salmerón U., P. 1970. Estudios bioestratigráfico preliminar de parte de la región meridional de la cuenca sedimentaria de Veracruz. Asoc. Mex. Geol. Petrol. Bol. 22(1–4):1–60.

Salvador, A. 1987. Late Triassic–Jurassic paleogeography and the origin of the Gulf of Mexico basin. Amer. Assoc. Petrol. Geol. Bull. 71(4):419–451.

Sánchez M. de O., R. 1979. Geología petrolera de la Sierra Madre de Chiapas. Asoc. Mex. Geol. Petrol. Bol. 31(1–2):67–97.

Santiago A. J., J. Carrillo B. & A.B. Martell. 1984. Geología petrolera de México. *In* D. Marmissolle–Daguerre (coordinador). Evaluación de Formaciones en México, México, D.F. PEMEX y Schlumbertger Offshore Serv. pp. 1–36.

Santillán, M. 1929a. Geología minera de la región comprendida entre Durango, Dgo. y Mazatlán, Sin., a uno y otro lado de la carretera en proyecto entre esas ciudades. UNAM. Inst. Geol. Bol. 48:1–46.

————. 1929b. Geología minera de las regiones norte, noreste y central del Edo. de Guerrero. UNAM. Inst. Geol. Bol. 48:47–102.

————. 1936. Synopsis of the geology of Mexico. Amer. Assoc. Petrol. Geol. Bull. 20(4):394–402.

———— & T. Barrera. 1930. Las posibilidades petrolíferas en la costa occidental de la Baja California, entre los paralelos 30 y 32 de latitud norte. UNAM. Inst. Geol. An. 5:1–37.

Sapper, C. 1896. Sobre la geografía física y la geología de la Península de Yucatán. Inst. Geol. Mex. Bol. 3:1–57.

Schlaepfer, C.J. 1968. Hoja México 14 Q–h(5), Distrito Federal y Estado de Mexico, con resumen de la geología. UNAM. Inst. Geol. Carta Geol. Méx. Ser. 1:100,000.

Schmidt, H.A. 1930. Geology of the Parral Area of the Parral District, Chihuahua, Mexico. American Inst. Min. Met. Eng. Techn. Publ. No. 304. 1–24.

Schmidt–Effing, R. 1980. The Huyacocotla Aulocogen in Mexico (Lower Jurassic) and the Origin of the Gulf of Mexico. *In* Proceedings of a Symposium: The Origin of the Gulf of Mexico in the Early Openning of the Central North Atlantic Ocean. Geol. Rundsch. 68(2):457–494.

Schuchert, C. 1935. Historical Geology of the Antillean–Caribbean Region. New York: John Wiley & Sons.

Schulze, G. 1953. Conglomerados terciarios continentales en la comarca lagunera de Durango y Coahuila y sus relaciones con fenómenos ígneos geomorfológicos y climatológicos. Inst. Nal. Invest. Rec. Min. Bol. 30:1–52.

Seewald, K. & D. Sundeen. (eds.). 1970. The Geologic Framework of the Chihuahua Tectonic Belt. West Texas Geol. Soc. & Univ. Texas–Austin, Symposium in Honor of Professor, R.K. DeFord. Midland, TX: West Texas Geol. Soc.

Segerstrom, K. 1961. Estratigrafía del Area Bernal Jalpan, Estado de Querétaro. Asoc. Mex. Geol. Petrol. Bol. 13(5–6):183–206.

Serna V., R., C. Acosta del C., J.J. Martínez B. & J. Nava, A. 1959. Reconocimiento geológico de la zona Alunítica de Romero, Guanajuato. Asoc. Ings. Min. Met. Geol. Mex. Min. y Metal. 9:93–123.

Shor, G. C., Jr. 1974. Continental margin of middle America. *In* C.A. Burk & C.L. Drake (eds.), The Geology of Continental Margins. New York: Springer–Verlag, pp. 502-509.

Silver, L. T., F.G. Stehli & C.R. Allen. 1963. Lower Cretaceous Pre–Batholitic rocks of Baja California, Mexico. Amer. Assoc. Petrol. Geol. Bull. 47 (12):2054–2059.

Singewald, Q.D. 1936. Evolution of the Coahuila Peninsula, Mexico, Part V. Igneous phenomena and geologic structure near Mapimi. Bull. Geol. Soc. Amer. 47(7):1153–1176.

Smith, Ch. I. 1970. Lower Cretaceous stratigraphy of Northern Coahuila, Mexico. Univ. Texas–Austin Bureau Econ. Geol. Rep. Invest. 65:1–101.

Staub, W. 1928. Uber die Verbreitung der oligocänen und der alterneogenen Schichten in der Golfregion des nordostlichen Mexiko. Eclogae Geol. Helv. 21(2):119–130.

Steele, D.R. 1986. Physical stratigraphy and petrology of the Cretaceous Sierra Madre Limestone, west Central Chiapas. UNAM. Inst. Geol. Bol. 1:1–101.

Stewart, J.H. 1988. Latest Proterozic and paleozoic southern margin of North America and the accretion of Mexico. Geology 16:186–189.

———, M. McMenamin & J.M. Morales-Ramírez. 1984. Upper Proterozoic and Cambrian rocks in the Caborca region, Sonora, Mexico; physical stratigraphy, biostratigraphy, paleocurrent studies and regional relations. U.S. Geol. Surv. Prof. Paper 1309. pp. 1–36.

Sutter, M. 1980. Tectonics of the external part of the Sierra Madre Oriental foreland thrust–and–fold belt between Xilitla and the Moctezuma River (Hidalgo and San Luis Potosi States). UNAM. Inst. Geol. Rev. 4(1):19–31.

———. 1984. Cordilleran deformation along the eastern edge of the Valles San Luis Potosi Carbonate Platform Sierra Madre Oriental fold–thrust belt, East Central Mexico. Bull. Geol. Soc. Amer. 95 (12):1387–1397.

———. 1987. Structural traverse across the Sierra Madre Oriental fold–thrust belt in East Central Mexico. Bull. Geol. Soc. Amer. 98(3):249–264.

Swanson, E. R., R.P. Keizer, J.I. Lyons & S.E. Clabaugh. 1978. Tertiary volcanism and caldera development near Durango City, Sierra Madre Occidental, Mexico. Bull. Geol. Soc. Amer. 89:l000–1012.

―――― & F.W. McDowell. 1984. Calderas of the Sierra Madre Occidental volcanic field, western Mexico. J. Geophys. Res. 89:8787–8799.

Taliefferro, N. 1933. An occurrence of upper Cretaceus sediments in Northern–Sonora, Mexico. J. Geol. 41(1):12–37.

Tardy, M. 1973. Sobre la Tectónica de la Sierra Madre Oriental en el sector de Parras, Coah. El Cabalgamiento de la Serie Parrense. Soc. Geol. Mex. Bol. 34(1–2):51–70.

――――. 1980. La Transversal de Guatemala y las Sierras Madre de México. *In* J. Auboin, R. Brousse & J.P. Lehman (eds.), Tratado de Geología, Tomo III, Tectónica, Tectonofísica y Morfología. Barcelona: David Serrat Trad. Editorial Omega. pp. 117–182.

――――, J. Signal & G. Glaçon. 1974. Bosquejo sobre la estratigrafía y paleontología de los Flysch Cretácicos en el sector transversal de Parras, Sierra Madre Oriental. UNAM. Inst. Geol. Rev. Ser. Divul. No. 2. pp. 1–72.

――――, J.F. Longoria, J. Martínez R., L.M. Mitre, M. Patino & R. Padilla. 1975. Observaciones generales sobre la estructura de la Sierra Madre Oriental: la aloctonía del conjunto Cadena Alta–Altiplano Central, entre Torreon, Coah. y San Luis Potosí, S. L. P. Mexico. UNAM. Inst. Geol. Rev. 1(1):1–11.

――――, C. Ramírez R. & M. Patiño. 1976. El frente de la Napa de Parras en el area de Aramberri, Nuevo León. Sierra Madre Oriental. Méx. UNAM. Inst. Geol. Rev. 1(2):1–12.

Thalmann, H.E. 1935. Die Miozäne Tuxpan–Stufe in Gebiete zwischen Rio Tuxpan und Rio Tecolutla (Staat Veracruz, Ost. México). Eclog. Geol. Helv. 28 (2):543–546.

Thorpe, R.S. 1977. Tectonic significance of alkaline volcanism in eastern Mexico. Tectonophysic 40:19–26.

Uchipi, E., 1973. Eastern Yucatan Continental Margin and western Caribbean Tectonics. Amer. Assoc. Petrol. Geol. Bull. 57:1075–1085.

Urrutia F., J. (ed.). 1981. Paleomagnetism and tectonics of Middle America and adjacent regions. Pt. 1. Geof. Int. 20(3):139–270.

――――. (ed.). 1983a. Paleomagnetism and tectonics of Middle America and adjacent regions. Part 2. Geof. Int. 22:87–110.

――――. 1983b. On the tectonic evolution of Mexico, paleomagnetic constraints. Amer. Geophys. Union Geodyn. Ser. 12:29–47.

――――. 1986a. Late Mesozoic–Cenozoic evolution of the northwestern Mexico magmatic arc zone. Geof. Int. 25(1):61–84.

――――. 1986b. Crustal thickness, heat flow arc magnetism and tectonics of Mexico: preliminary Report. Geof. Int. 25(4):559–573.

―――― & L. del Castillo. 1977. Un modelo del Eje Volcanico Mexicano. Soc. Geol. Mex. Bol. 38:18–28.

――――, J.D. Morán Z. & C.E. Cabral C. 1986. Paleomagnetism and Tectonics of Mexico. UNAM. Inst. Geof. Comun. Tecn. Ser. Invest. No. 16 pp. 1–22.

Van Andel, T.H. & G.G. Shore Jr. (eds.). 1964. Marine Geology of the Gulf of California, A Symposium. Amer. Assoc. Petrol. Geol. Mem. 3:1–408.

Valentine, W.G., 1936. Geology of the Cananea Mountains, Sonora, Mexico. Bull. Geol. Soc. Amer. 47:53–86.

Velasco, R. 1956. Geología del Mineral de Cananea, México. XX Cong. Geol. Internal. Mexico. Exc. A–l y C–4. pp. 9–78.

Viniegra O., F. 1965. Geología del Macizo de Teziutlán y la Cuenca Cenozoica de Veracruz. Asoc. Mex. Geol. Petrol. Bol. 17:101–163.

————. 1971. Age and evolution of salt basins of southeastern Mexico. Amer. Assoc. Petrol. Geol. Bull. 55(3):478–494.

————. 1981. El gran banco calcáreo yucateco. Rev. Ing. 1:20–44. (Also: Great Carbonate bank of Yucatan, southern Mexico. J. Petro. Geol. 3:247–278.

———— & C. Castillo T. 1975. Golden Lane fields, Veracruz, Mexico. Amer. Assoc. Petrol. Geol. Bull. Mem. 14:309–325.

Vinson, G.L. & J.J. Brineman. 1963. Nuclear Central America, hub of the Antillean Transverse Belt. *In* G. Childs (ed.), Backbone of the Americas, Tectonic History from Pole to Pole. Amer. Assoc. Petrol. Geol. Mem. 2:101–112.

Waite, L.E. 1986. Biostratigraphy and paleoenvironmental analysis of the Sierra Madre Limestone (Cretaceous), Chiapas. UNAM. Inst. Geol. Bol. 102(2):108– 245.

Waitz, P. 1926. Erupciones riolíticas ligadas con fracturas tectónicas entre Aguas Calientes y San Luis Potosí. Soc. Cient. Antonio Alzate. Mem. 46 (3–6):201–212.

Wall, J.R., G.E. Murray & G. Díaz T. 1961. Geologic occurrence of intrusive gypsum and its effect on structural forms in Coahuila. Amer. Assoc. Petrol. Geol. Bull. 45(9):1504–1522.

Walper, J.L. & C.L. Rowett. 1972. Plate tectonics and the origin of the Caribbean Sea and the Gulf of Mexico. Gulf. Coast. Assoc. Geol. Soc., Transact. 22:105–116.

Weber, R. 1972. La vegetación maestrichtiana de la Formación Olmos de Coahuila, México. Soc. Geol. Mex. Bol. 33(1):5–21.

Whal, D.F., Jr. 1976. Geología de la Faja del Salto, Durango, México. UNAM. Inst. Geol. Bol. 96:1–85.

Wilhelm, O. & M. Ewing. 1972. Geology and history of the Gulf of Mexico. Bull. Geol. Soc. Amer. 88:575–600.

Williams, H. 1950. Paricutin region Mexico. U.S. Geol. Surv. Bull. 965:165–273.

Wilson, II.II. 1987. The structural evolution of the Golden Lane, Tampico Embayment. J. Petrol. Geol. 10(1):5–40.

Wilson, I.F. & V. Rocha S. 1946. Los yacimientos de carbón de la región de Santa Clara, Mpio. de San Javier. Edo. de Sonora. Com. Direct. Invest. Rec. Min. (Mex.) Bol. 9:1–108.

————. 1957. Geología y depósitos minerales del Distrito Cuprífero del Boleo, Baja California, México. Inst. Nac. Invest. Rec. Min. Bol. 41:1–419.

———— & M. Veytia. 1949. Geología y yacimientos minerales de la región manganesífera de Lucifer, al noreste de Santa Rosalia, Baja California. Inst. Nac. Invest. Rec. Min. Bol. 25:1–68.

Wright, J.V. 1981. The Rio Caliente ignimbrite: analysis of a compound intraplinian ignimbrite from a major Late Quaternary Mexican eruption. Bull. Volcanol. 44:189–212.

Young, K. 1983. The Mesozoic. *In* M. Moullade & A.E.M. Nairn (eds.), The Phanerozoic Geology of the World. Vol. 2. Amsterdam: Elsevier. pp. 61–88.

Zwanzinger, J.A. 1978. Geología regional del sistema sedimentario Cupido. Asoc. Mex. Geol. Petrol. Bol. 30(1–2):1–56.

2

Historical Factors and Biological Diversity in Mexico

ALAN GRAHAM

For every complex problem there's a simple solution.
And it's always wrong.

—SOURCE UNKNOWN

The biological diversity of a region is a consequence of factors that promote the appearance of novel phenotypes, facilitate the accumulation of these phenotypes, and operate over a significant period of geologic time. The paleontological record provides some insight into the nature and timing of these events and their effect on the development (evolution), composition, and distribution of organisms and communities of organisms. In eastern and southeastern Mexico, the Sierra Madre Oriental has provided upland habitats for vegetation since the latest Cretaceous/ Paleocene. The Trans-Mexican Volcanic Belt (TVB) began development during the Early Tertiary but underwent its principal uplift and deformation during Miocene to Quaternary times. The Cenozoic was also a time of significant climatic change, with documented effect on the biota of southern Mexico. In particular, the drop in global temperatures beginning during the mid-Miocene is recorded in the composition and distribution of communities preserved in the Pliocene Paraje Solo formation near Coatzacoalcos in Veracruz. The presence of pollen of *Picea* (spruce), now restricted to the mountains to the north, *Abies* (fir), *Pinus* (pine), *Quercus* (oak), and *Liquidambar* (sweet gum) in the coastal lowland Paraje Solo depositional basin suggests cooler climates and a lowering of ecotones. Such physiographic and climatic fluctuations not only affected speciation rates through vicariance but provided a diversity of habitats for the perpetuation of new forms. In addition, southern Mexico is located at the confluence of two migration routes through Central America and the Antilles that have operated in varying degrees throughout the Tertiary. An extensive new source area to the south became increasingly available as South America moved closer to North America and the Panama land bridge became established about 2.4 Ma. By the end of the Tertiary, southern Mexico had access to temperate biota from the north, with introductions facilitated by the cooling climates of the late Eocene, middle to late Miocene, and Pleistocene; to tropical biota to the south, with introductions progressively increasing during the warm climates of the Paleocene, early and middle Eocene, Oligocene, and early Miocene (as well as by long-distance transport from both regions throughout the Cenozoic); and a landscape of sufficient diversity to accommodate the introduc-

Table 2.1. Comparison of diversity between microfossil floras and modern vegetation in Mexico and southern Central America

Modern			Miocene/Pliocene/Quaternary	
Region	Area (km²)	Species No.	Region	Identified Types
Mexico			Miocene	
Chiapas	74,000	8,250	Panama	
Veracruz	62,820	7,700	La Boca	54
Nueva Galicia	125,000	8,000	Cucaracha	21
			Culebra	55
Panama	77,082	7,345	Costa Rica	
			Uscari	46
Costa Rica	51,100	8,000		
			Pliocene	
			Mexico	
			Paraje Solo	124
			Quaternary	
			Panama	148

Figures for the Miocene and Pliocene floras are based on Graham (1976, 1987a, 1988a, b, 1989), Quaternary on Bartlett and Barghoorn (1973), Chiapas on Breedlove (1981), Veracruz on Gomez Pompa (pers. comm., 1988), Nueva Galicia on McVaugh (pers. comm., 1988), Panama on D'Arcy (1987), and Costa Rica on Burger (pers. comm. in Hartshorn, 1983, p. 119).

tions. The factors of topographic diversity, climatic change, and access to both temperate and tropical source areas account, in part, for the species richness observed in the modern communities of southern Mexico.

The vegetation of Mexico is characterized by endemism in the desert communities to the north (Medellin-Leal, 1982; Rzedowski, 1973, 1978, this volume) and by diversity in the tropical and subtropical communities to the south. The flora of the state of Veracruz (62,820 km²) is presently estimated at 7,700 species (Table 2.1), the state of Chiapas (74,000 km²) at 8,250 species, and Nueva Galicia (125,000 km²) at 8,000 species. These figures compare with approximately 2,700 species in the 115,719 km² of Ohio (Cooperrider, 1983) and 3,360 species in the 216,630 km² of the Carolinas (Radford et al., 1968) in the United States. Various models have been proposed to explain speciation and diversity (see reviews in Pianka, 1983; Ricklefs, 1979), and implicit in all of them is the assumption that the proposed factors (e.g., predation, competition, hybridization, introgression, mutation, reproductive isolation-vicariance, and founder effect) have operated over time to produce the species and species richness observed in the modern biota. The paleontological record constitutes documentation that these factors have, in fact, operated over time to produce changes in organisms and in communities of organisms. In addition, the fossil record provides evidence of other events difficult or impossible to observe from a single point in time, including the development and timing of physiographic diversity, e.g., uplift of the Trans-Mexican Volcanic Belt (TVB) during the Late Tertiary, climatic changes, e.g., the sharp drop in global temperatures during the late Eocene and middle to late Miocene, and isolation from vs. access to source areas during the Cenozoic, e.g. emergence of the Isthmian land bridge in the Plio-Pleistocene.

Tracing the Cenozoic history of vegetation and paleo- environments in Mexico is limited primarily to data from the Paraje Solo Formation of Veracruz (Graham, 1976) (Figs. 2.1 and 2.2), with additional information from the Oligo-Miocene Simojovel Group of Chiapas (Langenheim et al., 1967), the lower Miocene Uscari sequence of Costa Rica (Graham, 1987a,b), the lower Miocene Culebra (Graham, 1988a), Cucaracha (Graham, 1988b), La Boca (Graham, 1989) formations of Panama, and Quaternary studies by Bartlett and Barghoorn (1973) from Panama and by Leyden (1984) from Guatemala. The current studies of Alvarado & Delgado Rueda (1985), Gonzáles-Quintero (1980; González-Quintero & Fuentes Mata, 1980; González-Quintero & Montufar López, 1980; González-Quintero & Sanchez Martínez, 1980), Ludlow-Wiechers (1980; Ludlow-Wiechers & Ayala-Nieto, 1983), Martínez-Hernández et al. (1980), Palacios-Chavez (1985; Palacios-Chavez & Arreguin, 1980), Quiroz-García and Palacios-Chavez (1986a,b), Toledo (1976), and others will eventually add new and important data. For a more complete listing of vegetational history studies in Latin America, see the bibliographies in Graham (1973, 1979, 1982, 1986).

The Paraje Solo Formation was earlier regarded as middle Miocene in age by PEMEX geologists (pers. comm., 1965); the palynomorphs suggested an

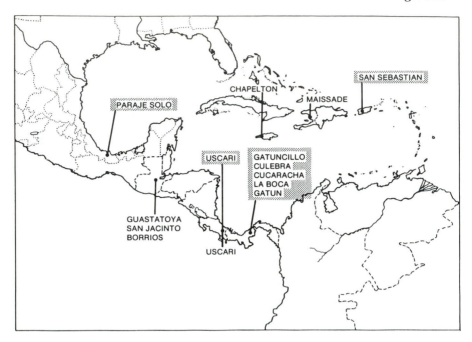

Figure 2.1. Distribution of Tertiary microfossil floras in the Gulf/Caribbean region. Stippling indicates the study of the assemblage has been completed and the results published.

age no older than uppermost Miocene (Graham, 1976), although the minimum age then could not be established precisely. More recent studies by Akers (1979, 1981, 1984) on planktic foraminifera and by Machain-Castillo (1985) on ostracodes from Late Tertiary formations of southeastern Mexico indicates a middle Pliocene age for the Paraje Solo Formation.

COMPARISON OF DIVERSITY IN TERTIARY FLORAS FROM MEXICO AND PANAMA

The composition of the Paraje Solo microfossil flora from southeastern Veracruz is presented in Table 2.2. The 106 taxa (plus 18 unknowns) recognized for the flora group into eight principal paleocommunities (Table 2.3), with 10 other communities represented by a few wide ranging genera (Graham, 1976). By contrast, the early Miocene Uscari, Culebra, Cucuracha, and La Boca microfossil floras from Costa Rica and Panama reflect a more simple vegetation (Tables 2.4 and 2.5). The Uscari assemblage consists of 40 identified palynomorphs and six unknowns (the difference between these numbers and the 25 taxa listed in column 2 [Uscari] of Table 2.4 is because subtypes of certain palynomorphs, e.g., monolet fern spores types 1–5, are recognized in the original study; see Graham, 1987a). The Culebra flora contains 41 identified taxa plus 14 unknowns, the Cucuracha flora 19 plus 2 unknowns, and the La Boca flora 39 plus 15 unknowns. These figures are

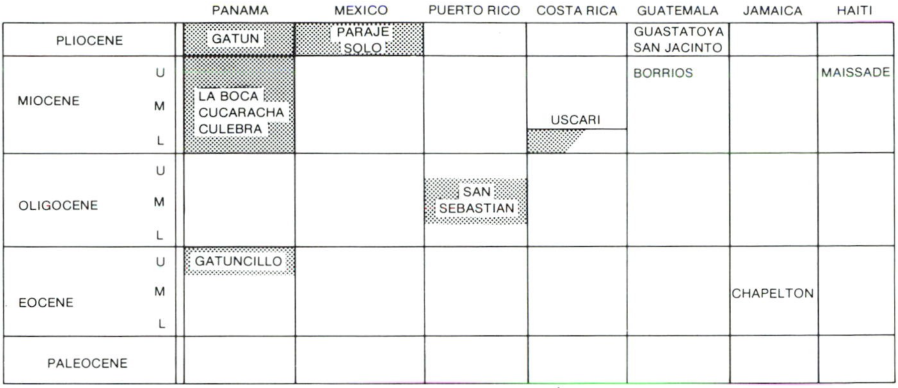

		PANAMA	MEXICO	PUERTO RICO	COSTA RICA	GUATEMALA	JAMAICA	HAITI
PLIOCENE		GATUN	PARAJE SOLO			GUASTATOYA SAN JACINTO		
MIOCENE	U					BORRIOS		MAISSADE
	M	LA BOCA CUCARACHA CULEBRA			USCARI			
	L							
OLIGOCENE	U			SAN SEBASTIAN				
	M							
	L							
EOCENE	U	GATUNCILLO						
	M						CHAPELTON	
	L							
PALEOCENE								

Figure 2.2. Age of Tertiary microfossil floras in the Gulf/Caribbean region. Position on the chart does not imply formal correlation between the formations. Stippling indicates study of the assemblage has been completed and the results published.

summarized in Table 2.1. The Paraje Solo flora has more than twice the number of recognized taxa as the most diverse of the Costa Rica/Panama fossil floras and six times as many as the Cucaracha assemblage. Several historical factors can be suggested to account, in part, for the differences.

Physiographic Diversity

The region of southeastern Veracruz state is influenced by two mountain systems. To the north and west is the Sierra Madre Oriental, a Mesozoic orogenic belt that began its development in latest Cretaceous/Paleocene times (Suter, 1984) and has served as source area for deposits in the Veracruz Basin since the end of the Cretaceous (Helu et al., 1977). The TVB is a younger system that began uplift during the Early Tertiary, with its principal period of intense vulcanism and uplift during the late Miocene, Pliocene, and Pleistocene (Demant, 1978; Nixon, 1982; Thorpe, 1977; Ferrusquía-Villafranca, this volume). The system includes the highest peaks in Mexico (Pico de Orizaba 5,650 m; Popocatepetl 5,450 m; Ixtaccihuatl 5,280 m). Collectively, the two mountain ranges have provided areas of significant physiographic relief throughout Middle to Late Tertiary time and were certainly prominent during the Pliocene when the Paraje Solo flora was being deposited. Habitats included coastal, brackish-water zones, low-lying freshwater lakes, swamps, and marshes, upland slopes, and middle to high-altitude mountains. Already present by the middle Pliocene were communities of manglar (*Rhizophora* [Fig. 2.3H]; *Laguncularia*; *Hibiscus*, probably *H. tiliaceous* [Fig. 2.3F]), marsh/swamp/aquatic vegetation (palms [Fig. 2.3B]; *Ceratopteris* and other ferns; *Utricularia*), various types of forests (*Cyathea*, [Fig. 2.3A]; *Alchornea*; *Bursera*; *Casearia*; *Cedrela*; *Guarea*; *Gustavia*; *Mortoniodendron*; *Symphonia*), pine-oak communities (*Pinus* [Fig. 2.3E]; *Quercus* [Fig. 2.3C]), deciduous forest (*Quercus*; *Liquidambar*; *Alnus* [Fig. 2.3I]; *Ilex* [Fig. 2.3K]; *Engelhardia* [Fig. 2.3J]; *Cleyera*; *Cyathea*), and high altitude needle-leaved forests (*Pinus*; *Abies* [Fig. 2.3D]; *Picea* [Fig. 2.3G]; Table 2.3). The relevance of this orogenic history to the origin of biological diversity in southern and southeastern Mexico is that "speciation rates increase in areas with high topographic complexity" (Cracraft & Prum, 1988; Lewin, 1988), and that fluctuating environments and an extensive array of habitats have been available in the region for at least four million years and gradually to a lesser extent for much longer.

The physiographic situation in southern Central America was different. The various tectonic models proposed for the Caribbean (Dickinson & Coney, 1980; Malfait & Dinkleman, 1972; Pindell & Dewey, 1982; Sykes et al., 1982; Wadge & Burke, 1983) are consistent in depicting the isthmian

Table 2.2. Revised composition of the Paraje Solo microfossil flora, Veracruz, Mexico

PSILOTACEAE: *Psilotum*
LYCOPODIACEAE: *Lycopodium*
SELAGINELLACEAE: *Selaginella*
CYATHEACEAE: *Alsophila, Cyathea, Cnemidaria (Hemitelia), Sphaeropteris/
    Trichipteris*
DRYOPTERIDACEAE: *Lomariopsis*
GLEICHINIACEAE: *Dicranopteris*
POLYPODIACEAE: Monolete fern spores types 1–6
PTERIDACEAE: *Ceratopteris; Pteris* types 1,2
PINACEAE: *Abies, Picea, Pinus* types 1, 2
PODOCARPACEAE: *Podocarpus*
ARACEAE: *Spathiphyllym*
CYPERACEAE
DIOSCOREACEAE: *Rajania*
GRAMINEAE
LILIACEAE: *Smilax*
PALMAE: cf. *Astrocaryum*, cf. *Attalea*, cf. *Brahea*, cf. *Chamaedorea*, cf.
    Maximiliana type
ACANTHACEAE: *Bravaisia, Justicia*
AMARANTHACEAE: *Irisene*
AMARANTHACEAE/CHENOPODIACEAE
ANACARDIACEAE: *Comocladia*
AQUIFOLIACEAE: *Ilex*
BETULACEAE: *Alnus*
BORAGINACEAE: *Tournefortia*
BURSERACEAE: *Bursera, Protium*
CHLORANTHACEAE: *Hedyosmum*
COMBRETACEAE: *Combretum, Terminalia, Laguncularia*
COMPOSITAE types 1–6
DICHAPETALACEAE: *Dichapetalum*
EUPHORBIACEAE: *Alchornea*, cf. *Bernardia*, cf. *Sapium*; cf. *Stillingia*, cf.
    Tetrorchidium, cf. *Tithymalus*
FAGACEAE: *Quercus*
FLACOURTIACEAE: *Casearia, Laetia*
GUTTIFERAE: *Symphonia*
HAMAMELIDACEAE: *Liquidambar*
JUGLANDACEAE: *Alfaroa/Engelhardia, Juglans*
LECYTHIDACEAE: *Gustavia*

Table 2.2. (cont.)

LEGUMINOSAE: *Acacia, Desmanthus, Mimosa* (cf. *M. pigra*)
LENTIBULARIACEAE: *Utricularia*
LORANTHACEAE: *Struthanthus*
LYTHRACEAE: *Cuphea*
MALPIGHIACEAE: cf. *Hiraea*, cf. *Malpighia*, cf. *Mezia* (?) type;
MALVACEAE: *Hampea/Hibiscus*
MELIACEAE: *Cedrela, Guarea*
MYRICACEAE: *Myrica*
MYRTACEAE: *Eugenia/Myrcia*
ONAGRACEAE: *Ludwigia*
PASSIFLORACEAE: *Passiflora*
POLYGALACEAE: cf. *Bredemeyera*, cf. *Securidaca*
POLYGONACEAE: *Coccoloba*
RANUNCULACEAE: *Thalictrum*;
RHIZOPHORACEAE: *Rhizophora*
RUBIACEAE: cf. *Alibertia, Borreria, Faramea, Terebrania*;
SALICACEAE: *Populus*
SAPINDACEAE: *Allophylus, Cupania, Matayba, Meliosma*, cf. *Paullinia* (e.g. *P. pinnata*), cf. *Paullinia* (e.g. *P. turbacensis*), *Serjania*
STERCULIACEAE: *Buettneria*
THEACEAE: *Cleyera*
THYMELIACEAE: *Daphnopsis*
TILIACEAE: *Mortoniodendron*
ULMACEAE: *Celtis, Ulmus*.

---

area as a series of low-lying peninsulas and volcanic islands throughout the Tertiary. Compared to southeastern Mexico, land area was less, extensive highlands were not present, and any potential altitudinal zonation was diminished by the buffering effect of the surrounding ocean. The paleocommunities were consequently fewer in number, consisting of a fringing zone of mangroves (*Pelliceria, Rhizophora*), floating or submerged freshwater communities (*Ceratopteris, Utricularia*), ferns and palm marshes, and versions of the low- to moderate-altitude cloud forest (*Cnemidaria, Cyathea, Pteris, Alchornea, Allophylus, Casearia, Combretum, Crudia, Cupania, Hiraea, Matayba, Sabicea, Sapium, Tetrorchidium*) (Table 2.5). Poorly represented to absent were elements of premontane to montane communities. This comparatively simple, although tectonically active, landscape persisted throughout the Cenozoic, with modern elevations reaching only 3,475 m in the western mountains (Volcan Baru) in Panama and 3,820 m (Cerro Chirripo) and 3,432 m (Volcan Irazu) in Costa Rica.

Climatic Changes

In addition to the dynamic physical environment, with concomitant diversity of habitats available in southeastern Mexico for much of the Tertiary

Table 2.3. Paleocommunities of the Paraje Solo microfossil flora.

Needle-leaved and scale-leaved forest (including pine and pine-oak forests): *Abies, Picea, Pinus, Alchornea, Alnus, Quercus,* cf. *Sapium, Coccoloba, Smilax,* cf. *Stillingia.* (Also *Cyathea* in atypically low altitude *Pinus strobus* forest near Tlapacoyan, 500 m, mixture of evergreen and deciduous forest elements; *Liquidambar* and *Myrica* in atypically low-altitude *Pinus oocarpa* forest on the eastern slopes of Volcan Santa Marta, 500 m, mixture of lowland and deciduous forest elements.)

Broad-leaved oak forest: *Alchornea, Coccoloba, Quercus,* cf. *Sapium*

Deciduous (oak-*Liquidambar*) forest: *Psilotum, Lycopodium, Sphaeropteris, ?Trichipteris, Selaginella, Alsophila, Cyathea, Pinus, Podocarpus, Alchornea, Alnus,* cf. *Chamaedorea, Cleyera, Dichapetalum, Alfaroa/Engelhardia, Eugenia, Guarea, Hampea, Hedyosmum, Ilex, Iresine, Juglans, Justicia* (widespread), *Liquidambar, Meliosma, Myrica, Populus* (riparian, widespread), *Quercus, Struthanthus, Tournefortia, Ulmus.*

High evergreen forest: *Lycopodium, Sphaeropteris/Trichipteris, Podocarpus, Alchornea, Allophyllus* (secondary), cf. *Astrocaryum, Bursera, Casearia, Cedrela,* cf. *Chamaedorea, Cupania, Faramea, Guarea, Gustavia, Hampea, Hibiscus,* cf. *Hiraea, Iresine, Matayba, Mortoniodendron, Myrica,* cf. *Paullinia,* cf. *Sapium, Spathiphyllum, Symphonia, Terminalia,* cf. *Tetrorchidium.*

High semievergreen forest: *Lycopodium, Selaginella, Cyathea, Alchornea,* cf. *Bernardia* (secondary), *Bursera, Casearia, Cedrela,* cf.*Chamaedorea, Cupania, Daphnopsis, Faramea, Hampea, Hibiscus, Ilex, Iresine,* cf. *Paullinia, Protium, Quercus, Rajania,* cf. *Sapium, Securidaca* (secondary), *Spathiphyllum,* cf. *Tetrorchidium, Ulmus.*

Low deciduous forest: cf. *Acacia,* cf. *Brahea, Bursera, Casearia, Celtis, Combretum, Cupania, Daphnopsis, Eugenia, Ilex,* cf. *Sapium, Comocladia.*

Swamp/aquatic vegetation: *Ceratopteris, Bravaisia, Ludwigia, Utricularia.*

Mangrove swamp: *Hibiscus, Laguncularia, Ludwigia, Mimosa, (M. pigra* type), *Rhizophora,* cf. *Sapium.*

Genera are placed according to typical occurrence; many range through several vegetation types. Fossil monolete fern spores of the Blechnaceae/Polypodiaceae/Pteridaceae and fossil pollen of Compositae,Cyperaceae, and Graminae, undifferentiated generically, are present and range throughout most present-day communities. The dominant or defining elements of the high evergreen forest are rare or absent, and most of the genera listed occur in several other communities.

period, the biota also experienced changes in climate. Study of ^{18}O from shells of marine invertebrates (Savin, 1977; Savin & Douglas, 1985; Savin et al., 1975) demonstrated that warm and relatively constant temperatures prevailed from the Early Cretaceous through the middle Eocene, with a short-lived drop at the Cretaceous-Tertiary boundary. As noted by Friis et

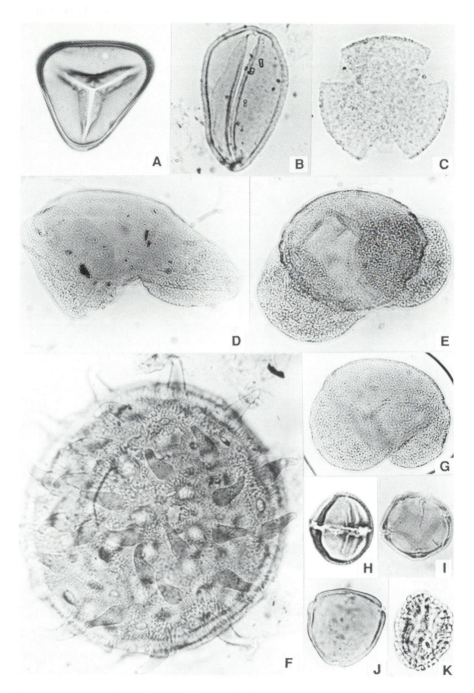

Figure 2.3. Plant microfossils from the Mid-Pliocene Paraje Solo Formation, state of Veracruz, Mexico. A,*Cyathea*; B,Palmae; C,*Quercus*; D,*Abies*; E,*Pinus*; F,*Hampea/Hibiscus*; G,*Picea*; H,*Rhizophora*; I,*Alnus*; J,*Engelhardia*; K,*Ilex*.

Table 2.4. Plant microfossils from lower Miocene formations in the Gulf/Caribbean region (Costa Rica and Panama)

Organism	Uscari	Culebra	Cucaracha	La Boca
FUNGI				
Microtherium type	X			
Ascomycete cleistothecium				X
PYRROPHYTA				
Operculodinium centrocarpum				X
Spiniferites				X
BRYOPHYTA				
Phaeoceros	X			
Muscae	X			
LYCOPSIDA				
Lycopodium	X	X	X	X
Selaginella	X	X	X	X
FILICINEAE				
cf. *Antrophyum*		X	X	X
Ceratopteris			X	
Cnemidaria	X			
Cyathea		X	X	X
Danaea		X		
cf. *Hymenophyllum*	X			
Lophosoria	X			
Lygodium		X		
Pityrogramma	X			
Pteris	X	X	X	X
Other monolete fern spores	X	X	X	X
Other trilete fern spores	X	X	X	X
GYMNOSPERMAE				
Podocarpus	X			
ANGIOSPERMAE				
Acacia		X		
cf.*Aguiaria*				X
Alchornea	X	X	X	X
Alfaroa/Engelhardia				X
Allophylus		X		
cf. *Attalea* type				X
cf. *Banisteriopsis*	X			
Bombacaceae	X			
Casearia		X		
cf. *Ceiba*				X
Chenopodac/Amaranthac		X		
Combretum/Terminalia		X		

Table 2.4. (cont.)

Organism	Uscari	Culebra	Cucaracha	La Boca
Compositae	X	X	X	
Crudia			X	X
Cryosophia type		X	X	X
Cupania		X		
Desmoncus type		X		
Dioscorea/Rajania		X		
cf. *Doliocarpus*		X		
Ericaceae	X			
Eugenia/Myrcia	X	X	X	
cf. *Glycydendrum*	X			
Gramineae		X		X
cf. *Guazuma*		X		
Hampea/Hibiscus		X		X
cf. *Hiraea*	X			
Ilex	X	X	X	X
Lisianthius	X			
Malpighiaceae		X		X
Manicaria type		X	X	X
Matayba		X		
Melastomataceae	X			X
Pelliceria				X
Psedobombax				X
Rhizophora	X	X	X	X
cf. *Rourea*		X		
Rubiaceae				X
Sabicea		X		
Sapium		X	X	
Sideroxylon		X		
Synechanthus type		X	X	X
Tetrorchidium		X		
Utricularia				X

al. (1987, p.11), Parrish (1987, p.51), and Frakes (1979) this interval was possibly the warmest in the history of the earth. A sharp drop occurred at the end of the Eocene, followed by more gradual cooling until the early Miocene when temperatures rose and then fell dramatically during middle and late Miocene times. A further lowering occurred during the late Pliocene, reflecting initiation of the Pleistocene glaciations. This pattern is for ocean bottom and surface water temperatures but is reflected in terrestrial Tertiary floras from the Northern Hemisphere (Hubbard & Boulter, 1983; Tanai & Huzioka, 1967; Upchurch & Wolfe, 1987; Wolfe, 1978; Wolfe & Hopkins, 1967). The Paraje Solo study revealed that the effects also extended at least to 18° N in the New World. The presence of *Picea* in southeastern Mexico during the middle Pliocene suggests cooler temperatures because spruce is presently represented by only two species (*P. chihuahuana*, *P. mexicana*) in the mountains of northern Mexico 1,000 km or more distant from the Paraje Solo locality. Cooler temperatures are also consistent with the poor representation of the tropical rain forest in the fossil flora and the presence of relict populations of a few temperate genera (*Myrica*, *Podocarpus*) isolated in the modern Veracruz lowlands. To the effects of topographic change in the region during the Tertiary can be added climatic changes documented to have significantly affected the composition and distribution of the plant and animal communities.

To the south the climatic picture was different. Data are not yet available for Pliocene vegetation in Central America, but some relevant information does exist for other fossil floras. The Culebra, Cucaracha, and La Boca floras are Lower Miocene in age, and all are similar in composition to the modern tropical vegetation of Panama. This is consistent with the global paleo-temperature curve, which shows warm conditions during the early Miocene. Although the effect of the Late Tertiary temperature drop on terrestrial vegetation at 9° N latitude cannot be measured directly, it is possible to extrapolate from data available for the middle (?) to late Eocene Gatuncillo Formation of Panama. Remnants of this vegetation were also being deposited during a time of general climatic cooling, but the communities were not significantly affected. The paleocommunities represented in the Gatuncillo assemblage are similar to those of present-day lowland

Table 2.5. Plant paleocommunities from the lower Miocene of the Gulf/Caribbean region (Costa Rica and Panama)

Floating or submerged fresh-water aquatic communities: *Ceratopteris, Utricularia.*

Mangrove swamp: *Pelliceria, Rhizophora, Hibiscus, Sapium.*

Tropical moist forest: *Lycopodium, Selaginella,* cf. *Antrophyum, Cyathea, Danaea, Lygodium, Pteris, Cryosophia* type, *Desmoncus* type, *Manicaria* type, *Synechanthus* type, *Acacia, Alchornea, Allophylus, Casearia, Combretum, Crudia, Cupania, Dioscorea,* cf. *Doliocarpus, Eugenia, Hampea, Matayba, Myrcia,* cf. *Rourea, Sabicea, Sapium, Terminalia, Tetrorchidium.*

Tropical wet forest: *Phaeoceros, Lycopodium, Selaginella,* cf. *Antrophyum, Cnemidaria, Danaea, Lygodium, Pteris, Cryosophia* type, *Desmoncus, Manicaria* type, *Alchornea, Allophylus,* cf. *Banisteriopsis, Casearia, Combretum, Crudia, Cupania, Dioscorea, Eugenia,* cf. *Glycydendrum, Hampea,* cf. *Hiraea, Matayba, Myrcia, Terminalia, Tetrorchidum.*

Premontane wet forest: *Lycopodium, Selaginella,* cf. *Anthophyum, Cyathea, Danaea, Lygodium, Pteris, Crysophia* type, *Manicaria* type, *Synechanthus* type, *Alchornea, Allophylus, Casearia, Combretum, Crudia, Cupania, Dioscorea,* cf. *Doliocarpus, Eugenia,* cf. *Guazuma, Matayba,* cf. *Rourea, Sabicea, Sapium, Terminalia, Tetrorchidium.*

Table 2.5. (cont.)

Premontane moist forest: cf. *Antrophyum, Lygodium, Allophylus, Combretum, Dioscorea, Eugenia,* cf. *Guazuma, Hampea, Ilex, Matayba,* cf. *Rourea, Terminalia.*

Lower montane moist forest: *Lycopodium, Selaginella, Cyathea,* cf. *Hymenophyllum, Lophosoria, Pityrogramma, Pteris, Manicaria* type, *Podocarpus, Engelhardia, Lisianthius.*

Tropical dry forest: *Lygodium, Allophylus, Casearia, Combretum,* cf. *Guazuma, Matayba,* cf. *Rourea.*

Lower montane wet forest: *Lycopodium, Selaginella, Cyathea, Pteris, Manicaria* type, *Sapium.*

Montane moist forest: cf. *Hymenophyllum, Lophosoria, Pityrogramma, Podocarpus, Engelhardia, Lisianthius.*

Premontane rain forest: *Synechanthus* type *Alchornea, Casearia, Hampea.*

Premontane dry forest: *Casearia, Combretum, Eugenia,* cf. *Guazuma.*

Placement is according to principal or most common occurrence(s), and most range through more than one community. Genera occurring in virtually all communities, or where placement is uncertain, are not included (e.g., the Ascomycete fungus *Microthyrium*). Also omitted are taxa identified only to family or higher taxonomic groups (Muscae, other monolete fern spores, Bombacaceae, etc.).

Panama, with all identified taxa present in the modern vegetation. If the Eocene assemblages experienced little effect of the cooling trend because of a more southern latitude, less physiographic diversity, and insular environment, it is likely that the effects of the Late Tertiary cooling at the same site and under similar physiographic conditions were also slight, at least compared to those in southern Mexico.

Source Areas

Throughout the Tertiary continuous land extended between the biotas of southeastern Mexico and temperate North America. As a consequence, when climatic cooling occurred beginning at the Mid-Miocene, temperate elements were able to migrate into the area from the north and, because of topographic diversity, were able to persist in a zone between 1,000 and 2,000 m along the eastern escarpment of the Mexican Plateau. Although some species were likely introduced during the Pleistocene and throughout the Cenozoic by long-distance dispersal, many were already present by the middle Pliocene. The Paraje Solo flora includes pollen of *Abies, Picea,*

Pinus, Ilex, Alnus, Quercus, Liquidambar, Juglans, Myrica, Populus, Celtis, and *Ulmus*. Although the exact time of introduction of these northern temperate elements is not known, they were not recovered from the Oligo-Miocene Simojovel Group of Chiapas (Langenheim et al., 1967), and the paleotemperature curve shows that earlier times were less suitable because of rising temperatures and more tropical conditions. Certainly from the middle Miocene onward the biota of southeastern Mexico periodically received elements from the north; and because of the physiographic diversity they were able to persist and augment the existing warm-temperate to tropical communities.

To the south through Central America, land surfaces were intermittent until about 2.4 Ma, when final connection between North and South America was established (Stehli & Webb, 1985). Prior to this time island "stepping stones" were available for migration through Central America, but they cannot be used uncritically to suggest opportunity for exchange of all tropical elements. Plate tectonic models imply that during the early stages of emergence these islands and peninsulas were relatively low-lying and that the even moderate altitudes present today are of recent origin. The paleobotanical data are consistent with this assumption. The five Tertiary floras from the region (middle[?] to late Eocene Gatuncillo, early Miocene Uscari, Culebra, Cucaracha, and La Boca floras) reflect only coastal and low- to moderate-altitude communities (Table 2.5), with no pollen from mid- to high-altitude communities being either blown or washed into the depositional basin.

Connections through the Antilles have been intermittent throughout the Cenozoic. The magmatic arc constituting the proto-Greater Antilles developed during the Late Cretaceous to Early Tertiary and moved eastward until it collided with the Florida-Bahama Platform during the late Eocene. The lesser Antilles arose during the Eocene/Oligocene (Coney, 1982). Regardless of whether the Greater Antilles were rafted in from great distances to the west, with the Cayman Trough representing the site of great fault motion or they are of more local origin, there was no continuous land connecting Mexico with South America via a single Antillean land mass any time during the Cenozoic. Opportunities for migration through the Antilles gradually increased after the Eocene as South America moved closer to North America and the Greater and Lesser Antilles chain became established.

Tectonic events and plate movements were not the only factors influencing the paleophysiography of Middle America during the Cenozoic. Vail and Hardenbol (1979) and Savin and Douglas (1985) discussed the effect of changing sea levels. They noted that the difference between the highest levels of the Late Cretaceous and the present was about 300 m; if all polar ice melted, sea levels would rise about 60 m; and at the maximum of Pleistocene glaciations, sea levels were lower by about 92 m. These figures are important in the context of present-day physiography. Across northern Costa Rica/southern Nicaragua there is a lowland region with maximum elevations of only 34 m, and across eastern Panama (Darien) the elevation is only 84 m. It is not possible at present to disentangle the effect of sea level changes from uplift and subsidence in this geologically complex region, but such fluctuations certainly contributed to the dynamic nature of the Central American landscape.

Thus prior to about 2.4 Ma the area of southeastern Mexico likely received tropical elements from the south during warm periods, as during the Paleocene through middle Eocene and the early Miocene, but physiographic conditions favored introduction of coastal and low-altitude-inhabiting species with effective means of long-distance dispersal. After about 2.4 Ma, introduction of a greater variety of elements from the south was facilitated by continuous land surfaces and increasing altitudes through Central America.

With the development of greater land surfaces, increasing topographic diversity, and continuous land established across eastern Panama, new taxa were added to the biological inventory of southern Central America. Now the estimated 7,345 species in the 77,082 km^2 of Panama and the 8,000 species in the 51,100 km^2 of Costa Rica, compare more closely to the 8,250 species in the 74,000 km^2 of Chiapas (Table 2.1). This trend is also reflected in Quaternary studies from Gatun Lake, Panama by Bartlett and Barghoorn (1973). They reported about 148 kinds of pollen and spores from these sediments, far more than the number recovered from any Tertiary deposits in Central America but comparable to the 124 types found in the Paraje Solo assemblage from southeastern Mexico.

The present configuration of the area shows southern Mexico located at the confluence of two major migration routes that have operated as described throughout much of the Tertiary. Thus a factor accounting for the diversity of the Mexican biota, in contrast to that of southern Central America, has been greater accessibility to more extensive source areas through longer periods of Cenozoic time. This accessibility provided both temperate and tropical supplements to the original Late Cretaceous/Early Tertiary plant and animal communities evolving *in situ* in the Sierra Madre Oriental and Transvolcanic region.

ACKNOWLEDGMENTS

The author gratefully acknowledges information provided by James Allan, Washington University, St. Louis, Missouri, United States. Research was supported by NSF grant 8500850, 8819771.

REFERENCES

Akers, W.H. 1979. Planktic foraminifera and calcareous nannoplankton bio-stratigraphy of the Neogene of Mexico. Part I. Middle Pliocene. Tulane Studies Geol. Paleontol. 15:1–32.

————. 1981. Planktic foraminifera and calcareous nannoplankton bio-stratigraphy of the Neogene of Mexico. Addendum to part I. Some additional mid-Pliocene localities and further discussion on the Agueguexquite and Concepcion superior beds. Tulane Studies Geol. Paleontol. 16:145–148.

————. 1984. Planktic foraminifera and calcareous nannoplankton bio-stratigraphy of the Neogene of Mexico. Part II. Lower Pliocene. Tulane Studies Geol. Paleontol. 18:21–36.

Alvarado, J.L. & M. Delgado Rueda. 1985. Flora apicola en Uxpanapa, Veracruz, México. Biotica 10:257–275.

Bartlett, A.S. & E.S. Barghoorn. 1973. Phytogeographic history of the Isthmus of Panama during the past 12,000 years (a history of vegetation, climate, and sea level change). *In* A. Graham (ed.), Vegetation and Vegetational History of Northern Latin America. Amsterdam: Elsevier. pp. 203–299.

Breedlove, D.E. 1981. Introduction to the flora of Chiapas. *In* D.E. Breedlove (ed.), Flora of Chiapas, Part 1. San Francisco, CA: California Academy of Sciences.

Coney, P.J. 1982. Plate Tectonic constraints on the biogeography of Middle America and the Caribbean Region. Ann. Missouri Bot. Gard. 69:432-443.

Cooperrider, T.S. 1983. Introduction. *In* T.S. Cooperrider (ed.), Endangered and Threatened Plants of Ohio. Ohio Biol. Surv. Biol. Notes No. 16. pp. 1–4.

Cracraft, J. & R. Prum. 1988. Patterns and processes of diversification: speciation and historical congruence in some neotropical birds. Evolution 42:603–620.

D'Arcy, W.G. 1987. Flora of Panama Checklist and Index. Vols. 1 & 2. St. Louis, MO: Missouri Botanical Garden.

Demant, A. 1978. Caracteristicas de Eje Neovolcanico Transmexicano y sus problemas de interpretación. UNAM. Rev. Inst. Geol. 2:172–187.

Dickinson, W.R. & P.J. Coney. 1980. Plate tectonic constraints on the origin of the Gulf of Mexico. *In* R.H. Pilger (ed.), The Origin of the Gulf of Mexico and the Early Opening of Central North Atlantic Ocean. Baton Rouge, LA: Lousiana State Univ.. pp. 27–36.

Frakes, L.A. 1979. Climates Throughout Geologic Time. Amsterdam: Elsevier.

Friis, E.M., W.G. Chaloner & P.R. Crane. 1987. Introduction to angiosperms. *In* E.M. Friis, W.G. Chaloner & P.R. Crane (eds.), The Origins of Angiosperms and their Biological Consequences. Cambridge: Cambridge Univ. Press. pp. 1–16.

Gonzáles-Quintero, L. 1980. Paleoecología de un sector costero de Guerrero, México (3000 años). Memorias III Coloquio Paleobotánica Palinología (México D.F., 1980). INAH, Colección Científica 86:133–157.

———— & M. Fuentes Mata. 1980. El Holoceno de la cuenca del valle de México. Memorias III Coloquio Paleobotánica Palinología (México D.F., 1980). INAH, Colección Científica 86: 113–132.

———— & A. Montufar López. 1980. Interpretación paleoecológica del contenido polínico de un núcleo cercano a Tula, Hidlago. Memorias III Coloquio Paleobotánica Palinología (México D.F., 1980). INAH, Colección Científica 86:113–132.

———— & F. Sanchez Martínez. 1980. Determinación palinológica del ambiente en que vivieron los mamuts en la cuenca de México. Memorias III Coloquio Paleobotánica Palinología (México D.F., 1980). INAH, Colección Científica 86:113–132

Graham, A. 1973. Literature on vegetational history in Latin America. *In* A. Graham (ed.), Vegetation and Vegetational History of Northern Latin America. Amsterdam: Elsevier. pp. 315–360.

————. 1976. Studies in Neotropical paleobotany. II. The Miocene communities of Veracruz, Mexico. Ann. Missouri Bot. Gard. 63:787–842.

————. 1979. Literature on vegetational history in Latin America. Supplement I. Rev. Palaeobot. Palynol. 27:29–52.

————. 1982. Literature on vegetational history in Latin America. Supplement II. Rev. Palaeobot. Palynol. 37:185–223.

————. 1986. Literature on vegetational history in Latin America. Supplement III. Rev. Palaeobot. Palynol. 48:199–239.

————. 1987a. Miocene communities and paleoenvironments of southern Costa Rica. Amer. J. Bot. 74:1501–1518.

————. 1987b. Tropical American Tertiary floras and paleoenvironments: Mexico, Costa Rica, and Panama. Amer. J. Bot. 74:1519–1531.

————. 1988a. Studies in Neotropical paleobotany. V. The lower Miocene communities of Panama—the Culebra formation. Ann. Missouri Bot. Gard. 75:1441–1466.

————. 1988b. Studies in Neotropical paleobotany. VI. The lower Miocene communities of Panama—the Cucaracha Formation. Ann. Missouri Bot. Gard. 75:1467–1479.

————. 1989. Studies in Neotropical paleobotany. VII. The lower Miocene communities of Panama—the La Boca Formation. Ann. Missouri Bot. Gard. 76:50–66.

Hartshorn, G.S. 1983. Plants. *In* D.H. Janzen (ed.), Costa Rican Natural History. Chicago, IL: The Univ. of Chicago Press. pp. 118–157.

Helu, P.C., R. Verdugo V., & R. Barcenas P. 1977. Origin and distribution of Tertiary conglomerates, Veracruz Basin, Mexico. Amer. Assoc. Petrol. Geol. Bull. 61:207–226.

Hubbard, R.N. & M.C. Boulter. 1983. Reconstruction of Palaeogene climate from palynological evidence. Nature 301:147–150.

Langenheim, J.H., B. Hackner & A. Bartlett. 1967. Mangrove pollen at the depositional site of Oligo-Miocene amber from Chiapas, Mexico. Bot. Mus. Leafl. 21:289–324.

Lewin, R. 1988. The stamp of history and ecology in Amazonia. Science 241:1619.

Leyden, B.W. 1984. Guatemalan forest synthesis after Pleistocene aridity. Proc. Natl. Acad. Sci. U.S.A. 81:4856–4859.

Ludlow-Wiechers, B. 1980. Palynological catalogue for the Veracruz flora. Abstract., 5th Internat. Palynol. Conf. (Cambridge, 1980). p. 235.

———— & M.L. Ayala-Nieto. 1983. Catalogo palinológico para la flora de Veracruz, 14. Familia Taxodiaceae. Biotica 8:309–314.

Machain-Castillo, M. 1985. Ostracode biostratigraphy and paleoecology of the Pliocene of the isthmian salt basin, Veracruz, Mexico. Tulane Studies Geol. Paleontol. 19:123–139.

Malfait, B. & M. Dinkleman. 1972. Circum-Caribbean tectonic and igneous activity and the evolution of the Caribbean plate. Bull. Geol. Soc. Amer. 83:251–272.

Martínez-Hernández, E., M. Sanchez López, & H. Hernández Campos. 1980. Palinología del Eocene en el noreste de México. Abstract, IV Coloq. Paleobotánica Palinología (México D.F. 1980). pp. 25–26.

Medellin-Leal, F. 1982. The Chihuhuan Desert. *In* G.L. Bender (ed.), Reference Handbook on the Deserts of North America. Westport, CT: Greenwood Press. pp. 331–381.

Nixon, G.T. 1982. The relationship between Quaternary volcanism in central Mexico and the seismicity and structure of subducted ocean lithosphere. Bull. Geol. Soc. Amer. 93:514–523.

Palacios-Chavez, R. 1985. Lluvia de polen moderno en el bosque tropical caducifolio de la estación de Biología de Chamela, Jalisco (México). Anales Esc. Nac. Ci. Biol. 29:43–55.

——— & M.L. Arreguin. 1980. Análisis polínico de algunos sitios de interés arqueológico en el Valle de San Juan de Río, Queretero. Memorias III Coloquio Paleobotánica Paleonología (México D.F., 1980). INAH, Colección Científica 86:179–183.

Parrish, J.T. 1987. Global palaeogeography and palaeoclimate of the late Cretaceous and early Tertiary. *In* E.M. Friis, W.G. Chaloner & P.R. Crane (eds.), The Origins of Angiosperms and their Biological Consequences. Cambridge: Cambridge Univ. Press. pp. 51–74.

Pianka, E.R. 1983. Evolutionary Ecology. 3rd ed. New York: Harper and Row.

Pindell, J. & J.F. Dewey. 1982. Permo-Triassic reconstruction of western Pangea and the evolution of the Gulf of Mexico/Caribbean region. Tectonics 1:179–211.

Quiroz-García, D. & R. Palacios Chavez. 1986a. Catalogo palinológica para la flora de Veracruz. No. 31. Familia Boraginaceae. Género *Rochefortia*. Biotica 11:47–50.

——— & ———. 1986b. Catalogo palinológica para la flora de Veracruz. No. 32. Familia Boraginaceae. Género *Tournefortia*. Biotica 11:51–66.

Radford, A.E., H.E. Ahles & C.R. Bell. 1968. Manual of the Vascular Flora of the Carolinas. Chapel Hill, NC: Univ. North Carolina Press.

Ricklefs, R.E. 1979. Ecology. 2nd ed. New York: Chiron Press.

Rzedowski, J. 1973. Geographical relationships of the flora of Mexican dry regions. *In* A. Graham (ed.), Vegetation and Vegetational History of Northern Latin America. Amsterdam: Elsevier. pp. 61–72.

———. 1978. Vegetación de México. Mexico City: Limusa.

Savin, S.M. 1977. The history of the Earth's surface temperature during the past 100 million years. Ann. Rev. Earth Planet. Sci. 5:319–355.

——— & R.G. Douglas. 1985. Sea level, climate, and the Central American land bridge. *In* F.G. Stehli & S.D. Webb (eds.), The Great American Biotic Interchange. New York: Plenum. pp. 303–324.

———, ——— & F.G. Stehli. 1975. Tertiary marine paleo-temperatures. Bull. Geol. Soc. Amer. 86:1499–1510.

Stehli, F.G. & S.D. Webb (eds.). 1985. The Great American Biotic Interchange. New York: Plenum.

Suter, M. 1984. Cordilleran deformation along the eastern edge of the Valles-San Luis Potosi carbonate platform: Sierra Madre Oriental fold-thrust belt, east-central Mexico. Bull. Geol. Soc. Amer. 95:1387–1397.

Sykes, L.R., W.R. McCann & A.L. Kafka. 1982. Motion of Caribbean plate during last 7 million years and implications for earlier Cenozoic movements. J. Geophys. Res. 87:10656–10676.

Tanai, T. & K. Huzioka. 1967. Climatic implications of Tertiary floras in Japan. 11th Pacific Science Congress (Tokyo, 1966), Symposium 25. pp. 77–87.

Thorpe, R.S. 1977. Tectonic significance of alkaline volcanism in eastern Mexico. Tectonophysics 40:T19–T26

Toledo, V.M. 1976. Los cambios climaticos del Pleistoceno y sus efectos sobre la vegetación tropical calida y humeda de México. M.S. Thesis. UNAM.

Upchurch, G.R. & J.A. Wolfe. 1987. Mid-Cretaceous to early Tertiary vegetation and climate: evidence from fossil leaves and woods. *In* E.M. Friis, W.G. Chaloner, & P.R. Crane (eds.), The Origins of Angiosperms and Their Biological Consequences. Cambridge: Cambridge Univ. Press. pp. 75–106.

Vail, P.R. & J. Hardenbol. 1979. Sea-level changes during the Tertiary. Oceanus 22:71–79.

Wadge, G. & K. Burke. 1983. Neogene Caribbean plate rotation and associated Central American tectonic evolution. Tectonics 2:633–643.

Wolfe, J.A. 1978. A paleobotanical interpretation of Tertiary climates in the Northern Hemisphere. Amer. Sci. 66:694–703.

———— & D.M. Hopkins. 1967. Climatic changes recorded by Tertiary land floras in northwestern NorthAmerica. 11th Pacific Science Congress (Tokyo, 1966), Symposium 25, pp. 67–76.

3

Diversity and Origins of the Phanerogamic Flora of Mexico

JERZY RZEDOWSKI

Mexico's phanerogamic flora is estimated at roughly 220 families, 2,410 genera, and 22,000 species. The highest convergence of diversity is found along a belt that originates in Chiapas, traverses Oaxaca, and continues to central Veracruz on one side and to Sinaloa and Durango on the other. Cloud and tropical evergreen forests are the most diverse per unit area; however, in absolute numbers of species, other vegetation types surpass them.

Approximately 10% of the genera and 52% of the species are endemic to Mexico. These figures rise to 17% and 72%, respectively, if a phytogeographically more natural area is considered as the point of reference, although it would extend the area of Mexico by about one-third. Endemism is most pronounced in the xerophilous scrubs and the grasslands; and at the species level it is also rich in other types of vegetation, with the exception of the evergreen tropical forests.

The above figures indicate that the country has been the site of origin and evolution of a great number of plant lineages: (1) In the arid and semiarid zones of northern Mexico plants have experienced intense evolution, giving way to a moderately rich and distinctive flora with specialized growth forms that are often unique. (2) The flora in the semihumid regions developed largely from elements that exist in other parts of the world; a considerable number of these elements have led locally to extensive secondary radiation resulting in an abundant and diverse flora. (3) The flora of the humid areas, especially of the warm-humid areas in the east and southeast of the country, is also varied; yet to date there is little evidence that Mexico could have been an important center in its evolution.

An analysis of the geographical affinities of Mexico's phanerogamic flora indicates that its links with the south are about four times more important than those with the north. This fact should not, however, be interpreted as meaning that such a large number of Mexican plants are derived directly from the south, as a good number of the elements common to Central and South America must have originated in Mexico or in other parts of the world, such as the Antilles, Africa, Eurasia, or North America.

Available fossil records show that the basic characteristics of Mexico's present phanerogamic flora were already well established by the mid-Tertiary or earlier.

The plant wealth of Mexico is of exceptional diversity, variety, and significance. This fact is reflected in multiple forms and levels. Its most important aspects are the following.

1. *Plant communities.* Practically all important vegetation types known to man are found in Mexico. Apart from Mexico, only India and Peru have a similar diversity of plant cover.
2. *Life forms.* The great variety and remarkable splendor of growth or biological forms of Mexico's flora, particularly in its arid zones, can be paralleled only by South African flora.
3. *Plant species.* Although the floristic richness of many parts of the world is not known accurately, Mexico, with a probable total of 30,000 plant species, is considered among the most important countries in the world in terms of this richness.
4. *Combination of northern and southern elements.* Two of the most significant attributes of Mexico's flora are that it includes a great number of components from both the southern and northern hemispheres and that both play an important role in the vegetation.
5. *Endemism.* The continental flora of Mexico behaves like an island flora and has a high proportion of exclusive taxa; this ratio is considerably higher if a more natural phytogeographical area extending somewhat beyond the present political boundaries is considered.
6. *Cultivated and semi-cultivated plants and weeds.* Together with northern Central America, Mexico has been an important center for the domestication of crops and, even today, maintains a great variety of selected and improved germplasm. Along with the development of agriculture and civilization, a considerable number of native weeds have evolved.

As a corollary it must constantly be kept in mind that fortuitous circumstances have bestowed upon this part of the planet an unusual profusion of plant resources; the commitment of inhabitants of Mexico should be a profound and unquestioning guarantee of its preservation through an appropriate balance of exploitation and conservation.

This chapter, building on earlier reports (Rzedowski, 1965, 1978), presents an updated outline of Mexico's floristic richness and endemism. It provides ideas on the evolution of plant lineages in the country and fresh insights and details on geographical affinities of the flora and its origin.

FLORISTIC RICHNESS

It has long been known that Mexico, together with Central America, constitutes one of the regions of greatest plant diversity. Attempts to quantify this diversity, however, have been hindered not only by the lack

Table 3.1. Ratio of the number of species and the number of genera (s/g coefficient) calculated for the family Compositae and for the whole of some floras and floristic lists

Location	Total phanerogamic flora			Compositae		
	Genera	Species	s/g	Genera	Species	s/g
California[1]	1,067	5,590	5.1	141	696	4.9
Arizona[2]	907	3,370	3.6	151	543	3.6
New Mexico[3]	941	3,728	4.0	138	564	4.1
Texas[4]	1,216	4,839	3.9	158	578	3.7
Baja California[5]	862	2,640	3.1	130	405	3.1
Sonoran Desert[6]	794	2,634	3.3	119	397	3.4
Est. Biol Chamela[7]	434	754	1.7	21	27	1.3
Valley of Mexico[8]	672	2,065	3.1	107	388	3.6
Est. Biol. Los Tuxtlas[9]	504	818	1.6	40	59	1.5
Tabasco[10]	852	2,147	2.5	62	101	1.7
Yucatan Peninsula[11]	828	1,907	2.3	59	107	1.8
Chiapas[12]	1,701	7,018	4.1	134	561	4.2
Guatemala[13]	1,799	7,078	3.9	140	595	4.3

[1]Munz & Keck (1959); [2]Kearney & Peebles (1951); [3]Martin & Hutchins (1980); [4]Correll & Johnston (1970); [5]Wiggins (1980); [6]Wiggins (1964); [7]Lott (1985); [8]Rzedowski & Rzedowski (1979, 1985, 1990); [9]Ibarra & Sinaca (1987); [10]Cowan (1983); [11]Sosa et al. (1985); [12]Breedlove (1986); [13]Standley, Williams et al. (1946–76).

of a comprehensive inventory of all known species but also by the fact that a significant number have yet to be described or discovered.

Earlier estimates, based mainly on Standley's (1920-1926) work, placed the number of vascular plant species in Mexico at close to or somewhat over 20,000 (Rzedowski, 1978). Data now available allow more accurate estimates. Figures from direct counts and a conservative classification (that of Engler and Diels, with slight modifications) suggest that the number of families in the known phanerogamic flora of Mexico is about 220. Without straying too far from the traditionally accepted criteria, the number of genera can be placed at approximately 2,410.

It was discovered, while estimating the number of species, that in the latitudes close to Mexico where the Compositae are prominent, the number of species/number of genera (coefficient s/g) ratio in this family (using moderately conservative generic concepts) is similar to the s/g ratio of the whole phanerogamic flora.

This point is illustrated in Table 3.1, which shows that most floras or reasonably complete floristic lists of a territory originating in California and extending to Texas and south to Guatemala, the s/g coefficient of the Compositae is close to or the same as that for all the flowering plants. The three regions which do not show this coincidence in Mexico (Yucatan Peninsula, Tabasco and the Biological Field Station at Chamela) are areas in which the proportion of species of the Compositae in the flora is below 6%.

Table 3.2. Approximate numbers of genera and species in best represented families in Mexican phanerogamic flora

Family	No. of genera	No. of species
Compositae	314	2,400
Leguminosae	130	1,800
Gramineae	170	950
Orchidaceae	140	920
Cactaceae	70	900
Rubiaceae	80	510

From these data it is possible to infer that for all of Mexico the s/g coefficient of the phanerogamic flora is probably similar to that of the Compositae. The ratio for the latter is approximately 7.8 (about 2,400 presently known species ÷ about 310 presently known genera). Thus the number of presently known native species of Mexican flowering plants can be calculated at 18,800 (7.8 X 2,410 estimated genera in Mexico).

Deriving the real number of all existing native species in the country from this figure is a more difficult task, and the resulting number must remain tentative. With reference to introduced plants, their proportion in some regional floras (e.g., Valley of Mexico) amounts to or exceeds 10%, but at a national level the percentage is lower, probably half this figure or less. If the number of species in this category is estimated at 800, the number of native known elements would be about 18,000.

Experience suggests that the knowledge of Mexico's phanerogamic flora is somewhere between 75 and 90% of its real total. Thus a figure of 20% above the 18,000 is a reasonable estimate. The provisional total of the phanerogamic species would then reach 21,600 and, estimating the richness of pteridophytes in some 1,200, the vascular plant species 22,800. However, given the tentative nature of these numbers, the margin of error may be as high as 8%.

The taxonomic groups best represented in the known flora are listed in Table 3.2; the approximate data show that six families make up about 40% of the total genera and species. The relative importance of these six families varies from region to region. The Compositae, Gramineae, and Cactaceae are better represented in the northern and central parts of the country, whereas Orchidaceae and Rubiaceae are more diverse in the southern half. The Leguminosae become more abundant in warmer climates. It is possible that the Orchidaceae will prove to be richer than the Gramineae in number, as there are many orchid species yet to be discovered and described whereas the grasses are relatively well known.

The geographical and ecological distribution of Mexico's plant diversity is interesting. Figure 3.1 shows broadly what is known and estimated as

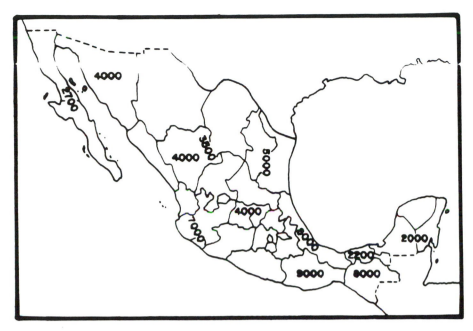

Figure 3.1. Known or estimated present floristic richness of some Mexican regions expressed in approximate numbers of phanerogamic species.

regards the floral richness of various regions in the country. The area of highest concentration of species extends from Chiapas to Oaxaca, from where it divides into two belts that are numerically somewhat less impressive, one extending toward central Veracruz and the other to Sinaloa and Durango. Significantly, the Yucatan Peninsula has a relatively poor flora.

Table 3.3 gives a general and rough idea of the contribution of different plant formations to Mexico's flora. The vegetation is classified into seven groups, each of which is assigned a percentage according to its estimated proportional, rather than its real, representation in the flora, because evidently many species thrive in more than one type of vegetation. For greater clarity, percentages are adjusted to a total of 100.

These estimates suggest that the coniferous and oak forests contribute to slightly less than one-fourth of the flora, whereas the xerophilous scrub and the grasslands contribute about 20%. The tropical forests comprise over one third of the whole flora, divided unequally into the humid forests on the one hand and the semihumid and dry forests on the other. Cloud forests make up about 10%. Ruderal and agrestal vegetation have lower percentages, as do the aquatic and semiaquatic communities.

These admittedly approximate values are especially significant when compared with the area each vegetation type covers in Mexico. By unit area, the cloud forest is by far the most diverse. It is followed in importance by the evergreen tropical forest, with xerophilous scrub and grasslands being last in the sequence.

Table 3.3. Estimated proportional participation of main vegetation types in Mexico's entire phanerogamic flora

Vegetation type	Approximate relative cover of each vegetation type in Mexico (%)	Estimated floristic richness in each vegetation type	
		No. of species	Percent of total flora
Xerophilous scrub and grassland	50	6,000	20
Conifer and oak forests	21	7,000	24
Cloud forests	1	3,000	10
Tropical evergreen forest	11	5,000	17
Tropical subdeciduous, deciduous, and thorn forests	17	6,000	20
Aquatic and subaquatic vegetation	—	1,000	3
Ruderal and agrestal vegetation	—	2,000	6

ENDEMISM

Taxa with restricted distributions hold special interest in studies of biogeography and organic evolution. Hemsley (1886–1888) was the first to study the endemism of Mexico's flora, outlining its main quantitative and qualitative characteristics. Today, advances in the understanding of this subject allow greater detail, but, as in the case of diversity, the need for a complete inventory and sufficient information on the complete geographical distribution of many taxa remains the cause of notable lack of accuracy.

The fact that the geographical distribution of organisms most frequently ignores political frontiers must not be forgotten; their areas are more often associated with the boundaries of natural regions, which in turn are determined by physiographic, climatic, edaphic, and other factors.

In order to be consistent with this fact and to report on true endemisms, it is necessary to extend the borders of the country (Fig. 3.2). Hence the term "Mega-Mexico 1" is used when parts of the Sonoran, Chihuahuan, and Tamaulipan arid zones that belong to the United States are included (Fig. 3.2B), "Mega-Mexico 2" when Central American territory as far as northern Nicaragua is included (Fig. 3.2C), and "Mega-Mexico 3" when both of these extensions are included (Fig. 3.2D).

Using these criteria, Table 3.4 offers quantitative information on the five endemic families, which are Canotiaceae, Fouquieriaceae, Plocospermataceae, Pterostemonaceae, and Simmondsiaceae. Table 3.5 provides the number of endemic genera.

At the species level, it is still not possible to make direct counts of similar values; the figures presented here are based primarily on counts made in the floras of areas bordering Mexico (California, Arizona, New Mexico, Texas and Guatemala), computing the species shared with Mexico. The

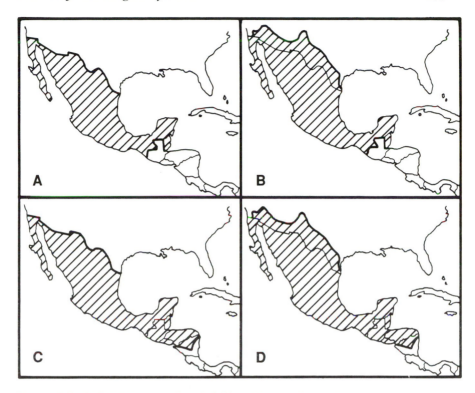

Figure 3.2. Reference areas for definition of endemism. A. Mexico. B. Mega-Mexico 1. C. Mega-Mexico 2. D. Mega-Mexico 3.

estimated 500 nonendemic species that do not occur in the mentioned neighboring areas or are not included in the corresponding floras, and the approximately 1,000 species that have been discovered in Mexico subsequent to the publication of the floras, were added to above total, which then was subtracted from 18,000. Table 3.6 presents the results.

This provisional estimate must be considered conservative because among the large number of species of Mexican flowering plants which have not yet been described (probably about 2,000), most evidently have a restricted distribution. Thus the absolute values of endemic plants will be higher; percentages should not vary markedly, as they will, no doubt, be compensated for by the numerous new records for Mexico of plants already known from other places.

The geographical distribution of endemisms does not follow the same patterns as those of diversity. Thus endemic taxa are concentrated, in the first instance, in regions with a dry climate. Other important areas include: (1) the Baja California Peninsula (20 genera and 25% of the species) as well as some of the distant islands, e.g., Isla Guadalupe (one genus and 21% of the species) and the Revillagigedo Archipelago (26% of the species); (2) some of the ecological islands and peninsulas such as the Sierra Madre chains, the Balsas river depression, high mountain peaks, gypsum soil

Table 3.4. Number of endemic families in Mexico's phanerogamic flora

Location	By Engler and Diels classification	By modern classifications
Mexico	0	1
Mega-Mexico 1	1	4
Mega-Mexico 2	0	2
Mega-Mexico 3	1	5

Table 3.5. Approximate number and percentage of endemic genera in Mexico's phanerogamic flora (2,410 genera)

Location	No.	%
Mexico	230	10
Mega-Mexico 1	310	13
Mega-Mexico 2	310	13
Mega-Mexico 3	400	17

Table 3.6. Approximate number and percentage of endemic species in Mexico's phanerogamic flora (18,000 species)

Location	No.	%
Mexico	9,300	52
Mega-Mexico 1	10,600	59
Mega-Mexico 2	11,500	64
Mega-Mexico 3	12,900	72

areas; and (3) areas that acted as refugia during the climatic changes of the Pleistocene.

In terms of general ecological links, the abundance of endemisms at family and generic ranks is positively and remarkably correlated with aridity and is concentrated in the xerophilous vegetation. At the species level (Table 3.7), however, coniferous and oak forests, which are essentially distributed along mountain chains with a cool, semihumid climate, rank almost as high as the xerophilous scrub and grassland. Tropical deciduous forests and cloud forests rank second, their relative wealth of endemisms becoming evident if Mega-Mexico 2 is taken as a unit of reference. The evergreen tropical forests, particularly those found west of the Isthmus of Tehuantepec, rank last, as even aquatic and semiaquatic vegetation and weeds have higher endemism.

Table 3.7. Estimated ratio of endemic species in the flora of vegetation types in Mexico, expressed as an approximate percentage of the total number of species in each unit of vegetation[a]

Vegetation type	Mexico	Mega-Mex. 1	Mega-Mex. 2	Mega-Mex. 3
Xerophilous scrub and grassland	60	90	60	90
Conifer and oak forests	70	75	80	85
Cloud forest	30	30	60	60
Tropical evergreen forest	5	5	20	20
Tropical deciduous, subdeciduous, and thorn forests	40	40	60	60
Aquatic and subaquatic vegetation	15	—	—	20
Ruderal and agrestal vegetation	20	—	—	30

[a]Values are based on few counts and, consequently, are broad estimates.

EVOLUTION OF PLANT LINEAGES IN MEXICO

The huge profusion and high degree of endemism of Mexico's flora, associated with its remarkable diversity, indicate that the country has been the place of origin and development of a great number of plant groups. This point is particularly spectacular in arid and semiarid areas where endemism not only affects high-ranking taxonomic groups but also life forms and is therefore responsible for the singularity of the flora. For example, the Cactaceae, albeit of South American origin, have reached their maximum diversity, abundance, and importance in Mexico with about 900 species, over 95% of which are restricted to Mega-Mexico 1.

The Fouquieriaceae, endemic to Mega-Mexico 1 and probably originated here, are notable for their growth forms, which are unique even among the xerophytes. The variants offered by species of *Agave*, a genus that at present is not restricted to Mexico, are no less remarkable. It diversified taxonomically and morphologically here and in all probability originated in Mexico as well. *Yucca, Dasylirion, Nolina, Krameria,* and several other genera present similar cases.

Other parts of the country have also been active centers of speciation: the mountainous cool, semihumid areas deserve special emphasis, as a surprisingly rich flora has evolved here, not only of herbaceous plants but also of shrubs and trees. Examples of this abundance are *Castilleja* (about 50 species), *Eryngium* (about 50 species), *Eupatorium* (sensu lato, about 220

Table 3.8. Approximate representation, at genus level, of main geographic elements in different vegetation types of Mexico

Vegetation type	Geographical elements		
	Meridional	Boreal	Endemic
Tropical evergreen forests	XXXXXXXX	.	.
Tropical subdeciduous forests	XXXXXXX	.	X
Tropical deciduous forests	XXXXXX	.	XX
Thorn forests	XXXXX	.	XXX
Xerophilous scrubs	XXX	X	XXXX
Grasslands	XXX	XX	XXX
Oak forests	XXX	XXX	XX
Conifer forests	XXX	XXX	XX
Cloud forests	XXXXX	XX	X

Source: Rzedowski (1978)
A point (.) indicates the presence of the element in question, but in insignificant proportions.

species), *Muhlenbergia* (about 100 species), *Pinus* (about 48 species), *Quercus* (about 165 species), *Salvia* (about 300 species, fide Ramamoorthy, 1984; Ramamoorthy & Elliott, this volume), *Sedum* (about 60 species), *Senecio* (*sensu lato*, about 180 species) and *Stevia* (about 70 species).

The cloud forests that cover less than 1% of national territory have been an important theater of speciation for the epiphytes; in this group the most diverse and prolific have been *Epidendrum, Peperomia,* and *Tillandsia*.

In the hot parts of the Pacific drainage area the outstanding feature is the variety and importance of the 60-odd species of *Bursera*, almost all endemic to Mega-Mexico 3, followed closely by *Acacia, Euphorbia, Ipomoea,* and many Malvaceae.

It has been known for a century or more that Mexico holds more genera, more species, and more individuals of the Compositae than any other country in the world, and it appears to be the largest of the families of flowering plants.

Within the Mexican Gramineae, apart from the already mentioned *Muhlenbergia, Bouteloua* is particularly outstanding. This taxon is now widely distributed on the American continent, but its diversity (about 40 species) is concentrated almost entirely in Mexico and its origin, like that of the eight derived satellite genera, must also be attributed to Mexico (Rzedowski, 1975; Stebbins, 1975). Similar cases are presented by *Achimenes* (Gesneriaceae), *Bouvardia* (Rubiaceae), *Cucurbita* (Cucurbitaceae), *Dalea* and *Marina* (Leguminosae), *Karwinskia* (Rhamnaceae), *Lamourouxia* (Scrophulariaceae), and *Lopezia* (Onagraceae), among others.

Mexico is also an important center of evolution for weeds. Unlike the situation in countries such as Canada, the United States, Argentina and Uruguay, in which almost all the weedy flora is composed of introduced species, it is remarkable that in most parts of Mexico native weeds strongly prevail. A great number of them have preserved their endemic character. Several representatives of genera such as *Argemone, Bidens, Euphorbia, Melampodium, Physalis, Sicyos,* and *Solanum* belong to this particular ecological group and are in an active state of evolution.

GEOGRAPHICAL AFFINITIES AND ORIGINS OF THE FLORA

Although Mexico has been an important center of intense evolution for certain plant lineages, it is evident that many components of its flora or their ancestors originated elsewhere in the world.

Research directed at establishing the places and dates of origin of the plant lineages and floras and at discovering how and when they migrated requires broad, solid support from the fossil records, but this source of information is only beginning to emerge for Mexico and the rest of Latin America. In some cases monographic and biosystematic studies of genera or other taxonomic groups may shed some light on the genesis of groups in question, but such studies are equally scarce for Mexico.

Consequently, ideas concerning the geographical origins of Mexico's flora and its various elements must still be largely based on what is known of the present distribution of species, genera, and other plant taxa. As is known and as will be seen further on, interpreting the distribution of organisms is not always easy and can sometimes be misleading. Nevertheless, it is a valid method, has survived the trial of time, and has contributed to the solution of many important problems.

In a classical and routine way, three main geographical elements have been recognized in the Mexican flora: southern, northern, and endemic or indigenous. Table 3.8 (Rzedowski, 1978) provides a preliminary evaluation of the proportional representation of these components at the generic level in Mexico's main vegetation types. From these estimates it is possible to deduce that the southern affinity of the flora overwhelmingly surpasses (by about four times) the northern one and is also far greater than the indigenous. A similar analysis done at the species level, however, would alter the latter ratio substantially.

Such an obvious affinity with the south consequently places most of Mexico within the Neotropical floristic realm and suggests that an important part of the Mexican flora must have originated in Central and South America, as has also been stressed by Raven and Axelrod (1974). This conclusion is supported by lineages of hot climates (*Byrsonima, Cecropia*), those of cool mountain climates (*Befaria, Brunellia*), and those of arid and semiarid regions (Cactaceae, *Nicotiana*).

On the other hand, many of the lineages at present distributed across the Neotropics might have originated in Mexico and migrated south or arrived in Mexico and the Neotropics from other parts of the world, with the group subsequently becoming extinct in important parts of its original area of

distribution. To illustrate the first possibility, there are numerous examples from the Compositae (e.g., *Flourensia, Galinsoga, Gutierrezia,
Montanoa, Tagetes*), the Gramineae (e.g., *Aegopogon, Bouteloua, Erioneuron,
Pappophorum*), and many other families. It could also be mentioned that the
spread northward of elements of Mexican origin, especially plants of arid
or semiarid climates, but of other environments as well, is equally important, as McVaugh (1952) has shown for the *Prunus serotina* and *Lobelia
cardinalis* complexes. Evidence for taxa arriving in Mexico and Neotropics
from other parts of the world is more difficult to come by, but the
importance of this source is no less significant. For example, the floristic
affinities between Mexico and Africa have not been adequately examined
but doubtless are more important than a superficial study would suggest.
Thus *Bursera*, which is so important in the flora of Mexico's Pacific drainage
area, is probably of African origin, as its closest relatives, *Aucoumea,
Boswellia*, and *Commiphora*, are concentrated there. The most interesting
fact is that the present geographical distribution of *Bursera* (southern
United States to the Guyanas and Peru) and the known fossil record
(western United States and England) suggest that this Gondwanan lineage
apparently did not reach Mexico via South America but probably migrated
via Laurasia, which for long periods of the Tertiary had a much warmer
climate than it has today. Likewise the familiar mesquite (*Prosopis*) is an
African descendant (Burkart, 1976; Sousa & Delgado, this volume); and
despite the great diversification of this group in South America, it may have
followed the same migration route in its displacement. This hypothesis is
in accordance with Thorne's (1973) thesis that the relatively scarce floristic
affinities between tropical America and Africa point to early (perhaps Pre-
Cretaceous) separation of western Gondwanaland. In such circumstances
South America becomes excluded from being an important bridge for
flowering plant exchanges between Africa and North America. It is also
plausible that an important contingent of genera of actual pantropical or
African-American distribution may have originated in that continent.
Probable genera include *Carpodiptera, Commiphora, Dichapetalum, Dioscorea,
Erblichia, Glinus, Hermannia, Oligomeris, Manilkara, Sesuvium*, and *Trianthema*.

The cloud forests of Mexico and Central America harbor an outstanding
group of genera in common with eastern Asia, many of which are represented in South America and others in the eastern United States. Some,
such as *Cleyera, Deutzia, Distylium, Engelhardtia, Microtropis*, and *Mitrastemon*,
have been collected only in Mexico, Central America, and Asia. Sharp
(1966) suggested an Asiatic origin for many of these elements and proposed
the Bering Strait and western North America as the migratory route,
although these plants are no longer found there except in fossil states.

As has been surmised by some (Berry, 1937; Raven & Axelrod, 1974), a
certain proportion of the plants that at present flourish in Mexico or in the
Neotropics in general have probable Laurasian ancestors of a subtropical
climate. The corresponding lineages have become extinct in many parts of
the Northern Hemisphere with the hardening of the climate in recent times.
It is possible that groups such as *Annona, Cedrela, Celastrus, Dendropanax*,

Meliosma, Persea, Saurauia, Stemmadenia, Symplocos, Ternstroemia, and many others, not easy to detect with existing knowledge, may have originated in Laurasia and later invaded South America (see also Wendt, this volume). Gentry (1982) has argued for a Gondwanan origin for many of these groups, but such an origin is hypothetical. In this connection it is important to recall that Miranda (1959) pointed out the phytogeographical significance of the relation of the flora of southeastern Mexico with those of Asia and Africa.

The insular condition and considerable geological age have made the Antilles an important nucleus for plant evolution, although, as in the case of Mexico, their flora has a definite neotropical affinity. In Mexico the direct influence of floristic elements characteristic of the Caribbean islands is manifest in the Yucatan Peninsula but less so in some other regions. However, it is probable that a significant contingent of lineages that are today widely distributed in Latin America originated in the Antilles. The following perhaps represent this source: *Bourreria, Calyptranthes, Exostema, Harpalyce, Hyperbaena, Jacquinia, Machaonia, Malpighia, Pisonia, Rhacoma, Rondeletia, Roystonea, Sabal,* and *Zamia.*

It may be concluded that the floristic contribution received by Mexico from the south, albeit significant in terms of numbers, was not as overwhelming as would seem to be indicated by the affinities with Central and South America presented in Table 3.8. It must be kept in mind that this floral continuity has profound ecological bases, mainly consisting in a great similarity of climatic conditions.

With reference to the chronology of past events, there is still not much to be said. Until relatively recently, it was believed that Mexico's present floristic complexity must have originated in large part during the Pleistocene or slightly earlier; there is no longer any doubt that it is much older. Paleopalynological findings (Graham, 1976; Palacios, 1985; Tomasini, 1980) have revealed that the northern element (e.g., *Abies, Acer, Carya, Cedrus, Cornus, Engelhardtia, Fagus, Fraxinus, Liquidambar, Liriodendron, Nyssa, Picea, Pinus, Platanus, Populus, Tilia, Ulmus,* etc.) was present in Veracruz and Chiapas during the Miocene and a smaller part even during the Oligocene. The southern element was already present there with representatives of hot climates (e.g., *Alibertia, Astrocaryum, Ayenia, Coccoloba, Enterolobium, Faramea, Matayba, Paullinia*) as well as from cool climates (e.g., *Brunellia, Calatola, Gunnera, Hoffmannia, Nothofagus, Phyllonoma, Podocarpus, Weinmannia*). In addition, elements (e.g., *Alfaroa, Chiranthodendron, Deppea*) existed that are presently restricted to Mexico and adjacent regions.

Information on the presence of fossil xerophytes is scarce, but it is well known that there is little likelihood for such fossils to appear other than occasionally. Hence it is difficult to estimate the age of the flora of the arid regions. Axelrod (1979) believed that the "Sonoran Desert" of northwestern Mexico and the southwestern United States has existed as such since the Pleistocene, indicating that its vegetation has evolved from enclaves of semiarid climate present since the early Miocene or perhaps late Eocene. However, there are no substantial reasons to support the idea that arid climates constitute such a recent phenomenon in Mexico; rather, its highly diversified xerophilous flora suggests a prolonged period of evolution that began perhaps during the Cretaceous itself. Eocene fossil remains of *Prosopis, Vauquelinia,* and *Agave* in the western United States and Miocene fossils of *Fouquieria, Pachycormus,* and *Condalia* in the same region (Axelrod, 1979) support this idea.

On the other hand, several authors (Axelrod, 1950; Raven, 1963; Wells & Hunziker, 1976) have insisted that *Larrea* arrived in Mexico from South America during the Quaternary or perhaps even later. This important element of North American xerophilous vegetation is probably southern in origin, but there is no concrete evidence as to when it arrived. There are clear indications that the exchange of plants between the arid zones of North and South America started in more ancient times. Species of *Condalia, Fagonia, Hoffmanseggia, Lycium, Menodora, Nicotiana, Prosopis, Selaginella,* and *Ziziphus* are examples of this phenomenon, as are several lineages of the Cactaceae family.

Consequently, from available paleobotanical information it can be concluded that, although the climatic fluctuations and physiographical changes that occurred during the Pleistocene may have contributed largely to the diversification of Mexico's flora, there is no doubt that its basic characteristics became well established by the Middle Tertiary and many of them possibly well before.

ACKNOWLEDGMENTS

I thank Helia Bravo, Eric Hágsater, Mario Sousa, Sergio Zamudio, and especially my wife, Graciela Calderón for their help in preparing this contribution. Financial support for this work was received from the Consejo Nacional de Ciencia y Tecnología (Mexican National Council of Science and Technology) and the Centro de Investigación y Desarrollo del Estado de Michoacán (Center of Research and Development of the State of Michoacan).

REFERENCES

Axelrod, D.I. 1950. Evolution of desert vegetation. Carn. Inst. Wash. Publ. 590:215–306.

———. 1979. Age and origin of Sonoran desert vegetation. Occas. Papers Calif. Acad. Sci. 132:1–74.

Berry, E.W. 1937. Tertiary floras of eastern North America. Bot. Rev. 3:31–46.

Breedlove, D.E. 1986. Listados Florísticos de México. IV. Flora de Chiapas. Mexico City: UNAM, Inst. Biol.

Burkart, A. 1976. A monograph of the genus *Prosopis* (Leguminosae subfam. Mimosoideae). J. Arnold Arb. 57:217–249; 450–485.

Correll, D.S. & M.C. Johnston. 1970. Manual of the Vascular Plants of Texas. Renner, TX: Texas Research Foundation.

Cowan, C.P. 1983. Listados Florísticos de México. 1. Flora de Tabasco. Mexico City: UNAM, Inst. Biol.

Gentry, A.H. 1982. Neotropical floristic diversity: phytogeographical connections between Central and South America, Plesitocene climatic fluctuations or an accident of Andean orogeny? Ann. Missouri Bot. Gard. 69:557–593.

Graham, A. 1976. Studies in neotropical paleobotany. II. The Miocene communities of Veracruz, Mexico. Ann. Missouri Bot. Gard. 63:787–842.

Hemsley, W.B. 1886–1888. Outlines of the geography and the prominent features of the flora of Mexico and Central America. *In* Biologia Centrali-Americana, Botany. Vol. IV. London: R.H. Porter. pp. 138–315.

Ibarra M., G. & S. Sinaca C. 1987. Listados Florísticos de México. VII. Estación de Biología Tropical Los Tuxtlas, Veracruz. Mexico City: UNAM, Inst. Biol.

Kearney, T.H. & R.H. Peebles. 1951. Arizona Flora. Berkeley, CA: Univ. of California Press.

Lott, E.J. 1985. Listados Florísticos de México. III. La Estación de Biología Chamela, Jalisco. Mexico City: UNAM, Inst. Biol.

Martin, W.C. & C.R. Hutchins. 1980. A Flora of New Mexico. Volumes 1 & 2. Vaduz, Germany: J. Cramer.

McVaugh, R. 1952. Suggested phylogeny of *Prunus serotina* and other wide ranging phylads in North America. Brittonia 7:317–346.

Miranda, F. 1959. Posible significación del porcentaje de géneros bicontinentales en América tropical. An. Inst. Biol. Univ. Nac. México 30:117–150.

Munz, P.A. & D.D. Keck. 1959. A California Flora. Berkeley, CA: Univ. of California Press.

Palacios Ch., R. 1985. Estudio palinológico y paleocológico de las floras fósiles del Mioceno Inferior y principios del Mioceno Medio de la región de Pichucalco, Chiapas. ENCB. Mexico City.

Ramamoorthy, T.P. 1984. Notes on the genus *Salvia* (Lamiaceae) in Mexico with three new species. J. Arnold. Arb. 65:135–143.

Raven, P.H. 1963. Amphitropical relations in the flora of North and South America. Quart. Rev. Biol. 29:151–177.

———— & D.I. Axelrod. 1974. Angiosperm biogeography and past continental movements. Ann. Missouri Bot. Gard. 61:539–673.

Rzedowski, J. 1965. Relaciones geográficas y posibles orígenes de la flora de México. Bol. Soc. Bot. Méx. 29:121–177.

————. 1975. An ecological and phytogeographical analysis of the grasslands of Mexico. Taxon 24:67–80.

————. 1978. Vegetación de México. Mexico City: Limusa.

———— & G.C. de Rzedowski (eds.), 1979. Flora Fanerogámica del Valle de México. Vol. 1. México City: Compañia Editorial Continental.

———— & ———— (eds.), 1985. Flora Fanerogámica del Valle de México. Vol. 2. Mexico City: ENCB, IPN/IE.

———— & ————. 1990. Flora Fanerogámica del Valle de México. Vol. 3. Pátzcaro, Michoacán: IE.

Sharp, A.J. 1966. Some aspects of Mexican phytogeography. Ciencia (Mexico) 24:229–232.

Sosa, V. & al. 1985. Etnoflora yucatanense: Lista florística y sinonimia maya. Xalapa, Veracruz: INIREB.

Standley, P.C. 1920–1926. Trees and shrubs of Mexico. Contr. U.S. Natl. Herb. 23:1–1721.

———, L.O. Williams et al. 1946–1976. Flora of Guatemala. Fieldiana Bot.Vols. 24 and 26.

Stebbins, G.L. 1975. The role of polyploid complexes in the evolution of North American grasslands. Taxon 24:91–106.

Thorne, R.F. 1973. Floristic relationships between tropical Africa and tropical America. *In* B.J. Meggers, E.S. Ayensu & D. Duckworth (eds.), Tropical Forest Ecosystems in Africa and South America. Washington, DC: Smithsonian Institution Press.

Tomasini O., A.C. 1980. Estudio palinológico del Oligoceno de Simojovel, Chiapas, México. Facul. de Ci., UNAM. Mexico City.

Wells, P.V. & J.H. Hunziker. 1976. Origin of the creosote bush (*Larrea*) desert of southwestern North America. Ann. Missouri Bot. Gard. 63:843–861.

Wiggins, I.L. 1964. Flora of the Sonoran Desert. *In* F. Shreve and I.L. Wiggins, Vegetation and Flora of the Sonoran Desert. Vols. 1 & 2. Stanford, CA: Stanford Univ. Press.

———. 1980. Flora of Baja California. Stanford, CA: Stanford Univ. Press.

II

SELECTED FAUNISTIC GROUPS OF MEXICO

4

Conservation-Oriented Analysis of Mexican Butterflies: Papilionidae (Lepidoptera, Papilionoidea)

JORGE LLORENTE-BOUSQUETS
AND ARMANDO LUIS-MARTÍNEZ

JORGE LLORENTE-BOUSQUETS
AND ARMANDO LUIS-MARTÍNEZ

Mexican Papilionoidea is composed of five families, over 20 subfamilies, about 50 tribes, 400 genera and just over 2,200 species. A conservation-oriented analysis of Mexican Papilionidae is presented here. Brief historic accounts of taxonomic and biogeographic aspects of the study of the family are given. Pertinent data from other families, mainly Pieridae and Nymphalidae, have been incorporated where appropriate to reinforce conclusions regarding areas of endemism and species richness. Several biogeographic patterns are identified. The southwestern part of the Nearctic region, mainly the United States and Mexico, are indicated as areas supporting relictual elements of the group. The various levels of diversity in the group (Lepidoptera) in the country, including many endemic taxa, suggest that Mexico has been an active center of speciation at both specific and transspecific levels. Fifty-seven species of Papilionidae are recognized in Mexico; species endemism is over 10%. Tropical deciduous and montane cloud forests are notably rich in Papilionidae. Among the areas richest in Papilionidae in Mexico are Los Tuxtlas (Veracruz) and Sierra de Juárez (Oaxaca), but areas richest in endemic taxa are border areas between the states of Morelos and Guerrero followed by Cañón del Novillo (Tamaulipas). Northern Mexico is generally poorer in butterflies than southern Mexico. The problems in the conservation biology of Mexican butterflies are discussed: habitat destruction resulting from man's activities and unrestricted commercial trade in Lepidopterans seem to be among the primary causes. The need for comprehensive studies is empahsized, and suggestions are offered for conservation and management of these diverse populations.

Mexico contains important biogeographic provinces that are species-rich, are high in endemics, contain primitive or plesiomorphic groups, and embrace zones of relictual biota. This chapter discusses these aspects in relation to butterflies, with the aim of evaluating the diversity of the Papilionidae of Mexico. After the Coleoptera, the Lepidoptera, of which

Table 4.1. Transspecific taxa in Mexican Papilionidae through time

Year	Tribe	Subtribe	Genus	Subgenus	Species Group
1758[1]	1	1	1	None	1
1836[2]	1	1	1	None	7
1879–1901[3]	2	2	2	None	12
1906[4]	2	2	2	None	16
1940[5]	2	2	2	None	—
1944[6]	4	5	4	None	—
1961[7]	5	5	6	Yes	17
1983[8]	5	6	9	Yes	17
1987[9]	5	6	6	Yes	—
This report	5	6	12	None	17

See text for details.
[1]Linnaeus; [2]Boisduval; [3]Godman & Salvin; [4]Rothschild & Jordan; [5]Hoffmann; [6]Ford; [7]Munroe; [8]Hancock; [9]Miller.

the Papilionidae is part, is the richest order, comprising roughly 200,000 species worldwide. Approximately 25,000 of them, including many paleoendemics and neoendemics, are found in Mexico, thus making it one of the countries with the highest diversity, along with Brazil and Indonesia. The Mexican Papilionidae family is composed of about 57 taxa. Its small size lends itself to an analysis of diversity. For comparative purposes, data from other families, particularly Pieridae and Nymphalidae, have been included in this study. Collins and Morris (1985) provided an excellent introduction to the biology and conservation of Papilionidae that has served as an invaluable background for the present survey. A brief historical review of the study of the family precedes discussion on the diversity of the family in Mexico.

Knowledge of diversity and of the nature of endemism are critical in conservation-oriented studies. Diversity, which has eluded definition, is conceptualized differently by authors with varying backgrounds. It should have greater biological significance, reflecting the evolutionary histories of organisms of an area; but it is usually equated with species richness (Rosenzweig, 1975; Wilson, 1988; other contributions in this volume). Discussions of diversity lay much emphasis on the term endemism, but there is some disagreement on its use. It is used here to suggest a restricted distribution. Taxa may be paleoendemic or neoendemic. In this chapter, those that are centered in Mexico or have their major distribution in the country are considered quasiendemic. Conservation biology generally stands to benefit from discussions of the interrelation of centers of endemism (Nelson, 1983; Patterson, 1983), as they may be useful when assigning priority to areas in need of conservation. In Mexico, conservation biology is intimately associated with preservation of the habitats of the

Table 4.2. Species and subspecies recognized in Mexican Papilionidae through time

Year	Species	Species indicated	Subspecies indicated
1758[1]	4	0	0
1836[2]	29	11	11
1879–1901[3]	49	41	41
1906[4]	53	43	55
1940[5]	52	52	65
1966[6]	57	51	62
1975[7]	57	47	57
1978[8]	59	58	71
1981[9]	44	41	52
1984[10]	62	62	70
1988[11]	57	57	82

See text for details
[1]Linnaeus; [2]Boisduval; [3]Godman & Salvin; [4]Rothchild & Jordan; [5]Hoffmann; [6]D'Almeida; [7]Tyler; [8]Díaz & De la Maza; [9]D'Abrera; [10]Beutelspacher; [11]LLorente & Luis.

monarch butterflies, *Danaus plexippus*, which migrates southward by the millions to overwinter in the central Mexican state of Michoacán, draping the firs and pines near Angangueo, among other places.

HISTORY

The history of taxonomic studies of this family in Mexico is summarized in Tables 4.1 and 4.2. Table 4.1 presents the gross taxonomy of Papilionidae accepted in ten relevant studies listed chronologically. The rank of subfamily has not been included as there has been no significant change at this level since the work of Rothschild and Jordan (1906). Recognition of transspecific taxa has increased over time except at the level of subgenera. In the case of monotypic groups (e.g., *Baroniini*), the intermediate subgroups (*Baroniini* and *Baroniina*) have been included. Table 4.2 lists, sequentially, species and subspecies recognized in 11 studies. The "recognized species" here include those whose original descriptions did not provide localities or areas of distribution in Mexico. "Species indicated" and "subspecies indicated" include those whose distribution in Mexico has been provided in the cited works. In cases where subspecies have not been designated by an author, one has been included for each indicated species. Figure 4.1 depicts the early surge in recognized and indicated species which, after a steady climb during the nineteenth century, stabilized at 58 ± 3 around the beginning of the 1960s. Although species concepts have varied among

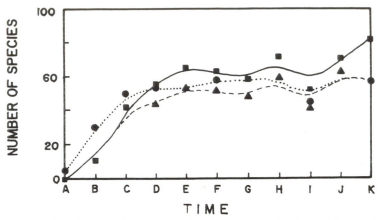

Figure 4.1. Species numbers of Papilionidae recognized over time. Circles, recognized species; Squares, indicated species; Triangles, indicated subspecies.

authors, recognition of specific groups (species or subspecies) has remained approximate (Fig. 4.1). The number of infraspecific taxa accepted by authors during the period 1975–1990 has stayed at 72 ± 8.

The difference in the number of species in the recognized or indicated categories in Table 4.2 reflects the changing perceptions of distribution of taxa across time (Fig. 4.1). Linnaeus (1758) and Boisduval (1836) provided general and vague distribution data. Godman and Salvin (1879–1901) assigned the collection localities to countries. The growth of knowledge of geography is clearly reflected in the contribution of Rothschild and Jordan (1906), who give altitudinal data as well as more precise localities. Two contributions that have added to our knowledge of Mexican Papilionidae are those of Hoffmann (1940) and Beutelspacher (1984). The former used administrative divisions and altitudinal zonation for providing distributional data, and the latter's data were accompanied by reliable distributional maps. This chapter, which builds on these earlier studies, incorporates the use of modern electronic techniques available today to analyze data from various sources (see below).

It is possible that 90–95% of species and subspecies of Mexico's Papilionidae have already been documented. Further exploration of little-studied areas, detailed analysis of geographic variation of disjunct populations of some species, and the study of the reproductive biology of others may reveal the remainder. There are few descriptive ecological studies of the Papilionidae in Mexico. Whereas knowledge of some species, e.g., *Baronia brevicornis*, may be deemed adequate, that of others (various endemic or quasiendemic species whose principal distribution is Mexican and those that have differentiated at subspecific levels mainly in Mexico) is meager and known only through museum samples or collections. Examples include *Parides alopius, Pyrrhosticta diazi, P. abderus, Priamides rogeri, P. erostratinus*, and *Pterourus alexiares*. In some cases information on their likely host plants is lacking. The above observations generally stress the need for field and laboratory studies in Mexican groups.

MATERIALS AND METHODS

Data were compiled from museum collections, catalogs, and literature. Collections in the Allyn Museum (Florida) and the Natural History Museum of New York, which house two of the most important collections from Mexico (Hoffmann and Escalante) provided valuable information. Their holdings cover over 50 years of sustained collections from many parts of Mexico. Historic collections of Welling, Hubbell, Wind, Holland, and Miller are housed in these museums as well. In Mexico, the collections of L. González and that of the Zoology Museum at UNAM were consulted. The data from the Holland Catalogue (MS) with information on northern Mexico, otherwise scarce, were included. The samples were all checked for taxonomic accuracy. Literature related to distribution, host plants, and life cycles of the Papilionidae of Mexico exceeding 100 entries was compiled.

A capture screen was designed for the dBASE III plus software program for analysis of data. Data relating to 15 areas (genera, species, subspecies, sex, day, month, year, state, municipality, locality, vegetation type, collector, altitude, number of individuals, and collection code or number of bibliographical reference) were compiled from the samples and bibliographical references. The ecological and geographical data were thus integrated. Lists were made for areas, states, and other parameters of distribution (vegetation types, altitude); and maps were generated from the data base. Areas of richness were determined from the above analysis. Phylogenetic factors (plesiomorphic and apomorphic groups and their interrelations) were taken into consideration in this exercise. The genealogical relations put forward by Durden and Rose (1978), Hancock (1983), Igarashi (1984), and Miller (1987) were also used.

RESULTS

Table 4.3 presents the classification of the Papilionidae of Mexico accepted here. It recognizes three subfamilies, five tribes, six subtribes, 12 genera, 57 species, and 82 subspecies. Of these groups, one monotypic subfamily, six species, and 28 subspecies are endemic; another 28 subspecies, which are restricted to the Mesoamerican region, are considered quasiendemics. The species and subspecies endemisms are 10.5% and 34.1% respectively. This fact, along with the high number of quasiendemics and the presence of important plesiomorphic elements and paleoendemics, suggests that the diversity of Papilionidae of Mexico is unique.

A preliminary analysis of the endemism of the Mexican Pieridae reveals similar percentages and biogeographical characteristics. The same can be seen for Hesperiidae, Nymphalidae, and Lycaenidae. The species ende-

Table 4.3. Classification of Mexican Papilionidae

Baronia brevicornis brevicornis Salvin	E
B. brevicornis rufodiscalis W. & M.	E
Parnasius phoebus ssp?	
Protesilaus marcellus Cra.	
P. philolaus Boi.	Q*
P. oberthueri R. & J.	Q*
P. epidaus epidaus Do.	Q*
P. epidaus fenochionis G. & S.	E
P. epidaus tepicus R. & J.	E
P. phaon Boi.	
P. branchus Do.	Q*
P. belesis belesis Bat.	Q*
P. belesis occidus Vaz.	E
P. thymbraeus thymbraeus Boi.	Q*
P. thymbraeus aconophos Gray	E
P. agesilaus neosilaus Hopffer	Q*
P. agesilaus fortis R. & J.	E
P. macrosilaus macrosilaus Gray	Q*
P. macrosilaus penthesilaus Felder	E
Eurytides marchandi marchandi Boi.	
E. marchandi occidentalis Maza et al.	E
E. lacandones lacandones Bat.	
E. calliste calliste Bat.	Q*
E. salvini Bat.	Q*
Battus philenor philenor Linnaeus	
B. philenor orsua G. & S.	E
B. philenor ssp?	
B. philenor acauda Oberthr	E
B. polydamas polydamas Linnaeus	
B. laodamas copanae Reakirt	Q
B. laodamas procas G. & S.	E
B. eracon G. & S.	E
B. belus varus Kollar	Q
B. belus chalceus R. & J.	E
B. lycidas Cramer	
Parides alopius G. & S.	Q*
P. montezuma Westwood	Q*
P. photinus Do.	Q*
P. photinus ssp.?	
P. erithalion sadyattes Druce	Q
P. erithalion polyzelus Felder	Q
P. erithalion trichopus R. & J.	E
P. lycimenes lycimenes Boi.	
P. lycimenes septentrionalis M. & D.	E
P. iphidamas iphidamas Fabricius	
P. sesostris zestos Gray	Q
P. childrenae Gray	
P. eurimedes mylotes Bat.	Q
Pterourus esperanza Beu.	E
P. palamedes leontis R. & J.	E

Table 4.3. (cont.)

P. glaucus glaucus Linnaeus	
P. alexiares alexiares Hopffer	E
P. alexiares garcia R. & J.	E
P. rutulus rutulus Lucas	
P. multicaudatus Kirby	
P. pilumnus Boi.	Q*
P. eurymedon Lucas	
Pyrrhosticta victorinus victorinus Do.	Q*
P. victorinus morelius R. & J.	E
P. victorinus ssp.	
P. diazi Rac. & Sbo.	E
P. garamas garamas Geyer	E
P. garamas ssp?	
P. abderus abderus Hopffer	E
P. abderus electryon Bat.	Q*
P. abderus baroni R. & J.	E
Heraclides thoas autocles R. & J.	
H. cresphontes Cra.	
H. ornythion Boi.	Q*
H. ornythion ssp.	
H. astyalus pallas Gray	Q
H. astyalus occidentalis Br. & Fau.	E
H. androgeus epidaurus G. & S.	Q
H. androgeus ssp?	
Troilides tolus tolus G. & S.	Q*
T. tolus mazai Beu.	E
Priamides pharnaces Do.	Q*
P. anchisiades idaeus Fabricius	
P. rogeri Boi.	Q*
P. erostratus erostratus Westwood	Q*
P. erostratus vazquezae Beu.	E
P. erostratinus Vas.	E
Papilio polyxenes asterius Stoll	
P. polyxenes coloro Wright	
P. bairdii Edwards	
P. zelicaon zelicaon Lucas	
P. zelicaon nitra.	
P. indra pergamus Edwards	
P. indra ssp?	

This classification is based on the contributions of Hancock (1983) and Miller (1987), and the taxa are phylogenetically arranged from plesiomorphic to apomorphic. Nevertheless, for the arrangement of genera in Papilionini a greater number of genera representing distinct groups of species are recognized, so the distinct wing patterns, the host plant, and the patterns of distribution stand out, reflecting the geneological relation in this tribe. Subgenera are not recognized. The same criteria have not been applied in Leptocircini, where the genus *Protesilaus* contains distinct groups of species that are relatively homogeneous among themselves. The availability of names is being examined by K. Brown (personal communication). The question mark (?) in front of names suggests the possibility of an unnamed subspecies as well as suggesting the need for additional studies to confirm its occurrence in

Figure 4.2. World and Mexican totals of Papilionoidea (left) and Papilionidae (right).

mism in Pieridae is over 10% and in the other families somewhat below this figure (see Fig. 4.8, below). Among the endemic species of the Pieridae are *Lieinix neblina*, *Euchloe guaymasensis*, *Eucheira socialis*, *Falcapica limonea*, *Heliochroma crocea*, *Prestonia clarki*, and *Neophasia terlooti*.

Figure 4.2 illustrates the representation of Papilionoidea in Mexico in relation to the rest of the world. Papilionidae consists of four subfamilies. Of them, Baroniinae and Papilioninae occur in Mexico. The presence of Parnasiinae is doubtful. Praepapilioninae, which is extinct, is known from the Colorado fossils of the Middle Eocene. Baroniinae, which is endemic to southern and southeastern Mexico, is monotypic and has more plesiomorphic characters than *Praepapilio*. Papilioninae is represented by three of the four subtribes (Miller, 1987): Leptocircini, Troidini, and Papilionini.

The generic and subgeneric concepts have differed among authors (Hancock, 1983; Miller, 1987; Miller & Brown, 1981). A less conservative taxonomy adopted here recognizes 10 genera in Papilioninae: *Protesilaus, Eurytides, Battus, Parides, Pterourus, Pyrrhosticta, Heraclides, Troilides, Priamides*, and *Papilio*.

Papilionidae consists of 580 species worldwide of which about 10% are found in Mexico (Fig. 4.2). Figure 4.3, which provides a comparison of species richness in various countries in North America (Nearctic region, Mesoamerica, and the Antilles), clearly suggests that Mexico is the richest. Collins and Morris (1985) have pointed out that Mexico holds tenth place in the world in terms of numbers of species and seventh in endemics (Figs. 4.4 and 4.5). However, the many quasiendemics recognized among Mexi-

Table 4.3. (cont.)

Mexico. The supraspecific classification precedes the species enumeration. There are three subfamilies, five tribes, six subtribes, 12 genera, 17 species groups in Papilioninae, 57 (+1?) species, 82 (+7?) subspecies.
PAPILIONIDAE. Baroniinae: Baroniini; *Baronia*. Parnasiinae: Parnasiini; *Parnassius*. Papilioninae: Leptocircini; *Leptocircina, Protesilaus, Eurytides*. Troidini; *Battina, Battus, Troidina, Parides*. Papilionini; *Pterourus, Pyrrhosticta, Heraclides, Troilides, Priamides, Papilio*.
E, endemic; Q, quasiendemic or mesoamerican (with an additional area in Central America). The asterisk indicates that species is considered in this category.

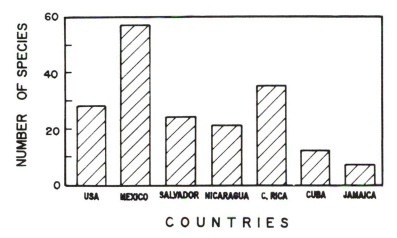

Figure 4.3. Numbers of species of Papilionidae in Mexico and adjacent countries. After Collins & Morris, 1985.

can taxa here would rank the country even higher with respect to the exclusivity of its lepidopteran fauna (Table 4.4). The presence of habitats of relictual biota of various plesiomorphic species and paleoendemics in Mexico (e.g., *Baronia brevicornis*, *Parides alopius*, and *Pterourus esperanza*) further supports this claim. Thus Mexico is among the richest and most diverse countries for butterflies, together with Indonesia, the Philippines, China, Brazil, Madagascar, and India. Table 4.4 and Figures 4.6 and 4.7 give the number of species, endemics, and quasiendemics and their respective percentages. In the Antilles the number of species is low, but endemism is relatively high. Forty-seven percent of the Papilionidae of Mexico are

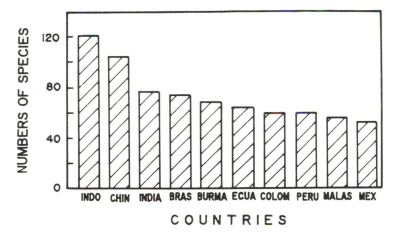

Figure 4.4. Numbers of species of Papilionidae for countries with the greatest diversity in the family. After Collins & Morris, 1985.

Table 4.4. Endemics and quasiendemics in Mexico and adjacent countries

Country	Total species	Endemics		Quasiendemics	
		No.	%	No.	%
United States	28	1	3.6	8	28.5
Mexico	57	6	10.5	21	36.8
Salvador	24	0	0	0	0
Nicaragua	21	0	0	0	0
Costa Rica	35	0	0	0	0
Cuba	12	4	33.3	2	16.7
Jamaica	7	3	42.8	0	0

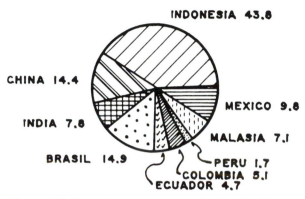

Figure 4.5. Endemism (percentages) of Papilionidae for countries with the greatest diversity in the family. After Collins & Morris, 1985.

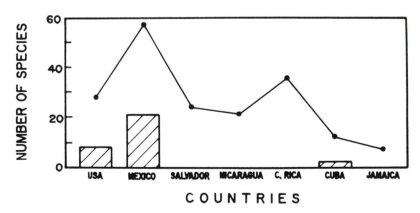

Figure 4.6. Numbers of species and endemics in Mexico and adjacent countries. Bars indicate endemics.

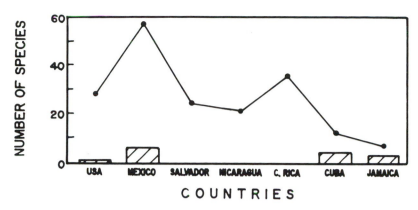

Figure 4.7. Numbers of species and quasiendemics in Mexico and adjacent countries. Bars indicate quasiendemics.

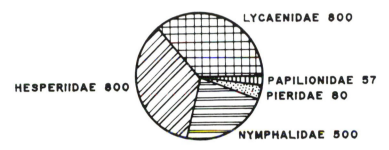

Figure 4.8. Approximate numbers of species in various families of Papilionoidea.

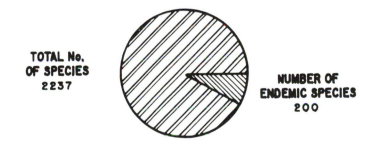

Figure 4.9. Approximate number of endemics in Papilionoidea in Mexico.

endemic and quasiendemic, making Mexico one of the most important countries in the region.

Figure 4.8 provides estimates of species numbers in families of Papilionoidea for Mexico from which a preliminary evaluation of endemism by family was obtained. The two estimates, integrated in Figure 4.9, give a total of 9.4% endemism for the Papilionoidea of Mexico. Like Papilionidae, other Lepidopteran families in Mexico have sets of genera or groups of plesiomorphic species that inhabit (like the paleoendemics) areas of relictual biota, principally xeric and mesic areas of medium and high mountains (500–1,500 m and 1,800–3,000 m above sea level, respectively). Examples include *Aegiale, Heliocroma, Prestonia, Eucheira, Lieinix, Anetia, Bolboneura, Cyclogramma, Manataria, Paramacera, Cyllopsis, Megisto, Pindis, Anemeca, Microtia, Caria, Apodemia, Emesis, Calephelis,* and *Eumaeus.* These supraspecific taxa have evolved mainly in the Mexican Transition Zone as defined by Halffter (1976, 1987).

PHYLOGENETIC CONSIDERATIONS

Durden and Rose (1978) were the first to report two fossil species of Praepapilioninae (*Praepapilio colorado* and *P. gracilis*). These fossils from the Middle Eocene (48 Ma) of Colorado present more apomorphic characters than the endemic Baroniinae of Mexico which indicates that generic and suprageneric differentiation in Papilionidae may date back to the Eocene-Paleocene (Miller, 1987). Differentiation of the families of Papilionoidea, then, may be placed at least in the Upper Paleocene.

The contributions of Ford (1944), Ehrlich (1958), and Munroe (1961) followed by the phylogenetic analysis and classifications of Hancock (1983), Igarashi (1984), and Miller (1987) form the basis for discussion of the evolutionary history of the Papilionidae. These studies suggest that the ancestral morphology of the group had its origins in *Baronia brevicornis.* The polarization of character states in the cladistic analysis of transspecific taxa in the Papilionidae has its plesiomorphic origins in *Baronia.* This genus, however, exhibits specializations.

Munroe (1961) considered the Parnasiinae and the Papilioninae as sister subfamilies. Only Papilioninae is found in Mexico, as the report of *Parnasius phoebus* ssp. in Tamaulipas is suspect. Its presence (if found) would be a marginal distribution of an apomorphic taxon of Parnasiinae. The genealogical relations of the Papilioninae tribes were little understood until the work of Hancock (1983) and Miller (1987). Hancock recognized three tribes and Miller four; all of Hancock's tribes and three of Miller's are found in Mexico (Tables 4.1, 4.3).

Of the ten generic groups recognized in Leptocircini by Miller (1987), only the most plesiomorphic occur in Mexico. In the case of *Protesilaus,* groups of plesiomorphic species of Mesoamerican evolution and derived species with a southern affinity are found. The *Eurytides* are apomorphic groups with a southern affinity except for, perhaps, *E. salvini* and *E. calliste,* which could be considered Mesoamerican.

The two subgroups of Troidini are the monotypic *Battina* and *Troidina.* The latter subtribe is composed of ten generic groups, of which only one— Parides, apomorphic within the subtribe—is present in Mexico. However, there are sets of plesiomorphic species of Mesoamerican evolution and derived groups with a southern affinity. *Battina* presents a similar case with two groups of species. In both subtribes taxa of Mesoamerican evolution are less numerous than those of a southern evolution.

The only phylogenetic analysis of Papilionini taxonomically less conservative in approach (Hancock, 1983) recognizes 11 generic groups. It is based on criteria that better express phylogenetic and biogeographic aspects in the classification of the tribe. These criteria, applied at world level, recognize over 20 generic groups in the Papilionini. Six generic groups of this tribe are found in Mexico (Table 4.3). In the cladogram of Hancock (1983), most plesiomorphic groups (*Pterourus*, *Pyrrhosticta*, *Heraclides*, *Troilides*, and *Priamides*) and one of the most apomorphic (*Papilio*) may be seen. These species groups are of Mesoamerican evolution (*Pterourus* and *Pyrrhosticta*), southern evolution (*Heraclides*, *Troilides*, and *Priamides*), and a section of *Papilio*, which is Nearctic.

BIOGEOGRAPHIC PATTERNS AND ORIGINS

Various lineages in Mexican Papilionidae display biogeographic patterns described by Halffter (1976, 1987); they include the following groups.

1. Paleoamerican (*Baronia*)
2. Mesoamerican (the several groups of species of Papilionini genera)
3. Nearctic (*Papilio*)
4. Neotropical (several Troidini and Leptocircini)

The lack of fossil evidence and genealogical studies impedes the interpretation of chronological relations of different clades. Nevertheless, paleo- and neoendemics in several communities are recognized, which is significant. *Baronia* is a good example. Its two subspecies occur in the lower deciduous tropical forests, one in southern Mexico (Balsas Basin) and the other in western Mexico (inland Chiapas), and suggest an old vicariant process for the communities. The levels of differentiation of these populations, however, do not warrant their recognition as species. *Parides alopius*, with a similar distribution, occurs in colder climates of higher altitudes and latitudes. The age of these elements in both communities may date back to the Paleocene. *Pterourus alexiares* sspp. and *P. esperanza* may represent elements of two groups of species that have converged into a community composed of groups of northern, southern, and local origins. These communities, which have remained as relict populations in montane cloud forests at least since the Oligocene, may have experienced vicariant processes comparable to *Pyrrhosticta abderus* and *P. victorinus*, during this time. Various groups of subspecies of neotropical affiliation may have differentiated during the Pleistocene, possibly in the tropical wet and humid forest refugia in southern and southeastern Mexico (e.g., *Eurytides marchandi occidentalis*, *Battus laodamas procas*, *Parides erithalion trichopus*, *Battus eracon*, and *Parides lycimenes septentrionalis*). The first three have their vicariants in the coastal areas of the Gulf of Mexico. Halffter (1976) has provided further examples from Papilionoidea and has also described other patterns. Additional examples may be seen in Dismorphini (LLorente, 1983; LLorente

& Luis, 1988). Some of the oldest paleoendemics found in montane cloud forests and tropical deciduous forests (sensu Rzedowski, 1978) may possibly have closer genealogical relations with relicts of the Greater Antilles (Cuba and Hispaniola). Examples include *Pterourus, Heraclides*, and *Parides* in Papilioninae, *Anetia* in Danainae, and *Prestonia, Heliocroma*, and *Apodemia* in other Papilionoidea groups.

It is possible that the time and degrees of differentiation of several Papilionidae of the region are related and may coincide with the geological ages of the areas. *Baronia* in the Cretaceous areas of southern and western Mexico and *Pterourus esperanza* in mountains may date back to the Eocene-Oligocene. The various biogeographical provinces of Mexico that link Nearctic and Neotropical regions, thus acting as a corridor or barrier for their biotic elements, and the presence of disjunct areas of extreme climates (xerics and mesics) seem to have generally provided for the high degree of speciation in these areas (Hancock, 1983; LLorente, 1983).

It is difficult to suggest a "center of origin" for any group with certainty, and the exercise is shrouded in controversy (Croizat et al., 1974; Nelson, 1978, 1983; Patterson, 1983). The problem is compounded if the group in question is of great age, as is the case with the families and subfamilies of Papilionoidea. The unpredictability of the differential extinction of species and clades often invalidates some criteria for locating centers of origin, e.g., more plesiomorphic groups, more diversified groups (Collins & Morris, 1985). Several areas of origin proposed for the Papilionidae by Shields and Dvorak (1979), Hancock (1983), and Collins and Morris (1985) are debatable. It is noteworthy that Miller (1987) did not find biogeographical patterns in the cladograms of areas as a function of a vicariant model based on tectonic plates, which may suggest that unrecognized extinction for some areas has made it difficult to interrelate centers of paleoendemics. For the present, Hancock's (1983) suggestion that the family and its more plesiomorphic groups may have originated in Laurasia seems reasonable (Collins & Morris, 1985). The present distribution of the Praepapilioninae, Baroniinae, and the primitive groups of Papilioninae (*Pterourus* and *Battus*) and Parnassinae (*Archon* and *Sericinus*) points to the southwest of the Nearctic region (United States and Mexico) as the most probable place of origin for the family.

The latitudinal gradients of species richness in the butterflies pointed out by Slansky (1972) and Scriber (1973a) have been observed in other continental areas (Collins & Morris, 1985). The higher diversity encountered toward the tropics has led to the characterization of the Papilionidae as "preeminently tropical" (Collins & Morris, 1985). The diversity that prompts such generalizations may be due not only to greater diversification there but also to a probable lower rate of extinction in these latitudes.

ENDEMISM AND RICHNESS

Of the 57 species, 25 have ample regional or continental distributions. Six species and 28 subspecies are endemic to the country (Table 4.3). The

Table 4.5. Distribution of endemics of Mexican Papilionidae in physiographic provinces

Taxa	1	2	3	4	5	6	7	8	9
Baronia brevicornis brevicornis		X							
B. b. rufodiscalis			X						
Protesilaus epidaus fenochionis	X	X				X			
P. e. tepicus					X				
P. belesis occidus	X				X				
P. thymbraeus aconophos	X	X					X		
P. agesilaus fortis	X	X					X		
P. macrosilaus penthesilaus	X			X	X	X	X		X
Eurytides marchandi occidentalis	X					X			
Battus philenor orsua								X	
B. p. acauda						X			
B. laodamas procas	X	X				X			
B. eracon		X				X			
B. belus chalceus						X			
B. e. trichopus	X	X				X			
Parides lycemenes septentrionalis				X			X		X
P. esperanza				X					
P. palamedes leontis					X				
P. alexiares alexiares					X				
P. a. garcia					X				
Pyrrhosticta victorinus morelius	X	X				X			
P. diazi		X							
P. abderus abderus				X	X				
P. a. baroni	X					X			
Heraclides astyalus occidentalis						X			
Troilides tolus mazai		X				X			
Priamides erostratus vazquezae		X				X			
P. erostratinus					X				
Total	10	11	1	4	8	13	4	1	2

1) Sierra Madre del Sur; 2) Balsas Basin; 3) Chiapan Interior; 4) Sierra de Juárez; 5) Sierra Madre Oriental; 6) Pacific Coastal Plain; 7) Yucatan Peninsula; 8) Islas Marias; 9) Gulf Coastal Province.

geographic areas with the highest number of endemic taxa are tropical deciduous forests of southern and western Mexico, the mesic areas of the Sierra Madre Oriental, the Sierra de Juárez, the Sierra Madre del Sur, and the Pacific Coastal Plains, particularly the last, which has 53% of the endemic subspecies.

Table 4.5 gives the distribution of endemic taxa in their recognized physiographic areas: Sierra Madre de Sur, Balsas Basin, Chiapan Interior, Sierra de Juárez, Sierra Madre Oriental, Pacific Coastal Plains, Yucatán Peninsula, Islas Marias, and Gulf coast.

The endemic elements of Papilionoidea of northern Mexico, particularly those associated with desert areas, are also found in southern areas of the

Figure 4.10 Areas of high endemism of Papilionidae in Mexico. 1. Durango-Sinaloa border. 2. Cañón del Novillo, Tamaulipas. 3. Sierra de San Juan, Nayarit. 4. Southern parts of Sierra Madre de Oriental. 5. Morelos. 6. Cañón de Zopilote, Guerrero. 7. Sierra de Atoyac, Guerrero. 8. Sierra de Juárez, Oaxaca. 9. Inland Chiapas.

United States. Some endemic taxa of Hesperiidae, Pieridae, Nymphalidae, and Lycaenidae can frequently be found associated with the Valley of Tehuacán-Cuicatlán, semiarid areas of Querétaro and Hidalgo, or in thorn scrub areas of Sonora-Sinaloa and in the Peninsula of Baja California. However, data pertaining to these taxa are too incomplete to provide a synthetic picture of their endemism.

Generally, areas of endemism of other Papilionoidea families coincide with those of Papilionidae; there are, as mentioned above, other zones of endemism in xeric areas in these families. These coincidences in various groups of Papilionoidea suggest biogeographic patterns that may have been reached through shared historical processes. Figure 4.10 shows areas of high endemism for the Papilionidae of Mexico that include representa-tive populations of quasiendemic taxa. Three notable biogeographic patterns of endemic and quasiendemic Papilionidae of Mexico, associated with a pattern of vicariant diversification, are discernible. These three areas can be divided into subpatterns.

1. Modern pattern along the coastal areas. This pattern is characterized by specifically or subspecifically differentiated populations along the coastal strips, e.g., *Protesilaus epidaus* (subspecific northern and south-

Table 4.6. Number of species of Papilionidae in Mexican states

State	No. of species
Baja California Norte	7
Baja California Sur	6
Sonora	7
Chihuahua	7
Coahuila	3
Nuevo León	23
Tamaulipas	24
Sinaloa	17
Durango	10
Zacatecas	2
San Luis Potosí	27
Nayarit	23
Jalisco	28
Aguascalientes	6
Guanajuato	9
Querétaro	7
Hidalgo	19
Colima	26
Michoacán	24
México	21
Distrito Federal	8
Tlaxcala	0
Guerrero	31
Morelos	24
Puebla	31
Veracruz	41
Oaxaca	40
Chiapas	41
Tabasco	28
Campeche	11
Yucatán	21
Quintana Roo	21

ern populations along the Pacific coast) or *Dismorphia amphiona* (LLorente, 1983).

2. Mesomontane pattern. In this pattern of elements restricted to montane cloud forests, the disjunct populations found in the large physiographic areas (e.g., Sierra Madre Occidental, Sierra Madre del Sur, Sierra Madre Oriental-Sierra de Juárez, and the mountains of Chiapas) are often subspecifically differentiated, e.g., members of *Pterourus alexiares*, *Pterourus esperanza*, and *Pyrrhosticta abderus*.

3. Xeric-relictual pattern. This pattern is made up of relictual elements associated with areas of tropical, deciduous forests and pine-oak

Table 4.7. Papilionidae in various areas of Mexico

Area	No. of species	Percent of species in the area	Percent of species in the country
Monterrey	16	69.5	28.0
Cañón del Novillo	21	87.5	36.8
Sierra de San Juan	20	87.0	35.1
Huasteca Potosina	16	59.3	28.0
Patla-Necaxa	26	84.0	45.6
San Nicolás Tolentino	16	76.2	28.0
Tepoztlán-Yautepec	20	83.3	35.1
Teocelo	20	48.8	35.1
Yanga-Tuxpango	25	61.0	43.9
Sierra de Atoyac	20	64.5	35.1
Sierra de Juárez	31	77.5	54.4
Los Tuxtlas	29	70.7	50.9
Boca del Chajul	26	63.4	45.6

forests of southern and western Mexico. Included are *Baronia brevicornis* and *Parides alopius*.

The ecological characterization of some of the historic elements in these patterns is similar to that described by Halffter (1976) in his patterns of dispersion (typical Neotropical pattern, Montane Mesoamerican pattern, and Paleoamerican pattern).

Table 4.6 lists the states of Mexico and the number of species found in each. The highest numbers of species are found in the southeastern states of Veracruz and Chiapas, with 41 each. Oaxaca, with 40, is a close second. These species account for 72% of the total in Mexico. The states of Guerrero and Puebla have the next highest number of species with 31 each. The northern states have fewer Papilionidae. They usually number less than ten as is the case in Baja California Norte, Sonora, Chihuahua, Durango, Zacatecas, and Coahuila. The northeastern states of Nuevo León and Tamaulipas have 23 and 24 species respectively. Numbers range from seven to 27 in the eastern states of San Luis Potosí, Querétaro and Hidalgo. In the central states (México, Morelos) there are more than 20. The Yucatán Peninsula has between 11 and 21 species. The states rich in species are also those whose vegetation is varied. It is of interest that this factor is independent of surface area. For example, Chihuahua is more than 30 times larger than Colima but has fewer than one-third its species.

The states with tropical forests and cloud forests are the most species-rich, particularly those with more mesic climates along the coastal strips, from Chiapas to Sinaloa on the Pacific side to Tabasco and Tamaulipas on the Gulf. Morelos, México, and Puebla in central Mexico with this type of vegetation have a considerable number of species. The paucity of

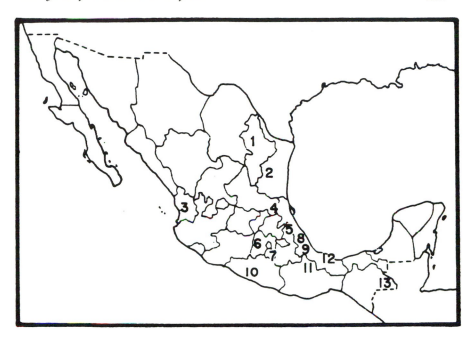

Figure 4.11. Papilionidae in Mexico: areas where the group has been studied. 1. Monterrey. 2. Cañón del Novillo. 3. Sierra de San Juan. 4. Huasteca Potosina. 5. Patla-Necaxa. 6. San Nicolás Tolentino. 7. Tepoztlán-Yautepec. 8. Teocelo. 9. Yanga-Tuxpango. 10. Sierra de Atoyac. 11. Sierra de Juárez. 12. Los Tuxtlas. 13. Boca del Chajul. See Table 4.7 for details.

Papilionidae in Zacatecas, Coahuila, and Tlaxcala may be due to the fact that these states have been generally undercollected. Table 4.7 lists the 13 best known geographic areas (Fig. 4.11) for the Papilionidae of Mexico.

The ten richest areas in terms of species and endemism of Papilionidae (in Mexico) are north to south and west to east: (1) Cañón del Novillo; (2) Durango-Sinaloa border; (3) Sierra de San Juan; (4) Sierra de Atoyac; (5) some parts of Morelos bordering Guerrero state; (6) Barranca de Patla; (7) Los Tuxtlas, Veracruz; (8) Sierra de Juárez; (9) inland Chiapas; (10) Boca de Chajul.

Most available studies on the biology of Papilionidae, such as mimicry, foraging, and gradients of diversity, do not include Mexican species or populations. The only reliable studies are those of Vázquez from the 1950s and the more recent efforts of Paul Spade. However, the lack of information for Mexican taxa is significant when compared with those of the United States or Central America. Often even their original host plants are not known. *Pterourus multicaudata, P. alexiares, P. pilumnus,* and *P. esperanza* are examples. These mostly Mexican species are poorly known, while *P. glaucus,* which is shared with the United States, is well known in scientific literature. There are no Mexican counterparts to studies of De Vries (1987) and Young (1985) and Muyshondt and Muyshondt (1975) concerning the biology of Papilioninae in Costa Rica and El Salvador. Studies of the Papilionidae of Mexico have until now focused more on aspects of variation, taxonomy, and biogeography. It is encouraging that, of late, Mexican Papilionidae has been attracting the attention of scholars, e.g., Leptocircini, *Baronia*, (K. Brown, unpublished data).

It is essential to have systematic information on the biology of the endemic and quasiendemic taxa, especially in areas of species richness or in relict areas. It is well known that various Papilionidae thrive in areas subject to perturbation by man. *Baronia brevicornis*, whose host plant is *Acacia cymbispina* ("cubata"), flourishes in disturbed and abandoned areas of the tropical deciduous forests. It has been observed that plantations of avocado (*Persea* spp.), orange and other citrus (*Citrus* spp.), and "chirimoya" (*Annona* spp.) promote population increases in *Pyrrhosticta*, *Heraclides*, *Priamides*, and *Protesilaus* in various parts of Mexico. This knowledge is key to the restoration of species through reforestation measures.

Aspects of ecology cannot be generalized for all the Papilionoidea from studies of Papilionidae, as the phytophagic interrelations differ from one family to another and are often monophagic or oligophagic. The reproductive strategies and life histories are also likely to be different so far as patterns are concerned. Some of the differences that have been noted include the foraging behavior of the groups: for example, the caterpillars of Papilionidae feed mostly on trees, whereas those of other family groups feed on creepers and annual plants. Adult Papilionidae rely on nectar for sustenance or feed on dissolved salts in moist earth, whereas other groups of Papilionoidea feed on decomposing (exuded or fermented) organic material. Of late, several genera of Pieridae and Nymphalidae have been the subjects of study, i.e., *Enantia*, *Eucheira*, *Hamadryas*, and *Bolboneura*, among others.

DIVERSITY AND ITS CONSERVATION

Data on maintenance of diversity in Mexican Papilionidae are presently lacking and urgently needed. It may be assumed safely that the diversity includes a set of delicate interrelations in specific habitats. Any disturbance in this balance of nature leads to possible extinctions. Destruction or alteration of habitat through deforestation, livestocking, intensive agriculture, urbanization, and industrialization are the greatest threats not only to the preservation of diversity of butterflies (Collins & Morris, 1985) but to whole biotic communities.

Commercial trade poses a significant threat to the diversity of Mexican butterflies. First, the trade does not discriminate among rare, restricted, threatened, or paleoendemic species. The problem is acute, as capture of samples frequently involves a long chain of local traders usually hired by middlemen working for collectors, businessmen, or museums. Uninformed children and adults are hired to indiscriminately collect such samples, as has been observed in Oaxaca, Chiapas, Veracruz, and Guerrero. An intense educational effort on the nature of this trade (actual worth of samples, rarity of species, and so on) which in part is international, may remedy the situation. The Appendix provides the contrast in prices. Second, the exporter has one or two itineraries he follows regularly for some years that include a minimum of two or three areas per itinerary (Escalante and González, pers. comm.), placing him at the top of a pyramid of major depredation. Alarmingly, some work, under the umbrella of "scientific" societies or conservation agencies, involves this type of trade depredation in covert ways, often with vulnerable or endangered species. To date, there is no effective legislation or regulation, illustrated by the fact that, following the law of supply and demand, the prices for butterflies in Mexico are much lower than those in other neotropical countries, which indicates a constant supply.

Habitat preservation is essential for further study of the biology of these organisms, as it may lead to a better understanding of the mechanisms by which diversity is preserved, as well as alternative ways to conserve these biota.

DISCUSSION AND CONCLUSIONS

Mexico has been an important theater of speciation for butterflies, as suggested by the specific and transspecific endemism in the group. Endemism in Mexican Papilionoidea is roughly 9%, which is a good index of comparison to the Lepidoptera in the country (25,000 species). Tropical and deciduous forests, semiarid zones, and in some cases pine-oak and montane cloud forests are centers of paleoendemism for butterflies and moths. The areas of high endemism for the Papilionoidea are tropical and montane cloud forests of the Balsas Basin, Sierra Madre del Sur, Pacific Coastal Plain, and Sierra Madre Oriental–Sierra de Juárez. The species-rich areas are tropical and montane cloud forests in southern, southeastern, and western Mexico. About 25% of Mexican Papilionidae are found in some of these areas (Fig. 4.11). Generally, areas with pronounced topographic, climatological, and vegetational heterogeneity have high numbers of Lepidoptera. The most species-rich areas, with the greatest number of endemics are montane cloud forests in the oldest mountains and in the coastal strips of southern Mexico between 500 to ,1600 m (Sierra de Juárez, Sierra de Atoyac, and Sierra de Chiconquiaco). Disturbed vegetations, rather than virgin ones, seem to harbor higher numbers of Papilionidae; but the degrees of disturbance are not known in terms of the proportion of each type of area, which might explain the maximum richness and permanence.

Despite what has been studied, there are gaps in our knowledge, which is generally evident in phylogenetic studies. Biological understanding of various plesiomorphic species are preliminary. Aspects of population dynamics of species and communities of Papilionidae are poorly known.

Anthropogenic factors contribute greatly to habitat alterations in species-rich areas, among which are intensive slash-and-burn agriculture, deforestation, cattle-raising, trade in wild biota, urbanization, and industrialization. These practices continue to take a heavy toll on a great genetic richness. Fortunately, there are areas that can be recovered and restored in the species-rich states of Guerrero, Oaxaca, and Chiapas. Conservation and study of diversity should go hand in hand, so the construction of a system of protected areas will be accompanied by ecological, biogeographic, taxonomic, and genetic research.

ACKNOWLEDGMENTS

We thank Drs. Lee D. Miller (Sarasota, Florida) and Frederick Rindge (Natural History Museum, New York) for facilities to study collections in their institutions. Dr. Jorge Soberon was an important early participant in the study. Isabel Vargas (Papilionidae) and Teresa Lilejhoult (Pieridae) helped variously in data collection and areographic studies. Alejandro Peláez and Jorge Moreno provided assistance with the use of dBASE III plus. Luis González Cota provided unpublished data and allowed access to his private collection. Isolda Luna, in addition to processing various versions of the Spanish text, reviewed them. Dr. K.S. Brown critically read and made valuable suggestions on an earlier version. Alejandro Martínez M. helped with photography. CONACyT of Mexico and PSPA of the Universidad Nacional Autónoma de Mexico and the Facultad de Ciencias of the same university provided financial assistance. Finally, we thank T.P. Ramamoorthy and Marguerite Elliott for help with the preparation of this chapter in English.

APPENDIX

Price list for Mexican butterflies (Papilionidae)[a]

Papilionidae	Cost				
	$0.50–2.00	$2–30	$30–150	$150–1,500	>$1,500
Baronia brevicornis (rare[b])	*				
Parnassius phoebus	*	*			
Protesilaus marcellus	*	*			
Protesilaus philolaus	*	*			
P. oberthueri					
P. epidaus	*				
P. phaon	*				
P. branchus	*	*			
P. belesis	*				
P. thymbraeus	*	*			
P. agesilaus	*				
P. macrosilaus	*				
Eurytides marchandi	*	*			
E. lacandones	*	*			
E. calliste		*			
E. salvini		*			
Battus philenor	*				
B. polydamas	*	*			
B. laodamas	*	*			
B. eracon		*			*
B. belus	*				
B. lycidas	*	*			
Parides alopius		*			
P. montezuma	*	*			
P. photinus	*				
P. erithalion	*	*			
P. lycimenes	*	*			
P. iphidamas	*				
P. sesostris	*	*			
P. childrenae		*			
P. eurymedes	*				
Pterourus esperanza (vulnerable[b])					*
P. palamedes	*	*			
P. glaucus[c]	*	*			
P. alexiares				*	*
P. rutulus	*	*			
P. multicaudatus[c]	*	*			
P. pilumnus		*			
P. eurymedon	*	*			
Pyrrhosticta victorinus[c]	*	*			
P. diazi					*
P. garamas[c]		*			
P. abderus		*	*		

Appendix (cont.)

Papilionidae	Cost				
	$0.50–2.00	$2–30	$30–150	$150–1,500	>$1,500
Heraclides thoas		*			
H. cresphontes[c]	*	*			
H. ornythion	*	*			
H. astyalus[c]	*				
H. androgeus	*	*			
Troilides tolus	*	*			
Priamides pharnaces	*				
P. anchisiades[c]	*				
P. rogeri	*	*			
P. erostratus	*		*		
P. erostratinus					
Papilio polyxenes	*	*			
P. bairdii		*			
P. zelicaon	*	*			
P. widia		*			

Asterisks in more than one of the price columns indicate price fluctuation due to demand in the market or to forms, sexes, or subspecies that have distinct prices.
[a]Compiled from De la Maza (1978), Collins & Morris (1985), and several private lists from European and United States businesses.
[b]Conservation category following IUCN.
[c]Usually grown in captivity.

REFERENCES

A selected bibiographical list of the principal works on the history of the study of the Papilionidae of Mexico follows here. References marked with an asterisk are the source of information used in the analysis. Others are cited in the text.

*Allyn, A.C., M. Rothschild & D.S. Smith. 1982. Microstructure of blue/green and yellow pigmented wing membranes in Lepidoptera, with remarks concerning the function of pterobilins. 1. Genus *Graphium*. Bull. Allyn Mus. 75:1–20.

*Anon. 1976. Zonas de interesante exploración lepidopterológica para 1976. Bol. Inf. Soc. Méx. Lep. 2(1):4–5.

*Balcázar, M.A. 1988. Fauna de mariposas de Pedernales, Municipio de Tacámbaro, Michoacán. (Lepidoptera: Papilionoidea y Hesperioidea). Thesis, Universidad Michoacana de San Nicolás de Hidalgo, Morelia, Michoacán.

*Barrera, A. & M.E. Díaz. 1977. Distribución de algunos lepidópteros de la Sierra de Nanchititla, México, con especial referencia a *Tisiphone maculata* Hopff. (Ins.: Lepid.). Rev. Soc. Méx. Lep. 3(1):17–28.

*Barrera, T. & L. Romero. 1986. Estudio faunístico de Lepidópteros (Superfamilia Papilionoidea) en un Bosque Mesófilo de Montaña en Cascada Los Diamantes, San Rafael, Estado de México. Thesis, UNAM.

*Bates, H.W. 1864–65. New species of butterflies from Guatemala and Panama collected by Osbert Salvin and F. du Cane Godman. Ent. Mon. Mag. 1(1):1–6, (2):31–35, (3):55–59, (4):81–85, (5):113–116, (6):126–131, (7):161–164, (8):178–180, (9):202–205.

*Beutelspacher, C.R.B. 1974. Reconsideración taxonómica de *Papilio tolus* G. & S. (Lep. Papilionidae) y descripción de una nueva subespecie. Rev. Soc. Méx. Hist. Nat. 35:149–157.

*———. 1975a. Una nueva especie de *Papilio* L. (Papilionidae). Rev. Soc. Méx. Lep. 1(1):3–6.

*———. 1975b. Notas sobre el Suborden Rhopalocera (Lepidoptera) de las Minas, Veracruz. Rev. Soc. Méx. Lep. 1(1):11–20.

*———. 1976. Nuevas formas de papiliónidos mexicanos (Papilionidae). Rev. Soc. Méx. Lep. 2(2):61–70.

*———. 1980. Mariposas Diurnas del Valle de México. Mexico City: La Prensa Médica Mexicana.

*———. 1981. Lepidópteros de Chamela, Jalisco, México. I. Rhopalocera. An. Inst. Biol. UNAM Serie Zool. 62(1):371–388.

*———. 1983. Mariposas diurnas de "El Chorreadero" Chiapas (Insecta: Lepidoptera). An. Inst. Biol. UNAM Serie Zool. 53(1):341–366.

*———. 1984. Mariposas de México, fascículo I. Introdución y generalidades, superfamilia Papilionoidea, familia Papilionidae. Ed. Científ. Mexico City: La Prensa Médica Mexicana.

*———. 1986. Una nueva subespecie mexicana de *Papilio erostratus* Westwood (Insecta, Lepidoptera, Papilionidae). An. Inst. Biol. UNAM Serie Zool. 56(1):241–244.

Boisduval, J.B.A. 1836. Histoire Naturalle des Insectes. Spécies général Lépidoptères. Libr. Paris: Encyclopedia de Roret.

*Brower, L.P. 1958b. Larval food plant specificity in butterflies of the *Papilio glaucus* group. Lep. News. 12:103–114.

*———. 1959a. Speciation in butterflies of the *Papilio glaucus* group I. Morphological relationship and hybridization. Evolution 13:40–63.

*——— & J.V.Z. Brower. 1962. The relative abundance of model and mimic butterflies in natural population of the *Battus philenor* mimicry complex. Ecology 43:319–323.

*———, J.V.Z. Brower, F.G. Stiles, H.J. Croze & A.S. Hower. 1964. Mimicry: differential advantage of color patterns in the natural environment. Science 144:183–185.

Brown, F.M. & B. Heineman. 1972. Jamaica and Its Butterflies. London: Classey.

Brown, K.S. 1984. Species diversity and abundance in Jaru, Rondonia, Brasil. News Lep. Soc. 3:1–2.

*Clench, H.K. 1965. A collection of butterflies from western Chihuahua, Mexico. Ent. News. 76(6):157–162.

*———. 1968. Butterflies from Coahuila, Mexico. J. Lep. Soc. 22(4):227–231.

Collins, N.M. & M.G. Morris. 1985. Threatened Swallowtail Butterflies of the World. Gland (Switzerland)/Cambridge: The IUCN Red data book.

*Comstock, J.A. 1958. Butterfly collecting in the Mexican tropics. J. Lep. Soc. 12(3–4):127–129.

*——— & L.G. Vázquez. 1960. Estudios de los ciclos biológicos en lepidópteros mexicanos. An. Inst. Biol. UNAM 31(1–2):349–448.

Croizat, L., G. Nelson & D.E. Rosen. 1974. Centers of origin and related concepts. Syst. Zool. 23(2):265–287.

D'Abrera, B. 1981. Butterflies of the Neotropical Region. Part I. Papilionidae and Pieridae. Melbourne: Landowne Editions & E.W. Classey.

*D'Almeida, R.F. 1966. Catálogo dos Papilionidae Americanos. Soc. Brasileira Ent.

*D'Almeida, R.C. 1977. Mariposas de Tabasco. Bol. Inf. Soc. Méx. Lep. 3(1):5–7.

*De la Maza, J.E. 1977. Reconsideración taxonómica de *Papilio garamas baroni* R. y J., 1906 (Lepidoptera: Papilionidae). Rev. Soc. Méx. Lep. 3(2):74–84.

*———— & A. Díaz. 1979. Notas y descripciones sobre la familia Papilionidae en México. Rev. Soc. Méx. Lep. 4(2):51–56.

*————, J.L. White & A.L. White. 1987. Observaciones sobre el polimorfismo femenino de *Baronia brevicornis* Salv. (Papilionidae: Baroniinae) con la descripción de una nueva subespecie del estado de Chiapas, México. Rev. Soc. Méx. Lep. 11(1):3–13.

*———— & R.E. de la Maza. 1976. Papiliónidos del Cañón del Novillo, Tamaulipas (Lepidoptera: Papilionidae). Rev. Soc. Méx. Lep. 2(1):24–31.

*———— & R.E. de la Maza. 1985a. La fauna de mariposas de Boca de Chajul, Chiapas, México (Rhopalocera). Parte I. Rev. Soc. Méx. Lep. 9(2):21–44.

*———— & R. E. de la Maza. 1985b. La fauna de mariposas de Boca de Chajul, Chiapas, México (Rhopalocera). Parte II. Rev. Soc. Méx. Lep. 10(1):1–24.

*————, R.E. de la Maza & R. de la Maza. 1982. Lepidópteros nuevos del estado de Guerrero, México (Papilionoidea). Rev. Soc. Méx. Lep. 7(1):2–14.

*————, R. Moreno & E. Fernández. 1975. Excursiones. Bol. Inf. Soc. Méx. Lep. 1(2):3.

*De la Maza, R.E. 1975. Notas sobre lepidópteros de Rancho Viejo y Tepoztlán, Morelos, México. Primera Parte: Papilionoidea. Rev. Soc. Méx. Lep. 1(2):42–61.

————. 1978. Los lepidópteros y su importancia como una explotación pecuaria. Thesis, UNAM.

*————. 1979. Notas sobre los papiliónidos de México (Lep). VII. Area de Monterrey a Cola de Caballo Nuevo León. Bol. Soc. Méx. Lep. 5(4):2–15.

*————. 1980a. Las poblaciones centroamericanas de *Parides erithalion* (Boisd.). (Papilionidae: Troidini). Rev. Soc. Méx. Lep. 5(2):51–74.

*————. 1980b. Notas sobre los Papilionidae en México (Lep). VIII. Area San Luis Potosí, S.L.P. Bol. Inf. Soc. Méx. Lep. 6(3):3–13.

*———— & A. Díaz F. 1978. Una nueva subespecie de *Parides lycimenes* Boisd. de México (Lepidoptera, Papilionidae). Rev. Soc. Méx. Lep. 4(1):7–14.

*———— & E. Olaya. 1979. Hallazgo de una población de *Papilio abderus* Hopf. en la Sierra de Alvarez, San Luis Potosí, México (Papilionidae). Bol. Inf. Soc. Méx. Lep. 5(2):9–12.

*———— & J.E. de la Maza. 1979. Confirmación de la existencia de *Parides lycimenes lycimenes* Boisd. en la región Lacandona, Chiapas, México (Papilionidae). Rev. Soc. Méx. Lep. 4(2):47–56.

*———— & J.E. de la Maza. 1981. Notas sobre los Papilionidae en México (Lep). IX. Sierra de Alvarez, S.L.P. Bol. Inf. Soc. Méx. Lep. 7(2):6–23.

*———— & R.R. de la Maza. 1978a. Notas sobre la familia Papilionidae en México. I. San Nicolás Tolentino, México. Bol. Inf. Soc. Méx. Lep. 4(2):3–7.

*———— & R.R. de la Maza. 1978b. Notas sobre la familia Papilionidae en México (Lep). IV. Area de Orizaba a Yanga, Veracruz. Bol. Inf. Soc. Méx. Lep. 4(5):15–30.

*———— & R.R. de la Maza. 1979. Notas sobre los papiliónidos en México. V. Zona de los Tuxtlas, Veracruz. Bol. Inf. Soc. Méx. Lep. 5(3):2–18.

*———— & R. Turrent D. 1978. Notas sobre la familia Papilionidae en México (Lep). III Area del Valle de México. Bol. Soc. Méx. Lep. 4(4):5–14.

*De la Maza, R.R. 1975. Colecta en el Sureste. Bol. Inf. Soc. Méx. Lep. 1(5): 2–5.

*————. 1976a. Una interesante aberración de *Parides alopius* (Godman & Salvin) (Papilionidae). Rev. Soc. Méx. Lep. 2(1):5–7.

*————. 1976b. Colecta del 14 al 23 de abril en los estados de Oaxaca y Chiapas. Bol. Inf. Soc. Méx. Lep. 2(2):6–7.

*———. 1976c. Colecta en Sierra de Juárez, Oaxaca. Bol. Inf. Soc. Méx. Lep.2(3):2–4.

*———. 1976d. Colecta en el Estado de Nuevo León, del 23 al 31 de julio de 1976. Bol. Inf. Soc. Méx. Lep. 2(4):2–3.

*———. 1987. Mariposas mexicanas, guía para su colecta y determinación. Mexico City: Fondo de Cultura Económica.

De Vries, P.J. 1987. The butterflies of Costa Rica and their Natural History. Princeton, NJ: Princeton Univ. Press.

*Díaz, A.F. 1975. Papiliónidos del Valle de Tepoztlán, Morelos. Bol. Inf. Soc. Méx. Lep. 1(3):5–7.

——— & J.E. de la Maza. 1978. Guía ilustrada de las mariposas mexicanas. Parte I, familia Papilionidae. Publ. Esp. Soc. Méx. Lep. 3:1–15.

*Domínguez, Y. & J.L. Carrillo. 1976. Lista de Insectos en la Colección Entomológica del Instituto Nacional de Investigaciones Agrícolas, SAG. Foll. Misc. Secr. Agr. Ganad. (México) No. 29.

*Doubleday, E. & W.C. Hewitson. 1846–1852. The Genera of Diurnal Lepidoptera. Vols.1 & 2. London: Green & Longman.

Durden, C.J. & H. Rose. 1978. Butterflies from the Middle Eocene: the earliest occurrence of fossil Papilionoidea (Lepidoptera). Pearce-Sellards Series (Texas Memorial Museum) 29:1–25.

Ehrlich, P.R. 1958. The comparative morphology, phylogeny and higher classification of the butterflies (Lepidoptera: Papilionoidea). Univ. Kans. Sci. Bull. 39(8):305–370.

*——— & P.H. Raven. 1967. Butterflies and Plants. Ecology, Evolution and Population Biology. Reading from Scientific American. San Francisco, CA: W.H. Freeman. pp. 131–138.

*Eisner, T.E., E. Plieske, M. Ikeda, D.F. Owen, L. Vazquez, H.R. Prez, J.G. Framclemont & J. Meinwald. 1970. Defense mechanisms of arthropods XXVII. Osmeterial secretions of papilionid caterpillars (*Baronia, Papilio, Eurytides*). Ann. Ent. Soc. Amer. 63(3):914–915.

*Esper, E.J.C. 1784–1801. Die auslndischen oder die ausserhalb Europa zur Zeit in den brigen Welttheilen vorgefundenen Schmetterlinge in Abbildungen nach der Natur mit Beschreibungen. Erlangen: Walther.

*Ferris, C. & J. Emmel. 1982. Discussion of *Papilio coloro* W. G. Wright (= *Papilio rudkini* F. & R. Chermock) and *Papilio polyxenes* Fabricius (Papilionidae). Bull. Allyn Mus. 76:13.

Ford, E.B. 1944. Studies on the chemistry of pigments in the Lepidoptera with reference to their bearing on systematics. 4. The classification of the Papilionidae. Trans. R. Ent. Soc. Lond. 94(2):201–223.

*Gibson, W. & J. L. Carrillo. 1959. Lista de Insectos en la Colección Entomológica de la Oficina de Estudios Especiales, S.A.G. Foll. Misc. Secr. Agric. Ganad. (Méx.) Vol. 9.

*Gilbert, L.E. & P.H. Raven (eds.). 1973. Coevolution of Animals and Plants. Austin, TX: Univ. of Texas Press.

*Godman, F.D. & O. Salvin. 1879–1901. Biologia Centrali Americana. Zoology: Insecta. Lepidoptera-Rhopalocera. London: Taylor & Francis.

*González, L.C. 1977. Reporte de la colecta en La Ceiba, Puebla. Bol. Inf. Soc. Méx. Lep. 3(3):6–7.

*———. 1978. Notas sobre la familia Papilionidae (Lepidoptera) en México. Barranca de Patla, Puebla y alrededores. Bol. Inf. Soc. Méx. Lep. 4(1):3–15.

*Guzmán, P. 1976. Algunas observaciones sobre lepidópteros de Chalma, Estado de México. Rev. Soc. Méx. Lep. 2(1):49–51.

Halffter, G. 1976. Distribución de los insectos en la Zona de Transición Mexicana: Relaciones con la entomofauna de Norteamérica. Folia Entomol. Mex. 35:1–64.

————. 1987. Biogeography of the Montane Entomofauna of Mexico and Central America. Annu. Rev. Entomol. 32:95–114.

Hancock, D.L. 1983. Classification of the Papilionidae (Lepidoptera): a phylogenetic approach. Smithersia 2:1–48.

*Hernández V., H., I. Martínez G. & S. Rodríguez N. 1981. Lepidópteros en la Colección Entomológica de la Dirección General de Sanidad Vegetal. Parte I. Fitófilo 84:15–17.

*Hodges, R.W. (ed.). 1983. Checklist of the Lepidoptera of America North of Mexico. London: Classey.

*Hoffmann, C.C. 1940. Catálogo sistemático y zoogeográfico de los lepidópteros mexicanos. Primera parte: Papilionoidea. An. Inst. Biol. UNAM 11(2):639–739.

*Holland, R. 1972. Butterflies of middle and southern Baja California. J. Res. Lep. 11(3):147–160.

*Howe, W.H. 1973. The Butterflies of North America. New York: Doubleday

*Hübner, J. & C. Geyer. 1796–1838. Sammlung Europischer Schmetterlinge. Augsburg. Published by author.

*———— & C. Geyer. 1808–1837. Zütrage zur Sammlung Exotischer Schmetterlinge. Vols. 1–5 Augsburg.

Igarashi, S. 1984. The classification of the Papilionidae mainly based on the morphology of their immature stages. Ty To Ga 34(2):41–96.

*Katthain, D.G. 1971. Estudio Taxonómico y datos ecológicos de especies del Suborden Rhopalocera (Insecta, Lepidoptera) en un área del Pedregal de San Angel, D.F. Mexico. Thesis, UNAM.

*Kendall, R.O. & W. McGuire. 1984. Some new and rare records of Lepidoptera found in Texas. Bull. Allyn Mus. 86:49.

*Lamas, G.M. 1983. How many butterfly species in your backyard? News Lep. Soc. 4:1–2.

*———— (ed.). 1985. Proceedings of the second symposium on Neotropical Lepidoptera, Arequipa, Peru, 1983. J. Res. Lep. (Suppl. 1):1–104.

*Latreille, P.A. & J.B. Godart. 1819–1824. Encyclopedie Méthodique. Histoire Naturelle des Insects. Vol. 9. Paris.

Linnaeus, C. 1758. Systema Naturae. Vol. 1. Regnum Animale. 10th ed. Holmiae: Impensis Salvii.

LLorente J. 1983. Sinopsis Sistemática y Biogeográfica de los Dismorphiinae de México con especial referencia al género *Enantia* Huebner (Lepidoptera: Pieridae). Folia Entomol. Mex. 58:1–207.

*————. 1988. Notas y comentarios sobre las mariposas de Cuba: 1. Algunos aspectos nomenclaturales y clasificatorios (unpublished).

———— & A. Luis. 1988. Nuevos Dismorphiini de México y Guatemala (Lepidoptera: Pieridae). Folia Entomol. Mex. 74:159–178.

*————, A. Garcés & A. Luis. 1986. Las mariposas de Jalapa-Teocelo, Veracruz. Teocelo 4:14–37.

*Luis, M. A. 1987. Distribución altitudinal y estacional de los Papilionoidea (Insecta: Lepidoptera), en la Cañada de los Dínamos, Magdalena Contreras, D.F. Thesis, UNAM.

Maes, J.M., J.P. Desmedt, V. Hellebuyk & J.C. Gantier. Catálogo de los Lepidoptera de Nicaragua. 1. Papilionidae. Rev. Nica. Ent. In Press.

Miller, J.S. 1987. Phylogenetic studies in the Papilioninae (Lepidoptera: Papilionidae). Bull. Am. Mus. Nat. Hist. 186(4):365–512.

*———— & P. Feeny. 1983. Effects of benzilisoquinoline alkaloids on the larvae of polyphagous Lepidoptera. Oecologia 58:332–339.

Miller, L.D. & F.M. Brown. 1981. A catalogue/checklist of the butterflies of America, North of Mexico. Mem. Lep. Soc. 2:1–280.

Munroe, E. 1961. The classification of Papilionidae (Lepidoptera). Can. Entomol. (Suppl.) 17:1–51.

Muysholndt, A. & A. Muyshondt, Jr. 1975. Notes on the duration of the pupal stage of some swallowtails of El Salvador (Lepidoptera: Papilionidae). Entomol. Rec. J. Var. 87(2):45–47.

Nelson, G. 1978. From Candolle to Croizat: comments on the history of biogeography. J. Hist. Biol. 11:269–305.

————. 1983. Vicariance and cladistics; historical perspectives with implications for the future. *In* R.W. Sims, J.H. Price & P.E.S. Whalley (eds.), Evolution, Time and Space, the Emergence of the Biosphere. New York: Academic Press. pp. 469–492.

Patterson, C. 1983. Aims and methods in Biogeography. *In* R.W. Sims, J.H.Price & P.E.S. Whalley (eds.), Evolution, Time and Space. The Emergence of the Biosphere. Vol. 23. The Systematic Association. New York: Academic Press. pp. 1–28.

*Pérez, H.R. 1969. Quetotaxia y morfología de la oruga de *Baronia brevicornis* Salv. (Lepidoptera, Papilionidae, Baroninae). An. Inst. Biol. UNAM Serie Zool. 40(2):227–244.

*————. 1971. Algunas consideraciones sobre la población de *Baronia brevicornis* Salv. (Lepidoptera, Papilionidae, Baroninae) en la región de Mezcala, Guerrero. An. Inst. Biol. UNAM Serie Zool. 42(1):63–72.

————. 1977. Distribución geográfica y estructura poblacional de *Baronia brevicornis* Salv. (Lepidoptera, Papilionidae, Baroninae) en la República Mexicana. An. Inst. Biol. UNAM Serie Zool. 48:151–164.

———— & R.S. Sánchez. 1986. Algunos aspectos demográficos de *Baronia brevicornis* Salv. (Lepidoptera: Papilionidae, Baroniinae) en dos localidades de México. An. Inst. Biol. UNAM Serie Zool. 57(1):191–198.

*Platt, A.P., R.P. Coppinger & L.P. Brower. 1971. Demonstration of the selective advantage of mimetic Limenitis butterflies presented to caged avian predators. Evolution 25(4):692–701.

*Powell, J.A. 1958. Additions to the knowledge of the butterfly fauna of Baja California Norte. Lep. News 12(1–2):26–32.

*Racheli, T. & V. Sbordoni. 1975. A new species of *Papilio* from Mexico (Lepidoptera, Papilionidae). Frag. Entom. 11(2):175–183.

*Rivera, L. 1975. Colecta de material entomológico en el Estado de Veracruz. Bol. Inf. Soc. Méx. Lep. 1(6):7–8.

*Rodríguez, S. 1982. Mariposas del Suborden Rhopalocera (Lepidoptera) de Acatlán de Juárez, Jalisco y alrededores. Thesis, UNAM.

Rosenzweig, M.L. 1975. On continental steady states of species diversity. *In* M.L. Cody & J.M. Diamond (eds.), Ecology and Evolution of Communities. Cambridge, MA: Belknap Press. pp. 121–140.

*Ross, G.N. 1967. A distributional study of the butterflies of the Sierra de Tuxtla in Veracruz, Mexico. Ph.D. dissertation, Louisiana State Univ..

*————. 1964a. Life history studies on Mexican butterflies. I. Notes on the early stages of four Papilionids from Catemaco, Veracruz. J. Lep. Soc. 3(1):9–18.

*————. 1964b. Life history studies on Mexican butterflies. III. Nine Rhopalocera (Papilionidae; Nymphalidae; Lycaenidae) from Ocotal, Chico, Veracruz. J. Lep. Soc. 3(4):207–229.

*————. 1975. An ecological study of the butterflies of the Sierra de Tuxtla in Veracruz, México. J. Res. Lep. 14(2):103–124.

Rothschild, W. & K. Jordan. 1906. A revision of the American Papilios. Novit. Zool. 13:412–752.

*Routledge, C. 1977. El Suborden Rhopalocera (Lepidoptera) del estado de Tabasco. Su lista, frecuencia, diversidad y distribución. Rev. Soc. Méx. Lep. 3(2):57–73.

Rzedowski, J. 1978. Vegetación de México. Mexico City: Limusa.

*Salvin, O. 1893. Description of a new genus and species of Papilionidae from Mexico. Trans. Ent. Soc. Lond. 41(4):331–332.

Scott, J.A. 1986. The Butterflies of North America. A Natural History and Field Guide. Stanford, CA: Stanford Univ. Press.

*Scriber, J.M. 1972. Confirmation of a disputed of *Papilio glaucus* (Papilionidae). J. Lep. Soc. 26(4):235–236.

*————. 1973a. Latitudinal Gradients in Larval Feeding Specialization of the World Papilionidae (Lepidoptera). Psyche 80(4):355–373.

*————. 1973b. Latitudinal gradients in larval feeding specialization of the world Papilionidae (Lepidoptera). A supplementary Table of Data. Published by author.

*————. 1978. The effects of larval feeding specialization and plant growth form on the consumption and utilization of plant biomass and nitrogen: an ecological consideration. Ent. Exp. Appl. 24:694–710.

*————. 1979a. The effects of sequentially switching foodplants upon biomass and nitrogen utilization by polyphagous and stenophagous *Papilio* larvae. Ent. Exp. Appl. 25:203–215.

*————. 1979b. Effects of leaf-water supplementation upon post-ingestive nutritional indices of forb-, shrub-, vine-, and tree-feeding Lepidoptera. Ent. Exp. Appl. 25:240–252.

*————. 1982a. Food plants and speciation in the *Papilio glaucus* group. Proc. 5th. Int. Symp. Insect-Plant Relationships. London: Wageningen. pp. 307–314.

*————. 1982b. The behavior and nutritional physiology of southern armyworm larvae as a fuction of plant species consumed in earlier instars. Ent. Exp. Appl. 31:359–369.

*————. 1983. Evolution of feeding specialization, physiological efficiency, and host races in selected Papilionidae and Saturniidae. *In* R.F. Denno & M.S. McClure (eds.), Variable Plants and Herbivores in natural and managed systems. New York: Academic Press. pp. 373–412.

*————. 1984a. Larval foodplant utilization by the world Papilionidae (Lepidoptera): latitudinal gradients reappraised. Tokurana (Acta Rhopaloc.) 6–7:1–50.

*————. 1984b. Host-Plant suitability. Chem. Ecol. Insects 7:159–202.

*———— & F. Slansky Jr. 1981. The nutritional ecology of immature insects. Annu. Rev. Ent. 26:183–211.

*———— & M. Finke. 1978. New foodplant and oviposition records for the eastern black swallowtail, Papilio polyxenes on an introduced and native Umbellifer. J. Lep. Soc. 32(3):236–238.

*———— & P. Feeny. 1979. Growth of herbivorus caterpillars in relation to feedings specialization and to the growth form of their plants. Ecology 60(4):829–850.

*————, G.L. Lintereur & M.H. Evans. 1982. Foodplant suitabilities and new oviposition record for *Papilio glaucus canadiensis* (Lepidoptera: Papilionidae) in northern Wisconsin and Michigan. Great Lakes Entomol. 15(1):39–46.

*———— & R.C. Lederhouse. 1982. Temperature as a factor in the development and feeding ecology of tiger swallowtail caterpillars, *Papilio glaucus* (Lepidoptera). Oikos 40(1):95–102.

*————, R.C. Lederhouse & L. Contardo. 1975. Spicebush, lindera benzoin, a little known foodplant of *Papilio glaucus* (Papilionidae). J. Lep. Soc. 29(1):10–14.

Seitz, A. (ed.). 1907–1935. Die Grossschmetterlinge der Erde. Vols. 1, 5, 9, and 13. Stuttgart: Alfred Kernen.

Serrano, F. & M.E. Serrano. 1972. Las mariposas del Salvador. Primera parte, Papilionidae. Comunicaciones (segunda época) 1:48–79.

Shields, O. & S.K. Dvorak. 1979. Butterfly distribution and continental drift between the Americas, the Caribbean and Africa. J. Nat. Hist. 13:221–250.

Slansky, F. 1972. Latitudinal gradients in species diversity of the New World swallowtail butterflies. J. Res. Lep. 11(4):201–218.

*————— & J.M. Scriber. 1982. Selected bibliography and summary of quantitative food utilization by immature insects. Entom. Soc. Am. Bull. 28(1):43–55.

*Southwood, T.R.E. 1978. The components of diversity. In Diversity Insect Faunas. Symposium of The Royal Entomological Society of London No. 9. Oxford: Blackwell Scientific Publication. pp. 19–40.

*Taylor, L.R. 1978. Bates, Williams, Hutchinson—a variety of diversities. In Diversity Insect Faunas. Symposium of The Royal Entomological Society of London No.9. Oxford: Blackwell Scientific Publication. pp. 1–18.

Tyler, H.A. 1975. The Swallowtail Butterflies of North America. Healdsburg, CA: Natural Publ. Naturegraph.

*Vane-Wright, R.I. 1978. Ecological and behavioral origins of diversity in butterflies. In Diversity Insect Faunas. Symposium of The Royal Entomological Society of London No. 9. Oxford: Blackwell Scientific Publications. pp. 56–70.

*Vázquez, L.G. 1942. Observaciones faunísticas de los lepidópteros de Izúcar de Matamoros, Puebla. An. Inst. Biol. UNAM 13(2):547–553.

*—————. 1947. Papilios nuevos de México. An. Inst. Biol. UNAM 18(1):249– 256.

*—————. 1953. Observaciones sobre papilios de México con descripciones de algunas formas nuevas. An. Inst. Biol. UNAM 23:257–267.

*—————. 1954. Notas sobre lepidópteros mexicanos. I. Papilionidae y Pieridae de la Mesa de San Diego, Puebla y sus alrededores. An. Inst. Biol. UNAM 25:391–416.

*————— & H.R. Prez. 1961. Observaciones sobre la biología de Baronia brevicornis Salvin (Lepidoptera: Papilionidae-Baroninae). An. Inst. Biol. UNAM 32:295–311.

*————— & H.R. Pérez. 1967. Nuevas observaciones sobre la biología de Baronia brevicornis Salvin. An. Inst. Biol. UNAM 37:(1–2):195–204.

*Vázquez, N.R. 1982. Mariposas diurnas del Altiplano Potosino en la Colección Entomológica del Instituto de Investigación de Zonas Desérticas. Resúmenes del Sexto Congreso Nacional de Zoología; U.A.S. Soc. Méx. Zool.

*Velázquez, C.A. 1976. Reporte de un viaje de colecta a los estados de Michoacán, Jalisco, Colima y Oaxaca. Bol. Inf. Soc. Méx. Lep. 2(4):6.

*Velázquez, N.V. de & C.A. Velázquez M. 1975. Viaje de colecta a Jalisco y Colima. Bol. Inf. Soc. Méx. Lep. 1(4):6–7.

*White, J.L. & A.L. White. 1980. Notas sobre los Papilionidae en México (Lep). VI. Area de la Huasteca Potosina. Bol. Inf. Soc. Méx. Lep. 6(1):10–35.

Wilson, E.O. (ed.). 1988. Biodiversity. Washington, DC: National Academic Press.

*Young, A.M. 1973. Notes on the life cycle and natural history of Parides arcas mylotes (Papilionidae) in Costa Rica premontane wet forest. Psyche 30(1–2):1–22.

*—————. 1977. Studies on the biology of Parides iphidamas (Papilionidae: Troidini) in Costa Rica. J. Lep. Soc. 31(2):100–108.

*—————. 1979. Oviposition of the butterfly Battus belus varus (Papilionidae). J. Lep. Soc. 33(1):56–57.

—————. 1985. Notes on the natural history of Papilio victorinus Doubl. (Papiliondae) in Northeastern Costa Rica. J. Lep. Soc. 38(3):237–242.

*—————, M.S. Blum & Z. Brian. 1986. Natural History and ecological chemistry of the Neotropical butterfly Papilio anchisiades (Papilionidae). J. Lep. Soc. 40(1):36–53.

5

The Native Bees of Mexico

RICARDO AYALA, TERRY L. GRISWOLD, AND STEPHEN H. BULLOCK

The rich bee fauna of Mexico encompasses 8 families and 153 genera, with approximately 1,589 currently recognized species. Due to gaps in collecting and the paucity of current revisions, the apifauna of Mexico is certainly much richer, perhaps well in excess of 2,000 species. By comparison, 3,745 species are recorded from the continental United States, whereas Panama has fewer than 400 species. The Apoidea of Mexico are of diverse origins, but little is known about the evolution of Mexico's fauna. The oldest fossil, Apoidea (96 million to 74 million years old), belongs to a genus of social bees, as does the only fossil bee known from Mexico (Middle Miocene).

Although the history of taxonomy on Mexican bees spans 230 years, most species have been described during the twentieth century. Three specialists Cockerell, Cresson, and Timberlake contributed more than half of the present total.

The species composition of the fauna by families is as follows: Andrenidae 30%; Anthophoridae 29%; Apidae 5%; Colletidae 5%; Halictidae 12%; Megachilidae 18%; Melittidae and Oxaeidae, each less than 1%. These percentages are probably biased toward Andrenidae and Anthophoridae, as most of the 58% of the genera that have been revised belong to these two families. Four of the 155 genera of bees and an additional eight subgenera are endemic to Mexico. There is no reliable estimate of endemism at the species level for any of the families except Andrenidae, for which it is 53%. Genera that attain their greatest diversity in Mexico include *Centris, Deltoptila, Exomalopsis, Mesoxaea, Mexalictus, Peponapis,* and *Protoxaea.* Five genera are represented by more than 50 species in Mexico: *Perdita* 226; *Andrena* 88; *Exomalopsis* 74; *Centris* 53; and *Megachile* 77. (All but *Megachile* have been revised; several additional genera will undoubtedly exceed this number when they have been revised.)

The arid and semiarid Madrean region (straddling the Mexico-United States border) has the richest apifauna in the world. The poorest regions of Mexico are probably the Yucatan Peninsula and montane regions above 3,000 m. These patterns are tentative, as analyses of sampling effort indicate large gaps, both geographical and seasonal, in our knowledge of the Mexican bee fauna.

Most Mexican bees are solitary or semisocial, pollen-gathering species (70% of the total genera and 90% of the species) compared with cleptoparasites (17% and 7% respectively) and highly social species (13% and 3% respectively). Virtually nothing is known concerning seasonal activity, nesting, or larval pollen diets of

179

bees in Mexico, all important factors in the ecology and evolution of the diversity of bees and possible mechanisms of reproductive isolation. Significant threats to the preservation of the bee diversity in Mexico include the use of insecticides, the destruction of nesting sites, the loss of the forest and wayside flora, and the advent of the Africanized bee.

The superfamily Apoidea compirses the insects commonly called bees. Although many people are familiar only with the domesticated honey bee, *Apis mellifera*, the Apoidea in reality is composed of over 20,000 species in the world (Hurd, 1979) classified into 11 families (McGinley, 1980; Michener, 1965; Michener & Greenberg, 1980; Rozen, 1965). All but three of these 11 families are found in Mexico; Ctenoplectridae is Paleotropical, Fideliidae is known only from Africa and Chile, and Stenotritidae is restricted to Australia. The greatest diversity of bees occurs in extratropical, warm, semidesert areas of the world, with the largest number of species recorded from the hot deserts and scrublands of California and Arizona (Linsley, 1958; Michener, 1979); there are no faunistic reports from adjacent xeric areas in Mexico. By contrast, the apifaunas of the humid tropics appear depauperate, with social Apidae predominating, though with greater diversity in the Neotropics than in the Paleotropics (Michener, 1979).

Bees range in size from 2 to 39 mm long (Michener, 1965) with a biomass from less than 1 mg to more than 1 g. Bees are hypothesized to have evolved from sphecoid wasps, with the distinction that the food provided their larvae is pollen and nectar or oils of angiosperms, whereas sphecoid wasps are predators of arthropods. Though some bees are similar in general appearance to wasps, they differ morphologically in that they have plumose hairs on the body and a posterior basitarsus wider than the other hind tarsal segments (Michener, 1965). The morphological diversity of bees is associated in part with structures for pollen collection (Michener, 1944; Michener & Brooks, 1984; Neff & Simpson, 1981; Roberts & Vallespir, 1978; Thorp, 1979) and nest construction in females, and with structures presumably related to mating in males. This often results in strong sexual dimorphism, presenting taxonomic problems in associating sexes.

Most bees are solitary creatures in which each female constructs her nest and provisions the individual cells with the food to be consumed by each of her progeny, without the cooperation of other females. In this case there is no separation into castes (workers and queens), and only rarely is there overlapping of generations within a nest. Each female undertakes all the activities of nesting and foraging during her life, which lasts several weeks to a few months (Linsley, 1958), rarely as long as 2 years. Highly eusocial behavior is found only among members of the family Apidae. Communal, quasisocial, or primitively eusocial behavior is found in most other families (Michener, 1969, 1974) and is especially prevalent in Halictidae. The families Halictidae, Megachilidae, and Anthophoridae include a considerable number of "cleptoparasitic" species, in which the female oviposits in the nests of other species of bees and her progeny consume the food designed for the host larva (Michener, 1974). Social parasites are present in Apidae. Most bees excavate their own nests in soil. Others (many Megachilidae, *Hylaeus, Ceratina, Xylocopa*) utilize existing cavities such as hollow stems, beetle burrows in dead wood, or abandoned wasp nests, or they excavate nests in wood (*Lithurge, Ceratina, Xylocopa*). Some groups of Megachilidae create exposed nests of resin or mud (sometimes mixed with pebbles or sticks) on stems or rocks. Bees vary widely in the degree of specialization in pollen collection, ranging from those that collect pollen from only a few related plant species (oligolectic) (Linsley, 1958) to ones that are broad generalists (polylectic). Though most bees are active during the warm middle of the day, a few are matinal, crepuscular, or nocturnal (Linsley, 1958), their activity patterns often related to the timing of pollen presentation in the flowers on which they forage.

The few known fossils shed little light on the history of bees. In fact, the oldest known fossil bee belongs to a genus clearly specialized morphologically (Michener & Grimaldi, 1988a,b): *Trigona prisca* (Meliponinae) from the Cretaceous in New Jersey (96 million to 74 million years old). The only known fossil bee from Mexico from the middle Miocene amber deposits of Chiapas is also a stingless bee (Wille, 1959) belonging to the extant group *Plebeia* (*Nogueirapis*) (Michener, pers. comm.).

Overall phylogenetic relations have yet to be elucidated for Apoidea. Even the standardized, comparative, and detailed morphological terminology needed for basic species descriptions was not available until relatively recently (Michener, 1944). Some specialized studies on particular morphological characters were published subsequently, but it has proved difficult to determine which states of these characters are derived (McGinley, 1980; Michener & Brooks, 1984; Rozen & McGinley,1974).

General treatments of Apoidea for the interested reader include morphology (Michener, 1944), life history (Linsley, 1958; Stephen et al., 1969), foraging and mating (Eickwort & Ginsberg, 1980), social behavior (Michener, 1974), biogeography (Michener, 1979), pheromones and glandular systems (Duffield et al., 1984), ecology and natural history of tropical bees (Roubik, 1989), and fossils (Wille, 1977; Zeuner & Manning, 1976).

Knowledge of native bees predates the arrival of Europeans. The management of stingless bees (Apidae: Meliponinae) was well developed by the native people in various regions of Mexico (Labougle & Zozaya, 1986). The importance of honey and wax as tribute, for trading and religious practices was recorded at the time of the Conquest (Gomarra, 1811; Sahagún, 1960). Recognition of different types of honey (Gomarra, 1811) suggests that more than one species was exploited. At present, the Mayas recognize 12 "species" of Meliponinae (González, 1984) from a total of 15 species present in this area (Ayala, unpublished). Management of meliponines has been recorded from Sinaloa, Guerrero, the Yucatan Peninsula, and Tabasco (Bennett, 1964; Dixon, 1987; González, 1984; Murillo, 1984). *Apis mellifera* was apparently introduced into New Spain from the Old World between 1520 and 1530, but only after 1920 did its management become intensive owing to commercial interest and to the development of standardized hives with interchangeable frames (Labougle & Zozaya, 1986), making it the most important bee species economically. The advent of *A. m. scutellata* (the "African bee") threatens both the honey bee industry (Labougle & Zozaya, 1986) and likely the native fauna (Roubik, 1978, 1983; Roubik et al., 1986).

A few indigenous names have survived to the present, most applying to groups of bees. Large bees are known as "jicotes" in Mesoamerica (from the Nahuatl word "xicotli" (Macazaga, 1982; Wille, 1983). The term "pipiol" (pipiolin, pipiyolin, pipiolo, or pipiola, from Nahuatl) was applied to diverse small bees. In the Purepecha language (or Tarasco) of Michoacan there are names designating bumble bees (uauap, kheri, and kaparhi) and others for species esteemed for their honey and roasted larvae (uauapu, tsitsis, turhipiti, and tinakua) (Velásquez, 1978).

Research in Mexico on bees is recent and has focused primarily on the flora exploited by Apidae (Villanueva, 1984, and citations therein). Bees have received little attention as pollinators of crop species (Búrquez & Sarukhán, 1980, 1984; Quiros & Marcias, 1978), even when only *Apis* is considered. The paucity of studies on the bees of Mexico is surprising, given their importance in the pollination of the native flora (Bawa et al., 1985; Frankie et al., 1983; Simpson & Neff, 1987), their exploitation by man for agricultural production (McGregor, 1976; Parker et al., 1987), and their importance to the understanding of the Nearctic fauna. Many revisionary studies have arbitrarily been limited in coverage to the United States. Unfortunately, even basic taxonomic knowledge of Mexican bees is still incomplete, in part because of large geographical and seasonal gaps in existing collections . Moreover, despite the description of many new species during this century, there are still few generic revisions for major components of the families Colletidae, Halictidae, and Megachilidae as well as for such speciose Neotropical groups as *Ceratina, Xylocopa, Euglossa,* and Meliponinae. A key to the genera of North American bees now in progress (Michener & McGinely, in prep.) will greatly aid in furthering work on Mexican bees.

Interest in the bee fauna of Mexico has come to focus in an informal working group, Programa Cooperativo Sobre la Apifauna Mexicana (PCAM). The group is comprised of Mexican and American workers interested in bee systematics, biology, and ecology. It has provided the opportunity not only to increase our understanding of the fauna through collecting expeditions but has greatly enhanced the interchange of information and ideas.

The work reported in this chapter summarizes present knowledge of the bee fauna of Mexico, the history of its taxonomy, the geographic and seasonal coverage of existing collections, and areas of major diversity for the better-known genera. We comment on factors influencing the evolution of bees and faunal diversity and on the importance of conserving this diversity. Given our present state of knowledge, a considerable part of the discussion is necessarily devoted to pointing to gaps in our understanding of Mexico's bee fauna. In view of the importance of bees in the natural ecosystems at risk, discussion of these gaps seems of value.

METHODS

There are no comprehensive studies for the Apoidea of Mesoamerica, so it was necessary to compile a list of all species recorded in Mexico. An exhaustive review of the literature resulted in a computerized checklist with taxonomic and historical information on all recognized Mesoamerican bees (Ayala & Griswold, unpub.). Species of Central America were included, as the only checklist for this region is for Panama (Michener, 1954), and distributions in Central America were needed to ascertain overall distributional patterns for Mexican bees.

The historical review relies principally on the sequence of publications of original species descriptions. No attempt was made to exclude species originally described based on material from outside Mexico; species now regarded as synonyms are not included. Historical periods were defined with respect to clustering of these dates, changes in active researchers, and the degree of taxonomic specialization of these workers.

Discussion of distributional patterns, areas of diversity, and evaluations of geographical and seasonal distribution of sampling are based on records found in revisionary works on Mesoamerican bees (59 genera, 810 species, Appendix) augmented (for distributional but not seasonal analysis) by unpublished records in the collection of the Bee Biology and Systematics Laboratory, Logan, Utah. A few revisions did not give specific locality records for a few of the most common species, as indicated in the Appendix. In these instances records for other species were included, as widespread species are not particularly useful for elucidating distributional patterns within Mexico. Revisions with limited records were excluded from the analyses. The resulting data base includes 6,162 records from 2,179 localities in Mexico.

Because there are no detailed studies of seasonality for any sites in Mexico, we compared the sum of the number of active species per month with intensive studies in Jamaica (Raw, 1985) and the lowlands of Guanacaste Province, Costa Rica (Heithaus, 1979a,b) as an index to seasonal adequacy of collections.

The terms "diversity" and "richness" are here used interchangeably to refer to the number of species of a given taxonomic group found in a given area or site. By "endemic" we mean that a taxon is encountered solely in the referred area; unless qualified, it is used as found only in Mexico. Mesoamerica is used to include all countries from Mexico to Panama. Zoogeographic regions are those used by Wallace (1876) plus the Madrean region (chaparral and deserts from northern California to Sonora and Chihuahua) (Michener, 1979). To discuss diversity and collecting effort within Mexico we refer to simple physiographic divisions (see Figs. 5.2 and 5.5 below) and a modification of the biotic provinces proposed by Smith

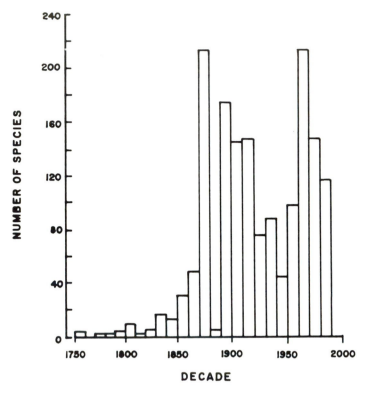

Figure 5.1. History of systematic work on Mexican bees: number of species described per decade.

and by Stuart as revised by Alvarez and de Lachica (1974) but taking into consideration the floristic provinces noted by Rzedowski (1978) (see Fig.5.7 below).

RESULTS

Historical Review

The history of bee systematics in Mexico is almost as old as that of modern systematics. Nearly 100 authors have described more than 1,400 species in the 230 years from Linnaeus up to the present (Fig. 5.1). The historical pattern is similar to that of America north of Mexico (McGinley, 1989) but with smaller numbers of species described.

The first works to include Mexican bees (1758-1819) were characterized by species descriptions in general studies on insects by such European naturalists as Linnaeus, Olivier, Fabricius, and Spinola based on limited collections brought back by early explorers. The 26 taxa described from Mexico during this period were placed in a few broadly defined genera.

For example, Linnaeus described all of his species of bees, including three from Mexico, in the single genus *Apis*.

The second period (1820–1889) represents a transition from general naturalists, who worked on diverse organisms, to the first specialists in Hymenoptera. This period also saw the first involvement by American workers. Cresson (1878) with 174 described Mexican taxa and Smith with 94 species were the first to direct their attention specifically to the Mexican bee fauna. Other important authors of the period include Dours, Guerin, Lepeletier, Radoszkowski, Sichel, and Say.

The years between 1890 and 1929 saw a considerable increase in both the number of specialists (*n* = 33), and the number of species described (*n* = 543), particularly in the prolific works of Cockerell (386 species) and Friese (42 species). The only catalog of the bees of Mexico (Cockerell, 1899), which included 311 species, many new, was published during this period, along with the first generic revisions.

Even greater involvement in the systematics of Mexican bees characterizes 1930 to the present; 44 authors have described 715 species. Five specialists have contributed over three-fourths of these taxa: Hurd, LaBerge, Michener, Snelling, and Timberlake. Timberlake's work focused on the speciose Panurginae and contributed 22% of the bee fauna presently known from Mexico. This period includes the first works expanding the study of Mexican bees beyond the taxonomic arena to studies on behavior (Janzen 1964, 1966), nesting biology (Linsley et al., 1955, 1956; Packer, 1985; Parker, 1977; Rozen & Ayala, 1987; Torchio, 1974), floral relations (Hurd & Linsley, 1964; Linsley et al., 1963) and pollination biology (Bullock et al., 1989; Devall & Thien, 1989). Important catalogs dealing with Mexican bees were also produced during this period: a catalog for New World Halictidae (Moure & Hurd, 1987) and one of Hymenoptera for America north of Mexico (Hurd, 1979) that recorded species extending into Mexico.

Despite the increase in published works and bee specialists, much has yet to be learned about the apifauna of Mexico. Forty-two percent of the genera known to occur in Mexico have yet to be revised, compared to only 19% for the United States. Only a tiny fraction of the species have been studied biologically or ecologically, and most of these studies are on populations outside Mexico. Furthermore, numerous seasonal and geographical gaps remain to be filled before biodiversity patterns can be fully analyzed.

Mexican Apifauna

The bee fauna of Mexico as presently known is composed of 8 families, 153 genera, and 1,589 species (Table 5.1; Appendix). These numbers are undoubtedly conservative, considering the lack of recent revisions for many large genera and the inadequate sampling of this topographically, climatically, and floristically diverse country. We anticipate the real number of species to well exceed 2,000.

Mexican bees show a number of broad geographical patterns. At the generic or tribal levels, these elements: cosmopolitan (*Lithurge, Megachile, Ceratina, Hylaeus, Colletes*), Holarctic (*Bombus, Osmia, Dufourea, Andrena, Lasioglossum, Dialictus*), Nearctic (*Ashmeadiella*), New World (Augochlorini), Madrean (*Perdita, Conanthalictus, Ancylandrena*), Pantropical (*Xylocopa,* Meliponinae), Neotropical (*Centris, Ptiloglossa,* Euglossini), Mesoamerican (*Protoxaea, Deltoptila, Crawfordapis*), and Amphitropical (*Epanthidium, Caupolicana, Martinapis, Protepeolus*).

Four genera are endemic to Mexico (see Fig. 5.3 below): *Paragapostemon* (Halictidae, Halictini), *Aztecanthidium* (Megachilidae, Anthidiinae), *Agapanthinus* and *Loxoptilus* (Anthophoridae, Eucerini). (*Pectinapis*, thought to be a Mexican endemic, has been found in Texas [LaBerge, 1989a; Zavortink, 1982].) *Mexalictus* and *Xenopanurgus* are largely restricted to Mexico, barely extending north of the border into the United States. Genera

Table 5.1. Comparison of Apifaunas in North America

Apifauna	Continental U.S.		Mexico		Costa Rica	Panama	
Colletidae	173	(5)	83	(8)	(6)	15	(5)
Hylaeinae	52	(1)	20	(1)	(1)	2	(1)
Xeromelissinae	—	—	2	(1)	(1)	1	(1)
Colletini	114	(1)	46	(1)	(1)	7	(1)
Paracolletini	1	(1)	2	(1)	—	—	—
Dissoglottini	—	—	4	(1)	—	2	(1)
Caupolicanini	6	(2)	9	(3)	(2)	4	(1)
Oxaeidae	4	(2)	10	(2)	—	—	—
Halictidae	530	(23)	196	(28)	(20)	88	(16)
Rophitinae	104	(7)	20	(5)	—	—	—
Nomiinae	26	(1)	16	(1)	—	—	—
Augochlorini	17	(5)	52	(9)	(8)	53	(7)
Halictini	383	(10)	108	(13)	(12)	33	(9)
Andrenidae	1303	(15)	471	(13)	(5)	4	(3)
Andreninae	448	(3)	89	(2)	(1)	2	(1)
Panurginae	855	(12)	382	(11)	(4)	2	(2)
Melittidae	33	(4)	10	(2)	—	—	—
Melittinae	12	(3)	1	(1)	—	—	—
Dasypodinae	21	(1)	9	(1)	—	—	—
Megachilidae	648	(21)	283	(26)	(13)	64	(7)
Lithurginae	6	(1)	5	(1)	(1)	—	—
Megachilinae	190	(2)	110	(2)	(2)	56	(2)
Osmiinae	311	(9)	75	(8)	(2)	1	(1)
Anthidiinae	141	(9)	93	(15)	(8)	7	(4)

Table 5.1. (cont.)

Apifauna	Continental U.S.		Mexico		Costa Rica	Panama	
Anthophoridae	1003	(48)	455	(56)	(32)	105	(23)
Nomadinae	544	(15)	66	(15)	(6)	14	(5)
Anthophorini	64	(2)	30	(3)	(2)	—	—
Centridini	20	(1)	56	(2)	(2)	32	(3)
Emphorini	37	(3)	27	(3)	(1)	2	(2)
Ericrocini	3	(2)	11	(4)	(4)	6	(3)
Eucerini	224	(17)	116	(18)	(9)	9	(5)
Exomalopsini	52	(2)	81	(4)	(4)	12	(2)
Melectini	17	(4)	5	(3)	—	—	—
Rathymini	—	—	1	(1)	(1)	—	—
Tetrapediini	—	—	4	(2)	(1)	1	(1)
Xylocopinae	42	(2)	58	(2)	(2)	29	(2)
Apidae	57	(4)	81	(18)	(17)	110	(18)
Meliponinae	—	—	29	(11)	(11)	48	(11)
Euglossinae	1	(1)	30	(4)	(4)	53	(5)
Bombinae	49	(2)	21	(2)	(1)	4	(1)
Apinae	1	(1)	1	(1)	(1)	1	(1)
TOTAL	3745	(122)	1589	(153)	(93)	387	(72)

Data are given as the number of species and subspecies for each country followed by the number of genera in parentheses. Figures for the United States include the continental area only. No complete list of Costa Rican species exists, so generic totals only are presented. *Sources:* United States: Hurd (1979), modified by more recent revisions; Mexico: unpublished checklist, Ayala & Griswold; Costa Rica: unpublished checklist, Griswold, plus recent collections by F. D. Parker; Panama: Michener (1954), modified by more recent revisions.

that have a major center of diversity in Mexico include *Centris, Deltoptila, Exomalopsis, Mexalictus, Xenoglossa,* and *Protoxaea.* The 10 largest genera in Mexico are *Perdita* (226 species), *Andrena* (88), *Megachile* (77), *Exomalopsis* (74), *Centris* (53), *Melissodes* (47), *Colletes* (46), *Heterosarus* (44), *Protandrena* (40), and *Coelioxys* (33). (It is notable that all of these genera except *Megachile, Colletes,* and *Coelioxys,* have been revised.)

Solitary, parasocial, and primitively eusocial pollen-collecting bees are the most diverse, with 68.4% of the total number of genera and 88.1% of the species, whereas the respective percentages for cleptoparasites are 19.5% cent and 8.5% and for highly eusocial bees 12.1% and 3.4%.

Representative Groups

The following comments on some representative groups for each family illustrate the patterns of diversity of bees in Mexico. (Parenthetical numbers throughout the text below refer to the number of species known to occur in Mexico; numbers presented as [x/y] refer to species in Mexico/ total for the genus, where there is a reliable estimate for the latter.)

Andrenidae

The family Andrenidae consists of small to moderate-sized solitary bees that construct their own nests in the soil. It is well represented in the Nearctic, diminishing rapidly in both genera and species as one moves southward into the tropics (Table 5.1; Appendix). It is one of the largest families of bees in Mexico and is certainly the best understood systematically, with all but one of its 13 Mexican genera revised. Andrenidae is characterized by a high degree of endemicity and a strong tendency toward oligolecty (LaBerge, 1986b; Moldenke, 1979).

The subfamily Panurginae is particularly well represented in Mexico with 11 genera. It is not well represented in the tropics (Michener, 1979), though it is diverse again in temperate South America. *Perdita* (194/612) is the most speciose genus of Apoidea in both the Nearctic and Mexico, with its greatest diversity in xeric regions such as the Sonoran (74) and Chihuahuan (52) Deserts. Subtropical areas along the Pacific Coast also have a diverse *Perdita* fauna including the subgenus *Callomacrotera*, which is endemic to coastal dunes. The small panurgine genera *Hypomacrotera* (3/ 4) and *Metapsaenythia* (1/3) are limited to northern deserts, and *Nomadopsis* (9/46) is limited to deserts plus mesic Baja California. *Calliopsis* is almost equally represented in the United States and Mexico (Table 5.2), occupying a diverse spectrum of habitats.

In contrast to such genera as *Perdita* and *Nomadopsis*, which are best represented in the United States (Appendix), the complex of panurgines, which includes *Heterosarus*, *Pterosarus*, *Protandrena*, and *Pseudopanurgus*, is best represented in Mexico (Table 5.2) and has the highest degree of endemism, at almost 70% of the species. These genera are better represented in subtropical to tropical zones of both coasts; over half of the andrenids from these regions belong to these genera. *Xenopanurgus* is endemic to the Altiplano, extending marginally into southeastern Arizona.

Andreninae is not as well represented in Mexico as is Panurginae and is limited to temperate environments. Two small genera of vernal bees, *Ancylandrena* and *Megandrena*, are endemic to the Sonoran and Mojave Deserts and adjacent cismontane southern California. *Megandrena* has not yet been recorded in Mexico, but should occur in Sonora on *Larrea*.

The primarily Holarctic genus *Andrena* (Andreninae) is represented almost equally in both hemispheres; in the New World it extends only as far south as Panama, whereas in the Old World it reaches South Africa (Michener, 1979). In the Nearctic and in Mexico, it is the second largest genus of bees (529 species in the United States and Canada, 95 species in Mexico) with its maximum diversity found in the West Coast and Western Montane Regions of the United States, and the least number of species, but

Table 5.2. Comparison of North American fauna for select genera

Fauna	Continental U.S.	Mexico	Costa Rica	Panama
Temperate element				
Andrena	442	88	1	2
Calliopsis	27	24	2	1
Perdita	657	226	—	—
Lasioglossum s. str.	29	29	5	2
Dolichostelis	5	5	1	—
Melissodes + *Svastra*	123	61	1	3
Bombus	42	19	6	4
Neotropical element				
Augochlorella	7	7	2	2
Osiris	—	5	8	9
Centris	20	53	31	23
Epicharis	—	3	8	7
Eufriesea	—	7	12	16
Eulaema	1	5	10	9
Mesoamerican element				
Mesoxaea + *Protoxaea*	4	10	—	—
Heterosarus	17	44	2	—
Pseudopanurgus	14	28	2	—
Protandrena	22	40	—	—
Exomalopsis	50	74	10	7

Sources as in Appendix.

highest percentage of endemism (78%) in Mexico and Central America (LaBerge, 1986b). In contrast to *Perdita*, it is most abundant in mesic parts of Mexico principally in areas above 1,500 m. *Andrena* only marginally enters the tropics as far south as Panama (Table 5.2).

At the subgeneric level several faunal elements in *Andrena* are apparent. The most diverse component consists of temperate, spring-flying groups that enter Mexico only in the parts of Baja California Norte that have a Mediterranean climate; these groups account for 11 of the 25 subgenera recorded from Mexico. (Thirteen additional subgenera that are found in San Diego County, California, will likely be added to this list when the fauna of mesic northern Baja California is better known. The vernal montane fauna of northeastern Mexico, when sampled, should also significantly increase the number of species and probably subgenera known from Mexico.) An endemic element is represented by the monotypic subgenus *Celetandrena*. This resident of the Valley of Mexico is apparently a matinal bee oligolectic on the cucurbit *Sicyos* (LaBerge & Hurd, 1965). *Andrena toluca*, also from the Transverse Volcanic Belt, is a New World disjunct of the otherwise Eurasian subgenus *Charitandrena* (LaBerge, 1969). *Callandrena*

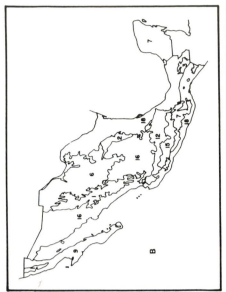

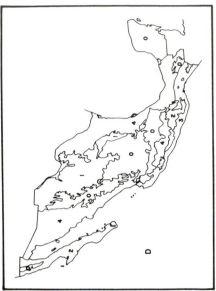

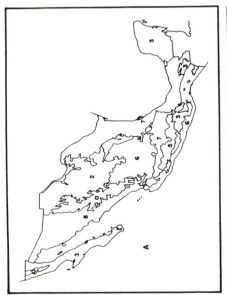

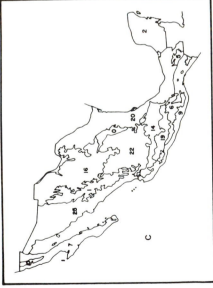

is the most abundant subgenus in Mexico, with nearly half the species of *Andrena* and 21 of 29 endemics. It seems likely that the switch from spring season, forest flora relations to later season Asteraceae (LaBerge, 1986b) provided the opportunity to expand into the predominantly summer/fall flowering regime of the altiplano. Most of the 38 Mexican species of *Andrena* (*Callandrena*) are endemic to the altiplano or marginally extend either into the mountains of southeastern Arizona and southwestern New Mexico or down into lower elevation subtropical zones.

Anthophoridae

The tribe Eucerini has several speciose genera in Mexico, some polylectic, others specialists. *Peponapis* (10/30) (Fig. 5.2A) and *Xenoglossa* (6/7) are oligolectic on Cucurbitaceae, collecting pollen during the early morning sometimes before light (Hurd & Linsley, 1964; Hurd et al., 1971). Both seem to have centers of abundance in Mexico (Hurd & Linsley, 1970), with recent extensions to the north and northeast following the dispersal of cultivated squashes by man (Hurd et al., 1971). The amphitropical *Martinapis* is known only from deserts and semideserts of southwestern North America and Argentina where it is a polylege active only during the early morning and evening (Zavortink & LaBerge, 1976). *Pectinapis* (5/5) is known from the mountains of the Eje Neovolcanico plus isolated records from Texas (LaBerge, 1957, 1989a; Zavortink, 1982) (Fig. 5.3A), whereas *Loxoptilus* (2/2) is known solely from the basin of the Rio Balsas and the lowlands of Nayarit (LaBerge, 1970) (Fig. 5.3B). The monotypic genus *Agapanthinus* is known only from San Jose Island in the Gulf of California (Fig. 5.3B).

The species of *Diadasia* (Emphorini) are oligolectic, with five species groups each specializing on members of a different plant family: Malvaceae, Cactaceae, Convolvulaceae, Asteraceae, and Onagraceae (Linsley & MacSwain, 1958). Species of all groups except the last are present in Mexico. *Diadasia* probably occupies all of the dry areas of Mexico; the genus has not been revised. *Melitoma*, which is oligolectic on *Ipomoea* (Hurd, 1979; Michener, 1954), is unrevised but appears to be widespread in Mexico.

Centridini is a clearly Neotropical tribe of three genera. *Ptilotopus* extends north only as far as Panama (Snelling, 1984), *Epicharis* (3) into Mexico, and *Centris* (53) into the United States where it is limited to the warm Southwest, except for two species, which range east into Kansas and Florida. Subgeneric diversity is greatest in Mexico (Panama 8 subgenera; Costa Rica 8; Mexico 11; United States 5). All *Epicharis* and most *Centris* apparently utilize floral oils in their provisions (Neff & Simpson, 1981) and possess specialized combs composed of modified setae on the legs to aid in

---

Figure 5.2. Species richness by physiographic region for *Peponapis* (A), Centridini (B), *Exomalopsis* (C), and Oxaeidae (D).

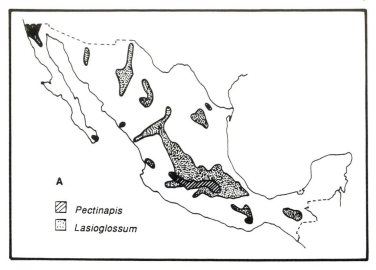

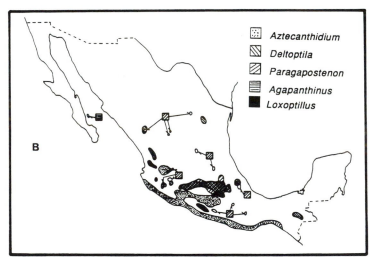

Figure 5.3. Distribution of genera of bees. A. *Pectinapis* and *Lasioglossum*.
B. Endemic genera: *Loxoptilus, Aztecanthidium, Paragapostemon, Agapanthinus,*
and the Mesoamerican *Deltoptila.*

their collection. Neff and Simpson (1981) described three types of these
structures for Mesoamerican *Centris*. Most species have a "typical" comb
pattern (as found in *Epicharis*) related to collecting oils from the epithelial
elaiophores of Malpighiaceae. (Some South American *C.* [*Paracentris*] and
C. [*Hemisiella*] have a modification of this typical pattern in which the
posterior setae are more enlarged, thus allowing collection of oils from
flowers with trichome elaiophores as well as those with epithelial
elaiophores. According to Neff [pers. comm.] Mexican members of these
two subgenera lack this modification even though they utilize flowers with

trichome elaiophores.) The monotypic Mexican subgenus *Exallocentris* has dense pads of finely branched setae in place of the combs The function is unknown but is assumed to be related to oil collection. Finally, species of the subgenus *Xerocentris* have the combs reduced or absent. They do not collect oils. Oil collection is considered ancestral in *Centris*, with the genus postulated to have its origin in lowland tropical regions, with secondary radiations into montane and xeric habits accompanying shifts to flowers with trichome elaiophores and non-oil flowers, respectively (Neff & Simpson, 1981). These radiations resulted in the rich fauna of Mexico, in the northwest dominated by exclusively nectar- and pollen-gathering species and in the northeast and south by those that also collect oils.

It is possible that *Centris* finds its greatest diversity in Mexico, though it is uncertain, as there has been no revision of the South American forms. As many as 14 species of *Centris* may be encountered at a single site in tropical regions (Ayala, 1988). There are few species in the high mountains of Mexico. It is unclear why, as *Centris* is well represented in the Andes to elevations over 3,000 meters (Neff, pers. comm.). *Centris* is also poorly represented on the Yucatan and Baja California peninsulas (Fig. 5.2B).

Exomalopsis (Exomalopsini) (68) belongs to a tribe with much endemism in South America (Michener, 1979; Michener & Moure, 1957). It is present in all of the states of Mexico but with maximum diversity in the Altiplano and southern Sonora (Fig. 5.2C). (This datum may be an artifact related to intensity of collecting effort.) Most of the species appear to have restricted distributions (Timberlake, 1980a), creating distinct regional faunas. This apparent endemism needs to be checked against more widespread collecting throughout Mexico. Another exomalopsine, *Ancyloscelis*, whose Mexican species are oligolectic on *Ipomoea*, ranges widely from the southern United States to Argentina (Michener & Moure, 1957). It has not been well studied in Mexico; a number of undescribed species exist, concentrated in the tropical Pacific lowlands and Balsas Basin. *Paratetrapedia* and *Monoeca*, the other two exomalopsines in Mexico, and the distantly related *Tetrapedia* (Tetrapediini), collect oils (Neff & Simpson, 1981). They constitute a significant component of the fauna in the wet tropics (Michener, 1979). None has been revised.

The tribe Anthophorini is represented in Mexico by three genera. *Anthophora* and *Habropoda* are temperate elements with their greatest New World diversity in the United States. *Deltoptila*, on the other hand, is endemic to Mesoamerica (Fig. 5.3B), where it is restricted to montane environments, probably with greatest diversity in Mexico.

The subfamily Nomadinae consists exclusively of cleptoparasitic species. Known and suspected hosts were summarized by Alexander (1990). Of the three most speciose genera, *Triepeolus* and *Epeolus* have not been revised, and only two small subgenera of *Nomada* have been revised, so nothing can be said of their distributions. (All are certainly more diverse than their numbers recorded in Appendix indicate. For example, there are four times the recorded number of *Triepeolus* species in the Logan collection alone.)

Large (*Xylocopa*) and small (*Ceratina*) carpenter bees   constitute the subfamily Xylocopinae in Mexico. *Xylocopa* includes the largest bees in Mexico, some attaining a length of more than 3 cm. Both genera have more species in tropical than in temperate regions. Neither has been revised with the exception of the subgenus *X.* (*Notoxylocopa*). *X.* (N.) *tabaniformis* has been divided into a number of allopatric subspecies that correspond well with physiographic regions in Mexico (O'Brien & Hurd, 1965). Janzen (1964, 1966) reported on the behavior of two Mexican *Xylocopa*.

Apidae

The largely social Apidae is the family best represented in lowland tropical environments and the only one whose proportional representation increases in tropical climes (Roubik, 1989). Only the tribe Bombini, *Bombus* and its cleptoparasite *Psithyrus*, and the introduced honey bee *Apis mellifera* occur in temperate climates. *Bombus* is most diverse in the Palearctic and dominates the bee fauna north of 60° latitude. The species in Mexico are found principally in mountainous areas forested with *Pinus*, *Quercus*, and *Abies*, with as many as six sympatric species recorded (Labougle, 1990). In tropical and arid climates it is absent or represented by a single species.

Meliponinae and Euglossinae are the only subfamilies or tribes of bees with greater diversity in Central America than in Mexico (Table 5.1). Stingless bees (Meliponinae) are Pantropical, found predominantly in lowland areas, with their greatest diversity in the Neotropics (Michener, 1979). Up to 26 species co-occur at typical sites in evergreen tropical forests of Mexico (Palenque in Chiapas) (Ayala, unpublished), whereas eight species are present at a single site in drought-deciduous tropical forests (Chamela, Jalisco) (Ayala, 1988). One species of *Melipona* does enter pine and pine-oak forests up to 3,000 m. Latitudinally, Meliponinae penetrates as far north as 29° in Sonora (Bennett, 1964) and 24°30' in Nuevo León along the Gulf Coast (Ayala, in prep., Neff, pers. com.). The only revision of Mexican meliponines (Schwarz, 1949) is out of date; a number of endemic new species exist (Ayala in prep.). The complex biology of meliponines has been reviewed by Wille (1983) and Roubik (1989).

Orchid bees (*Eulaema*, *Eufriesea*, *Euglossa*, and their cleptoparasites *Exaerete* and *Aglae*; Euglossinae) are restricted to the Neotropics. Their numbers decrease with increasing latitude (Tables 5.1 and 5.2) despite the great disparity in land mass between Central American countries and Mexico. *Euglossa* has been incompletely revised, but euglossines do not appear to have levels of endemism in Mexico comparable to that in Central America.

Colletidae

The two largest genera of colletids, *Colletes* (Colletini) and *Hylaeus* (Hylaeinae), are virtually cosmopolitan (though *Colletes* does not reach

Australia) and are widespread throughout Mexico. The size of their faunas in Mexico is unknown, as neither has been revised for Mesoamerica. *Eulonchopria* (Paracolletini) (2/4), an amphitropical member of a tribe otherwise restricted to South America and Australia, is presently known in Mexico only from tropical and subtropical regions of the Pacific Coast, marginally extending into the Transverse Volcanic Belt. It may occur on the lower slopes of the Sierra Madre Occidental, as it is known to occur in southeastern Arizona. Long-distance dispersal is thought unlikely for this or other amphitropical bees. Simpson and Neff (1985) suggested short hops between arid pockets as a reasonable alternative but noted the absence of members in modern arid pockets. Collections of *Eulonchopria* from arid sites in Nicaragua, Venezuela, and Ecuador have recently been found (Griswold, unpub.), supporting their hypothesis.

Diphaglossinae, which is limited to the Americas, is well represented in Mexico, with four of the nine genera present. Diphaglossines reach their greatest diversity in the Transverse Volcanic Belt. They are poorly represented on the Yucatan Peninsula and are not known from the northwest (Sinaloa and Sonora) but likely occur there because they are present in southern Arizona. The monotypic genus *Crawfordapis* is endemic to the mountains of Mesoamerica, ranging from Panama to north of the Isthmus of Tehuantepec (Roubik & Michener, 1985). Species of *Mydrosoma* are rarely collected. All four Mexican species are endemic, two of them known from single sites on the Pacific slope (Michener, 1986). Only *Ptiloglossa* is common in tropical areas.

Halictidae

The family Halictidae, made up of the three subfamilies Rophitinae, Nomiinae, and Halictinae, constitutes an important component of bees in tropical as well as temperate parts of Mexico. The solitary, ground-nesting rophitines are found only in temperate parts of Mexico, particularly along the northern border; only *Dufourea* is known to extend southward in the altiplano to central Mexico. Five of the seven North American genera are presently known from Mexico (Appendix), but the other two genera should be found when northwestern Mexico has been better collected.

Nomiinae also occurs primarily in the temperate zone, but *Nomia* (*Curvinomia*) extends south to the margins of the tropics. Nomiines nest in the ground, often gregariously. One has been used commercially as a pollinator of alfalfa (Parker et al., 1987).

The strong contribution of the two halictine tribes Augochlorini and Halictini makes Mexico the richest in Halictidae of any country in North America at the generic level (Table 5.1). The tribe Augochlorini is found exclusively in the Americas, principally in the Neotropics. Specific patterns for the species-rich augochlorines are difficult to analyze because *Augochlorella* is the only one of the nine Mexican genera that has been revised, but at least *Augochloropsis* is an important component of the fauna in Mexico (Eickwort, pers. comm.). *Caenaugochlora* (7/14) is apparently oligolectic on *Cucurbita* (Moure & Hurd, 1987) and is found from Mexico to Ecuador, in Mexico occupying lowland and tropical environments as well as the mountains. The polylectic members of *Augochlorella* (6/16) range even more widely, from Canada to Argentina, and in Mexico are widely distributed apparently without respect to climate, altitude or vegetation type. Most species are widespread, but *A. maritima* is restricted to the Pacific coastal zone of mainland Mexico. (*A. maritima* has in the past been considered a subspecies of the widespread *A. neglectula* [Ordway, 1966] but is a valid species. In addition to the characters cited in the revision [Ordway, 1966], the shape of the tegula is unique.)

Lasioglossum (s. str.) is a holarctic member of the cosmopolitan tribe Halictini extending southward at high elevations to northern Panama. Of its 34 American species, 29 are found in Mexico (McGinley, 1986) primarily in cold to temperate climates (Fig. 5.3A). Nearly half are endemic. *Mexalictus* (2/3) is known only from the Altiplano northward to the Huachuca and Santa Catalina Mountains of southeastern Arizona (Eickwort, 1978). *Agapostemon* is found throughout Mexico and includes both widespread species and narrow endemics (Roberts, 1972). Species seem to be of both North American and South American stock (Roberts, 1972). All appear to be polyleges. Two Mesoamerican genera related to *Agapostemon*— *Dinagapostemon* and *Paragapostemon*—have rarely been collected; perhaps females are matinal, crepuscular or both (Roberts & Brooks, 1987). The monotypic *Paragapostemon* is endemic to the Mexican plateau (Fig. 5.3B). Both Mexican species of *Dinagapostemon* are endemic. *Dialictus* is the most diverse halictid genus in Mexico (Eickwort, pers. comm.), but it has not been revised. The unrevised parasitic genus *Sphecodes* is also much more speciose than the 12 recorded species imply (Appendix).

Megachilidae

The subfamily Megachilinae is the richest group of Megachilidae in Mexico, with *Megachile* and its cleptoparasite *Coelioxys* well represented in both temperate and tropical regions, though few subgenera have as broad an ecological tolerance. Neither genus has been revised.

Lithurge is the only North American representative of the subfamily Lithurginae, with almost equal representation in the southern United States and northern Mexico (Snelling, 1983). *Lithurge planifrons* has the widest distribution, extending from southern Arizona to Costa Rica. *Lithurge bitorulosa* is endemic to the west coast of Mexico from Jalisco to Oaxaca. All North American species provision their nests with Cactaceae pollen (Snelling, 1983).

Although there are no genera or subgenera of Osmiinae endemic to Mexico, osmiines are particularly rich in the Madrean Region and include a number of groups endemic to the arid regions of the southwest United

States and northern Mexico: *Hoplitis* (*Eremosmia*), *H.* (*Isosmia*), *H.* (*Dasyosmia*). Groups with their greatest diversity in this region are *Ashmeadiella* and *Proteriades*. The high degree of plant specificity found in Osmiini, coupled with the ability to adapt to xeric conditions, seems to have contributed to the diversity at both generic-subgeneric and species levels. With the exception of *Heriades* and one or two species of *Ashmeadiella*, osmiines are absent from tropical environments.

The Mexican component of the subfamily Anthidiinae is rich. Fifteen genera are known in Mexico compared to nine in the United States (Table 5.1). Though more species are currently known in the United States, the number of Mexican species may well be greater; most of the Mexican genera have not been revised (Appendix). Anthidiinae shows the greatest diversity of distributional patterns of any Megachilidae. They include widespread temperate elements such as the cosmopolitan (except for Australia) genus *Anthidium*, the Holarctic genera *Stelis*, *Protostelis*, and *Dioxys*, and the Nearctic subgenera *Trachusa* (*Heteranthidium*), *Dianthidium* s.s., and *Paranthidium* s.s. Neotropical genera (*Hypanthidium*, *Hoplostelis*) and amphitropical disjuncts (*Epanthidium* and *Anthodioctes* [*Nananthidium*]) are both apparently restricted to tropical deciduous forests. *Anthidiellum* and *Dolichostelis* contain both Nearctic and Neotropical species, but neither extends into South America. Groups limited to Mesoamerica include *Paranthidium* (*Rapanthidium*), restricted to the highlands, and four Mexican endemics: *Aztecanthidium* (Fig. 5.3B), *Dianthidium* (*Deranchanthidium*), *D.* (*Mecanthidium*), and *Trachusa* (*Ulanthidium*). (The last two subgenera barely extend beyond the boundaries of Mexico into southeastern Arizona and southwestern New Mexico, respectively.)

Melittidae

The family Melittidae is restricted to temperate regions of the world. Only two of the five North American genera are known from Mexico. *Hesparapis* is most abundant in the southwestern United States but does extend in more xeric areas to south of Mexico City. *Melitta* is confined in Mexico to mesic northern Baja California.

Oxaeidae

The small family Oxaeidae, with only four genera, is restricted to the Americas (Michener, 1979). *Oxaea* is a genus of the lowland tropics of South America and apparently Mexico (Michener & McGinley, in prep.). *Protoxaea* (10/10), including the questionably distinct *Mesoxaea*, is restricted to Mexico and the southwestern United States (Hurd & Linsley, 1976), with its greatest diversity in the warm lowlands of Mexico except for the Yucatan Peninsula (Fig. 5.2D). Species of *Protoxaea* and *Oxaea* are narrowly polylectic

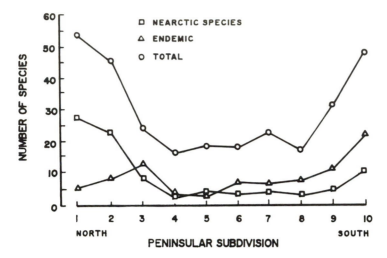

Figure 5.4. Species richness along the Baja California Peninsula for the entire fauna and Nearctic and endemic elements, by 1° latitude subdivisions. (The Nearctic element includes only species found also in the United States but not in mainland Mexico.)

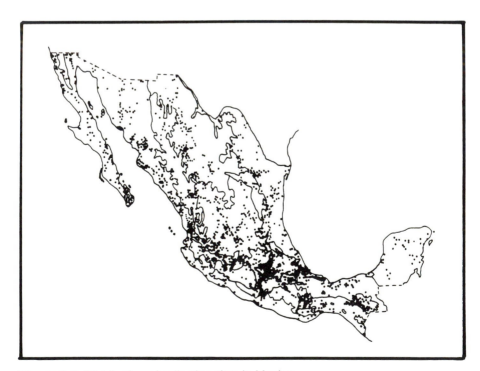

Figure 5.5. Distribution of collecting sites in Mexico.

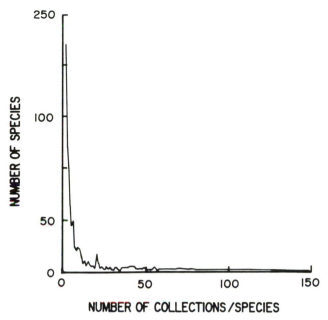

Figure 5.6. Number of reported collections per bee species.

matinal bees, which seemingly prefer plants with poricidal anthers and no nectar, e.g., *Cassia* and *Solanum*, as their pollen sources (Camargo et al., 1984; Hurd & Linsley, 1976), but use many other plants as well.

DISCUSSION

Biogeographical Patterns

The rich bee fauna of Mexico is in part the result of its unique position at the juncture of the Nearctic and Neotropical regions, with strong contributions of both temperate and tropical elements plus a unique Mesoamerican component (Tables 5.2, 5.3; Fig. 5.3). This is evidenced by the larger number of genera when compared with the continental United States, despite the apparently smaller number of species (Table 5.1). Twenty biotic provinces are tentatively identified (Fig. 5.7). Any conclusions on overall diversity or biogeographical patterns must remain tentative, however, as the distribution of collecting shows major gaps (Fig. 5.5), and the number of reported collections per species for the Mexican fauna is poor (Fig. 5.6).

Xeric Regions

The deserts of North America are rich in bees (Michener, 1979; Moldenke, 1979), with particularly large complements of Panurginae (Andrenidae),

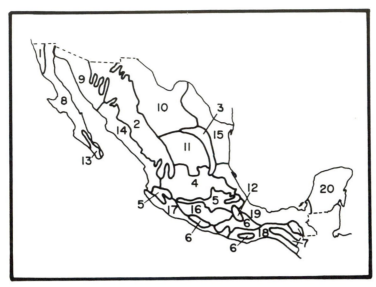

Figure 5.7. Biotic provinces of Mexico. 1, California; 2, Sierra Madre Occidental; 3, Sierra Madre Oriental; 4, Transverse Volcanic Belt; 5, Altiplano Sur; 6, Sierra Madre del Sur; 7, Serranias Transistmicas; 8, Baja California; 9, Sonora; 10, Chihuahua; 11, Altiplano Norte (Saladin); 12, Tehuacan; 13, San Lucas (Cape); 14, Sinaloa; 15, Balsas; 16, Tamaulipas; 17, Guerrero; 18, Tehuantepec; 19, Veracruz; 20, Yucatan. From Alvarez & de Lachica (1974), following Smith (1941) and Stuart (1964).

Dufoureinae (Halictidae), Osmiinae (Megachilidae), Eucerini (Anthophoridae), and Exomalopsini (Anthophoridae). There are five desert areas in Mexico: Baja California, Sonora, Chihuahua, Altiplano Norte, and Tehuacan (Fig. 5.7). Mesic elements such as *Andrena* (Andrenidae) and Bombini (Apidae) are poorly represented and many neotropical groups are marginal or absent (Diphaglossinae, Colletidae; Augochlorini, Halictidae; Euglossinae and Meliponinae, Apidae). Numerous genera and subgenera are endemic to North American deserts or are largely restricted to xeric regions, including two small genera of Andreninae (*Ancylandrena* and *Megandrena*), several Eucerini (*Agapanthinus, Martinapis*—also found in xeric South America—and *Simanthedon*), the melittid *Xeralictoides*, and four of the seven genera of Rophitinae (*Conanthalictus, Michenerula, Sphecodosoma*, and *Xeralictus*). These patterns are based largely on records from the United States. The fauna of the Mexican portions of the deserts is poorly known (Table 5.3); genera found just north of the Mexico–United States border, e.g., *Megandrena, Conanthalictus, Xeralictus*, and *Xeralictoides*, have yet to be recorded in Mexico.

There appears to be a significant faunal distinction between the Sonoran Desert (including Baja California) and the Chihuahuan Desert. Of the 181 Sonoran and 169 Chihuahuan species analyzed, only 38 are common to both. The small sampling effort in the deserts of Mexico (Table 5.3; Fig. 5.7) likely exaggerates the differences in their faunas; yet bees should be expected to mirror the distinctive nature of their floras (Rzedowski, 1973). The isolated Tehuacan Desert (Fig. 5.7) needs to be studied before its relationship to other desert areas can be analyzed.

The absence of strong ties between desert regions also suggests isolated development of their faunas rather than fragmentation of a "proto-desert." This idea is concordant with the view of Axelrod (1979) that deserts are of recent origin, with their distinctive vegetation evolved from mesic types as a response to increasing local aridity. The bee fauna also reflects the greater affinity of the Sonoran flora to subtropical and tropical zones to the south and the greater similarity of the Chihuahuan flora to montane floras (Rzedowski, 1973).

Mesic Regions

Mesic habitats include the southward extension of California's mediterranean climate into northwestern Baja California, the higher elevations of the Mexican Plateau (except for interior deserts), and the southward extension of the Cordillera into Central America (Fig. 5.7). With the exception of the Transverse Volcanic Belt, they have not been well collected by comparison to xeric regions (Table 5.3). *Paragapostemon, Mexalictus,* and *Deltoptila* are endemic here. *Colletes, Hylaeus, Andrena, Dufourea,* Halictini, Osmiinae, *Anthidium, Anthophora,* and Bombini are groups well represented in these environments in the north, less so in the south. The discontinuous nature of montane habitats to the south, with low elevation barriers isolating the Sierra Madre del Sur and the mountains of Chiapas, may account for the attenuation of montane elements, as has been shown for carabid beetles (Ball, 1968). Only one of the 13 montane species of *Andrena (Callandrena)* extends southward into the Sierra Madre del Sur; none is known to reach the mountains of Chiapas. *Andrena (Cnemidandrena),* the montane *Bombus huntii,* and seven of nine montane-restricted *Lasioglossum* are apparently absent from both southern montane regions. These indications of the effects of barriers and small areas need to be tested by more extensive collecting (Table 5.3).

Subtropical Regions

The faunas of both coasts and the Balsas Basin are a rich mixture of Nearctic, Mesoamerican, and Neotropical species. Such groups as Oxaeidae and *Centris* have their greatest diversity here (Fig. 5.2B,D). Sinaloa (Fig. 5.7) appears to be a particularly rich area (Table 5.3) perhaps due to the broad contact zone with the rich Sonoran Desert fauna. Tamaulipas (Fig. 5.7), although sharing many of the same generic elements, appears to have a

Table 5.3. Evident species richness and measures of sampling effort for Mexican biotic provinces[a]

Region and Province	Species	Mexican endemics	Endemics unique to province	Number		Species/site (ave. No.)
				Collecting sites	Species-site[b] collections	
Mesic						
California	66	15	10	55	119	2.2
Sierra Madre Occidental	129	43	13	119	363	3.1
Sierra Madre Oriental	40	13	1	34	70	2.1
Transverse Volcanic Belt	165	75	10	166	915	5.5
Altiplano Sur	204	123	41	334	523	1.6
Sierra Madre del Sur	80	39	9	60	168	2.8
Serranias Transistmicas	46	10	3	62	124	2.0
Xeric						
Baja California	111	33	20	113	296	2.6
Sonora	109	21	13	75	262	3.5
Chihuahua	169	47	25	147	585	4.0
Altiplano Norte (Saladin)	85	32	10	46	186	4.0
Tehuacan	36	17	2	17	47	2.8
Subtropical						
San Lucas (Cape)	54	24	15	70	157	2.2
Sinaloa	171	57	15	196	665	3.4
Balsas	135	79	27	149	576	3.9
Tamaulipas	45	11	1	47	106	2.3

Table 5.3. (cont.)

Region and Province	Species	Number				
		Mexican endemics	Endemics unique to province	Collecting sites	Species-site[b] collections	Species/site (ave. No.)
Tropical						
Guerrero	84	44	7	47	235	5.0
Tehuantepec	72	26	4	81	180	2.2
Veracruz	61	21	5	108	222	2.1
Yucatan	19	2	0	19	44	2.3

[a]Based on revised genera; see Appendix.
[b]Species-site is defined as a single collection of a species.

depauperate fauna (Table 5.3). Inadequacy of the collecting effort cannot be overlooked, but the disparity between the rich west coast and the relatively poor east coast appears real for this and tropical climates to the south. The fauna of the San Lucas region of Baja California is unique (see below).

Tropical Regions

The bee fauna of the tropical regions of Mexico is probably the least well known of any in Mexico. It does not appear to be as rich as the fauna of tropical Central America. The Yucatan Peninsula (Fig. 5.7)in particular has a depauperate fauna (Table 5.3). Characteristic taxa of this region include Augochlorini, *Tetrapedia, Paratetrapedia, Monoeca, Epicharis*, Euglossinae, and Meliponinae. Dominant bee groups in this region have yet to be revised and collecting has been sparse, so little can be said about distributional patterns within this region.

Baja California

Although the bees of Baja California are still not well known, they appear to constitute a distinctive fauna, with approximately 70% of the species not found elsewhere in Mexico. Of 197 species (in 30 recently revised genera) recorded in the peninsula, 57 (29%) are endemic, comparable to the 23% figure given for flowering plants (Wiggins, 1980). In addition to this high degree of endemism, there is a large contribution from the mediterranean region to the north; 16% of the peninsular species sampled are restricted to this climate in northwestern Baja California. Of the nonendemic species only 52% occur elsewhere in Mexico, whereas 96% are found in the United States, almost all of them in California. The relatively low percentage of bees in common with those on the mainland may be due in part to disparities in sampling effort, but it seems largely due to the distinctive vegetation of the peninsula. In addition to the distinctive mediterranean zone and the San Lucas region with its high degree of endemism, two subdivisions of the Sonoran Desert, Vizcaino and Magdalena Plain, are restricted to the peninsula (Turner & Brown, 1982). The poorly collected Vizcaino region should have a distinctive spring bee fauna owing to its predominantly winter rainfall regime compared to the primarily summer rainfall of other parts of the Sonoran Desert (Turner & Brown, 1982).

The applicability of the "peninsular effect," where species richness is expected to decrease as one proceeds from the base to the tip of a peninsula, has been called into question for at least some groups of organisms in Baja California (Brown, 1987). Plotting the distribution of butterflies onto ten 1° latitude segments of the peninsula did not yield a southward attenuation of species. Rather, both the northern base and the southern San Lucas were species-rich, with the middle of the peninsula depauperate. Analysis of bee distributions shows a similar bimodal pattern (Fig. 5.4), though the apparent causes seem divergent. Brown (1987) suggested the San Lucas region (Fig. 5.7) as a pool of primarily Neotropical species compared to the principally Nearctic-derived fauna of the northern base. Bees similarly show a strong Nearctic component in the northern part of the peninsula (Fig. 5.4) in large part due to the southward extension of the mediterranean zone, but there does not seem to be a significant Neotropical component in the San Lucas region. Rather, the diversity of bees in the south appears to result largely from a high degree of endemism (Fig. 5.4) plus a contingent of Nearctic species reaching their farthest southward extension here. Together these species account for 64% of those recorded from the southernmost subdivision. The paucity of Neotropical bees in the San Lucas region is surprising given the past geologic and vegetational connection with subtropical portions of the mainland (Axelrod, 1979), the relative proximity of the Sinaloan coast (Fig. 5.7) with its rich component of Neotropical bees, and the vagility of bees.

Vicariance

There are no biogeographical analyses for any Mexican bees. A review of cladograms for a wide spectrum of other insects in North America (Noonan, 1988) suggested four zones of disjunction in Mexico separating allopatric sister taxa: (1) Sierra Madre Occidental and Sierra Madre Oriental separating western and eastern lowland areas; (2) Central Gulf Coast, separating northeastern Mexico from coastal regions south; (3) a transverse central Mexico zone separating north and north-central Mexico from a south-central area; and (4) Pacific slope of southwestern Mexico separating zones to the north and south. However, vicariance accounted for only a small fraction of geographical patterns of the insect groups studied. Noonan (1988) concluded that four other factors—dispersal across barriers, cyclic presence-absence of barriers, drastic range shifts particularly during the Pleistocene, and extinction—have contributed to current distributional patterns. Inspection of the only species level cladistic analyses of bees to date (McGinley, 1986; Roberts & Brooks, 1987) shows no support for vicariance between sister species at these or any other zones in Mexico. All four pairs of Mesoamerican sister species in *Lasioglossum* (McGinley, 1986) are sympatric or have extensively overlapping distributions. This finding is perhaps not surprising given the vagility of bees. The presence in the Antilles of 16 of the 27 tribes and subfamilies recorded from the mainland Neotropics (Michener, 1979) suggests that bees are able to occasionally but not readily (Eickwort, 1988) colonize across even large barriers. Of the 11 not recorded in the Greater Antilles, Hylaeinae has since been found in Jamaica (Snelling, 1982); Xeromelissinae and Paracolletini are primarily found in southern South America, only marginally entering Mesoamerica

(Michener, 1979); Xeromelissinae and Panurginae are small, weak-flying bees less likely to have crossed the water gap.

Areas of Major Diversity

The diversity of bees in Mexico appears intermediate between that of the United States and countries to the south (Table 5.1). Although the Madrean region is considered the richest in the world, this statement is based largely on that portion lying within the United States. Much of this diversity is due to the rich bee fauna of desert regions, particularly in the Panurginae; almost 900 species have been found in the United States' portions of the Sonoran and Chihuahuan Deserts (Moldenke, 1979). Because five of the seven subdivisions of the Sonoran Desert (desert portions of Baja California included) lie entirely within Mexico (Shreve & Wiggins, 1964) and over 80% of the Chihuahuan Desert is in Mexico (MacMahon & Wagner, 1985), many more species may be expected from the xeric regions of Mexico (Moldenke, 1979). When these deserts and other poorly sampled biotas have been adequately surveyed, the faunal richness of Mexico may approach that of the United States and even surpass it on a per area basis.

Temperate, more mesic environments such as parts of the altiplano are probably the second richest faunistically (Table 5.3; Fig. 5.7), followed by areas with tropical deciduous forests. As an example of the latter, 228 species have been recorded from Chamela, Jalisco (Ayala, 1988). The Yucatan Peninsula appears to be an exception. Limited collections from this region indicate that the area has a depauperate bee fauna. Because southern Florida is also poor (Graenicher, 1930), we suggest that shallow soils restrict the number of species able to nest. The recent age of these areas geologically may also be a factor.

Other vegetation types have not been analyzed, but according to Michener (1979), it is possible that pine, pine-oak, and tropical evergreen forests have relatively low diversities. These generalities must remain tentative until intensive sampling has been done particularly in chaparral and montane regions throughout Mexico (Fig. 5.5).

Linsley (1958) stated that the diversity of bees in a region is normally well represented at specific localities. Therefore one expects to find a high frequency of sympatry, with ecological separation by host plants, nest sites and seasonality. This finding leads to the suggestion that intensive local studies may yield better information than more dispersed collecting because rare populations and temporal variation are better sampled. In addition, it permits closer observation on nesting and interactions with host plants. However, studies by Moldenke (1976) suggested that the high bee diversity of deserts is poorly represented at specific sites. In any case, multiyear studies are probably essential even when intensive inventories are conducted. Roubik (1989) suggested that only half of the bee species visiting a particular flower can be collected in a year of transect sampling.

Unfortunately, there is no published fauna for any locality in Mexico, except Chamela (Ayala, 1988), let alone adequate descriptions of their ecology.

Endemism

As previously noted, Baja California has a high degree of endemism (around 30%) concentrated in the north and in the San Lucas region (Fig. 5.4). The Pacific Coast of mainland Mexico has a unique component of endemic bees such as *Perdita* (*Callomacrotera*) and *Augochlorella maritima* restricted to the coastal strand. Other areas with strong endemism include southern parts of the Altiplano, the Chihuahuan Desert, and Balsas Basin (Table 5.3; Fig. 5.7). To what extent this endemism is correlated with sampling effort is unclear.

Brown (1982) identified five areas of endemism for forest butterflies coinciding with paleoecological tropical forest refuges in southern Mexico. Evidence suggesting comparable centers of endemism for Neotropical bees is lacking. However, there has not been adequate sampling of these presumed paleoecological refuges, nor have most genera of Neotropical bees been revised.

Evaluation of Existing Collecting

Geographical patterns presented here are limited by inadequate analysis of existing material in collections and by large geographical and seasonal gaps in sampling. Analysis of recorded collecting localities gives an indication of regions that require greater attention (Fig. 5.5; Table 5.3). Among poorly collected areas are the Bolsón de Mapimí; the Sierras Madre Occidental, Oriental, and Sur; the Pacific Coast from Jalisco to Guatemala; the low part of the Rio Balsas Basin; Tabasco; and the Yucatan Peninsula. Sampling has been most complete for the Federal District (0.43 localities per square kilometer), and the states of Morelos (0.28), Mexico (0.09), Hidalgo (0.08), and Puebla (0.08).

Another index of the adequacy of collecting is the number of times a species has been recorded. Most species of Mexican bees have been reported from only one or two localities (Fig. 5.6).

Analysis of the data on seasonality of collections (Fig. 5.8) shows that most collections have been made between June and September, with a dearth of sampling from November to February and in May. Recent collections made in Jalisco, Michoacan, and Morelos during October and November (Ayala, Bullock, Griswold, and Parker, pers. obs.) demonstrate that it is not due to a depauperate fall fauna. Only *Perdita*, a northern genus, has many records for the first 3 months of the year (Fig. 5.8A). For comparison, we looked at the seasonality of complete faunas studied in

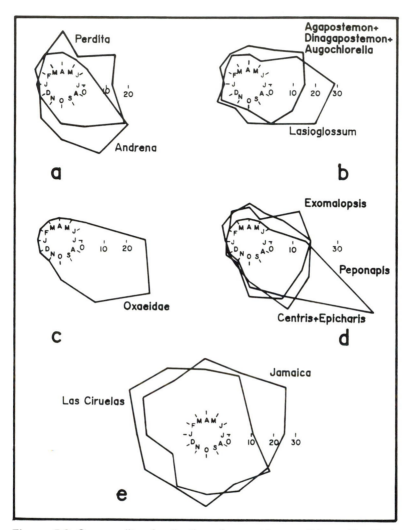

Figure 5.8. Seasonality of collections (percent by month). a. *Perdita* (*n* = 569); *Andrena* (*n* = 168). b. Halictidae: *Lasioglossum* (*n* = 299), *Agapostemon* plus *Dinagapostemon* plus *Augochlorella* (*n* = 135).   c. Oxaeidae (*n* = 213). d. Anthophoridae: *Exomalopsis* (*n* = 394), *Peponapis* (*n* = 464), *Centris* and *Epicharis* (*n* = 346).   e. Percent of all species active by month for the faunas of lowland Guanacaste Province, Costa Rica, and Jamaica.

other neotropical areas (Heithaus, 1979a; Raw, 1985). These seasonal patterns, shown as the percentage of the fauna present each month (Fig. 5.8E), demonstrated no outstanding fluctuations at either site, though there were notable seasonal changes in the composition of the faunas. Patterns certainly are more seasonal and diverse in the rough topography of many tropical areas of Mexico and even more distinct in temperate climates.

MAINTENANCE OF DIVERSITY

Reproductive Isolation and Other Mechanisms

Although there are many interesting biogeographic and morphologic patterns among bee species that may be related to speciation processes, the lack of basic data on genetics is notable. Genetic determination of sexes has not been studied experimentally except in Apidae. However, males develop from unfertilized eggs, and it is probable that in general the female controls fertilization; there are reports of parthenogenetic females in *Apis* (Mackensen, 1943), *Ceratina* (Daly, 1966), and *Osmia* (Stephen et al., 1969). It is generally considered that most species are monandric. In the few species whose isoenzymes have been studied, heterozygosity tends to be low (Moritz, 1986).

Pheromones must have important effects on mating patterns through their roles in intersexual attraction, species recognition, and territoriality (Eickwort & Ginsberg, 1980), as well as in the suppression of reproduction among females (Greenberg & Buckle, 1981) and sibling recognition (Greenberg, 1979). There are as yet no critical studies of speciation models in the Apoidea, but opportunities for sympatric speciation seem to exist because of such factors as temporal or seasonal shifts in flight times, changes in host plants, changing energetic budgets (Moldenke, 1979), sibling matings, and parasite and predator pressures. LaBerge (1986b) offered interpretations of geography and speciation in *Andrena* based on host and habitat factors.

Visual mimicry has not been reported as a major factor in bee diversification, perhaps due to their small size and rapid flight; but opportunities seem to exist in tropical climates where abundant, year-round populations of social bees and stinging social wasps (Polistinae, Vespidae) occur. Observations of *Dianthidium* (*Mecanthidium*) *macrurum* in Colima tend to support this statement. Not only does the bees' coloration closely resemble that of the local *Polistes*, but its slow, leg-dangling flight mimics that of the wasp, in sharp contrast to the tight, rapid flight of most anthidiines (Griswold, pers. obs.). Other examples include *Mydrosoma saussurei* with *Melipona beechii* (Michener, 1986), *Neocorynura* with *Polybia* (Eickwort, pers. comm.) and *Paratetrapedia* and *Tetrapedia* with *Trigona* s.l. (Parker, pers. comm.). The only well-studied example of Müllerian mimicry is in *Eulaema* (Apidae: Euglossini) (Dressler, 1979), but the partially sympatric bumble bees *Bombus diligens* and *B. brachycephalus* appear to be part of a mimetic complex with the anthophorid *Deltoptila aurulentocauda* (Labougle, 1990).

Another factor in species richness may be the availability of nesting sites, although actual data are remarkably scarce (Michener et al., 1958). Many groups nest in the ground, so their capacity to excavate tunnels and chambers may influence their distribution and abundance on different substrates. Water-impermeable substances derived from glandular secretions or plants are commonly applied to the cells (Batra & Schuster, 1977;

Roubik & Michener, 1980; Rozen, 1984a,b; Stephen et al., 1969), thus providing protection from either inundation or desiccation.

Animal–Plant Interactions

Feeding specialization in bees has been suggested as a factor in speciation (Linsley & MacSwain, 1958), although the evolutionary interaction of bees and plants does not necessarily imply mutualistic coevolution, as bees often take advantage of plants (Bullock et al., 1989; Roubik, 1989).

The bases for feeding specialization have not been adequately studied. There is some evidence that the nutritional value of pollen of different plants is not equivalent for the larvae of a particular bee species (Levin & Haydack, 1957). Also, oils characteristic of a given pollen species may condition the individual larva so that as an adult it searches for similar host plants (Dobson, 1987). However, this and other bases for feeding specialization have not been thoroughly studied, nor have sufficient comparisons been made among solitary species. In evolutionary terms, the problem of cause-effect remains, e.g., in the observation that long-lived bees tend to exploit a greater diversity of plants.

Phenology is an important determinant of diversity through its effect on separation of competing populations and reproductive isolation. Studies of tropical bee faunas show a notable percentage of species active throughout the year and a long mean flight period compared to that in temperate regions (Heithaus, 1979a; Raw, 1985). The degree of pollen-feeding specialization may affect flight periods through the flowering seasonality of potential host plants. Thus seasonal changes in the fauna of a particular site might be expected, as well as intersite differences in fauna due to factors that influence plant seasonality (e.g., climate, flora, and life forms) (Beatley, 1974; Bullock & Solis, 1990; Carabias & Guevara, 1985; Castillo & Carabias, 1982; Kummerow et al., 1981; Monasterio & Sarmiento, 1976).

The existence of abundant generalists in the social group Meliponinae might result in a monopolization of resources by a few species (Heithaus, 1979b; Roubik, 1989). This fact may help explain why in the tropics the richness of the apifauna does not compare with that of plants, but that it occurs to the extent of largely excluding solitary bees remains unproved.

CONSERVATION NEEDS

The bee fauna of Mexico faces serious threats to its survival from destruction of habitat, agricultural practices, and potential competition from the invasion by the Africanized honey bee. Protection of the native flora—not only natural habitats but also the dicotyledons in ruderal vegetation—may provide the only resources for bees in more populated areas. Nesting sites must also be protected. This measure should include modification of agricultural practices to preserve ground-nesting sites, preservation of old adobe walls, and, for cavity nesters, wise fire control practices. The effect of broad-spectrum insecticides on nontarget organisms such as bees must also be addressed. Utilization of integrated pest management methods to reduce pesticide use and appropriate timing of necessary applications can minimize effects on bee populations. Some might suggest that such inconspicuous creatures are scarcely worth the trouble, but as with all things in nature no element lives in isolation. Native bees form a critical component of natural ecosystems. In their role as pollinators, they are keystone species vital to the longevity of natural communities (Soulé, 1990). As such, they deserve special attention. It is essential, therefore, that we gain an understanding of their distribution and ecology if we are to have the tools necessary to develop effective management schemes to conserve threatened habitats.

We need to heighten the awareness of people to bees, to their rich diversity in form and habits. Ultimately, the survival of our native bees is dependent upon the educated will of the people, their understanding of the importance of bees not only in home and commercial agricultural production but in natural ecosystems as well.

ACKNOWLEDGMENTS

We are grateful to many: to the members of PCAM (Programa Cooperativo sobre la Apifauna Mexicana), who provided initial stimulus for this work; to E. Naranjo, R. Dirzo, G. Eickwort, C. Michener, J. Neff, F. Parker, and J. Rozen, who took the time to review various stages of the manuscript; to Jordi Bosch, who kindly assisted in translation from the original Spanish draft to English; to Marianne Cha Filbert, who spent endless hours entering data and proofing the work; to Frank Parker, whose prodigious collecting in Costa Rica clarified relations between its fauna and that of Mexico; and lastly to the many collectors whose nameless efforts made this possible— may their numbers increase.

APPENDIX

North America bee genera, known species for geographical regions, and status of knowledge for Mexico

Genera	No. of Species Cont. U.S.	Mex.	Cent. Amer.	Mexico Rev.	Dist.	Dates	Source
Andrenidae							
Andreninae							
Ancylandrena	4	1		X	X	X	Zavortink (1974)
Andrena	442	88	7	X	P	P	LaBerge & Hurd (1965); LaBerge (1967, 1969, 1971, 1973, 1977, 1978, 1980, 1986a, 1987, 1989b); LaBerge & Ribble (1972, 1975); Donovan (1977); Ribble (1967, 1968a, 1974); Bouseman & LaBerge (1979); Thorp (1969)
Megandrena	2						
Panurginae							
Anthemurgus	1						
Calliopsis	27	24	2	X	X	P	Shinn (1967)
Heterosarus	17	44	3	X	P	P	Timberlake (1975)
Hypomacrotera	4	3					
Metapsaenythia	2	1		X	X	X	Timberlake (1969b)
Nomadopsis	45	9		X	P		Rozen (1958, 1959)
Panurginus	19	3					
Perdita	657	226	1	X	X	X	Timberlake (1954, 1956, 1958, 1960, 1962, 1964, 1968, 1971, 1977, 1980b)
Protandrena	22	40		X	P	P	Timberlake (1976)
Pseudopanurgus	14	28	5	X	X	X	Timberlake (1973)
Pterosarus	46	3					
Xenopanurgus	1	1		X	X	X	Shinn (1964)
Anthophoridae							
Anthophorinae							
Anthophorini							
Anthophora	53	23	5				
Deltoptila		5	2	X	X	X	LaBerge & Michener (1963)
Habropoda	11	2					
Centridini							
Centris	20	53	37	X	P	P	Snelling (1966, 1974, 1984)
Epicharis		3	9	X	X	X	Snelling (1984)

Appendix (cont.)

Genera	Cont. U.S.	Mex.	Cent. Amer.	Mexico Rev.	Dist.	Dates	Source
Ptilotopus			1				
Emphorini							
Diadasia	32	20	4				
Melitoma	3	5	5				
Ptilothrix	2	2	1				
Ericrocini							
Aglaomelissa			1				
Ctenioschelus		1	1				
Ericrocis	2	2		X	X	X	Snelling & Zavortink (1984)
Mesocheira		1	1				
Mesoplia	1	7	3				
Eucerini							
Agapanthinus		1		X	X	X	LaBerge (1957)
Anthedonia	2	2		X	X	X	LaBerge (1955)
Cemolobus	1						
Florilegus	1	1	2	X			Urban (1970)
Gaesischia	1	2	1	X	X	X	Michener et al. (1955); LaBerge (1958a); Urban (1968b)
Idiomelissodes	1	1		X	X		LaBerge (1956); Zavortink (1975b)
Loxoptilus		2		X	X	X	LaBerge (1957)
Martinapis	2	2		X	X	X	Zavortink & LaBerge (1976)
Melissodes	102	47	7	X	X	P	LaBerge (1956, 1961, 1963a)
Melissoptila	1	1	3	X			Urban (1968a)
Pectinapis	1	5		X	X	X	LaBerge (1970, 1989a)
Peponapis	6	10	6	X			Hurd & Linsley (1966)
Simanthedon	1	1		X	X	X	Zavortink (1975a)
Svastra	21	14	2	X	X	P	Laberge (1956, 1958b, 1963b)
Syntrichalonia	1	1		X	X		Zavortink (1975b)
Tetralonia	59	12		X	P		Timberlake (1969a)
Thygater	1	5	5	X	X		Urban (1967)
Xenoglossa	5	7	1	X			Hurd & Linsley (1967b)
Xenoglossodes	19	2	3				
Exomalopsini							
Ancyloscelis	2	3	4				
Exomalopsis	50	74	19	X	P	P	Timberlake (1980a)
Monoeca		1	2				
Paratetrapedia		6	10				
Melectini							
Brachymelecta	1						
Melecta	12	1					

Appendix (cont.)

Genera	No. of Species Cont. U.S.	Cent. Mex.	Cent. Amer.	Mexico Rev.	Dist.	Dates	Source
Xeromelecta	3	2					
Zacosmia	2	2					
Rathymini							
Rathymus		1	*				
Tetrapediini							
Tetrapedia		4	2				
Coelioxoides		1	1	X	X	X	Roig-Alsina (1990)
Nomadinae							
Epeoloides	2						
Epeolus	50	15	15				
Hexepeolus	2	*					
Holcopasites	17	6		X	X	X	Hurd & Linsley (1972)
Melanomada	6	1		X	X	X	Snelling & Rozen (1987)
Neolarra	19	2		X	X	X	Shanks (1978)
Neopasites	5						
Nomada	317	20	10	P	P	P	Broemeling (1988); Bromeling & Moalif (1988)
Odyneropsis	*	*	1				
Oreopasites	7	*					
Osiris		5	13	X	X	X	Shanks (1986, 1987)
Paranomada	3	1					
Protepeolus	1	1		X			Eickwort & Linsley (1978)
Townsendiella	3	1					
Triepeolus	108	11	4				
Triopasites	4	1					
Xylocopinae							
Ceratinini							
Ceratina	21	22	31				
Xylocopini							
Xylocopa	21	36	24	P	P	P	O'Brien & Hurd (1965)
Apidae							
Apinae							
Apis	1	1	1				
Bombinae							
Bombini							
Bombus	42	19	11	X	X		Labougle (1990)
Psithyrus	7	2	4				
Euglossini							
Aglae	1						
Eufriesea		7	18	X	X		Kimsey (1982)
Euglossa		15	45				

Appendix (cont.)

Genera	No. of Species Cont. U.S.	Cent. Mex.	Cent. Amer.	Mexico Rev.	Dist.	Dates	Source
Eulaema	1	5	11				
Exaerete		3	4	X	X	P	Kimsey (1979)
Meliponinae							
Cephalotrigona		2	1				
Lestrimelitta		1	1				
Melipona		4	12				
Nannotrigona		1	1				
Oxytrigona		1	2				
Paratrigona		1	4				
Partamona		1	2				
Plebeia		4	8				
Scaptotrigona		4	6				
Trigona		8	14				
Trigonsica		2	5				
Colletidae							
Colletinae							
Colletini							
Colletes	114	46	12				
Paracolletini							
Eulonchopria	1	2	1	X	X	X	Michener (1963, 1985)
Diphaglossinae							
Caupolicanini							
Caupolicana	3	5		X	X	P	Michener (1966)
Crawfordapis		1	1	X	X		Michener (1966); Roubik & Michener (1985)
Ptiloglossa	3	3	8				
Dissoglottini							
Mydrosoma		4	2	X	X	X	Snelling (1980); Michener (1986)
Hylaeinae							
Hylaeus	52	20	14				
Xeromelissinae							
Chilicola		2	1	X	X	X	Eickwort (1967); Toro & Michener (1975)
Halictidae							
Halictinae							
Augochlorini							
Augochlora	5	19	28				
Augochlorella	7	7	3	X	X	P	Ordway (1966)
Augochloropsis	3	10	11				
Caenaugochlora		7	7				
Chlerogella			1				

Appendix (cont.)

Genera	No. of Species Cont. U.S.	Cent. Mex.	Cent. Amer.	Mexico Rev.	Mexico Dist.	Mexico Dates	Source
Megalopta		1	3				
Neocorynura		4	8				
Pereirapis		1	3				
Pseudaugochloropsis	1	2	2				
Temnosoma	1	1	1				
Halictini							
Agapostemon	12	14	9	X	P	P	Roberts (1972)
Agapostemonoides			1				
Caenohalictus		*	1				
Dialictus	183	22	30				
Dinagapostemon		2	5	X	X	X	Roberts & Brooks (1987)
Evylaeus	68	19	4				
Habralictus		1	2				
Halictus	9	4	3				
Hemihalictus	1						
Lasioglossum	29	29	9	X	X	P	McGinley (1986)
Mexalictus	1	2		X	X	X	Eickwort (1978)
Microsphecodes			1				
Paragapostemon		1		X	X		Roberts & Brooks (1987)
Paralictus	5						
Ptilocleptis		1	1	X	X	X	Michener (1978)
Rhinetula			1				
Sphecodes	71	12	4				
Sphecodogastra	4	1					
Nomiinae							
Nomia	26	16		P	P	P	Cross (1958); Ribble (1965, 1968b)
Rophitinae							
Conanthalictus	13	*					
Dufourea	76	16					
Michenerula	1	1					
Micralictoides	9						
Protodufourea	2	1					
Sphecodosoma	2	2		X	X	X	Timberlake (1961)
Xeralictus	1						
Megachilidae							
Anthidiinae							
Anthidiellum	7	6	2				
Anthidium	26	26	5				
Anthodioctes		3	2				
Aztecanthidium		3		X	X	X	Michener & Ordway (1964); Snelling (1987)
Dianthidium	33	19					

Appendix (cont.)

Genera	Cont. U.S.	Mex.	Cent. Amer.	Mexico Rev.	Dist.	Dates	Source
Dioxys	9	3		X	X	X	Hurd (1958)
Dolichostelis	5	5	1	X			Parker & Bohart (1979)
Epanthidium		1		X	X	X	Stange (1983)
Hoplostelis		1	1				
Hypanthidioides			2				
Hypanthidium		4	4				
Paranthidium	4	3	2				
Protostelis	7	1					
Saranthidium		2	2				
Stelis	35	6					
Trachusa	15	10		P	P	P	Snelling (1975); Brooks & Griswold (1988)
Lithurginae							
Lithurge	6	5	1	X	X	P	Snelling (1983, 1986)
Megachilinae							
Coelioxys	47	33	28				
Megachile	143	77	28				
Osmiinae							
Anthocopa	32	12					
Ashmeadiella	58	26					
Chelostoma	10	1					
Heriades	12	7	2				
Hoplitis	27	5					
Osmia	135	11					
Proteriades	35	12					
Protosmia	1	1					
Xeroheriades	1						
Mellitidae							
Dasypodinae							
Hesperapis	20	9					
Xeralictoides	1						
Melittinae							
Dolichochile	1						
Macropis	7						
Melitta	4	1					
Oxaeidae							
Mesoxaea	3	7		X	X	X	Hurd & Linsley (1976)
Protoxaea	1	3		X	X	X	Hurd & Linsley (1976)

Rev., recent revision of Mexican species; Dist., distributional data given; Dates, seasonal records given; *, new record based on one or more undetermined species; X, complete data; P, partial data.

REFERENCES

Alexander, B. 1990. A cladistic analysis of the nomadine bees. Syst. Entomol. 15:121–152.

Alvarez, T. & F. de Lachica. 1974. Zoogeografía de los vertebrados de México. *In* I. Bernal & J.L. Lorenzo. (eds.), El Escenario Geográfico. Recursos Naturales. México: Panorama Histórico y Cultural. Vol. II. México City: INAH.

Axelrod, D.I. 1979. Age and origin of Sonoran Desert vegetation. Occas. Papers Calif. Acad. Sci. 132:1–74.

Ayala, R. 1988. Abejas silvestres de Chamela, Jalisco, México. Fol. Entomol. Méx. 77:395–493.

Ball, G.E. 1968. Barriers and southward dispersal of the holarctic boreo-montane element of the family Carabidae in the mountains of Mexico. Anales Esc. Nac. Ci. Biol. 17:91–112.

Batra, S.W.T. & J.C. Schuster. 1977. Nests of *Centris, Melissodes*, and *Colletes* in Guatemala. Biotrop. 9:135–138.

Bawa, K.D., S.H. Bullock, D.R. Perry, R.E. Coville & M.H. Grayum. 1985. Reproductive biology of tropical lowland rain forest trees. II. Pollination systems. Amer. J. Bot. 72:346–356.

Beatley, J.C. 1974. Phenological events and their environmental triggers in Mohave desert ecosystems. Ecology 55:856–863.

Bennett, F.C. 1964. Stingless bee keeping in western Mexico. Geogr. Rev. 54:85–92.

Bouseman, J.K. & W.E. LaBerge. 1979. A revision of the bees of the genus *Andrena* of the Western Hemisphere. Part IX. Subgenus *Melandrena*. Trans. Amer. Entomol. Soc. 104:275–389.

Broemeling, D.K. 1988. A revision of the *Nomada* Subgenus *Nomadita* of North America. Pan-Pac. Entomol. 64:321–344.

Broemeling, D.K. & A.S. Moalif. 1988. A revision of the *Nomada* Subgenus *Pachynomada*. Pan-Pac. Entomol. 64:201–227.

Brooks, R.W. & T.L. Griswold. 1988. A key to the species of *Trachusa* subgenus *Heteranthidium* with descriptions of new species from Mexico. J. Kans. Entomol. Soc. 61:332–346.

Brown, J.W. 1987. The peninsular effect in Baja California: an entomological assessment. J. Biogeog. 14:359–365.

Brown, K.S., Jr. 1982. Paleoecology and regional patterns of evolution in Neotropical forest butterflies. *In* G. T. Prance (ed.), Biological Diversification in the Tropics. New York: Columbia Univ. Press. pp 255–308.

Bullock, S.H., C. Martínez del Rio & R. Ayala. 1989. Bee visitation rates to trees of *Prockia crucis* differing in flower number. Oecologia 78:389–393.

Bullock, S.H. & J.A. Solís M. 1990. Phenology of canopy trees of a tropical deciduous forest in Mexico. Biotrop. 22:22–35.

Búrquez, A. & J. Sarukhán. 1980. Biología floral de poblaciones silvestres y cultivades de *Phaseolus coccineus* L. I. Relaciones planta-polinizador. Bol. Soc. Bot. Méx. 39:5–27.

——— & ———. 1984. Biología floral de poblaciones silvestres y cultivadas de *Phaseolus coccineus* L. II. Sistemas reproductivos. Bol. Soc. Bot. Méx. 46:3–12.

Camargo, J.M.F., A. Gottsberger & I. Silberbauer-Gottsberger. 1984. On the phenology and flower visiting behavior of *Oxaea flavescens* (Klug in São Paulo, Brazil). Beitr. Biol. Pflanzen 59:159–179.

Carabias-Lillo, J. & S. Guevara S. 1985. Fenología de una selva tropical humeda y en una comunidad derivada; Los Tuxtlas, Veracruz. *In* A. Gomez-Pompa & S. DelAmo R. (eds.), Investigaciones sobre la regeneración de selvas altas en Veracruz, México. 2:27–66. Xalapa, VER: INIREB/Mexico City: Alhambra Mexicana.

Castillo, S. & J. Carabias-Lillo. 1982. Ecología de la vegetación de dunas costeras: fenología. Biótica 7:55a–568.

Cockerell, T.D.A. 1899. Catálogo de las abejas de México. México: Biblioteca Agricola de la Secretaría del Fomento de México.

Cresson, E.T. 1878. Catalogue of North American Apidae with descriptions of new species. Trans. Amer. Entomol. Soc. 7:61–214.

Cross, E.A. 1958. A revision of the bees of the subgenus *Epinomia* in the New World. Univ. Kans. Sci. Bull. 38:1261–1301.

Daly, H.V. 1966. Biological studies on *Ceratina dallatorreana*, an alien bee in California which reproduces by parthenogenesis. Ann. Entomol. Soc. Amer. 59:1138–1154.

Devall, M.S. & L.B. Thien. 1989. Factors influencing the reproductive success of *Ipomoea pes-caprae* around the Gulf of Mexico. Amer. J. Bot. 76:1821–1831.

Dixon, C.V. 1987. Beekeeping in southern Mexico. Conf. Latin Amer. Geogr. Yearbook 13:66–71.

Dobson, H.E.M. 1987. Role of flower and pollen aromas in host-plant recognition by solitary bees. Oecologia 72:618–623.

Donovan, B.J. 1977. A revision of North American bees of the subgenus *Cnemidandrena*. Univ. Calif. Publ. Entomol. 81:1–107.

Dressler, R. 1979. *Eulaema bombiformis, E. meriana*, and Mullerian mimicry in related species. Biotrop. 11:144–151.

Duffield, R.M., J.W. Wheeler & G.C. Eickwort. 1984. Sociochemicals of bees. *In* W.J. Bell & R.T. Cardé (eds.), Chemical ecology of insects. London: Chapman & Hall. pp. 387–428.

Eickwort, G.C. 1967. Aspects of the biology of *Chilicola ashmeadi* in Costa Rica. J. Kans. Entomol. Soc. 40:42–73.

———. 1978. *Mexalictus*, a new genus of sweat bees from North America. J. Kans. Entomol. Soc. 51:567–580.

———. 1988. Distribution patterns and biology of west indian sweat bees. *In* J.K. Liebherr (ed.), Zoogeography of Caribbean Insects. Ithaca, NY: Cornell Univ. Press. pp. 231–253.

——— & H.S. Ginsberg. 1980. Foraging and mating behavior in Apoidea. Ann. Rev. Entomol. 25:421–446.

——— & E. G. Linsley. 1978. The species of the parasitic bee genus *Protepeolus*. J. Kans. Entomol. Soc. 51:14–21.

Frankie, G.W., W.A. Haber, P.A. Opler & K.S. Bawa. 1983. Characteristics and organization of the large bee pollination system in the Costa Rican dry forest. *In* C.E. Jones and R.J. Little (eds.), Handbook of Experimental Pollination Biology. New York: Van Nostrand Reinhold. pp. 411–447.

Gomarra, F.L. de. 1811. Conquista de México. Madrid: Vedra T.

González, A.J. 1984. Acerca de la regionalización de la nomenclatura de las abejas sin aguijón en Yucatán. Rev. Geogr. Agric. (Mex.). 5–6:190–193.

Graenicher, S. 1930. Bee fauna and vegetation of the Miami region of Florida. Ann. Entomol. Soc. Amer. 23:153–174.

Greenberg, L. 1979. Genetic component of bee odor in kin recognition. Science 206:1095–1097.

——— & G.R. Buckle. 1981. Inhibition of worker mating by queens in a sweat bee, *Lasioglossum zephyrum*. Insectes Soc. 28:347–352.

Heithaus, E.R. 1974. The role of plant-pollinator interactions in determining community structure. Ann. Missouri Bot. Gard. 61:675–691.

———. 1979a. Community structure of Neotropical flower visiting bees and wasps; diversity and phenology. Ecology 60:190–202.

———. 1979b. Flower-feeding specialization in wild bee and wasp communities in seasonal Neotropical habitats. Oecologia 42:179–194.

Hurd, P.D., Jr. 1958. American bees of the genus *Dioxys* Lepeletier and Serville. Univ. Calif. Publ. Entomol. 14:275–302.

———. 1979. Superfamily Apoidea. *In* K.V. Krombein, P.D. Hurd, Jr., D. R. Smith & B.D. Burks (eds.), Catalog of Hymenoptera in America north of Mexico. Washington: Smithsonian Inst. Press. pp. 1741–2209.

——— & E.G. Linsley. 1964. The squash and gourd bees—genera *Peponapis* Robertson and *Xenoglossa* Smith—inhabiting America north of Mexico. Hilgardia 35:375–477.

——— & ———. 1966. The squash and gourd bees of the genus *Peponapis*. Ann. Entomol. Soc. Amer. 59:835–851.

——— & ———. 1967. Squash and gourd bees of the genus *Xenoglossa*. Ann. Entomol. Soc. Amer. 60:988–1007.

——— & ———. 1970. A classification of the squash and gourd bees *Peponapis* and *Xenoglossa*. Univ. Calif. Publ. Entomol. 62:1–39.

——— & ———. 1972. Parasitic bees of the genus *Holcopasites* Ashmead. Smithsonian Contr. Zool. 114:1–41.

——— & ———. 1976. The bees family Oxaeidae with a revision of the North American species. Smithsonian Contr. Zool. 220:1–75.

———, E.G. Linsley & T.W. Whitaker. 1971. Squash and gourd bees (*Peponapis*, *Xenoglossa*) and the origin of the cultivated *Cucurbita*. Evolution 25:218–234.

Janzen, D.H. 1964. Notes on the behavior of four subspecies of the carpenter bee, *Xylocopa* (*Notoxylocopa*) *tabaniformis*, in Mexico. Ann. Entomol. Soc. Amer. 57:296–301.

———. 1966. Notes on the behavior of the carpenter bee *Xylocopa fimbriata* in Mexico. J. Kans. Entomol. Soc. 39:633–641.

Kimsey, L.S. 1979. An illustrated key to genus *Exaerete* with descriptions of male genitalia and biology. J. Kans. Entomol. Soc. 52:735–746.

———. 1982. Systematics of bees of the genus *Eufriesea*. Univ. Calif. Publ. Entomol. 95:1–195.

Kummerow, J., G. Montenegro & D. Krause. 1981. Biomass, phenology and growth. *In* P.C. Miller (ed.), Resource use by chaparral and matorral. New York: Springer Verlag.

LaBerge, W.E. 1955. Bees of the genus *Anthedonia* Michener in North America. J. Kans. Entomol. Soc. 28:132–135.

———. 1956. A revision of the bees of the genus *Melissodes* in North America and Central America. Univ. Kans. Sci. Bull. 37:911–1194.

———. 1957. The genera of bees of the tribe Eucerini in North and Central America. Amer. Mus. Novitates. 1837:1–44.

———. 1958a. Notes on *Gaesischia* Michener, LaBerge and Moure, with description of a new species. Pan-Pac. Entomol. 34:195–201.

———. 1958b. Notes on the North and Central American bees of the genus *Svastra* Holmberg. J. Kans. Entomol. Soc. 31:266–273.

———. 1961. A revision of the bees of the genus *Melissodes* in North and Central America. Part III. Univ. Kans. Sci. Bull. 42:283–663.

————. 1963a. New species and records of little-known species of *Melissodes* from North America. Univ. Nebr. St. Mus. Bull. 4:227–242.

————. 1963b. Records of the genus *Svastra* in North America with the description of a new species. J. Kans. Entomol. Soc. 36:52–56.

————. 1967. A revision of the bees of the genus *Andrena* of the Western Hemisphere. I. *Callandrena*. Univ. Nebr. St. Mus. Bull. 7:1–316.

————. 1969. A revision of the bees of the genus *Andrena* of the Western Hemisphere. II. *Plastandrena, Aporandrena, Charitandrena*. Trans. Amer. Entomol. Soc. 95:1–47.

————. 1970. A new genus with three new species of Eucerine bees from Mexico. J. Kans. Entomol. Soc. 43:321–328.

————. 1971. A revision of the bees of the genus *Andrena* of the Western Hemisphere. IV. *Scrapteropsis, Xiphandrena* and *Rhaphandrena*. Trans. Amer. Entomol. Soc. 97:441–520.

————. 1973. A revision of the bees of the genus *Andrena* of the Western Hemisphere. VI. *Trachandrena*. Trans. Amer. Entomol. Soc. 99:235–371.

————. 1977. A revision of the bees of the genus *Andrena* of the Western Hemisphere. VIII. Subgenera *Thysandrena, Dasyandrena, Psammandrena*. Trans. Amer. Entomol. Soc. 103:1–143.

————. 1978. *Andrena (Callandrena) micheneriana*, a remarkable new bee from Arizona and Mexico. J. Kans. Entomol. Soc. 51:592–596.

————. 1980. A revision of the bees of the genus *Andrena* of the Western Hemisphere. X. Subgenus *Andrena*. Trans Amer. Entomol. Soc. 106:395–525.

————. 1986a. A revision of the bees of the genus *Andrena* of the Western Hemisphere. XI. Minor subgenera and subgeneric key. Trans. Amer. Entomol. Soc. 111:441–567.

————. 1986b. The zoogeography of *Andrena* Fabricius of the Western Hemisphere. *In* G.K. Clambey & R.H. Pemble (eds.), The Prairie: Past, Present and Future. Proc. Ninth North Amer. Prairie Conf. pp. 110–115.

————. 1987. A revision of the bees of the genus *Andrena* of the Western Hemisphere. XII. Subgenus *Leucandrena, Ptilandrena, Scoliandrena*, and *Melandrena*. Trans. Amer. Entomol. Soc. 112:191–248.

————. 1989a. A review of the bees of the genus *Pectinapis*. J. Kans. Entomol. Soc. 62:524–527.

————. 1989b. A revision of the bees of the genus *Andrena* of the Western Hemisphere. Part XIII. Subgenera *Simandrena* and *Taeniandrena*. Trans. Amer. Entomol. Soc. 115:1–56.

———— & P.D. Hurd, Jr. 1965. A new subgenus and species of matinal *Andrena* from the flowers of *Sicyos* in Mexico. Pan-Pac. Entomol. 41:186–193.

———— & C.D. Michener. 1963. *Deltoptila*, a Middle American genus of anthophorine bees. Bull. Univ. Nebr. St. Mus. 4:211–225.

———— & D.W. Ribble. 1972. A revision of the bees of the genus *Andrena* of the Western Hemisphere. Part V. *Gonandrena, Geissandrena, Parandrena, Pelicandrena*. Trans. Amer. Entomol. Soc. 98:271–358.

———— & ————. 1975. A revision of the bees of the genus *Andrena* of the Western Hemisphere. Part VII. Subgenus *Euandrena*. Trans. Amer. Entomol. Soc. 101:371–446.

Labougle, J.M. 1990. *Bombus* of Mexico and Central America. Univ. Kans. Sci. Bull. 54:35–73.

———— & J.A. Zozaya. 1986. La apicultura en México. Ciencia y Desarrollo 69:17–36.

Levin, M.D. & M.H. Haydak. 1957. Comparative value of different pollens in the nutrition of *Osmia lignaria*. Bee World 38:221–226.

Linsley, E.G. 1958. The ecology of solitary bees. Hilgardia 27:534–599.

————. 1961. The role of flower specificity in the evolution of solitary bees. XI Intern. Cong. Entomol., Wien, 1960. Sonderdruck aus den Verlandlungen BO. I.

———— & J.W. MacSwain, 1958. The significance of floral constancy among bees of the *Diadasia*. Evolution 12:219–223.

————, ————, & P.H. Raven. 1963. Comparative behavior of bees and Onagraceae. I. Oenothera bees of the Colorado Desert. Univ. Calif. Publ. Entomol. 33:1–24.

————, ———— & R.F. Smith. 1955. Biological observations on *Xenoglossa fulva* Smith with some generalizations on biological characters of other eucerine bees. Bull. So. Calif. Acad. Sci. 54:128–141.

————, ———— & ————. 1956. Biological observations on *Ptilothrix sumichrasti* (Cresson) and some related groups of emphorine bees. Bull. S. Calif. Acad. Sci. 55:83–101.

Macazaga, O.C. 1982. Diccionario de Zoología Náhuaatl. Mexico City: Editorial Innovación.

Mackensen, O. 1943. The occurrence of parthenogenetic females in some strains of honey bees. J. Econ. Entomol. 36:465–467.

MacMahon, J.A. & F.H. Wagner. 1985. The Mojave, Sonoran and Chihuahuan Deserts of North America. *In* M. Evenari & I. Noy-Meir (eds.), Hot Deserts and Arid Shrublands. Amsterdam: Elseveir. pp. 105–202.

McGinley, R.J. 1980. Glossal morphology of the Colletidae and recognition of the Stenotritidae at the family level. J. Kans. Entomol. Soc. 53:539–552.

————. 1986. Studies of Halictinae. I. Revision of New World *Lasioglossum* Curtis. Smithsonian Contrib. Zool. 429:1–294.

————. 1989. A catalog and review of immature Apoidea. Smithsonian Contrib. Zool. 494:1–24.

McGregor, S.E. 1976. Insect Pollination of Cultivated Crop Plants. USDA-ARS Agric. Handbook No. 496, Washington, D.C: Government Printing Office.

Michener, C.D. 1944. Comparative external morphology, phylogeny, and classification of the bees. Bull. Amer. Mus. Nat. Hist. 82:151–326.

————. 1954. Bees of Panama. Bull. Amer. Mus. Nat. Hist. 104:1–176.

————. 1963. The bee genus *Eulonchopria*. Ann. Entomol. Soc. Amer. 56: 844–849.

————. 1965. A classification of the bees of the Australian and South Pacific regions. Bull. Amer. Mus. Nat. Hist. 130:1–362.

————. 1966. The classification of the Diphaglossinae and North American species of the genus *Caupolicana*. Univ. Kans. Sci. Bull. 46:717–751.

————. 1969. Comparative social behavior of bees. Ann. Rev. Entomol. 14:299–342.

————. 1974. The Social Behavior of the Bees. Cambridge, MA: Belknap Press.

————. 1978. The parasitic groups of Halictidae. Univ. Kans. Sci. Bull. 51:291–339.

————. 1979. Biogeography of the bees. Ann. Missouri Bot. Gard. 66:277–347.

————. 1985. A fourth species of *Eulonchopria* and a key to the species. J. Kans. Entomol. Soc. 58:236–239.

————. 1986. A review of the tribes Diphaglossini and Dissoglottini. Univ. Kans. Sci. Bull. 53:183–214.

———— & R.W. Brooks. 1984. Comparative study of the glossae of bees. Contrib. Amer. Entomol. Inst. 22:1–73.

———— & L. Greenberg. 1980. Ctenoplectridae and the origin of long-tongued bees. Zool. J. Linn. Soc. 69:183–203.

———— & D.A. Grimaldi. 1988a. The oldest fossil bee, Apoid history, evolutionary status, and antiquity of social behavior. Evolution 85:6424–6426.

———— & D.A. Grimaldi. 1988b. A *Trigona* from late Cretaceous amber of New Jersey. Amer. Mus. Novit. 2917:1–10.

———, W.E. LaBerge & J.S. Moure. 1955. Some American Eucerini bees. Dusenia 6:213–230.

———, R.B. Lange, J.J. Bigarella & R. Salamuni. 1958. Factors influencing the distribution of bees' nests in earth barks. Ecology 39:207–217.

——— & J.S. Moure. 1957. A study of the classification of the more primitive non-parasitic anthophorine bees. Bull. Amer. Mus. Nat. Hist. 112:395–452.

——— & E. Ordway. 1964. Some Anthidine bees from Mexico (Hymenoptera: Megachilidae). J. New York Entomol. Soc. 72:70–78.

Moldenke, A.R. 1976. California pollination ecology and vegetation types. Phytologia 34:305–361.

———. 1979. The role of host-plant selection in bee speciation processes. Phytologia 43:433–460.

Monasterio, M. & G. Sarmiento. 1976. Phenological strategies of plant species in the tropical savanna and the semi-deciduous forest of the Venezuelan llanos. J. Biogeog. 3:325–356.

Moritz, R.F.A. 1986. The genetics of bees other than *Apis mellifera*. *In* T.E. Rinderer (ed.), Bee Genetics and Breeding. pp. 121–154.

Moure, J.S. & P.D. Hurd, Jr. 1987. An Annotated Catalog of the Halictid Bees of the Western Hemisphere. Washington, DC: Smithsonian Inst. Press.

Murillo, M.R.M. 1984. Uso y manejo actual de las colonias de *Melipona beecheii* Bennett en el estado de Tabasco. Biótica 9:423–428.

Neff, J.L. & B.B. Simpson. 1981. Oil-collecting structures in the Anthophoridae: morphology, function, and use in systematics. J. Kans. Entomol. Soc. 54:95–123.

Noonan, G.R. 1988. Biogeography of North American and Mexican insects, and a critique of vicariance biogeography. Syst. Zool. 37:366–384.

O'Brien, L.B. & P.D. Hurd, Jr. 1965. Carpenter bees of the subgenus *Notoxylocopa*. Ann. Entomol. Soc. Amer. 58:177–196.

Ordway, E. 1966. Systematics of the genus *Augochlorella* north of Mexico. Univ. Kansas Sci. Bull. 46:509–624.

Packer, L. 1985. Two social halictine bees from southern Mexico with a note on two bee hunting philanthine wasps. Pan-Pac. Entomol. 61:291–298.

Parker, F.D. 1977. Biological notes on some Mexican bees. Pan-Pac. Entomol. 53:189–192.

——— & G.E. Bohart. 1979. *Dolichostelis*, a new genus of parasitic bees. J. Kans. Entomol. Soc. 52:138–153.

———, S.W.T. Batra & V.J. Tepedino. 1987. New pollinators for our crops. Agric. Zool. Rev. 2:279–304.

Quiros, C.F. & A. Marcias. 1978. Natural cross pollination and pollinator bees of the tomato in Celaya, central Mexico. Hortscience 13:290–291.

Raw, A. 1985. The ecology of Jamaican bees. Rev. Bras. Entomol. 29:1–16.

Ribble, D.W. 1965. A revision of the banded subgenera of *Nomia* in America. Univ. Kans. Sci. Bull. 65:277–359.

———. 1967. The monotypic North American subgenus *Larandrena* of *Andrena*. Bull. Univ. Nebr. St. Mus. 6:27–42.

———. 1968a. Revisions of two subgenera of *Andrena*: *Micrandrena* Ashmead and *Derandrena*, new subgenus. Bull. Univ. Neb. St. Mus. 8:237–394.

———. 1968b. A list of recent publications on the alkali bee, *Nomia melanderi*, with notes on related species of bees. Univ. Wyo. Agr. Exper. Sta. Sci. Monog. 11:1–18.

———. 1974. A revision of the bees of the genus *Andrena* of the Western Hemisphere. Subgenus *Scaphandrena*. Trans. Amer. Entomol. Soc. 100:101–189.

Roberts, R.B. 1972. Revision of the bee genus *Agapostemon*. Univ. Kans. Sci. Bull. 49:437–590.

———— & R. W. Brooks. 1987. Agapostemonine bees of Mesoamerica. Univ. Kans. Sci. Bull. 53:357–392.

———— & S. R. Vallespir. 1978. Specialization of hairs bearing pollen and oil on the legs of bees. Ann. Entomol. Soc. Amer. 71:619–627.

Roig-Alsina, A. 1990. *Coelioxoides* Cresson, a parasitic genus of Tetrapediini. J. Kans. Entomol. Soc. 13:279–287.

Roubik, D.W. 1978. Competitive interactions between neotropical pollinators and Africanized honeybees. Science 201:1030–1032.

————. 1983. Experimental community studies: time-series tests of competition between African and Neotropical bees. Ecology 64:971–978.

————. 1989. Ecology and Natural History of Tropical Bees. Cambridge: Cambridge Univ. Press.

———— & C.D. Michener. 1980. The seasonal cycle and nests of *Epicharis zonata*, a bee whose cells are below the wet-season water table. Biotrop. 12:56–60.

———— & ————. 1985. Nesting biology of *Crawfordapis* in Panama. J. Kans. Entomol. Soc. 57:662–671.

————, J.E. Moreno, C. Vergara & D. Wittmann. 1986. Sporadic food competition with the African honey bee: projected impact on neotropical social bees. J. Tropical Ecol. 2:97–111.

Rozen, J.G., Jr. 1958. Monographic study of the genus *Nomadopsis* Ashmead. Univ. Calif. Pub. Entomol. 15:1–202.

————. 1959. A new species of *Nomadopsis* and notes on some previously described ones. Proc. Entomol. Soc. Wash. 61:255–259.

————. 1965. The biology and immature stages of *Melitturga clavicornis* (Latreille) and of *Sphecodes albilabris* (Kirby) and the recognition of the Oxaeidae at the family level. Amer. Mus. Novit. 2224:1–18.

————. 1978. The relationships of the bee subfamily Ctenoplectrinae as revealed by its biology and mature larva. J. Kans. Entomol. Soc.51:637–652.

————. 1984a. Comparative nesting biology of the bee tribe Exomalopsini. Amer. Mus. Novit. 2798:1–37.

————. 1984b. Nesting biology of diphaglossine bees. Amer. Mus. Novit. 2786:1–33.

———— & R. Ayala. 1987. Nesting biology of the squash bee *Peponapis utahensis*. J. New York Entomol. Soc. 95:28–33.

———— & R.J. McGinley. 1974. Systematics of Ammobatine bees based on their mature larvae and pupae. Amer. Mus. Novit. 2551:1–16.

Rzedowski, J. 1973. Geographical relationships of the flora of Mexican dry regions. *In* A. Graham (ed.), Vegetation and vegetational history of Northern Latin America. Amsterdam: Elsevier. pp. 61–72.

————. 1978. Vegetación de México. Mexico City: Editorial Limusa.

Sahagún, Fray B. de. 1960. Historia General de las Cosas de la Nueva España. Mexico: Editorial Purrúa.

Schwarz, F.H. 1949. The stingless bees (Meliponidae) of Mexico. An. Inst. Biol. Mex. 20:357–370.

Shanks, S.S. 1978. A revision of the cleptoparasitic bee genus *Neolarra*. Wasmann J. Biol. 35:212–246.

————. 1986. A revision of the Neotropical bees genus *Osiris*. Wasmann J. Biol. 44:1–56.

————. 1987. Two new species of *Osiris*, with a key to the species from Mexico. Wasmann J. Biol. 45:1–5.

Shinn, A.F. 1964. The bee genus *Xenopanurgus*. Entomol. News 75:73–78.

————. 1967. A revision of the bee genus *Calliopsis* and the biology of *C. andreniformis*. Univ. Kans. Sci. Bull. 46:753–936.

Shreve, F. & I.L. Wiggins. 1964. Vegetation and Flora of the Sonoran Desert. Vol. I. Stanford, CA: Stanford Univ. Press.

Simpson, B.B. & J.L. Neff. 1985. Plants, their pollinating bees, and the great American interchange. *In* F.G. Stehli & S.D. Webb (eds.), The Great American Biotic Interchange. New York: Plenum. pp. 427–452.

——— & ———. 1987. Pollination ecology in the arid Southwest. Aliso 11:417–440.

Snelling, R.R. 1966. The taxonomy and nomenclature of some North American bees of the genus *Centris* with descriptions of new species. Los Angeles Co. Mus. Contr. Sci. 112:1–33.

———. 1974. Notes on the distribution and taxonomy of some North American *Centris*. Los Angeles Co. Mus. Contr. Sci. 259:1–40.

———. 1975. Range extension of two *Heteranthidium*, with description of *H. cordaticeps* male. Proc. Entomol. Soc. Wash. 77:87–90.

———. 1980. The bee genus *Bicornelia*. Contrib. Sci. Nat. Hist. Mus. Los Angeles Co. 327:1–6.

———. 1982. The taxonomy of some Neotropical *Hylaeus* and descriptions of new taxa. Bull. S. Calif. Acad. Sci. 81:1–25.

———. 1983. The North American species of the bee genus *Lithurge*. Contr. Sci. Nat. Hist. Mus. Los Angeles Co. 343:1–11.

———. 1984. Studies on the taxonomy and distribution of American centridine bees. Contr. Sci. Nat. Hist. Mus. Los Angeles Co. 347:1–69.

———. 1986. The taxonomic status of two North American *Lithurge*. Bull. S. Calif. Acad. Sci. 85:29–34.

———. 1987. A revision of the bee genus *Aztecanthidium*. Pan-Pac. Entomol. 63:165–171.

——— & J.G. Rozen, Jr. 1987. Contributions toward a revision of the New World nomadine bees. 2. The genus *Melanomada*. Contr. Sci. Nat. Hist. Mus. Los Angeles Co. 384:1–12.

——— & T.J. Zavortink. 1984. A revision of the cleptoparasitic bee genus *Ericrocis*. Wasmann J. Biol. 42:1–26.

Soulé, M.E. 1990. The real work of systematics. Ann. Missouri Bot. Gard. 77:4–12.

Stange, L.A. 1983. A synopsis of the genus *Epanthidium* Moure with the description of a new species from northeastern Mexico. Pan-Pac. Entomol. 59:281–297.

Stephen, W.P., G.E. Bohart & P.F. Torchio. 1969. The biology and external morphology of bees, with a synopsis of the genera of northwestern America. Corvallis, OR: Agric. Exp. Sta., Oregon State Univ.

Thorp, R.W. 1969. Systematics and ecology of bees of the subgenus *Diandrena*. Univ. Calif. Pub. Entomol. 52:1–146.

———. 1979. Structural, behavioral, and physiological adaptations of bees for collecting pollen. Ann. Missouri Bot. Gard. 66:788–812.

Timberlake, P.H. 1954. A revisional study of the bees of the genus *Perdita* F. Smith, with special reference to the fauna of the Pacific Coast. Part I. Univ. Calif. Pub. Entomol. 9:345–432.

———. 1956. A revisional study of the bees of the genus *Perdita* F. Smith, with special reference to the fauna of the Pacific Coast. Part II. Univ. Calif. Pub. Entomol. 11:247–350.

———. 1958. A revisional study of the bees of the genus *Perdita* F. Smith, with special reference to the fauna of the Pacific Coast. Part III. Univ. Calif. Pub. Entomol. 14:303–410.

———. 1960. A revisional study of the bees of the genus *Perdita* F. Smith, with special reference to the fauna of the Pacific Coast. Part IV. Univ. Calif. Pub. Entomol. 17:1–156.

―――. 1961. A review of the genus *Conanthalictus*. Pan-Pac. Entomol. 37:145–160.

―――. 1962. A revisional study of the bees of the genus *Perdita* F. Smith, with special reference to the fauna of the Pacific Coast. Part V. Univ. Calif. Pub. Entomol. 28:1–124.

―――. 1964. A revisional study of the bees of the genus *Perdita* F. Smith, with special reference to the fauna of the Pacific Coast. Part VI. Univ. Calif. Pub. Entomol. 28:125–388.

―――. 1968. A revisional study of the bees of the genus *Perdita* F. Smith, with special reference to the fauna of the Pacific Coast. Part VII. Univ. Calif. Pub. Entomol. 49:1–196.

―――. 1969a. A contribution to the systematics of North American species of *Synhalonia*. Univ. Calif. Pub. Entomol. 57:1–76.

―――. 1969b. *Metapsaenythia*, a new panurgine bee genus. Entomol. News 80:89–92.

―――. 1971. Supplementary studies on the systematics of the genus *Perdita*. Univ. Calif. Pub. Entomol. 66:1–63.

―――. 1973. Revision of the genus *Pseudopanurgus* of North America. Univ. Calif. Pub. Entomol. 72:1–58.

―――. 1975. The North American species of *Heterosarus* Robertson. Univ. Calif. Pub. Entomol. 77:1–64.

―――. 1976. Revision of the North American bees of the genus *Protandrena* Cockerell. Trans. Amer. Entomol. Soc. 102:133–227.

―――. 1977. Descriptions of new species of *Perdita* in the collection of the California Academy of Sciences. Proc. Calif. Acad. Sci. 41:281–295.

―――. 1980a. Review of North American *Exomalopsis*. Univ. Calif. Publ. Entomol. 86:1–151.

―――. 1980b. Supplementary studies on the systematics of the genus *Perdita*. Part II. Univ. Calif. Pub. Entomol. 85:1–65.

Torchio, P.T. 1974. Notes on the biology of *Ancyloscelis armata* Smith and comparisons with other anthophorine bees. J. Kans. Entomol. Soc. 47:54–63.

Toro, H. & C.D. Michener. 1975. The subfamily Xeromelissinae and its occurrence in Mexico. J. Kans. Entomol. Soc. 48:351–357.

Turner, R.M. & D.E. Brown. 1982. Sonoran desert scrub. *In* D.E. Brown (ed.), Biotic Communities of the American Southwest-United States and Mexico. Desert Plants 4:181–221.

Urban, D. 1967. As especies do genero *Thygater* Holmberg, 1884. Bol. Univ. Fed. Parana (Zool.) 2:177–307.

―――. 1968a. As especies do genero *Melissoptila* Holmberg, 1884. Rev. Brasil. Entomol. 13:1–94.

―――. 1968b. As especies de *Gaesischia* Michener, LaBerge e Moure, 1955. Bol. Univ. Fed. Parana (Zool.) 3:79–129.

―――. 1970. As especies do genero *Florilegus* Robertson, 1900. Bol. Univ. Fed. Parana (Zool.) 3:245–280.

Velásquez, G.P. 1978. Diccionariio de la Lengua Purépecha. México City: Fondo do Cultura Económica.

Villanueva, R. 1984. Plantas de importancia apicola en el ejido Plan del Rio, Veracruz, México. Biotica 9:279–340.

Wallace, A.R. 1876. The Geographical Distribution of Animals. London: Macmillan.

Wiggins, I.L. 1980. Flora of Baja California. Stanford, CA: Stanford Univ. Press

Wille, A. 1959. A new fossil stingless bee (Meliponini) from amber of Chiapas, Mexico. J. Paleont. 33:849–852.

―――. 1977. A general review of the fossil stingless bees. Rev. Biol. Trop. 25:43–46.

————. 1983. Biology of the stingless bees. Ann. Rev. Entomol. 28:41–64.

Zavortink, T.J. 1974. A revision of the genus *Ancylandrena*. Occas. Papers Calif. Acad. Sci. 109:1–36.

————. 1975a. A new genus and species of eucerine bee from North America. Proc. Calif. Acad. Sci. 40:231–242.

————. 1975b. Host plants, behavior, and distribution of the eucerine bees *Idiomelissodes duplocincta* (Cockerell) and *Syntrichalonia exquisita* (Cresson). Pan-Pac. Entomol. 51:230–242

————. 1982. Occurrence of the bee genus *Pectinapis* in the United States. Wasmann J. Biol. 40:18.

———— & W.E. LaBerge. 1976. Bees of the genus *Martinapis* Cockerell in North America. Wasmann J. Biol. 34:119–145.

Zeuner, F.E. & F.J. Manning. 1976. A monograph of fossil bees. Bull. Brit. Mus. (Nat. Hist.) Geol. 27:149–268.

6

Notes on Mexican Ichthyofauna

HÉCTOR ESPINOSA-PÉREZ,
PATRICIA FUENTES-MATA,
MA. TERESA GASPAR-DILLANES,
AND VIRGILIO ARENAS

Mexico's Ichthyofauna is estimated at about 2,122 species of fishes in 779 genera representing 206 families and 41 orders. About 384 species are considered exclusively freshwater. Roughly 375 marine fishes have been registered in continental environments, and the rest are strictly oceanic. Distributional notes on representative families in different areas of the country are given. Endemism, in both marine and freshwater species, is high. Loss of diversity of fish fauna of the country is discussed. Brief notes on introduced species and their possible role in the extinction of indigenous species are provided.

Mexico's fish fauna, comprising the most numerous vertebrate group in the country, is diverse. It is composed of about 2,122 species in 779 genera representing 206 families and 41 orders. This remarkable diversity reflects a variety of aquatic ecosystems from coral reefs to desert streams, embracing major rivers, cenotes, lakes, and coastal lagoons on the continental side, gulfs, islands, bays, and peninsulas in coastal areas, and a vast oceanic area within the limits of the Exclusive Economic Zone (EEZ). As claimed by Mexico, the EEZ covers 2,946,825 km^2, 70% of which is concentrated in the Pacific Ocean. The Mexican Continental Shelf has 375,000 km^2, of which 65% is in the Gulf of Mexico.

The composition of fish species in these ecosystems is greatly influenced by the different physiographic, geologic, and climatic conditions. These fish groups include taxa of South and North American origins (in freshwater ecosystems) as well as of Indopacific and eastern Atlantic origins (in the oceanic and coastal parts). In addition, there are a great number of autochthonous fish genera and species represented in the country. The two coasts share over 100 coastal lake systems of fluvial origin that are either ecotonic or transitional. Species richness is not associated with the size of the country's aquatic environments but with their heterogeneity and tropical latitudes. Mexico's richness is highlighted when its total 2,122

species is compared with 2,268 species described for the combined area of
Canada and the United States (Robins et al., 1980). Miller's (1986) estimate
of almost 500 species of freshwater fishes for Mexico constitutes 60% of
those for North America.

TAXONOMIC STUDIES

Interest in Mexican fishes predates pre-Hispanic times when different
species were named according to their color, habitat type, and use. Species
of *Chirostoma*, currently called "Charales" and "Pescado Blanco" (Alvarez,
1981) were known by such names as "Amiloth," "Iztamichin" and
"Papalomichin."

Linnaeus (1758) treated fish species of wide distributions, some of which
extended into Mexico's waters. Studies of Cuvier and Valenciennes (1828-
1849) included several species from Mexico and those by Haeckel (1848) a
few species native to the state of Veracruz. Among the early Mexican
students was Bustamente y Septien (1837), who described *Cyprinus viviparus*
(= *Girardinichthys viviparus*), a goodeid from the Valley of Mexico. A
significant work of the last century by Baird and Girard (1854) included
most of the Nearctic fishes of Mexico. The contributions of Günther (1859-
1870), Steindachner (1875a,b, 1879), and Bean (1883, 1887) were notable.
Jordan and Evermann (1896-1900) attributed over 500 species to Mexico.
Meek (1902, 1904, 1907) described over 50 species of Mexican freshwater
fishes. Studies by Regan (1906-1908) and Vaillant and Bocourt (1878) were
important for understanding the fishes of Mexico. Contributions by Hubbs
(1918, 1924, 1932, 1952, 1960) and de Buen (1947a,b) influenced Alvarez
(1946, 1949a,b, 1950, 1952, 1959, 1960, 1964, 1970, 1981), a distinguished
Mexican ichthyologist. Others who have contributed significantly are
Cuesta-Terrón (1925), Beltrán (1934), Martín del Campo (1938), Berdegué
(1956), and Maldonado-Koerdell (1956). More recently, Miller (1982, 1986)
has been one of the notable students of Mexico's fish fauna.

The study described in this chapter is based on the collections at the
Instituto de Biología UNAM (IBUNAM-P), the Secretaría de Marina of
Mexico (SM-P), and the Universidad Autónoma de Nuevo León (UANL),
as well as the published literature (Castro, 1983; Castro-Aguirre, 1978;
Hoese & Moore, 1977; Miller, 1986; Miller & Smith, 1986; Smith & Miller,
1986; Randall, 1984; Thompson et al., 1979). Estimates of diversity were
made following the data presented in Cohen (1970), Robins et al. (1980),
Berra (1981), and Nelson (1984).

AN OVERVIEW OF MEXICO'S ICHTHYOFAUNA

Moyle and Cech (1988) estimated that about 21,700 living fish species have
been described worldwide, but this number may reach 28,000 (Nelson,
1984). It is remarkable that only 58% of modern fish species are marine

Table 6.1. Mexico's ichthyofauna: order, number of families, genera, and species

Order	Families (No.)	Genera (No.)	Species (No.)
Myxiniformes	1	2	2
Petromyzontiformes	1	2	4
Chimaeriformes	2	2	2
Hexanchiformes	2	4	6
Heterodontiformes	1	1	3
Lamniformes	7	23	62
Squaliformes	3	15	28
Rajiformes	7	23	69
Acipenseriformes	1	2	3
Lepisosteiformes	1	2	4
Elopiformes	3	3	4
Notacantiformes	1	1	1
Anguilliformes	14	39	79
Clupeiformes	2	19	61
Gonorynchiformes	1	1	1
Cypriniformes	2	28	102
Characiformes	1	5	7
Siluriformes	3	11	40
Gymnotiformes	1	1	1
Salmoniformes	6	15	22
Stomiiformes	9	18	35
Aulopiformes	9	21	39
Myctophiformes	2	21	46
Ophidiiformes	3	14	27
Gadiformes	5	9	11
Batrachoidiformes	1	3	12
Lophiiformes	4	9	23
Gobiesociformes	1	5	27
Cyprinodontiformes	8	52	184
Atheriniformes	1	10	40
Lampriformes	4	6	8
Beryciformes	6	12	25
Zeiformes	1	1	1
Gasterosteiformes	2	2	2
Syngnathiformes	4	11	32
Dactylopteriformes	1	1	1
Synbranchiformes	1	2	3
Scorpaeniformes	7	29	115
Perciformes	68	302	851
Pleuronectiformes	4	29	88
Tetraodontiformes	5	21	46
Total	206	779	2,122

(Moyle & Cech, 1988) even though saltwater covers 70% of the earth's surface. Forty-one percent of the fishes are freshwater inhabitants, with 1% moving on a regular basis between the two environments (Cohen, 1970). The greatest number of species is found in the tropical and subtropical regions. The richest zones, in order of importance, are the eastern Indo-Pacific (including Indian Ocean, Polynesia, and Oceania), the Caribbean and the Antilles (from Florida to Brazil). The higher biomass in the temperate regions and the poles has facilitated the development of big fishing industries, which exploit a relatively small number of species.

The fishes of the world were grouped into 50 orders and 445 families by Nelson (1984). As many as 41 orders (82%) and 206 families (46.3%) are represented in Mexican waters. These figures are based on a provisional list (Espinosa et al., in press.) that records 779 genera and 2,122 species for Mexico (Table 6.1). Approximately, 102 genera and 384 of these species are considered strictly freshwater (Table 6.2). However, Miller (1986) estimated the number of continental fishes in the country to be around 500, which includes species from brackish waters, coastal lagoons, and estuaries. Table 6.3 presents families of marine fishes found in continental waters. The generic and species representations for the country, respectively, are 19.8% and 10.2% of the world total.

The principal zoogeographical continental provinces of Mexico are Nearctic and Neotropical. Parts of Mexico considered Nearctic in this chapter, following Alvarez and De Lachica (1974), include the Mexican altiplano, the mountain ranges that border it, the Baja California Peninsula, the Sonoran Desert region, and the coastal plain of Tamaulipas south to about 24° latitude (Rio Soto La Marina). The high valleys of the Sierra Madre del Sur and the Chiapas-Guatemala altiplano in Central America are considered to be the extreme southern limits of this province. The river basins in this province are located in the Baja California Peninsula, northwestern Sinaloa, Chihuahua, northern Mexico (including the Bravo and Salado Rivers), Cuatro Cienegas, El Tunal, and the basins of the Nazas-Aguanaval and Mezquital, which comprise the Armeria River and the central part of the Lerma River. In these river basins 47 genera and 152 species have been recorded. The following families, genera, and species are common in these areas (Table 6.2): Petromizontidae (*Lampetra spadicea, L. geminis* endemic to Lerma-Santiago); Acipenseridae (*Scaphirhynchus platorynchus*, which is now extinct in Rio Bravo [Miller, 1986]), Salmonidae (*Oncorhynchus chrysogaster, O. gairdneri*); Catostomidae (*Carpiodes carpio, Catostomus* with several endemic species to the Yaqui-Mayo River, *Ictiobus*, and *Moxostoma*); Cyprinidae (*Algansea* spp., *Campostoma* spp., *Dionda* spp., *Cyprinella* spp., *Notropis* spp., and *Gila* spp.); Goodeidae (*Allodontichthys, Allophorus, Allotoca, Chapalichthys, Characodon, Ameca, Goodea, Girardinichthys, Neophorus, Ataeniobius, Hubbsina, Ilyodon, Skiffia, Xenoophorus, Xenotaenia,* and *Zoogonecticus*); Centrarchidae (*Lepomis* spp. and *Micropterus salmoides*); and Percidae (*Etheostoma* spp. and *Percina macrolepida*).

Areas of Mexico in the Neotropical province include the Mexican lowlands, which border the Nearctic region. Bodies of water are more

Table 6.2. Families of strictly freshwater Mexican fishes with numbers of genera and species

Family	Distribution	Genera (No.)	Species (No.)
Petromyzontidae	Nearctic	1	2
Acipenseridae	Nearctic	1	1
Salmonidae	Nearctic	1	4
Catostomidae	Nearctic	6	20
Cyprinidae	Nearctic	17	78
Goodeidae	Nearctic	17	37
Centrarchidae	Nearctic	2	5
Percidae	Nearctic	2	5
Characidae	Neotropical	5	7
Gymnotidae	Neotropical	1	1
Pimelodidae	Neotropical	1	6
Ariidae	Neotropical	2	2
Gobiesocidae	Neotropical	1	2
Hemirhamphidae	Neotropical	1	1
Anablepidae	Neotropical	1	1
Synbranchidae	Neotropical	2	3
Bythitidae	Neotropical	1	1
Batrachoididae	Neotropical	1	1
Belonidae	Neotropical	1	1
Gerreidae	Neotropical	1	1
Lepisosteidae	Transitional nearctic	2	4
Ictaluridae	Transitional nearctic	4	12
Poeciliidae	Transitional neotropical	11	75
Cichlidae	Transitional neotropical	2	41
Cyprinodontidae	Shared	7	30
Atherinidae	Shared	3	31
Sciaenidae	Shared	1	1
Mugilidae	Shared	2	2
Clupeidae	Shared	1	2
Eleotrididae	Shared	2	4
Gobiidae	Shared	2	3
Total		102	384

abundant in this province; the important river basins are those of the Lower Lerma, those along the coasts of Guerrero and Michoacán, the Balsas and Papaloapan Rivers, the Isthmus of Tehuantepec, the Grijalva-Usumacinta system, and the Yucatán. The families, genera, and species of this region (Table 6.2) include Characidae (*Astyanax fasciatus*, *Brycon guatemalensis*, *Bramocharax caballeroi* [which is endemic to Catemaco Lake], and *Hyphessobrycon compresus*); Gymnotidae (*Gymnotus cylindricus*); Pimelodidae (*Rhamdia guatemalensis*, *R. reddelli* [blind fish, endemic to the Papaloapan River]); Ariidae (*Cathorops aguadulce* and *Potamarius nelsoni*, endemic to the

Table 6.3. Marine fishes registered in Mexican continental waters temporarily or permanently, including coastal lagoons, estuaries, and brackish waters

Family	Genera	Species	Family	Genera	Species
Heterodontidae	1	1	Centropomidae	1	10
Orectolobidae	1	1	Serranidae	5	11
Carcharhinidae	4	8	Grammistidae	1	1
Sphyrnidae	1	3	Echeneididae	1	1
Pristidae	1	2	Carangidae	7	16
Torpedinidae	1	1	Lutjanidae	3	15
Rhinobatidae	1	2	Lobotidae	1	1
Rajidae	1	1	Gerreidae	5	14
Dasyatidae	4	8	Haemulidae	6	18
Myliobatidae	2	2	Sparidae	5	6
Elopidae	1	1	Sciaenidae	14	34
Megalopidae	1	1	Kyphosidae	1	1
Albulidae	1	1	Ephippidae	2	3
Anguillidae	1	1	Chaetodontidae	1	1
Muraenidae	1	1	Pomacentridae	1	2
Ophichthidae	3	5	Mugilidae	2	7
Clupeidae	5	9	Sphyraenidae	1	2
Engraulididae	4	22	Polynemidae	1	2
Chanidae	1	1	Uranoscopidae	1	1
Ariidae	5	16	Clinidae	2	2
Salmonidae	1	2	Chaenopsidae	1	1
Synodontidae	1	2	Blenniidae	2	2
Gadidae	1	7	Dactyloscopidae	1	1
Ophidiidae	1	1	Eleotrididae	4	10
Batrachoididae	3	4	Gobiidae	18	27
Antennariidae	2	2	Acanthuridae	1	1
Ogcocephalidae	1	1	Trichiuridae	1	1
Gobiesocidae	1	2	Scombridae	1	2
Hemiramphidae	3	7	Stromateidae	2	2
Belonidae	2	4	Bothidae	5	7
Atherinidae	6	20	Cynoglossidae	1	2
Gasterosteidae	1	1	Soleidae	3	7
Syngnathidae	4	9	Balistidae	2	2
Dactylopteridae	1	1	Ostraciidae	1	1
Scorpaenidae	1	3	Tetraodontidae	2	3
Triglidae	1	2	Diodontidae	2	2
Cottidae	1	1			
Total				179	375

Grijalva-Usumacinta system); Gobiesocidae (*Gobiesox fluviatilis* and *G. mexicanus*); Hemirhamphidae (*Hyporhamphus mexicanus*), Anablepidae (*Anableps dowi*); Synbranchidae (*Synbranchus marmoratus* and *Ophisternon infernalis*); Bythitidae (*Typhliasina pearsei*, endemic to the Yucatán Peninsula); Batrachoididae (*Batrachoides goldmani*); Belonidae (*Strongylura hubbsi*); and Gerreidae (*Diapterus mexicanus*, endemic to the Grijalva-Usumacinta system). These families make up about 27 species.

The division of these two provinces is both ecological and climatic. Mountain systems, which define their limits, allow Neotropical and Nearctic elements to intergrade, resulting in a complex zone of transposition of faunas that has been called the "Central American–Mexican transition zone" (Darlington, 1957). Freshwater fish have a clearly marked transition zone extending along the length of the coastal plain from Rio Soto La Marina (75% North American forms) to Papaloapan (95% Mesoamerican forms). On the Pacific coast the fauna is poor with the exception of the Lerma River basin. A few primary northern species (*sensu* Myers, 1940), some secondary ones, and some marine invaders in the south (Miller, 1982) are found here.

The families, genera, and species found in this transition zone include Ictaluridae (*Amiurus melas, Ictalurus* [nine species], *Prietella phreatophila* endemic to the Bravo River, and *Pylodictis olivaris*) and Lepisosteidae (*Lepisosteus oculatus* and *Atractosteus spatula*) in the transitional Nearctic. In the transitional Neotropical region are found Poeciliidae (*Gambusia* [about 19 spp.], *Poecilia* [about 12 species], *Poeciliopsis* [17 species], and *Xiphophorus* [13 species]) and Cichlidae (*Cichlasoma* [about 40 species]). These families make up about 132 species (Table 6.2). The shared families are Cyprinodontidae (*Cualac tessellatus, Cyprinodon* [18 species], *Profundulus* spp., *Lucania, Megupsilon, Fundulus,* and *Rivulus*), Atherinidae (*Chirostoma* [19 species], *Atherinella* [nine species], *Poblana alchichica, P. ferdebueni,* and *P. letholepis*), Sciaenidae (*Aplodinotus grunniens*), Mugilidae (*Agonostomus monticola* and *Joturus pichardoi*), Eleotrididae (*Gobiomorus dormitor, G. maculatus, G. polylepis,* and *Leptophilypnus*), and Gobiidae (*Gobionellus atripinnis, Sicydium gymnogaster,* and *S. multipunctatum*). These families comprise about 75 species (Table 6.2).

The wide distribution of marine fishes makes their geographic assignment difficult. Those captured in Mexican waters are considered Mexican for the purposes of this study. On Mexico's Pacific coast three subprovinces may be recognized: San Dieguina or Californian (northern frontier of Baja California and Bahia Magdalena), Sea of Cortes (Gulf of California to Mazatlan), and Panamic extending from the southern limits of the Sea of Cortes to the southern frontier of the State of Chiapas (Briggs, 1974; Ekman, 1953). The three provinces on the Atlantic coast include the Carolinian (from the northern frontier of the state of Tamaulipas to Cabo Rojo between Tuxpan and Nautla in Veracruz), the Caribbean, extending from here to the southern frontier of state of Quintana Roo, and the Antillean (which includes the island region and the coral reefs (Greenfield & Johnson, 1981). A total of 1,304 species of fish have been recorded for both regions.

Studies by Osburn and Nichols (1916), Meek and Hildebrand (1923–1928), Breder (1928a,b), Fowler (1941), Hildebrand (1946), Berdegué (1956), Ricker (1959a,b), Ramírez and Paez (1965), Amezcua-Linares (1977), and Fuentes-Mata and Gaspar-Dillanes (1981) have provided data for our understanding of the fish fauna along the Pacific coast: 507 species are known from the Californian Region.

The common families and genera include Scorpaenidae (*Scorpaenoides xyris, Scorpaena guttata, Sebastes atrovirens, S. auriculatus, S. aurora,* and *S. caurinus*); Triglidae (*Bellator xenisma* and *Prionotus stephanophrys*); Cottidae (*Artedius corallinus, A. creaseri, A. lateralis, Leptocottus armatus, Leiocottus hirundo,* and *Paricelinus hopliticus*); Carangidae (*Caranx caninus, Seriola lalandi, S. rivoliana, Chloroscombrus orqueta, Selene peruvianus, Trachinotus symmetricus, T. rhodopus,* and *Uraspis secunda*), some Elasmobranchies; Alopiidae (*Alopias superciliosus,* and *A. vulpinus*); Carcharhinidae (*Carcharhinus leucas, C. brechyurus, C. obscurus,* and *C. longimanus*); Sphyrnidae (*Sphyrna lewini,* and *S. zygaena*); Rhinobathidae (*Rhinobatus productus*); Dasyatidae (*Urolophus halleri, Gymnura marmorata,* and *Dasyatis brevis*); Myliobatidae (*Myliobatis californica*); and Rajidae (*Raja inornata, R. rhina, R. binoculata, R. stellulata*).

Miller and Lea (1972) listed 554 species for a larger zone of the same region. In the Sea of Cortes there are about 404 species. Thompson et al. (1979) have claimed that the reef species alone, excluding pelagic and deep-sea species, make up 271 species. Walker (1960) has suggested that this number could reach 800 species without taking into account the deep-sea fishes. Briggs (1964, 1967) has mentioned at least 62 Transpacific species. A brief account of these species can be found in Castro-Aguirre et al. (1970). The main families for this zone in terms of richness of species and genera include Serranidae (*Epinephelus panamensis, E. acanthistius, E. labriformis, Mycteroperca jordani, M. rosacea, Paranthias colonus, Paralabrax maculatofasciatus,* and *P. auroguttatus*); Sciaenidae (*Cynoscion xanthulus, Cynoscion reticulatus, Totoaba macdonaldi, Pareques viola, Bairdiella armata, Larimus pacificus,* and *Isopisthus remifer*); Pomacentridae (*Eupomacentrus rectifraenum, Abubdefduf troschelii,* and *Chromis atrilobatus*); Labridae (*Halichoeres semicinctus, H. dispilus, Bodianus diplotaenia, Decodon melasma, Pseudojulis notospilus,* and *Thalassoma lucasanum*); Clinidae (*Labrisomus xanti, Malacoctenus gigas, M. hubbsi, Starksia hoesei,* and *Paraclinus sini*); and Gobiesocidae (*Gobiesox adustus, G. papillifer, G. pinniger, G. schultzi, Arcos erythrops, Pherallodiscus funebris, Tomicodon boehlkei, T. eos, T. zebra, T. humeralis,* and *T. myersi*).

The Panamic zone, as defined by Ekman (1953), is a continuum extending from the equator to California. In areas restricted, and relevant, to this study (Mexico), it is estimated that there are 505 species. The main families in this area include Ariidae (*Arius liropus, A. seemani, Ariopsis coerulescens, A. gilberti,* and *Bagre panamensis*); Bothidae (*Citharichthys gilberti, Cyclopsetta querna, Etropus crossotus, Hippoglossina bollmani, Syacium ovale,* and *S. latifrons*); Carangidae (*Caranx caballus, C. marginatus, Chloroscombrus orqueta, Oligoplites saurus, Selene brevoorti,* and *Trachinotus kennedyi*); Clupeidae (*Harengula thrissina, Lile stolifera, Opisthonema libertate,* and *Pleisteostoma lutipinnis*); Engraulididae (*Anchoa scofieldi, A. ischana, Anchoviella analis,* and *Anchovia macrolepidota*); Gerreidae (*Diapterus peruvianus, Eucinostomus dowii,*

Table 6.4. Amphiamerican genera and typical species registered on Mexican coasts

Family	Atlantic coast	Pacific coast
Dasyatidae	*Urolophus jamaicensis*	*U. halleri*
Clupeidae	*Opisthonema oglinum*	*O. libertate*
Engraulididae	*Cetengraulis edentulus*	*C. mysticetus*
	Anchovia clupeoides	*A. macrolepidota*
	Anchoviella perfasciata	*A. analis*
	Anchoa mitchilli	*A. scofieldi*
Batrachoididae	*Batrachoides surinamensis*	*B. pacifici*
Gobiesocidae	*Gobiesox strumosus*	*G. adustus*
Triglidae	*Prionotus tribulus*	*P. stephanophrys*
Centropomidae	*Centropomus nigrescens*	*C. parallelus*
Serranidae	*Diplectrum formosum*	*D. pacificum*
Carangidae	*Chloroscombrus chrysurus*	*C. orqueta*
Gerreidae	*Eucinostomus gula*	*E. dowii*
	Diapterus auratus	*D. peruvianus*
	Eugerres plumieri	*E. axillaris*
Haemulidae	*Anisotremus surinamensis*	*A. interruptus*
	Anisotremus virginicus	*A. taeniatus*
	Haemulon parrai	*H. scudderi*
Sparidae	*Calamus lecosteus*	*C. brachysomus*
Sciaenidae	*Cynoscion nebulosus*	*C. xanthulus*
	Cynoscion arenarius	*C. reticulatus*
	Micropogon altipinnis	*M. undulatus*
	Menticirrhus americanus	*M. panamensis*
	Larimus effulgens	*L. breviceps*
	Bairdiella ronchus	*B. armata*
Dactyloscopidae	*Dactyloscopus tridigitatus*	*D. amnis*
Chaenopsidae	*Emblemaria pandionis*	*E. hypacanthus*
Gobiidae	*Gobiomorus dormitor*	*G. maculatus*
	Eleotris pisonis	*E. picta*
	Awaous tajasica	*A. trasandeanus*
	Gobionellus hastatus	*G. microdon*
Bothidae	*Citharichthys spilopterus*	*C. gilberti*
Soleidae	*Trinectes maculatus*	*T. fonsecensis*
	Achirus lineatus	*A. mazatlanus*

E. melanopterus, and *Eugerres lineatus)*; Gobiidae *(Awaous trasandeanus, Dormitator latifrons, Gobionellus microdon, Eleotris pictus, Gobiosoma etheostoma, Microgobius miraflorensis,* and *Gobiomorus maculatus)*; Lutjanidae *(Lutjanus peru, L. guttatus, L. novemfasciatus, L. argentiventris, Hoplopagrus guentheri,* and *Rabirrubia inermis)*; Sciaenidae *(Larimus effulgens, Menticirrhus panamensis, Ophioscion strabo,* and *Umbrina xanti);* and Triglidae *(Prionotus ruscarius* and *P. quiescens).* Several amphiamerican genera and sibling species of reef fishes have been recorded by Thompson et al. (1979). Castro-

Aguirre (1978) has provided data for coastal sand-bottom fishes and some pelagic species. In Table 6.4 we integrated this information with our observations.

In the Carolinian subprovince on the Atlantic coast, about 555 species have been estimated. This figure contrasts with the earlier 497 reported by Hoese and Moore (1977) and the 161 reported by Walls (1975). Studies by Castro-Aguirre et al. (1986), who listed 100 species for the Tuxpam-Tampachoco system, and Gómez-Soto and Contreras-Balderas (1988), who reported 78 species for the Laguna Madre, as well as unpublished data (Espinosa et al., in prep.), which estimate 100 species for the Laguna de Tamiahua, have contributed to the estimate provided here. The best represented families in the region are Carangidae (*Caranx ruber, C. fusus, C. bartholomaei, C. lugubris, C. hippos, C. latus, Trachinotus carolinus, T. falcatus, T. goodei, Oligoplites saurus, Seriola rivoliana, S. fasciata, S. dumerili, S. zonata, Decapterus punctatus*, and *Chloroscombrus chrysurus*); Haemulidae (*Haemulon aurolineatum, H. striatum, H. melanurum, H. parrai, H. plumieri, Anisotremus virginicus, A. surinamensis, Conodon nobilis*, and *Pomadasys crocro*); Mullidae (*Upeneus parvus, Mullus auratus, Pseudupeneus maculatus*, and *Mulloidichtys martinicus*); Centropomidae (*Centropomus undecimalis, C. ensiferus, C. poeyi, pectinatus*, and *C. parallelus*); Gerreidae (*Ulaema lefroyi, Eucinostomus argenteus, E. gula, Diapterus auratus, D. rhombeus, Eugerres plumieri, E. brasilianus*, and *Gerres cinereus*); Serranidae (*Pikea mexicana, Paranthias furcifer, Hemanthias leptus, Diplectrum formosum, D. bivittatum, Serranus phoebe, S. atrobranchus, Centropristes striata, C. ocyura, Epinephelus nigritus, E. flavolimbatus, E. niveatus, E. inermis, E. itajara, E. guttatus, Mycteroperca rubra, M. phenax*, and *M. microlepis*); Sparidae (*Calamus campechanus, C. bajonado, C. nodosus, C. leucosteus, C. arctifrons, Lagodon rhomboides, Archosargus probatocephalus, Diplodus holbrooki*, and *Stenotomus caprinus*); and Tetraodontidae (*Lagocephalus laevigatus, Canthigaster rostrata, Sphoeroides dorsalis, S. spengleri, S. pachygaster, S. testudineus, S. nephelus*, and *S. parvus*).

In the Caribbean region about 627 species have been recorded. This province is one of the richest in Mexico, possibly due to complex reef developments in the area. It shares over 30 species with the eastern Atlantic (Briggs, 1967). The most representative data are perhaps those of Zaneveld (1983), who gives a figure of 683 species for a wider area of the same region. Fisher (1978) estimated over 800 species in the Western Central Atlantic ("fishing area 31"), but part of this area covers what has here been defined as Antillean. The most representative families of the region are Ariidae (*Ariopsis felis* [abundant], *Bagre marinus*, and *Cathorops spixii*); economically important Lutjanidae (*Lutjanus* [red snapper]: *L. analis, L. apodus, L. griseus, L. jocu, L. mahogani, L. synagris, L. vivanus; Ocyurus chrysurus, Rhomboplites aurorubens*, and *Pristipomoides macrophthalmus*); Chaetodontidae (*Centropyge argi, Chaetodon capistratus, Ch. ocellatus, Ch. sedentarius, Holocanthus ciliaris, H. isabelita, Pomacanthus arcuatus*, and *P. paru*); Sciaenidae (*Cynoscion nebulosus, Equetus acuminatus, E. lanceolatus, E. punctatus, Larimus fasciatus, Micropogonias furnieri, M. undulatus, Pogonias cromis, Stellifer lanceolatus*,

Umbrina coroides, Bairdiella batabana, and *B. ronchus);* Sphyraenidae *(Sphyraena* ["barracuda"]: *S. barracuda, S. borealis, S. guachancho, S. picudilla);* Serranidae *(Epinephelus guttatus, E. adscensionis, E. flavolimbatus, E. niveatus, Petrometopon cruentatum,* and *Serranus phoebe);* and Clupeidae *(Harengula clupeola, H. humeralis, H. pensacolae, Jenkinsia lamprotaenia, J. stolifera, Sardinella anchovia,* and *Opisthonema oglinum).*

The Antillean province, so called because of its proximity to that region, has 221 species, which, in comparison with the 323 noted by Böhlke and Chaplin (1968), may be considered the poorest zone. Taxa peripheral to the island and reef areas have not been considered in this report. If included, they increase the number of species for the area. On the other hand, several species found here have been included in the Caribbean region. The best represented families and genera in this zone are Scaridae *(Nicholsina usta, Sparisoma rubripinne, S. chrysopterum, S. viride, Scarus iserti, S. vetula,* and *S. guacamaia);* Pomacentridae *(Stegastes dorsopunicans, S. planifrons, S. variabilis, Abudefduf saxatilis, A. taurus, Chromis multilineata,* and *Ch. cyanea);* Haemulidae *(Haemulon steindachneri, H. chrysargyreum, Orthopristis,* and *Anisotremus);* Labridae *(Bodianus rufus, B. pulchellus, Halichoeres bivittatus,* and *Thalassoma bifasciatum);* Blenniidae *(Ophioblennius atlanticus, Parablennius marmoreus, Entomacrodus nigricans,* and *Labrisomus nuchipinnis);* Clinidae *(Malacoctenus triangulatus* and *Paraclinus fasciata);* and Acanthuridae *(Acanthurus chirurgus, A. bahianus,* and *A. coeruleus).*

ORIGIN AND EVOLUTION OF THE GROUP IN MEXICO

The complex geological history of Mexico, which has played a pivotal role in the evolution of its biota (see other contributions in this volume), has had an equally important role in the evolution of fishes in the country. Aquatic habitats are notoriously unpredictable. Their drying leads to local extinction of taxa. Their division results in uninhabitable intervening dry areas, in turn resulting in fragmentation and genetic isolation of populations, possibly leading to local speciation events. The high number of autochthonous taxa in Mexican fishes suggests significant local evolution, as in the Petromyzontidae *(Lampetra),* one of the more primitive groups of vertebrates.

Most of the continental fishes of Mexico have an oceanic origin (Miller & Smith, 1986). Examples include species complexes of *Chirostoma* in the Lerma River (see Fig. 6.5 below), or the species in estuaries. Their radiation and evolution "on land" are intimately related to the geological history of Mexico as has been discussed by Miller and Smith (1986) for the freshwater fishes in what they called the "Mesa Central," which includes parts of various morphotectonic provinces of Mexico (see Ferrusquía-Villafranca, this volume), including the Sierra Madre Occidental, Sierra Madre Oriental, Trans-Mexican Volcanic Belt, and Sierra Madre del Sur. Miller and Smith (1986) have discussed examples of evolution through fragmentation

Table 6.5. Estimated percentages of endemic fishes in Mexican drainages and coasts

Region	%
Continental waters	
Lerma-Santiago	66
Usumacinta-Grijalva	36
Panuco	40
Rio Balsas	35
Rio Ameca	32
Rio Papaloapan	21
Rio Coatzacoalcos	13
Rio Conchos	21
Rio Tunal	62
Cuatro Cienegas Depression	50
Chichankanab lake	85
Media Luna lake	65
Marine waters	
Gulf of California	20
Mexican Caribbean	15
Gulf of Tehuantepec	15+
Gulf of Mexico (north)	15+

and vicariance. Fossils belonging to Goodeidae from these parts have been analyzed by them and others (Alvarez & Arreola, 1972; Alvarez & Moncayo, 1976). Additional cases of vicariance may be seen in Atherinidae (*Chirostoma*) and Petromyzontidae (*Lampetra spadicea* and *L. geminis*). Applegate and Espinosa-Arrubarena (1982) suggested estuarine environments during the Cretaceous for parts of this region.

Examples of population fragmentation and differentiation in northern Mexico have been noted in Cyprinidae by Uyeno and Miller (1971). Smith and Miller (1986)·considered this point when discussing the relations between extant species and fossils in the Río Bravo region.

The evolution of the present-day Baja California Peninsula during the Miocene has isolated populations of Nearctic taxa from the mainland. Populations of *Fundulus* are classic examples, and those in Baja California have remained in isolation since the Miocene. The Pacific islands of continental and oceanic origin are recent.

In the southeast, bodies of water—cenotes, hummocks (petenes), and sinkholes (aguadas)—dating back to the Eocene-Miocene have been identified (Beltrán, 1959) in the Yucatán Peninsula. It is possible that the freshwater fauna here have special characteristics of vicariance and sympatry. Hubbs (1936, 1939) has discussed their origin and dispersal. No fossil records exist for this region.

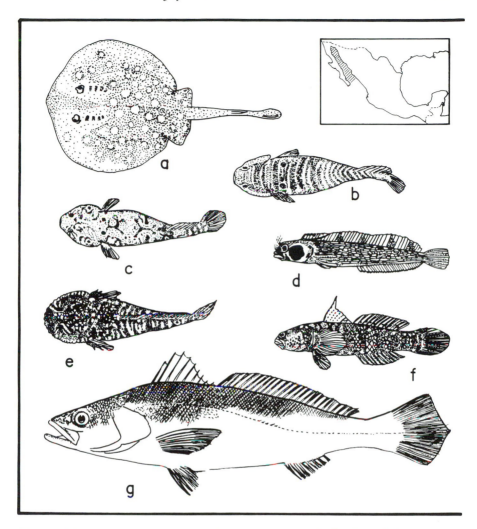

Figure 6.1. Some endemic fishes from the Gulf of California. A. *Urolophus concentricus.* B. *Tomicodon humeralis.* C. *Gobiosoma boehlkei.* D. *Acanthemblemaria crockeri.* E. *Gobiesox pinniger.* F. *Gobiosoma chiquita.* G. *Totoaba macdonaldi.*

The Atlantic coast has a great number of islands with well developed coral reefs and an extensive continental shelf, sandy beaches, and numerous rivers with deltas and estuaries. The Gulf of Mexico has been subject to recent geological changes as well as to effects of glaciations during the Quaternary. The emergence of the Yucatán Peninsula during this period may have resulted in the present-day distribution patterns seen among the following amphiamerican genera: *Epinephelus* (Serranidae), *Gerres* and *Eucinostomus* (Gerreidae), *Conodon* (Haemulidae), *Scorpaena* (Scorpaenidae) and *Epinnula* (Gempylidae) (Briggs, 1974; Castro-Aguirre, 1978).

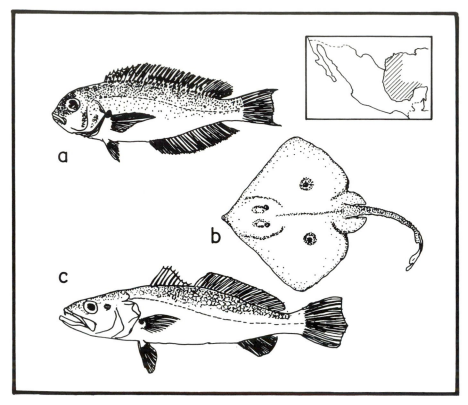

Figure 6.2. Some endemic fishes from the Gulf of Mexico.   A. *Caulolatilus intermedius.* B. *Raja texana.* C. *Cynoscion arenarius.*

ENDEMISM

Different river basins are notable for varying compositions of fish species, with many harboring exclusive fish faunas. Endemism is high in Goodeidae and Atherinidae in the Lerma-Santiago basin. The Grijalva-Usumacinta system supports endemics of Poeciliidae (*Gambusia* and *Priapella*) and Cichlidae. As many as ten species of *Cichlasoma* (Cichlidae) are endemic to this system. The Panuco River provides examples of endemics in Poeciliidae (*Xiphophorus*) and Cichlidae (*Cichlasoma*).   Table 6.5 lists principal river basins and percentages of endemic species known for each. Miller (1986) has discussed the origins and causes of endemism in each of these regions. The families with the highest number of endemic species are Petromyzontidae, Clupeidae, Cyprinidae, Cichlidae, Cyprinodontidae, Goodeidae, Atherinidae, and Poeciliidae.

There are no reliable data on endemic marine fishes. It has been suggested that approximately 20% of the species in the Gulf of California are endemic (Walker, 1960). Briggs (1974) recognized a Mexican Province in the region (here considered part of the Panamic) where species endemism exceeds 15%. It must be emphasized that few studies have been

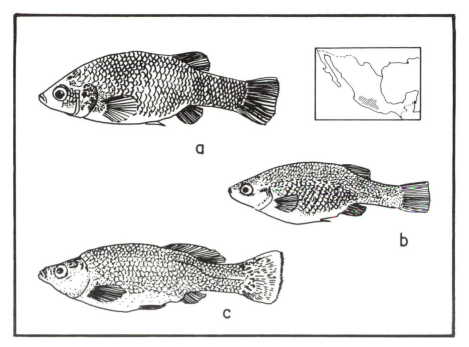

Figure 6.3. Family Goodeidae, endemic to the Lerma River. A. *Xenoophorus captivus erro*. B. *X. captivus captivus*. C. *Ilyodon xantusi*.

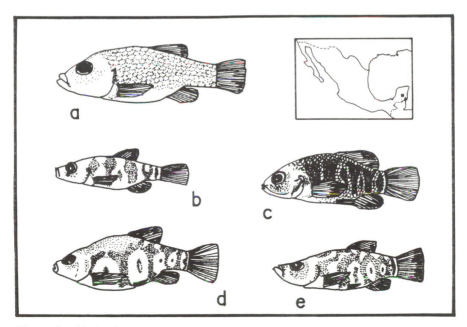

Figure 6.4. Endemic Pupfishes from Chichankanab Lake in the Yucatan Peninsula. A. *Cyprinodon maya*. B. *C. simus*. C. *C. verecundus*. D. *C. beltrani*. E. *C. labiosus*.

Table 6.6. Exotic species in Mexico

Family	Species	Sites in Mexico
Clupeidae	*Dorosoma petenense*	Colorado and Bravo Rivers
Salmonidae	*Oncorhynchus gairdneri*	Michocán, Chiapas, Durango, Valley of Mexico
	Salvenilus fontinalis	Sonora, Michoacán, Valley of Mexico
	Oncorhynchus tshawytscha	Northern Mexico
Cyprinidae	*Carassius auratus*	Several states
	Ctenopharyngodon idella	Several states
	Cyprinus carpio	Several states
	Hypophtalmichthys molitrix	Several states
	Megalobrema amblycephala	Several states
	Aristichthys nobilis	Several states
	Mylopharyngodon piceus	Several states
	Barbus conchonius	Unknown
	Barbus titleya	Unknown
	Gila orcutti	Unknown
	Notemigonus crysoleucas	Unknown
Catostomidae	*Carpiodes cyprinus*	Unkown
Cobitidae	*Misgurnus anguillicaudatus*	Valley of Mexico
Ictaluridae	*Ictalurus furcatus*	Yaqui River
	Ictalurus melas	Sonora, Chihuahua
	Ictalurus punctatus	Jalisco, Sonora Nayarit, Tampico.
Percichthidae	*Morone chrysops*	Bravo River
	Morone saxatilis	Colorado & Bravo Rivers
Centrarchidae	*Amboplites rupestris*	Chihuahua
	Lepomis auritus	Coahuila
	Lepomis cyanellus	Baja California N., Sonora
	Lepomis gulosus	Aguascalientes, Tampico.
	Lepomis macrochirus	Valley of Mexico, several states
	Lepomis microlophus	Unknown
	Micropterus dolomieui	Unknown
	Micropterus salmoides	Patzcuaro, Michoacán, several states
	Pomoxis annularis	Several states
	Pomoxis nigromaculatus	Unknown
Cichlidae	*Oreochromis mossambicus*	Several states
	Oreochromis niloticus	Several states
	Oreochromis aureus	Several states
	Oreochromis urolepis hornorum	Several states

Table 6.6. (cont.)

Family	Species	Sites in Mexico
	Tilapia rendalli	Several states
	Tilapia zillii	Northern Mexico
Characidae	*Collosoma macropomun*	Zacatepec, Morelos
Cyprinodontidae	*Fundulus zebrinus*	Bravo River
Poeciliidae	*Poecilia reticulata*	Several states

carried out on the Gulf of Tehuantepec, where species endemism may exceed 15%. Figure 6.1 illustrates some endemic fishes from this region: *Urolophus concentricus, Gobiesox pinniger, Tomicodon humeralis, T. boehlkei, Acanthemblemaria crockeri,* and *Totoaba macdonaldi.* Degrees of endemism in the Gulf of Mexico and the Caribbean are not known. However, Böhlke and Chaplin (1968) estimated that endemism here may exceed 20%. Hoese and Moore (1977) suggested that endemism may be high in the northern Gulf of Mexico, and their estimate exceeded 15% (Table 6.5). Figure 6.2 illustrates some of the endemic fishes from this region: *Raja texana, Caulolatilus intermedius,* and *Cynoscion arenarius.* Endemisms in Goodeidae, principally in the Lerma-Santiago basin (Fig. 6.3), are exceptional. Speciation in *Cyprinodon* (Cyprinodontidae, pupfishes), with five species (Fig. 6.4) in Chichankanab Lake in the Yucatán Peninsula, is remarkable. The highly restricted distribution of genera and species suggests local speciation. Species with reduced distributions indicate that geographic, geological, and ecological characteristics have contributed to this endemism.

LOSS OF DIVERSITY

The study and rational use of the country's fishes cannot be separated from their conservation and maintenance. Fishes, more than any other group, reflect the paradox of conservation when they are not thought of as a resource. Human activities are among the principal factors that contribute to loss of diversity. Water from many rivers and river basins has been pumped for human and agricultural needs, that has resulted in local extinction of species, some of which were probably never documented. This problem is more aggravated at present. Exotic fishes introduced into the country have displaced indigenous species or caused extinctions in native taxa usually through predation, competition, and habitat alteration. In this respect, commercial use of native species should be considered first before introducing exotic ones. The dispersal of exotic species in the country should be weighed carefully.

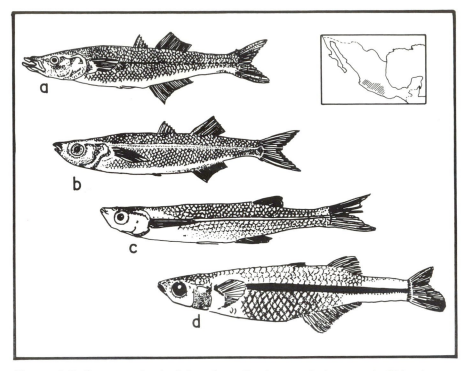

Figure 6.5. Some endemic fishes from the Lerma drainage.   A. *Chirostoma sphyraena*, B. *C. grandocule*. C. *C. chapalae*. D. *C. humboldtianum*.

The construction of dams has led to irreversible changes in estuaries, affecting numerous marine and continental fishes. The population decrease in *"Totoaba" (Totoba macdonaldi)* (Fig. 6.1) in the Gulf of California may be due to the local ecological changes caused by the Hoover dam (Arvizu & Chavez, 1972). The Balsas River provides other examples where dams have altered the delta in dramatic ways and affected the ichthyofauna. Indiscriminate use of insecticides has contributed to a reduction or loss of small populations, especially in tidal waters. About 100 species form the basis of the fisheries industry in the country, but in many cases their biology or their potential use are not known. No species has become extinct owing to activities of the fisheries industry, but special measures are needed to protect about 250 species that are caught along with shrimp and discarded as "waste."

EXOTIC SPECIES

Several exotic species, mostly Asian and African, have been introduced by the fishing industry. Table 6.6 provides a list of these species which now form part of the Mexican ichthyofauna. Native fishes have been transplanted to areas where they had not been reported before. Welcomme (1988) has presented a list of introduced species, and Contreras-Balderas and Escalante-Cavazos (1984) have provided details of these introductions and their impact on indigenous species.

ACKNOWLEDGMENTS

We thank Drs. José Luis Castro-Aguirre and Robert Rush Miller for critically reviewing an earlier version of this chapter. Silvia Toral Almazan of the Secretaria de Marina provided access to the collection housed there. We are especially grateful to T.P. Ramamoorthy and Marguerite Elliott, who reviewed several versions of this chapter, and helped in various ways with its preparation. CONACyT (Proyecto PCCNCNA-031542) provided partial financial support.

REFERENCES

Alvarez, J. 1946. Revisión del género *Anoptichthys* con descripción de una especie nueva (Pisc., Characidae). An. Esc. Nac. Ci. Biol. Mex. 4(2–3):263–282.

——. 1949a. Una nueva especie de peces Cyprinodontiformes de la Laguna de Chichankanab, Quintana Roo. Rev. Soc. Mex. Hist. Nat. 10(1–4):235–240.

——. 1949b. Ictiología dulceacuícola mexicana. I. Resumen de los estudios ictiológicos. Rev. Soc. Mex. Hist. Nat. 10(1–4):309–327.

——. 1950. Claves Para la Determinación de Peces en Las Aguas Continentales Mexicanas. Mexico City: Sec. de Marina.

——. 1952. *Dicerophallini*, nueva tribu de Poeciliidae en Chiapas. Ciencia (Mexico) 12(3–4):40–42.

——. 1959. Contribución al conocimiento del género *Neoophorus* (Pisc., Goodeidae). Ciencia (Mexico) 19(1–3):13–22.

——. 1960. Cincuenta años de ictiología en México. Rev. Soc. Mex. Hist. Nat. 21(1):46–91.

——. 1964. Ictiología Michoacana IV. Contribución al conocimiento biológico y sistemático de las lampreas de Jacona, Mich., Mexico. An. Esc. Nac. Ci. Biol. Mex. 13(1–4):107–144.

——. 1970. Peces mexicanos (Claves). Ser. Inv. Pesq. Inst. Nacl. Inv. Biol. Pesq. Mex. 1:1–166.

——. 1981. Platicas Hidrobiológicas. Mexico City: C.E.C.S.A.

—— & J. Arreola. 1972. Primer goodeido fósil procedente del Plioceno Jalisciense, (Pisces, Teleostomi). Soc. Ci. Nat. Jalisco A.C. 6:6–15.

—— & M.E. Moncayo. 1976. Contribución a la Paleoictiología de la Cuenca de México. Prehist. Epoca 7a. 4:191–292.

Alvarez, T. & F. de Lachica. 1974. Zoogeografía de los Vertebrados de México. *In* El Escenario Geográfico. Recursos Naturales. Mexico City. INAH.

Amezcua-Linares, F. 1977. Generalidades ictiológicas del sistema lagunar costero de Huizache-Caimanero, Sinaloa, México. An. Centro Ci. del Mar y Limnol. UNAM 4(1):1–26.

Applegate, S.P. & L. Espinosa-Arrubarena. 1982. Field Guide for the Excursion. Lithographic Limestone-like Deposits in Tepexi de Rodríguez, Puebla, Mexico. Mexico City: UNAM, Inst. Geol.

Arvizu, J. & H. Chavez. 1972. Sinopsis sobre la biología de la Totoaba, *Cynoscion macdonaldi* Gilbert, 1890. FAO Fish. Synopsis 108:1–24.

Baird, S.F. & C. Girard. 1854. Descriptions of new species of fish collected in Texas, New Mexico and Sonora, by Mr. John H. Clark of the U.S. and Mexican Boundary Survey, and in Texas by Capt. Stewart Van Vliet, U.S.A. Proc. Acad. Nat. Sci. Philadelphia 1854:1–50.

Bean, T.H. 1883. Catalogue of the collection of fishes exhibited by the U.S. Natl. Mus. Rep. upon exhibit of United States made at London. Bull. U.S. Natl. Mus. 27:387–510.

———. 1887. Descriptions of five new species of fishes sent by Prof. A. Dugés, from the province of Guanajuato, Mexico. Proc. U.S. Natl. Mus. 10:370–375.

Beltrán, E. 1934. Lista de peces mexicanos. Mexico City: Sria. de Agric. Inst. Biotécnico.

———. 1959. Los recursos naturales del sureste y su aprovechamiento. I. Situación, problemas, perspectivas. Mexico City: IMERNAR.

Berdegué, J. 1956. Peces de importancia comercial en la costa noroccidental de México. Mexico City: Secr. Marina. Com. Pisc. Rural.

Berra, T.M. 1981. An Atlas of Distribution of the Freshwater Fish Families of the World. Lincoln, NB: Univ. of Nebraska Press.

Böhlke, J.E. & C.C.G. Chaplin. 1968. Fishes of the Bahamas and adjacent tropical waters. Wynnewood, PA: Livingston.

Breder, C.N. 1928a. Elasmobranchii from Panama to Lower California Scientific results of the Second Oceanographic Expedition of the "Pawnee," 1926. Bull. Bingham Oceanogr. Coll. 2(1):1–13.

——— 1928b. Scientific results of second oceanographic expedition of the "Pawnee," 1926. Nemathongnathi, Apodes, Isospondyli, Synentognathi, and Thoracostraci from Panama to Lower California. Bull. Bingham Oceanogr. Coll. 2(2):1–25.

Briggs, J.C. 1964. Additional transpacific shore fishes. Copeia 1964(4):706–708.

———. 1967. Relationship of the tropical shelf regions. Stud. Trop. Oceanogr. Miami. 5:569–578.

———. 1974. Marine Zoogeography. New York: McGraw–Hill.

Bustamente y Septién, M. 1837. Descripción del Mextlapique *(Cyprinus viviparus)*. Mosaico Mex. 2:116.

Castro, J.I. 1983. The Sharks of North American waters. College Station, TX: Texas A & M Univ. Press.

Castro-Aguirre, J.L. 1978. Catálogo sistemático de los peces marinos que penetran a las aguas continentales de México, con aspectos zoogeográficos y ecológicos. Dir. Gral. Inst. Nal. de Pesca Mex. Ser. Cient. 19:1–296.

———, J. Arvizu-Martinez & J. Paez-Barrera. 1970. Contribución al conocimiento de los peces del Golfo de California. Rev. Soc. Mex. Hist. Nat. 31:107–180.

———, R. Torres-Orozco B., M. Ugarte & A. Jimenez. 1986. Estudios ictiológicos en el sistema estuarino lagunar Tuxpam-Tampamachoco, Veracruz. I. Aspectos ecológicos y elenco sistemático. An. Esc. Nac. Ci. Biol. Mex. 30:155–170.

Cohen, D.M. 1970. How many recent fishes are there? Proc. Calif. Acad. Sci. (ser. 4) 38:341–345.

Contreras-Balderas, S. & M.A. Escalante-Cavazos. 1984. Distribution and known impacts of exotic fishes in Mexico. *In* W.R. Courtenay Jr. & J.R. Stauffer, Jr. (eds.), Distribution, Biology, and Management of Exotic Fishes. Baltimore, MD: The Johns Hopkins Univ. Press. pp. 102–130.

Cuesta-Terron, C. 1925. La fauna ictiológica y malacológica comestible del Lago de Chapala, Jal., y su pesca. Mem. Soc. Cient. Antonio Alzate 44:39–67.

Cuvier, G. & A. Valenciennes. 1828–49. Histoire Naturelle des Poissons. Paris. Vols. 1–22.

Darlington, P.J. 1957. Zoogeography: the geographical distribution of animals. New York: John Wiley & Sons.

de Buen, F. 1947a. Investigaciones sobre ictiología mexicana. I. Catálogo de los peces de la región neártica en suelo mexicano. Anales Inst. Biol. UNAM 18(1):257–292.

———. 1947b. Investigaciones sobre ictiología mexicana. III. Zoogeografíade los peces de agua dulce, con estudio especial de la región neártica. An. Inst. Biol. UNAM 18(1):304–348.

Ekman, S. 1953. Zoogeography of the Sea. London: Sidwick & Jackson.

Espinosa-P., H., M.T. Gaspar-D., & P. Fuentes-M. Listados Faunisticos: Peces Dulciacuicolas Mexicanos. Mexico City: UNAM. Inst. Biol. In press.

Fisher, W. (ed.). 1978. FAO Species identification sheets for fishery purposes. Western Central Atlantic (Fishing area 31). Vols. 1–7. Italy: U.N. FAO.

Fowler, H.W. 1941. The fishes of the groups Elasmobranchii, Holocephali, Isospondyli, and Ostariophysi obtained by the United States Bureau of Fisheries Steamer "Albatross" in 1907 to 1910, chiefly in the Philippine Islands and adjacent seas. *In* Contributions to the Biology of the Philippine Archipielago and adjacent regions. 13. Bull. U.S. Natl. Mus. Vol. 100.

Fuentes-Mata, P. & M.T. Gaspar-Dillanes. 1981. Aspectos biológicos y ecológicos de la ictiofauna de la desembocadura del Río Balsas, Mich.-Gro. Thesis, UNAM.

Gómez-Soto, A. & S. Contreras-Balderas. 1988. Ictiofauna de la Laguna Madre, Tamaulipas, México. Mem. IX Cong. Nal. Zool. 2:8–17.

Greenfield, D.W. and R.K. Johnson. 1981. The blennioid fishes of Belize and Honduras, Central America, with comments on their systematics, ecology, and distribution (Blenniidae, Chaeposidae, Labrisomidae, Tripterygiidae). Fieldiana, Zool. (New Ser.) 8:1–106.

Günther, A. 1859–70. Catalogue of the fishes in the British Museum. Vols. 1–8 London: Taylor and Francis.

Haeckel, J. 1848. Eine neue Gattung von Poecilien nit rochenartigen Amklammerungs-Organe. Sitzber. K. Akad. Wiss. Wien. Math. Narwiss. Cl. 1:289–303.

Hildebrand, S.F. 1946. A descriptive catalog of the shore fishes of Peru. Bull. U.S. Natl. Mus. 189:1–530.

Hoese, H.D. & R. H. Moore. 1977. Fishes of the Gulf of Mexico. Texas, Louisiana, and Adjacent Waters. College Station, TX: Texas A & M Univ. Press.

Hubbs, C.L. 1918. *Colpichthys, Thyrynops* and *Austromenidia* new genera of Atherinoid fishes from the New world. Proc. Acad. Nat. Sci. Philadelphia 1918:305–308.

———. 1924. Studies of the fishes of the order Cyprinodontes I–IV. Misc. Publ. Mus. Zool. Univ. Mich. 13:1–31.

———. 1932. Studies of the fishes of the order Cyprinodontes. XI. *Zoogeneticus zonistius*, a new species from Colima, Mexico. Copeia 1932(2):68–71.

———. 1936. Fishes of the Yucatán Peninsula. Carnegie Inst. Wash.457:157–287.

———. 1939. Fishes from the caves of Yucatán Peninsula. Carnegie Inst. Wash. 491:261–295.

———. 1952. Antitropical distribution of fishes and other organisms. Symposium on Problems of Bipolarity and Pantemperate faunas. Proc. 7th. Pac. Sci. Cong. 3:324–329.

———. 1960. The marine vertebrates of the outer coast. *In* The biogeography of Baja California and Adjacent Seas. Pt. II. Marine Biotas. Syst. Zool. 9(3–4):134–147.

Jordan, D.S. & B.W. Evermann. 1896–1900. The fishes of North and Middle America. Bull. U.S. Nat. Mus. 47(1–4):1–3313.

Linnaeus, C. 1758. Systema Naturae. Vol. 1. Regnun Animale. 10 ed. Holmieae: Impensis Salvii.

Maldonado-Koerdell, M. 1956. Peces fósiles de México. III. Nota preliminar sobre los peces del turoniano superior de Xilitla, San Luis Potosí (México). Ciencia (Mexico) 16(1–3):31–35.

Martin del Campo, R. 1938. Nota acerca de los peces del Lago de Catemaco. An. Inst. Biol. UNAM 9:225–226.

Meek, S.E. 1902. A contribution to the ichthyology of Mexico. Publ. Field. Columbian Mus. Ser. Zool. 3:63–128.

———. 1904. The freshwater fishes of Mexico, north of the Isthmus of Tehuantepec. Publ. Field Columbian Mus. Zool. Ser. 5:1–252.

———. 1907. Notes on freshwater fishes from Mexico and Central America. Publ. Field Columbian Mus. Ser. Zool. 7:135–157.

——— & S.F. Hildebrand. 1923–28. The marine fishes of Panama. Publ. Field Mus. Nat. Hist. Zool. ser. 15(1–4):1–1045.

Miller, D.J. & R.N. Lea. 1972. Guide to the coastal marine fishes of California. Fish Bull. 157:1–249.

Miller, R.R. 1982. Pisces. *In* S.H. Hurlbert & A. Villalobos-Figueroa (eds.), Aquatic Biota of Mexico, Central America and the West Indies: being a compilation of taxonomic bibliographies for the fauna and flora of inland waters of Mesoamerica and the adjacent Caribbean region. San Diego, CA: San Diego State Univ.

———. 1986. Composition and derivation of the freshwater fish fauna of Mexico. An. Esc. Nac. Ci. Biol. Mex. 30:121–153.

——— & M.L. Smith. 1986. Origin and geography of the fishes of Central Mexico. *In* C.H. Hocutt & E.O. Wiley (eds.), The Zoogeography of North American Freshwater Fishes. New York: John Wiley & Sons. pp. 487–517.

Moyle, P.B. & J.J. Cech. 1988. Fishes: An Introduction to Ichthyology. Englewood Cliffs, NJ: Prentice Hall.

Myers, G.S. 1940. Freshwater Fishes and West Indian Zoogeography. Washington, DC: Annual Report, Smithson. Inst. [for 1937]:339–364.

Nelson, J.S. 1984. Fishes of the World. New York: John Wiley & Sons. 2nd ed.

Osburn, R.C. & J.T. Nichols. 1916. Shore fishes collected by the "Albatross" expedition in Lower California with descriptions of new species. Bull. Amer. Mus. Nat. Hist. 35:139–181.

Ramírez, E. & J. Paez. 1965. Investigaciones ictiológicas en las costas de Guerrero. I. Lista de peces marinos de Guerrero colectados en el período 1961–1965. An. Inst. Nal. Invest. Biol. Pesq. 1:329–360.

Randall, J.E. 1984. Caribbean reef fishes. Neptune City, NJ: T.F.H. Publications.

Regan, C.T. 1906–08. Pisces. *In* F.D. Godman & O. Salvin (eds.), Biologia Centrali Americana 8:1–201.

Ricker, K.E. 1959a. Mexican shore and pelagic fishes collected from Acapulco to Cape San Lucas during the 1957 cruise of the "Marijean." Mus. Contrib. Inst. Fish. Univ. Brit. Colum. 3:1–18.

———. 1959b. Fishes collected from the Revillagigedo Islands during 1954–1958 cruises of the "Marijean." Mus. Contrib. Inst. Fish. Univ. Brit. Colum. 4:1–10.

Robins, C.R., R.M. Bailey, C.E. Bond, J.R. Brooker, E.A. Lachner, R.N. Lea and W.B. Scott. 1980. A list of common and scientific names of fishes from the United States and Canada. Amer. Fish. Soc. (Spec. Publ.) 12:1–174.

Smith, M.L. & R.R. Miller. 1986. The evolution of the Rio Grande Basin as inferred from its fish fauna. *In* C.H. Hocutt & E.O. Wiley (eds.), The Zoogeography of North American Freshwater Fishes. New York: John Wiley & Sons. pp. 457–485.

Steindachner, F. 1875a. [No title.] Ichthy. Beirtrage No. 3 Sitzb Akad. Wiss. 72:1–68.

———. 1875b. [No title.] Ichthy. Beirtrage No. 4 Sitzb Akad. Wiss. 72:1–65.

———. 1879. Uber einige Neue und Seltene Fisch Arte, kk. Zoologischen Husseen zu Wien, Stutgart un Warschau. Denkschr. Akad. Wiss. Wien 41:1–52.

Thompson, D.A., L.T. Findley & A.N. Kerstitch. 1979. Reef Fishes of the Sea of Cortez. New York: John Wiley & Sons.

Uyeno, T. & R.R. Miller. 1971. Comparative study of chromosomes of the North American killifish genera, family Cyprinodontidae. Abst. 51st. Ann. Mtg., Amer. Soc. Ich. Herpt. Los Angeles.

Vaillant, L.L. & F. Bocourt. 1878. Studes sur les poissons. *In* Mission Scientifique au Mexique. 4:1–155.

Walker, B.W. 1960. The distribution and affinities of the marine fish fauna of the Gulf of California. Syst. Zool. 9(3):123–133.

Walls, J.F. 1975. Fishes of the Northern Gulf of Mexico. Neptune City, NJ: T.F.H. Publications.

Welcomme, R.L. 1988. International introductions of inland aquatic species. FAO Fish. Tech. Paper 294:1–318.

Zaneveld, J.S. 1983. Caribbean Fish Life. Leiden: E.J. Brill/DR. W. Backhuys.

7

Herpetofauna of Mexico: Distribution and Endemism

OSCAR FLORES-VILLELA

There are over 950 species of reptiles and amphibians in Mexico (9.8% of the world's herpetofauna), of which 55% are endemic to Mexico; there are more species in Mexico than in the United States and Canada or any of the Central American countries. The degree of endemism in the larger groups of amphibians and reptiles is significant. Among the amphibians, endemism is highest in the salamanders (Plethodontidae and Ambystomatidae) and frogs (Hylidae, Leptodactylidae, and Ranidae). Among the reptiles, endemism is highest in lizards (especially Iguanidae, Anguidae, Teiidae, Xantusiidae) and snakes (notably Colubridae, Elapidae and Viperidae).

The tropical highlands of southern and central Mexico have the highest number of endemic taxa followed by the tropical lowlands of the Pacific coast. Endemism among reptiles is high in the extratropical arid zones of northwestern Mexico, partly due to the large number of endemic species in the islands of the Sea of Cortes. The high number of endemics, especially in the highlands of the country, is probably due to vicariance and may have been associated with climatic changes during the Pleistocene.

Mexico's herpetofauna is an important biological resource, as over 40 species are economically useful. Conservation of large areas of cloud, oak, and pine-oak forests, presently subject to intense exploitation, is essential as they harbor most of the amphibian and reptile endemics.

The diversity of herpetofauna of Mexico, which constitutes one of the more important elements of the country's vertebrate fauna, is impressive: 13 of the 37 amphibian and 29 of the 50 reptilian families are found in Mexico, as are 285 amphibian and over 693 reptile species. Patterns and percentages of endemism at generic and species levels are significant. This chapter reviews briefly the history of studies of this group in Mexico, discusses the distribution of amphibians and reptiles in the country and the world, and details their richness and endemism in Mexico.

Numerous students have contributed to our understanding of Mexico's herpetofauna. Smith and Smith (1973–79), in their Synopsis of the

Table 7.1. Amphibian and Reptile species in Mexico

Species	No. of Species	
	Hernández[1] (1570–77)	Dugés (1896)
Amphibians		
Anura	7	24
Caudata	1	12
Gymnophiona	1	1
Unidentified species	3	—
Total	12	37
Reptiles		
Amphisbaenia	1	2
Sauria	26	59
Serpentes	31	102
Crocodylia	1	4
Testudines	—	14
Total	59	181

[1]From Smith (1970) and Flora-V. (1982).

Table 7.2. Herpetofaunistic richness of Mexico

Order or Suborder	Families	Genera	Species & subspecies
Gymnophiona	1	2	2
Caudata	4	16	65
Salientia (*Anura*)	7	25	163
Testidines	8	18	49
Amphisbaenia	1	1	3
Sauria	10	47	394
Serpentes	8	80	486
Loricata (*Crocodylia*)	2	2	3
Total	41	191	1,165

Source: Smith & Taylor (1950).

Herpetofauna of Mexico, reviewed the noteworthy contributions of past students. The works by Francisco Hernández and A.A. Dugés are of historic interest. The former, the court physician of Philip II of Spain, in his "Historia Natural de la Nueva España" mentioned nine amphibians and 59 reptiles (Table 7.1). Dugés (1896), who provided data on the distribution of many species, listed 37 amphibians and 181 reptiles (Table 7.1). The more recent studies of interest are those of Smith and Taylor (1945, 1948, 1950). They listed Mexican species (Table 7.2) and provided keys to their identi-

Table 7.3. Herpetofaunistic richness of Mexico

Order or suborder	Families	Genera	Species	Species & subspecies
Reptiles				
Amphisbaenia	1	1	3	4
Crocodylia	1	2	3	3
Lacertilia (*Sauria*)	12	48	321	561
Serpentes	7	84	320	583
Testudines	7	17	38	59
Total	28	152	685	1,210
Amphibians				
Anura	7	26	177	202
Caudata	4	15	88	98
Gymnophiona	1	2	2	2
Total	12	43	267	302
Total amphibians & reptiles	40	195	952	1,512

Source: Smith & Smith (1976).

fication. Smith and Taylor (1966) reprinted their earlier work with some recombinations. This work was followed by that of Smith and Smith (1973, 1976a,b, 1977, 1979); their 1976 contribution, which presents an overview of the herpetofauna of Mexico, is reproduced in Table 7.3 here.

Figure 7.1 shows the increase in the number of species of amphibians and reptiles recorded for Mexico over the years. During the first 300 years, the number of species tripled from 68 to 218. Over a period of a little more than 54 years following 1896 this number quadrupled; and it has continued to rise, at a slower rate, from 1950 to the present. Several more species are in the process of being described by various students (pers. comm.; Flores-V., unpublished data).

METHODS

Lists of endemic and nonendemic species have been made from the data of Smith and Smith (1976a,b). The information thus prepared and the geographic zonal distribution of endemic taxa have been updated with the help of various students of Mexican herpetofauna, curators of museums, and the most recently published literature.

The country has been divided into natural geographic zones based on their climatic and vegetational characteristics. West (1964) divided Mexico into five natural regions, which have been further divided to produce the nine discussed in this chapter (see Appendix). The presence of a species in a zone, when its distribution is close to the limits of two or more regions,

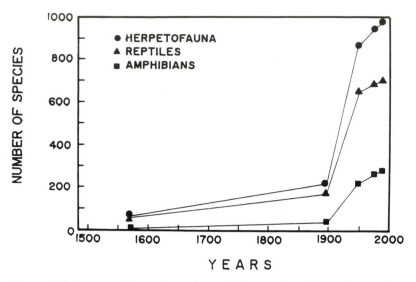

Figure 7.1. Increase in number of amphibian and reptile species registered for Mexico from 1570 to date, based on data in **Tables 7.1 to 7.4.**

has been determined on the basis of the vegetation types, where the species occur, and their characteristics.

The distribution of nonendemics has been analyzed using data from field guides (Behler & King, 1979; Conant, 1975; Stebbins, 1966) for northern species and checklists (Henderson & Hoevers, 1975; Stuart, 1963) and the maps in Duellman (1970) and Lee (1980) for southern species.

The criteria proposed by Savage (1966, 1982) for historic units of the herpetofauna are adopted here. Tables 7.4, 7.6 and 7.7 (see below) present the group's richness in Mexico, their percentages of the world total, number of species that share territories, endemic species, and distribution by natural regions.

BRIEF REVIEW OF WORLD DISTRIBUTION OF AMPHIBIANS AND REPTILES

Excepting Antarctica, amphibians and reptiles are found in all parts of the world (Darlington, 1957; Halliday & Adler, 1986; Porter, 1972; Savage, 1973). Their diversity is greater in tropical latitudes. The distribution of these ectothermic organisms is subject to environmental factors: for example, they are rare in cold parts of the world.

The occurrence of amphibians is greatly influenced by humidity and salinity. They are scarce in arid and sandy zones. Species of *Rana*, *Bufo*, and *Scaphiopus*, which do occur in arid areas (Darlington, 1957), are exceptions. The most widespread amphibians are frogs and toads. They are found on all continents; and some, e.g., the European *Rana temporaria* and the American *R. sylvatica*, have even reached the Arctic Circle. Frogs are the only amphibians found on islands such as Madagascar, New Guinea, Australia, New Zealand, and the islands of the western and southwestern Pacific as well as in the Caribbean. They are found on most of the continental islands: British Isles, Japan, Taiwan, Sri Lanka, Sumatra, Java, and even in the Seychelles. Most families and subfamilies of the frogs (86%) are tropical.

The salamanders (order Caudata) have a different distribution, being restricted mostly to temperate zones of the Northern Hemisphere. One species, *Hynobius keyserlingii*, reaches the Arctic Circle. The family Plethodontidae (which comprises almost half the living species of salamanders) is found in the New World tropics. Salamanders are found on the continental masses of the Northern Hemisphere and on some of its islands: British Isles, Taiwan, and Japan.

Among amphibians, the caecilians (order Gymnophiona) have the most restricted ecological and geographical distribution. They are circumtropical and restricted mostly to lowlands and submountainous areas as well as to some regions that border temperate zones in Asia and southern South America. They are found in a few continental islands and the Seychelles. Their general distribution ranges from southern China, India, and Sri Lanka to the Philippines to the east and Central Africa to the west, and from southern Mexico to southern South America.

The reptiles with the widest distribution are squamates (Squamata), mainly lizards and snakes (Bellairs, 1975; Darlington, 1957; Halliday & Adler, 1986; Porter, 1972). The Amphisbaenians (suborder Amphisbaenia) in this order are the most restricted in their geographical and ecological distribution. They occur in subtropical regions of North America and the Caribbean islands, in South America reaching Patagonia, in central, southern, and extreme northeastern Africa, the Iberian Peninsula, Arabia, and extreme southwestern Asia.

Snakes are virtually cosmopolitan in distribution. Only two species are known to occur near the Arctic Circle (*Vipera verus* in Europe and *Thamnophis sirtalis* in North America). In the Southern Hemisphere, they exist in Santa Cruz, Argentina, extreme southern Africa, Australia, and Tasmania; they are also found in almost all the continental and oceanic islands in the tropical and warm temperate zones. There are no snakes in Iceland, Ireland, or New Zealand. Sea snakes are found in the warm waters of the Indian and Pacific oceans and in the Indo-Australian seas.

The suborder Sauria (lizards) has almost the same distribution as the snakes but is more diverse. Only one species is known to reach the Arctic Circle, *Lacerta vivipara*, in Europe. Lizards reach as far south as Tierra del Fuego in the Americas, southern parts of Africa, Australia, Tasmania, and New Zealand. They are also found in Ireland and in more oceanic islands than those where snakes are found owing to the fact that their habits allow them to establish resident populations in islands more easily.

Table 7.4. Herpetofaunistic richness of Mexico and its percentage with respect to the world total

Taxa	No. of families			No. of genera			'No. of species		
	Mex.[1]	World[2]	%	Mex.[1]	World[2]	%	Mex.[1]	World[2]	%
Amphibia									
Anura	8	23	34.7	27	303	8.9	194	3494	5.5
Caudata	4	9	44.4	17	61	27.8	89	357	24.9
Gymnophiona	1	5	20.0	1	34	2.9	2	155	1.2
Total	13	37	35.1	45	398	11.3	285	4014	7.1
Reptilia									
Amphisbaenia	1	4	25.0	1	21	4.7	3	135	2.2
Sauria	10	17	58.8	49	362	13.5	330	3317	9.9
Serpentes	7	14	50.0	87	419	20.7	319	2268	14.0
Testudines	9	12	75.0	19	73	26.0	38	223	17.0
Crocodylia	2	3	66.6	2	8	25.0	3	22	13.6
Total	29	50	58.0	158	883	17.8	693	5965	11.6
TOTAL	42	87	48.2	203	1281	15.8	978	9979	9.8

[1]Flores-V., unpublished data.
[2]Duellman (1979) and Frost (1985).

Turtles are widely distributed, but their southern and northern limits are narrower than those of the lizards and snakes. On the American continent they range as far north as British Columbia; in the Old World they occur in northern Germany and northeastern Asia; in the Southern Hemisphere they reach the extreme south of Africa, northern Argentina, and northern and eastern Australia. They are not present in Tasmania or New Zealand. Sea turtles are found mainly in tropical waters, although they can reach temperate and occasionally cold seas.

Crocodiles are circumtropical. *Alligator* is the only genus found within the borders of temperate zones in areas with warm climates; some populations of *A. mississippensis* possibly hibernate in the most northern part of their distribution. In the New World, they occur in the tropical Americas including the southeastern United States and Caribbean islands. In the Old World they extend from Africa to Australia through Madagascar, southern Asia, southeast China, and the Indo-Malayan Archipelago.

QUANTIFICATION OF THE MEXICAN HERPETOFAUNA

Table 7.4 presents the number of families, genera, and species of amphibians and reptiles in Mexico in their respective orders. (The suborders of Squamata are treated here as orders as are crocodiles and tortoises.)

With regard to Mexico's representation on a world scale, it is instructive to highlight the percentages of the families, genera, and species of the amphibians and reptiles that occur in the country. The families in each order fluctuate between a low of 20% for Gymnophiona and a high of 75% for Testudines. In the more diverse groups such as the frogs, lizards, and snakes, the percentages are 34.7%, 58.8%, and 50.0%, respectively; 36.1% of the amphibian families and 58% of the reptilian are represented in the country. The percentage of families of the reptiles and amphibians combined is 48.2% in Mexico, or roughly half the families in the world.

Percentages at the generic level fluctuate between 2.9% in Gymnophiona and 27.8% in Caudata. The most diverse groups are the anurans with 8.9%, the lizards with 13.5%, and the snakes with 20.7%. The high endemism in snakes (17 endemic genera) contributes to their greater proportion. In total, 11.3% of the amphibian genera and 17.8% of the reptilian genera occur in Mexico. Together these groups constitute 15.8% of the world's genera of amphibians and reptiles.

The percentages in species range from 1.2% in Gymnophiona to 24.9% in Caudata. In the most diverse groups, anurans make up 5.5%, lizards 9.9%, and snakes 14.0%. A total of 7.1% of the world's amphibians and 11.6% of the reptiles occur in Mexico. Together 9.8% of the world's herpetofaunal species occur in this country.

The reptiles, which constitute 70.8% of the total species in Mexico, are richer in number than amphibians. Saurians represent 33.7% and snakes 32.6%. Among the amphibians, frogs and toads represent 19% and salamanders 9.1%. Caecilians, *Bipes*, tortoises, and crocodiles make up the remaining 4.8%. The lizards and snakes are richer than the other groups: the former with 49 genera and 330 species and the latter with 87 genera and 319 species. Many genera (12 among lizards and 36 among snakes) are monotypic.

As a corollary to this section, it is worth noting the observations made by Smith and Smith (1976b) with respect to Mexico's richness in amphibians and reptiles:

> We have repeatedly stated our conviction that Mexico houses the greatest diversity of herpetofauna of any area of equal size on earth . . . . The great rainforests of South America and Africa harbor considerably greater assemblages of anurans, but are limited in reptilian diversity. The only country likely to approach the herpetodiversity of Mexico is Indonesia, with its unparalleled abundance of tropical insular isolation and an equally precarious existence of endemic species. No continental area, however, even remotely approximates Mexico in diversity, at least in comparison of comparable area. Certainly smaller areas exceed the species-subspecies:area ratio of Mexico, but areas of equal size do not do so.

Further on they concluded:

> the herpetofauna of Mexico is at least twice as rich as that of any other area of the globe, with the exception of the East Indies and West Indies . . . . In no continental area does there exist a comparable diversity.

Table 7.5. Fossil records of present day Mexican Herpetofauna

Taxa	Age	State	Present Distribution
Amphibia			
Anura			
Bufonidae			
Bufo			
alvarius	Pleistocene	Sonora	+
campi	Pliocene	Chihuahua	x +
cf *horribilis-marinus*	Pleistocene	Yucatán	+
cf *valliceps*	Pleistocene	Yucatán	+
sp.	Pleistocene	México	-
Leptodactylidae			
Leptodactylus			
cf *labialis-fragilis*	Pleistocene	Tamaulipas	+
Syrrhophus campi	Pleistocene	Tamaulipas	+
Ranidae			
Rana			
pipiens-berlandieri?	Pleistocene	Tamaulipas	+
sp.	Pleistocene	México	+
sp.	Pleistocene	México	+
Rhinophrynidae			
Rhinophrynus dorsalis	Pleistocene	Tamaulipas	+
Caudata			
Ambystomatidae			
Ambystoma mexicanus			
mexicanum	Pleistocene	México	-
sp.	Pleistocene	México	?
sp.	Pleistocene	México	+
Reptilia			
Sauria			
Iguanidae			
Ctenosaura cf *similis*	Pleistocene	Yucatán	+
Deltatmena premaxillaris	Pleistocene	Yucatán	x
Laemanctus			
cf *alticoronatus* or *serratus*	Pleistocene	Yucatán	+
Paradipsosaurus			
mexicanus	Eocene-Oligocene	Guanajuato	x
Prynosoma josecitensis	Pleistocene	Nuevo León	+
P. orbiculare	Pleistocene	Nuevo León	+
Sceloporus jarrovi	Pleistocene	México	+
cf *variabilis*	Pleistocene	Tamaulipas	+
sp.	Pleistocene	Tamaulipas	+
Teiidae			
Cnemidophorus gularis	Pleistocene	Tamaulipas	+
Xantusiidae			
Lepidophyma			
arizeloglyphus	Pleistocene	Yucatán	x -
sp.	Pleistocene	Tamaulipas	+

Table 7.5. (cont.)

Taxa	Age	State	Present Distribution
Serpentes			
Boidae			
Constrictor-Boa			
cf *constrictor*	Pleistocene	Yucatán	+
sp.	Pliocene-Pleistocene	Baja Calif. Sur	-
Colubridae			
Coluber-Masticophis			
mentovarius	Pleistocene	Yucatán	+
Drymarchon cf *corais*	Pleistocene	Yucatán	+
Drymobius cf *margaritiferus*	Pleistocene	Yucatán	+
Elaphe cf *flavirufa* or *triaspis*	Pleistocene	Yucatán	+
Lampropeltis intermedius	Pliocene	Michoacán	x +
Pituophis sp.	Pliocene-Pleistocene	Baja Calif. Sur	+
Spilotes cf *pullatus*	Recent Pleistocene	Yucatán	+
Thamnophis			
macrostema	Pleistocene	México	+
scalaris	Pleistocene	México	+
sp.	Pleistocene	México	+
sp.	Pleistocene	Jalisco	+
Trimorphodon tau	Pleistocene	Jalisco	+
Viperidae			
Crotalus			
scutulatus	Pleistocene	México	+
sp.	Pliocene-Pleistocene	Baja Calif. Sur	+
Testudines			
Emydidae			
Chrysemys-Trachemys			
scripta	Pleistocene	Sonora	x
Pseudemys ?			
sp.	Pliocene-Pleistocene	Jalisco	x
sp.	Pleistocene	Tabasco	+
Terrapene culturatus	Pleistocene	Jalisco	x
Kinosternidae			
Kinosternon			
hirtipes	Pleistocene	Jalisco	+
integrum	Pleistocene	Jalisco	+
scorpioides integrum-			
integrum	Pleistocene	Aguascalientes	+
cf *cruentatun* o			
creaseri	Pleistocene–Recent	Yucatán	+
sp.	Pleistocene	México	+
sp.	Pliocene	Michoacán	+
sp.	Pleistocene	San Luis Potosí	+
sp.	Pleistocene	San Luis Potosí	+
cf *K.* sp.	Pleistocene	México	+
Testudinidae			
Geochelone			

Table 7.5. (cont.)

Taxa	Age	State	Present Distribution
sp.	Pleistocene	Aguascalientes	x
sp.	Pliocene-Pleistocene	Baja Calif. Sur	x
Gopherus-Scaptochelys			
agassizi	Pleistocene	Sonora	+
auffenbergi	Pleistocene	Aguascalientes	x
Gopherus			
flavomerginatus (sic)	Pleistocene	Aguascalientes	-
pargensis	Pleistocene	Aguascalientes	x
sp.	Pleistocene	Baja Calif. Sur	x
cf *G.* sp.	Pliocene-Pleistocene	Chihuahua	x -
Testudo			
sp.	Pleistocene	México	x
sp.	Pleistocene	Puebla	x
Trionychidae			
Trionyx sp.	Cretaceous	Coahuila	x -
Crocodylia			
Crocodylidae			
Crocodylus cf *moreleti*	Pliocene-Pleistocene	Baja Calif. Sur	-
C. sp.	Pliocene-Pleistocene	Jalisco	x -

+, fossil record within the recent distributional range; x, extinct species or genus; -, fossil record not within the recent distributional range of the species or genus; ?, not enough information to relate to recent distributional range.
Source: Barrios-R. (1985) and Flores-V., unpublished data.

It is, however, possible that the herpetofauna of Colombia may be richer but as yet is only poorly known.

ORIGINS AND DISTRIBUTIONS OF MEXICAN HERPETOFAUNA

Fossil documentation of amphibians and reptiles in Mexico is incomplete. Most reports are from the Pliocene and Pleistocene of which many are late Pleistocene. Table 7.5 summarizes known fossils belonging to recent groups of amphibians and reptiles in Mexico.

Various authors have provided explanations regarding the origin and distribution of Mexico's herpetofauna (Savage, 1960, 1966, 1973, 1982; Smith, 1949). Notable contributions on distribution patterns have been made by Stuart (1964) and Savage (1966, 1982). Duellman (1966) presented a detailed review of the ecological distribution of the herpetofauna.

The origins of Mexico's herpetofauna may be traced to four historical source units (Savage, 1982). The Old Northern Element is comprised of genera that are extratropical but are represented by tropical forms in the Americas. These genera or their ancestors had a continuous circumpolar

Table 7.6. Number of species Mexico shares with adjacent areas

Taxa	Nonendemic Species	North America	Central America
Amphibia			
Anura	88	37	57
Caudata	22	6	16
Gymnophiona	1	0	1
Total	111	43	74
Reptilia			
Sauria	133	70	61
Serpentes	161	76	99
Testudines	21	19	16
Crocodylia	3	1	3
Total	318	166	179
TOTAL	429	209	253

distribution but, when forced southward, fragmented into geographic isolates owing to increased cooling and aridity through the Cenozoic. Examples are Rhinophrynidae, Xantusiidae, and Dermatemydidae. The second source unit is the South American Element, whose evolutionary history was centered in South America during the Cenozoic. Its contribution to Mexican faunal diversity is recent. *Leptodactylus*, *Physalaemus*, and *Micrurus* species are examples of this category.

The Young Northern Element is composed of genera *Phrynosoma*, *Sceloporus*, and *Cnemidophorus*, which are found in xeric extratropical areas. The most important historical source unit is the Middle American Element, which is represented, among others, by *Dermophis*, *Tomadactylus*, and *Ptychohyla* (amphibians) and *Basiliscus*, *Ungaliophis*, and *Imantodes* (reptiles). This unit is primarily Mesoamerican and has its allies in the region or in tropical North America but is mostly endemic to Mexico and Central America. It may have evolved in isolation until the end of the Eocene.

The other components of Mexico's herpetofauna are possibly associated with what Savage (1960) has called complexes, and which he viewed as subdivisions of the Ancient Northern Element. They are Eastern American, Western American, and Southeastern American. Some examples of genera in these complexes are *Ambystoma*, *Acris*, *Chrysemys* (Morafka, 1977).

Savage (1982) recognized four groupings in the Mesoamerican groups based on geographical limits: broadly tropical (*Bufo*, *Hyla*, *Micrurus*), South American staggered in its northern distribution (*Centrolenella*, *Ameira*, *Caiman*), tropical Mesoamerican (*Bolitoglossa*, *Claudius*, *Toluca*), and North American extratropical (*Salvadora*, *Micruroides*, *Crotalus*).

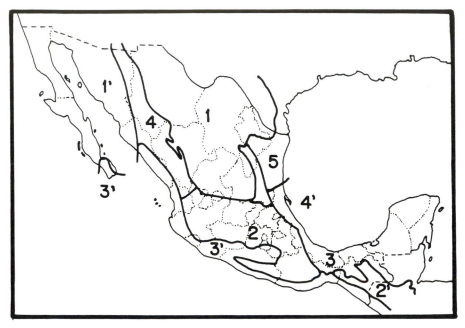

Figure 7.2. Natural regions. Extratropical arid areas (1 and 1'), tropical highlands (2 and 2'), tropical lowlands (3 and 3'), extratropical highlands (4 and 4'), subhumid extratropical areas (5). See Appendix for more detail. Modified from West (1964).

DISTRIBUTION PATTERNS

Table 7.6 gives a breakdown of the 429 (111 amphibians and 318 reptiles) nonendemic taxa that Mexico shares with adjacent territories. Mexico and North America share 209 species and Mexico and Central America 253. Mexico and Central America share a larger number of species, most notably among frogs, salamanders, and snakes. Four of the salamander families that occur in North America do not extend into Mexico. Moreover, most of the genera of plethodontid family found in Mexico also occur in northern Central America. Among frogs, the families Hylidae and Leptodactylidae have the highest number of species common to Mexico and Central America, particularly among genera *Hyla* and *Eleutherodactylus*, which are, according to Savage (1982), Mesoamerican and South American. It is difficult to assess this difference among snakes.

Some families are found in Mexico but not in North or Central America. Two amphibian families, Centrolenidae and Caeciliadae, do not occur in North America but are diverse in Central and South America. The Pelobatidae, Ambystomatidae, Sirenidae, and Salamandridae are not present in Central America and are considered typically North American, with the exception of Salamandridae, which are also well represented in the Old World. Of the reptiles, those that are not present in North America north of Mexico are Bipedidae (endemic to Mexico), Xenosauridae (Mexico,

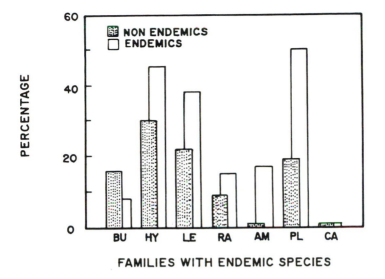

Figure 7.3. Proportion of endemisms among amphibian families in Mexico. BU, Bufonidae; HY, Hylidae; LE, Leptodactylidae; RA, Ranidae; AM, Ambystomatidae; PL, Plethodontidae; CA, Caeciliadae.

Guatemala, and southern China), and Dibamidae (Mexico and the Indo-Malayan archipelago) with relictual distributions; Typhlopidae (southern Mexico and the Antilles); the Loxocemidae (endemic to Mexico and Central America); and Dermatemydidae (a relict in Mexico and northern Central America). The Bipedidae, Dibamidae, and Anniellidae do not occur in Central America. Dibamidae are found only in northeastern Mexico and Anniellidae on the west coast of the United States. Trionychidae has a northern distribution in the New World. Interestingly, only 23 species are common to North America, Mexico, and Central America.

Areas of endemism that are important in conservation policies can be delimited through an analysis of distribution patterns of endemic taxa.

West (1964) divided Mexico into five large, natural regions based on climate and vegetation; these regions have been modified here (Fig. 7.2): arid extratropical areas (regions 1 and 1'), cold tropical highlands (regions 2 and 2'), tropical lowlands (regions 3 and 3'), extratropical highlands (regions 4 and 4') and extratropical subhumid areas (region 5). The Appendix gives a more detailed description of each of these regions.

Figures 7.3 and 7.4 present the proportion of endemic species in families for amphibians and reptiles. Amphibian families, with the exception of the Bufonidae and Caeciliadae, have a higher proportion of endemic species. Among reptiles, eight of the 19 families with endemic species have a higher proportion of nonendemics.

Table 7.7 gives the number of endemic species in the orders of amphibians and reptiles present in each of the regions. The tropical highlands in the center of the country (Fig. 7.2, zone 2) have the highest number of

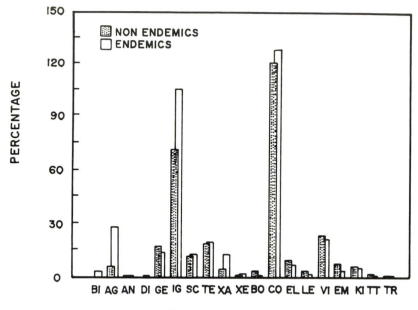

FAMILIES WITH ENDEMIC SPECIES

Figure 7.4. Proportion of endemisms among reptile families in Mexico. BI, Bipedidae; AG, Anguidae; AN, Anniellidae; DI, Dibamidae; GE, Gekkonidae; IG, Iguanidae; SC, Scincidae; TE, Teiidae; XA, Xantusiidae; XE, Xenosauridae; BO, Boidae; CO, Colubridae; EL, Elapidae; LE, Leptotyphlopidae; VI, Viperidae; EM, Emydidae; KI, Kinosternidae; TT, Testudinidae; TR, Trionychidae.

endemic species for both reptiles and amphibians. This region also affords the greatest number of species, in both groups, whose distribution is restricted to the highlands. In the case of amphibians, of the 123 species here, 98 are restricted to these ecosystems; and among the reptiles, 107 of the 163 are restricted. Finally, of the 286 species of the two groups here, 205 (71.6%) are restricted to zone 2.

The tropical lowlands of the Pacific coast (zone 3') are next in importance. Of the 29 endemic amphibians, 14 are restricted to these confines. Of the 124 endemic reptiles, 71 are known only from this area. The Pacific coast has a total of 153 species, and 85 (55.5%) are found locally.

Reptiles also have a great number of endemic species in the arid areas of the northwest (zone 1'), which is not the case with the amphibians. Of the 63 species of the former, 43 (68.2%) are local. The islands of the Sea of Cortes may be among the factors contributing to the high endemism here.

Taxonomically diverse genera in the amphibian Hylidae, Leptodactylidae, Ambystomatidae, and Plethodontidae and the reptilian Anguidae, Iguanidae, and Colubridae are noteworthy and have a certain degree of zonation in their distribution. The Hylidae are well represented in the cold tropical areas (zone 2) with 38 endemic species, of which 30 are

Table 7.7. Distribution of endemic species by natural region

Taxa	No. of species by regions 1–5[a]								
	1	1'	2	2'	3	3'	4	4'	5
Amphibia									
Anura	5/0	6/0	72/52	9/5	11/5	26/12	7/2	5/2	1/0
Caudata	0/0	0/0	51/46	3/3	8/7	3/2	2/1	7/3	1/0
Gymnophiona	0/0	0/0	0/0	0/0	1/1	0/0	0/0	0/0	0/0
Total	5/0	6/0	123/98	12/8	20/13	29/14	9/3	12/5	2/0
Reptilia									
Amphisbaenia	0/0	1/0	0/0	0/0	0/0	3/2	0/0	0/0	0/0
Sauria	15/4	45/36	75/50	8/7	20/16	60/36	6/1	14/2	6/1
Serpentes	8/0	15/6	87/57	7/5	15/11	57/31	12/4	15/6	2/0
Testudines	5/3	2/1	1/0	0/0	2/2	4/2	0/0	1/0	0/0
Total	28/7	63/43	163/107	15/12	37/29	124/71	18/6	30/8	8/1
TOTAL	33/7	69/43	286/205	27/20	57/42	153/85	27/8	42/13	10/1

Data are given as: number of species recorded for the region/number of species endemic to the region.
[a]Regions 1–5 are outlined in **Figure 7.2.**
Source: West (1964).

local. The latter number represents 66.6% of all endemic Hylidae, and for the most part they are represented by *Hyla*. The Leptodactylidae are also well represented in this area. There are 18 endemic species, of which 14 are local, representing 36.8% of all endemic species of the family. *Eleutherodactylus* and *Tomodactylus* are well represented in this area. The Ambystomatidae has 16 species (94.1% of them are endemic species of axolotls) in zone 2, all of which are restricted to this area. They belong to *Ambystoma* and *Rhyacosiredon*, the latter endemic to Mexico. The last amphibian family well represented in zone 2 is the Plethodontidae, with 35 endemic species (30 of which, or 60%, are local). Genera of this family that are best represented here are *Pseudoeurycea* and *Thorius*; the latter is endemic to Mexico. The amphibians whose distribution is restricted to zone 2 and that belong to the families mentioned comprise 51.7% of all amphibians endemic to the country.

Among the three reptile families that are well represented in the cold tropical areas is the Anguidae, with 17 endemics, 14 of which are local, constituting 50% of all endemic Anguidae. *Abronia* and *Mesaspis* have the highest number of species in this zone. The Iguanidae can be considered well represented in zones 1', 2, and 3'. In zone 1' (extratropical arid areas in western Mexico) 16 of the 20 species present are local and make up 15.2% of all endemic iguanids. In zone 2, 24 of the 40 species are restricted and represent 22.8% of the total number of endemics of this family. Zone 3' has 38 endemic species of which 25 are local and this figure represents 23.8% of all endemic iguanids; 61.9% of the endemic iguanids of Mexico have areas of distribution that are limited to only one of the areas mentioned. Only genera *Sauromalus* and *Uta* can be considered to be well represented in zone 1', as all the endemic species of these genera are restricted to this zone. The other iguanid genera have distributions that are too heterogeneous to be characterized. Finally, the colubrids are well represented in zones 2 and 3'. In zone 2 there are 50 restricted species (of 72, representing 39.3% of all endemic colubrids). Zone 3' has 40 colubrids of which 23 (18.1% of the total number of endemics of the family) are local. In total, the colubrid species locally occurring in these two areas represent 57.4% of all endemic colubrids. In zone 2 the genera that are well represented are *Geophis*, *Tantilla*, and *Toluca*; the last is exclusive to zone 2 and endemic to Mexico.

The 410 species of the seven families analyzed are endemic to the country, i.e., 75.6% of all endemics in Mexico. Of the zones analyzed, the poorest were 4 and 5, with 27 and 10 endemic species, respectively.

POSSIBLE REASONS FOR THIS RICHNESS

As has been seen, the percentage of endemism is high in the mountainous areas of central Mexico and on the Pacific coast. The first of these regions is associated with pine, pine-oak, and cloud forests and the second with deciduous and semideciduous tropical forests. Any explanations on the origin of this richness would be speculative, as reliable data on the distribution of endemic taxa and phylogenetic relations between them presently are lacking.

Wake and his colleagues (Wake, 1987; Wake & Lynch, 1976) have drawn interesting conclusions about the evolution of the Plethodontidae in the cloud forests of Mesoamerica based on evolutionary ecological studies of these salamanders: most of the salamanders in the tropics tend to occur in medium to high altitudes. Endemism is high in cloud forests. The variation in the number of species in different geographical areas seems to be due to climatic and topographical factors; species abundance is not influenced by altitude. In northern Central America most species are partially or totally arboreal. The salamander communities tend to segregate altitudinally in the tropics; and within each altitudinal belt, the species tend to be separated by body size and trophic behavior, habitat, and microhabitat.

Mexico's rugged topography has resulted in a great variety of habitats and microhabitats that are subject to variable environmental conditions. Consequently, there are different ecological conditions that allow the establishment of distinct animal populations isolated in small areas. Amphibians and reptiles are organisms of low vagility with a few exceptions, e.g., sea turtles and sea snakes; their tolerance levels of climatic and ecological factors are generally low, and they are susceptible to environmental changes. The rugged topography and the varied environments have contributed to the differentiation and radiation of these isolated populations into species, making Mexico exceptionally rich in amphibians and reptiles.

The formation of the Central American isthmus (Rosen, 1978; Savage, 1982) and the climatic and vegetational changes during the last several thousand years (Axelrod, 1975; Duellman, 1960, 1966; Toledo, 1982) have influenced the evolution of Mexico's biota.

Savage (1982) suggested the following sequence of events for the extant diversity of herpetofauna and its distribution in the region based on the evidence that most of the present-day amphibians and reptiles are known in the fossil records of Mesoamerica from the Cretaceous-Paleocene. With the establishment of the land bridge between North and South America during the Paleocene, a dispersal of South American groups toward Central America occurred. With the disappearance of this land bridge during the Eocene, faunas of Central and South America became isolated and differentiated. A second, later dispersal led to the establishment of northern groups in the north-central region of Mexico. The formation of mountains and climatic changes during the Eocene-Pliocene further isolated these and the Mesoamerican groups from the northern elements mainly in the southeastern United States. According to Savage, these groups (including plethodontids, frogs, anguids, and various colubrids) evolved together for the rest of the Cenozoic. The orographic formations of the Oligocene in Mexico and Central America led to further vicarious speciation. The formation of these mountain systems fragmented the more or less homogeneous Mesoamerican herpetofauna into three groups (Savage, 1982): those of the eastern lowlands, those of the western lowlands and those of the highlands. Examples of the dichotomy between the highland and lowland groups can be seen in species of the *Rana palmipes* (Hillis & de Sá, 1988) and the *Bufo valliceps* groups (Blair, 1972). When land connections were reestablished between North and South America, the two continents further exchanged faunistic elements.

The history of the distribution of the fauna since the end of the Pliocene and the Pleistocene has been largely influenced by climatic and vegetational changes. Toledo (1982) suggested that the climatic fluctuations during the Pleistocene affected the vegetation of the lowlands in the following three ways: (1) during the cold and dry cycles the vegetation of the lowlands was dominated by pine and oak forests; (2) during the cold and wet cycles, communities of cloud and oak forests were established; and (3) during the hot and dry cycles, deciduous and semideciduous forests possibly dominated. The distribution of most of present-day amphibians and reptiles, which were already present in the regions (Table 7.5) of North America, Mexico, and Central America, is due to the factors discussed.

It is suggested that the climatic changes during the Pleistocene may have contributed to high diversity and endemism in these regions. The species in these areas may be recent, neoendemic, or paleoendemic. The distribution of the paleoendemic species may have been reduced by the Pleistocene climatic changes.

The highlands of central Mexico became isolated from the central plateau and the southern United States as a consequence of increased aridity and low temperatures from about the middle of the Tertiary (Axelrod, 1975; Morafka, 1977; Rosen, 1978). The present distribution of many groups supports the north-south division between the southern part of the United States/northern Mexico and the central highlands. Examples include the frogs of the *Rana pipiens* complex (Hillis et al., 1983), genera of the xantusids (*Xantusia* and *Klauberina* in the north and *Lepidophyma* in the south [Crother et al., 1986]), and the species of *Phrynosoma* (Montanucci, 1987). The climatic fluctuations possibly led to numerous events of extinction and isolation as in *Terrapene* and *Chelydra* (Milstead, 1967; Van Devender and Tessman, 1975). Extinctions may have led to the present-day low endemism in northwestern Mexico.

The formation of the Isthmus of Tehuantepec isolated the southern Mexican highlands, and it seems to have acted as a barrier and corridor for biotic elements at different times during the Cenozoic. Several faunistic discontinuities have been recorded for different groups between the highlands of Central Mexico and those of Central America (for various elements of the herpetofauna, see Bezy & Sites, 1987; Duellman, 1966; Good, 1988; Halffter, 1987; Hillis and de Sá, 1988; Savage, 1982). These isolated allopatric and parapatric groups experienced further speciation due to climatic fluctuations during the Pleistocene. Good (1988) suggested other events of vicariance in the highlands of Mexico. Examples may be found in *Hyla, Eleutherodactylus, Rana, Ambystoma, Rhyacosiredon, Pseudoeurycea, Chiropterotriton, Abronia,* and *Geophis.* Wake's studies (1987) of the Plethodontidae suggested a similar course of speciation.

The highland herpetofaunas are locally distributed, leading to a higher degree of endemism than those in humid, tropical forest lowlands (Duellman, 1966). In Omiltemi, Guerrero, for example, of 37 herpetological taxa recorded, 13 (35.1%) are endemic (Muñoz-Alonso, 1988), most of which are associated with pine and cloud forests. In the Los Tuxtlas area of Veracruz, of 149 taxa recorded (Pérez-Higareda et al., 1987), 19 (12.7%) are endemic; some of them are *Pseudoeurycea werleri,* and there are two species of *Abronia* associated with cloud forests. The cloud and pine-oak forests in the Los Tuxtlas region may be relatively recent, as they may have become isolated in the higher elevations of Los Tuxtlas during the last 40,000 years when the massif was formed. The existing relations between some elements of this fauna and of the highlands in the extreme east of Oaxaca can be explained by dispersal during times of cold and wet climate; later, the fauna in both places may have become isolated, leading to allopatric speciation as in species of *Abronia* whose closest congeners are found in the highlands of the extreme east of Oaxaca (Campbell, 1984).

The biogeography of the Pacific lowlands is more difficult to interpret (Toledo, 1982), although the herpetofauna of this area and that of the coastal plain of the Gulf present distributions that are more or less continuous. Endemism is higher in the first group. This point may also be true for the extreme north of the Yucatan Peninsula (Duellman, 1966; Lee,1980). The endemic species in the peninsula may have evolved *in situ* due to isolation resulting from climatic fluctuations during the Pleistocene (Lee, 1980). There are also significant differences in climate and vegetation types (Savage, 1982). Some of the most characteristic elements of the Pacific areas are *Syrrhophus, Bipes,* some species of the *Phyllodactylus,* and several species of *Anolis.* Generally, these groups of species exhibit allopatric or parapatric distributions along the Pacific coast, possibly as a result of climatic changes. A similar phenomenon has been observed by Toledo-Manzur (1982) in *Bursera* (Burseraceae) among which vicariance has been noted in the Eastern Depression of the Balsas and northwestern regions of this zone.

Although it is widely recognized that the humid tropical forests are the richest ecosystems with high endemism (Myers, 1986; Prance, 1982) it is not true of the humid tropical forests of Mexico (Toledo, 1982). Similarly, the amphibian and reptile species of Mexico's tropical forests are poor in endemism. Although zone 3 (Table 7.7) has 57 species endemic to the country, 42 of which are restricted to this region, not all the species are associated with the vegetation of the humid tropical forests. In fact, only a few species are endemic to this vegetation type; others are found in the cloud forests of Los Tuxtlas (a zone that falls geographically into this region but whose fauna has some elements that bear affinities to the fauna of the highlands of Oaxaca) or are endemic to the Yucatan Peninsula, where the humid tropical forest is not the characteristic vegetation type.

Duellman (1966) observed that the herpetofauna of the humid lowlands of Mesoamerica behave similarly and that the species have more or less continuous distributions. Of the 149 reptiles and amphibians that occur in Los Tuxtlas, Veracruz (Pérez-Higareda et al., 1987), at least 117 have a distribution that goes beyond the region, spreading in many cases as far as Central America or some other region of Mexico.

HIGH DIVERSITY AREAS OF MEXICAN HERPETOFAUNA

Precise data on the distribution of endemic species (Flores V., in prep.) will identify areas of high species richness. The distribution of endemic species in the zones identified in this study is preliminary. Data on wide-ranging and narrow endemics are important and will help in the selection of priority areas for conservation. Most endemic species of the same genus have allopatric or parapatric distributions. Preliminary data suggest that the distribution of several endemic species of different groups overlap. Areas of high concentration of endemic species have been considered as "refugia" (Toledo, 1982). It is possible that some of these sites correspond more to areas of endemism than to "refugia" (Cracraft, 1985).

Toledo (1982) has identified the following "refugia" for Mexico: Lacandon forests, Soconusco area, Sierra de Los Tuxtlas, and Sierra de Juárez and Cordoba. Emerging data on endemism of Mexican herpetofauna do not support Toledo's conclusions regarding "refugia." Cracraft (1985) observed that not all "refugia" are necessarily areas of high centers of endemism.

IDENTIFICATION OF AREAS IN NEED OF CONSERVATION

The areas for conservation require careful evaluation for their species richness. The Mexican states with the highest number of endemics in decreasing order are Oaxaca, Chiapas, Veracruz, Guerrero, Michoacán, Jalisco, Puebla, Sinaloa, San Luis Potosí, and Nayarit (Flores-V. & Gerez, 1988). The ecosystems in these states with the highest vertebrate diversity (Flores-V. & Gerez, 1988) are oak, pine-oak, and cloud forests. To date, 468 species of Mesoamerican vertebrate endemics have been recorded in these forests with amphibians and reptiles accounting for 57.2%. In the xerophyllous underbrush and chaparrals (which include approximately 14 types of underbrush) 230 terrestrial vertebrates have been recorded, of which 51.3% are reptiles. In the lowland deciduous forests, 229 species of Mesoamerican endemic vertebrates have been noted, 38.8% of which are reptiles and amphibians. Tropical evergreen forests have 197 Mesoamerican endemic vertebrates, of which 51.7% are amphibians and reptiles. Finally, coniferous forests have 121 vertebrates, and amphibians and reptiles account for 67.7% of them. Knowledge of several factors, including biodiversity of states and ecosystems, is necessary during the selection of priority zones for conservation.

HERPETOFAUNA AS A NATURAL RESOURCE

Several species of Mexico's herpetofauna are important as a resource to man. Generally, two groups of species are exploited in Mexico (Flores-V., 1978, 1980; Lazcano-Barrero et al., 1986): (1) those that are (or were), exploited on a large scale and (2) those that are exploited at a regional level. In the first group there are three species of crocodiles, six species of sea turtles, and at least five species of frogs. The second group includes at least 13 freshwater turtles and terrapins, approximately 25 species of lizards and serpents, and eight species of frogs and salamanders.

In Mexico, the main reason for exploiting these species is to obtain their skin to meet the growing demand of the skin trade or for consumption. Amphibians and reptiles are also kept as pets or are used in the arts and crafts industry, among other uses.

For example, the exploitation of crocodile skins in Mexico was 101,703 tons in 1938 and the production of skin had decreased to 15,035 tons by 1965, mainly owing to overexploitation. The use of turtle products was modest until the middle 1960s, then increasing rapidly by 1967–1968. Subsequently their exploitation has abated (Flores-V., 1980).

There are few endemics that are subject to exploitation, but among them are some axolotl species (Ambystomatidae), some turtle species of *Kinosternon*, and several species of rattlesnakes and semi-aquatic *Thamnophis*. It is important to point out that several of the species exploited are in danger of extinction through overexploitation and habitat destruction. The loss of habitat is especially notable in the case of amphibians. The ever-increasing need for water for human needs is drying up several areas where amphibians once were common. Again man's economic interest is a direct threat to these taxa. The introduction of carp (for human consumption) into Mexican waters is proving detrimental to the amphibians. The disappearance of amphibians in areas where carp has been introduced may be due to predation by the latter.

CONSERVATION NEEDS

Conservation measures are needed to preserve this qualitatively and quantitatively impressive diversity. The primary identifiable threat to them come from habitat loss. Additional national parks where representative diversity is available should be set aside. Studies should be encouraged in the already existing ones, so their diversity can be documented.

Many groups are taxonomically still incompletely known. Distribution data for many groups continues to be fragmentary. Field-oriented studies to produce information to cover taxonomy and distribution are urgently needed as they can help in the formulation of conservation policies. In addition, phylogenetic studies of Mexican groups or groups centered in Mexico are scarce and thus are required to highlight the unique taxonomic status of many groups and the nature of the endemism of many of them. Groups in great need of study are iguanids (*Anolis*), colubrids (various genera, e.g., *Thamnophis*), leptodactilids (*Eleutherodactylus*), and kinosternids (*Kinosternon*).

Historical biogeography of the groups can contribute useful information for conservation; for example, the study of areas of endemism could be the key to a solid conservation plan. It is important to take into account the ecosystems with the highest diversity, those that suffer most from human pressure, and those whose rates of destruction are high. By putting together all these factors a significant contribution can be made to conservation planning in Mexico.

ACKNOWLEDGMENTS

I am most grateful to Dr. H. Smith whose review of an earlier version of this chapter resulted in significant changes and improvements. Part of the work was carried out while I was a Resident Museum Specialist at Carnegie

Museum at Pittsburgh where I enjoyed the collaboration and assistance of Dr. C.J. McCoy. I thank E. Ezcurra, J. Llorente B., M. Montellano, V. Sánchez C., J.M. Savage, J. Soberón, J. Therbourgh, and V. Toledo for reviewing various drafts of this paper. I appreciate the help of many who helped with taxonomic and nomenclatural aspects. They include R.L. Bezy, R.A. Brandon, J.A. Campbell, J.R. Dixon, W.E. Duellman, J.S. Frost, D.R. Frost, D. Hahn, D.M. Hillis, J.B. Iverson, A.G. Kluge, J.D. Lynch, C.J. McCoy, C.W. Myers, J.A. Roze, J. Villa, D.G. Wake, R.G. Webb, and L.D. Wilson. C.S. Lieb helped variously. Anelio Aguayo L., Héctor Arita W., R. Crombie, D. Good, J. Hanken, Efrain Hernández G., Mónica Herzig, P. Miramontes, Antonio Muñoz A., Adolfo Navarro, and Gonzalo Pérez H. provided valuable information. Finally, I thank T.P. Ramamoorthy and Marguerite Elliott for their valuable help with the preparation of this chapter in English.

APPENDIX[1]

Arid Extratropical Lands, Regions 1 and 1'

Region 1 includes the Mesa del Norte, which comprises northern, central, and southeastern Chihuahua; Coahuila, excluding the extreme southeast; Durango, excluding its southernmost part; western San Luis Potosí; and extreme northern Jalisco. This arid region is characterized by extensive areas of internal river basins and classical desert topography, with elevations of 1,000 to 2,000 m above sea level with annual mean precipitation of 200–300 mm. The vegetation consists of thorny underbrush of species of agave (usually *Agave lechugilla*), *Larrea divaricata*, species of *Yucca*, and several cacti. *Prosopis juliflora* occurs in sandy soils, and in some areas there are small forests of *Yucca* and thickets of *Opuntia*.

Region 1' comprises the peninsula of Baja California, excluding the northwest and south; Sonora, excluding the central, eastern, and northeastern parts; and northwestern Sinaloa. West (1964) considered the Baja California Peninsula a subregion separate from Sonora and Sinaloa. In this chapter only one region is recognized in order to provide easy access to the information provided. The region is characterized by its low, arid areas with BWh (semiarid, warm to very cold) and BSh (semiarid, warm to moderately cold) climates with annual rainfall below 200 mm in the central part of Baja California and the Desierto de Altar in Sonora and below 400 mm in northern Sinaloa. The vegetation of these arid areas is composed of xerophytic shrubs or semishrubs: *Cercidium* in the more arid areas; *Prosopis* and *Pithecellobium* in deep alluvia along streams; the pitaya (several genera of *Cereus* group); and a great number of deciduous and evergreen shrubs. In the Peninsula there are also species of *Washingtonia* and *Erythea* palms. Endemic to the central desert are *Idria columnaris* and *Pachycormus discolor*.

Cold Tropical Highlands, Regions 2 and 2'

Region 2 comprises the Mesa Central; part of the Sierra Madre Oriental; and the Sierra and Mesa del Sur. It includes southeastern Nayarit; the southern tip of Zacatecas; northern, central, and eastern Jalisco; south Aguascalientes; Guanajuato; northern Michoacán; Querétaro, excluding the northernmost region; Hidalgo, excluding the extreme northeast; the state of Mexico, excluding its southwest tip; the Federal District; Tlaxcala; Puebla, excluding the northern and southwestern extremes; the extreme west of the central part of Veracruz; the extreme north of Morelos; and the Sierra del Sur in Oaxaca, Guerrero, and Michoacán. The Mesa Central is volcanic and has a peculiar hydrography; the Transverse Neovolcanic Belt is the southern limit of the Meseta and has the most recent volcanic formations. The

[1]Modified from West (1964).

northern limit near the Mesa Central has older (Tertiary) volcanic formations.

One of the most salient characteristics of the Meseta is its great number of flat, dry basins, occupied during the Pleistocene by large lakes caused by the lack of normal drainage through vulcanism. The Sierra Madre del Sur and the Mesa del Sur have mountain massifs of old crystalline rock. The vegetation of the arid areas (the Valleys of Mezquital, Tehuacan, and others) have columnar cacti, yuccas, and other xerophytic shrubs. The vegetation of the wet regions is composed of temperate forests mainly of *Abies, Alnus, Juniperus, Pinus, Pseudotsuga,* and *Quercus.* The vegetation of the Sierra and Mesa del Sur is composed of forests of *Pinus* and *Quercus* as well as cloud forests, principally of *Fagus, Liquidambar, Podocarpus* and *Tilia.*

Region 2' comprises the Sierra Madre in Chiapas, which has characteristics similar to those of the wet areas of the region.

Tropical Lowlands, Regions 3 and 3'

Region 3 covers the lowlands of the Gulf of Mexico and of the Caribbean coast. It includes a small part of the following states: extreme south of Tamaulipas; eastern San Luis Potosí; northeastern Hidalgo; northern Puebla; north-northeastern Oaxaca; Veracruz, excluding the extreme west of the center; central, northern, and northeastern Chiapas; Tabasco; Campeche; Yucatán; and Quintana Roo. It features lowlands with altitudes from 0 to 1,200 m above sea level, abundant rain with a relatively short dry season (climate Am to Af) and high temperatures throughout the year. The vegetation comprises tropical evergreen forests in southern Veracruz, underbrush in the arid parts of Yucatán, and medium subevergreen forests in two-thirds of the Yucatán Peninsula.

Region 3' includes the lowlands of the Pacific coast, Balsas Basin, and Valley of Chiapas. It comprises central and southern Sinaloa; extreme southern Baja California; western Nayarit; western and extreme southern Jalisco; Colima; central and western Michoacán, excluding the Sierra de Coalcomán; northern and southern Guerrero; central and southern Morelos; southwestern Puebla; southern Oaxaca; and southern Chiapas.

It has lower precipitation levels (1,000–2,000 mm) than region 3 and a long markedly dry season that lasts 5–6 months. Tropical deciduous and semideciduous forests predominate. Plants with high concentrations of resin and tannin are abundant, probably as a result of the long, dry spells. In the Balsas Basin this type of plant is also abundant, but the dominant species are those of the legume family. In the Valley of Chiapas,which is not as dry as the Balsas depression, the vegetation is composed of tropical shrubs, low trees, cacti, and agaves.

Extratropical Highlands, Regions 4 and 4'

Region 4 corresponds to the Sierra Madre Occidental. It includes northeastern Sonora; eastern and southwestern Chihuahua; northeastern Sinaloa; and eastern Durango. Above 2,200 m, the volcanic mesas give way to pine-oak forests. In the northernmost zones there is a marked winter with little precipitation, whereas in the remaining parts the seasonal temperature variation is lower and is similar to the tropical highlands of the south.

Region 4' corresponds to the higher areas of the Sierra Madre Oriental. It includes the extreme southeast of Coahuila; southern Nuevo León; extreme southwestern Tamaulipas; and central and north-northeastern San Luis Potosí. It has pine-oak forests, and in the higher elevations (>3,600 m) there are patches of alpine vegetation.

Subhumid Extratropical Areas, Region 5

Region 5 comprises central and eastern Nuevo León; Tamaulipas, excluding the northwest and southwest; and a small part of northeastern San Luis Potosí. Climatically, it forms a transition zone between the humid tropical areas of the Gulf and Caribbean and the subtropical subhumid areas of the southeastern United States. The vegetation is composed of underbrush and small trees such as *Acacia* and *Cordia* occasionally intermingled with *Opuntia*, *Yucca*, organ cacti, and various herbs. This community differs from the desert flora of the adjacent Mesa del Norte. Gallery forests of *Carya, Juniperus*, and *Salix* occur along the lowland streams. In the isolated low mountains there are sporadic thickets of *Quercus* and *Pinus*.

REFERENCES

Axelrod, D.I. 1975. Evolution and biogeography of Madrean-Tethyan Sclerophyll vegetation. Ann. Missouri Bot. Gard. 62(2):280–334.

Barrios-Rivera, H. 1985. Estudio Análitico del Registro Paleovertebradológico de México. Thesis, UNAM.

Behler, J.L. & F.W. King. 1979. The Audubon Society Field Guide to North American Reptiles and Amphibians. New York: Knopf.

Bellairs, A. 1975. Los Reptiles. Historia Natural Destino. 10:145–420. Barcelona: España.

Bezy, R.L. & J.W. Sites. 1987. A preliminary study of allozyme evolution in the lizard family Xantusiidae. Herpetologica 43(3):280–292.

Blair, W.F. 1972. *Bufo* of North and Central America. *In* W.F. Blair (ed.), Evolution in the Genus *Bufo*. Austin, TX: Univ. Texas Press. pp. 93-101.

Campbell, J.A. 1984. A new species of *Abronia* (Sauria: Anguidae) with comments on the herpetogeography of the highlands of southern Mexico. Herpetologica 40(4):373-381.

Conant, R. 1975. A Field Guide to Reptiles and Amphibians of the Eastern and Central North America. 2nd Ed. Boston: Houghton Mifflin.

Cracraft, J.L. 1985. Historical biogeography and patterns of differentiation within the South American avifauna: areas of endemism. *In* P.A. Buckley, M.S. Foster, E.S. Morton, R.S. Ridgely, & F.G. Buckley (eds.), Neotropical Ornithology. Am. Ornith. Union Ornithol. Monogr. (36):49–84.

Crother, B.I., M.M. Miyamoto & W.R. Presch. 1986. Phylogeny and biogeography of the lizard family Xantusiidae. Syst. Zool. 35(1):37–45.

Darlington, P.J. 1957. Zoogeography: The Geographical Distribution of Animals. New York: John Wiley & Sons.

Duellman, W.E. 1960. A distributional study of the amphibians of the isthmus of Tehuantepec, Mexico. Univ. Kansas. Mus. Nat. Hist. Pub. 13(2):19–72.

――――. 1966. The Central American herpetofauna: an ecological perspective. Copeia 1966(4):700–719.

――――. 1970. The Hylid Frogs of Middle America. Vols. 1 & 2. Lawrence, KA: Monograph of the Museum of Natural History, Univ. Kansas.

――――. 1979. The numbers of the amphibians and reptiles. Herp. Rev. 10(3):83-84.

Dugés, A.A. 1896. Reptiles y batracios de los Estados Unidos Mexicanos. Naturaleza 2(2):479-485.

Flores-Villela, O.A. 1978. Contribución al Conocimiento de los Anfibios y Reptiles de Importancia Económica. Mem. II Congr. Nal. Zool. Monterrey N.L. 1:343–356.

――――. 1980. Reptiles de Importancia Económica en México. Thesis, UNAM.

――――. 1982. Contribución a la Historia de la Herpetología en México. Ponencia en la Reunión Latinoamericana de Historiadores de la Ciencia. Puebla, México.

―――― & P. Gerez. 1988. Conservación en México: Sintesis sobre Vertebrados Terrestres, Vegetación y Uso del Suelo en México. Mexico City: INIREB-Conservation International.

Frost, D. (ed.). 1985. Amphibian Species of the World, A Taxonomic and Geographical Reference. Lawrence, KA: Association of Systematic Collections.

Good, D.S. 1988. Phylogenetic Relationships Among Gerrhonotine Lizards (Sauria: Anguidae): An Analysis of External Morphology. Univ. Calif. Pub. Zool. 121:1-139.

Halffter, G. 1987. Biogeography of the Montane Entomofauna of Mexico and Central America. Ann. Rev. Entomol. 32:95–114.

Halliday, T.R. & K. Adler (eds.). 1986. The Encyclopedia of Reptiles and Amphibians. New York: Facts on a File Publications.

Henderson, R.W. & L.G. Hoevers. 1975. A checklist and key to the amphibians and reptiles of Belize, Central America. Cont. Biol. Geol. Milwaukee Pub. Mus. (5):1-63.

Hillis, D.M., J.S. Frost & D.A. Wright. 1983. Phylogeny and biogeography of the *Rana pipiens* Complex; a biochemical evaluation. Syst. Zool. 32(2):132-143.

Hillis, D.M. & R. de Sá. 1988. Phylogeny and taxonomy of the *Rana palmipes* group (Salientia: Ranidae). Herp. Monogr. 2:1–26.

Lee, J.C. 1980. An ecogeographic analysis of the herpetofauna of the Yucatan Peninsula. Misc. Pub. Univ. Kansas Mus. Nat. Hist. (67):1-75.

Lazcano-Barrero, M.A., O. Flores-V., M. Benabib-N., J. Hernández-G., M. Chávez-P. & A. Cabrera-A. 1986. Estudio y conservación de los Anfibios y Reptiles de México: Una Propuesta. Cuad. de Divulg. INIREB. No. 25.

Milstead, W.W. 1967. Fossil box turtles (*Terrapene*) from Central North America, and box turtles from eastern Mexico. Copeia 1967(1):168–179.

Montanucci, R.R. 1987. A phylogenetic study of the horned lizards, genus *Phrynososma*, based on skeletal and external morphology. Contr. Sci. Los Angeles Co. Mus. Nat. Hist. 390:1–36.

Morafka, D.J. 1977. A Biogeographic Analysis of the Chihuahuan Desert through its Herpetofauna. Biogeographica Vol. 9. The Hague: Dr. W. Junk B. V. Publ.

Muñoz-Alonso, L.A. 1988. Estudio herpetofaunístico del Parque Ecológico Estatal de Omiltemi, Mpio. de Chilpancingo, Guerrero. Thesis, UNAM.

Myers, N. 1986. Tropical deforestation and a mega-extinction spasm. *In* M. Soulé (ed.). Conservation Biology, The Science of Scarcity and Diversity. Sounderlan, MA: Saunders. pp. 394–409.

Pérez-Higareda, G., R.C. Vogt & O.A. Flores-V. 1987. Lista Anotada de los Anfibios y Reptiles de la Región de Los Tuxtlas, Veracruz. Mexico City: UNAM. Inst. Biol.

Porter, K.R. 1972. Herpetology. Philadelphia: Saunders.

Prance, G.T. (ed.). 1982. Biological Diversification in the Tropics. New York: Columbia Univ. Press.

Rosen, D.E. 1978. Vicariant patterns and historical explanation in biogeography. Syst. Zool. 27(2):159–188.

Savage, J.M. 1960. Evolution of a Peninsular Herpetofauna. Syst. Zool. 9(3):184–212.

———. 1966. The origins and history of the Central American herpetofauna. Copeia 1966(4):719–766.

———. 1973. The geographic distribution of frogs: patterns and predictions. *In* J.L. Vial (ed.), Evolutionary Biology of Anurans. Columbia,MO: Univ. Missouri Press. pp. 351–445.

———. 1982. The enigma of the Central American herpetofauna: dispersal or vicariance?. Ann. Missouri Bot. Gard. 69(3):464–547.

Smith, H.M. 1949. Herpetogeny in Mexico and Guatemala. Chicago Acad. Sci. Nat. Hist. Mus. 20:1-3.

———. 1970. The first herpetology of Mexico. Herpetology 3(1):1-16.

——— & R.B. Smith. 1973. Synopsis of the Herpetofauna of Mexico. Vol 2. Analysis of the Literature Exclusive of the Mexican Axolotl. Augusta, WV: Eric Lundenberg.

——— & ———. 1976a. Synopsis of the Herpetofauna of Mexico. Vol. 3. Source Analysis and Index for Mexican Reptiles. North Bennington, VT: John Johnson.

——— & ———. 1976b. Synopsis of the herpetofauna of Mexico. Vol. 4. Source Analysis and Index for Mexican Amphibians. North Bennington, VT: John Johnson.

——— & ———. 1977. Synopsis of the Herpetofauna of Mexico. Vol. 5. Guide to Mexican Amphisbaenians and Crocodylians. North Bennington, VT: John Johnson.

——— & ———. 1979. Synopsis of the Herpetofauna of Mexico. Vol. 6. Guide to Mexican Turtles. North Bennington, VT: John Johnson.

——— & E.H. Taylor. 1945. An annotated checklist and key to the snakes of Mexico. Bull. U.S. Natl. Mus. 187:1–239.

——— & ———. 1948. An annotated checklist and key to the amphibia of Mexico. Bull. U.S. Natl. Mus. 194:1–118.

——— & ———. 1950. An annotated checklist and key to the reptiles of Mexico. Bull. U.S. Natl. Mus. 199:1–253.

——— & ———. 1966. Herpetology of Mexico. An Annotated Checklist and Key to the Amphibians and Reptiles. A Reprint of Bulletins 187, 194 and 199 of the U.S. National Museum, with a List of Subsequent Taxonomic Innovations. Ashton, MD: Eric Lundenberg.

Stebbins, R.C. 1966. A Field Guide to Western Reptiles and Amphibians. Boston, MA: Houghton Mifflin.

Stuart, L.C. 1963. A checklist of the Herpetofauna of Guatemala. Misc. Pub. Mus. Zool. Univ. Michigan 122:1-150.

————. 1964. Fauna of Middle America. *In* R. Wauchope (ed.), Handbook of Middle American Indians. 1:316-362. Austin, TX: Univ. Texas Press.

Toledo-Manzur, C.A. 1982. El género *Bursera* (Burseraceae) en el Estado de Guerrero. Thesis, UNAM.

Toledo, V. M. 1982. Pleistocene Changes of Vegetation in Tropical Mexico.*In* G.T. Prance (ed.), Biological Diversification in the Tropics. New York: Columbia Univ. Press. pp. 93–111.

Van Devender, T.R. & N.T. Tessman. 1975. Late Pleistocene snapping turtles (*Chelydra serpentina*) from southern Nevada. Copeia 1975(2):249–253.

Wake, D.B. 1987. Adaptative radiation of salamanders in Middle American cloud forest. Ann. Missouri Bot. Gard. 74:242–264.

———— & J.F. Lynch. 1976. The distribution, ecology and evolutionary history of Plethodontid salamanders in tropical America. Nat. Hist. Mus. Los Angeles Co. Sci. Bull. 25:1-65.

West, R.C. 1964. The natural regions of Middle America. *In* R. Wauchope (ed.), Handbook of Middle American Indians. 1:363-383. Austin, TX: Univ. Texas Press.

8

A Geographic, Ecological, and Historical Analysis of Land Bird Diversity in Mexico

PATRICIA ESCALANTE PLIEGO,
ADOLFO G. NAVARRO SIGÜENZA,
AND A. TOWNSEND PETERSON

Patterns of distribution and diversity in the Mexican avifauna are summarized herein. We outline the history of ornithological exploration in Mexico, from pre-Hispanic times to present-day efforts. Using a data base describing the distribution of 773 resident land bird species in 35 geographical regions and 23 habitats, we attempt to detect patterns of diversity and endemism. The regions with greatest species richness are the southeastern lowlands, the Yucatan Peninsula, and the Isthmus of Tehuantepec, whereas the Pacific islands, the southern tip of Baja California, and parts of the Balsas Basin have low numbers of species. Patterns of endemism are different: species endemic to Mexico are concentrated in the Sierra Madre Occidental, Transvolcanic Belt, and Sierra Madre Oriental, as well as on several of the oceanic islands. Habitats richest in species are semideciduous tropical forest, lowland rain forest, and pine-oak forest, whereas seacoast, chapparal, and urban habitats contain few species. Habitats richest in Mexican endemics are desert scrub, pine-oak forest, tropical deciduous forest, and cloud forest. Patterns of similarity among geographic regions and among habitats are summarized, as is the historical biogeography of the Mexican avifauna, considering the relative importance of Mexico as a center of diversification versus Mexico as a mixture of north temperate and southern tropical avifaunas. Finally, based on patterns of diversity, endemism, and extinctions, we outline several recommendations of priority regions and habitats for conservation efforts.

The physical environments of Mexico include a wide array of mountain ranges and intervening high plateaus, lowlands, and coastal plains. Consequently, the country contains a wide variety of habitats, many of which are subdivided by geographic barriers or occur only in small isolated patches. This geographic and ecological diversity is reflected in the great variety of animals and plants living in the country.

Many attempts have been made to analyze and summarize the diversity of various taxonomic groups in Mexico (Toledo, 1988), including plants (Rzedowski, 1978), scarabid beetles (Halffter, 1976), and mammals (Ramírez Pulido & Müdespacher, 1987). For birds, however, attempts to synthesize and understand patterns of diversity in Mexico have been surprisingly few. Grinnell (1928), Ridgway and Friedmann (1901–46), Friedmann et al. (1950), Goldman (1951), and Miller et al. (1957) summarized distributional data for Mexican birds. Griscom (1950) analyzed the distribution and origin of the Mexican avifauna, but concentrated chiefly on historical patterns and did not analyze patterns of diversity within the country. Phillips (1961) discussed patterns of distribution and diversity in Mexican birds but focused largely on contrasts between resident and migrant distributions.

Our goal in this chapter is to summarize patterns of distribution and diversity in the avifauna of Mexico. We first outline the historical development of ornithology in Mexico, then summarize avian distribution and diversity in terms of geography and ecology, and analyze patterns of endemism in the avifauna as a whole. Finally, we interpret these patterns in terms of the biogeography, history, and evolution of the Mexican avifauna and discuss goals and strategies for preserving avian diversity in Mexico.

HISTORY OF ORNITHOLOGY IN MEXICO

The study of birds in Mexico began well before the Spanish conquest. The Indian cultures of that time had an impressive knowledge of their environment, including complex taxonomic systems based on the same principles of hierarchical grouping that are employed in folk taxonomies of present-day cultures throughout the world (Berlin, 1973; Raven et al., 1971). Birds were an important part of these taxonomic systems, which included most of the flora and fauna in the region.

The arrival of the Europeans in the New World during the fifteenth and sixteenth centuries initiated an effort that continues today to catalog scientifically the faunal diversity of Mexico. The Spanish naturalists, most of them soldiers or priests, began describing the species of the New World, mainly by recording original descriptions from Indians. Many of the descriptions contain mistakes and misunderstandings. Nevertheless, priceless works such as those by Bernardino de Sahagún, Juan de Torquemada, Gonzalo Fernández de Oviedo, and Tomás López Medel were produced during this era (Trabulse, 1985).

The most important work during the sixteenth century is that of Francisco Hernández, physician to the king of Spain, who devoted most of his life to describing the natural environment of Mexico based on an expedition made in 1570–77. Many of the Hernández descriptions are the earliest known for birds from the New World, e.g., those of vultures, toucans, and some birds of prey. Unfortunately, most of the zoological works of Hernández were lost in a fire in the library of El Escorial in Spain in 1641. Those that survived have been only recently published (Trabulse, 1985). Alvarez del Toro (1985) listed the 228 species of birds recognized in the Hernández manuscripts.

An early attempt to compile extensive data about the flora and fauna of part of Mexico was made by Miguel Venegas during an expedition in 1757 to California and Baja California. His *Noticias de la California* includes many descriptions of birds and other features of the natural history of the area. This work was later revised and corrected by Miguel del Barco, who added the results of a 1773 expedition (Trabulse, 1985). This work contains a relatively thorough (for the time) description of the avifauna of Baja California.

During the 1780s the Jesuit Francisco Javier Clavijero wrote extensively about the natural history of Mexico and California, treating both plants and animals (Beltrán, 1982). His work included some elements of the taxonomic system developed in Sweden by Carl Linneaus. Other Jesuits and scientists, such as Ferrer de Valdecebro, José Antonio Alzate, and Velázquez de León produced important works about the birds of Mexico during this period as well (Beltrán, 1982; Trabulse, 1985).

The Spanish interest in science and natural history stimulated the Royal Botanical Expedition to Mexico (1787–1803) led by Martín Sessé. Although the expedition concentrated primarily on plants, it provided descriptions and illustrations of many Mexican birds.

The geographic expedition of Alejandro Malaspina from Acapulco to Mexico City in 1789–94 also yielded important ornithological information. Antonio Pineda, the head naturalist of the expedition, collected specimens of plants and animals, including many birds that were later sent to Spain for study (González-Claverán, 1988).

The first part of the nineteenth century was a time of political instability in Mexico, a poor environment for scientific development (Trabulse, 1985). In Europe, however, the ideas of Buffon, Lamarck, and Cuvier began the evolutionary approach to understanding biological diversity. Many European scientists, chiefly German and French, visited Mexico during this period. One, the famous Alexander von Humboldt, made extensive collections of plants and animals in Mexico that were later studied and described by German scientists (Minguet, 1985; Von Mentz de Boege, 1982).

The French-led creation of the Mexican Empire during the 1860s brought both increased European scientific interest and participation by Mexicans. The Mission Scientifique au Mexique, a survey of natural resources, produced the descriptions of many new bird taxa by scientists of the Musée d'Histoire Naturelle de Paris.

The formation of the first scientific societies in Mexico, such as the Sociedad Mexicana de Historia Natural in 1868, brought together most of the scientists in the country and catalyzed the development of biology and natural history. Important ornithological works of this time included a folio of descriptions and illustrations of the hummingbirds of Mexico (Montes de Oca, 1874) and the description of the quetzal *Pharomachrus mocinno* (de la LLave, 1871).

During 1879–84, the Comisión Geográfica Exploradora de la República Mexicana employed Fernando Ferrari-Pérez, a skilled naturalist, to collect and identify animal specimens as a part of a general geographic survey of the country. A great quantity of specimens, including many birds, were sent by ship to be studied in New York and later exhibited in New Orleans, but the ship burned in Cuba in 1884 and the entire collection and field notes were lost. Later, in 1884–85, another expedition was made to obtain more specimens. Many of these specimens were later studied and identified as new taxa by Robert Ridgway of the United States National Museum (Ferrari-Pérez, 1886).

During the late nineteenth century, Osbert Salvin and F. Ducane Godman, British naturalists, went to Mexico and Central America to make a faunistic survey of the region as a whole. They collected extensively in several countries and acquired material from other collectors (Godman, 1915). In the end, they produced one of the most important faunistic works for the New World, including much new information about distribution and many descriptions of new species in the four volumes about birds in the series *Biologia Centrali-Americana* (Salvin & Godman, 1879–1904). Other important collectors and ornithologists from the United States or Europe of this period were John Xantus, A.J. Grayson, Ferd Bischoff (Lawrence, 1874), and F. Sumichrast (Lawrence, 1875).

In order to make available to nonscientists the new information on Mexican birds, Alfonso L. Herrera, a Mexican naturalist, published *Ornitología Mexicana* (Herrera, 1898–1914), based largely on Salvin and Godman (1879–1904). It was one of the first efforts by a Mexican scientist to summarize the diversity of Mexican birds.

Another important set of expeditions to Mexico was that of E.W. Nelson and E.A. Goldman, members of the Biological Survey of the Fish and Wildlife Service of the United States. They made several trips throughout the country during 1892-1906, exploring many new sites and habitats (Goldman, 1951; Nelson, 1922). They collected a total of 12,400 bird specimens (Goldman, 1951), many of them new taxa later described by Nelson (see Sánchez-León, 1969, for a complete reference list of Nelson's descriptions). This collection has since served as the main reference for taxonomic and biogeographic works dealing with the birds of Mexico (e.g., Friedmann et al., 1950; Hellmayr, 1918-49; Miller et al., 1957; Ridgway & Friedmann, 1901-46).

Progress in Mexican ornithology between 1910 and 1960 was analyzed by Phillips (1960). Detailed studies of geographic variation in Mexican birds (e.g., Johnson, 1980; Miller, 1941; Pitelka, 1951) dominated studies of bird diversity in Mexico during the latter part of the twentieth century. This topic still deserves much attention (Phillips, 1960), especially given the improved knowledge of many poorly known areas and the modern techniques now available.

Much of the information published on Mexican birds, and also data from specimens in U.S. museums (especially the Robert T. Moore Collection in Occidental College, California) were summarized by Friedmann et al. (1950) and Miller et al. (1957) making available much information on distribution and within-species geographic variation in Mexican birds. This reference still serves as the main source of distributional data for the birds of Mexico. The information became more accessible to nonscientists with Blake's (1953) and Peterson and Chalif's (1973) field guides, which led to many distributional discoveries by bird-watchers (e.g., Terrill & Terrill, 1986; Williams, 1987). Detailed information about the distribution and status of game birds of Mexico was published by Leopold (1977).

Finally, during the twentieth century, scientists have produced many studies of regional bird faunas, giving a more accurate picture of the geographic diversity of the birds of Mexico. Avifaunal summaries have been published for a number of states, the most recent and complete of which are listed in Appendix 1. Further information about regional surveys and descriptions of Mexican bird taxa can be found in Sánchez León (1969) and Gómez and Terán (1981). Additional distributional information on Mexican bird species is summarized in the A.O.U. (1983) and Phillips (1986).

METHODS

For the purposes of this chapter, we used the term diversity to refer to species richness; we did not take into account aspects of relative abundance or evenness. We followed the biological species concept, as outlined by the A.O.U. (1983), in defining speices-level diversity.

We restricted our analysis to species that breed in Mexico, including both permanent and summer residents, but excluding the numerous migrants and vagrants from Canada and the United States. Because current knowledge of breeding distributions of marine birds is still incomplete, we excluded several groups (Procellariiformes, Pelecaniformes, Scolopacidae, Haematopodidae, Laridae) from our quantitative analyses but discussed their distribution and diversity in the text (see Everett & Anderson, in press, for a careful treatment of patterns of distribution and diversity in these groups). Because paleontological information on Mexican birds is all but nonexistent (Simpson, 1943), we restricted our discussion to recent species.

We considered a species endemic to a region if it occurs there and nowhere else. In this sense, we treated both species endemic to Mexico ("endemics"), and species endemic to particular parts of the country ("narrow endemics"). Because some habitats and bird species overlap narrowly into adjacent countries, we include another category,

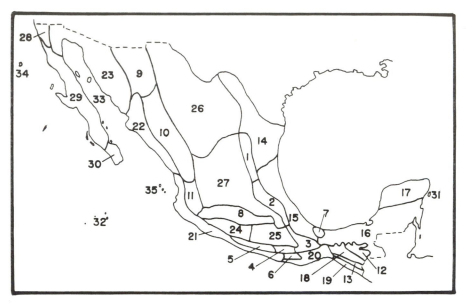

Figure 8.1. Map of Mexico showing the 35 biotic provinces used for the analyses in this chapter. See **Table 8.2** for province names.

"quasiendemics" (e.g., many species restricted to pine-oak woodland occur in the Chisos Mountains in western Texas and/or in extreme southeastern Arizona).

Although the optimal description of geographic faunal diversity would be based on patterns of occurrence on a fine grid system (e.g., Crowe & Crowe, 1982), such analyses require accurate distributional data that are not presently available for Mexican birds. So, for our analyses of geographic distribution, we used a modified version of the map of the biotic provinces of Mexico of Smith (1941). The modifications are mostly subdivisions of geographically extensive regions in order to provide more detail, and additions of several offshore islands as provinces. In all, we considered avian distribution among 35 biotic provinces (Fig. 8.1). Nomenclature for and information on habitat use were taken from Blake (1953), Rzedowski (1978), and A.O.U. (1983).

The principal sources for data on avian distribution were Friedmann et al. (1950) and Miller et al. (1957). Additional data were obtained from the more recent surveys of biologically important regions listed in Appendix 1 and from our personal observations. Nomenclature and arrangement of bird species are from A.O.U. (1983) for North and Central America and from Meyer de Schauensee (1970) for South America. Given the lack of a more recent synthesis, subspecific taxonomy follows Friedmann et al. (1950) and Miller et al. (1957).

Patterns of similarity among biotic provinces or habitats were summarized as follows. The taxon–region or taxon–habitat matrix was trans-

Table 8.1. Taxonomic diversity and endemicity in the Mexican breeding avifauna, including seabirds

Taxonomic rank	No.	True endemics		True and quasiendemics	
		No.	%	No.	%
Order	21	0	0	0	0
Family	72	0	0	0	0
Genus	397	9	2.3	14	3.5
Species	769	101	13.1	125	16.3

formed into a region–region or habitat–habitat matrix of values of Jaccard's index of similarity using programs written in BASIC. This index of faunal similarity incorporates elements of both faunal similarity and faunal size (Jackson et al., 1989), a problem that must be borne in mind when interpreting the results. The similarity matrices were then used to form phenograms using unweighted pair-group method of analysis (UPGMA) algorithms in a program developed by R.E. Strauss. Standard errors were estimated at each cluster node from all possible pairs of between-group distances, where the groups are the two clusters that are joined at the node calculated as the standard deviation of the distances divided by the number of distances.

TAXONOMIC DIVERSITY

Avian diversity is concentrated in several centers in the world, including Amazonia, northern South America, central Africa, southeast Asia, and others. Each continent, to some degree, has a unique fauna with varying degrees of diversification and endemism (Van Tyne & Berger, 1976). Mexico, located in the transition between the Nearctic (North America) and Neotropical (South and Central America) regions, contains a unique mixture of the two avifaunas.

In total, 769 bird species breed in Mexico (Table 8.1); an additional 257 species occur in Mexico as migrants, winter residents, or accidentals, for a total of 1007 species of birds recorded from the country. This figure is higher than that for the United States and Canada (>800; National Geographic Society, 1983), although Mexico has only about 11% as much land area.

Endemism in the Mexican avifauna is concentrated at the level of genera, species, and subspecies (Table 8.1). Fourteen (3.5%) of 397 genera, and 125 of 769 (16.3%) species are endemic or quasiendemic to the country.

Of the 72 families of birds breeding in Mexico (see summary in Appendix 2), a total of 43 (59.7%) are families of cosmopolitan distribution (e.g., Ardeidae) or are widespread in North and South America (e.g., Vireonidae). Another seven (9.7%) are primarily of Nearctic distribution, reaching their southernmost limits in Mexico (e.g., Sittidae). Twenty-two (30.6%) are

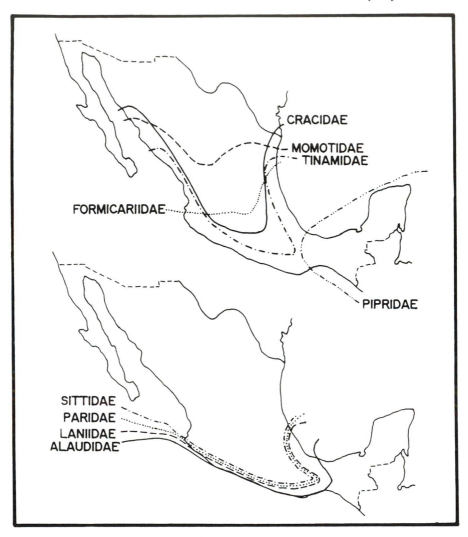

Figure 8.2. Range limits of some families of primarily South American (bottom) and North American (top) distribution.

primarly Neotropical, reaching their northernmost limits in Mexico (e.g., Furnariidae). Figure 8.2 shows several examples of northern and southern range limits in Mexico.

DIVERSITY AND GEOGRAPHY

Species richness in the 35 geographic provinces is summarized in Table 8.2 and Figure 8.3. The highest number of species is found in the southeastern lowlands of Mexico: the Petén, Sierra de los Tuxtlas, and Middle East

Table 8.2. Avian diversity and endemism in 35 biotic provinces in Mexico

Province[a]	No. of species	True endemics	All endemics	Narrow endemics[b]
1. Sierra Madre Or. N	139	14 (10.0)	20 (14.4)	1
2. Sierra Madre Or. S	191	25 (13.0)	33 (17.3)	1
3. Nudo de Zempoaltépetl	122	22 (18.0)	29 (23.8)	2
4. Sierra Madre Sur—Oax	101	22 (21.8)	31 (30.7)	0
5. Sierra Madre Sur—Gro	182	23 (12.6)	33 (18.1)	0
6. Sierra de Miahuatlán	110	17 (15.5)	23 (20.9)	1
7. Sierra de los Tuxtlas	245	1 (0.4)	1 (0.4)	1
8. Transvolcanic Belt	165	27 (16.4)	37 (22.4)	2
9. Sierra Madre Occ. N	103	5 (4.9)	13 (12.6)	0
10. Sierra Madre Occ. M	171	33 (19.3)	46 (26.9)	2
11. Sierra Madre Occ. S	161	26 (16.1)	35 (21.7)	0
12. Sierra Norte de Chiapas	105	5 (4.7)	15 (14.3)	1
13. Sierra Madre de Chiapas	100	1 (1.0)	9 (9.0)	0
14. East Coast N	144	4 (2.7)	5 (3.5)	0
15. East Coast M	233	6 (2.5)	7 (3.0)	2
16. Petén	301	7 (2.3)	9 (3.0)	0
17. Yucatán	192	7 (3.6)	9 (4.7)	1
18. Lowlands C Chiapas	110	7 (6.3)	8 (7.2)	0
19. Lowlands S Chiapas	117	6 (5.1)	9 (7.6)	1
20. Lowlands Isthmus	190	11 (5.8)	12 (6.3)	1
21. West Coast S	170	20 (11.8)	23 (13.5)	0
22. West Coast M	185	23 (12.4)	26 (14.0)	1
23. West Coast N	125	4 (3.2)	7 (5.6)	0
24. Balsas Basin W	113	15 (13.3)	21 (18.6)	0
25. Balsas Basin E	81	11 (13.6)	15 (18.5)	0
26. Chihuahuan Desert N	95	0	2 (2.1)	0
27. Chihuahuan Desert S	149	8 (5.3)	14 (9.3)	0
28. Baja California N	122	0	3 (2.5)	0
29. Baja California M	81	3 (3.7)	4 (4.9)	0
30. Baja California S	83	3 (3.6)	4 (4.8)	0
31. Cozumel-Mujeres Is.	74	4 (5.4)	4 (5.4)	2
32. Revillagigedo Islands	19	4 (21.1)	4 (21.1)	4
33. Islands Gulf of California	32	2 (6.3)	3 (9.4)	1
34. Guadalupe Island[c]	13	0	0	0
35. Tres Marias Islands	47	9 (19.1)	9 (19.1)	0

Numbers in parenthesis are percents.
[a]See Fig. 8.1 for location of the provinces (by number)
[b]Includes quasiendemic species.
[c]Excludes two extinct allospecies.

Coast. Each of these areas supports more than 230 resident bird species. This group of provinces could be called the "tropical southeastern lowlands." Next in numbers are the Yucatán, Southern Sierra Madre Oriental, lowlands of the Isthmus of Tehuantepec, and Middle West Coast.

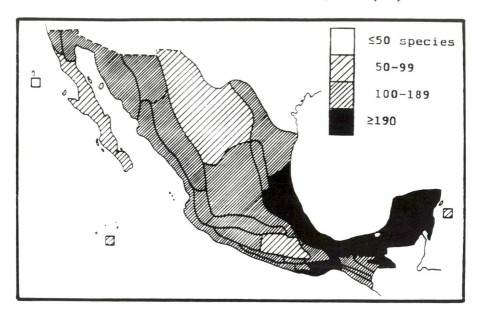

Figure 8.3. Geographic distribution of avian diversity in Mexico.

On the other hand, several provinces support a low diversity, such as Guadalupe Island, Revillagigedo Islands, islands of the Gulf of California, and Tres Marias Islands. A thorough avifaunal survey is not yet available for the islands of the Gulf of California. These islands are rich in breeding populations of seabirds (Everett & Anderson, in press), that were not included in our analyses. The low land bird diversity reflects low native habitat diversity, small land area, and limited opportunity for colonization. Mainland regions with low diversity are the Eastern Balsas Basin and Middle and Southern Baja California.

An analysis of overall avifaunal similarity among regions yielded some interesting results (Fig. 8.4). As might be expected, regions dominated by similar habitats grouped together. Hence, the montane regions (regions 1–6, 8–13) form one group, the lowland areas (regions 14–22, 31) another, and the desert areas (regions 26–30) still another. More interesting are patterns of association among individual regions. The Balsas Basin falls into the desert group instead of with the coastal areas, even though it is geographically isolated from the deserts to the north. Similarly, the Northern West Coast region groups with the deserts, reflecting the influence of the Sonoran Desert to the north. Within the coastal group, there is a north–south division in which the Northern East Coast and Middle and Southern West Coast associate in one group and the lowlands of the Isthmus of Tehuantepec, the Yucatán, and Chiapas in another. The Sierra de los Tuxtlas groups with lowland areas, reflecting the absence of higher-altitude areas to support pure patches of upland habitats such as cloud

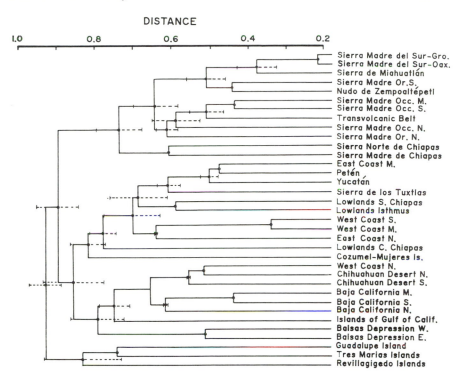

Figure 8.4. Results of a UPGMA analysis of avifaunal similarity among geographic regions in Mexico. Error bars represent standard errors.

forest. Finally, the islands far removed from the mainland (Tres Marías, Revillagigedo, and Guadalupe Islands) fall outside the main groups of continental areas, reflecting high levels of endemism, impoverished avifaunas, and possibly faunal size effects (Jackson et al., 1989).

The distribution of Mexican endemics and quasiendemics takes on a strikingly different pattern (Table 8.2, Fig. 8.5). The Middle Sierra Madre Occidental is the region with the highest number of endemic species, followed by the Transvolcanic Belt, Southern Sierra Madre Occidental, Southern Sierra Madre Oriental, and Sierra Madre del Sur in Guerrero. Patterns are similar for proportional levels of endemism, except that the Revillagigedo Islands, Tres Marias Islands, and Balsas Basin also emerge as areas of high endemism. Hence, avian endemism is concentrated in the western, central, and southern highlands of Mexico.

Areas of low endemism are Sierra de los Tuxtlas (despite the presence of such well-marked subspecies as *Chlorospingus opthalmicus wetmorei*, *Atlapetes brunneinucha apertus*, and *Geotrygon lawrenceii carrikeri*), Sierra Madre de Chiapas, Guadalupe Island (not including the extinct storm-petrel *Oceanodroma macrodactyla* or the extinct caracara allospecies *Polyborus plancus lutosus*), northern Baja California, and the northern Chihuahuan

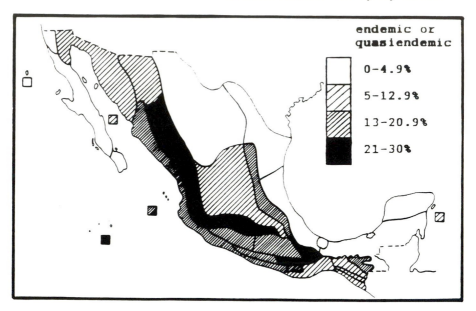

Figure 8.5. Geographic distribution of avian endemism (including quasiendemics) in Mexico.

Desert. The latter two regions share most of their species with the arid lands of the adjacent southwestern United States.

The number of species restricted to one or only a few contiguous regions (narrow endemics) is an indicator of the degree of isolation provided by, and the amount of evolutionary change that has occurred in, those areas. Of the 125 species endemic or quasiendemic to Mexico, 24 are restricted to only one province (Table 8.2). The various islands contain the highest levels of narrow endemism (four in Revillagigedo Islands and two in Cozumel-Mujeres Islands and Guadalupe Island). On the mainland, the Nudo de Zempoaltépetl, Transvolcanic Belt, Middle Sierra Madre Occidental, and Middle East Coast each contain two narrow endemics.

Insular regions—like the islands of the Gulf of California, Tres Marías and Revillagigedo Islands, and some Caribbean islands such as Alacranes off the coast of Yucatán—support a great diversity of breeding seabirds (Everett & Anderson, in press). Specifically, the islands of the Gulf of California and coastal Baja California support most of the world breeding populations of seabirds, e.g., *Sterna elegans*, *Synthliborhamphus craveri*, and *Larus livens* (A.O.U., 1983). *Oceanodroma macrodactyla*, now extinct, bred only on Guadalupe Island. Guadalupe Island also supports the only Mexican breeding populations of the albatross *Diomedea immutabilis* (Dunlap, 1988). The Revillagigedo Islands support the only Mexican breeding populations of the petrel *Puffinus auricularis*.

Table 8.3. Avian diversity and endemism in 23 habitat types in Mexico

Habitat	Number of species	True endemics	All endemics	Species restricted to habitat
Mixed coniferous forest	77	5 (6.5)	10 (13.0)	1 (1.3)
Pine forest	118	12 (10.2)	20 (17.0)	1 (0.9)
Pine-oak forest	218	34 (15.6)	43 (19.7)	2 (0.9)
Oak forest	136	15 (11.0)	20 (14.7)	1 (0.7)
Pinyon-juniper woodland	70	2 (2.9)	4 (5.7)	0
Cloud forest	182	23 (12.6)	30 (16.5)	20 (11.0)
Semideciduous tropical forest	240	10 (4.2)	12 (5.0)	1 (0.4)
Tropical rain forest	225	4 (1.8)	5 (2.2)	25 (11.1)
Tropical decicuous forest	211	38 (18.0)	41 (19.4)	3 (1.4)
Tropical grassland	84	0	2 (2.4)	3 (3.6)
Temperate grassland	52	3 (5.8)	3 (5.8)	2 (3.9)
Mangroves	64	4 (6.3)	4 (6.3)	4 (6.3)
Desert scrub	206	37 (18.0)	45 (21.8)	8 (3.9)
Chaparral	47	0	0	1 (2.1)
Mesquite	70	8 (11.4)	12 (17.1)	0
Lakes and ponds	61	1 (1.6)	1 (1.6)	0
Marshes	90	5 (5.6)	5 (5.6)	14 (15.6)
Rivers and streams	50	0	0	0
Cultivated areas	189	9 (4.8)	10 (5.3)	0
Coastal scrub	126	23 (18.3)	25 (19.8)	0
Gallery forest	76	8 (10.5)	12 (15.8)	0
Seacoast	20	1 (5.0)	1 (5.0)	2 (10.0)
Cities and towns	44	3 (6.8)	3 (6.8)	0

Numbers in parentheses are percents.

DIVERSITY AND HABITATS

Species richness in different habitats is summarized in Table 8.3. The habitats containing the most species are semideciduous tropical, tropical rain, pine-oak, and tropical deciduous forests, and desert scrub, each with more than 200 species. The habitats containing the fewest species are seacoast (20), cities and towns (44), chaparral (47), and rivers and streams (50).

Many species endemic or quasiendemic to Mexico are found in desert scrub (45 species), pine-oak forest (43), tropical deciduous forest (41), and cloud forest (30). In terms of proportion of avifauna endemic, the ranking is desert scrub (21.8%), coastal scrub (19.8%), pine-oak forest (19.7%), and tropical deciduous forest (19.4%). Chaparral and rivers and streams hold no endemic species; only one endemic species is found in lakes and ponds and seacoast.

Ecological restriction is high in some habitats and low in others (Table 8.3). Habitats with many ecologically restricted species are tropical rain

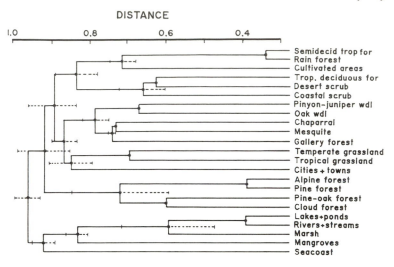

Figure 8.6. The results of a UPGMA analysis of avifaunal similarity among habitat types in Mexico. wdl, woodland; for, forest. Error bars represent standard errors.

forest (25), cloud forest (20), and marshes (14). Proportionally, the ranking is desert scrub (15.6%), tropical rain forest (11.1%), and cloud forest (11.0%). In 14 habitats, fewer than 2% of the species are ecologically restricted.

A phenetic analysis of avifaunal similarity among habitats (Fig. 8.6) shows clusters of similar or contiguous habitats, such as wetlands, high-altitude forest, and xeric woodlands. Interestingly, the oak and pinyon-juniper woodlands group with other xeric woodlands, and not with the relatively similar pine-oak forest. Also, desert scrub and coastal scrub group with the tropical forest habitats, reflecting their relative continuity with the dry tropical deciduous forest. Finally, cloud forest groups with the high-altitude forests, instead of with tropical forest habitats.

HISTORICAL BIOGEOGRAPHY OF THE BIRDS OF MEXICO

The biota of Mexico is one of the richest in the world, to the point that it has been called a "megadiversity country" (Mittermeier, 1988; Toledo, 1988). Indeed, Mexico is a center of diversity for many taxonomic groups, including cacti, oaks (*Quercus*), colubrid snakes, iguanid lizards, and bats (see other chapters in this volume).

Griscom (1950) analyzed the origin and diversification of the Mexican avifauna. He identified seven elements in the avifauna: pelagic seabirds, a West Indian element, a wide-ranging element (temperate and tropical, tropical only), a South American element, preglacial relicts, an old North American element (Middle North America, Sonoran endemics), and a modern Holarctic element. Griscom's conclusions about the origins of the avifauna, however, depend to some extent on presumed origins of the various families.

In this analysis, we also will treat the origin and diversification of the Mexican avifauna. Instead of attempting to understand the origin of each taxon, we discuss two nonexclusive hypotheses explaining Mexico's avian diversity: (1) Mexico contains an avifauna that is a mixture of North American, Caribbean, and South American elements (Blake, 1953; Edwards,1972); and (2) extensive diversification has taken place within Mexico (Griscom, 1950). We give examples illustrating the action of the two processes, and assess the importance of each.

Faunal Mixture

Several taxa that are diverse in South America reach their northernmost limits in Mexico, often with much reduced diversity. In many cases, the Mexican populations are closely related to, or conspecific with, South American forms. We interpret these situations as representing cases of northward range expansion without subsequent diversification. Examples include the hummingbird genus *Phaethornis*, which has 22 species in South America, two of which extend north into Mexico, and the antwren genus *Myrmotherula*, which has 30 species in Central and South America, one of which reaches Mexico. Other examples include the jacamars *(Galbula)*, the puffbirds *(Malacoptila)*, the yellow-finches *(Sicalis)*, the flower-piercers *(Diglossa)*, and the manakins *(Pipra)*. Patterns in other groups are less clear owing to limited diversification in Mexico or Central America, e.g., *Crypturellus, Synallaxis, Tangara, Saltator,* and *Cacicus* (Meyer de Schauensee, 1970).

Similarly, several groups that are diverse in North America have their southern range limits in Mexico, apparently having expanded their ranges southward without extensive subsequent diversification. Examples include the warblers *(Dendroica;* 21 species in the United States and Canada, four reaching Mexico and Central or South America); nuthatches *(Sitta;* four in North America, two reaching Mexico, and one apparently relictual population of a third on Guadalupe Island); and the shrikes *(Lanius;* two in North America, one reaching Mexico). Of the ten tits *(Parus)* in North America, three have ranges that extend into northern Mexico, and two more are distributed largely in Mexico (A.O.U., 1983). A Caribbean influence is suggested by the presence of the grassquits *(Tiaris;* three species in the Caribbean, one reaching Mexico), the elaenias *(Elaenia;* two of the three Caribbean species occurring in eastern Mexico), the stripe-headed tanager *(Spindalis zena)*, and the distinctive Caribbean subspecies of the yellow warbler *(Dendroica petechia;* A.O.U., 1983).

Hence, it appears that many taxa have expanded their ranges into Mexico from the north, the south, and possibly from the Caribbean. Some of these taxa have subsequently speciated, but several remain conspecific with parent forms. Clear examples of such colonization appear more numerous from South America than from North America or the Caribbean.

Center of Diversification

That extensive speciation has occurred in Mexico is indicated by groups with centers of diversity in the country. Several higher taxa appear to have undergone their early radiation chiefly in Mexico and Central America. All of the genera of New World quail (Griscom, 1950) and New World jays (Pitelka, 1951) and most genera of wrens and motmots occur in Mexico, with diversity decreasing substantially north and south. Other groups appear to be still in the process of diversification. All representatives of the quail (*Callipepla*: four species), jays (*Aphelocoma*: three species) and ground-sparrows (*Melozone*: three species) breed in Mexico. Several other genera have their highest diversity in Mexico, including the thrashers (*Toxostoma*: 9 of 10 species), towhees (*Pipilo*: 5 of 6), sparrows (*Aimophila*: 10 of 12), woodpeckers (*Melanerpes*: 8 of 21), wrens (*Thryothorus*: 7 of 21; *Campylorhynchus*: 8 of 11), vireos (*Vireo*: 14 of 25), orioles (*Icterus*: 14 of 21), and hummingbirds (*Eupherusa*: 3 of 4).

Hence, assuming that taxa with centers of diversity in Mexico have indeed diversified *in situ*, several groups have radiated extensively in Mexico. Other groups appear to have radiated secondarily in Mexico (e.g., *Otus, Columba, Catharus, Myiarchus, Turdus, Tyrannus, Geothlypis*). Additionally, many groups have speciated to a limited extent in Mexico, producing sister-species pairs (e.g., *Corvus, Cyanocorax, Atlapetes, Ortalis, Colinus, Zenaida*).

Synthesis

Ample evidence exists that each hypothesis—that of a faunal mixture and that of internal diversification—can explain aspects of the avian diversity of Mexico. Beyond representing just a mixture of North and South American elements, the Mexican avifauna contains many elements that are autochthonous, having diversified extensively within the country.

BIOLOGICAL DIVERSITY AND CONSERVATION

A thorough understanding of geographic and ecological patterns of biological diversity is critical for conservation efforts. Without accurate information on centers of diversity and endemism, identifying priority habitats and regions is difficult (Soulé, 1986). In the next section, we discuss problems regarding the preservation of avian diversity in Mexico and identify habitats and regions that should be emphasized in any plan to preserve the avian diversity of Mexico.

Problems

Already, six species or allospecies of Mexican birds are extinct: Guadalupe storm-petrel (*Oceanodroma macrodactyla*), California condor (*Gymnogyps californianus*), Guadalupe caracara (*Polyborus plancus lutosus*), Socorro dove (*Zenaida graysoni*), imperial woodpecker (*Campephilus imperialis*), and slender-billed grackle (*Quiscalus palustris*; A.O.U., 1983). Half of these species were insular forms; the other three inhabited the mountains of northern Baja California (*G. californianus*), Sierra Madre Occidental (*C. imperialis*), and marshes at the headwaters of the Río Lerma (*Q. palustris*).

In many parts of the country, birds of diverse types are hunted for food, a practice that in some instances may compromise the health of the populations. For example, *Aphelocoma unicolor* is a common food item for people living within its range (A.T.P., pers. observ.), and this pressure may be responsible for the low population levels of this species in several areas.

Another pressure on populations of many species is the pet trade. Birds of all sorts are captured throughout the country and either sold in markets as pets or exported to collectors in the United States and Europe. Mortality is high at all stages of the process, and only a small proportion of birds captured ever reach pet owners. In one day in the Mercado Libertad in Guadalajara, A.T.P. counted 411 individuals of 26 native Mexican bird species for sale at prices ranging up to $100. In the famous Mercado de Sonora in Mexico City, many thousands of animals are sold each year. Iñigo Elias (1986), focusing on raptors, counted 644 individuals of 15 species in the Mercado de Sonora, including such rare species as *Herpetotheres cachinnans* and *Falco mexicanus.* Popular and expensive species as the parrots *Amazona* and birds of prey are probably most seriously threatened by the pet trade.

Although nonnative species of animals have been introduced or have spread into mainland Mexico (e.g., *Rattus norvegicus*, *Mus musculus*, *Passer domesticus*, *Sturnus vulgaris*), no serious ill effects on birds are yet apparent. On the offshore islands, however, the problem is alarming. On Guadalupe Island, two endemic forms (*Oceanodroma macrodactyla* and *Polyborus plancus lutosus*) and the only Mexican population of *Sitta canadensis* have been extirpated. Similarly, on Socorro Island in the Revillagigedo group, the endemic dove *Zenaida graysoni* is extinct in the wild (Jehl & Parkes, 1983). Other insular forms (e.g., *Puffinis auricularis*) are now threatened by introduced predators (Everett & Anderson, in press). In all cases, every indication is that introduction of feral cats, rats, pigs, and goats contributed significantly to the problems.

The most serious threat to avian diversity in Mexico and the tropics in general is habitat destruction (Soulé, 1986). Already, rain forest in southern Veracruz, Tabasco, and Chiapas; cloud forest in Guerrero, Veracruz, and Puebla; marshes in central Mexico; and highland forests in northwestern and central Mexico have been seriously disturbed. In addition to destruction of habitats, vast areas of habitat are exploited for various purposes (e.g., food, wood), often leading to serious declines in habitat quality. This disturbance has caused numerous local avian extinctions. With Mexico's human population growing so rapidly (3.4% per year in 1974; Enciclopédia de México, 1985), and with serious economic and bureaucratic problems, this problem will worsen in coming years.

Recommendations

We offer three recommendations to help ensure the preservation of avian diversity in Mexico.

1. Regulation of hunting and live-trapping of wild birds
2. Establishment of a series of carefully planned and positioned reserves (see discussion below)
3. Better protection of habitat and animals in existing reserves, including removal of introduced predators and herbivores from insular reserves

The motivation behind the first and third suggestions has been discussed above. The reasoning behind the second follows: the planning of a reserve system in terms of geographic position and habitat depends critically on the goals of the effort. Comparing Figures 8.3 and 8.5, it is evident that patterns of diversity and endemism in Mexico are not concordant. If the objective is to preserve avian diversity in Mexico alone, then semideciduous tropical, tropical rain, pine-oak, and tropical deciduous forests should be priority habitats, and the Petén, Sierra de los Tuxtlas, Middle East Coast, and Yucatán should be priority regions. However, if conservation in Mexico is to be part of a global plan to preserve biological diversity, areas of high endemism such as the Sierra Madre Occidental, Transvolcanic Belt, Sierra Madre Oriental, and Sierra Madre del Sur, and habitats such as desert scrub, pine-oak forest, tropical deciduous forest, and cloud forest should also be emphasized. In addition, areas of high narrow-sense endemism (e.g., Revillagigedo Islands) should be protected. Ideally, both high-diversity and high-endemism areas would be incorporated into conservation strategies for Mexico.

ACKNOWLEDGMENTS

We would like to thank the following people for critical review and incisive discussion of the manuscript: Hector E. Colón, Robert Dickerman, John W. Fitzpatrick, Debra Moskovits, Douglas F. Stotz, and J. Fernando Villaseñor. A.T.P. is greatly indebted to his companions in field work—D. Scott Baker, Hesiquio Benítez Díaz, and Noé Vargas Barajas—for helpful discussion, careful reading, and patience while parts of the manuscript were being written. A.G.N.S. thanks Claudia Abad for her patience and support while the manuscript was written. Thanks also to Larry Heaney for helpful advice, and to Amy Peterson, Miriam Torres Chávez, and Judith Vega Morales for editorial and bibliographic assistance, and to Leslie Marcus and R.E. Strauss for providing computer programs for the faunal similarity analyses. A.T.P.'s field work in Mexico was supported by grants from the National Science Foundation Dissertation Improvement Grant Program, the National Geographic Society, the Chapman Fund of the American Museum of Natural History, Sigma Xi, the University of Chicago, and the Field Museum of Natural History. Finally, thanks to Jorge LLorente-Bousquets for his encouragment and assistance at all stages of the work.

APPENDIX 1

Avifaunal summaries for various Mexican states

State	References
Baja California	Grinnell (1928)
Coahuila	Urban (1959); Ely (1962)
Colima	Brattstrom & Howell (1956); Schaldach (1963, 1969)
Chiapas	Alvarez del Toro (1980)
Distrito Federal	Wilson & Ceballos-Lascuráin (1987)
Durango	Webster & Orr (1952)
Guerrero	Griscom (1934); Martín del Campo (1948); Blake (1950b); Navarro (1986)
Michoacán	Blake & Hanson (1942); Villaseñor (1988)
Morelos	Davis & Russell (1953)
Nayarit	Grant & Cowan (1964); Escalante (1988)
Nuevo León	Martín del Campo (1959)
Oaxaca	Blake (1950a); Rowley (1966, 1984); Binford (1988)
Querétaro	Navarro & Hernández (in press)
Sonora	Van Rossem (1945)
Tamaulipas	J. Phillips (1911); Martin (1955)
Veracruz	Wetmore (1943); Lowery & Dalquest (1951); Andrle (1967)
Yucatán	Paynter (1955)
Zacatecas	Webster & Orr (1952); Webster (1958)

APPENDIX 2

Taxonomic summary of the avifauna of Mexico

Order	Family	Breeding	Nonbreeding	Endemic[a]
Tinamiformes	Tinamidae	4	0	0
Gaviiformes	Gaviidae	0	3	0
Podicipediformes	Podicipedidae	4	0	0
Procellariiformes		9	9	2
	Diomedidae	1	2	0
	Procellariidae	3	5	1
	Hydrobatidae	5	2	1
Pelecaniformes		12	3	0
	Phaethontidae	1	1	0
	Sulidae	4	0	0
	Pelecanidae	1	1	0
	Phalacrocoracidae	4	0	0
	Anhingidae	1	0	0
	Fregatidae	1	1	0
Ciconiiformes		20	2	0
	Ardeidae	15	2	0
	Threskiornithidae	3	0	0
	Ciconiidae	2	0	0
Phoenicopteriformes	Phoenicopteridae	1	0	0
Anseriformes	Anatidae	9	26	0
Falconiformes		50	5	0
	Cathartidae	5	0	0
	Accipitridae	34	4	0
	Falconidae	11	1	0
Galliformes		23	0	6
	Cracidae	6	0	1
	Phasianidae	17	0	5
Gruiformes		18	2	0
	Rallidae	15	0	0
	Heliornithidae	1	0	0
	Eurypygidae	1	0	0
	Aramidae	1	0	0
	Gruidae	0	2	0

Appendix 2 (cont.)

Order	Family	Breeding	Nonbreeding	Endemic[a]
Charadriiformes		27	56	3
	Burhinidae	1	0	0
	Charadriidae	4	5	0
	Haematopodidae	2	0	0
	Recurvirostridae	2	0	0
	Jacanidae	1	0	0
	Scolopacidae	2	32	0
	Laridae	13	17	2
	Alcidae	2	2	1
Columbiformes	Columbidae	24	0	1
Psittaciformes	Psittacidae	21	0	6
Cuculiformes	Cuculidae	11	1	0
Strigiformes		26	2	2
	Tytonidae	1	0	0
	Strigidae	25	2	2
Caprimulgiformes		12	1	1
	Caprimulgidae	10	1	1
	Nyctibiidae	2	0	0
Apodiformes		59	4	12
	Apodidae	8	1	1
	Trochilidae	51	3	11
Trogoniformes	Trogonidae	9	0	2
Coraciiformes		10	1	0
	Momotidae	6	0	0
	Alcedinidae	4	1	0
Piciformes		26	3	7
	Bucconidae	2	0	0
	Galbulidae	1	0	0
	Ramphastidae	3	0	3
	Picidae	20	3	4
Passeriformes		394	120	83
	Furnariidae	7	0	0
	Dendrocolaptidae	12	0	1
	Formicariidae	9	0	0
	Tyrannidae	58	13	5
	Cotingidae	2	0	0

Appendix 2 (cont.)

Order	Family	Breeding	Nonbreeding	Endemic[a]
	Pipridae	4	0	0
	Alaudidae	1	0	0
	Hirundinidae	10	3	1
	Corvidae	22	0	8
	Paridae	5	0	2
	Remizidae	1	0	0
	Aegithalidae	1	0	0
	Sittidae	3	0	0
	Certhiidae	1	0	0
	Troglodytidae	29	1	10
	Cinclidae	1	0	0
	Muscicapidae	27	7	4
	Mimidae	11	7	6
	Motacillidae	0	3	0
	Bombycillidae	0	1	0
	Ptilogonatidae	2	0	1
	Laniidae	1	0	0
	Vireonidae	20	23	5
	Emberizidae	159	58	38
	Fringillidae	8	4	2

[a]Includes quasiendemics.

REFERENCES

Allen, J.A. 1893. The geographical origin and distribution of North American birds, considered in relation to faunal areas of North America. Auk 10:97–150.

Alvarez del Toro, M. 1980. Las aves de Chiapas, 2nd ed. Tuxtla Gutiérrez, Chiapas: Univ. Autón. Chis.

———. 1985. Las aves. *In* Francisco Hernández: Obras Completas VII. Mexico City: UNAM. pp. 237–240.

Andrle, R.M. 1967. Birds of the Sierra de Tuxtla in Veracruz, Mexico. Wilson Bull. 79:163–187.

A.O.U. 1983. Check-list of North American Birds, 6th ed. Lawrence, KS: American Ornithologist's Union.

Beltran, E. 1982. Contribución de México a la Biología: Pasado, Presente y Futuro. Mexico City: C.E.C.S.A.

Berlin, B. 1973. Folk systematics in relation to biological classification and nomenclature. Ann. Rev. Ecol. Syst. 4:259–271.

Binford, L.C. 1988. A distributional survey of the birds of the Mexican state of Oaxaca. Orn. Monogr. 43:1–405.

Blake, E.R. 1950a. Report on a collection of birds from Oaxaca, Mexico. Fieldiana Zool. 31:395–419.

———. 1950b. Report on a collection of birds from Guerrero, Mexico. Fieldiana Zool. 31:375–391.

————. 1953. Birds of Mexico. A guide for field identification. Chicago, IL: Univ. Chicago Press.

———— & H.C. Hanson. 1942. Notes on a collection of birds from Michoacán, Mexico. Field Mus. Nat. Hist. Zool. Ser. 22:513–551.

Brattstrom, B.H. & T.R. Howell. 1956. The birds of the Revillagigedo Islands, Mexico. Condor 58:107–120.

Crowe, T.M. & A.A. Crowe. 1982. Patterns of distribution, diversity and endemism in African birds. J. Zool. Lond. 198:417–442.

Davis W.B. & R.J. Russell. 1953. Aves y mamíferos del Estado de Morelos. Revista. Soc. Mex. Hist. Nat. 14:77–145.

De la LLave, P. 1871. El *Pharomachrus mocinno*, su descripción y fundamento de este género de aves. La Naturaleza (Ser. 1) 7:17–18.

Dunlap, E. 1988. Laysan albatross nesting on Guadalupe Island, Mexico. American Birds 42:180–181.

Edwards, E.P. 1972. A field guide to the birds of Mexico. Sweet Briar, VA: Ernest P. Edwards.

Ely, C.A. 1962. The birds of southeastern Coahuila, Mexico. Condor 64:34–39.

Enciclopédia de México. 1985. Todo México: Compendio Enciclopédico, 1985. Mexico: Enciclopédia de México.

Escalante, P. 1988. Aves de Nayarit. Tepic, Nayarit: Univ. Autón. Nayarit.

Everett, W.T. & D.W. Anderson. In press. Status and conservation of the breeding seabirds on offshore Pacific islands of Baja California and the Gulf of California. ICBP Tech. Bull.

Ferrari-Pérez, F. 1886. Catalogue of animals collected by the Geographical and Exploring Comission of the Republic of México. Proc. U.S. Natl. Mus. 9:125–199.

Friedmann, H., L. Griscom & R.T. Moore. 1950. Distributional check-list of the birds of Mexico. Part I. Pac. Coast Avif. 29:1–202.

Godman, F.D. 1915. Biologia Centrali-Americana: Introductory volume. London: Taylor and Francis.

Goldman, E.A. 1951. Biological investigations in Mexico. Smiths. Misc. Coll. 115:1–476.

Gomez A., G. & R. Terán O. 1981. Contribución para el estudio de los vertebrados terrestres mexicanos. Thesis, UNAM.

González-Claverán, V. 1988. La expedición científica de Malaspina en Nueva España, 1789–1794. Mexico City: El Colegio de México.

Grant, P.R. & I. McT. Cowan. 1964. A review of the avifauna of the Tres Marías Islands, Nayarit, Mexico. Condor 66:221–228.

Grinnell, J. 1928. A distributional summation of the ornithology of Lower California. Univ. Calif. Publ. Zool. 32:1–300.

Griscom, L. 1934. The ornithology of Guerrero, Mexico. Bull. Mus. Comp. Zool. 75:365–422.

————. 1950. Distribution and origin of the birds of Mexico. Bull. Mus. Comp. Zool. 103:341–382.

Halffter, G. 1976. Distribución de los insectos en la Zona de Transición Mexicana: relaciones con la entomofauna de Norteamérica. Folia Entom. Mex. 35:1–64.

Hellmayr, C.E. (and various coauthors). 1918–49. Catalogue of the birds of the Americas. Field Mus. Nat. Hist. Zool. Ser. 13. Parts 1–11.

Herrera, A.L. 1898–1914. Ornitología Mexicana. La Naturaleza (Ser. 2) 3:129–229, 267–358, 407–547, 563–680; ser. 3. 1(A):1–232.

Iñigo Elias, E. 1986. The trade in diurnal birds of prey in Mexico. Birds Prey Bull. 3:128–140.

Jackson, D.A., K.M. Somers, & H.H. Harvey. 1989. Similarity coefficients: measures of co-occurrence and association or simply measures of occurrence? Amer. Natur. 133:436–453.

Jehl, J.R. & K.C. Parkes. 1983. "Replacements" of landbird species on Socorro Island, Mexico. Auk 100:551–559.

Johnson, N.K. 1980. Character variation and evolution of sibling species in the *Empidonax difficilis-flavescens* complex (Aves: Tyrannidae). Univ. Calif. Publ. Zool. 112:1–151.

Lawrence, G.N. 1874. The birds of western and northwestern Mexico, based upon collections made by Col. A. J. Grayson, Capt. J. Xantus, and Ferd Bischoff, now in the museum of the Smithsonian Institution, at Washington, DC. Mem. Boston Soc. Nat. Hist. 2:265-319.

———. 1875. Birds of southwestern Mexico collected by Sumichrast for the United States National Museum. Bull. U.S. Nat. Mus. 5:5–56.

Leopold, A.S. 1977. Fauna silvestre de México. Mexico City: IMERNAR.

Lowery, G.H. Jr. & W.W. Dalquest. 1951. Birds from the state of Veracruz, Mexico. Univ. Kansas Publ. Mus. Nat. Hist. 3:531–649.

Martin, P.S. 1955. Zonal distribution of vertebrates in a Mexican cloud forest. Amer. Natur. 89:347–362.

Martin del Campo, R. 1948. Contribución para el conocimiento de la fauna ornitológica del estado de Guerrero. An. Inst. Biol. México 11:241–266.

———. 1952. Aves en la historia antigua de México. Biol. Soc. Mex. Geogr. Estad. 70:243–249.

———. 1959. Contribución al conocimiento de la ornitología de Nuevo León. Universidad (Monterrey, N.L.) 16–17:121–180.

Meyer de Schauensee, R. 1970. A guide to the birds of South America. Philadelphia: Academy of Natural Science.

Miller, A.H. 1941. Speciation in the avian genus *Junco*. Univ. Calif. Publ.Zool. 44:173–434.

———, H. Friedmann, L. Griscom & R.T. Moore. 1957. Distributional check-list of the birds of Mexico, part 2. Pac. Coast Avif. 33:1–436.

Minguet, C. 1985. Alejandro de Humboldt: historiador y geógrafo de la América española (1799–1804). Vols. 1 & 2. Mexico City: UNAM.

Mittermeier, R.A. 1988. Primate diversity and the tropical forest. *In* E.O. Wilson (ed.), Biodiversity. Washington, DC: National Academy Press. pp. 145–154.

Montes de Oca, R. 1874. Ensayo ornitológico de la familia Trochilidae, o sea de los colibríes o chupamirtos de México. La Naturaleza Ser. 1. 3:15–31, 59–66, 99–106, 159–167, 203–211, 299–304.

National Geographic Society. 1983. Field guide to the birds of North America. Washington, D.C: NGS.

Navarro S., A. 1986. Distribución altitudinal de las aves en la Sierra de Atoyac, Guerrero. Thesis, UNAM.

——— & B. Hernández B. in press. Estado actual del conocimiento de las aves del Estado de Querétaro. *In* S. Zamudio & P. Rojas (eds.), Los Recursos Naturales del Estado de Querétaro. Mexico City: Inst. Ecol.

Nelson, E.W. 1922. Lower California and its natural resources. Mem. Nat. Acad. Sci. 16:1–194.

Paynter, R.J. Jr. 1955. The ornithogeography of the Yucatan Peninsula. Peabody Mus. Nat. Hist. Bull. 9:1–347.

Peterson, R.T. & E. Chalif. 1973. A field guide to Mexican birds. Boston, MA: Houghton-Mifflin.

Phillips, A.R. 1960. La ornitología Mexicana en los últimos cincuenta años. Rev. Soc. Mex. Hist. Nat. 21:375–389.

————. 1961. Emigraciones y distribución de aves terrestres en México. Rev. Soc. Mex. Hist. Nat. 22:295–311.

————. 1986. The known birds of North and Middle America, Part I. Denver, CO: A.R. Phillips.

Phillips, J. 1911. A year's collecting in the state of Tamualipas, Mexico. Auk 28:67–89.

Pitelka, F. 1951. Speciation and ecologic distribution in American jays of the genus *Aphelocoma*. Univ. Calif. Publ. Zool. 50:194–464.

Ramirez-Pulido, J. & C. Müdespacher. 1987. Estado actual y perspectivas del conocimiento de los mamíferos de México. Ciencia (Mexico) 38:49–67.

Raven, P.H., B. Berlin & D.E. Breedlove. 1971. The origins of taxonomy. Science 174:1210–1213.

Ridgway, R. & H. Friedmann. 1901–46. The birds of North and Middle America. Bull. U.S. Nat. Mus. 50.

Rowley, J.S. 1966. Breeding records of birds of the Sierra Madre del Sur, Oaxaca, Mexico. Proc. West. Found. Vert. Zool. 1:107–204.

————. 1984. Breeding records of land birds in Oaxaca, Mexico. Proc. West. Found. Vert. Zool. 2:74–221.

Rzedowski, J. 1978. La vegetación de México. Mexico City: Limusa.

Salvin, O. & F.D. Godman. 1879–1904. Biologia Centrali-Americana: Aves. Vols. 1–4. London: Taylor & Francis.

Sanchez-León, V.M. 1969. Los recursos naturales de México. IV. Estado actual de las investigaciones de fauna silvestre y zoología cinegética. Mexico City: IMERNAR.

Schaldach, W.J. Jr. 1963. The avifauna of Colima and adjacent Jalisco, Mexico. Proc. West. Found. Vert. Zool. 1:1–100.

————. 1969. Further notes on the avifauna of Colima and adjacent Jalisco, Mexico. An. Inst. Biol. México 40:299–316.

Simpson, G.G. 1943. Turtles and the origin of the fauna of Latin America. Amer. J. Sci. 241:413–429.

Smith, H.M. 1941. Las provincias bióticas de México, según la distribución geográfica de las lagartijas del género *Sceloporus*. An. Esc. Nac. Ci. Biol. 2:103–110.

Soulé, M.E. 1986. Conservation biology: the science of scarcity and diversity. Sunderland, MA: Sinauer.

Terrill, S.B. & L.S. Terrill. 1986. Common paraque (*Nyctidromus albicollis*) record from Sonora, Mexico. Amer. Birds 40:430.

Toledo, V.M. 1988. La diversidad biológica de México. Ciencia y Desarollo 14:17–30.

Trabulse, E. 1985. Historia de la Ciencia en México. Mexico City: CONACYT-Fondo de Cultura Económica.

Urban, E.K. 1959. Birds from Coahuila, Mexico. Univ. Kansas Publ. Mus. Nat. Hist. 11:443–516.

Van Rossum, A.J. 1945. A distributional survey of the birds of Sonora, Mexico. Occ. Pap. Mus. Zool. Louisiana State Univ. 15:265–269.

Van Tyne J. & A.J. Berger. 1976. Fundamentals of ornithology, 2nd ed. New York: John Wiley & Sons.

Villaseñor-Gomez, J.F. 1988. Aves costeras de Michoacán, México. Thesis, Escuela de Biologia, Universidad Michoacana de San Nicolás de Hidalgo. Morelia, Michoacán.

Von Mentz de Boege, B.M. 1982. México en el siglo XIX visto por los alemanes. Mexico City: UNAM.

Webster, J.D. 1958. Further ornithological notes from Zacatecas, Mexico. Wilson Bull. 70:243–256.

———— & R.T. Orr. 1952. Notes on Mexican birds from the states of Durango and Zacatecas. Condor 54:309–313.

Wetmore, A. 1943. The birds of southern Veracruz, Mexico. Proc. U.S. Nat. Mus. 93:215–340.

Williams, S.O. III. 1987. The changing status of the wood duck (*Aix sponsa*) in Mexico. Amer. Birds 41:372–375.

Wilson, R. & H. Ceballos-Lascurain. 1987. The birds of Mexico City. Mexico City: Published by authors.

9

Ecological Diversity in Scrub Jays (*Aphelocoma coerulescens*)

A. TOWNSEND PETERSON
AND NOÉ VARGAS-BARAJAS

Geographic variation in the habitat use of scrub jays (*Aphelocoma coerulescens*) in Mexico and the United States is reviewed. Scrub jays live in a wide variety of habitats, including oak woodland, pinyon-juniper woodland, riparian brush, desert woodlands, tropical thorn scrub, alpine pine-spruce forest, and mangrove swamps. Two contrasting interpretations of scrub jay ecological requirements are compared: (1) within-locality studies have identified the nuts of oaks or pinyons as critical elements in scrub jay biology, whereas (2) a geographic overview suggests that the presence of particular food types does not limit the distribution of scrub jays. The conflict between these views is due to a typological assumption of uniformity across the species' geographic range implicit in the geographic overview. One example of the translation of ecological diversity into morphological diversity (in this case beak shape) is described, and other factors leading to morphological and ecological diversity in scrub jays are discussed.

Most studies of biological diversity deal with the number of species present in a locality or region (see other chapters in this volume). Although within-species morphological variation is well documented (e.g., Johnson, 1980; Miller, 1941), few studies have treated within-species ecological variation (e.g., Collins, 1983; Dow, 1969).

Nevertheless, studies of ecological-geographic diversity are important because ecological variation influences processes such as population differentiation and speciation. Ecological-geographic diversity can, through the process of natural selection, lead to local adaptation and hence differentiation if gene flow is not overwhelming (Slatkin, 1987). If ecological specialization is sufficiently strong, populations can become reproductively isolated as well as differentiated, and speciation can take place (Mayr, 1963). Hence, under certain circumstances, ecological diversity is translated into morphological diversity, taxonomic diversity, or both.

The concept of ecological niches has led to many advances in the understanding of the roles of particular species in biotic communities. The original formulation of the niche concept (Grinnell, 1904, 1917) consisted of the range of values of environmental factors, including other species, that is necessary and sufficient to allow a species to carry out its life history (James et al., 1984). Later workers (e.g., Hutchinson, 1958, 1968; James et al., 1984; MacArthur, 1972) modified the niche concept, making it more specific and mathematical. Typical studies of avian niches (e.g., Titus and Mosher, 1981) examine a species' use of habitats within single localities and attempt to determine environmental factors critical to the species' life history.

Because most bird species have extensive geographic distributions, a species often encounters a variety of environments in different parts of its range and hence carries out its life history in different environmental contexts. Thus, a definition of the niche that a species occupies despite different climates, habitats, and so on would be invaluable for understanding the ecology of that species. Some problems with this approach are discussed below.

SCRUB JAYS

Scrub jays (*Aphelocoma coerulescens*) range from southern Washington to Oaxaca and Guerrero in Mexico (lat. 16°–46° N), and from peninsular Florida and central Texas to Santa Cruz Island off southern California (long. 80°–124° W; Pitelka, 1951). Within this broad geographic range, scrub jays show a high degree of geographic variation in terms of morphology and coloration. A broad north-south trend of increasing size exists in interior North America, but the trend is reversed with size, decreasing from north to south along the Pacific coast of the United States and Baja California (Pitelka, 1951). Coloration varies widely as well, from all-over dusky gray-blue to contrasting purple and white, making populations from different regions easily recognizable.

Scrub jays use a wide variety of habitats (Pitelka, 1951). Their habitat use in Mexico is of special interest because: (1) Mexico contains the southern limit of the woodlands that they normally inhabit, placing the jays in unusual ecological settings; and (2) the ecology and even the distribution of scrub jays in Mexico are still poorly known. The purpose of this study is to provide a summary of ecological variation in scrub jays in the United States and Mexico and to use that information to explore niche characteristics, possible competitive interactions or exclusion, and patterns of natural selection.

Information on the ecology and habitat use of scrub jays was gathered from the literature (especially Pitelka, 1951), and during field work from 1986 to 1989. Seventy-five sites in the United States and Mexico were visited as part of a study of genetic differentiation in the species. At most sites, jays were observed and collected, vegetation characteristics noted, and photographs taken. In this review, no attempt is made to treat the diversity of specific habitat types existing within a specific vegetation type

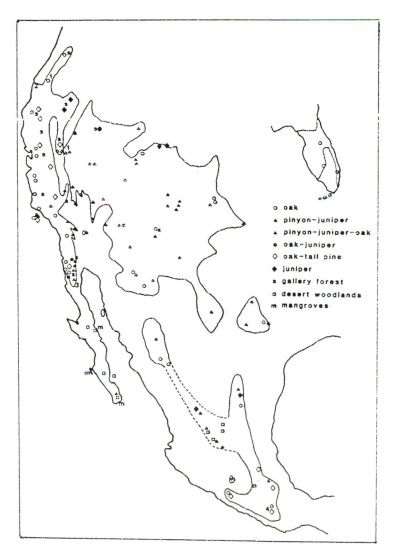

Figure 9.1. Geographic variation in habitat use of scrub jays. Dashed lines indicate an area of local and somewhat irregular distribution in the southern Sierra Madre Occidental and Chihuahuan Desert. For the sake of simplicity, adjacent sites with similar habitats have been designated with one symbol. Note that "gallery forest" in some areas includes occasional oaks, tall pines or both. Numbers indicate unusual habitats, as follows: (1) mountain mahogany woodland, (2) chaparral-like habitats lacking oaks, (3) alpine pine-spruce forest, and (4) palmetto scrub and open thorn scrub near the upper margin of the tropical deciduous forest. Sources: Pitelka (1951,unpubl. data), Dixon (1952), Fitton & Scott (1984), Brown & Horvath (1989), personal observation, and personal communications with other ornithologists.

(e.g., "oak" vegetations in this chapter include both chaparral and oak woodland), nor are plant species replacements considered (e.g., *Pinus monophylla* versus *P. quadrifolia* in northern Baja California).

HABITAT USAGE

Geographic variation in habitat use by scrub jays is summarized in Figure 9.1. A wide variety of vegetation types is inhabited across the species' range, from oak (*Quercus* spp.) and pinyon (*Pinus* spp.) woodlands to riparian brush, juniper (*Juniperus* spp.) woodlands, yellow pine (*P. ponderosa*)-oak forest, desert woodlands (*Acacia*, mesquite, various Agavaceae), tropical thorn forest, alpine pine-spruce forest, and mangrove swamps. An additional record exists for "cactus-acacia grassland" in a locality in Nayarit but probably refers to a locality in Jalisco (Peterson, in prep.).

Within localities, however, scrub jays are often remarkably habitat-restricted. In Florida, scrub jays are limited to oak-palmetto scrub and do not use other available habitats, e.g., oak forest, pine woodland, palmetto grasslands, or mangroves (Woolfenden & Fitzpatrick, 1984). Similar degrees of habitat restriction exist in scrub jays in coastal California and the Great Basin (Peterson, pers. observ.). Scrub jays live in more diverse habitats near the limits of their range, such as in Oregon, northern California, southern Mexico, and Baja California. Scrub jays in Baja California represent an extreme in this respect, living in habitats ranging from coastal mangrove swamps up through desert woodlands to pinyon-oak woodland (Pitelka, 1951; Peterson, pers. observ.).

ECOLOGICAL LIMITATIONS

In studies at single localities, investigators have concluded that scrub jays are heavily dependent on acorns or pinyon seeds (Bent, 1946; Woolfenden & Fitzpatrick, 1984). In Florida and California, acorns constitute a major proportion of the scrub jays' diet, and the jays invest a considerable amount of time in harvesting and storing acorns (Koenig & Heck, 1988; Woolfenden & Fitzpatrick, 1984). Nevertheless, Koenig and Heck (1988) demonstrated that scrub jays are physiologically less specialized for eating acorns than another acorn-eating species, the acorn woodpecker (*Melanerpes formicivorus*).

Scrub jays living in pinyon-juniper woodlands appear to be similarly dependent on pinyon seeds. The jays spend large amounts of time harvesting and storing pinyon seeds (Vander Wall & Balda, 1981) and actively defend pinyon trees against other jay species, e.g., *Gymnorhinus cyanocephalus* and *Nucifraga columbiana* (Peterson, pers. observ.). Although Vander Wall and Balda (1981) argued that scrub jays are relatively unspecialized for eating pinyon seeds, behavioral observations and morphometric studies in progress indicate that such specializations are indeed present (Peterson, in prep.).

However, scrub jays live in habitats lacking acorns and pinyon seeds in areas from northern California east to southern Wyoming (juniper woodland), Baja California (mangrove swamp and desert woodlands), and southern Mexico (desert riparian woodland, tropical thorn forest). Thus, from one viewpoint (Grinnell, 1917), scrub jays are not limited ecologically by the presence or absence of acorns or pinyon seeds. Although no detailed studies of scrub jay biology in these habitats have been conducted, it appears that sources of food comparable to acorns or pinyon seeds are apparently lacking (Peterson, pers. observ.).

Another view of a species' ecological requirements can be obtained by looking at the habitats outside the species' range (Grinnell, 1917). Most of the range limits make intuitive sense. Nowhere in their extensive range are scrub jays found commonly in closed-canopy forest, so their absence from the extensive forest belt separating the Florida and Texas populations is not surprising. Scrub jays seem to be limited in the north by the absence of open, xeric woodlands and by excessive snow cover. Similarly, excepting Baja California populations, scrub jays seem limited in several areas (e.g., southern Arizona, New Mexico, and western Texas) by the presence of hot deserts. Puzzlingly, scrub jays do not cross the dry lowlands of the Isthmus of Tehuantepec, even though they live in desert scrub and thorn scrub habitats just west of the Isthmus in Puebla and Oaxaca.

More interesting, however, are two gaps in the species' range in apparently suitable habitat: the Transvolcanic Belt and the northern Sierra Madre Oriental (Pitelka, 1951). Despite extensive field work in these areas by ourselves and others, no records of scrub jays exist. Scrub jays do not breed in the state of Michoacán (L. and F. Villaseñor, pers. comm.), in the Chisos Mountains of western Texas (Wauer, 1973), or the Sierra del Carmen of northern Coahuila (Miller, 1951, pers. observ.) despite the presence of ample habitat. Interestingly, gray-breasted jays (*A. ultramarina*) in the Sierra del Carmen have an unusually wide ecological range, overlapping completely not only the scrub jay's usual range but also that of the (absent) steller's jay (*Cyanocitta stelleri*) (Miller, 1951; pers. observ.).

A complication in the interpretation of these patterns is that the degree of habitat restriction of scrub jays varies geographically. Whereas in Florida and coastal California the jays are restricted to oak habitats, and in the Great Basin they are largely restricted to pinyon-juniper woodlands, elsewhere (e.g., Baja California, northern Oregon, southern Mexico) the jays range widely into other habitats. A typical explanation of this variation involves the presence and absence of competitors in other habitats. For example, in taller, marginal habitats in Florida, scrub jays are subject to aggression and nest destruction by blue jays (*C. cristata*), which leads to lowered reproductive success (Woolfenden & Fitzpatrick, 1984). Interactions between steller's jays and scrub jays in the mosaic of woodland and forest in central California appear similar (pers. observ.). Behavioral observations (Edwards, 1986) and experimental removal of gray-breasted jays in areas of sympatry (New Mexico and Arizona) indicate that the dominant gray-breasted jays probably limit the ecological distribution of scrub jays (David Ligon, pers. comm.). Similar arguments could be made that the large, aggressive magpie-jays (*Calocitta* spp.) limit scrub jays ecologically in southern Mexico, although scrub jays range widely into dry tropical habitats in southern Mexico, a habitat commonly inhabited by magpie-jays. The wide ecological range of scrub jays in Baja California may be due to the absence of other jays, e.g., *Cyanocitta stelleri*, *Cyanocorax* (*Cissilopha*) spp., *Calocitta* spp., on the peninsula (Pitelka, 1951).

However, habitats exist that are not occupied by other jays on the mainland but are inhabited by scrub jays in Baja California. In some cases (e.g., mangroves in Florida, deserts in much of the range of the species), scrub jays do not use these habitats, indicating that competition with other jay species alone does not limit habitat use by scrub jays. (This argument assumes, of course, that other, noncorvid species are not acting as competitors in these habitats. A detailed analysis of the interactions among corvid species is in progress by N.V.B.)

Many studies (e.g., James, 1971; Titus & Mosher, 1981) have attempted to determine the ecological requirements, or "niche gestalt," of species. However, species are not genetically uniform across their ranges; rather, at some scale most species are genetically structured (Slatkin, 1987). It is possible that the habitat use and limitation of species vary geographically because populations in different regions have adapted differentially to the locally available habitat mosaics. Put another way, the niche concept has an implicit typological assumption that a species encounters its environment armed only with a single species-specific set of ecological requirements.

Scrub jays are strongly differentiated morphologically and genetically among regions (Pitelka, 1951; Peterson, in prep.). An alternative to Grinnellian's interpretation of scrub jay ecology is that the jays have also differentiated in their ecological requirements. For example, scrub jays in Baja California may have "figured out," in an evolutionary sense, how to live in desert and mangrove habitats. Scrub jays in Florida, California, and the Great Basin may indeed be critically dependent on features of their habitats. Different degrees of habitat restriction without apparent changes in the presence of competitors occur even within the scrub jay populations of Baja California. (Although an argument can be made in the case of scrub jays that various populations should be considered separate species [Phillips, 1986], the problem with typological assumptions is general and applies at any taxonomic level.)

ECOLOGICAL AND MORPHOLOGICAL DIVERSITY: AN EXAMPLE

Because scrub jays are highly genetically structured among regions (Peterson, in prep.), and experience a wide variety of selective environments (see above), they might be expected to be locally adapted to different habitats. Preliminary work on geographic variation and phylogenetic origins of different beak forms in scrub jays indicates that beak shape variation (1) probably has a genetic basis, and (2) has functional implications in handling different types of food; (3) beak shape is closely correlated with habitat type, with hooked beaks occurring in oak habitats and pointed beaks in pinyon habitats; and (4) there have been multiple evolutionary derivations of the beak forms under similar environmental conditions (Peterson, in prep.). Scrub jays of Florida, California, Baja California, and southern Mexico have strongly hooked beaks, whereas those of the interior of North America generally have straight, pointed beaks. This apparently adaptive morphological variation probably developed in parallel with ecological specialization.

DIVERSITY OF SCRUB JAYS IN MEXICO

In Figure 9.1, it is evident that scrub jays use a wide variety of habitats in Mexico. In addition to ecological diversity, Mexican populations of scrub jays show a high degree of diversity in morphology (Pitelka, 1951). At least three factors probably have acted to produce these patterns.

1. Colonization history. The Baja California scrub jays, for example, probably represent a reinvasion of Mexico from California and for that reason are phylogenetically (and morphologically) distinct from the jays of mainland Mexico (Pitelka, 1951).
2. Isolation, such as that caused by historical geographic shifts of habitats during climatic fluctuations. For example, scrub jay distribution patterns in the mountains of northern Mexico probably were established less than 11,500 years ago (Wells, 1974). Hence, differentiation in morphology among populations in the Sierra Madre Oriental and Occidental is relatively weak (Pitelka, 1951).
3. Current ecological variation. Mexican scrub jays live in an exceptionally diverse array of habitats, which places the species in a variety of selective environments, as discussed above.

SUMMARY

This review demonstrates that scrub jays show considerable ecological diversity across their geographic range. Although patterns of habitat restriction and use vary widely among regions, certain populations of scrub jays may be critically dependent on acorns, pinyon seeds, or similar foods. The role of interspecific competition in limiting the distribution and habitat use of scrub jays is unclear and is likely to be clarified only by in-depth geographic comparisons and experimental manipulations. Ecological variation, colonization patterns, and historical geographic habitat shifts probably have interacted to produce the morphological and ecological diversity existing among Mexican scrub jays.

ACKNOWLEDGMENTS

We thank David Ligon for information about jay interactions in the southwestern United States, and David Willard for incisive discussion of topics treated in this chapter. Dale Clayton, Robert Dickerman, Patricia Escalante, John Fitzpatrick, Keith Karoly, Debra Moskovits, Adolfo Navarro, and Tom Schulenberg provided valuable comments on drafts of the manuscript. Amy Peterson, Hesiquio Benítez Díaz, D. Scott Baker, and Fernando Villaseñor provided thousands of hours of companionship and assistance in the field. Héctor Colón translated the abstract. Amy Peterson gave expert editorial assistance. A.T.P.'s field work was supported by the National Science Foundation Dissertation Improvement Grant Program, the National Geographic Society, the Chapman Fund of the American Museum of Natural History, Sigma Xi, the University of Chicago, and the Field Museum of Natural History. We especially thank Frank Pitelka for providing extensive unpublished data and for untiring interest and assistance.

REFERENCES

Bent, A.C. 1946. Life histories of North American jays, crows, and titmice. Bull. U.S. Nat. Mus. 191.
Brown, J.L. & E.G. Horvath. 1989. Geographic variation of group size, ontogeny, rattle calls, and body size in *Apheloma ultramarina*. Auk 106:124–128.
Collins, S.L. 1983. Geographic variation in habitat structure of the Black-throated green warbler (*Dendroica virens*). Auk 100:382–389.
Dixon, K.L. 1952. Scrub jay in Bexar County, Texas. Condor 54:208.
Dow, D.D. 1969. Home range and habitat of the cardinal in peripheral and central populations. Can. J. Zool. 47:103–114.
Edwards, T.C. 1986. Ecological distribution of the gray-breasted jay: the role of habitat. Condor 88:456–460.
Fitton, S.D. & O.K Scott. 1984. Wyoming's juniper birds. Western Birds 15:85–90.
Grinnell, J. 1904. The origin and distribution of the chestnut-backed chickadee. Auk 21:364–382.
――――. 1917. Field tests of theories concerning distributional control. Amer. Natur. 51:115–128.
Hutchinson, G.E. 1958. Concluding remarks. Cold Spring Harbor Symp. Quant. Biol. 22:415–427.
――――. 1968. When are species necessary? *In* R.C. Lewontin (ed.), Population biology and evolution. Syracuse, NY: Syracuse Univ. Press. pp. 177–186.
James, F.C. 1971. Ordinations of habitat relationships among breeding birds. Wilson Bull. 83:215–236.

————, R.F. Johnston, N.O. Wamer, G.J. Niemi & W.J. Boecklen. 1984. The Grinnellian niche of the wood thrush. Amer. Natur. 124:17–30.

Johnson, N.K. 1980. Character variation and evolution of sibling species in the *Empidonax difficilis-flavescens* complex (Aves: Tyrannidae). Univ. Calif. Publ. Zool. 112:1–151.

Koenig, W.D. & M.K. Heck. 1988. Ability of two species of oak woodland birds to subsist on acorns. Condor 90:705–708.

MacArthur, R.A. 1972. Geographical ecology. Princeton, NJ: Princeton Univ. Press.

Mayr, E. 1963. Animal species and evolution. Cambridge, MA: Harvard Univ. Press.

Miller, A.H. 1941. Speciation in the avian genus *Junco*. Univ. Calif. Publ. Zool. 44:173–434.

————. 1951. The avifauna of the Sierra del Carmen of Coahuila, Mexico. Condor 57:154–178.

Phillips, A.R. 1986. The known birds of North and Middle America. Vol. 1. Denver, CO: Denver Mus. Nat. Hist.

Pitelka, F. 1951. Speciation and ecologic distribution in American jays of the genus *Aphelocoma*. Univ. Calif. Publ. Zool. 50:195–464.

Slatkin, M. 1987. Gene flow and the geographic structure of natural populations. Science 236:787–792.

Titus, K. & J.A. Mosher. 1981. Nest-site habitat selected by woodland hawks in the central Appalachians. Auk 98:270–281.

Vander Wall, S.B. & R.P. Balda. 1981. Ecology and evolution of food-storage behavior in conifer-seed-caching corvids. Z. Tierpsychol. 56:217–242.

Wauer, R.H. 1973. Birds of Big Bend National Park and vicinity. Austin, TX: Univ. Texas Press.

Wells, P.V. 1974. Post-glacial origin of the present Chihuahuan Desert less than 11,500 years ago. *In* R.H. Wauer and D.H. Riskind, (eds.), Transactions of the symposium on the biological resources of the Chihuahuan Desert region, United States and Mexico. Washington, DC: U.S. Govt. Printing Office. pp. 67–83.

Woolfenden, G.E. & J.W. Fitzpatrick. 1984. The Florida scrub jay: demography of a cooperative-breeding bird. Princeton, NJ: Princeton Univ. Press.

10

Patterns of Mammalian Diversity in Mexico

JOHN E. FA AND LUIS M. MORALES

There are 449 mammal species, mostly bats and rodents, known for Mexico. This chapter describes this diversity. Computer-assisted analyses of species chorology are used to identify regions of richness and endemism. Greater species numbers are concentrated in the tropical regions, but endemisms are more important in the Trans-Mexican Neovolcanic Belt and the Sierra Madre del Sur. Mexico is unique in containing more total species and endemics per area than any other country in the Americas. Percent endemism is around 32%, a figure above the "normal" endemism expected for a country of this area. Gradients of diversity are examined with respect to variation in latitude, elevation, and aridity. Explanations of the country's high mammalian diversity are given by analyzing the influence of geographic barriers, habitat heterogeneity, and the mix of temperate and tropical forms converging in Mexico.

Pattern is a word used to refer to characteristic (often mathematical) features of a set of results, such as trends and repetitions in a data set. They are derivations, analogs, or generalizations based on initial observations and, as such, are not without bias (Eldredge, 1981). They reflect the type of questions asked as well as the way they are formulated. Patterns of species diversity in space and time have explanations in terms of a process or processes. "Diversity" in the present context is defined as species richness. It provides an umbrella term that includes the genetic variability within species and their populations. Species numbers reflect the population dynamics of the individual species; and these numbers, in turn, relate to the variability of environments in space and time.

Descriptions of biodiversity have centered primarily on a pervading axiom in biogeography—that species numbers are directly proportional to land area (MacArthur, 1972; MacArthur & Wilson, 1967; Preston, 1962). Most explanations of number of species have concentrated on this point (Connor & McCoy, 1979), in most cases because of lack of better information on other intervening variables. However, the number and complexity of species templates—habitats—contained in an area can be responsible for the correlation between area and species numbers. As the area increases, the number of habitats, each with an associated set of species, also increases (Williams, 1964).

Here we describe the distribution of Mexican mammals by empirically documenting patterns of geographic variation in diversity and assemblages. We also attempt to explain the observed trends in species richness in terms of predictable ecological and historical processes. Examining the biodiversity of an area as large and complex as Mexico necessarily obviates the problems of defining and quantifying spatial variation in abundance and distribution of species. This point is not trivial as measures of species richness are highly dependent on the spatial and temporal scale of sampling (Auerbach & Shmida, 1987; Bond, 1983; Ricklefs, 1987; Whittaker, 1972). Area and distance influence results because large regions contain more species than small ones (Wilson & Shmida, 1984) and resolution of factors controlling species richness patterns becomes more limited. Thus this chapter is by no means definitive but provides new approaches to understanding Mexican mammalian diversity. It therefore conveniently centers all interpretations on the use of inventory diversity data: the size of the regional species pool, which is in itself a function of the interaction between alpha, beta, and gamma diversities (Cowling et al., 1989). Clarification of the patterns will no doubt still be open to further debate as refined evaluations of the different hypotheses become available. Despite this fact and within the limitations imposed by the level of resolution, the generality of the chapter's approach can assist in an overall appreciation of biogeographic topics common to different specialities and applicable to most plants and animals.

STUDIES OF MEXICAN MAMMALS

According to Ramirez-Pulido & Britton (1981) the study of Mexican mammals can be divided into four historical periods (Fig.10.1). The first coincides with the exploration of the western United States (1831–1881). Seventy-seven categories were described largely by such explorers as Henri de Saussure and Spencer Fullerton Baird. During the second period, from 1887 to 1909, well-funded expeditions from U.S. museums and government agencies resulted in the recognition of 460 new forms, 164 of which (35%) were classified by Clinton Hart Merriam alone. Together with E.W. Nelson and E.A. Goldman, Merriam identified 227 taxa, almost half the total for this period. The Mexican Revolution (1910–19) and the First World War (1914–18) disrupted scientific exploration in the country, but the third period (1922–42) saw a renewal of exploration from U.S. scientists in Mexico. One hundred and eighty-four categories were described, most by Nelson and Goldman. A fourth period, recognized byRamirez-Pulido and Britton as a stage of synthesis, started in 1943 and extends to the

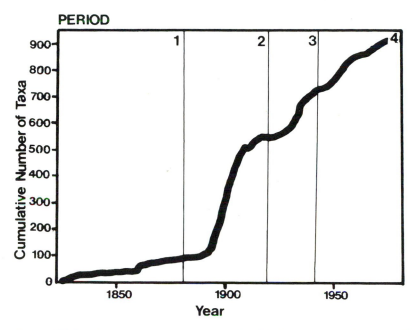

Figure 10.1. Historical periods for describing Mexican mammalian taxa. See text for definition of periods. Adapted from Ramírez-Pulido & Britton (1981).

present. Families, genera, and species complexes have been revised but another 190 new categories have been described. In all, 73% of the mammalian fauna native to Mexico were described by 18 authors (Ramirez-Pulido & Britton, 1981). Most work on mammals in Mexico has concentrated on systematics and taxonomy. There is still little information on ecology and behavior and even less on inventory diversity, even though advances have been made from Mexico City-based collections at the Instituto de Biología, Universidad Autónoma Metropolitana, and Instituto Politécnico. Nonetheless, of more than 3,000 publications on Mexican mammals (Ramirez-Pulido & Müdespacher, 1987) only 2% are studies undertaken in natural communities.

THE MEXICAN MAMMAL FAUNA

The total terrestrial Mexican mammal fauna is composed of 449 species of 10 orders, 32 families, and 149 genera (Ramirez-Pulido & Britton, 1981; Ramirez-Pulido & Müdespacher, 1987) (Table 10.1). Among species, 160 are monotypic and 283 polytypic—an overall total of 1,070 subspecies and 1,230 species and subspecies. Rodents and bats are the best represented, with 215 and 133 species, respectively. Both orders account for a total of 79% of the Mexican mammalian fauna.

Table 10.1. Mexican mammals

Order	Family	Subfamily	Genera	Species (No.)	Sub-species (No.)
Marsupialia	Didelphidae		*Caluromys*	1	2
			Chironectes	1	1
			Didelphis	2	3
			Marmosa	2	6
			Philander	1	1
Insectivora	Soricidae		*Cryptotis*	6	12
			Megasorex	1	0
			Notiosorex	1	2
			Sorex	14	10
	Talpidae		*Scalopus*	1	2
			Scapanus	1	2
Chiroptera	Emballonuridae	Emballonurinae	*Balantiopteryx*	2	2
			Centronycteris	1	1
			Peropteryx	2	2
			Rhynochonycteris	1	0
			Saccopteryx	2	0
		Diclidurinae	*Diclidurus*	1	0
	Noctlionidae		*Noctilio*	1	1
	Mormoopidae		*Mormoops*	1	1
			Pteronotus	4	4
	Phyllostomidae	Phyllostominae	*Chrotopterus*	1	1
			Lonchorhina	1	1
			Macrophyllum	1	0
			Macrotus	2	2
			Micronycteris	4	1
			Mimon	2	1
			Phylloderma	1	1
			Phyllostomus	1	1
			Tonatia	2	0
			Trachops	1	1
			Vampyrum	1	0
		Glossophaginae	*Anoura*	1	1
			Choerniscus	1	0
			Choeronycteris	1	0
			Glossophaga	4	6
			Hylonycteris	1	2
			Leptonycteris	2	0
			Musonycteris	1	0

Table 10.1. (cont.)

Order	Family	Subfamily	Genera	Species (No.)	Sub-species (No.)
		Carolliinae	*Carollia*	3	1
		Sturnirinae	*Sturnira*	2	3
		Stenoderminae	*Artibeus*	8	13
			Centurio	1	1
			Chiroderma	2	3
			Uroderma	2	3
			Vampyressa	1	1
			Vampyrodes	1	1
			Vampyrops	1	0
		Desmodontinae	*Desmodus*	2	1
			Diphylla	1	0
	Natalidae		*Natalus*	1	2
	Thyropteridae		*Thyroptera*	1	1
	Vespertilionidae	Vespertilioninae	*Eptesicus*	3	6
			Euderma	1	0
			Idionycteris	1	0
			Lasionycteris	1	0
			Lasiurus	5	6
			Myotis	20	24
			Nycticeus	1	2
			Pipistrellus	2	5
			Plecotus	2	2
			Rhogeesa	5	0
		Nyctophyllinae	*Antrozous*	2	4
			Bauerus	1	0
	Molossidae		*Eumops*	5	6
			Molossops	1	1
			Molossus	5	4
			Nyctinomops	4	2
			Promops	1	1
			Tadarida	1	2
Primates	Cebidae	Alouattinae	*Alouatta*	2	1
		Atelinae	*Ateles*	1	1
Xenarthra	Myrmecophagidae		*Cyclopes*	1	1
			Tamandua	1	2
	Dasypodidae		*Dasypus*	1	2

Table 10.1. (cont.)

Order	Family	Subfamily	Genera	Species (No.)	Sub-species (No.)
Lagomorpha	Leporidae	Palaeolaginae	*Romerolagus*	1	0
		Leporinae	*Lepus*	5	18
			Cabassous	1	1
			Sylvilagus	8	28
Rodentia	Sciuridae	Sciurinae	*Ammospermophilus*	4	6
			Cynomys	2	1
			Sciurus	12	24
			Spermophilus	10	25
			Tamias	6	7
			Tamiasciurus	1	0
		Petauristinae	*Glaucomys*	1	4
	Castoridae		*Castor*	1	3
	Geomyidae		*Geomys*	3	2
			Orthogeomys	4	18
			Pappogeomys	9	43
			Thomomys	2	79
			Zygogeomys	1	2
	Heteromyidae	Perognathinae	*Perognathus*	8	32
			Chaetodipus	14	49
		Dipodomyinae	*Dipodomys*	10	39
		Heteromyinae	*Heteromys*	7	2
			Liomys	4	12
	Muridae	Sigmodontinae	*Baiomys*	2	12
			Habromys	4	2
			Hodomys	1	4
			Megadontomys	1	3
			Nelsonia	1	3
			Neotoma	15	49
			Nyctomys	1	4
			Onychomys	3	12
			Oryzomys	7	32
			Osgoodomys	1	2
			Otonyctomys	1	0
			Ototylomys	1	2
			Peromyscus	45	101
			Reithrodontomys	11	39

Table 10.1. (cont.)

Order	Family	Subfamily	Genera	Species (No.)	Sub-species (No.)
			Rheomys	2	1
			Scotinomys	1	1
			Sigmodon	7	24
			Tylomys	2	4
			Xenomys	1	0
		Arvicolinae	*Microtus*	5	12
			Ondatra	1	2
			Pitymys	2	0
	Erethizontidae		*Coendou*	1	2
			Erethizon	1	1
	Agoutidae		*Agouti*	1	1
	Dasyproctidae		*Dasyprocta*	2	2
Carnivora	Canidae		*Canis*	2	12
			Urocyon	1	8
			Vulpes	1	3
	Ursidae		*Ursus*	4	4
	Procyonidae	Bassariscinae	*Bassariscus*	2	13
			Potos	1	2
		Procyoninae	*Procyon*	3	9
			Nasua	2	3
	Mustelidae	Mustelinae	*Eira*	1	1
			Galictis	1	1
			Mustela	1	9
		Melinae	*Taxidea*	1	1
		Mephitinae	*Conepatus*	3	10
			Mephitis	2	6
			Spilogale	2	11
		Lutrinae	*Enhydra*	1	1
			Lutra	1	1
	Felidae		*Felis*	4	16
			Lynx	1	6
			Panthera	1	5
Perissodactyla	Tapiridae		*Tapirus*	1	0

Table 10.1. (cont.)

Order	Family	Subfamily	Genera	Species (No.)	Sub-species (No.)
Artiodactyla	Tayassuidae		*Tayassu*	2	8
	Cervidae	Odocoileinae	*Mazama*	2	2
			Odocoileus	2	19
	Antilocapridae		*Antilocapra*	1	3
	Bovidae	Bovinae	*Bison*	1	1
		Caprinae	*Ovis*	1	3
TOTAL				449	1070

Numerical Significance

Comparisons of the area of a variety of countries with recorded mammalian species show two main trends: (1) Diversity is higher in the tropics than in temperate countries; and (2) diversity is higher in continents than in islands, divisible also into oceanic and continental islands (Eisenberg, 1981). Data for North and South American countries (Toledo, 1989) (Fig. 10.2A) show Mexico, together with Cuba and the Lesser Antilles, departing from the overall species–area relation. As with species richness, endemisms are known to increase from arctic to temperate latitudes but are more numerous on islands than on continents. The degree of endemism also increases with the size of the area (Major, 1988). Percentage species endemism of Mexico's mammalian fauna is around 32%, well below the diagonal line in a nomogram for determining "normal" endemism (Fig. 10.2B). The index of endemicity (Bykov, 1979) for Mexican mammals is 1.23 (derived from the factual percentage endemism divided by the normal percentage endemism read off from the nomogram), which points at a greater than normal level of endemism.

Figure 10.2. A. Species–area relationships for all mammals. B. Nomogram for normal endemism in Latin American countries. Key to countries: 1, Tobago; 2, Netherland Antilles; 3, Cuba; 4, Trinidad; 5, Lesser Antilles; 6, Puerto Rico; 7, Jamaica; 8, Bahamas; 9, El Salvador; 10, Belize; 11, Costa Rica; 12, Hispaniola; 13, Panama; 14, French Guiana; 15, Guatemala; 16, Honduras; 17, Nicaragua; 18, Surinam; 19, Uruguay; 20, Guyana; 21, Ecuador; 22, Paraguay; 23, Chile; 24, Venezuela; 25, 26, Colombia; 27, Peru; 28, Mexico; 29, Argentina; 30, Brazil. Data from Toledo (1989).

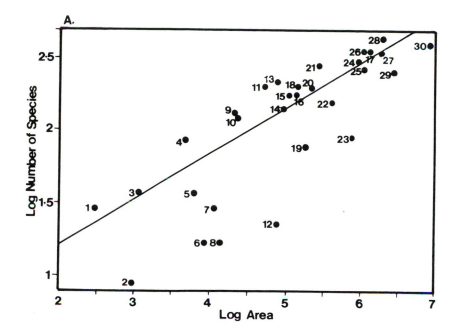

A.

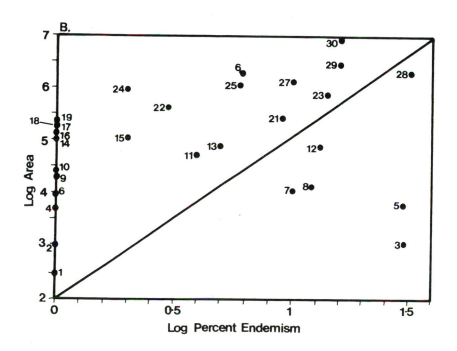

B.

Taxonomic Diversity

Straddling Two Domains: Nearctic and Neotropical

The Mexican fauna and flora are greatly influenced by the fact that the country is at the interface of temperate and tropical climates. The boundary between the Nearctic and Neotropical biotas is interdigitated with species that have spread northward or southward in a series of interchange events. The delimiting line does not correspond precisely, however, to a present or past natural geographic barrier. Although the ancient separation and the narrowest point of the present land connection between North and South America is the Isthmus of Panama, the Neotropical biota ranges well north of this point and begins to be replaced by Nearctic forms only in southern Mexico. The zone between the two floras and faunas is confluent. Some authors draw the irregular boundary between Nearctic and Neotropics through the spine of the transverse group of peaks in the Trans-Mexican Neovolcanic Belt (Rzedowski, 1978). Darlington (1957) and Hershkovitz (1972) delimited the Neotropical region from the Nearctic by the Tropic of Cancer on the basis of broad environmental aspects. Ferrusquía-Villafranca (1977) argued that the division is too vague and encompasses "tropical" animals and plants of vastly different ecological requirements. The boundary between Neotropical and Nearctic mammals, based on the postulated geographic origin of each family, also conforms to a demarcating line of natural regions of different physical and biotic conditions. As he suggested, the realms are distinguishable by physiography, climate, and vegetation. Fa (1989) also showed that mammalian distribution patterns closely parallel vegetation distribution patterns.

Without doubt, the demarcation of the Neotropical and Nearctic is associated with environmental tolerance of various mammalian groups. Baker (1967) showed that climatic control may be important in the observed distribution of Pacific Coast mammals of the entire continent. In some groups, such as the cricetid rodents, distribution may respond to environmental control (Patterson & Pascual, 1972). The boundary between regions seems to correlate with climatic and habitat factors (Fa, 1989). Climate limits the altitude of southern biota in the Trans-Mexican Neovolcanic Belt, but the biota can still penetrate some of the peaks at Colima and Cofre de Perote and Pico de Orizaba. Nearctic elements, on the other hand, are less limited by climate than tropicals and can penetrate farther south. However, Brown and Gibson (1983) argued that even though organisms are limited primarily by climate and habitat there is in fact a gradual replacement of Neotropical species by Nearctic species throughout regions in Mexico that are subtropical. Likewise, a few tropical mammals such as the collared peccary (*Tayassu tajacu*, Tayassuidae) and coati (*Nasua nasua*, Procyonidae) range as far north as the southern United States. Climate may not be totally responsible, as features such as habitat structure as well as interspecific interactions may also be important.

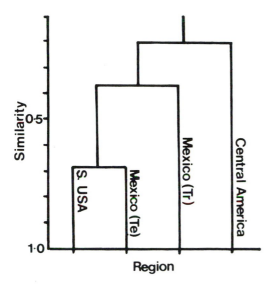

Figure 10.3. Dendrogram showing similarity between mammals found in the four main regions from the southern United States through tropical (Tr) and temperate (Te) Mexico to Central America.

The present patterns of mix between tropical and temperate faunas have explanations in the past interchange of forms across the narrow isthmus between North and South America since the Tertiary. During this period land mammals entered South America from the north and formed the basic stock there. Central America was a North American peninsula separated by the Bolivar Geosyncline Strait. Regional tectonic uplift at the end of the Tertiary connected North and South America by the Late Pliocene, and extensive faunistic interchange took place. These biotas are not highly discrete units composed of closely interdependent species but are composed of forms that originated before the land bridge. Patterson and Pascual (1972) suggested that the interchange caused minor competition and replacement of some South American forms by northern immigrants, which caused the extinction of carnivorous marsupials and slow-moving herbivores. Southern immigrants such as edentates and caviomorph rodents found "empty niches" in the north, as is true for most of the mammals involved in the interchange. The net result has been the enrichment of both faunas. Ferrusquía-Villafranca (1977) implied early ecological balance of northern and southern faunas because of coexistence throughout the Pleistocene.

Distribution of Orders in a Middle American Context

Faunal comparisons of the Middle American regions show that they are influenced by the affinities to the Nearctic or Neotropical domains

Table 10.2. Distribution of living rodent genera across main geographic areas

Family/Genus	Southern United States	Nearctic Mexico	Neotropical Mexico	Central America
Sciuridae				
Ammospermophilus	x	x	x	—
Cynomys	x	x	—	—
Eutamias	x	x	—	—
Glaucomys	x	x	x	—
Marmota	x	x	—	—
Microsciurus	—	—	—	x
Sciurus	x	x	x	x
Spermophilus	x	x	—	—
Syntheosciurus	—	—	—	x
Tamias	x	—	—	—
Tamiasciurus	x	—	—	—
Geomyidae				
Cratogeomys	x	x	—	—
Geomys	x	x	—	—
Orthogeomys	—	—	x	x
Pappogeomys	—	x	x	—
Thomomys	x	x	—	—
Zygogeomys	—	—	x	—
Heteromyidae				
Dipodomys	x	x	x	—
Heteromys	—	—	x	x
Liomys	—	x	x	x
Microdipodops	x	—	—	—
Perognathus	x	x	x	—
Castoridae				
Castor	x	x	—	—
Muridae				
Baiomys	x	x	x	x
Clethrionomys	x	—	—	—
Lagurus	x	—	—	—
Megadontomys	—	x	—	—
Microtus	x	x	x	x
Neacomys	—	—	—	x
Neatomys	—	—	—	x
Nelsonia	—	—	—	x
Neofiber	x	—	—	—
Neotoma	x	x	x	—
Nyctomys	—	—	x	x
Ondatra	x	—	—	—
Onychomys	x	x	—	—
Oryzomys	x	x	x	x
Osgoodomys	—	x	—	—

Table 10.2. (cont.)

Family/Genus	Southern United States	Nearctic Mexico	Neotropical Mexico	Central America
Otonyctomys	–	–	x	–
Ototylomys	–	–	x	x
Peromyscus	x	x	x	x
Phenacomys	x	–	–	–
Reithrodontomys	x	x	x	x
Rheomys	–	–	–	x
Rhipidomys	–	–	–	x
Scotinomys	–	–	x	x
Sigmodon	x	x	x	x
Thomasomys	–	–	–	x
Tylomys	–	–	x	x
Xenomys	–	x	–	–
Zygodontomys	–	–	–	x
Zapodidae				
Zapus	x	–	–	–
Erethizontidae				
Coendou	–	–	x	x
Erethizon	x	x	–	–
Hydrochaeridae				
Hydrochoerus	–	–	–	x
Dasyproctidae				
Cuniculus	–	–	x	x
Dasyprocta	–	–	–	x
TOTAL	31	29	25	29

Source: Adapted from Ferrusquía-Villafranca (1977).
x, present; –, absent.

(Ferrusquía-Villafranca, 1977). Similarities between generic composition from the southern United States to Central America indicate that faunas change gradually from mostly Nearctic forms in the north to tropical ones toward the south (Fig. 10.3). The summary of terrestrial forms for Middle America provided below describes the Mexican mammalian fauna in a broader context.

Marsupialia. Marsupials are some of the oldest most primitive mammals of the region. *Didelphis*, the most widespread genus, is the only one in the Nearctic domain. Within the tropical zone there are seven other didelphid genera and one caenolestid, all restricted to South America. There is a southern increase in marsupial diversity, with five genera occurring in tropical Mexico, seven in Central America, and eight in northern South America. The Cenozoic fossil record shows that marsupials have South American affinities appearing in other areas only during the Pleistocene (Ferrusquía-Villafranca, 1977).

Insectivora. Insectivores are represented by three families (Solenodontidae, Talpidae, and Soricidae). Solenodons are endemic to the Antilles, and talpids are restricted to the Nearctic with a few species just entering Mexico. Shrew species are more numerous in North America, with only one genus (*Cryptotis*) extending into northern South America. Because insectivores are absent from the Tertiary record of Mexico and Central America and South America, Soricids are known to have migrated south after originating in the Nearctic.

Xenarthra. Edentates except *Dasypus* are restricted to the tropical domain and increase in diversity southward. Four genera are found in Mexico (one, *Cabassous*, was recently discovered in Chiapas [Cuarón et al., 1989]), seven in Central America and ten in northern South America.

Lagomorpha. All Middle American lagomorphs have North American affinities. Three genera are found in the Nearctic domain, and two range into the tropics. *Sylvilagus* is the most widespread lagomorph, occurring in all continental Middle America. *Lepus* is largely a Nearctic form that just enters the tropical domain in the Mexican state of Oaxaca. The other genus, *Romerolagus*, has a restricted range in the Trans-Mexican Neovolcanic Belt.

Rodentia. Rodents are the most abundant mammals in Middle America, with 15 families (43%) and 94 genera. Table 10.2 indicates the genera found within each main region. Overall generic diversity decreases slightly from the southern United States to Central America. Sciurid genera are more abundant in the Nearctic, with nine, in contrast to six in the tropics. Six of the Nearctic genera are endemic, whereas only *Sciurullus* is restricted to northern South America. *Sciurus* is the most abundant genus in Middle America, whereas *Syntheosciurus* and *Microsciurus* occur in Central America and northern South America. *Ammospermophilus* and *Tamiasciurus* have their southernmost range in Mexico.

The most abundant rodents in Middle America are the murids. The sigmodontine group (the complex-penised) are dominant in tropical Middle America, whereas the peromyscines (simple-penised) dominate in Nearctic Middle America (Ferrusquía-Villafranca, 1977). Geomyid and heteromyid rodents are represented by four genera each in Nearctic Middle America and in the tropical regions. Three geomyid genera are endemic to the Nearctic, the fourth genus just reaching the northern edge of tropical biomes. Heteromyids, which are largely northern forms, have a genus (*Heteromys*) restricted to the tropical domain and the only one to inhabit northern South America. Both zapodids and castorids are found in the Nearctic and are represented by a single genus each.

Primates. There are no primates in Nearctic Middle America. Four genera (*Allouata, Cebus, Saimiri, Ateles*) occur in the tropics. Only two genera are known for Mexico.

Carnivora. Carnivores are represented by five families and 25 genera. Two felid genera occur in Middle America and only one (*Felis*) in the tropics. Eight genera of mustelids are more numerous in tropical Mexico than in any other part of Middle America, but at least three genera (*Eira, Galictis, Taxidea*) are in their southernmost limit. *Mustela* is the only genus that inhabits all of Middle America. The Canids, in contrast to Mustelids, occur throughout Middle America, but *Urocyon* is present throughout. *Canis* is absent from northern South America, and its southernmost limit is northern Mexico. The affinities of the family are clearly North America or at least Northern Hemisphere. Procyonids, however, include five genera that are more abundant in Central America than in any other part of Middle America. *Procyon* occurs in all regions, and *Nasua* is absent from the Antilles. Ursids are restricted to *Ursus* in the Nearctic and *Tremarctos* in South America.

Perissodactyla and Artiodactyla. Perissodactyls and artiodactyls are demonstrably northern immigrants into Middle America. The tapir, the only living perissodactyl, itself regarded a living fossil (Brown & Gibson, 1983), is now restricted to tropical Middle America. Peccaries (*Tayassu*) and *Odocoileus* are found throughout Middle America.

PATTERNS

Chorology

Basic Data and Measurements

Given the absence of adequate inventory information for most of the country, it is necessary at present to rely on gamma diversity projections to understand patterns of mammalian species richness in Mexico. This method describes the concept of species turnover in similar habitats along geographic gradients (Cody, 1975, 1983). It is identical to Whittaker's (1977) delta diversity. Gamma diversity is independent of habitat differences and invokes the concept of ecologically equivalent species, i.e. species that occupy the same habitat in different geographical localities (Shmida & Wilson, 1985). These data are traditionally utilized by biogeographers to explain patterns in species richness across geographic space (Simpson, 1964).

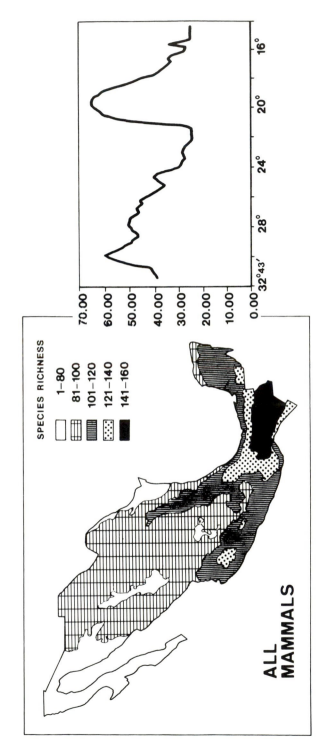

Figure 10.4. Choropleth map and latitudinal gradient of species richness patterns for all Mexican mammalian species.

The basic data presented in this chapter consist of records of species of recent mammals present in 20 x 20 km quadrants covering the entire country. Further details of the method are given in Fa (1989). The method corresponds to the superimposition of a grid of 20 x 20 km not oriented with respect to physiographic features or other known zoogeographic factors. A list of species of recent mammals occurring in Mexico was compiled from Hall (1981). Species nomenclature was based on Hall (1981) unless modified subsequently by Ramirez-Pulido et al. (1988). Only continuous mainland areas and species occurring in them were included in the analyses. Insular distributions were treated separately.

Each species range map was digitized and transferred to a data base. A raster program for screening all maps into a one-zero grid per species was performed by computer. All grids were then overlaid and numbers of species counted per grid square. The total number of quadrants used was 2,000. The resulting cumulative grids were then employed for drawing the final choropleth maps.

Continental Species Richness

Species richness increases from the north of the country toward the Chiapan interior (Fig. 10.4). Lowest species diversity is found along the Baja California peninsula and eastern foothills of the Sierra Madre Occidental. This trend in species richness primarily reflects the origin of the families concerned (Fig. 10.5). Tropical orders have more species within the Neotropical area of the country, whereas Nearctic forms are found more abundantly in the northern and central parts. Primates, edentates and perissodactyls are restricted to tropical regions of the Yucatan Peninsula and the tropical coastal areas. Species richness for lagomorphs, insectivores, and chiropterans is higher around central Mexico, in the Trans-Mexican Neovolcanic Belt. Rodent species are more numerous around central Mexico stretching from the Mexico-United States border to the Chiapan Highlands. As with bats and insectivores, species richness is highest around 20° latitude.

Insular Faunas

Twenty-four species (6%) of the mammal fauna occur on the larger Trés Marías Islands adjacent to the continent and on islands in the Sea of Cortes and the Mexican Caribbean. The Sea of Cortes islands are by far the most significant in terms of numbers of species present and endemism. One endemic rabbit species and two endemic subspecies are present together with 11 species and 40 rodent subspecies. The Trés Marías mammal fauna is unique in that it includes an endemic raccoon and an endemic rabbit with

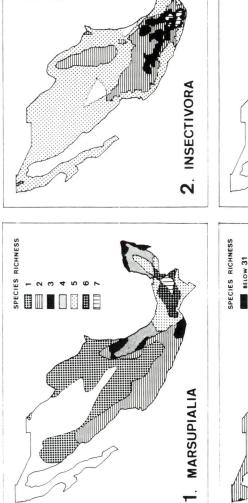
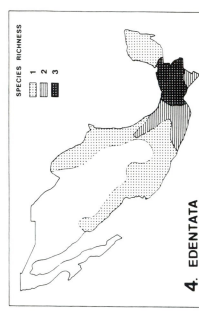

Figure 10.5. (Part one) Choropleth maps of species richness patterns for all Mexican mammalial orders.

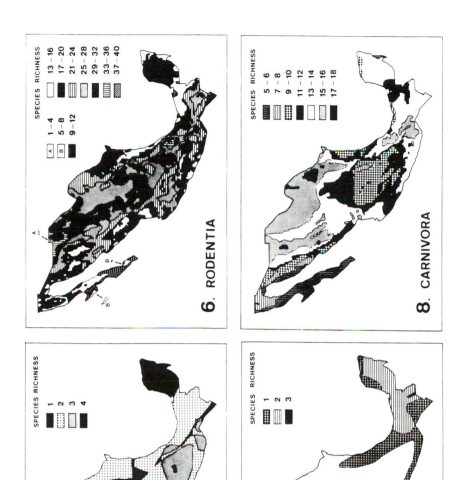

Figure 10.5. (Part two)

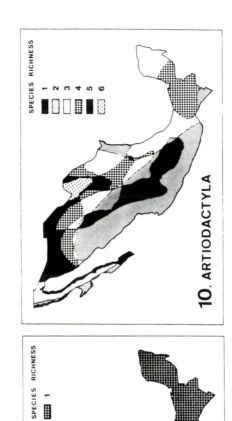

SPECIES RICHNESS

1
2
3
4
5
6

10. ARTIODACTYLA

SPECIES RICHNESS

1

9. PERISSODACTYLA

Figure 10.5. (Part three)

only two rodents present (one full endemic and another endemic subspecies).

Endemics

A taxon is considered endemic if confined to a particular area through historical, ecological, or physiological reasons. Endemics can be restricted to large or small areas but are regarded as old (paleoendemics) or new (neoendemics); the endemic taxon can be of any rank, though it is usually at family level or below. Discussion on endemism has focused on relative frequency of neo- and paleoendemics as well as differences in rates of evolution and phylogenetic persistence abilities of neo- and paleoendemics (Cain, 1974; Stebbins & Major, 1965). These classes have been subdivided even further based upon cytological evidence (Favager & Contandriopoulos, 1961). More elaborate discussions of endemisms recognize four types of endemism (neo-, paleo-, insular, and ecologic). Despite the heuristic utility of such categories, it is obvious that the classes are neither absolute nor mutually exclusive. Ecologic endemics can be classified as neo- and paleoendemics, for example; and insular endemics may exist that are also ecological isolates.

Six genera and 142 species are endemic to Mexico (Table 10.3). Twenty-three species (16%) are restricted to islands; the remainder are continental endemics. Over three-fourths of all endemics are made up of 106 rodent species, 14 chiropterans, 11 insectivores, 8 lagomorphs, 4 carnivores, and 1 marsupial. Most endemics are found within the Trans-Mexican Neovolcanic Belt and the Sierra Madre del Sur region (Fig. 10.6): 26 to 30 endemic species occur in this region. The area of importance for mammalian endemics stretches along the Sierra Madre Occidental and into the Tehuantepec Isthmus. All endemic genera occur exclusively in the Trans-Mexican Neovolcanic Belt, Sierra Madre Occidental and Sierra Madre del Sur. Ramirez-Pulido & Müdespacher (1987) came to the same conclusion by arbitrarily denoting endemism regions on the basis of geophysical regions. Their main region for endemics stretches from the northern Nayarit coast to the Tehuantepec Isthmus and across the Trans-Mexican Neovolcanic Belt. They (1987) calculated that 45.3% of the country's endemic species are restricted to this region. Figure 10.7 shows that up to 25% of Mexico's endemic mammals are found within the Colima area, stretching into Michoacan and Jalisco. In particular, the Trans-Mexican Neovolcanic Belt contains all the country's endemic genera and 52.2% of the endemic species. Most endemics are rodents; and 50% of Mexican rodents occur in this belt. They are largely restricted to the belt's main axis, with centers of endemism along Colima, Michoacan, Toluca/Valley of Mexico/Sierra Nevada, and Cofre de Perote-Pico de Orizaba, all of which are above 2,000 m in temperate pine-oak forest habitats (Fa, 1989; Fa & Morales,

Table 10.3. Endemic Mexican mammals
_____

ISLAND ENDEMICS

Lagomorpha
Lepus insularis
Sylvilagus mansuetus
S. graysoni

Rodentia
Ammospermophilus insularis
Perognathus anthonyi
Neotoma anthonyi
N. bryanti
N. bunkeri
N. martinensis
N. varia
Peromyscus caniceps
P. dickeyi
P. guardia
P. interparietalis
P. madrensis
P. pembertoni
P. pseudocrinitus
P. sejugis
P. slevini
P. stephani
Reithrodontomys spectabilis

Carnivora
Nasua nelsoni
Procyon insularis
P. pygmaeus

CONTINENTAL ENDEMICS

Marsupialia
Marmosa canescens

Insectivora
Cryptotis goldmani
C. magna
C. mexicana
Notiosorex gigas
Sorex emarginatus
S. macrodon
S. milleri
S. oreopolus
S. sclateri
S. stizodon
S. ventralis

Table 10.3. (cont.)

Chiroptera
  Artibeus hirsutus
  Choeronycteris harrisoni
  Myotis carteri
  M. findleyi
  M. fortidens
  M. milleri
  M. peninsularis
  M. planiceps
  M. vivesi
  Plecotus mexicanus
  Rhogeesa alleni
  R. gracilis
  R. mira
  R. parvula

Lagomorpha
  Lepus callotis
  L. flavigularis
  Romerolagus diazi
  Sylvilagus cunicularius
  S. insonus

Rodentia
  Cynomys mexicanus
  Sciurus alleni
  S. colliaei
  S. nayaritensis
  S. oculatus
  Spermophilus odocetus
  S. annulatus
  S. atricapillus
  S. madrensis
  S. perotensis
  Tamias bulleri
  Geomys tropicalis
  Orthogeomys cuniculus
  O. lanius
  Pappogeomys alcorni
  P. bulleri
  P. fumosus
  P. gymnurus
  P. merriami
  P. neglectus
  P. thylorhinus
  P. zinzeri
  Zygogeomys trichopus
  Dipodomys gravipes
  D. insularis
  D. nelsoni

Table 10.3. (cont.)

D. phillipsii
Heteromys gaumeri
Liomys irroratus
L. pictus
L. spectabilis
Perognathus arenarius
P. artus
P. dalquesti
P. goldmani
P. lineatus
P. pernix
Chaetodipus nelsoni
Habromys chinanteco
H. lepturus
H. simulatus
Hodomys alleni
Megadontomys thomasi
Nelsonia neotomodon
Neotoma angustapalata
N. goldmani
N. nelsoni
N. palatina
N. phenax
Oryzomys caudatus
O. fulgens
O. nelsoni
O. peninsulae
Osgoodomys banderanus
Peromyscus alstoni
P. bullatus
P. eva
P. furvus
P. hooperi
P. megalops
P. mekisturus
P. melanocarpus
P. melanophrys
P. melanotis
P. ochraventer
P. perfulvus
P. polius
P. simulus
P. spicilegus
P. winkelmanni
P. yucatanicus
P. zarhynchus
Reithrodontomys burti
R. chrysopsis
R. hirsutus
Rheomys mexicanus

Table 10.3. (cont.)

Sigmodon alleni
S. leucotis
S. mascotensis
Tylomys bullaris
T. tumbalensis
Xenomys nelsoni
Microtus fulviventer
M. oaxacensis
M. umbrosus
Pitymys quasiater
Dasyprocta mexicana

Carnivora
Spilogale pygmaea

1991). The area is also important for subendemics, in that it has up to 76 subspecies of Mexican endemics and 23 of non-Mexican species (Fa, 1989).

Criteria of "endemicity" are also components of rarity of a fauna or flora (Usher, 1986). Rarity can be illustrated by the area each species covers. Species numbers plotted against the logarithm of the area (in square kilometers) show a generally steady increase in numbers of species from rare to common, though there is a relatively high number of uncommon species, up to 28% (Fig. 10.8A and B). A closer examination of the uncommon species reveals that the rarer taxa are the Mexican endemics, with those in the Trans-Mexican Neovolcanic Belt having the more restricted ranges (Fig. 10.8C).

Geographic Gradients

Species numbers correlate closely with variation in physical characteristics of the earth's surface, e.g., as latitude, elevation, aridity, salinity, water depth, and others (Brown, 1988). Numbers vary gradually and incrementally with differences in these physical variables. Many of these gradients are nonrandom patterns and represent important diagnostic trends of any country's fauna or flora. The gradients discussed are chosen as significant in that they describe biotic variation from northern to southerly latitudes, aridity, and elevation. Although the three chosen gradients are interrelated because they reflect a pronounced physical gradient in solar radiation, temperature, seasonality, and availability of some resources they serve as descriptors of Mexico's mammalian fauna.

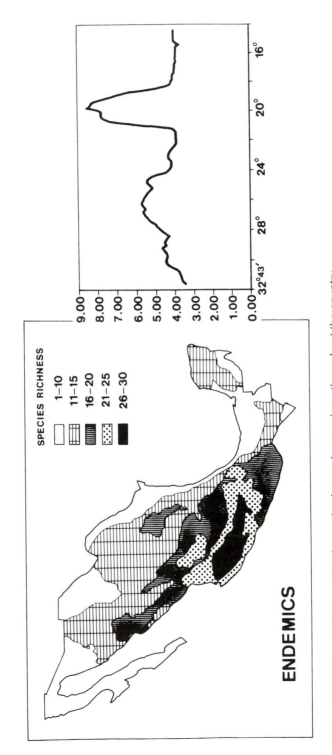

Figure 10.6. Choropleth map showing endemic species numbers throughout the country.

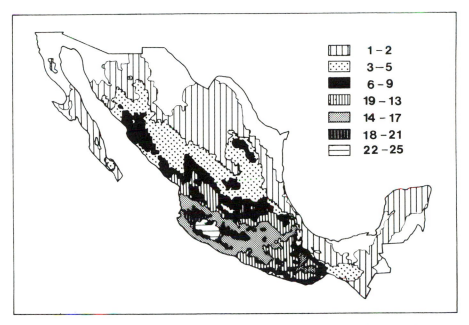

Figure 10.7. Choropleth map showing percentage of the country's total number of endemic species found throughout the country. This map represents a measure of representativeness of endemics within Mexico.

Latitudinal Gradients

The tendency for numbers of species to increase from high to low latitudes has been long perceived for terrestrial macroorganisms. For North American mammals, a number of studies from Simpson's (1964) early pioneering work reported increased changes in density of species toward the Equator. Methods for studying mammalian latitudinal gradients have differed from the use of quadrat methods to latitudinal bands (McCoy & Connor, 1980). Using Simpson's quadrats, Wilson (1974) found that quadrupedal mammals did not exhibit a latitudinal density gradient but, rather, a concentration at midlatitudes. According to Wilson, bats showed a nonlinear increment toward tropical latitudes that accounted for the overall pattern attributed to mammals in general. McCoy and Connor (1980) produced patterns different from those of Wilson (1974) for quadrupedal mammals but similar results for gradients of bat density. They (1980) concurred that observed latitudinal trends for overall mammalian density could be ascribed mostly to the increase in bat species density from the North American temperate zone to the Central American tropics.

Within the country, latitudinal gradients, expressed as the average number of species per quadrat, show no clear decline in numbers with higher latitude but point to a definite increase in species numbers around 20° latitude (Fig. 10.9). This peak at 20° appears in the Rodentia, Insectivora,

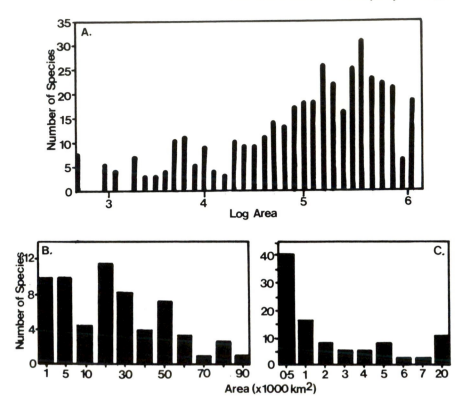

Figure 10.8. Species ranges of all Mexican mammals (A), Mexican endemics (B), and Trans-Mexican Neovolcanic Belt endemics (C).

and Chiroptera but not in the other orders and was similarly reported by McCoy and Connor (1980). The difference in the relation between species numbers of all mammals and degrees north below 38° N compared with the above is attributed by McCoy and Connor (1980) to the quadrupedal function. Above 38° N, numbers of species decrease regularly with increasing latitudes. Below this latitude, species numbers fluctuate, becoming reduced between 25° and 30° N and below 17° N. This observation is influenced by quadrupeds, which affect the all-mammals curve because bats generally increase in the number of species with decreasing latitude, although slight drops in diversity occur between 25° and 30° N and below 15° N. Further investigations for both North and South America by Willig and Selcer (1989) presented latitude, rather than longitude, biome richness, or area, as the best predictor of bat species density. This conclusion is explicable in terms of the eurytopic nature of bats, so that habitat diversity does not explain diversity. Organisms of lower vagility are nonetheless more restricted by habitat and hence exhibit no clear correlation with latitude.

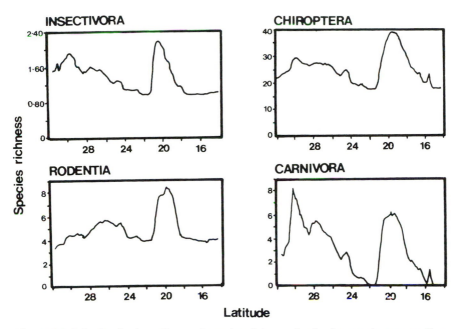

Figure 10.9. Latitudinal gradients of species richness for the four main mammalian orders in Mexico.

Elevational Gradients

Increasing elevation mirrors latitudinal gradients of species richness (Kikkawa & Williams, 1971; Terborgh, 1977; Yoda, 1967). However, even though altitudinal effects are pronounced in some cases according to Kendeigh (1974) their impact as a confounding factor in the delimitation of biotic provinces is minimal. The "life belts" concept of Dice (1943) was used to describe the effects of altitude on mammal distribution. This idea remains intuitively correct, as relations between organisms and climate are evolutionarily constrained: Taxa that attain their greatest species richness at temperate or arctic latitudes also reach their greatest diversity at intermediate to high elevations on tropical and temperate mountains.

Aridity Gradients

Although it is often difficult to isolate the separate effects of the antagonistic influences of aridity and elevation (primarily temperature), variation in moisture availability correlates with more or fewer species numbers. Species numbers of Mexican mammals show that there is correspondence between evapotranspiration levels and diversity, as the highest diversity occurs in the moist and dry tropical forests and in the tropical high land regions.

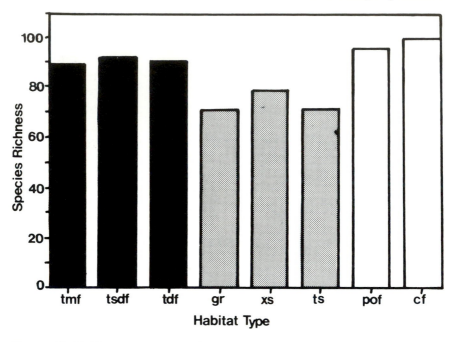

Figure 10.10. Mean average species numbers by habitat types. Key to habitat types. tmf, tropical moist forest; tsdf, tropical semi-deciduous forest; tdf, tropical deciduous forest; gr, grasslands; xs, xerophytic scrub; ts, thorn scrub; pof, pine-oak forest; cf, cloud forest. Modified from Fa (1989).

Provinciality

Geoecological Assemblages

Specific ecological assemblages of species tell of a connection between particular organisms and the habitat they occupy. The related idea of provinces is often defined on the basis of assemblages, but here assemblages are interpreted strictly ecologically. Although the concept is of great importance in paleogeography, where it is used to infer ancient ecological environments, it can be used in the present context to discuss the relation between mammal faunas and ecological entities. Ecological associations are often best identified in terms of vegetation. Distribution of many mammals is correlated with the variety and abundance of vegetation, which in turn, depend largely on physiographic and climatic factors. Different mammals vary in their responses to vegetation changes. For example, as Schmidly (1977) mentioned, carnivores and bats (except for the genus *Myotis*) are less affected by vegetation than other mammals in the Chihuahuan Desert. Other species, (e.g., rodents) are heavily influenced by vegetation type and cover, so that vegetation communities of low cover (e.g., desert vegetation) contain lower mammal biomass and lower species numbers. Estimates of mean species numbers of all vegetation types in the

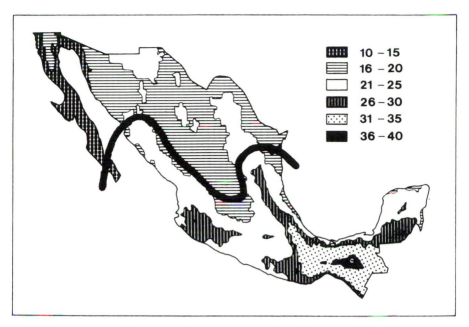

Figure 10.11. Distribution (percent) of total mammalian fauna in relation to the tropical and temperate areas of the country. The dividing line between tropics and northern temperate regions is shown by the dark solid line.

Trans-Mexican Neovolcanic Belt (Fa, 1989) have shown that arid habitats are less rich in species than are tropical and temperate environments (Fig. 10.10). Species assemblages within each habitat type vary somewhat in relation to the evolutionary associations between species and the environment. The distribution of 119 recorded mammals in the Chihuahuan Desert shows that 44% enter the succeeding vegetation type (desert scrub), but only 21% are restricted to the desert vegetation zone (Schmidly, 1977). A rapid faunal change occurs between montane forest and the desert, but there is little appreciable change between grassland and desert vegetation. For mammals inhabiting the montane plant communities, the intervening desert presents an impenetrable barrier.

Species richness in terms of the percent distribution of all species in the country increases below an imaginary line that divides the tropics from temperate regions (Fig. 10.11). From 36 to 40% of all species are found in the tropical regions, whereas up to 20% of the Mexican mammal fauna occur in the arid and semiarid areas.

Provinces

Areas can be divided into a hierarchy of regions reflecting patterns of faunal and floral similarities from realms or kingdoms, regions, subre-

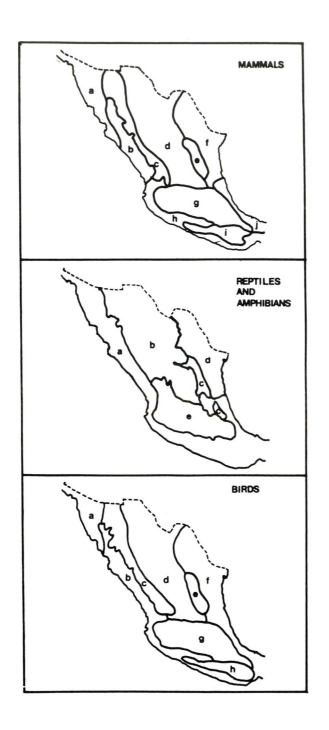

gions, provinces, and districts. Formal biogeographic provinces imply that the biota within each region is more homogeneous than between adjacent areas. The province concept has been based on the presence or absence data of taxa alone, either through identifying regions of endemism or statistically, as geographical assemblages. Provinces can be viewed empirically or as a response to particular processes. The latter can be either maintenance-based or evolutionary and longterm.

Fundamental areas of species assemblages can also be viewed as regions of endemisms whose characteristic taxa are confined to a particular region by the same biogeographic barriers. Such emphasis on endemism has led to historical and evolutionary explanations in which ideas about historical and barrier mechanisms become confused with explanations based on pure ecology (Rosen, 1988). Distinguishing geographical regions as distinct biotic provinces based on endemics was carried out for mammals by Goldman and Moore (1946). They distinguished 18 biotic provinces. The larger units, which they denominated provinces, are regional areas embracing natural general groupings. A major unit area may therefore be similar in topography throughout its extent and restricted to a single life zone, or it may be composed of both mountains and plains with arid and humid sections and including several life zones. Goldman and Moore (1946) argued that although the areas are not all of coordinate rank they could designate a "nearly balanced" concept of major centers of distribution in the country.

The provinces described for mammals are not the same for other groups of animals and plants, but the similarity existing between those detected for reptiles and amphibians, birds, and mammals (Fig. 10.12) points to the generality of ecological factors that limit the range of assemblages. Species assemblages largely follow the dividing lines of the main physiographic regions of the country. It is the geographic barriers that denote the variety of physiographic regions in the country influencing the biological attributes of groups of species and denoting complexes of unique forms. The effects of major climatic units (e.g., deserts) limit the assemblage of taxa that can survive within each, but more modest barriers and the influence of historical distributions operate at a smaller scale and may further separate assemblages even within climatically identical regions. A more statistical

---

Figure 10.12. Comparison of biotic provinces for mammals (Goldman & Moore, 1946), birds (Moore, 1945), and reptiles and amphibians (Morafka, 1977) for northern and central Mexico. Key to provinces (nomenclature from original sources): Mammals: a, Sonoran; b, Sinaloan; c, Sierra Madre Occidental; d, Chihuahuan desert; e, Sierra Madre Oriental; f, Tamaulipan; g, Transverse Volcanic; h, Coastal Pacific; i, Sierra Madre del Sur; j, Veracruzian. Reptiles and Amphibians: a, Veracruzian; b, Chihuahuan desert; c, Sierra Madre; d, Tamaulipan; e, Transvolcanic. Birds: a, Sonora; b, Sinaloa; c, Sierra Madre Occidental; d, Chihuahua desert; e, Sierra Madre Oriental; f, Tamaulipas; g, Transverse Volcanic; h, Sierra Madre del Sur.

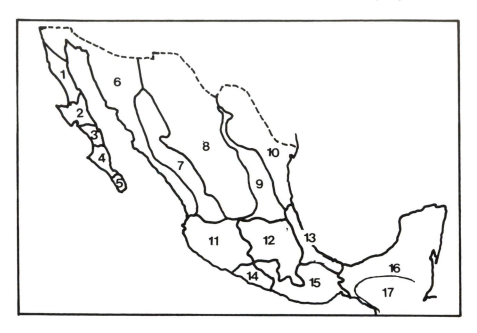

Figure 10.13. Mammalian primary areas as determined by maximal IFC regions: Key to areas: 1, San Pedro Mártir Mountains; 2, Sierra de la Bruja; 3, Sierra de Santa Lucia; 4, Sierra de la Giganta; 5, Sierra San Lazaro; 6, Sonoran Desert; 7, Sierra Madre Occidental; 8, Chihuahuan Desert; 9, Sierra Madre Oriental; 10, Tamaulipan Coastal Plain; 11, Western Pacific lowlands and Western Trans-Mexican Neovolcanic Belt; 12, Central Trans-Mexican Neovolcanic Belt; 13, Veracruzian Coastal Plain; 14, Southern Pacific Coastal Plain; 15, Sierra Madre del Sur; 16, SouthernTropical Regions; 17, Chiapan Highlands. *Sensu* Hagmeier & Stults, 1964).

detection of species assemblages of mammals employed here and following Hagmeier and Stults's (1964) procedures also demonstrates that the emergent pattern is essentially similar to those previously described (Fig. 10.13). Additionally, the IFC (Index of Faunistic Change) methodology further categorizes primary areas in the center of the country and in Baja California, denoting the influence of other factors. This is caused by the higher resolution of the methods employed herein (Fa, 1989).

INTERPRETATIONS OF SPECIES RICHNESS

Overall Patterns

All explanations of species richness can be divided into historical or equilibrial processes (Brown & Gibson, 1983). Historical arguments discuss the processes of past climatic changes in promoting speciation and equilibrial ones the role of such factors as productivity and habitat heterogeneity. Both types of reasoning must rely on the ultimate assumption that it is the nature of the physical environment that has promoted large species numbers in Mexico. Our treatment of hypotheses here steers away from mechanistic ones that deal with such factors as competition and predation to concentrate on arguments that have been used commonly to account for geographic patterns of species diversity. Thus they are at only one level of explanation. Here we are seeking to account for the possible primary effects of physical variables on species diversity.

Habitat Heterogeneity

For terrestrial mammals, habitat factors—spatial heterogeneity, habitat productivity, and environmental stability—can account for species numbers (Brown & Gibson, 1983). The habitat heterogeneity explanation has fallen out of favor (McGuinness, 1984) primarily because of the difficulty quantifying how the environment varies and relating it to the number of species. Yet the species-area relation still relies on an explanation based on secondary variates (Williamson, 1989) based on the nature of the habitats where species are found. According to Williamson (1989), the causes of species–area effects are simple. In small areas passive sampling effects are important, but in large areas the effect of habitat heterogeneity takes over.

In Mexico the tropical areas contribute the largest number of species. This greater diversity in the tropics is accounted for by the strong latitudinal gradient that exists from the poles to the equator. However, explanations of species diversity can be sought also in the nature of species. Stenotopy may be enhanced in the tropics by virtue of environmental constancy, which in turn, increases the effectiveness of barriers—thus producing more species (Emlen, 1973).

Refugia

The vicariant nature of the Mexican flora and fauna and its relation to Pleistocene climatic changes are well recognized (Martin & Hurrell, 1957; Ramamoorthy & Lorence, 1987). The biogeographical consequences of allopatric speciation (Mayr, 1942) result from once-continuous species distributions becoming fragmented into disjunct populations. Temperature and rainfall fluctuations during the Pleistocene caused enlargement or reduction of vegetation types in Mexico (Ohngemach, 1973; Toledo, 1982) within a sequence of cold-wet and warm-dry climates during the last 40,000 years (Emiliani et al. 1975; Lorenzo, 1969; Heine, 1973; Toledo, 1982; White, 1962).

Paleoenvironmental reconstruction in New Mexico, Texas (Galloway, 1970; Harris, 1988; Messing, 1986; Smartt, 1977), central Mexico (Heine, 1973; Ohngemach, 1973) and the Mexican tropics (Sarukhán, 1968, 1977;

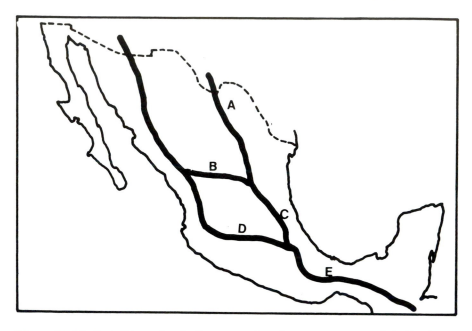

Figure 10.14. Probable routes for the expansion of montane mammals in Mexico. A. Northern route along eastern escarpment of the Rocky Mountains in New Mexico and Texas. B. Trans-plateau route through mountains across the Mesa del Norte from Durango; C. Southern route along the Sierra Madre Oriental; D. Transverse route across the Trans-Mexican Neovolcanic Belt. E. Southern route into the Sierra Madre del Sur and Chiapan Highlands.

Toledo, 1982) indicate that pine-oak forests and other temperate vegetation dominated the Sierras region, the Trans-Mexican Neovolcanic Belt, and all highlands to the south, with subtropical vegetation restricted to the coastal areas. These areas would have formed corridors for temperate species expansion along the Sierra Madres and the Trans-Mexican Neovolcanic Belt during maximal extension during the Quaternary glacials (Fig. 10.14) (Schmidly, 1977). Examples exist of a number of biotic elements that have extended their range southward during this time by using Pleistocene corridors across northern Mexico and southern Texas (Dressler, 1954; Lundelius, 1967; Sharp, 1953). Similarity of insect faunas (Halffter, 1976) and mosses (Delgadillo, 1987) indicates that the Trans-Mexican Neovolcanic Belt highlands and the Sierras were originally connected (see McDonald, this volume). For montane mammals, late Pleistocene climatic changes and concomitant changes in the environment have had a direct effect upon speciation and patterns of distribution of living species. Evidence that many northern species extended their ranges southward into Mexico during the Wisconsin pluvial cycle is provided by current distribution of relict species (Russell, 1969). For example, *Glaucomys volans* and *Microtus* spp. may have invaded MiddleAmerica recently along woodland corridors (Axelrod, 1975; Braun, 1988; Martin & Hurrell, 1957), leaving behind relict populations of these species in the highlands of Mexico and Central America (Handley, 1971) as aridity increased. Other examples, e.g., *Sorex milleri*, a living monotypic species endemic to higher elevations in the Sierra del Carmen of Coahuila and Sierra Madre Oriental, suggest the isolation and speciation of a *Sorex cinereus* after expansion southward. Findley (1955) suggested that the southern segment of *S. cinereus*, which ranged southward during the Wisconsin pluvial stage, became isolated during post-Wisconsin time and gave rise to *S. milleri*. Two other species, *Scalopus montanus* and *Cynomys mexicanus*, have ranges restricted to northeastern Mexico, and both species are geographic isolates. Like these species, *Reithrodontomys chrysopsis* (Hall, 1981) and *Peromys cus* (Neotomodon) *alstoni* (Williams *et al.*, 1985), both strongly associated with pine-fir-"zacaton" habitat, occur disjunct in pockets of favorable habitat found at higher elevations of the Trans-Mexican Neovolcanic Belt. *Neotomodon* is found in Late Pleistocene deposits in the Valley of Mexico (Ferrusquía-Villafranca, 1977) at lower altitudes than at present, pointing to a bridge between Trans-Mexican Neovolcanic Belt volcanoes for pine-fir-"zacaton" species.

For many desert species, the cooling climate decreased available habitat and fragmented the species into isolated populations in desert refugia.

Barriers

Although ecological processes are likely to have been responsible for maintaining or increasing high local diversity, they still depend on a pool of available species (Vermeij, 1978). This point directly linked to the porosity of barriers in regard to allowing or disallowing the mix of species.

Barriers may be abiotic (geological or climatic) or biotic (ecological) in origin. Physical barriers may vary from tectonic, eustatic, climatic, or oceanographic (TECO barriers) (Rosen, 1988) to small scale topographic. The effectiveness of a particular kind of barrier for a given taxon depends on both the physical and biotic environment provided by the barrier and the biological characteristics of the organism. Animals that inhabit temporary or highly fluctuating environments are also likely to have resistant life-history stages that enable them to survive unfavorable conditions and serve as effective dispersal agents. Consequently, organisms from temporary or fluctuating environments are better dispersors and are less limited by barriers than species from more permanent or constant environments. By studying frontier lines between mammal subspecies, Rapaport (1982) estimated that 52% of the cases are segregated by large rivers and mountains and less so by vegetational changes. This figure is clearly a rough approximation of only heuristic value. However, Rapaport's (1982) figures showing lines of demarcation or frontiers between species of North Ameri-

can mammals of the same genus with contiguous areas indicate an abundance of barriers of various types (physical or biological) in Mexico. The
country's orogenic complexity and climatic variety present a mosaic of
conditions clearly demarcated by areas of separation.

CONCLUSIONS

This chapter has concentrated on a global view of the patterns of mammalian species diversity in Mexico. In an American context Mexico has the
largest mammalian fauna, whereas in a world scenario it has possibly the
highest number, rivaled only by Zaire (Eisenberg, 1981). This diversity is
composed largely of rodents and bats, but a considerable number of orders
are represented in the country.

Geographic variation in species richness is a complex phenomenon that
shows repeated trends with general explanations (Schall & Pianka, 1978).
Patterns of diversity of Mexican mammals have been described briefly by
presenting how species numbers vary in space and by analyzing gradients
in relation to latitude, elevation, and aridity. These gradients vary relatively continuously with geographic variation in physical environmental
variables, (e.g., temperature, solar radiation, soil water potential) but
reflect simultaneous variation in several of these physical factors. We have
concentrated on these gradients because they are characteristic of a large
number of organisms and represent nonrandom patterns. Simple regression techniques are sufficient to demonstrate highly significant relations
between number of species and the physical environmental variables
(Brown, 1988). However, patterns of spatial richness cannot be fully
understood if the dynamics of past events are not taken into consideration.
The creation of refugia by the contraction and expansion of some habitats
during past glaciations as well as the interchange of taxa between North
and South America during geological times have shaped the extant fauna.

Explanations for the high mammalian species richness in Mexico have
received a broad treatment herein. Three main hypotheses have been
evaluated related to habitat heterogeneity and processes that promote
speciation by isolating mechanisms (refugia and barriers). Complete
mechanistic explanations must obviously await further study, but the
evaluation of these hypotheses in the present context confirms that
such processes correlate with diversity on biogeographic spatial scales. In
so doing, it steers away from the circular argument that the high diversity
of forms in Mexico is due to the complexity of environments. What is
clearly apparent is that Mexican mammalian fauna is abundant in endemics,
a substantial number of which are microendemics of small ranges: more
than one-fourth of the fauna. The contribution of the tropical regions in
supporting large numbers of species is recognized, although it is expressed
in terms of total numbers rather than unique species. This evaluation
coincides with Toledo's (1982) appraisal of diversity in the Mexican tropics,
where the humid tropical forests are considered low in general. This

tendency is also observed for amphibians and reptiles (see Flores-Villela, this volume).

The analyses of provinciality initiated here show that there is a correlation between natural regions of particular climate, vegetation, and topography and species assemblages. Other published distributional data for Mexican reptiles and amphibians, birds, and mammals coincide in terms of their general boundaries, with the major units identified in this chapter. No comparisons of the defined primary areas are made, as this subject is treated elsewhere (Fa, in prep.).

Although the material herein is rather dogmatic in its restriction to a limited number of hypotheses to describe mammalian richness in Mexico, it is appreciated that a better understanding of ecological and evolutionary processes will make important inroads in Mexican biogeography. The data base used here is adequate as it is without regard to artificial political boundaries and the methodology employed is more objective than that used in previous studies. Finer resolution of what is now a large, diffuse subject is essential. There is no doubt that more effort must be placed on understanding geological history, past and present physical environments, phylogenetic history, the dynamics of species origination and extinction processes and the ecological relations of species with both their physical environment and other organisms (Brown, 1988). Such a systems approach offers the best chance of saving the greatest number of taxa by focusing on species-rich areas as the most efficient and cost-effective way to retain maximal biological diversity in the minimal area (Fa, 1989; Scott et al., 1987).

ACKNOWLEDGMENTS

We are most grateful to the countless number of people who now remain anonymous who have contributed to the distribution data which form the database of this work. We thank Alberto Vega for drawing figures 5-7 and Ms. M. Williamson for editorial help.

REFERENCES

Auerbach, M. & A. Shmida. 1987. Spatial scale and determinants of plant species richness. Trends Ecol. Evol. 2:238–242.
Axelrod, D.J. 1975. Evolution and biogeography of Madrean-Tethyan sclerophyll vegetation. Ann. Missouri Bot. Gard. 62:280–334.
Baker, H.R. 1967. Distribution of recent mammals along the Pacific coast of the Western Hemisphere. Syst. Zool. 16:28–37.
Bond, W.J. 1983. An alpha diversity and the richness of the Cape Flora: a study in southern Cape fynbos. In F.J. Kruger, D.T. Mitchell & J.U.M. Jarvis [eds.], Mediterranean-type Ecosystems: The Role of Nutrients. Berlin: Springer. pp. 225–243.

Braun, J.K. 1988. Systematics and biogeography of the southern flying squirrel, Glaucomys volans. J. Mammalogy 69:422–426.

Brown, J.H. 1988. Species diversity. *In* A.A. Myers & P.S. Giller (eds.), Analytical Biogeography: An Integrated Approach to the Study of Animal and Plant Distributions. London: Chapman & Hall. pp. 57–90.

Brown, J.H. & A.C. Gibson. 1983. Biogeography. St. Louis, MO: Mosby.

Bykov, B.A. 1979. On a quantitative estimate of endemism. Botan. Mater. Gerb. Inst. Botan. Akad. Nauka Kazakhskoi SSR 11:3–8.

Cain, S.A. 1974. Foundations of Plant Geography. New York: Hafner Press.

Cody, M.L. 1975. Towards a theory of continental diversities: bird distribution over Mediterranean habitat gradients. *In* M.L. Cody & J. Diamond (eds.), Ecology and Evolution of Communities. Cambridge, MA: Harvard Univ. Press. pp. 214–257.

Cody, M.L. 1983. Continental diversity patterns and convergent evolution of bird communities. *In* F.J. Krugeer, D.T. Mitchell & J.U.M. Jarvis (eds.), Mediterranean-type Ecosystems—The Role of Nutrients. Berlin: Springer. pp. 357–402.

Connor, E.F. & E.D. McCoy. 1979. The statistics and biology of the species-area relationship. Amer. Naturalist 113:791–833.

Cowling, R.W., G.E. Gibbs Russell, M.T. Hoffman & C. Hilton-Taylor. 1989. Patterns of plant species diversity in southern Africa. *In* B.J. Huntley (ed.), Biotic Diversity in Southern Africa: Concepts and Conservation. Cape Town: Oxford Univ. Press. pp. 19–50.

Cuarón, A.D., I.J. March & P.M. Rockstroh. 1989. A second armadillo (*Cabassous centralis*) for the faunas of Guatemala and Mexico. J. Mammalogy 70:870–871.

Darlington, P.J. 1957. Zoogeography: The Geographical Distribution of Animals. New York: John Wiley & Sons.

Delgadillo, C. 1987. Moss distribution and the phytogeographical significance of the Neovolcanic Belt of Mexico. J. Biogeo. 14:69–78.

Dice, L.R. 1943. The Biotic Provinces of North America. Ann Arbor, MI: Univ. of Michigan.

Dressler, R.L. 1954. Some floristic relationships between Mexico and the United States. Rhodora 56:81–96.

Eisenberg, J. 1981. The Mammalian Radiations. Chicago, IL: Chicago Univ. Press.

Eldredge, N. 1981. Discussion of M.D.F. Udvardy's paper: The riddle of dispersal: dispersal theories and how they affect biogeography. *In* G. Nelson & D.E. Rosen (eds.), Vicariance Biogeography: A Critique. New York: Columbia Univ. Press. pp. 34–38.

Emiliani, C., S. Gartner, B. Lidz, K. Eldridge, D.K. Elvey, T.C. Huang, J. J. Stipp & M.F. Swanson. 1975. Paleoclimatological analysis of late quaternary cores from the northeastern Gulf of Mexico. Science 189:1083–1086.

Emlen, J.M. 1973. Ecology—An Evolutionary Approach. Reading, MA: Addison-Wesley Publishing.

Fa, J.E. 1989. Conservation-motivated analysis of mammalian biogeography in the Trans-Mexican Neovolcanic Belt. Nat. Geog. Res. 5:296–315.

——— & L.M. Morales. (1991). Mammals and protected areas in the Trans-Mexican Neovolcanic Belt. *In* M.A. Mares & D. Schmidly (eds.), Topics in Latin American Mammalian Biology: Ecology, Conservation and Education. Norman, OK: Oklahoma Univ. Press. pp. 196–227.

Favager, C. & J. Contandriopoulos. 1961. Essai sur l'endemisme. Ber. Schweiz. Bot. Ges. 71:384–408.

Ferrusquía-Villafranca, I. 1977. Distribution of Cenozoic vertebrate faunas in Middle America and problems of migration between North and South America. *In* I. Ferrusquía-Villafranca (ed.), Conexiones Terrestres entre Norte y SudAmerica: Simposio Interdisciplinario sobre Paleogeografía Mesoamericana. Mexico City: UNAM. Inst. Geol. pp. 193–321.

Findley, J.S. 1955. Taxonomy and distribution of some American shrews. Univ. Kansas Publ. Mus. Nat. Hist. 7:613–618.

Galloway, R. W. 1970. The full glacial climate in the southwestern United States. Ann. Assoc. Amer. Geographers 60:245–256.

Goldman, E.A. & R.T. Moore. 1946. Biotic provinces of Mexico. J. Mammalogy 26:347–360.

Hagmeier, E.M. & C.D. Stults. 1964. A numerical analysis of the distributional patterns of North American mammals. Syst. Zool.13:125–155.

Halffter, G. 1976. Distribución de los insectos en la zona de transición mexicana: relaciones con la entomofauna de Norteamérica. Folia Entomol. Mex. 35:1–64.

Hall, E.R. 1981. The Mammals of North America. 2nd. ed. New York: John Wiley & Sons.

Handley, C.O. 1971. Appalachain mammalian geography—recent epoch. *In* P. C. Holt, R.A. Patterson & J.P. Hibbard (eds.), The Distributional History of the Biota of the Southern Appalachians. Part III. Blacksburg, VA: Research Division Monographs, Virginia Polytechnic Univ. pp. 1–306.

Harris, A.H. 1988. Late Pleistocene and Holocene *Microtus* (*Pitymys*) (Rodentia: Cricetidae) in New Mexico. J. Vert. Paleontol. 8:307–313.

Heine, K. 1973. Variaciones más importantes del clima durante los ultimos 40,000 años en México. Comunicaciones 7:51–56.

Hershkovitz, P. 1972. The recent mammals of the Neotropical region: A zoogeographic and ecologic review. *In* A. Keast (ed.), Evolution, Mammals and Southern Continents. Albany, NY: State Univ. of New York Press. pp. 311–431.

Kendeigh, S.C. 1974. Ecology with Special Reference to Animals and Man. Englewood Cliffs, NJ: Prentice-Hall.

Kikkawa, J. & E.E. Williams. 1971. Altitudinal distribution of land birds in New Guinea. Search 2:64–69.

Lorenzo, J.L. 1969. Condiciones periglaciares de las altas montañas de México. Paleoecologia 4:1–25.

Lundelius, E.L. 1967. Late-Pleistocene and Holocene faunal history of central Texas. *In* P.S. Martin & H.E. Wright, Jr. (eds.), Pleistocene Extinctions. New Haven, CT: Yale Univ. Press. pp. 287–319.

MacArthur, R.H. 1972. Geographical Ecology; Patterns in the Distribution of Species. New York: Harper & Row.

——— & E.O. Wilson. 1967. The Theory of Island Biogeography. Princeton, NJ: Princeton Univ. Press.

Major, J. 1988. Endemism: a botanical perspective. *In* A.A. Myers & P.S. Giller (eds.), Analytical Biogeography: An Integrated Approach to the Study of Animal and Plant Distributions. London: Chapman & Hall. pp. 117–148.

Martin, P.C. & B.E. Hurrell. 1957. The pleistocene history of temperate biotas in Mexico and Eastern United States. Ecology 38:486–490.

Mayr, E. 1942. Systematics and the Origin of Species. New York: Columbia Univ. Press.

McCoy, E.D. & E.F. Connor. 1980. Latitudinal gradients in the species diversity of North American mammals. Evolution 34:193–203.

McGuinness, K.A. 1984. Equations and explanations in the study of species-area curves. Biol. Rev. 59:423–440.

Messing, H.J. 1986. A late Pleistocene-Holocene fauna from Chihuahua, Mexico. Southw. Naturalist. 31:277–288.

Moore, P.T. 1945. The Transverse Volcanic Biotic Province of central Mexico and its relationship to adjacent provinces. Trans. San Diego Soc. Nat. Hist. 10:217–236.

Morafka, D.J. 1977. Is there a Chihuahuan Desert? A quantitative evaluation through a herpetofaunal perspective. *In* R.H. Wauer & D.H. Riskind (eds.), Transactions of the Symposium on the Biological Resources of the Chihuahuan Desert Region, United States and Mexico. U.S. Department of the Interior, National Park Service and Proceedings Series, No. 3. pp. 437–454.

Ohngemach, D. 1973. Analisis polínico de los sedimentos del Pleistoceno reciente y del Holoceno en la region Puebla-Tlaxcala. Comunicaciones 7:47–49.

Patterson, B. & R. Pascual. 1972. The fossil mammal fauna of South America. *In* A. Keast (ed.), Evolution, Mammals and Southern Continents. Albany, NY: Univ. of New York Press. pp. 247–309.

Preston, F.W. 1962. The canonical distribution of commoness and rarity: Part I. Ecology 43:185–215.

Ramamoorthy, T.P. & D.H. Lorence. 1987. Species vicariance in Mexican flora and a description of a new species of *Salvia*. Bull. Mus. Hist. Nat. (Paris). Adansonia 2:167–175.

Ramirez-Pulido, J. & M.C. Britton. 1981. An historical synthesis of Mexican mammalian taxonomy. Proc. Biol. Soc. Wash. 94:1–17.

——— & C. Müdespacher. 1987. Estado actual y perspectivas del conocimiento de los mamíferos de México. Ciencia (Mexico) 38:49–67.

———, R. López-Wilchis, C. Müdespacher, & I.E. Lira. 1988. Los Mamíferos de México: Lista y Bibliografía de los Mamíferos de México. Mexico City: UAM.

Rapaport, E.H. 1982. Areography: Geographical Strategies of Species. Oxford: Pergamon Press.

Ricklefs, R. 1987. Community diversity: relative roles of local and regional processes. Science 235:167–171.

Rosen, B.R. 1988. Biogeographic patterns: a perceptual overview. *In* A.A. Myers & P.S. Giller (eds.), Analytical Biogeography: An Integrated Approach to the Study of Animal and Plant Distributions. London: Chapman & Hall. pp. 23–55.

Russell, R.J. 1969. Revision of pocket gophers of the genus *Pappogeomys*. Univ. Kansas Mus. Nat. Hist. 51:337–371.

Rzedowski, J. 1978. La Vegetación de México. Mexico City: Limusa.

Sarukhán, J. 1968. Estudio Sinecológico de las Selvas de *Terminalia amazonica* en la Planicie Costera del Golfo de México. M.S. thesis, Colegio de Posgraduados, Chapingo, México.

———. 1977. Algunas consideraciones sobre los paleoclímas que afectaron los ecosistemas de la planicie costera del Golfo. *In* Reunión sobre Fluctuaciones Climáticas. (ed. CONACyT). Mexico City: CONACyT. pp. 197–209.

Schall, J.J. & E.R. Pianka. 1978. Geographical trends in numbers of species. Science 201:679–686.

Schmidly, D. 1977. Factors governing the distribution of mammals in the Chihuahuan Desert region. *In* R.H. Wauer & D.H. Riskind (eds.), Transactions of the Symposium on the Biological Resources of the Chihuahuan Desert Region, United States and Mexico, U.S. Department of the Interior, National Park Service and Proceedings Series, No. 3. pp. 163–192.

Scott, J.M., B. Csuti, J.D. Jacobi, & J.E. Estes. 1987. Species richness: A geographic approach to protecting future biological diversity. Bioscience 37:782–788.

Sharp, A.J. 1953. Notes on the flora of Mexico: world distribution of the woody dicotyledonous families and the origin of the modern vegetation. J. Ecology 41:374–380.

Shmida, A. & M.V. Wilson. 1985. Biological determinants of species diversity. J. Biogeo. 12:1–20.

Simpson, G.G. 1964. Species density of North American recent mammals. Syst. Zool. 13:57–73.

Smartt, R.A. 1977. The ecology of late Pleistocene and recent *Microtus* from south-central and southwestern New Mexico. Southw. Naturalist 22:1–19.

Stebbins, G.L. & J. Major. 1965. Endemism and speciation in the California flora. Ecol. Monographs 35:1–35.

Terborgh, J. 1977. Bird species diversity on an Andean elevational gradient. Proc. Int. Ornithol. Congress 17:1005–1012.

Toledo, V.M. 1982. Pleistocene changes of vegetation in tropical Mexico. *In* G. Prance (ed.), Biological Diversification in the Tropics. New York: Columbia Univ. Press. pp. 93–111.

————. 1989. La diversidad biológica de México. Ciencia y Desarrollo 81:17–30.

Usher, M.B. 1986. Wildlife conservation evaluation: attributes, criteriaand values. *In* M.B. Usher (ed.), Wildlife Conservation Evaluation. London: Chapman & Hall. pp. 3–44.

Vermeij, G.J. 1978. Biogeography and Adaptation; Patterns of Marine Life. Cambridge, MA: Harvard Univ. Press.

Whittaker, R.H. 1972. Evolution and measurement of species diversity. Taxon 21:213–251.

————. 1977. Evolution of species diversity in land communities. Evol. Biol. 10:1–67.

White, S.E. 1962. Late Pleistocene glacial sequence for the west side of Iztaccihuatl, Mexico. Bull. Geol. Soc. Amer. 73:935–958.

Williams, C.B. 1964. Patterns in the Balance of Nature and Related Problems in Quantitative Ecology. New York: Academic Press.

Williams, S.L., J. Ramirez-Pulido & R.J. Baker. 1985. *Peromyscus alstoni*. Mammalian Species 242:1–4.

Williamson, M.A. 1989. Relationship of species number to area, distance and other variables. *In* A.A. Myers & P.S. Giller (eds.), Analytical Biogeography: An Integrated Approach to the Study of Animal and Plant Distributions. London: Chapman & Hall. pp. 91–116.

Willig, M.R. & K.W. Selcer. 1989. Bat species density gradients in the New World: A statistical assessment. J. Biogeo. 16:189–195.

Wilson, J.W. 1974. Analytical zoogeography of North American mammals. Evolution 28:124–140.

Wilson, M.V. & A. Shmida. 1984. Measuring beta diversity with presence-absence data. J. Ecology 72:1055–1064.

Yoda, K. 1967. A preliminary survey of the forest vegetation of eastern Nepal. II. General description, structure and floristic composition of sample plots chosen from different vegetation zones. J. Coll. Arts Sci. Chiba Univ. (Natural Sci. Ser.) 5:99–140.

III
SELECTED FLORISTIC GROUPS OF MEXICO

11

Diversity in the Mexican Bryoflora

CLAUDIO DELGADILLO M.

Most work on Mexican bryophytes has been performed by foreign scientists who have emphasized taxonomy, floristics, and phytogeography. The estimated bryoflora of Mexico, consisting of about 1,200 mosses and 800 liverworts and hornworts, is apparently as rich as in other tropical areas of the world but richer than in temperate areas of larger size. This situation is perhaps due to a large number of habitats and substrates in different types of vegetation, diverse topographic and climatic gradients, and long histories of migration and dispersal from and into Mexico. The degree of specific endemism is unknown, but a preliminary figure suggests 15% for mosses, which is regarded as low for the size and the environmental diversity of the country; there is only a handful of endemic moss genera. In addition to migration from other continental areas, bryophyte diversity is due to speciation processes in many sites where selective pressures are numerous, e.g., the alpine zone. However, continued floristic interchange has ensured that many of the newly differentiated taxa have migrated to other continental areas. Long-distance dispersal has had little effect on the bryophyte diversity of Mexico. To conserve our bryophytes for scientific and pragmatic use we must preserve their habitats. Their presence adds stability to our ecosystem.

Plants are part of our history, religion, and everyday life. Mexicans are well known for their interest in plant domestication, use, and scientific study, as is evident in archeological remains, colonial documents, and present-day scientific literature. However, contrary to expectations, bryophytes have been the subject of little concern among the general public or among past or contemporary generations of Mexican botanists.

Reference to Mexican bryophytes goes back to the mid-sixteenth century. De la Cruz (1552) cited and roughly illustrated two mosses that were supposed to be useful, in mixture, for the treatment of headaches and throat fever. Except for the citation of *Aitonia* (*Plagiochasma*) by Hernández (1651), Mexican bryophytes were apparently ignored until the mid-nineteenth century when several important works on mosses and liverworts were published, including studies by Müller (1848–51) and Gottsche (1863).

Liverworts, including hornworts, have received little attention since Gottsche's publication; however, several collectors obtained and distributed specimens that were described and reported upon in various floristic

and taxonomic treatments, the most important being that by Fulford (1963–
76). In contrast, interest in mosses flourished thanks to foreign scientists
who received and studied numerous collections from Mexico. Toward the
end of the nineteenth century this activity produced the first comprehensive study of Mexican mosses (Bescherelle, 1872), which listed about 375
species and varieties; by the early 1950s there was a second listing that
included about 836 species and varieties of mosses (Crum, 1951).

Current bryological work abroad is concerned with the preparation of
a Manual of Mexican Mosses, having been coordinated by A.J. Sharp since
the early 1970s and recently joined by H.A. Crum. This project has enlisted
the cooperation of about 26 bryologists throughout the world (Sharp, 1977),
who have published, in addition to their contributions to the Manual,
numerous works dealing with the immediate taxonomic and nomenclatural problems of Mexican mosses.

In Mexico, the first bryological group was established in 1973. Claudio
Delgadillo joined the staff of the Instituto de Biología of the Universidad
Nacional Autónoma de México (UNAM, the National University of Mexico)
and developed a project on the phytogeographical significance of the
Trans-Mexican Volcanic Belt (also known as the Neovolcanic Belt). Because of the nature of the project—floristic and phytogeographic—this
group has contributed distributional data, new records, range extensions,
and a better understanding of phytogeographical patterns in Mexico.

Differential activity in the study of Mexican bryophytes has made mosses
a well-known plant group relative to liverworts and hornworts. To date
there are only about 578 published reports on Mexican bryophytes; most are
concerned with the taxonomy, floristics, and distribution of mosses, but their
geographic and taxonomic coverages are generally limited. Furthermore,
even in the case of mosses, few areas may be considered well collected, and
many have hardly been visited by bryologists. The former category includes
the Yucatan Peninsula, the valley of Mexico, and the states of Veracruz,
Tamaulipas, and Oaxaca; among the latter are Aguascalientes, Guanjuato,
Querétaro, and Tabasco. It is clear, then, that although publication of the
Manual of Mexican Mosses will be an important landmark, our knowledge
of Mexican bryophytes will remain inadequate.

Considering the short history of Mexican bryology, this chapter offers
preliminary analysis on the nature of bryophyte diversity in Mexico.
Diversity is used here in the sense of species richness, i.e, the number of
species per unit area. Because no floristic catalog is yet available, when
necessary, estimated figures based upon bibliographic data and results of
this writer's research are offered, with emphasis on mosses.

HABITAT AND SPECIES DIVERSITY

Bryophytes are conspicuous in certain vegetation and substrate types. They
are frequent in the tropical lowland forests and comparatively abundant in
the alpine zone of our highest mountains. Between these two extremes they

are also common in *Quercus* and *Abies* forests, in the deciduous forests of eastern Mexico, and in other types of vegetation. Bryophytes are known from desert areas to aquatic environments, but in these extreme situations they have neither numerous taxa nor large biomass, except in a few protected habitats. Mosses dominate most bryophytic habitats, but in certain areas and microhabitats liverworts take over; hornworts are fairly common at intermediate and high elevations along water courses, but they are seldom the main feature of bryophytic vegetation.

Substrates within each type of vegetation are not uniformly occupied with bryophytes. In well-preserved tropical rain forests of southern Mexico, bryophytes are not common on soil or rocks, and they may even be scarce in exposed sites where they are outcompeted by herbaceous angiosperms. However, their number is comparatively high on trunks, buttresses, and branches of shrubs and trees; a few are found on leaves, woody vines, or tree-fern trunks. Pristine examples of such habitats were readily available in the valley of Uxpanapa (Delgadillo, 1976) before it was heavily disturbed by man.

In other areas, such as the Yucatan Peninsula, the soil and rocks are perhaps equally important as substrates for bryophytes. Although this moss flora is considered depauperate (Delgadillo et al., 1982), its interesting feature is that many of the rock and soil inhabitants may truly be regarded as calciphilous. Certain species may live on a humus layer overlying the rock, but such others as *Barbula agraria* Hedw., *Luisierella barbula* (Schwaegr.) Steere, and *Neohyophila sprengelii* (Schwaegr.) Crum live directly on the rock.

The alpine area of Mexico is particularly rich in bryophytes (Delgadillo, 1987a). The main substrates here are rocks and volcanic ash; the former shelters species of *Grimmia* and *Andreaea*, whereas species of *Leptodontium*, *Aongstroemia*, and *Pohlia* are frequent on the latter. The *Pinus* and *Abies* forests immediately below the alpine area share many of these taxa, but the trees themselves are apparently poor substrates for bryophytes: The only conspicuous bryophyte on *Pinus hartwegii* is *Leptodontium viticulosoides* (P. Beauv.) Wijk & Marg.

The bryophyte flora of the deciduous forests of eastern Mexico is incompletely known; but among mosses there are about 202 species and varieties recorded, and most are epiphytic (Delgadillo, 1979). In addition to a complex phytogeographical history and a large number of tree species, moss diversity in the deciduous forests is due to their altitudinal location in rainshadow areas where fog prevails during the cooler months of the year.

Substrate and habitat diversity are not limiting factors to bryophyte diversity in Mexico as may be inferred from the examples cited above. In fact, Mexico is already recognized as the site of a rich flora (Delgadillo, 1987b; Rzedowski, 1962) which includes an estimated 1,200 moss species and 800 liverworts (Delgadillo, 1982; Rzedowski, 1978). A large number of species are not unexpected for a country located in the area of contact between holarctic and neotropical floristic realms, with diverse geological and vegetational histories, a rough topography, and a variable climate. The

Table 11.1. Moss diversity in certain tropical and temperate areas of the world

Area	Mosses	Surface (km²)
Southern Africa[1]	591	1,659,000
West Africa[2]	964	ca. 4,280,000
Southeastern Africa[3]	1,227	ca. 5,000,000
Polynesia[4]	1,427	—
North America (U.S. and Canada)[5]	1,170[A]	11,526,622
USSR[6]	706[B]	13,445,000
Bolivia[7]	1,222	1,098,581
Colombia[8]	750	1,138,000
Ecuador[9]	781	270,670
Venezuela[10]	626	912,050
Mexico[11]	**1,200**	**1,969,269**

[1]Magill & Schelpe (1979); [2]Schultze-Motel (1975); [3]Kis (1985); [4]Miller et al. (1978); [5]Crum et al. (1973); [6]Anderson (1974a); [7]Hermann (1976); [8]Florschütz-de Waard & Florschütz (1979); [9]Steere (1948); [10]Pursell(1973); [11]Estimate, this report.
[A]Not including infraspecific categories.
[B]Acrocarpus mosses only.

question is whether the estimated size of the bryophyte flora is reliable and acceptable in view of the size of the country.

Table 11.1 lists several geographical areas for which the number of moss species is known. Strikingly, the moss flora of Canada and the continental United States is roughly equivalent in size to that of Mexico, but their combined surface is more than five times as large. The available records for the Soviet Union include the acrocarpus mosses only; but with the addition of the pleurocarpic species, the total number of taxa may not be substantially different from that of North America.

In terms of the number of species, the moss flora of Mexico is equivalent to that of southeastern Africa (as defined by Kis, 1985) and larger than that of western Africa (Schultze-Motel, 1975), but those areas are more than twice as large. It should be noted, however, that the bryological exploration of Africa is still in progress, and these preliminary figures may undergo drastic changes in future years. The same condition applies to most Latin American countries. These figures show that even though the catalog of Mexican mosses is incomplete, the number of species is at least similar to that of other tropical areas and greater than that of temperate regions with larger areas.

PHYTOGEOGRAPHY AND DIVERSITY

Bryophytes exhibit the same patterns of distribution as vascular plants; in an evolutionary sense, bryophyte species are long-lived and thus can disperse in time to distant places with little or no morphological differentiation. It is not to say, as is commonly believed, that they are evolutionary deadends (Anderson, 1974b; Wyatt, 1985; Wyatt & Anderson, 1984) but that the high degree of genetic variability observed in various taxa is not necessarily associated with morphology (Szweykowski, 1984).

Because of their evolutionary longevity, for many species the events of dispersal have resulted in broad ranges and patterns of distribution peculiar to a given area. In Mexico the patterns exhibited by the bryophyte flora relate it to western and eastern North America, central and northern South America, and to a lesser extent to eastern Asia, Europe, and tropical Africa. The Mexican bryoflora contains, in addition, a large number of endemics. The proportion of these elements varies in different parts of Mexico, but it is of phytogeographical interest that they are all generally represented in any given locality.

The following paragraphs outline the relative importance of the geographical elements in various parts of Mexico. For this purpose, floristic data derived from the author's own research on mosses are used.

The flora of the Peninsula of Baja California includes more than 118 moss species. The northern part of the peninsula is floristically related to western North America and in particular to the flora of California and the western United States. Part of the flora is also shared with areas of northwestern Mexico, but there are many species with a broad distribution in Mexico or the world. The southern part of the Baja California Peninsula has a stronger floristic affinity to mainland Mexico; and although many of its moss species are also found in the United States, they are not as widely distributed in the West as those of northern Baja California (Bowers et al., 1976).

A phytogeographical study of the state of Zacatecas (Delgadillo & Cárdenas, 1987) showed that its 115 species and varieties of mosses can be grouped into four elements. The element of wide distribution is known from several continents and includes the cosmopolitan species as well as different intercontinental disjunctions; the so-called Mexican element is distributed from the southwestern United States to Guatemala, with only three species distributed elsewhere in El Salvador, Costa Rica, Panama, Jamaica, and the Virgin Islands. The Mesoamerican element is distributed from Mexico to northern South America, and the endemic element includes five species restricted to Mexico. The relations of the Zacatecas moss flora to the Caribbean flora is virtually nil, as there are only three taxa whose distribution extends to that area. Similar distributions have been observed in the Tehuacan Valley in Puebla (Delgadillo & Zander, 1984).

As one progresses southward toward more mesic areas, the proportion of the moss flora with neotropical affinities increases. In the Yucatan Peninsula the Caribbean element that links the peninsular moss flora to that of the West Indies forms more than 60% of the known species; the element of wide distribution is still important, and endemic taxa are nearly absent from the peninsular moss flora (Delgadillo, 1984). The deciduous forests of eastern Mexico, on the other hand, also maintain strong ties with the Caribbean area, but perhaps because of a complex vegetational history and a wide latitudinal distribution there are other important elements in their moss flora. The Mesoamerican element, with a large number of species in Central America, and the element of wide distribution are well represented in the deciduous forests; this moss flora shows, in addition, strong ties with eastern North America and to a lesser extent with eastern Asia and Africa. The relations with eastern North America were identified by Sharp (1939) and Crum (1951) and those with eastern Asia were noted by Sharp (1966) and pointed out for the entire country by Sharp and Iwatsuki (1965). The relations with the African moss flora are not well understood owing partly to incomplete information on the moss floras of Mexico and Africa and partly to taxonomic and floristic data that have modified specific concepts and geographical ranges that had been accepted in the deciduous forests publication (Delgadillo, 1979). It is true, however, that the Mexican-African connection exists in the deciduous forest moss flora and in other areas. This fact was recognized by Crum (1956) and has been substantiated by the publication of lists of neotropical bryophytes that are also found in Mexico and Africa (Buck & Griffin, 1984; Gradstein et al., 1983; Reese, 1985).

The alpine moss flora has a more balanced proportion of elements showing different affinities; the Mesoamerican and widely distributed elements are the most important in this flora, but those distributed northward or southward are also significant in terms of number of species. The endemic element is the largest thus far observed among the areas studied, but most taxa are not restricted to the alpine area of the Trans-Mexican Volcanic Belt (Delgadillo, 1987a).

There are few regional studies on the tropical rain forests of Mexico. Because most of this floristic information is scattered as incidental reports, it is impractical for the purpose of this report to produce a species list for southern Mexico. The following statements are drawn from information in Pursell and Reese (1970) and Delgadillo (1976).

The mosses of the tropical rain forest are characteristic of the tropical American lowlands. Many of these species are widely distributed in the West Indies, Central America, and northern South America; their ranges may extend northward along the coastal plain to the southern United States, or they may enter Florida via the West Indies. In addition to this pattern there are species distributed in other continents, but their relative importance cannot be predicted with our limited data; it is of interest, however, that the endemic element seems to be nearly absent among the mosses from the tropical areas of Mexico.

We may conclude from the above sample cases that, despite a paucity of knowledge about a large part of the country, the richness of moss species in Mexico is partly due to continued contact with other floras and migration from other areas.

ENDEMISM

It has already been stated that with the sole exception of the alpine moss flora of Central Mexico, the examples cited above contain few or no endemic species. This point should not be misleading, as it does not apply to the entire country.

A preliminary survey of the endemic species and varieties of mosses indicates that there are about 180 in Mexico, comprising 15% of the estimated moss flora. This figure is at best a rough estimate because it includes recently described taxa that, in time must be tested against previously described species in other areas. It is not surprising to discover that there are morphologically equivalent plants in Mexico and South America that are known under different names in each area. For example, *Pleuridium kieneri* Bartr. and *Astomiopsis amblycalyx* C. Müll. were considered separate entities until Snider (1987) concluded that they are conspecific.

In addition to newly described species, the endemism list contains the names of taxa that have not undergone critical taxonomic and nomenclatural evaluation. For these reasons the 15% figure is used only as a reference for further discussion. Considering the size and environmental diversity in Mexico, moss endemism is low; but in comparison with other areas it is not totally out of proportion. Bartram (1949) concluded that the endemism in the Guatemalan moss flora was about 11% of the total and, at the same time, conceded that when the adjacent regions were more thoroughly explored many of the endemics would show wider distributions. Lawton (1971) stated that endemism in the Pacific Northwest area of North America (Washington, Oregon, Idaho, western Montana, Wyoming, British Columbia, and Alberta) was about 4–5% and Schofield (1980) pointed out that it reaches about 23% in North America north of Mexico. In the West Indies, Crosby (1969) detected a 30% endemism, which seems rather high for the area; however, Robinson (1986) observed that many species once regarded as West Indian are now known more widely and that "the evidence of endemism in the geologically young Lesser Antilles seems particularly weak at this time." Figures for other areas were cited by Schofield (1985).

Generic endemisms are mostly restricted to small taxa or recent segregates that have not been detected outside Mexico. Among mosses, *Acritodon* (Robinson, 1964), *Curviramea* (Crum, 1984), *Hymenolomopsis*, and *Bryomangiana* (Thériot, 1931) may be cited; the last genus belongs to this category only if one accepts it as taxonomically distinct from *Astomiopsis*.

The reason for this relative poverty in endemism is perhaps twofold. One possible explanation is related to the continued floristic interchange that seems to have taken place from Mexico to various continental areas and to the Caribbean islands and (vice versa); recently differentiated taxa were soon dispersed to adjacent areas. Indirect evidence in favor of this explanation is the floristic relation that the country maintains at present. A second hypothesis is concerned with the relative youth of the floristically better-known areas in Mexico such as those along the Trans-Mexican Volcanic Belt; the time elapsed since these areas have become available for plant occupation has not been sufficient to allow for the differentiation of genera among mosses. Older underexplored areas and younger areas with high selective pressures are the sites where additional specific endemics may be discovered, but it is doubtful if the former will yield numerous endemics of higher rank as they have had time to disperse or migrate to other areas.

SOURCES OF BRYOPHYTE SPECIES DIVERSITY

New taxa in a flora may be introduced by dispersal from other areas or may result from various processes of speciation *in situ*. Bryophyte dispersal may take place by means of spores or by other asexual propagules whether by stepwise migration or by events of long-distance dispersal. Aspects of these processes have been discussed and summarized by Crum (1972), Delgadillo and Pérez-Bandín (1982), Edwards (1980), Equihua (1987), Mogensen (1983) and van Zanten and Pócs (1981). The reader is referred to these references for background information.

The patterns of distribution mentioned in a preceding section suggest that most taxa have reached Mexico by stepwise migration from several sources. It is generally recognized, for instance, that the Central American bridge has served as a pathway of migration of tropical plants to and from South America at various times and that the mountains have favored the floristic displacement of high-altitude taxa from South America. Similar floristic movements seem to have occurred along the western mountains during the Pleistocene, resulting in the introduction of numerous temperate bryophytes into Mexico, e.g., the moss genus *Grimmia*.

It has been stated that the mountains of eastern Mexico have various important phytogeographical elements which certainly reflect the latitudinal displacements of the moss flora that have occurred since Tertiary times in this area. The present mixture of tropical and temperate elements in the deciduous forests is indirect evidence that migration has taken place from eastern North America, the Caribbean islands, and Central America. In other areas, such as the drylands of northern Mexico, the movement of floras has not been dramatic, but because they have been open and in close connection with similar areas of the southwestern United States it is assumed that the floristic interchange has also been continuous. The similarity of the moss floras of the valleys of Zacatecas with those of Texas (Magill, 1976) and New Mexico (Mahler, 1978) attest to this connection.

Long-range disjunctions often show that the bryophytes are able to colonize areas from distant sites. Although long-range dispersal is regarded as an infrequent process, it is also accepted as a likely explanation for the presence of bryophytes in oceanic islands and for certain continental disjunctions. In Mexico the bryophyte flora of Guadalupe Island is apparently derived from the western United States, Baja California, and the Mexican mainland; several disjunctive taxa in Mexico and South America probably arrived by long-range dispersal, as no intermediate stations have thus far been detected. A conspicuous example of this disjunction is *Zygodon pichinchensis* (Tayl.) Mitt., a large moss with spiny leaves that seldom forms sporophytes.

Stepwise migration and certain events of long-range dispersal may be considered the major forces promoting the bryological diversity in Mexico. Intercontinental disjunctions, however, are not necessarily explained by long range dispersal; and though spectacular, they seem to have had little effect on the number of taxa in the Mexican bryoflora.

Diversity may also be increased by speciation processes in a given area. Polyploidy as well other forms of genetic variability are not always expressed and recognized as separate taxonomic entities. However, in some cases polyploids may show distinct characteristics that merit taxonomic recognition. Hybrid (moss) populations, on the other hand, although sometimes treated as nomenclaturally distinct, have complete sterility barriers at the generic level (Anderson, 1972) that prevent their permanence in nature. The role of polyploidy and hybridization as major sources of variation and of new taxa is unknown, but it is possible that most taxonomic differentiation is due to the added effect of comparatively simple mutations; these would accumulate rapidly in highly selective environments, e.g., alpine areas, giving rise to ecotypes and new species. Similar behavior may not be expected among lowland bryophytes where continuous interchange with other areas may swamp new mutations. These hypotheses are still to be tested in Mexico.

CONSERVATION

Like all other organisms, including man, bryophytes should be conserved not for romantic reasons. Bryophytes play a role in ecological succession, substrate transformation, seed germination, establishment of insect populations, and as a link in the trophic chains of the ecosystem in which stability of the parts is a prerequisite to sustain the integrity of the whole. Individuals interested in economic plants are referred to Ando (1982) and Ando and Matsuo (1984) for accounts on the use of the bryophytes and a bibliography on the subject. For our purposes, suffice it to say that, among other uses, bryophytes are important in horticulture, as indicators of pollution and in the production of biologically active compounds (antitumor or cytotoxic). However, their scientific relevance in ecological and cytological research alone plainly justifies their conservation and study.

From my observations, bryophytes are abundant in numbers of species or in biomass in various areas of Mexico. Along the Trans-Mexican Volcanic Belt, beside the alpine areas, bryophytes exhibit exuberant growth along the canyons that climb down the mountains and on wooded slopes of major volcanoes where mesophyllous forests contain mixtures of temperate and tropical taxa. Such areas are likely candidates for preservation. Although most of the high volcanoes are already protected as national parks, in practice they are under continuous disturbance by large numbers of visitors, cattle, and man-made fires; furthermore, the "protected" limits seldom include canyons and gorges that once were covered by forests and now are heavily cut for timber or affected by hydraulic construction. This type of devastation is evident in the mountains surrounding Mexico City, especially on the northwestern and eastern flanks of Iztaccíhuatl and on Ajusco. The lower slopes of Nevado de Colima in Jalisco and the area of Milcumbres in Michoacan are under pressure from lumber dealers, tree diseases, and human populations.

In southern Mexico the state of Tabasco still harbors a rich bryoflora in areas bordering Chiapas, but the entire state has remained underexplored except for a few collections that do not represent its wealth. In the neighboring Yucatan Peninsula most areas have sustained great damage from centuries of occupation by man. In southern Quintana Roo, the municipal limits of Othón P. Blanco, where *Pinus caribaea* Morelet grows (Chavelas, 1981), should be declared a protected area similar to the Sian Ka'an reserve in eastern Quintana Roo. From the standpoint of bryophytes, both are of floristic and phytogeographical interest.

The lower elevations in southwestern Chiapas were originally covered by tropical rain forests; numerous collections by A.J. Sharp from Mapastepec and El Triunfo attest to the once rich bryophyte flora. If we were to preserve portions of the Lacandona region, we would maintain a representation of the tropical bryoflora that has gone unstudied in Mexico.

The drylands of northern Mexico are apparently depauperate in terms of bryophytes and may not be regarded as important for conservation. However, certain ecological adaptations linked to the economy of water and reproduction make desert bryophytes unique biological objects. For instance, desert mosses absorb water from dew or humid night air for their brief period of photosynthesis following dawn (Richardson, 1981). Such mosses as *Tortula* spp., *Pleurochaete squarrosa* (Brid.) Lindb., and *Desmatodon convolutus* (Brid.) Grout are frequent inhabitants of Mexican deserts; they have been shown to possess efficient capillary systems made up of leaf cell papillae (Proctor, 1979), which allow plants to moisten rapidly. For reproduction, bryophytes may have to shorten their life cycle and produce sporophytes toward the end of the rainy season. *Phascum hyalinotrichum* Card. & Thér., *Pleuridium* spp., and *Pterygoneurum subsessile* (Brid.) Jur. seem to follow this pattern in Zacatecas, with optimum sporophyte yield per gametophyte. This strategy is worth preserving for further study in Zacatecas and elsewhere in Mexican drylands where there are open spaces occupied by analogous examples.

The list of critical examples is endless, but the point seems clear. In order to preserve bryophytes we must preserve their habitats, i.e., the green cover of higher plants, because when both groups are gone the luxury of their sight and their myriad potential uses will also be gone.

ACKNOWLEDGMENTS

The author is indebted to Beranardina Bello and A. Cárdenas for comments and logistic support; Drs. T.P. Ramamoorthy and Robert Bye (both of the

National University of Mexico), A.J. Sharp (University of Tennessee), and W.R. Buck (New York Botanical Garden), who read and criticized an earlier version of the chapter.

REFERENCES

Anderson, L.E. 1972. Cytological studies of natural intergeneric hybrids and their parental species in the moss genera, *Astomum* and *Weissia*. Ann. Missouri Bot. Gard. 59:382–416.

———. 1974a. Bryology 1947–1972 Ann. Missouri Bot. Gard. 61:56–85.

———. 1974b. Taxonomy and evolution of bryophytes: introduction. J. Hattori Bot. Lab. 38:1–11.

Ando, H. 1982. Bryophytes as useful plants. Bryol. Times 14:1.

——— & A. Matsuo. 1984. Applied bryology. Adv. Bryol. 2:133–229.

Bartram, E.B. 1949. Mosses of Guatemala. Fieldiana, Bot. 25:1–442.

Bescherelle, E. 1872. Prodromus bryologiae mexicanae ou enumeration des mousses du Mexique avec description des especes nouvelles. Mem. Soc. Nat. Sci. Nat. Cherbourg 16:144–256.

Bowers, F.D., C. Delgadillo M. & A.J. Sharp. 1976. The mosses of Baja California. J. Hattori Bot. Lab. 40:397–410.

Buck, W.R. & D. Griffin III. 1984. *Trachyphyllum*, a moss genus new to South America with notes on African-South American bryogeography. J. Nat. Hist. 18:63–69.

Chavelas, P.J. 1981. El *Pinus caribaea* Morelet, en el estado de Quintana Roo, Mexico. Inst. Nal. Invest. For. Nota Técnica 10:1–8.

Crosby, M.R. 1969. Distribution patterns of West Indian mosses. Ann. Missouri Bot. Gard. 56:409–416.

Crum, H.A. 1951. The Appalachian-Ozarkian element in the moss flora of Mexico with a check-list of all known Mexican mosses. Ph.D. thesis, Univ. of Michigan, Ann Arbor.

———. 1956. Notes on *Hypnodon*, a genus of Orthotrichaceae new to North America. Bryologist 59:26–34.

———. 1972. The geographic origins of the mosses of North America's eastern deciduous forest. J. Hattori Bot. Lab. 35:269–298.

———. 1984. Notes on tropical American mosses. Bryologist 87:203–216.

———, W.C. Steere & L.E. Anderson. 1973. A new list of mosses of North America north of Mexico. Bryologist 76:85–130.

de la Cruz, M. 1552. Libellus de medicinalibus indorum herbis. [Latin translation by Juan Badiano.] Mexico City: IMSS 1964.

Delgadillo M., C. 1976. Estudio botánico y ecológico de la región del Río Uxpanapa, Veracruz. No. 3. Los Musgos. Publ. Inst. Invest. Rec. Bióticos 1(2):19–28.

———. 1979. Mosses and phytogeography of the *Liquidambar* forest of Mexico. Bryologist 82:432–449.

———. 1982. Current knowledge of the Mexican moss flora and its temperate element. Beih. Nova Hedwigia 71:455–459.

———. 1984. Mosses of the Yucatan Peninsula, Mexico. III. Phytogeography. Bryologist 87:12–16.

———. 1987a. Moss distribution and the phytogeographical significance of the Neovolcanic Belt of Mexico. J. Biogeo. 14:69–78.

———. 1987b. The Meso-American element in the moss flora of Mexico. Lindbergia 12:121–124.

———— & A. Cárdenas S. 1987. Musgos de Zacatecas, México III. Síntesis y fitogeografía. Bol. Soc. Bot. Méx. 47:13–24.

————, ————, & A.J. Sharp. 1982. Mosses of the Yucatan Peninsula, Mexico. I. Bryologist 85:253–257.

———— & E. Pérez-Bandín. 1982. Spore liberation in mosses. I. Problems and perspectives of wind tunnel experiments. Cryptogamie, Bryol. Lichénol. 3:39–49.

———— & R. H. Zander. 1984. The mosses of the Tehuacan valley, Mexico, and notes on their distribution. Bryologist 87:319–322.

Edwards, S.R. 1980. Spore discharge in *Calymperes*. J. Bryo. 11:95–97.

Equihua, C. 1987. Diseminación de yemas en *Marchantia polymorpha* L. (Hepaticae). Cryptogamie, Bryol. Lichénol. 8:199–217.

Florschütz-de Waard, J. & P.A. Florschütz. 1979. Estudios sobre criptógamas colombianas III. Lista comentada de los musgos de Colombia. Bryologist 82:215–259.

Fulford, M. 1963–76. Manual of the leafy hepaticae of Latin America. I–IV. Mem. New York. Bot. Gard. 11:1–535.

Gottsche, C.M. 1863. Die mexicanske Levermosser. Efter Prof. Liebannus Samling. Kgl. Dansk. Vid. Selsk. Skrift. 6:97–380.

Gradstein, S.R., T. Pócs, & J. Vana. 1983. Disjunct Hepaticae in tropical America and Africa. Acta Bot. Hung. 29:127–171.

Hermann, F.J. 1976. Recopilación de los musgos de Bolivia. Bryologist 79:125–171.

Hernández, F. 1651. Rerum Medicarum Novae Hispaniae Thesaurus seu Plantarum Animum Mineralium Mexicanorum. Rome: Typographeio Vitalis Muscardi.

Kis, G. 1985. Mosses of southeast tropical Africa. Inst. Ecol. Bot. Hungarian Acad. Sci. Vácrátót.

Lawton, E. 1971. Moss Flora of the Pacific Northwest. Hattori Bot. Lab. Nichinan.

Magill, R.E. 1976. Mosses of Big Bend National Park, Texas. Bryologist 79:269–295.

———— & E.A. Schelpe. 1979. The bryophytes of southern Africa. An annotated checklist. Mem. Bot. Surv. S. Africa 43:1–39.

Mahler, W.F. 1978. Preliminary checklist of mosses of New Mexico. Bryologist 81:593–599.

Miller, H.A., H.O. Whittier & B.A. Whittier. 1978. Prodromus florae muscorum Polynasiae with a key to genera. Bryophyt. Bibl. 16:1–334.

Mogensen, G.S. 1983. The Spore. *In* R.M. Schuster (ed.), New Manual of Bryology. Vol. 1. Hattori Bot. Lab. Nichinan. pp. 325–342.

Müller, C. 1848–1851. Synopsis muscorum frondosorum omnium hucusque cognitorum. Vols. 1 & 2. Berlin: Sumptibus Alb. Foerstner.

Proctor, M.C.F. 1979. Structure and ecophysiological adaptation in bryophytes. *In* G.C.S. Clark & J.G. Duckett (eds.), Bryophyte Systematics. London: Academic Press. pp. 479–509.

Pursell, R.A. 1973. Un censo de los musgos de Venezuela. Bryologist 76:473–500.

———— & W.D. Reese. 1970. Phytogeographic affinities of the mosses of the gulf coastal plain of the United States and Mexico. J. Hattori Bot. Lab. 33:115–152.

Reese, W.D. 1987. Tropical lowland mosses disjunct between Africa and the Americas, including *Calyptothecium planifrons* (Ren. & Par.) Argent, new to the western hemisphere. Acta. Amaz. (suppl.) 15:115–121.

Richardson, D.H.S. 1981. The biology of mosses. Oxford: Blackwood Sci. Publ.

Robinson, H. 1964. New taxa and new records of bryophytes from Mexico and Central America. Bryologist 67:446–458.

————. 1986. Notes on the bryogeography of Venezuela. Bryologist 89:8–12.

Rzedowski, J. 1962. Contribuciones a la fitogeografía florística e histórica de México. I. Algunas consideraciones acerca del elemento endémico en la flora mexicana. Bol. Soc. Bot. Méx. 27:52–65.

————. 1978. Vegetación de México. Mexico City: Limusa.

Schofield, W.B. 1980. Phytogeography of the mosses of North America (north of Mexico). *In* R.J. Taylor & A.E. Leviton (eds.), The Mosses of North America. Pacific Division, AAAS. pp. 131–170.

————. 1985. Introduction to Bryology. New York: Macmillan.

Schultze-Motel, W. 1975. Katalog der Laubmoose von west Afrika. Willdenowia 7:473–535.

Sharp, A.J. 1939. Taxonomic and ecological studies of eastern Tennessee bryophytes. Amer. Midl. Naturalist 21:267–354.

————. 1966. Some aspects of Mexican phytogeography. Ciencia (Mexico) 24:229–232.

————. 1977. The preparation of a manual of Mexican mosses. Bull. Bryol. X. Taxon 26:151–153.

———— & Z. Iwatsuki. 1965. A preliminary statement concerning mosses common to Japan and Mexico. Ann. Missouri Bot. Gard. 52:452–456.

Snider, J.A. 1987. A revision of the moss genus *Astomiopsis* (Ditrichaceae). Bryologist 90:309–320.

Steere, W.C. 1948. Contribution to the bryogeography of Ecuador. I. A review of the species of Musci previously reported. Bryologist 51:65–167.

Szweykowski, J. 1984. What do we know about the evolutionary process in bryophytes? J. Hattori Bot. Lab. 55:209–218.

Thériot, I. 1931. Quelques nouveautes bryologiques pour la Mexique. Rec. Trav. Crypt. L. Mangin 7–10.

Van Zanten, B.O. & T. Pócs. 1981. Distribution and dispersal of bryophytes. Adv. Bryol. 1:479–562.

Wyatt, R. 1985. Species concepts in bryophytes: input from population biology. Bryologist 88:182–189.

———— & L.E. Anderson. 1964. Breeding systems in bryophytes. *In* A.F. Dyer & J.G. Duckett (eds.), The experimental biology of bryophytes. London: Academic Press. pp. 39–64.

12

Mexican Pteridophytes: Distribution and Endemism

RAMON RIBA

Mexican Pteridoflora is composed of about 110 genera and between 1,000 and 1,100 species. The affinities of this flora to those of tropical America are strong. The tropical south is richer in terms of species numbers than the arid north but poorer in endemics. Diversity is highest among tropical cloud forest pteridophytes. Distribution of pteridophytes in various vegetation types is briefly reviewed. A provisional list of pteridophytes endemic to Mexico is presented.

It is estimated that over 110 genera and about 1,000–1,100 species of pteridophytes occur in Mexico (Mickel & Beitel, 1988; Riba, pers. obs.) which is about 5% of the vascular plant flora of Mexico, estimated at 21,600 species (Rzedowski, this volume). Most species are found in the tropical latitudes. The cloud forests are the richest in number of species followed by lowland tropical forests, pine-oak forests, tropical deciduous forests, and xeric and aquatic vegetation. Endemism is high in the Sierra Madre del Sur Morphotectonic Province as defined by Ferrusquía-Villafranca (this volume).

Notable contributions of the nineteenth century to our knowledge of Mexican pteridoflora are those by Martens and Galeotti (1842), Liebmann (1848), and Fée (1857). The work of Rovirosa (1909), which followed these contributions, provided fairly accurate notes on the distribution of ferns in Chiapas, Tabasco, and Veracruz. A gap of nearly 30 years separates this work from that of Conzatti (1939) who recognized 18 families, 88 genera, and 631 species. Later floristic contributions of a more regional scope were those by Matuda (1956a, b) and Knobloch and Correll (1962). The former listed 24 genera and 88 species for the state of Mexico, and Knobloch and Correll recognized 31 genera and 126 species. Several revisionary and taxonomic studies that have appeared during the last 30 years have included taxa of Mexican pteridophytes. A review of about 280 publications (Riba & Butanda, 1987) included eight references each to the states of Chiapas and Oaxaca and 18 to the state of Veracruz. Mexican pterido-

Table 12.1. Genera and species of pteridophytes in selected countries in the Americas

Country	Genera	Species
United States and Canada[1]	61	406
Mexico[2]	108	900+
Greater Antilles[3]	98	900
Lesser Antilles[4]	68	323
Belize[5]	47	132
Central America[6,7,8]	102	900+
Colombia[9]	105	1,150
Venezuela[10]	109	1,059
Surinam[11]	63	298

[1]Lellinger (1985a); [2]This report; [3]Tryon & Tryon (1982); [4]Proctor (1977); [5]Standley & Record (1936); [6]Goméz-Pignataro (1976); [7]Stolze (1976, 1981, 1983); [8]Lellinger (1985a); [9]Murillo (personal communication); [10]Smith (1981); [11]Kramer (1978).

phytes have also been the subject of active study (Mickel & Beitel, 1988; Smith, 1981). As early as 1946 Conzatti observed that knowledge of Mexico's flora was incomplete. It is still true for Mexico's pteridoflora, which is not well documented for several parts of the country, e.g., the northeast, the Pacific coast, and the Central region.

This chapter presents generic and species numbers of pteridophytes in some of the relevant temperate and tropical American floras (Table 12.1) and in some regions and states of Mexico (Table 12.2); percentages of species that Mexico shares with the adjacent United States, Guatemala, and the Caribbean islands of Jamaica and Lesser Antilles are also presented (Table 12.3).

The pteridophytes are estimated at 9,000 species, about 50% (4,500), of which are in Asia (Malaysia and the Pacific islands). In the Americas about 3,250 species are suspected, of which about 3,000 are concentrated in tropical America (Tryon & Tryon, 1982). This concentration of species in the tropical latitudes is clearly illustrated by Panama, with 687 species (Lellinger, 1985b), and Costa Rica, with 900 (Lellinger, 1985b). Of the 127 genera in the Americas, 95 (75%) are represented in Mexico. The country shares 45 (48%) genera with the United States and Canada and 77 (81%) with the Antilles and Central and South America.

Mexican pteridophytes are most closely related to those of tropical Central America, a natural consequence of the physiographic and climatic continuity. The pteridoflora of Chiapas and Guatemala, territories considered by Rzedowski (1978) to constitute one phytogeographic unit, is similar. This similarity may extend further south to Nicaragua. A major center of endemism for ferns in Central America south of Nicaragua is Costa Rica. The incidence of endemism (in terms of numbers) among

Table 12.2. Genera and species of pteridophytes in various regions and States of Mexico

State or region	Genera	Species
Baja California (Pen.)[1]	17	64
Sonoran Desert[2]	14	54
Chihuahua[3]	31	126
Nuevo León[4]	29	99
Nueva Galicia[5]	70	250
Veracruz[6]	90	508
Mexico[7]	30	88
Guerrero[8]	86	295
Oaxaca[9]	101	690
Tabasco[10]	32	104
Yucatan[11]	14	31
Quintana Roo[12]	13	20
Chiapas[13]	109	650

[1]Wiggins (1980); [2]Shreve & Wiggins (1964); [3]Knobloch & Correl (1962); [4]Aguirre & Arreguin (1988); [5]Mickel (personal communication); [6]Riba (this report); [7]Matuda (1956b); [8]Lorea (personal communication); [9]Mickel & Beitel (1988); [10]Cowan (1983); [11]Standley (1930); [12]Sousa & Cabrera (1983); [13]Smith (1981); Breedlove (1986); Riba et al. (1987).

Mexican pteridophytes parallels those of other groups in the country and is high in the Sierra Madre del Sur Morphotectonic Province (see Daniel, Hunt, and Flores-Villela, among others in this volume). However, endemism (in terms of percentages) is higher in the arid north. A preliminary estimate suggests that the number of endemic pteridophytes in Mexico will exceed 190 (see Appendix).

DISTRIBUTION AND DIVERSITY OF PTERIDOPHYTES

The neotropical affinity of the Mexican Pteridoflora is obvious. Many Central and South American taxa have their northern limits in Mexico, some even reaching central Mexico. The number of species in the country is about the same as in Costa Rica and 25% higher than in Guatemala or Panama (Lellinger, 1985b; Stolze, 1976, 1981, 1983). A comparison of regional floristic treatments (Table 12.2; Fig. 12.1) illustrates the diversity (species numbers) of pteridophytes in Mexico. The flora of Baja California (Wiggins, 1980) lists 64 species, of which four are endemic. The flora of the Sonoran Desert including Baja California and parts of Sonora as well as southern California and Arizona (Shreve & Wiggins, 1964) recorded 54 species of pteridophytes, 14 of which (24%) are restricted to that area. Of the 126 species Knobloch and Correll (1962) recognized for Chihuahua, seven are restricted to that state; but the number of endemic species rises

Table 12.3. Numbers of species Mexico shares with some countries in the Americas

Country or region	Species (No.)[a]	Shared species (No.)	%
U.S. and Canada[1]	406	127	14
Jamaica[2]	579	239	24
Lesser Antilles[3]	323	142	14
Guatemala[4]	690	535	54

[1]Lellinger (1985a); [2]Proctor (1985); [3]Proctor (1977); [4]Stolze (1976, 1981, 1983).
[a]Estimated number of species in Mexico is approximately 1,000.

to 39 (37%) if adjacent areas are included. Fourteen of the 650 species in Chiapas (Breedlove, 1986; Riba et al., 1987; Smith, 1981), are endemic. However, the state shares 31 species with Belize and Guatemala, which would make the number of species that are quasiendemic to Chiapas 45 (7% of the total). Oaxaca is probably the richest state in fern species with 690 species (Mickel & Beitel, 1988). Twenty-eight species are endemic to the state, which shares 42 species with adjacent areas. In the Yucatan Peninsula, where there are no endemics, diversity is low especially in the northern flatlands. Sousa and Cabrera (1983), for example, cited only 20 species for Quintana Roo. Standley (1930) recorded 31 species for Yucatan, but the number for the Peninsula are higher as number of taxa is higher in the southern portion. The tropical perennial forest vegetation may harbor additional species, and their number may reach 100 or more.

A number of vegetation types has been recognized in Mexico (Rzedowski, 1978[1]). Pteridophytes (including arborescent ferns) are not dominant elements of any vegetation type. Terrestrial ferns grow in disturbed areas as part of secondary growth, and numerous species with wide distributions occur in the fringes of several vegetation types. A brief account of the pteridophytes in these communities is provided below.

Tropical Evergreen Forest

Among the common pteridophytes are *Tectaria heracleifolia*, *T. incisa*, *T. transiens*, *Adiantum tetraphyllum*, and *A. princeps*. *Asplenium resiliens*, *A. monanthes*, and *A. myriophyllum* grow on fallen trees or on their bases where litter is abundant. Hemiepiphytes such as *Polybotrya polybotryoides*, *Olfersia cervina*, and *Bolbitis bernoullii* are infrequent.

Among epiphytes in the south and southeast, there are numerous species of Hymenophyllaceae, notably those of *Trichomanes*, at low to middle elevations (e.g., *T. capillaceum*, *T. collariatum*, *T. diversifrons*, and *T. krausii*). At higher altitudes species of *Hymenophyllum*, (*H. elegantulum*, *H.*

[1]See this contribution for definitions and distributions of vegetation types in Mexico.

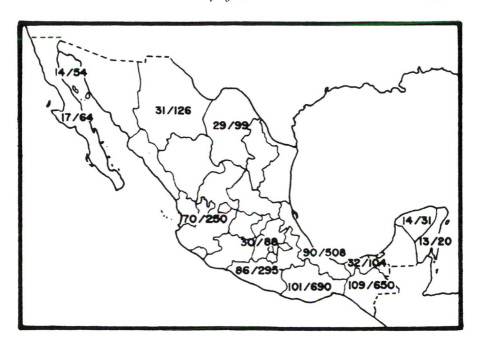

Figure 12.1. Mexican pteridophytes: numbers of genera and species by regions.

trapezoidale, and *H. polyanthos*) are seen. The last, incidentally, has been reported from near sea level.

In disturbed areas the understory trees bear epiphytes such as *Pleopeltis macrocarpa*, *Phlebodium decumanum*, and several species of *Pecluma* (*Polypodium pectinatum-plumula* species complex). These species, whose rhizomes stay dormant during the dry season and grow turgid when moisture becomes available, have adapted to low humidity levels in these areas.

The common species in the Los Tuxtlas area of Veracruz are *Danaea nodosa*, *Adiantum trapeziforme*, *A. tetraphyllum*, *Asplenium auritum*, *A. serratum*, *A. laetum*, *Diplazium lonchophyllum*, and *Didymochlaena truncatula* (in the Moreno peninsula). Several species of *Thelypteris* (*T. torresiana*, *T. griesbreghtii*, and *T. rhachiflexuosa*) are found here. *Schizaea elegans* is occasional in flooded places. In open shady stretches, *Selaginella delicatissima*, *S. martensii*, *S. mollis*, and *S. schizobasis* are common, and in exposed sites *Lycopodium cernuum* is seen. *Cyathea schiedeana*, *C. bicrenata* and *C. princeps* grow along roadsides and disturbed areas.

Tropical Deciduous Forest

Generally, pteridophytes are scarce, and they tend to be xerophytic owing to low humidity levels. The characteristic species are *Bommeria pedata*, *Doryopteris pedata* var. *palmata*, *Cheilanthes myriophylla*, *Ch. microphylla* var.

fimbriata, Ch. *kaulfussii*, Ch. *angustifolia*, and *Cheiloplecton rigidum*. *Tectaria mexicana* is found in relatively undisturbed areas.

Tropical Cloud Forest

Pteridophytes are most abundant and diverse in the vegetation of tropical cloud forests owing to a favorable combination of temperature, moisture, and fragmented topography. They may not seem obvious except for the arborescent ferns in some places. Lira and Riba (1984) reported, among the large life forms, *Cyathea salvinii*, *Nephelea tryoniana* (= *Cyathea tryoniana*, Riba in edit., the only Mexican locality), *C. princeps*, *C. mexicana*, *C. microdonta*, and *C. bicrenata* in the cloud forests of the Sierra de Santa Marta, which they called "*Liquidambar* forest and deciduous forests." *Hymenophyllum undulatum*, *Trichomanes capillaceum*, *T. hymenoides*, *T. krausii*, *Grammitis asplenifolia*, *Blechnum falciforme*, *Campyloneurum xalapense*, *Anthrophyum ensiforme*, *Polypodium fraternum*, *P. echinolepis*, and *Vittaria dimorpha* are frequent epiphytes on these tree ferns.

The terrestrial elements include *Selaginella martensii*, *S. pulcherrima*, *S. schizobasis*, *Anemia phyllitidis*, *Thelypteris hispidula*, *Th. meniscioides*, *Th. pilosohispida*, and *Adiantum trapeziforme*. In exposed barren places *Lycopodium cernuum* and *L. clavatum* frequently grow together.

The cloud forests of the Sierra de Chiconquiaco, Veracruz, north of Xalapa, is composed of elements of Lauraceae and *Brosimum* (Gómez-Pompa, 1966). Here *Tectaria heracleifolia* and *Lygodium venustum* grow in disturbed areas, and *Pteris longifolia*, *P. podophylla*, *Thelypteris albicaulis*, and *Ctenitis melanosticta* occur in the canyons. Common species in these forests are *Phlebodium pseudoaureum*, *Pityrogramma tartarea*, and *Hemionitis palmata*; *Blechnum occidentale* and *B. unilaterale* are found in exposed areas. Arborescent cyatheaceous ferns (*Cyathea princeps*, *C. mexicana*, *C. fulva*, and *C. costaricensis*) are frequent and form small communities.

Oak and Pine Forests

Common epiphytes in oak forests are *Pleopeltis astrolepis*, *P. macrocarpa*, *P. angusta*, *Polypodium alfredii*, *P. lepidotrichum*, and *P. madrense*. In Sierra de Juárez, Oaxaca, *Plecosorus speciosissimus* and *Dryopteris wallichiana* are exceptional among the terrestrial species in oak or pine-oak forests (at 2,800 m). In Zacualtipan, Hidalgo, *Woodwardia spinulosa*, *W. martinezii*, *Plagiogyria semicordata* (along river banks), *Lycopodium cernuum*, and *L. serratum* (presently known only from this locality in Mexico) are frequent at 2,000 m; *Botrychium* sp. is sporadic. *Equisetum hyemale* var. *affine* occurs along banks of streams. *Asplenium monanthes*, *Elaphoglossum latifolium* (with a wide altitudinal range), *Selaginella martensii*, and *S. pallescens* are common in coniferous forests.

In pine-oak and fagaceous forests (at 2,000 m) in Hidalgo, *Cyathea divergens* var. *tuerckheimii* and *C. fulva* are frequent, along with other terrestrial elements such as *Woodwardia spinulosa, W. martinezii, Lycopodium thyoides, Elaphoglossum latifolium, Adiantum princeps,* and *Lophosoria quadripinnata.*

Pteridophytes are abundant in the lower portions of Volcán Tacaná (at 1,500–1,700 m) in Chiapas. *Cibotium regale, Polypodium lindenianum, Elaphoglossum* spp., *Campyloneuron xalapense, Pleopeltis munchii, P. angusta, Asplenium* spp., and diverse Selaginellas are common in disturbed forests. *Jamesonia alstonii* grows at high altitude and is known only in the States of Chiapas (Volcán Tacaná, 3,600 m) and Oaxaca (Cerro Pelón at 2,750 m).

Grasslands, Thorn, and Desert Scrubs

Pteridophytes are rare. Xerophytes such as *Selaginella lepidophylla* and *S. marginata* occur in grasslands, thorn, and desert scrub vegetation. Examples of xeromorphics include *Pellaea atropurpurea* and *Cheilanthes lozanii* and other typical cheilanthoids. Mostly species of *Cheilanthes* and *Pellaea* are found in these vegetation types. Selaginellas are occasional in well-conserved localities.

An interesting xeric pteridoflora with a high number of endemics is found in northwestern Mexico, in the Baja California Peninsula and Sonora. Among the endemics are *Pellaea andromedaefolia, P. mucronata, Woodsia plummerae,* and *Adiantum jordanii*; species with wider distribution include *Equisetum laevigatum, Cheilanthes castanea,* and *Pellaea ternifolia.*

Other Pteridophytes

Some pteridopytes occur in salty, sodium, or gypsum soils (Valdés & Flores, 1983). *Notholaena bryopoda* of northern Mexico and *Selaginella gypsophylla* in Nuevo León are apparently restricted to gypsum soils, and *S. lepidophylla,* with a wider distribution, occurs in other soil types. Other gypsophilous species are *Cheilanthes cochisensis* and *Ch. integerrima* in Chihuahua and Nuevo León, *N. aschenborniana* in Coahuila and Nuevo León, and *N. neglecta* in Nuevo León. Among the halophytes, *Acrostichum aureum* and *A. daneaefolim* are common in mangroves and associated vegetation. *Pityrogramma trifoliata, Ophioglossum engelmanii,* and *O. nudicaule* are facultative halophytes.

In wet tropical areas, subject to seasonal flooding, *Blechnum serrulatum, Lygodium volubile, Schizaea elegans, S. poeppigiana,* and *Actinostachys germannii* occur; the last two are restricted to a small area in the Lacandon forest in Chiapas. Finally, among the aquatic pteridophytes there are several species of *Isoëtes* and *Marsilea.* Floating ferns include *Azolla* and *Salvinia.*

Pteridophytes with wider distribution, e.g., *Pteridium aquilinum,* behave as weeds in grasslands and disturbed areas between 800 and 2,000 m or more in altitude. *Selaginella lepidophylla, Equisetum hyemale* var. *affine, Thelypteris tetragona, Tectaria heracleifolia, Polypodium polypodioides* (including varieties), *Notholaena sinuata*, *N. aurea*, and *Mildella intramarginalis* are among a few others that cannot be assigned to any particular vegetation type.

CONCLUSIONS

The above briefly reviews diversity and distribution of pteridophytes in Mexico in various vegetation types. For comparative purposes, the numbers of genera and species in selected countries in the Americas and in several states of Mexico have been given. However, phytogeographic conclusions from these figures may be drawn only with caution, as: (1) the limits of the geographic entities are political and not natural, and (2) generic and species concepts of authors of the works cited in Tables 12.1, 12.2, and 12.3 differ from one other.

A note concerning conservation of these plants is in order. In many parts of the country population pressure has endangered many species. Among them are the rare *Schaffneria nigripes*, *Holodyctium ghiesbreghtii*, *Onocleopis hintonii*, and *Asplenium modestum*, which has not been seen for over 30 years. The arborescent ferns continue to be exploited, without control, for "maquique" (the thick layer of adventitious roots) used as substrate for epiphytes in greenhouses. The major threat to taxa of Mexican pteridoflora and other groups, however, comes from habitat loss. Measures are being discussed and implemented to conserve natural habitats, and the botanical gardens and institutions can play a useful role by maintaining living collections of endangered taxa. Also of urgent need is completion of the inventory of these plants in areas not hitherto well studied (northeastern Mexico, Yucatán Peninsula).

APPENDIX

Provisional list of pteridophytes endemic (restricted to the territorial limits) to Mexico

Species	States
Adiantum *amblyopteridium* Mick. & Beit.	Oaxaca
A. galeottianum Hook.	Guerrero, Oaxaca
A. oaxacanum Mick. & Beit.	Oaxaca
A. shepherdii Hook.	Guerrero, México, Michoacán
A. trichochlaenum Mick. & Beit.	Guerrero, Oaxaca, Chiapas
***Anemia** *brandegeea* Davenp.	Sinaloa
A. colimensis Mick.	Colima
A. familiaris Mick.	Oaxaca, Chiapas
A. intermedia Copeland	Nayarit
A. karwinskyana (Presl) Prantl	Jalisco, Michoacán, México, Oaxaca, Chiapas
A. munchii Christ	Oaxaca, Veracruz, Chiapas
***Antrophyum** *chlorosporum* Mick. & Beit.	Oaxaca
Aspidotis *meifolia* (D. C. Eaton) Pic. Ser.	Chihuahua, Coahuila, Nuevo León, Tamaulipas, San Luis Potosí
Asplenium *blepharodes* D.C. Eaton	Baja California
A. hesperium Mick. & Beit.	Sinaloa, Nayarit, Jalisco, Guerrero, Oaxaca, Chiapas
A. insolitum A.R. Sm.	Oaxaca, Chiapas
A. lamprocaulon Fée	Oaxaca, Chiapas
A. minimum Martens & Galeotti	Jalisco, Tamaulipas, Veracruz, Oaxaca, Chiapas
A. modestum Maxon	Chihuahua
A. munchii A.R. Sm.	México, Jalisco, Guerrero, Oaxaca, Morelos, Chiapas
A. oligosorum Mick. & Beit.	Oaxaca
A. potosinum Hieronymus	San Luis Potosí, Veracruz, Oaxaca
A. pringlei Davenp.	Chihuahua, Nayarit, Jalisco
A. soleirolioides A.R. Sm.	Guerrero, Oaxaca, Chiapas
A. stolonipes Mick. & Beit.	Oaxaca
A. tryonii Correll	Chihuahua
A. yelagagense Mick. & Beit.	Oaxaca
Blechnum *danaeaceum* (Kunze) C. Chr.	Veracruz
B. wardiae Mick. & Beit.	Veracruz, Oaxaca, Chiapas
Bolbitis *hastata* (Fourn.) Hennipman	Veracruz, Oaxaca, Chiapas

Appendix (cont.)

Species	States

B. umbrosa (Liebmann) Ching | Veracruz

Bommeria *ehrenbergiana* (Klotzsch) Under. | Nuevo León, Hidalgo, San Luis Potosí, México, Veracruz, Puebla, Oaxaca

B. subpaleacea Maxon | Chihuahua, San Luis Potosí, Hidalgo, Guanajuato, México, Puebla, Oaxaca

Cheilanthes *albida* Baker | Central Mexico

C. allosuroides Mett. | Baja California, Sonora to Oaxaca

C. angustifolia HBK | Nayarit, Jalisco, Michoacán, Guerrero, Oaxaca, Chiapas

C. beitelii Mick. | Jalisco, Hidalgo, Veracruz, Oaxaca

C. brandegeii D. C. Eaton | Baja California
C. chipinquensis Morton & Lelli. | Nuevo León
C. complanata A.R. Sm. | Chiapas
C. x coruscans Mick. & Beit. | Oaxaca
C. crassifolia (Moore & Houlston), Mick. & Beit. | Oaxaca
C. cuneata Link | Sinaloa, Nayarit, Jalisco, Michoacán, Morelos, México, Guerrero, Oaxaca, Veracruz, Chiapas

C. decomposita (Mart. & Gal.) Fée | Sinaloa, Nayarit, Jalisco, Veracruz, Oaxaca

C. elongata Gooding | Veracruz
C. lemmonii (D. C. Eaton) Domin
  var. *australis* (R. Tryon) Mick. & Beit. | Puebla, Oaxaca
C. lerstenii Mick. & Beit. | Nayarit, Jalisco, Guerrero, Oaxaca, Chiapas

C. longipila Baker | Jalisco, Guerrero, Oaxaca
C. lozanii (Maxon) R. Tryon | Baja California, Sonora to Chiapas

C. membranacea (Davenp.) Maxon | Oaxaca, Chiapas
C. mickelii Reeves | Oaxaca, Chiapas
C. moncloviensis Baker | Coahuila
C. pallens (Weath. in R. Tryon), Mick. & Beit. | Chihuahua, Durango, Jalisco, Guanajuato, Hidalgo, Puebla

C. palmeri Eaton | Jalisco
C. peninsularis Maxon | Baja California
C. pringlei Davenp. | Sonora, Chihuahua, Oaxaca
C. purpusii Reeves | San Luis Potosí
C. sonorensis Gooding | Sonora

Appendix (cont.)

Species	States
Cheiloplecton rigidum (Sw.) Fée var. *lanceolatum* C.C. Hall ex Mick. & Beit.	Puebla, Oaxaca
Cibotium schiedei Schldl. & Cham.	Veracruz, Oaxaca
Cnemidaria apiculata (Hook.) Stolze	Veracruz, Oaxaca
*Ctenitis baulensis A.R. Sm. *C. bullata* A.R. Sm. **C. chiapasensis* (Christ) A.R. Sm. *C. ursina* A.R. Sm.	Chiapas Chiapas Chiapas Chiapas
Diplazium entecnum Mick. & Beit. *D. hellwigii* Mick. & Beit.	Oaxaca Oaxaca
Dryopteris munchii A.R. Sm. *D. polyphylla* Copel. *D. rosea* (Fourn.) Mick. & Beit. *D. rossii* C. Chr. *D. tremula* Christ	Chiapas Veracruz Jalisco, Guerrero, Veracruz, Oaxaca, Chiapas Baja California, Sinaloa, Nayarit, Jalisco, Michoacán, Guerrero, México, Morelos, Distrito Federal, Oaxaca Michoacán
Elaphoglossum alan-smithii Mick. *E. decursivum* Mick. **E. ipshookense* Mick. **E. leonardii* Mick. **E. lepidopodum* Mick. *E. parduei* Mick. *E. pringlei* (Davenp.) C. Chr. *E. rufescens* (Liebmann) Moore *E. seminudum* Mick. *E. vestitum* (Schldl. & Cham.) Schott ex Moore	Oaxaca, Chiapas Oaxaca Oaxaca Oaxaca Oaxaca Oaxaca Michoacán, Oaxaca Veracruz Veracruz, Oaxaca Hidalgo, Veracruz, Puebla, Oaxaca
Equisetum ferrissii Clute	Baja California
Eriosorus flexuosus (HBK) Copel. var. *galeanus* A.F. Tryon	Guerrero
Grammitis basiatenuata (Jenm.) Proc. var. *valens* Mick. & Beit. *G. hellwigii* Mick. & Beit. *G. oidiophora* Mick. & Beit.	Oaxaca, Chiapas Oaxaca Oaxaca

Appendix (cont.)

Species	States

G. zempoaltepetlensis Mick. & Beit.   Oaxaca, Chiapas

Hemionitis *elegans* Davenp.   Nayarit, Jalisco, Morelos, Oaxaca

Holodyctium *ghiesbreghtii* (Fourn.) Maxon   Tamaulipas, San Luis Potosí, Hidalgo, Veracruz, Oaxaca

Hypolepis *eurichlaena* Mick. & Beit.   Oaxaca
H. melanochlaena A.R. Sm.   Chiapas
*H. microchlaena Mick. & Beit.   Oaxaca
H. thysanochlaena Mick. & Beit.   Oaxaca, Chiapas
H. trichochlaena Mick. & Beit.   Oaxaca

Isoëtes *mexicana* Underwood   Chihuahua, Durango, Hidalgo
I. pringlei Underwood   Jalisco, México

Lomariopsis *mexicana* Holttum   Hidalgo, Veracruz, Oaxaca, Tabasco, Chiapas

Lycopodium *cuernavacense* Underw. & Lloyd   Morelos, Chiapas
L. orizabae Underwood & Lloyd   Veracruz, Chiapas

Marattia *laxa* Kunze   Veracruz, Puebla, Oaxaca, Chiapas

Notholaena *aliena* Maxon   Coahuila
N. angusta R. Tryon   Hidalgo
N. aurantiaca D. C. Eaton   Jalisco
N. bryopoda Maxon   Coahuila, San Luis Potosí
N. delicatula Maxon ex Weatherby   Coahuila, Nuevo León
N. galeottii Fée   San Luis Potosí, Jalisco, Hidalgo, Distrito Federal, Morelos, Puebla, Guerrero, Oaxaca, Chiapas
*N. leonina Maxon   Nuevo León
N. lumholtzii Maxon ex Weatherby   Sonora
*N. palmeri Baker   San Luis Potosí
N. peninsularis Maxon ex Weatherby   Baja California
N. pilifera Tryon   Morelos
N. rigida Davenp.   Nuevo León, Tamaulipas
N. rosei Maxon   Jalisco, Oaxaca
N. weatherbyana Tryon   Chihuahua

Pecluma *ferruginea* (Mart. & Gal.) Price   Sinaloa, Nayarit, Jalisco, Colima, Guerrero, México, Morelos, Hidalgo, Oaxaca, Chiapas

Appendix (cont.)

Species	States
Pellaea *notabilis* Maxon	Tamaulipas, Nuevo León
P. *x oaxacana* Mick. & Beit.	Oaxaca
P. *pringlei* Davenp.	Sinaloa, Nayarit, Michoacán, Guerreo, Morelos
P. *seemannii* Hook.	Baja California, Chihuahua, Sonora, Sinaloa, Nayarit, Colima, Jalisco, México, Morelos, Michoacán, Guerrero, Oaxaca, Chiapas
Phanerophlebia *nobilis* (Schldl. & Cham.) Presl	Jalisco, Michoacán, México, Distrito Federal, Puebla, Oaxaca
Phlebodium *araneosum* (Mart. & Gal.) Mick. & Beit.	Jalisco, Michoacán, Guerrero, México, Distrito, Federal, Morelos, San Luis Potosí, Hidalgo, Puebla, Oaxaca
Plagiogyria *truncata* Mick. & Beit.	Oaxaca
Pleopeltis *conzattii* (Weatherby) R. Tryon & A. Tryon	Guerrero, Oaxaca
P. *polylepis* (Roem. ex Kze.) Moore	Durango, Jalisco, Michoacán, Guerrero, Guanajuato, México, Distrito Federal, Morelos, San Luis Potosí, Tamaulipas, Veracruz, Puebla, Oaxaca
Polypodium *arcanum* Maxon	Jalisco, Hidalgo, Puebla, Veracruz, Oaxaca
P. *californicum* Klf.	Baja California, Oaxaca
P. *chiapense* Evans & Smith	Oaxaca, Chiapas
P. *collinsii* Maxon	Veracruz, Oaxaca, Chiapas
P. *conterminans* Liebm.	Puebla, Veracruz, Chiapas
*P. *diplotrichum* Mick. & Beit.	Oaxaca
P. *eatonii* Baker	Veracruz, Chiapas
P. *eperopeutes* Mick. & Beit.	Veracruz, Oaxaca, Chiapas
P. *fallacissimum* Maxon	Coahuila
P. *glaberulum* Mick. & Beit.	México, Oaxaca
P. *guilleminianum* Fourn.	México
P. *guttatum* Maxon	Baja California, Nuevo León, Tamaulipas, Coahuila, Chihuahua
P. *hahnii* Fourn.	Veracruz
P. *lepidotrichum* (Fée) Maxon	Hidalgo, Veracruz, Oaxaca
P. *lesourdianum* Fourn.	Veracruz

Appendix (cont.)

Species	States
P. madrense J. Smith	Sinaloa, Durango, Hidalgo, Michoacán, México, Morelos, Oaxaca, Puebla
P. martensii Mett.	Jalisco, México, San Luis Potosí, Veracruz, Puebla, Oaxaca
P. maxonii C. Chr.	San Luis Potosí
P. puberulum Schldl. & Cham.	Veracruz, Oaxaca, Chiapas
P. pyrrholepis (Fée) Maxon	Veracruz, Oaxaca, Chiapas
P. rosei Maxon	Sinaloa, Jalisco, México, Michoacán, Guerrero, Morelos
P. subpetiolatum Hook.	Chihuahua, Colima, Jalisco, Michoacán, México, Distrito Federal, Puebla, Oaxaca
P. sursumcurrens Copel.	Veracruz
*P. tricholepis Mick. & Beit.	Oaxaca
P. villagranii Copel.	Hidalgo
Polystichum *distans* Fourn.	Hidalgo, Veracruz, Puebla, Oaxaca, Chiapas
P. drepanoides Fourn.	Guerrero, Chiapas
P. rachichlaena Fée	México, Distrito Federal, Morelos, Puebla, Oaxaca, Chiapas
Pteris *chiapensis* A.R. Sm.	Guerrero, Chiapas
P. pulchra Schldl. & Cham.	San Luis Potosí, Hidalgo, Veracruz, Guerrero, Oaxaca, Chiapas
Selaginella *arsenei* Weatherby	San Luis Potosí, Querétaro, Hidalgo, Guerrero, Chihuahua
S. californica Spring	Baja California (?)
S. carnerosana Reeves	Coahuila
S. chiapensis A.R. Sm.	Chiapas
*S. corrugis Mick. & Beit.	Oaxaca
S. cuneata Mick. & Beit.	Oaxaca
S. delicatissima Lind. ex A. Br.	Nuevo León, Tamaulipas, San Luis Potosí, Jalisco, Hidalgo, Veracruz, Colima, Michoacán, Morelos, Guerrero, Oaxaca
S. disticha Mick. & Beit.	Guerrero, Oaxaca, Chiapas
S. extensa Underwood	Tamaulipas, San Luis Potosí, Veracruz, Jalisco, Hidalgo, Puebla
S. finitima Mick. & Beit.	Veracruz, Oaxaca, Chiapas
S. gypsophylla A.R. Sm. & T. Reeves	Nuevo León

Appendix (cont.)

Species	States
S. landii Greenman & Pfeiffer	Nayarit, Jalisco, México, Morelos, Puebla, Oaxaca
S. lindenii Spring	Oaxaca, Tabasco, Chiapas
S. lineolata Mick. & Beit.	Nayarit, Jalisco, Coahuila, Guerrero, Oaxaca
S. macrathera Weatherby	Chihuahua
S. mixteca Mick. & Beit.	Oaxaca
S. mosorongensis Hieron.	Veracruz, Oaxaca
S. novoleonensis Hieron.	Tamaulipas, Nuevo León, Chihuahua
S. parishi Underwood	Coahuila, Zacatecas
S. pulcherrima Liebm. ex Fourn.	Veracruz, Chiapas
S. rzedowskii F. Lorea-Hernández	Guerrero
S. schaffneri Hieron.	San Luis Potosí, Jalisco, México
S. schiedeana A. Braun	San Luis Potosí, Veracruz, Chiapas
S. subrugosa Mick. & Beit.	Oaxaca
S. tarda Mick. & Beit.	Michoacán, México, Oaxaca
S. tropidophora Mick. & Beit.	Oaxaca
Stigmatopteris *chimalapensis* , Mick. & Beit.	Oaxaca
Thelypteris *albicaulis* (Fée) A.R. Smith	Jalisco, Michoacán, Guerrero, México, Morelos, Veracruz, Oaxaca, Chiapas
T. cheilanthoides (Kze.) Proc. var. *mucosa* A.R. Sm.	Oaxaca
T. lanosa (C. Chr.) A.R. Sm.	Veracruz
T. munchii A.R. Sm.	Chiapas
T. rhachiflexuosa R. Riba	Veracruz
*T. tablana (Christ) A.R. Sm.	Chiapas
Trichomanes *bucinatum* Mick. & Beit.	Hidalgo, Oaxaca
Woodsia *pusilla* Fourn.	México
Woodwardia *martinezii* Maxon	Hidalgo, Veracruz
W. x semicordata Mick. & Beit.	Puebla, Veracruz, Oaxaca

Binomials lacking reliable information (distribution or synonymy) are excluded.
*denotes taxa known only from type collections.

REFERENCES

Aguirre-Claverán, R. & M.L. Arreguín Sánchez. 1988. Claves de familias, géneros, especies y variedades de pteridofitas del estado de Nuevo León. Anales Esc. Nac. Ci. Biol. 32:9–61.

Breedlove, D.E. 1986. Listados florísticos de México. IV. Flora de Chiapas. Mexico City: UNAM. Inst. Biol.

Conzatti, C. 1939. Flora Taxonómica Mexicana. Pteridofitas o Helechos. Vol. 1. Mexico City: Oaxaca de Juárez.

————. 1946. Flora Taxonómica Mexicana. Pteridofitas o Helechos. Revised edition with additions. Mexico City: Sociedad Mexicana de Historia Natural.

Cowan, C.P. 1983. Listados florísticos de México. I. Flora de Tabasco. Mexico City: UNAM. Inst. Biol.

Fée, A.L.A. 1857. IX Catalogue Métodique des Fougéres et des Lycopodiacées du Mexique I–IV. Lithograph. pp. 1–48.

Gómez-Pignataro, L.D. 1976. Contribuciones a la Pteridología centroamericana. I. Enumeratio filicum nicaraguensium. Brenesia 8:41—57.

Gómez-Pompa, A. 1966. Estudios botánicos en la región de Misantla, Veracruz. Mexico City: IMERNAR.

Knobloch, I.W. & D.S. Correll. 1962. Ferns and Fern Allies of Chihuahua, Mexico. Austin, TX: Texas Research Foundation.

Kramer, K.U. 1978. The Pteridophytes of Suriname: an enumeration with keys to the ferns and fern allies. Uitgaven Natuurw. Studiekring Suriname Ned. Antillen 93:1–198.

Lellinger, D.B. 1985a. A Field Manual of the Ferns and Fern Allies of the United States and Canada. Washington, DC: Smithsonian Institution Press.

————. 1985b. The distribution of Panama's Pteridophytes. In W.G. D'Arcy & Mireya D. Correa A. (eds.), The Botany and Natural History of Panama. Monogr. Syst. Bot. 10:43–47.

Liebmann, F.M. 1848. Mexicos Bregner. In Systematisk Critisk, Plantgeographisk Undersögelse. Kongel Danske Vidensk. Selskr. Forhandl. 1:151–332; 353–362.

Lira, R. & R. Riba. 1984. Aspectos fitogeográficos y ecológicos de la flora Pteridofita de la Sierra de Santa Marta, Veracruz, México. Biotica 9(4):451–467.

Martens, M. & H. Galeotti. 1842. Memoire sur les fougéres du Mexique et considerations sur la géographie botanique de cette contrée. Mem. Acad. Sci. Belg. 15:1–99.

Matuda, E. 1956a. Los helechos del Valle de México y alrededores. Anales Inst. Biol. Univ. Nac. México 27:49–168.

————. 1956b. Los Helechos del Estado de México. Toluca, State of Mexico: Gobierno del Estado de México. Dirección de Agricultura y Ganadería.

Mickel, J.T. & J.M. Beitel. 1988. Pteridophyte flora of Oaxaca, Mexico. Mem. New York Bot. Gard. 46:1–568.

Proctor, R.G. 1977. Pteridophytes. In R.A. Howard (ed.), Flora of Lesser Antilles. Vol. 2. Cambridge, MA: Harvard Univ. Press.

————. 1985. Ferns of Jamaica. A Guide of Pteridophytes. London: Brit. Mus. (Nat. Hist.).

Riba, R. & A. Butanda. 1987. Bibliografía comentada sobre Pteridofitas de Mexico. Mexico City: Consejo Nacional de la Flora de México.

————, L. Pacheco & E. Martínez S. 1987. New records of Pteridophytes from the state of Chiapas, Mexico. Amer. Fern J. 77(2):69–71.

Rovirosa, J.N. 1909. Pteridografía del sur de México o Sea Clasificación y Descripción de los Helechos de Esta Región Precedida de un Bosquejo de la Flora General. Mexico City: Imprenta de I. Escalante.

Rzedowski, J. 1978. Vegetación de México. Mexico City: Limusa.

Shreve, F. & I.L. Wiggins. 1964. Vegetation and Flora of the Sonoran Desert. Vol. 1. Stanford, CA: Stanford Univ. Press.

Smith, A.R. 1981. Pteridophytes *In* D.E. Breedlove (ed.), Flora of Chiapas. II. San Francisco, CA: California Academy of Sciences.

Sousa S., M. & E. Cabrera C. 1983. Listados Florísticos de México. II. Flora de Quintana Roo. Mexico City: UNAM. Inst. Biol.

Standley, P.C. 1930. Flora of Yucatan. Field Mus. Publ. Bot. 3:157–492.

———— & S.J. Record. 1936. The forests and flora of British Honduras. Field Mus. Publ. Bot. Ser. Vol. 12.

Stolze, R.G. 1976. Ferns and fern allies of Guatemala. I. Ophioglossaceae through Cyatheaceae. Fieldiana, Bot. 39:1–130.

————. 1981. Ferns and fern allies of Guatemala II. Polypodiaceae. Fieldiana, Bot. 6:1–522.

————. 1983. Ferns and fern allies of Guatemala III. Marsileaceae, Salviniaceae and the fern allies. Index. Fieldiana, Bot. 12:1–91.

Tryon, R.M. & A.F. Tryon. 1982. Ferns and allied plants, with special reference to tropical America. New York: Springer Verlag.

Valdés, J. & H. Flores. 1983. Las pteridofitas en la flora halófila y gipsófila de México. Anales Inst. Biol. Univ. Nac. México. Ser. Bot. 54:173–188.

Wiggins, I.L. 1980. Flora of Baja California. Stanford, CA: Stanford Univ. Press.

13

Genus *Pinus*: A Mexican Purview

BRIAN T. STYLES

Pinus (Pinaceae, Coniferales), an exceptionally natural taxon almost entirely confined to the Northern Hemisphere, is represented in Mexico by about half of its 100
or so species. The 49 taxa recognized in this report are arranged according to the
system of Little and Critchfield. Distribution of various species in the country is
discussed. Observations concerning endemism are made. Six areas of species
diversity and richness of pines in Mexico, which is a major center of diversity for the
genus, are identified. The importance of pines in Mexican forest economy is
highlighted, and the conservation biology of pines in Mexico is considered.

The genus *Pinus* (Pinaceae—Coniferales) forms a natural group of plants
comprised of some 90–120 species (Critchfield & Little, 1966; Debazac,
1964; Eguiluz, 1977; Gaussen, 1960; Harrison, 1966; Krüssmann, 1971;
Loock, 1977; Mirov, 1967; Pilger, 1926; Silba, 1984). The constituent taxa
show more features in common with each other than they do with any other
genus in the Pinaceae. In fact during the last 50 years only one species, *P.
krempfii* Lecomte, a rare, localized endemic in the high mountains of
Vietnam, has been placed in a separate genus (*Ducampopinus* A. Chev.)
because it bears paired, broad, flat leaves that lack a fascicle sheath at
maturity; the leaves strongly resemble those of *Podocarpus*. The mature
female cone is, however, typical of *Pinus*, and it has been properly included
in this genus in all of the more recent revisions (Critchfield & Little, 1966;
Mirov, 1967).

The genus is distinguished from other members of the Coniferales by its
needle-like leaves, borne singly or in fascicles of one to six, on short shoots,
with a sheath of bud scales at the base when young. The female cone
consists of few to many woody, ovuliferous scales, each of which is
subtended by a small, insignificant bract. At maturity the latter either fuses
with the ovuliferous scale or completely or partially disappears. All pines
are woody, and many form large forest trees, some up to 70 m. They
frequently grow in stands, forming extensive forests composed of a single
species or a mixture of several.

HISTORICAL BACKGROUND TO THE STUDY

Interest in pines (particularly the tropical species) at Oxford University's Forestry Institute (OFI) began in 1963. The Eighth Commonwealth Forestry Conference held in Nairobi, Kenya in 1962 expressed concern at the loss of timber for constructional and other industrial purposes through the destruction of tropical forests and suggested that the shortage can be overcome by soft wood conifer plantations, particularly comprising species of pines as they are well represented in the tropical latitudes. The Latin American *Pinus caribaea* was recommended, as it was already used as a plantation tree in the tropics. The British Government provided funds for exploration and seed collection. OFI appointed A.F. Lamb, a student of *P. caribaea* (Lamb, 1973), to oversee the project, which was based at Oxford. The project, with the support of local government forestry departments, successfully collected seeds in different parts of Central America with voucher specimens and ecological observations.

Subsequent Commonwealth Forestry Conferences have recommended continuation of the work on *Pinus caribaea* but at the same time urging the inclusion of a number of other pine species such as *Pinus oocarpa, P. patula* subsp. *tecunumanii, P. pseudostrobus* and its infraspecific taxa, *P. maximinoi*, and *P. chiapensis*. These species were added to the OFI list (Annual Reports of C.F.I., 1975–86). Consequently, interest in Mexican pines developed, and cooperation with the Mexican Government through the Instituto Nacional de Investigación Forestal y Agropecuaria (INIFAP) and the Secretaria de Desarrollo Urbano y de Ecología (SEDUE) was formally agreed. Collaborative links were established between OFI and university forestry departments including that of the Instituto de Silvicultura y Manejo de Recursos Renovables at Linares, part of the Universidad Autónoma de Nuevo León, and the Departamento de Bosques at the Agricultural University at Chapingo. OFI has received botanical support from the staff of the Herbario Nacional at the Departamento de Botánica (UNAM) and the Escuela Nacional de Ciencias Biológicas (ENCB) for about two decades.

The Mexican government's forestry department (INIFAP) has for a long time been involved in international collaborative research projects (particularly with the United Nations' Food and Agricultural Organization, or FAO) to collect forest tree seed and has assisted the OFI in its provenance work. It has been responsible for sampling parts of the ranges of the *Pinus patula* complex, *P. pseudostrobus* (and its infraspecific taxa), and *P. oocarpa*. Loock (1977) mentioned that some of the first pine seed ever used for forestry trials in South Africa was received from Mexico in 1903. Consignments of supposedly single species bought from foreign seed merchants, however, frequently turned out to be mixtures of several species. Loock mentioned that the first seeds of *P. leiophylla* received under this name turned out to contain a mixture of *P. montezumae, P.pseudostrobus, P. douglasiana,* and *P. patula*. The need for correctly determined, site-identified material for such widespread international trials is great.

With the full cooperation of INIFAP and SEDUE, further collections of seeds and botanical material have been made by OFI staff of the taxa mentioned above as well as others, extending the cover over their entire distribution range. Liaison has also been maintained with other organizations interested in the collection and conservation of Mexican pine germplasm, e.g., the Centro de Genética Forestal at Chapingo and The Central America and Mexico Coniferous Resources Co-operative (CAMCORE) based at Raleigh, North Carolina (United States).

An enormous amount of botanical material and ecological data have thus been gathered since the early 1970s, and a number of projects on individual species or groups of species have been started especially at Oxford on the pines of importance to forestry (Stead & Styles, 1984; Styles, 1976). Accounts of the genus *Pinus* for Flora Mesoamericana, Flora de México, and Flora Neotropica are in progress (Styles, in prep.). The latter will include the pines not only of Mexico but also of Central America (all taxa growing here are common to Mexico) and the four or five that occur on some of the Caribbean islands.

This chapter focuses on the progress made so far on these studies. It details some of the problems that have been encountered in the taxonomy of pines and reasons for the disagreement on the delimitation of taxa.

ORIGINS OF THE GENUS *PINUS*

According to Florin (1963) the Pinaceae and the genus *Pinus* originated in the Northern Hemisphere during the Jurassic Period of the Mesozoic era. In fact, soon after this time the two major pine groups Haploxylon and Diploxylon had already differentiated. To determine more exactly where pines first arose is difficult, although Laurasia seems most likely. So far as the Mexican pines are concerned, their ancestors have existed in North America since the lower Cretaceous where fossil cones similar to those of *P. devoniana* var. *cornuta* (comb. nov., in edit.), have been found (Chaney, 1954). These pines did not, it seems, form extensive forests but were more widespread than they are now. They then migrated southward owing to climatic and other changes during glaciations and fluctuating physical barriers. According to Miller (1977), they reached Mexico from the United States, apparently not later than the Mid-Tertiary, and at the latest by the Mid-Cenozoic. Mirov (1967) suggested that they entered Mexico along the Sierra Madre Occidental from the North American Cordillera during the Late Cretaceous or at the beginning of the Tertiary. Martin and Harrell (1957) proposed that there was a later invasion of Mexico by pines during the Middle Tertiary from the Appalachian uplands of the eastern United States, round the Gulf of Mexico, and along the Sierra Madre Oriental, together with angiosperm genera such as *Quercus*, *Alnus*, *Carpinus*, *Ostrya*, *Magnolia*, and the well-known *Liquidambar styracifolia*.

Pines are pioneer species and probably spread quickly on the lava flows resulting from volcanic activity that occurred throughout the greater part of Mexico. This spread allowed them to radiate into the central plateaus, and thence gradually to the Isthmus of Tehuantepec. This area, according to Schuchert (1935), was above water during the Lower and Middle Eocene and remained so until the Oligocene and Miocene. During the Pliocene land contact between Mexico and America was reestablished, but it appears that these lowlands were difficult areas for pines to cross.

No fossil pines have been found in Mexico, Central America (including the Caribbean), or South America. It may be concluded from this and the fact that Central and South America were separated before the Upper Cretaceous that the pines arrived in Mexico from the United States in a north–south direction.

At the present time pines have an almost exclusive distribution in the Northern Hemisphere. As Critchfield and Little's (1966) excellent maps showed, the genus is widespread throughout North America and Eurasia, just extending into northern Africa. It is thus found mainly in temperate areas, but major centers of diversity occur where it crosses the Tropic of Cancer southward into Mexico, Central America, and the Caribbean islands. In Southeast Asia and South China particularly, there is another major center of genetic diversity. Only one species, *Pinus merkusii*, crosses the equator to 2° S in Sumatra. Other than here, pines are not known to have penetrated the Southern Hemisphere. Some three species are known to grow within the Arctic Circle. It is the most widespread genus of all woody plants in North America, as well as in Mexico and northern Central America.

INFRAGENERIC CLASSIFICATION

The genus *Pinus* is clearly defined on a number of distinct morphological characters and is readily separable from the other genera in the Pinaceae. The subdivision of the genus into natural groupings is difficult, however, and is still the source of much argument.

In the system proposed by Shaw (1914), the author used "evolutionary" characters, primarily the symmetry and woodiness of the cone, its degree of serotiny, characteristics of the seed, and the presence or absence of a wing. Pilger's (1926) classification included the two major subgroups proposed by Koehne (1893), Haploxylon and Diploxylon, to separate the "soft" or "white" pines from the "hard." However, even these divisions are not 100% diagnostic. Pilger rearranged some of Shaw's groupings and introduced a number of new names. His classification is unnatural because he placed too much reliance on needle number per fascicle. Duffield (1952) made a major contribution to the classification of the genus, although he did not publish names for the groups he proposed. He based his proposals on hybridization work carried out at the Institute of Forest Genetics, Placerville, California. As a result of this work a number of Shaw's groups were again rearranged or species transferred between them. The system proposed by Gaussen (1960) has received little attention from botanists. In a curious "phylogenetic" classification, he used such characters as the form of the seedling—particularly the vascular strands in the cotyledons—the position of the resin ducts in the needles, and the size of pollen grains (a character that is given great weight). The outcome is a complex arrangement that differs markedly from those proposed by previous investigators. A new (illegitimate) system of nomenclature is also used, further complicating an already complex situation.

The infrageneric classification proposed by Little and Critchfield (1969) is today the most acceptable to students of the group. It is based on both morphological differences and evidence from breeding and hybrid experiments cited by Duffield (1952). For many of the temperate and North American species, genetic and cytological relations are reasonably well known, which is certainly not true for the more tropical taxa. For this reason one or two of Little and Critchfield's subgroupings are considered heterogeneous, unwieldy, and in need of further refinement.

Most of the classifications of *Pinus* proposed since those noted above have added little to our understanding of the supposed evolutionary relationships within the group. Mention must be made of Murray's report (1983). This author departed radically from all previous workers and introduces a series of new taxonomic categories. "Genitor" and "subgenitor" were intercalated between genus and subgenus. A whole new array of subgroupings were erected with a confusing system of nomenclature. Many of the Subsections proposed by Little and Critchfield (1969) were radically modified and given different taxonomic ranks.

It is not surprising, therefore, in a genus with such an extensive worldwide distribution that there is little agreement on the exact number of the species extant. Shaw's (1914) exceptionally conservative monograph acknowledged only 66, whereas Critchfield and Little's (1966) geographic distribution maps show 94. Gaussen's (1960) treatment totaled 120, and Mirov (1967) recognized some 103 full species.

There is little doubt, based on the number of new pines being described each year, that the species concept in *Pinus* is becoming much narrower. It is ascribed mainly to the use of chemical information, where if a population (or populations) differ from the species as a whole regarding the level of a certain compound, or the enzyme is absent the population *must* be a different taxon. Evidence from work done at Oxford suggests that such information must be treated with extreme caution. Chemical constituents can be as variable as any other characters.

Widespread taxa, as so many plant taxonomists realize, often show geographical variations that may or may not be worthy of infraspecific rank. Given the known history of migration of the pines throughout the world, it is considered that the variation in many of them is clinal in nature, a point that is certainly true of some of the Mexican taxa.

Perhaps the most important reason for the disagreement about species' limits lies in the morphology of pine plants themselves; taxonomists are

Table 13.1. *Pinus* in Mexico

Major subgroup **HAPLOXYLON**
   Subgenus *Strobus* Sect. *Strobus* Subsect. *Strobi*: *Pinus chiapensis* (Mart.)
      Andresen, *P. strobiformis* Engelm., *P. ayacahuite* Ehrenb. ex Schlecht. (and
      infraspecific taxa), *P. lambertiana* Douglas ex Taylor & Philips.
   Subgenus *Strobus* Sect. *Parrya* Subsect. *Cembroides*: *P. culminicola* Andresen
      & Beaman, *P. catarinae* Robert-Passini, *P. cembroides* Zucc. subsp.
      cembroides and *subsp. *orizabensis* Bailey, *P. lagunae* (Robert-Passini)
      Passini, *P. discolor* Bailey, *P. johannis* Robert, *P. juarezensis* Lanner, *P.
      monophylla* Torr. & Frem., *P. x quadrifolia* Parl. ex Sudw., *P. remota* (Little)
      Bailey & Hawks., *P. nelsonii* Shaw, *P. pinceana* Gordon, *P. maximartinezii*
      Rzedowski.
   Subgenus *Strobus* Sect. *Parrya* Subsect. *Rzedowskianae*: *P. rzedowskii*
      Madrigal & Caballero.

Major subgroup **DIPLOXYLON**
   Subgenus *Pinus* Sect. *Pinea* Subsect. *Leiophyllae*: *Pinus leiophylla* Schiede &
      Deppe ex Schlecht. & Cham., *P. chihuahuana* Engelm. *P. lumholtzii* Robins.
      & Fern.
   Subgenus *Pinus* Sect. *Pinus* Subsect. *Australes*: *Pinus caribaea* var. *hondurensis*
      (Sénécl.) Barr. & Golf.
   Subgenus *Pinus* Sect. *Pinus* Subsect. *Ponderosae*: *Pinus arizonica* var.
      stormiae Mart., *P. jeffreyi* Grev. & Balf., *P. ponderosa* Laws., *P. engelmannii*
      Carr., *P. durangensis* Mart., *P. cooperi* C.E. Blanco, *P. montezumae* Lamb.,
      P. hartwegii Lindl., *P. devoniana* Lindl.(and infraspecific taxa), *P. maximinoi*
      H.E. Moore, *P. praetermissa* Styles & McVaugh, P. pseudostrobus Lindl.
      subsp. *pseudostrobus*, subsp. *apulcensis* (Lindl.) Stead and var. *oaxacana*
      (Mirov) Harrison, *P. douglasiana* Mart., *P. teocote* Schiede & Deppe ex
      Schlecht. & Cham., *P. herrerae* Mart., *P. lawsonii* Roezl ex Gord.
   Subgenus *Pinus* Sect. *Pinus* Subsect. *Sabinianae*: *Pinus coulteri* D. Don
   Subgenus *Pinus* Sect. *Pinus* Subsect. *Contortae*: *Pinus contorta* Dougl. ex
      Loud. subsp. *murrayana* (Balf.) Critchf.
   Subgenus *Pinus* Sect. *Pinus* Subsect. *Oocarpae*: *Pinus radiata* D. Don *var.
      binata* Lemm. and *var. *cedrosensis* (Howell) Axelrod, *P. attenuata*
      Lemm., *P. muricata* D. Don var. *muricata*, *P. patula* Schiede & Deppe ex
      Schlecht. & Cham. *subsp. *patula* and subsp. *tecunumanii* (Eguiluz &
      Perry) Styles, *P. greggii* Engelm. ex Parl., *P. oocarpa* Schiede ex
      Schlecht. subsp. *oocarpa* and *subsp. *trifoliata* (Mart.) Styles (*ined.*), *P.
      pringlei* Shaw ex Sargent, *P. jaliscana* Perez de la Rosa.

Classification after Little and Critchfield (1969). With modifications.
Taxa prefixed by an asterisk (*) have distributions restricted to Mexico.

forced to base their decisions on the variation found in needles and mature
female cones. Needles show comparatively little variation compared with
the leaves of angiosperms and may be greatly subjected to extremes of
environment. This fact means that the cones are the chief source of
taxonomic information. Until we know more about the development of
these important structures, even they must be treated with caution (Styles,

1976). Whereas in angiosperms some 25–30 characters may show variations that can be used in the taxonomy of a group, in pines these traits are frequently reduced to fewer than 15.

GENUS *PINUS* IN MEXICO

Mexico appears to have a larger concentration of pines than any other country in the world. This total approaches 45.5% of the known species. The INIF (1983) considered that there are about 30.0 million hectares (ha) of forest or wooded land in Mexico of which 21.0 million consists of coniferous forest, the dominant species being pines. Guerrero, one of the most highly forested and still commercially relatively unexploited States, may provide up to 10.9% of the country's annual yield of saw timber, most of it being obtained from pine species (Anonymous, 1971). In the State of Chiapas alone, Zamora (1981) estimated that the area of forest cover is about 3.5 million ha, which includes 1.4 million ha of pine and associated species such as *Quercus*; 45% of the pine present consists of a single species, *Pinus oocarpa*. Only the three tropical lowland states of Campeche, Tabasco, and Yucatán have no native pines. Others, such as Nuevo León, Jalisco, Michoacán, Chihuahua, México, Puebla, Coahuila and Hidalgo have more than 20 species each, not counting infraspecific taxa.

No other woody group of plants of equivalent size has been so intensively studied in Mexico as the genus *Pinus*. Monographic accounts have been published by Shaw (1909), Martínez (1948), FAO (1962), Loock (1977), and Eguiluz (1982, 1988). Accounts for regional floras have been prepared by Pineda et al. (1973), Ochoa (1973), Rzedowski (1979), Madrigal, (1982), Carvajal (1986), Narave F. and Taylor (in press), McVaugh (in press). Taxonomic and ecological treatments of individual species or groups of related species have appeared regularly, emanating from both forestry and botanical institutions (Bailey, 1979a,b, 1983, 1987; Passini, 1982; Barrett, 1972; Stead & Styles, 1984; Vela, 1976; Zamora, 1981; Zamora & Velasco, 1978).

Thus the literature on Mexican pines is voluminous, and only a small fraction of it can possibly be accounted for in this chapter. Much useful information is contained in purely forestry publications, which are often neglected by botanists.

In Table 13.1 the pines that occur in Mexico are arranged according to the infrageneric classification proposed by Little and Critchfield (1969), with minor amendments to take account of the new species described. The richness and diversity of the genus in Mexico are exemplified by the fact that representatives of only six of these authors' Subsections do not occur in the Republic. *Pinus rzedowskii*, which was formerly placed in *Balfourianae*, now seems better accommodated in its own new Subsection *Rzedowskianae* (Carvajal, 1986).

Perhaps the one overriding feature that pervades all recent publications is that there appears to be little agreement about the taxonomy of some of the more variable species of pines in the Mexican flora. Not only are many varieties and other infraspecific taxa being regularly described, but some well-known species are being split and divided at an alarming rate. Taxa formerly described at the level of variety or subspecies have been elevated to the rank of full species. This situation is reminiscent of the European seed collector and plantsman Roezl (1857), who alone described over 80 species of pine from the Valley of Mexico and a little further north, only one of which (*P. lawsonii*) survives today.

A conspectus of the Subsections of *Pinus* and details on the most important species in the Mexican flora follows.

Haploxylon: White Pines

Considerable uncertainty still exists regarding the number of species of white or soft pines that occur in Mexico. There is also much debate about the classification of the infraspecific variation that at least one of them exhibits. According to Little and Critchfield's (1969) system all our representatives are placed in the Section *Strobus* Subsection *Strobi*, which is one of the three subsections in this large group. The New World species (together with those from Eurasia) form a morphologically coherent group with five-needled fascicles and elongate, thin-scaled cones that open to shed their seeds at maturity. The seeds are typically winged, but a variety within *Pinus ayacahuite* has been described with a short wing that is sometimes vestigial. In addition to the above-mentioned, three other species are recognized in this Subsection, but this number could easily be increased when further work has been completed on relatives that grow in northwest Mexico. All of them are special to foresters and the lumber industry because of their excellent but resinous timber.

Although the white pines are well known for the ease with which they hybridize (several species from North America have been crossed with those from Asia), one member of the group *Pinus lambertiana*, the sugar pine of the United States, appears to be isolated genetically and has not been crossed with any of its immediate relatives. It is, nevertheless, as Critchfield (1986) stated, an archetypal white pine with long, unarmed cones and winged seeds that has been successfully hybridized with only two species from eastern Asia. These species do not resemble it or each other in terms of the morphological characteristics used to classify white pines. *Pinus lambertiana* is a taxon that has a mainly North American distribution in the western United States but that just enters northern Baja California in the Sierra Pedro Mártir. Here, as elsewhere, it forms an enormous tree up to 70 m tall; its cones may also reach up to 70 cm in length, some of the longest known in the genus.

Two of the best known taxa form important elements in the coniferous forests of Central and South Mexico: *Pinus chiapensis* and *P. ayacahuite*. The former occurs in the more tropical areas of the states of Puebla, Chiapas, Oaxaca, Guerrero, and Veracruz, between 700 and 1,800 m altitude, growing on rich, fertile soils with a higher than average rainfall of 1,500 mm per annum. Originally, Martínez (1940) described it as a variety of the eastern North American *P. strobus* but was much puzzled by the enormous gap in its range of distribution (some 1,200 miles) when it was first discovered. In one of the best computer-based analyses on a group of Mexican pines, Andresen (1964, 1966) showed that it should be elevated to the rank of a full species.

The middle altitudes occupied by this pine mean that it is becoming more and more under human pressure. Many of the stands are being greatly depleted as deforestation occurs to make way for agriculture. There is now a definite need for efforts to be made to conserve some of the best stands of this restricted taxon (Donahue, in press; Zamora and Velasco, 1977).

Pinus ayacahuite has a large distribution throughout Central and South Mexico. It forms an elegant, tall forest tree with a crown of many regularly whorled branches bearing large pendulous, resinous cones. It tends to replace *P. chiapensis* at higher altitudes (2,300–2,800 m) and must be considered fully frost-hardy in these areas. Its range of distribution has recently been considerably extended in Central America through the work of OFI staff, who have shown that it occurs as far south as southwestern Honduras close to the Nicaraguan border in hitherto unknown, but extensive, pure stands. Although considered to be ecologically and geographically allopatric, an area of overlap in the distribution of this pine and *P. chiapensis* has been reported in Oaxaca (Donahue & Arizmendi, 1986), but no morphologically intermediate trees were noted. Several infraspecific taxa have been described in *P. ayacahuite*, and their status is at present being investigated.

The status of the fourth taxon at specific level accepted in this group, *Pinus strobiformis*, is polemical. Many students of the group (Eguiluz 1977; Loock, 1977; Martínez, 1948) also list *P. flexilis* and *P. reflexa* as white pine representatives in the forests of the mountainous areas of the northern states of Mexico. This southwestern pine extends from the United States along the Sierra Madre Occidental in the west as far as Nayarit and also along the Sierra Madre Oriental to the state of Tamaulipas. *Pinus flexilis* has its range of distribution totally confined within the United States (Andresen & Steinhoff, 1971; Steinhoff & Andresen, 1972). *Pinus reflexa* is considered to be a synonym of the latter species.

The decision of Critchfield and Little (1966) to exclude *Pinus flexilis* and *P. reflexa* as separate species from the Mexican pine flora is followed here. However, good, fully correlated botanical material must be collected and examined in the large, as yet little explored northern states of the Republic.

Pinyon (Nut) Pines

The Pinyon (nut) pines representing Little and Critchfield's (1966) Subsection *Cembroides* among the Haploxylon pines, forms a fairly large, natural

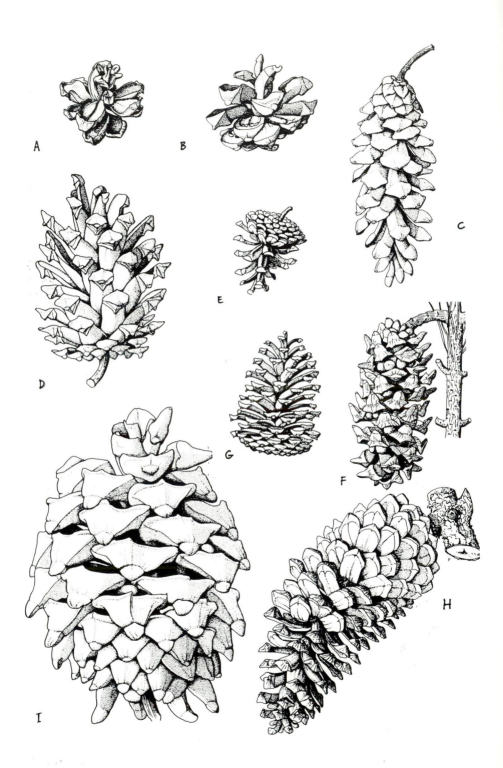

A

B

C

D

E

F

G

H

I

assemblage of some 14 species. In Mexico they are known as "piñoneros" on account of their edible nuts, which are known colloquially as piñones. They share a number of morphological characters in common, including needles borne singly or in fascicles of two to four, with deciduous sheaths (except *P. nelsonii*) and non-decurrent leaf-bract bases. Cone scales are relatively few, and the seed is always wingless. These pines are essentially trees or shrubs of arid environments in northern and central Mexico and some are also present in several States of the southern United States. A few are localized in high mountains, and most have a mainly upland distribution. One species, *P. cembroides*, plays a major role in the coniferous forest cover of arid areas, and Cuanalo (1979) estimated that as many as 862,000 ha throughout the Repubic may be occupied by it and its relatives. For one State alone, Nuevo León, Rzedowski (1966) suggested that there may be about 63,000 ha covered.

The group has been the subject of considerable taxonomic, ecological, and chemical research; and the classification of the variation revealed has led to conflicting results regarding the distinctness or otherwise of several of the taxa accepted here. Their taxonomic rank is also disputed (Bailey & Hawksworth, 1988). Although it is true that several species are distinct and must be considered relicts e.g., *Pinus pinceana*, *P. nelsonii*, and the more recently discovered *P. maximartinezii*, the various segregate species (Table 13.1) of the archetypal *P. cembroides* are treated by taxonomists in different ways. Silba (1985, 1986) considered many of them as subspecies of the latter or as mere varieties of *P. culminicola*. An interesting hybrid origin of *P. quadrifolia* was proposed by Lanner (1974b). Natural hybridization between several other taxa has been reported (Bailey et al., 1982; Lanner, 1974a; Zavarin et al., 1980). So great is the interest in these pines that symposia to consider various aspects of their taxonomy, ecology, pests and diseases, and economic uses have been held at the Facultad de Silvicultura y Manejo de Recursos Renovables at Linares, Nuevo León (Flores et al., 1985) and in Mexico City (Passini, 1988).

Only *Pinus cembroides* has an extensive range in Mexico, occurring in 16 States, as far south as Puebla and Veracruz. It has been the subject of intense investigation by Passini (1982). On the other hand, *P. culminicola*, *P. catarinae*, *P. lagunae*, *P. remota*, and *P. maximartinezii* are restricted to single or a few, small, disjunct populations. The latter, described by Rzedowski in 1964, is one of the most remarkable pines in the entire Mexican conifer flora on account of its large cones and seeds (Fig. 13.1). It is known only in one area in Zacatecas, and the size of the population is uncertain. It is considered one of the most highly threatened of all pine species. Rzedowski (1964), in his original description, suggested it was closest to *P. pinceana* in the pinyon group. This idea has been substantiated by Zavarin and Snajberk (1987), who discovered that the main constituents of the monot-

Figure 13.1. Variation in morphology and size of the cones of some Mexican pines. A. *Pinus cembroides*. B. *P.* quadrifolia. C. *P. chiapensis*. D. *P. rzedowskii*. E. *P. lumholtzii*. F. *P. nelsonii*. G. *P. caribaea*. H. *P. montezumae*. I. *P. maximartinezii*.

erpenes in the wood of both are limonenes, whereas those of *P. nelsonii* are α- and β- pinene.

The remarkable and isolated *Pinus rzedowskii* was described by Madrigal and Caballero (1969). Its morphological characteristics do not allow it to fit well into the infrageneric classification proposed by Little and Critchfield (1969). Although it shows most of the features of the Haploxylon group of pines, the female cone scales have protruding dorsal umbos and the seeds are winged (Fig. 13.1), making it somewhat intermediate between the two major groups. The authors classified their discovery in Shaw's (1914) Subsection *Paracembra* in the group *Balfourianae*. Later monographers (Eguiluz, 1985) followed this decision, but Carvajal (1986) placed it in a Subsection of its own—*Rzedowskianae*. Investigation of its xylem oleoresin has unfortunately failed to shed further light on its relationship (Perry & Madrigal, 1986).

Diploxylon: Hard Pines

Pinus Subsection *Leiophyllae* belongs with the Diploxylon group of pines. The intricate and reticulate evolutionary history of the genus is well illustrated in this group. Although the species contained in it show most of the characters of the hard pines, both the needle sheaths of the fascicles and the decurrent leaf-bract bases are deciduous and not persistent. Mature vegetative branchlets can strongly resemble those of Haploxylon pines. The cones and needle sections are, however, typical of Diploxylon in all respects. Workers studying the Mexican pines are again divided on the number of species recognizable in the group. *Pinus lumholtzii* is distinct and is confined locally to the western and northern slopes of the Sierra Madre Occidental. *Pinus leiophylla* and *P. chihuahuana* are closely related and, if treated as conspecific, have one of the most extensive ranges of any pine in Mexico. The two probably represent the north–south extremes of a geographical cline, a situation that is more common in the genus than hitherto realized. All taxa form trees, frequently with twisted boles and much-branched crowns. Epicormic shoots, not commonly seen in other species, are a feature of members of this group.

Pinus caribaea is the only member of the Mexican pine flora that is placed in Subsection *Australes*, the southern yellow pines (Little & Critchfield, 1969). It is a natural group of some 11 species that occur in the eastern and southeastern United States, the Caribbean, and Central America. Many of the species have enormous importance in forestry, and some are widely cultivated as plantation trees in the tropics and subtropics. The group has two or three needles per fascicle with internal or medial resin canals. The cones are symmetrical and fall whole without leaving basal scales on the twig. Most species have cone scales armed with prickles. *P. caribaea* has been monographed by Barrett and Golfari (1962), who proposed three varieties based primarily on geographical distribution. The mainland populations are included within variety *hondurensis*.

Standley and Steyermark (1958) first suggested that this species should be looked for in Mexico, and for many years its presence here was hotly debated. Lückhoff (1964) mentioned that he flew over the southern part of Yucatán and saw scattered stands of a pine but did not collect specimens or identify the species. Styles and Stead (unpublished) stated in the 1980 Annual Report of the CFI that they had identified trees of this taxon by means of field glasses across the Rio Hondo in Quintana Roó when they were studying its distribution in northern Belize. Its presence here was later definitely confirmed by Chavelas (1981), who collected botanical specimens and mapped its limited distribution. This isolated population is obviously of the greatest interest to foresters and geneticists, being at the northwest extreme of the species' range. It is therefore an obvious candidate for conservation efforts.

Pinus Subsect. *Ponderosae* is one of the larger, more heterogeneous infrageneric groupings proposed by Little and Critchfield (1969). All species belong to the New World and most of the Mexican "hard" pines are placed here. Some of them have little in common with other members of the group, and the whole Subsection seems to be in considerable need of refinement.

These pines have needles with mainly medial resin canals that occur in fascicles of two to six; needles vary in length from a few centimeters to over 45 cm. Similarly, the cones can be small (4 cm) or measure over 30 cm. The cones of most species, but by no means all, leave a few basal scales attached to the branchlet on falling.

Within this Subsection Martínez (1948) recognized a number of smaller subgroups that are probably more natural: "*Pseudostrobus*," "*Teocote*," "*Montezumae*," "*Rudis*," "*Michoacana*," and "*Ponderosa*." The relationships of the species within and between these groups is complex, and several are in need of taxonomic revision. Stead and Styles (1984) have studied the widespread Pseudostrobus group, and Favela Lara (1991) has revised the high altitude *Rudis* complex and its affinities with the *Montezumae* group. A change of the well-known name *Pinus michoacana* to the earlier *P. devoniana* in the large-coned Michoacana group has also been proposed (Rushforth, 1987). Martínez's *Ponderosa* group is chiefly confined to the northwestern States of México, Sonora, Sinaloa, Durango, Chihuahua, and Nuevo León. The species include *P. jeffreyi*, *P. engelmannii*, *P. ponderosa*, and *P. arizonica*, which have a distribution mainly in the United States. They are not well known in Mexico. Many more botanical collections are needed before this group can be adequately revised taxonomically, especially from the botanically undercollected states of Coahuila, Chihuahua, and Sonora.

The small-coned *Teocote* group of three species recognized by Martínez seems to have little in common with the other species in Little and Critchfield's Subsection *Ponderosae*, and their needles have mostly internal resin canals. Both *Pinus teocote* and *P. lawsonii* readily cross with *P. patula* of the closed-cone group, perhaps suggesting affinities with it (Critchfield, 1967).

Pinus Subsection *Sabinianae* has one of the smallest ranges of any equivalent group of pines in the New World. It is comprised of three species with a mainly Californian distribution, one of which, *Pinus coulteri*, just crosses the Mexican frontier into North Baja California. It has a short bole and long drooping branches but is remarkable and well known for its large, heavy, woody cones, some of which can weigh up to 2 kg. The umbos of the cone scales are protuberant. It is now a rare tree, occurring in chaparral in a few, small disjunct populations on steep, almost inaccessible granitic or volcanic slopes. Fires, which frequently occur in the area, are not able to reach and pass through these remaining stands, and the species probably finds a safe refuge in such fire-resistant habitats (Minnich, 1986).

Coulter pine is known to hybridize with *Pinus jeffreyi* (Subsection *Ponderosae*) in California (Zobel, 1951); but in Mexico, where the two species occur almost sympatrically, no intermediates have been reported (Minnich, 1986). Their respective habitats are, in any case, rather different.

Pinus Subsection *Contortae* is one of the new Subsections proposed by Little and Critchfield (1966) to include four species with distributions almost exclusively in North America. The needles of these pines are short and occur two per fascicle. Cones are also small and serotinous with a persistent prickle. One of these pines, *Pinus contorta*, however, also just penetrates Mexico as a single population in northern Baja California in Sierra San Pedro Mártir. It is named subsp. *murrayana*, following the monographic study by Critchfield (1957). As the taxon is at the southern limit of the natural distribution of the parent species in Mexico, it is another population that will prove of particular interest to geneticists and tree breeders for future research programs. As most of the remaining trees now occur within the confines of a national park, their future is fairly secure.

Little and Critchfield's (1966) Subsection *Oocarpae*, the "closed cone pines," is one of the more natural of the infrageneric groups now recognized. Some seven or eight species are involved, all of which occur in the New World and which, without exception, appear as elements in the Mexican flora (Barnes & Styles, 1983). One or two of the species, however, are represented only as infraspecific taxa and occur as isolated, rather sparse, discontinuous populations within our area. Others have enormous geographical ranges throughout the whole of Latin America as far as the southern limit of the genus in Nicaragua. Critchfield (1967) stated that the closed cone pine group is of unusual interest because it includes one of the most readily crossable combinations of species so far encountered among the pines. Morphologically, it is fairly diverse, with the needles mostly three to six per fascicle and with medial, septal, and internal resin canals. The serotinous cones are oblique or asymmetrical and long persistent on the tree after the seed is shed.

The considerable interest shown in all species of the closed cone pine group by foresters, geneticists, and botanists makes these trees one of the best studied groups of all pines. Their great silvicultural importance is discussed below. Millar (1986), in an impressive review of the taxonomy of the Californian members of the group, pointed out that the three species (*Pinus radiata, P. muricata, P. attenuata*) that just reach northern Baja California or the off-shore Mexican islands form a more closely knit group than the others. In chemistry and morphology, they resemble each other in that their terpenes contain α- and β- pinenes. They possess medial resin canals in the needles and bear asymmetrical cones that are strongly serotinous. Although these traits do occur scattered throughout the other mainland members of the group, the species do not show true serotiny but, rather, are tardy in their cone opening. They have cones that are only weakly asymmetrical, and their resin canals are variable in position. *Pinus oocarpa* and the *P. patula* complex have enormous distributions in Mexico and Central America. In fact, the former taxon has one of the most extensive distributions of any tropical pine, ranging from the north of Mexico in the state of Sonora (lat. 27° N) to as far south as Nicaragua (lat. 12° N). Similarly, *P. patula* and its infraspecific taxon stretch from the Tamaulipas/Nuevo León border to northern Nicaragua. The other members have more localized distributions in western and central Mexico.

As mentioned earlier, the whole group is of great economic importance in forestry operations, both within their natural areas of distribution and in *ex situ* plantings in various parts of the tropics. Several of the species have been the object of intensive experimental provenance research projects, especially by OFI, CAMCORE, and INIFAP staff.

Distribution of Species within Mexico

Eguiluz (1985) pointed out that only 35% of the land surface in Mexico lies below 500 m sea level; most of this land being chiefly coastal. Quoting Rzedowski (1978), Eguiluz stated that over one-half of the country lies above 1,000 m. Only *Pinus caribaea* var. *hondurensis* can be described as a truly lowland, tropical tree. *Pinus hartwegii* and *P. culminicola* occur at the other altitudinal extreme, between 3,500 and 4,000 m at the timberline. Other species, e.g., *P. oocarpa*, have enormous altitudinal ranges from 200 m at the Isthmus of Tehuantepec to 2,500 m above sea level in Central Mexico, i.e., from semitropical lowlands to temperate upland forests, which suffer frequent frosts. Most of the species are thus confined to mountain chains or plateaus. Eguiluz (1985) also identified six principal areas of distribution (Fig. 13.2).

1. Sierra Madre Occidental, mainly formed of Cretaceous rocks.
2. Sierra Madre Oriental.
3. Eje Neovolcanico and the Mesa Central. This area is considered of great significance in the evolutionary history of the genus, as it connects the two former ranges. Topographical (volcanic) disturbance has provided many microhabitats, allowing hybridization, adaptive radiation, and speciation to occur.

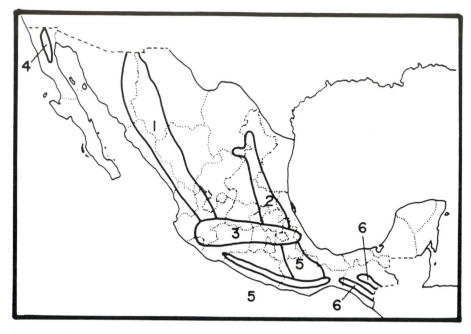

Figure 13.2. Principal areas of distribution of Mexican pines: 1. Sierra Madre Occidental, mainly formed of Cretaceous rocks. 2. Sierra Madre Oriental. 3. Eje Neovolcanico and the Mesa Central. 4. Sierra de Juárez and San Pedro Mártir. Baja California. 5. Sierra Madre del Sur and Macizo de Oaxaca. 6. Sierra de San Cristóbal and Sierra Madre de Chiapas.

4. Sierra de Juárez and San Pedro Mártir, Baja California. These areas have allowed penetration of the species with a mainly North American distribution into Mexico.
5. Sierra Madre del Sur and Macizo de Oaxaca.
6. Sierra de San Cristóbal and Sierra Madre de Chiapas.

Each of these areas is rich in its own particular group of species (Table 13.2 and Fig. 13.2) but the first, third, and the southern Sierras (areas 5 and 6) are richest in number of species and infraspecific taxa. *Pinus oocarpa, P. leiophylla, P. montezumae,* and *P. cembroides* are widely distributed and occur in several of them.

Endemism Among Mexican Pines

The use of the term "endemic" has been much abused in the literature and without careful qualification will become redundant. The species and infraspecific taxa in Table 13.1 prefixed by an asterisk (*) are considered to be restricted to Mexico; but some of these pines, although confined within the territorial boundaries of the Republic, occupy large areas within it. The

Table 13.2. Centers of species diversity of pines in Mexico

Area	*Pinus* species
1	*P. praetermissa; P. jaliscana; P. pringlei; P. engelmannii; P. herrerae; P. lumholtzii; P. durangensis*
2	*P.* patula subsp. *patula; P. pseudostrobus* subsp. *apulcensis; P. pinceana; P. nelsonii*
3	*P. devoniana* and infraspecific taxa; *P. montezumae* and infraspecific taxa; *P. lawsonii; P. hartwegii*
4	*P. jeffreyi; P. monophylla; P. lambertiana; P. lagunae; P. quadrifolia; P. contorta* var. *murrayana; P. coulteri; P. attenuata*
5 & 6	*P. chiapensis; P. patula* subsp. *tecunumanii; P. ayacahuite; P. maximinoi*

term should preferably be used only to refer to those pines (and indeed any other organisms) that have a localized, limited area of geographical distribution or occur infrequently in habitats exhibiting specialized ecological conditions (species with limited distributions).

Among the pines, several of the Haploxylon group qualify here, including the segregates of *Pinus cembroides* viz *P. remota, P. discolor, P. lagunae, P. catarinae, P. johannis,* and *P. culminicola,* all of which are known only as a single or a few small disjunct populations in northern Mexico. More significant perhaps are the three distinct species also in Subsection *Cembroides: P. nelsonii, P. pinceana,* and *P. maximartinezii.* The first two do not form forests either as pure stands or in mixture but tend to occur as rather isolated trees or in small groups growing with *Pinus cembroides.* Their numbers are probably decreasing. *Pinus maximartinezii* is known only in a single habitat and is threatened. *Pinus rzedowskii* is also restricted in distribution, but little is known about its ecology.

Sympatry Among Mexican Pines

Pines are wind-pollinated plants and perforce frequently grow in dense stands and form large forests. Sometimes such forests can be monospecific, composed of a single species, e.g., *Pinus oocarpa* or *P. pseudostrobus.* On the other hand, they have been found growing in a close mixture with up to seven species composing the forest. Hybridization between sympatrically occurring species seems to be rare, although Styles et al. (1982) have shown that *P. caribaea* var. *hondurensis* and *P. oocarpa* do cross naturally in Honduras. Critchfield (1967) also mentioned that *P. patula* can be easily crossed with *P. oocarpa*. Other artificial crosses cited by Critchfield include *P. patula* X *P. greggii* and *P. patula* X *P. teocote*. There must be many others (Lanner, 1974a).

Many pines in Mexico exhibit clinal variation (associated with geography or altitude), with pairs of species appearing different morphologically at their extremes but with some apparent intermediacy within parts of their range. Such examples are *Pinus chihuahuana* and *P.leiophylla*, *P. jaliscana* and *P. pringlei*, *P. strobiformis* and *P. ayacahuite*, *P. patula* subsp. *patula* and subsp. *tecunumanii*. Others such as *P. montezumae* and *P. hartwegii/rudis* complex might represent a cline associated with altitude.

MAINTENANCE OF DIVERSITY AND DISPERSAL

The hard pines in the group Diploxylon and most of the white pines (Subgenus *Strobus*) have winged seeds that are small, light, and easily dispersed by the wind. All of the pinyon pines and a few white or "soft" pines in Haploxylon have large, wingless seeds. Some of these seeds have been shown to be dispersed by birds, resulting in an interesting symbiosis and mutualism (Lanner & Vander Wall, 1980; Ligon, 1978; Vander Wall & Balda, 1977). The birds responsible are pinyon jays (*Gymnorhinus cyanocephalus*) and Clark's nutcracker (*Nucifraga columbiana*), which collect the pine seeds and carry them often over considerable distances, burying them in the soil in the form of caches. These birds thus form important dispersal agents, and several pines show distribution areas coincident with those of the birds. They have been proved to be responsible for reforestation of burned areas in the southern United States. Of the pine species studied, only *P. remota* occurs in Mexico and may therefore be dispersed by similar means. It is probable that a similar symbiosis exists between other Mexican pinyons and members of the crow family, but it has yet to be established.

PINES AS RENEWABLE RESOURCES

Mexican pines as a forest resource are of great significance and importance to the economy of the country. The timber of many species is used for industrial and commercial purposes, e.g., saw logs, and the number of end uses of such timber is large. It is also one of the best sources of pulp for craft paper and cardboard. Its importance lies in the fact that the xylem of most "hard" pine species produces long fibers, giving the final product extra strength. Among those currently being heavily exploited are *Pinus patula*, *P. oocarpa*, *P. pseudostrobus*, and *P. herrerae*. Frequently mixtures of trees are used without regard to species.

Branches from felled trees are also used for fuel wood, although the resin often causes a great deal of smoke, and broad-leaved species are preferred for cooking. Among poor communities, pine stakes are used as "ocotes" for lighting houses. Charcoal is also increasingly being made from almost any species of *Pinus*. Pine resin forms the basis of the turpentine industry, which is a major source of revenue in several states particularly Oaxaca, Chiapas, Jalisco, Nuevo León, and Michoacán. If the process of resin tapping is carried out carefully, the tree does not appear to suffer any ill effects. Frequently, however, large strips of bark are crudely removed from the tree, allowing entry of pathogens that gradually weaken and kill it. The preferred species appear to be *P. oocarpa*, *P. montezumae*, *P. teocote*, and *P. pseudostrobus*.

The pinyon pines are also an important source of local revenue in northern Mexico, producing edible nuts that may be eaten whole or used as a decoration for confectionary. This local industry is a seasonal one. When the piñones are ripe (late summer to autumn) whole communities move to the forest areas to harvest them. The most important species is *Pinus cembroides*, but nuts of *P. pinceana* and *P. nelsonii* are also gathered. Rzedowski (1964) related that he discovered *P. maximartinezii* only when he noticed its seeds on sale in a local market in Zacatecas; they appeared to be larger than those of the well known *P. cembroides*.

The trunks of *Pinus cembroides* are often too short and too crooked and knotted to form useful saw timber, but they are used in local craft industries and as firewood. As they often form a closed cover on severe sites such as hillsides with shallow soil over rock, such trees help to prevent soil erosion and assist in watershed conservation.

CONSERVATION MEASURES REQUIRED

The importance of pines to the forest industry in Mexico means that timber exploitation is increasing, and in many areas it is indiscriminate. Areas formerly covered with pine forests have now been completely decimated, and the resulting eroded hill and mountain sides are a common feature in the landscape throughout the entire Republic. Saw-mills require and use only trees with the best form, so forests are frequently "creamed" of the straightest, small-crowned specimens, a practice that will ultimately lead to the genetic impoverishment of a species. In forestry terms it means that the superior provenances (trees best adapted to a particular site and set of environmental conditions) are frequently the first to be removed. The trees left to regenerate are often of poor form, stagheaded (or rogue), and as seed trees produce genetically inferior progeny.

Vela (1976) pointed out that *P. patula* is a good example of a pine suffering from exploitation and consequent genetic erosion. It is one of the first species in the genus to be grown as an *ex situ* plantation crop, and demand for its seed is still great.

The plantation method will probably prove the best means for genetic conservation of this taxon, as well as of others. Sadly, though, many other Mexican species are not known commercially outside the Republic and may be destroyed. Their protection in designated national parks is curently the best method for their conservation.

Other reasons for the diminution of a species include indiscriminate destruction of the forest environment for agricultural production, particularly where population pressure is great and the soil is richer. *Pinus chiapensis* in southern Mexico, where many populations of this tree are highly threatened, is a good example. Zamora and Velasco (1977) discussed the problem in detail, but luckily CAMCORE (North Carolina) is making seed collections throughout its range, again for *ex situ* conservation. The species is being depleted because of slash and burn agricultural practices, particularly in the state of Chiapas.

Although many species of pinyon pines are known only from small disjunct populations, at present none is under threat of extinction except *Pinus maximartinezii*. It is the most vulnerable of all the Mexican taxa at species level. This pine is known from a small area in Zacatecas and is possibly restricted to a forest of only 8 km² (Eguiluz, unpublished). Although the forest is fairly isolated, the large nuts are collected annually by local people, who lop off whole branches from trees to obtain the cones. In doing so they frequently destroy young, developing cones. Another major factor that affects regeneration of the tree (as with those of many other species) is the frequency of man-induced fires. Seedlings are destroyed by too frequent burning and by domestic animals that trample the habitat. The whole area is highly degraded.

Requests for financial support have been made to the World Wide Fund for Nature to help conserve this species. Because it is not a pine of major economic importance emphasis must be placed on its great scientific interest as a relict species with potential as a source of high protein food and also as an attractive tree that could be used as an ornamental in parks, arboreta, and botanical gardens in the tropics. Because it is isolated within the pinyon group with its enormous cones and seeds, its "curiosity value" is also great. It is suggested that both *in situ* and *ex situ* stands should be established, the former with the help of the local population (Hughes & Wardell-Johnson unpublished; Eguiluz, unpublished).

The single population of *Pinus caribaea* var. *hondurensis* in Quintana Roó, the only representative of the species in Mexico, was mentioned earlier. Although the stand consists of a small number of poorly formed trees, efforts should be made to preserve this important gene pool. Similar efforts should be made for further Mexican pine gene pools.

ACKNOWLEDGMENTS

I would like to thank many colleagues formerly and presently at the Oxford Forestry Institute, particularly Peter McCarter and Colin Hughes, who over the last 10 years have accompanied me on annual trips to collect pines in Mexico, allowing me to visit almost every State in the Republic. Mexican botanists have always shown me great kindness on my visits to the National Herbarium at UNAM, and I would particularly like to thank Bióloga Teresa Germán and her husband, Biólogo Luis Pinzón, for the hospitality they have always shown me. Miss Christine Brotherton helped me in numerous ways during the preparation of this chapter, and I wish to record my indebtedness for her help during the period. Mrs. Cynthia Styles kindly typed the manuscript. The British Council kindly provided funds for me to attend the symposium.

REFERENCES

Annual Reports of the Commonwealth Forestry Institute for the Years 1975–86. Oxford: Oxford Univ.

Anonymous 1971. Resumen del Inventario Forestal del Estado de Guerrero. Inventario Nacional Forestal. Nota 6, 2–1. Num. 2.

Andresen, J.W. 1964. The taxonomic status of *Pinus chiapensis*. Phytologia 10:417–421.

———. 1966. A multivariate analysis of the *Pinus chiapensis—monticola—strobus* phylad. Rhodora 68:1–24.

——— & R.J. Steinhoff. 1971. The taxonomy of *Pinus flexilis* and *P. strobiformis*. Phytologia 22:57–70.

Bailey, D.K. 1979a. Pinyons of the Chihuahuan Desert Region. Phytologia 44(3):129–133.

———. 1979b. New pinyon records from Northern Mexico. Southw. Naturalist 24(2):389–390.

———. 1983. A new allopatric segregate from and a new combination in *Pinus cembroides* Zucc. at its southern limits. Phytologia 54(2):89–100.

———. 1987. A study of *Pinus* subsect. *Cembroides* I: The single needles pinyons of the Californias and the Great Basin. Notes from Roy. Bot. Gard. Edinburgh 44(2):275–310.

——— & F.G. Hawksworth. 1988. Phytogeography and Taxonomyof the Pinyon Pines. *In* M.F. Passini, T. Cibrian, & T. Eguiluz P. (eds.), 11 Simposio nacional sobre pinos piñoneros. Chapingo, MEX: Centro de Genetrica Forestal A.C. División de Ciencias Forestales. ppl 41–64.

———, K. Snajberk & E. Zavarin. 1982. On the question of natural hybridization between *Pinus discolor* and *Pinus cembroides*. Biochem. Syst. and Ecol. 10(2):111–119.

Barnes, R.D. & B.T. Styles. 1983. The closed-cone pines of Mexico and Central America. Commonw. Forest. Rev. 62(2):81–84.

Barrett, W.H.G. 1972. Variación de caracteres morfológicos en poblaciones naturales de *Pinus patula* Schlecht. et Cham. en Mexico. Idia Suppl. Forestal 7:9–35.

——— & L. Golfari. 1962. Descripción de dos nuevas variedades del Pino del Caribe. Carib. For. 23(2):59–71.

Carvajal, S. 1986. Notas sobre la Flora fanerogámica de Nueva Galicia III (*Pinus*). Phytologia 59(2):127–147.

Chaney, R.W. 1954. A new pine from the Cretaceous of Minnesota and its paleoecological significance. Ecology 35(2):145–151.

Chavelas, P.J. 1981. El *Pinus caribaea* Morelet en el Estado de Quintana Roó, México. Inst. Nac. Invest. For. Nota Tecnica No. 10.

Critchfield, W.B. 1957. Geographic variation in *Pinus contorta*. (Maria Moors Cabot Lodge Foundation, Publ. 3.) Cambridge, MA: Harvard Univ. Press.

———. 1967. Crossability and relationships of the closed-cone pines. Silv. Genet. 16:89–97.

———. 1986. Hybridization and classification of the white pines (*Pinus* Section *Strobus*). Taxon 35(4):647–656.

——— & E.L. Little. 1966. Geographic Distribution of the Pines of the World. U.S. Department of Agriculture, Forest Service. Miscellaneous Publication 991. Washington, DC.

Cuanalo, C.P. 1979. El pino piñonero *Pinus cembroides* en los bosques de México. Durango. Mexico City: INIF.

Debazac, E.F. 1964. Manuel des Coniféres. Nancy. École Nationale des Eaux et Forêts.

Donahue, J. Conservation efforts for endangered whte pines in Southern Mexico. *Pinus chiapensis* and *P. ayacahuite*. IX Congreso Forestal Mundial. Mexico City. In press.

——— & M. Arizmendi. 1986. Natural range of *Pinus chiapensis* and *P. ayacahuite*: found to overlap in Ixtlan, Oaxaca, Mexico. CAMCORE Technical Note No. 1.

Duffield, J.W. 1952. Relationships and species hybridization in the genus *Pinus*. Zeitschrift für Forstgenetik 1(4):93–97.

Eguiluz, T. 1977. Los Pinos del Mundo. Publicaciones Especiales 1. Chapingo, Mexico: Escuela Nacional de Agricultura, Departamento de Enseñanza, Investigación y Servicio en Bosques.

———. 1982. Clima y distribución del género *Pinus* en Mexico. Ciencia Forest. 7(38):30–44.

———. 1985. Orígen y evolución del género *Pinus* (con referencia especial a los pinos mexicanos). Dasonomía Mex. 3(6):5–31.

———. 1988. Distribución natural de los Pinos en México. Nota Tecnica No. 1. Chapingo, Mexico: Centro Genetico Forestal, A.C.

FAO. 1962. Seminario y viaje de estudio de coniferas latinoamericanas. INIF. SAG. Publicacion Especial 1. Rome: FAO.

Favela, L.S. 1991. Taxonomía de *Pinus pseudostrobus* Lindl., *P. montezumae* Lamb. y *P. hartwegii* Endl. reporte Cientifico No. 26. Linares, NL: Facultad de Ciencias Forestales.

Flores Lara, J.E., C.M. Cantú Ayala & J.S. La Fuente (eds.). 1985. Memorias Iᵉʳ Simposium Nacional Sobre Pinos Piñoneros. Linares, Nuevo León: Universidad Autónoma de Nuevo León.

Florin, R. 1963. The distribution of conifers and taxad genera in time and space. Acta Horti Bergiani 20(4):121–312.

Gaussen, H. 1960. Les Gymnospermes actuelles et fossiles. Fasc. IV. Chp. Xl. Generalités, Genre *Pinus*. Laboratoire Forestiére. Tome 2, Sect. 1. Vol. 1, pt. 2. Toulouse, France: Faculté des Sciences.

Harrison, S.G. 1966. *In* W. Dallimore & A.B. Jackson (eds.), *Pinus*. pp. 442–575. Handbook of Coniferae & Ginkgoaceae. 4th ed. London: Edward Arnold Ltd.

Hughes, C.E. & G. Wardell Johnson. Unpublished OFI report (Mimeograph).

Instituto Nacional de Investigaciones Forestales. Subsecretaría Forestal. SARH, 1983. Superficies Forestales de la República Méxicana (Millones de hectareas). Boletin Divulgativo No. 65. INIF, SR SARH.

Koehne, E. 1893. Deutsche Dendrologie. Stuttgart: F. Enke.

Krüssmann, G. 1971. Handbuch der Nadelhölze. Berlin and Hamburg. Paul Parey.

Lamb, A.F.A. 1973. Fast Growing Timber Trees of the Lowland Tropics No. 6. *Pinus caribaea*. Vol. 1. Oxford: UK Commonwealth Forestry Institute.

Lanner, R.M. 1974a. Natural hybridization between *Pinus edulis* and *Pinus monophylla* in the American S. West. Silv. Genet. 23(4):108–116.

————. 1974b. A new pine from Baja California and the hybrid origin of *Pinus quadrifolia*. Southw. Naturalist (19)1:75–95.

———— & S.B. Vander Wall. 1980. Dispersal of timber pine seed by Clark's nutcracker. J. Forestry 78(10):637–639.

Ligon, J.D. 1978. Reproductive interdependence of piñon jays and piñon pines. Ecol. Monog. 48:111–126.

Little, E.L. & W.B. Critchfield. 1969. Subdivisions of the Genus *Pinus* (Pines). USDA. Misc. Publ. 114. Washington, DC.

Loock, E.E.M. 1977. The Pines of Mexico & British Honduras. Bulletin No. 35. 2nd ed. Pretoria: Union of S. Africa, Dept. of Forestry,

Lückhoff, H.A. 1964. The natural distribution, growth and botanical variation of *Pinus caribaea* and its cultivation in South Africa. Ann. Univ. van Stellenbosch. (Ser. A.) 39(1):1–153.

Madrigal, X. 1982. Claves para la identificación de las Coníferas Silvestres del Estado de Michoacán. Mexico City: Boletin Divulgativo INIF. No. 58.

———— & D.M. Caballero. 1969. Una nueva especie mexicana de *Pinus*. Boletin Tecnico 26. Mexico City: INIF.

Martin, P. & B.E. Harrell. 1957. The Pleistocene history of temperate biotas in Mexico and the eastern United States. Ecology 38(3):468–480.

Martínez, M. 1940. Las Pinaceas mexicanas: descripción de algunas especies y variedades nuevas. An. Inst. Biol. Univ. Nac. Méx. 11:57–84.

————. 1948. Los Pinos Mexicanos. 2nd ed. Mexico City: Ediciones Botas.

McVaugh, R. *Pinus. In* W. Anderson (ed.), Flora Novo-Galiciana. Ann Arbor, MI: Univ. of Michigan Press. In press.

Millar, C.I. 1986. The Californian closed cone pines (Subsection *Oocarpae* Little & Critchfield). a taxonomic history and review. Taxon. 35(4):657–670.

Miller, C.N. Jr. 1977. Mesozoic Conifers. Bot. Rev. (Lancaster) 43(2):217–280.

Minnich, R.A. 1986. Range extensions and corrections for *Pinus jeffreyi* and *P. coulteri* (Pinaceae) in northern Baja California. Madroño 33:144–146.

Mirov, N.T. 1967. The Genus *Pinus*. New York: The Ronald Press.

Murray, E. 1983. Unum minutum monographum generis Pinorum (*Pinus* L.). Kalmia 13:11–24.

Narave Flores, H.V. & K. Taylor. Pinaceae. *In* A. Gomez-Pompa (ed.), Flora de Veracruz. Xalapa, VER: Inst. Ecol./Riverside, CA: Univ. of California Press. In press.

Ochoa, G.S. 1973. Clave para la identificación de Coniferas del Estado de Hidalgo. Inventario Nacional Forestal. Nota 43–4. No. 22.

Passini, M.F. 1982. Les forêts de *Pinus cembroides* au Mexique. Mission archeologique et ethnologique francaise au Mexique. Etudes Mesoamericaines 11–5. Cahier No. 9. Paris: Editions recherche sur les civilisations.

————, T. Cibrian, & P. Eguiluz (eds.). 1988. 11 Simposio nacional sobre pinos piñoneros. Chapingo, MEX: Centro de Genetica Forestal A.C. División de Ciencias Forestales.

Perry, J.P. & S.X. Madrigal. 1986. Turpentine Composition of *Pinus rzedowskii* xylem oleoresin. Naval Stores Review 96(4):18–19.

Pilger, R. 1926. The genus *Pinus. In* A. Engler & K. Prantl (eds.), Die Natürlichen Pflanzenfamilien. Vol. 13. Gymnospermae. Leipzig: W. Engelmann. pp. 331–342.

Pineda, R.A., G.S. Ochoa & T.J. Lopez. 1973. Clave dicotómica de identificación de especies de Pino en el estado de Chiapas. Nota INIF. 4.3–3. No. 21:1–22.

Roezl, B. 1857. Catalogue des graines de Conifères Mexicaines en vente chez B. Roezl et Cie, Horticulteurs à Napoles près Mexico. Pour Automne 1857 et Printemps 1858. Mexico City: M. Murguia.

Rushforth, K. 1987. Conifers. London: Christopher Helm.

Rzedowski, J. 1964. Una especie nueva de pino piñonero en el Estado de Zacatecas (México). Ciencia (Mexico) 23(l):17–20.

———. 1966. Vegetación del Estado de San Luis Potósí. Acta Ci. Potos. 5(1–2):5–291.

———. 1978. Vegetación de México. Mexico City: Limusa.

———. 1979. *Pinus. In* J. Rzedowski & G. Rzedowski (eds.), Flora Fanerogámica del Valle de México I. Mexico City: Compañia Editorial Continental. S.A. pp. 65–69.

Schuchert, C. 1935. Historical Geology of the Antillean-Caribbean Region. New York: John Wiley & Sons.

Shaw, G.R. 1909. The Pines of Mexico. Boston, MA: Arnold Arboretum Publication No. 1.

———. 1914. The genus *Pinus*. Boston, MA: Arnold Arboretum Publication No. 5.

Silba, J. 1984. An International Consensus of Coniferae l. Phytologia Memoirs 7. Plainfield, NJ: Moldenke. pp. 47–59.

———. 1985. The infraspecific taxonomy of *Pinus culminicola* Andr. & Beam. (Pinaceae). Phytologia 54(7):489–491.

———. 1986. Encyclopaedia Coniferae. Phytologia Memoirs 8. Corvallis, OR: Moldenke & Moldenke. pp. 1–217.

Standley, P.C. & J.A. Steyermark. 1958. *Pinus. In* P. Standley & L. Williams (eds.), Flora of Guatemala. Fieldiana, Bot. 24:40–56.

Stead, J.W. & B.T. Styles. 1984. Studies in Central American pines: a revision of the "Pseudostrobus" group (Pinaceae). Bot. J. Linn. Soc. 89:249–275.

Steinhoff, R.J. & J.W. Andresen. 1972. Geographic variation in *Pinus flexilis* and *P. strobiformis* and its bearing on their taxonomic status. Silv. Genet. 20(5–6):159–167.

Styles, B.T. 1976. Studies of variation in C. American pines. I. The identity of *Pinus oocarpa* var. *ochoterenae* Martínez. Silv. Genet. 25:109–118.

———, J.W. Stead & K.J. Rolph. 1982. Studies of variation in C. American pines. II. Putative hybridization between *Pinus caribaea* var. *hondurensis* and *P. oocarpa*. Turrialba 32:229–242.

Vander Wall, S.B. & R.P Balda. 1977. Co-adaptations of the Clark's nutcracker and the Piñon pine for efficient seed harvest and dispersal. Ecol. Monog. 47:89–111.

Vela, G.L. 1976. *Pinus patula*, una importante especie mexicana de Pino. Ciencia Forest. 1(1):10–20.

Zamora, C. 1981. Algunos aspectos sobre *Pinus oocarpa* en el Estado de Chiapas. Ciencia Forest. 6(32):25–53.

——— & F. Velasco. 1977. *Pinus strobus* var. *chiapensis*, una especie en peligro de extinción en el Estado de Chiapas. Ciencia Forest. 8(2):3–23.

——— & F.V. Velasco. 1978. Contribución al estudio de los Pinos en el Estado de Chiapas. Bol. Tec. Inst. Nac. Invest. For. Méx. No. 56.

Zavarin, E., & K. Snajberk. 1987. Monoterpene differentiation in relation to morphology of *Pinus culminicola, P. nelsonii, P. pinceana* and *P. maximartinezii*. Biochem. Syst. & Ecol. 15(3):307–312.

———, K. Snajberk & R. Debry. 1980. Terpenoid and morphological variability of *Pinus quadrifolia* and its natural hybridization with *P. monophylla* in N. Baja California and adjoining United States. Biochem. Syst. & Ecol. 8:225–235.

Zobel, B. 1951. The natural hybrid between Coulter & Jeffrey pines. Evolution 5:405–413.

14

The Commelinaceae of Mexico

DAVID R. HUNT

The classification and distribution of the predominantly tropical family Commelinaceae are outlined. Of approximately 40 genera and 650 species, 12 genera and about 100 species occur in Mexico. Ninety percent of these species belong to the tribe Tradescantieae, and 50% of the total are endemics. The principal genera are *Tradescantia* (37 species), *Callisia* (13 species), *Tripogandra* (13 species) and *Gibasis* (11 species), occurring in a range of vegetation types from high broad-leaved forest to coniferous forest and various types of matorral, and from sea-level to 3,700 m.

In terms of the number of species, the richest states are Chiapas and Oaxaca. Chiapas has several Mesoamerican elements not known further to the northwest. As in other families, Oaxaca seems to be a refugium for relictual species, including four of five genera regarded as *Tradescantia* precursors (the other being endemic to Ecuador). However, it is north of the Trans-Mexican Volcanic Belt that the predominantly temperate genera *Tradescantia* and *Gibasis* show their greatest morphological diversity.

The morphological diversity in the northern sections of *Tradescantia* is not matched by cytological diversity, all the species having $n = 6$ large metacentric chromosomes. South of the volcanic belt, however, the picture is different. The now extensive chromosome evidence suggests a tropical origin for the group and subsequent migration into more temperate and seasonal habitats.

The commelinaceae are one of the smaller families highlighted in this volume, with a total of about 650 species mainly in the warmer parts of the world. They are also of relatively slight economic importance, although many are grown as ornamentals. Mexico has a relatively high concentration of species. In all, there are about 100 native and naturalized species, last revised by the late Professor Matuda over 30 years ago (Matuda, 1956).

The Commelinaceae are monocotyledons characterized by having distinct calyx and corolla and a superior syncarpous ovary with few ovules and operculate seeds. It is generally agreed that the family is allied to a number of predominantly South American families (Mayacaceae, Eriocaulaceae, Rapateaceae, Xyridaceae). The Commelinaceae differ from all them in gross morphology and in having a distinct, closed leaf sheath, operculate seeds, and some other characters. In Engler's classification they

Table 14.1. Primary tribal characters for Tradescantieae and Commelineae

System	Tradescantieae	Commelineae
Meisner (1842) Clarke (1881)[a]	Fertile stamens 6 (plus Callisia)	Fertile stamens 3–2
Brückner (1926)	Flowers actinomorphic Embryotega dorsal Stomata four-celled	Flowers zygomorphic Embryotega lateral Stomata six-celled[b]
Woodson (1942) Rohweder (1956)	Cymes paired (or "Tradescantia" type)	Cymes simple

[a]Clarke had a third tribe, Pollieae, for genera with indehiscent fruits.
[b]Includes *Tinantia, Cochliostema, Geogenanthus.*

were associated with the Bromeliaceae and other families in an order called "Farinosae," but modern opinion based on embryological and anatomical evidence suggests they are more closely related to the grasses, sedges, and their allies (Dahlgren et al., 1985).

The family, which contains about 40 genera and 650 species in all, is divided into two unequal subfamilies: (1) Cartonematoideae with two genera and a handful of species; and (2) Commelinoideae with all the rest:

1. **Cartonematoideae**: Raphide canals absent or next to the veins of the lamina; glandular microhairs lacking; flowers yellow, actinomorphic (*Cartonema*, about 6 species, Australia; *Triceratella*, one species, Zimbabwe)
2. **Commelinoideae**: Raphide canals present and never next to the veins of the lamina; glandular microhairs almost invariably present; flowers pink or blue to white, rarely yellow to orange, actinomorphic or zygomorphic (38 genera, about 630 species, mostly tropics and subtropics)

Traditionally, the bulk of the subfamily Commelinoideae has been divided into two principal tribes, Tradescantieae and Commelineae, though the relative scope of the two tribes has varied considerably according to the weight placed on particular characters (Brückner, 1926; Clarke, 1881; Meisner, 1842). Latterly, the tribe Tradescantieae has been restricted to genera that have the peculiar bifacial inflorescence unit of *Tradescantia* itself, leaving the less-studied tribe Commelineae as a kind of "trashcan" for the rest (Rohweder, 1956; Woodson, 1942). (Table 14.1).

In a new classification of the family based on a wider range of characters Faden and Hunt (1991) attached less significance to the specialized inflorescence unit and emphasized anatomical and palynological features.

1. Tribe **Tradescantieae**. Stomata with 4(-2) subsidiary cells, or if 6 then the terminal pair equal to or larger than the second lateral pair; pollen exine lacking spines (except *Tripogandra*), mostly with cerebroid tectum; flowers mainly actinomorphic; filament hairs (when present) moniliform (except *Tripogandra* spp.)
2. Tribe **Commelineae**. Stomata with 6 subsidiary cells, the terminal pair smaller than the second lateral pair; pollen exine spinulose, tectum perforate; flowers actinomorphic or zygomorphic; filament hairs (when present) usually not moniliform

These two major tribes are about equal in numbers of species, but the Commelineae are predominantly Old World. The Tradescantieae has roughly equal-sized components, one endemic to the Old World and one to the New. About one-third of the family occurs in the New World.

Mexico has the relatively high complement of 12 genera and 100 species. These figures are about one-half the New World total and one-sixth the world total. Ninety percent of the Mexican species belong to the tribe Tradescantieae. Fifty percent of the total species are endemics, a figure that corresponds with the estimate for the Mexican Flora as a whole (Rzedowski, this volume). In other words, one-fourth of the New World Commelinaceae are endemic to Mexico.

MEXICAN COMMELINEAE

The rest of this chapter is concerned largely with the Tradescantieae only, although two brief points about the New World Commelineae seem pertinent. Apart from *Murdannia nudiflora*, which is a weedy species adventive in tropical America, *Commelina* itself is the only genus in Mexico and is represented by rather few species, some of which are pantropical and may be adventive in America.

There has evidently been some New World radiation in the group, with the parallel evolution of two indehiscent-fruited species. These have been assigned to two satellite genera, *Phaeosphaerion* (syn. *Athyrocarpus*) and *Commelinopsis*, both of which are to be considered as subgenera (Faden & Hunt, 1987). The species involved are virtually indistinguishable from capsular-fruited species when not in fruit.

MEXICAN TRADESCANTIEAE

Faden and Hunt (1991) divided the Tradescantieae into seven subtribes (Table 14.2), four confined to the Old World and three to the New, implying an early separation of the tribe into these two components. The American component consists of one predominantly South American group (*Dichorisandra* and its satellites) and one North American (*Tradescantia* and

Table 14.2. Subtribes within Tradescantieae (Commelinaceae)[a]

Cymes not fused in pairs:
  Fruit a berry, sepals petaloid ....Palisotinae
  Fruit a capsule:
    Plants scandent ....Streptoliriinae
    Plants rarely scandent:
      Seed-embryotega apical or basal ....Cyanotinae
      Seed-embryotega dorsal or lateral:
        Inflorescences all lateral ....Coleotrypinae
        Inflorescences terminal or terminal and lateral:
          Ovules biseriate; anther-dehiscence sometimes
            poricidal; $x = 19$ ....Dichorisandrinae
          Ovules uniseriate; anther-dehiscence
            longitudinal; $x = 4$–8, 17 ....Thyrsantheminae

Cymes fused in pairs (or if free then contracted, stipitate, with
  geniculate cyme-axis) ....Tradescantiinae

---

[a]Selected characters only.
Source: Faden & Hunt (1991).

its allies). (The circumscription of this subtribe Tradescantiinae, by Faden and Hunt, corresponds to that of tribe Tradescantieae sensu Woodson, as emended by Rohweder.) In addition, there is the small but heterogeneous group, Thyrsantheminae, nearly all of whose species are Mexican.

The subtribe **Dichorisandrinae** has a mixture of primitive and specialized characters, and may be more closely related to the Old World Palisotinae than to the other New World subtribes. It is strictly tropical, and only the relatively widespread and weedy *Dichorisandra hexandra* gets into Mexico, just extending into Chiapas.

The subtribe **Thyrsantheminae**, established by Faden and Hunt (1991), contains six miscellaneous genera that do not have any of the diagnostic features of either the Dichorisandrinae or Tradescantiinae. The subtribe may well be artificial or paraphyletic. All of these plants have derived features, some more than others, but all might be considered to be relicts of clades that diverged from Tradescantioid stock before *Gibasis* and the other genera of the Tradescantiinae—and so to be, in a sense, precursors (Hunt, 1978). Five of the six genera occur in Mexico, and three of them are endemic. The sixth genus, *Elasis*, is endemic to Ecuador. The most primitive-looking is *Thyrsanthemum* itself, which is endemic to Mexico, with three or more species. It has a large thyrsiform inflorescence.

In *Gibasoides*, it is as if the main axis of the thyrse has been contracted, and we are left with an umbel of stalked cymes, resembling *Gibasis*. But the cymes are not congested (at least when fully developed), and several other *Gibasis*-features are lacking.

A similar type of inflorescence is found in *Tinantia*, a genus of annuals, with about 9 of its 13 species in Mexico. *Tinantia* spp. have zygomorphic flowers, superficially resembling those of *Commelina*. *Tinantia erecta* is a widespread weed throughout tropical America, with only two cymes per inflorescence.

The other two Mexican genera are both monotypic and from high elevations, and both are stemless geophytes. *Matudanthus nanus* is endemic to Oaxaca and has stalked, axillary flowers, 1-2 per axil. *Weldenia candida* is found in subalpine pastures from Zacatecas to Guatemala and has a very dense thyrsiform inflorescence with long-tubed flowers.

All inflorescence types found in the Commelinaceae are considered to be derived from a thyrse of simple scorpioid cymes of the general type seen in *Thyrsanthemum* (Brenan, 1966). The most conspicuous unifying feature of the subtribe Tradescantiinae is the contracted cyme. This feature is not unique in the family, as the cymes of the Old World genus *Cyanotis* are also contracted, but in most of the Tradescantiinae the inflorescence is reduced to a pair of such cymes that are virtually sessile and fused back-to-back to form a bifacial unit. This structure is unique, and it is essential to be able to recognize it if one is using an identification key to the family.

In its typical expression, the structure is readily recognized because each of the two cymes is subtended by a conspicuous bract. In various groups, the bracts are obsolete, but it is usually possible to detect that the little "umbel" consists of two cymes. Also, often two flowers open simultaneously in the inflorescence, indicating that it is double. In *Gibasis* we have what may be another precursory condition, where the individual cymes are separate and individually stalked but sharply bent at the junction between the fertile axis and the subtending stalk or stipe. Since the last world monograph of the Commelinaceae by Clarke (1881), the genus *Tradescantia* has been gradually dissected into smaller and smaller units. The same is also true of the other Linnaean genus *Callisia*, which is a much smaller group overall. The subtribe as a whole is capable of division into about 24 such units, but because the total number of species is only about 130 the average genus-size would be amongst the lowest in any group of angiosperms (Hunt, 1975). There is no reason why this should be, other than oversplitting, and nearly all the segregates may be regarded as sections or subgenera within larger units.

This arrangement means the submergence of some generic names used by Matuda (1956) and other students such as the late Dr. H.E. Moore. At present, I recognize four genera only, which are fairly readily distinguished.

1. *Gibasis* (including *Aneilema* sensu Matuda, in part): cymes free, stipitate, two to several in a pseudoumbel, with the cyme-axis reflexed at the junction with the stipe. 11/12? species (11 of the 12[?] species occur in Mexico)

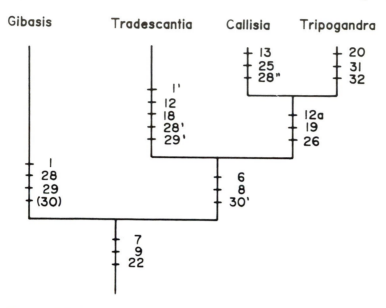

Figure 14.1. Cladogram: Tradescantiinae.

2. *Tradescantia* (including *Campelia, Cymbispatha, Rhoeo, Separotheca, Setcreasea, Zebrina*): cymes fused in bifacial pairs; inflorescence bracts usually well developed, paired; flowers regular; 37/c. 70
3. *Callisia* (including *Aploleia, Cuthbertia, Hadrodemas, Leptorhoeo, Phyodina, Spironema*): cymes fused in bifacial pairs; inflorescence bracts mostly obsolete; flowers regular (rarely stamens 3-1); 13/16
4. *Tripogandra* (including *Neodonnellia*): cymes fused in bifacial pairs; inflorescence bracts obsolete; flowers zygomorphic (stamens dimorphic); 13/22

Figure 14.1 is a simple model of the possible evolution of the Tradescantiinae. The numbered synapomorphies are as follows: 1, habit annual; 6, cyme-pairs fused; 7, cyme-axis condensed; 8, cymes sessile; 9,cyme-axis geniculate; 12, inflorescence-bracts spathaceous; 12a, inflorescence-bracts obsolete; 13, pedicels obsolete; 18, petal-filament tube present; 19, petal color pink/white only; 20, outer stamens staminoidal or absent; 22, anther-connective broad, versatile; 25, stigma penicilliform; 26, seed-hilum punctiform; 28, embryotega dorsal; 29, karyotype symmetric; 30, chromosomes large; 31, pollen verrucate; 32, self-compatible.

ECOLOGICAL AND REGIONAL DIVERSITY

The diversity of the Commelinaceae in Mexico can be further expressed in terms of habit or life-form. Table 14.3 is an analysis of the 100 species

Table 14.3. Strategies of Mexican Commelinaceae

Genus	Perennials[a]				Annual	Total species
	M	S	R	T		
Dichorisandra	1					1
Thyrsanthemum				3		3
Gibasoides				1		1
Tinantia					9	9
Matudanthus				1		1
Weldenia				1		1
Gibasis	4			6	1	11
Tradescantia	3	7	6	20	1	37
Callisia		10		1	2	13
Tripogandra	3	1		9		13
Murdannia					1	1
Commelina				9		9
Total	11	18	6	42	23	100

[a]M = mesophyte; S = leaf-succulent; R = rhizomatous; T = tuberous.

according to whether they are mesophytic perennials (thin-leaved chama-ephytes) of nonseasonal habitats, or variously adapted to more or less severe seasonal drought or cold. The species of seasonal habitats may be xerophytes (leaf-succulent chamaephytes), rhizomatous (hemi-cryptophytes), tuberous (geophytes), or annual (therophytes). The high proportion of geophytes and annuals is a reflection of the strongly seasonal distribution of rainfall in much of Mexico. Most Commelinaceae have succulent stems, but the leaf-succulent habit of many *Callisia* and some *Tradescantia* species is an additional adaptation to survival on rocks and other sites where surface water disappears rapidly even in areas of moderate rainfall. Various Mexican species of *Tradescantia* (and virtually all those in the United States) are shortly rhizomatous. The persistent stem base permits more flexibility in dormancy than occurs in the geophytes, and the habit seems suited to areas where the climate is more temperate and periods of drought less predictable.

Ecologically, many of the Commelinaceae are "weedy" in the sense that they are widespread and often found as colonists of clearings, abandoned cultivation, and so on. Many of the annuals, in particular, are good colonists, especially of damp situations. Most species, however, are characteristic of one or other of the three principal vegetation types of Mexico, i.e., "selva" (mainly tropical forests), "bosque" (mainly oak-pine forests), and "matorral" (scrub forests).

Table 14.4 is an attempt to assign all the species to one of these vegetation types. It suggests that half are characteristic of forest (though this figure includes many of the saxicoles and weeds, as well character-

Selected Floristic Groups of Mexico

Table 14.4. Distribution of Mexican Commelinaceae in the major vegetation types

Genus	Selva	Bosque	Matorral	Total
Dichorisandra	1			1
Thyrsanthemum	2	1		3
Gibasoides		1		1
Tinantia	7	1	1	9
Matudanthus		1		1
Weldenia		1		1
Gibasis	4	3	4	11
Tradescantia	14	10	13	37
Callisia	9		4	13
Tripogandra	6	5	2	13
Murdannia	1			1
Commelina	7	1	1	9
Total	51	24	25	100

Each species is assigned to the type of which it is considered most characteristic, but some species occur in two or all three types. Selva = tropical forests; Bosque = temperate forests; Matorral = scrub forests.

istic forest floor elements), and about one-fourth each in the oak-pine and matorral zones.

Figure 14.2 summarizes the available distributional data on a state-by-state basis. As yet, some states are clearly under-recorded, which merely reflects that botanical effort has hitherto been concentrated in certain states. Nevertheless, it comes as no surprise that, in numbers of species, the richest states are Oaxaca and Chiapas.

The total number of taxa for the two states is 65 (including five geographical varieties). Although the states are adjacent, 19 (38%) of the taxa recorded in Oaxaca do not occur in Chiapas, and 16 (34%) of those in Chiapas do not occur in Oaxaca. I do not know if figures for other families are comparable, but those for Commelinaceae seem to demonstrate rather forcibly the well known importance of the isthmus as a floristic boundary.

Chiapas has several Mesoamerican elements not known further to the northwest or extending into Tabasco and Veracruz but not (so far as is known) into Oaxaca. The higher parts of the state can also be said to be part of a local center of diversity in adjacent Guatemala. These Mesoamerican elements, not recorded from Oaxaca, include *Dichorisandra hexandra*, *Tinantia standleyi*, *Tradescantia huehueteca*, *T. deficiens*, *T. gracillima*, *T. zebrina* sspp. *mollipila* and *flocculosa*, *Callisia warszewicziana*, *C. gentlei* ssp. *macdougallii*, *Tripogandra grandiflora*, *Commelina standleyi*.

As for other families, Oaxaca seems to be a refugium for relictual species, including four of five genera of subtribe Thyrsantheminae regarded as *Tradescantia* precursors. It must be significant that *Thyrsanthemum*, *Gibasoides*,

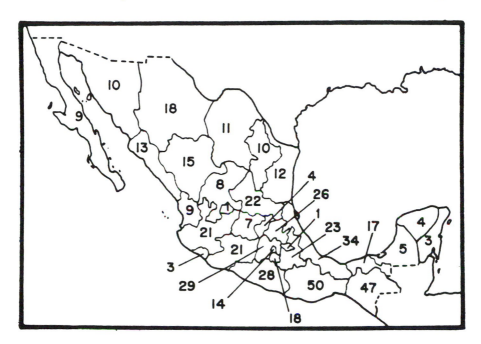

Figure 14.2. Mexican Commelinaceae: taxa recorded per state (1987).

Matudanthus, and *Weldenia* all occur in Oaxaca. The species of particular interest are *Thyrsanthemum floribundum*, possibly the most "primitive" of the Mexican Commelinaceae (it extends to Guerrero, Puebla, Michoacán and Hidalgo but is less common and variable in these states), *Gibasoides laxiflora* and *Matudanthus nanus* (both endemic), and *Weldenia candida*. Other interesting Oaxaca endemics are *Callisia laui* and *Gibasis oaxacana*.

If the higher parts of Chiapas and Oaxaca might be considered rather ancient centers of diversity, the Balsas Depression and adjacent Pacific coastal zone, taking in Michoacán, Guerrero, and Oaxaca, is perhaps another center but more modern, as most of the endemics can be regarded as relatively derived species. The following are all endemic to the zone (those marked "G" extending to Chiapas and Guatemala), and all are regarded as derived species: *Thyrsanthemum goldianum, T. macrophyllum, Gibasis triflora* (G), *Tradescantia llamasii, T. plusiantha, T. orchidophylla, T. mirandae, Callisia soconuscana* (G), *Tripogandra guerrerensis, T. kruseana, T. saxicola*.

On the Atlantic side, on the mostly younger sedimentary and calcareous rocks, we find the xerophytes of the genus *Tradescantia* showing their greatest morphological diversity: *Tradescantia sillamontana, T. rozynskii, T. pinetorum, T. wrightii, T. gypsophila, T. maysillesii, T. nuevoleonensis, T. potosina, T. cirrifera, T. brevifolia, T. buckleyi, T. pallida, T. hirta*. Other endemics of northeastern Mexico are *Tinantia pringlei, Gibasis karwinskyana, Callisia micrantha*, and *C. navicularis*. Further north, in the United States,

Table 14.5. Mexican Commelinaceae: phytogeographic elements

Genus	Am	Rq	Ms	Mx[a]	Total
Dichorisandra	1				1
Thyrsanthemum		3			3
Gibasoides		1			1
Tinantia	1		6	2	9
Matudanthus		1			1
Weldenia		1			1
Gibasis	1	1		9	11
Tradescantia	1		10	26	37
Callisia	4	3	2	4	13
Tripogandra	2		3	8	13
Murdannia	1				1
Commelina	6		2	1	9
Total	17	10	23	50	100

Am, "weedy" species; Rq, relictual and endemic species; Ms, moist tropical ("Mesoamerican") element; Mx, temperate endemic ("Mexican") element.
[a]Some of these elements extend into the southwestern United States (i.e., Mega-Mexico 1 of Rzedowski (see Rzedowski, this volume).

Tradescantia is represented by a whole series of species belonging to an endemic group based on *T. virginiana*.

To summarize the distributional picture, we can visualize the Mexican Commelinaceae as consisting of four phytogeographic components (Table 14.5): (1) the adventive and pan-American "weedy" species (Am); (2) the more or less isolated and endemic relictual genera and species (Rq); (3) the indigenous moist tropical ("mesoamerican") element (Ms); and (4) the indigenous temperate endemic ("Mexican") element, with species-rich areas in the southwestern and northeastern of the country (Mx).

According to Raven and Axelrod (1974), in their classic paper on angiosperm biogeography, "all the genera [of Commelinaceae] north of Panama, except perhaps *Thyrsanthemum* and a few others, may have been derived from South America in Neogene time and subsequently," i.e., during the Miocene, Pliocene, and Pleistocene—the past 27 million years. This statement needs some qualification. Of the four components I have identified, all but the first (Am) are restricted to or show their greatest diversity north of Panama. It seems possible that more than a few of the Tradescantieae could have originated in Central America, perhaps on the so-called Nicaraguan bank, or Nicaraguan Rise. Continental North America is less likely, but several of the principal subgroups in *Tradescantia, Callisia,* and *Gibasis* are endemic north of the isthmus and presumably evolved in North America.

Table 14.6. Chromosome constitution of *Tradescantia* sect. *Cymbispatha*: NF list

| Species | 2n | Karyotype | | NF |
		Metacentric	Acrocentric	
T. cymbispatha	14	—	14	14
T. plusiantha	12	2	10	14
T. poelliae	36	6	30	42
T. standleyi	16	12	4	28
T. commelinoides	14	14	—	28
	22	20	2	42
	28	28	—	56
	30	26	4	56

NF, "nombre fondamental."

KARYOTYPIC EVIDENCE FOR EVOLUTIONARY DISPERSAL

Evidence from cytology generally supports Raven and Axelrod's hypothesis for the evolutionary dispersal of the Tradescantiinae, subject to the qualification made above. Since the 1960s cytological research at Kew by Professor Keith Jones and colleagues has greatly altered and advanced our understanding of chromosome evolution. Prior to this time, the North American *Tradescantia virginiana* group, much used as a classroom example, with few, large, metacentric chromosomes, were thought to be chromosomally primitive and species with more numerous, smaller chromosomes to be derived by a process of fission. The *T. virginiana* karyotype pattern is very widespread in the temperate sections of the genus but is now believed to be derived. It would certainly be improbable—if the family as a whole originated in Gondwanaland and diversified northward, spreading from the tropics into the temperate zone—that *Tradescantia* would have originated at the present northern limit of its range and diversified southward through Mexico.

The probable course of karyotype evolution in the subtribe was first demonstrated by Jones (1974) in *Gibasis pellucida* (*G. schiedeana*). Two morphologically similar but cytologically distinct races of this species have been collected by this writer), one a diploid with $2n = 10$ and the other a tetraploid, but with $2n = 16$, not the expected $2n = 20$. Jones demonstrated a Robertsonian relationship between these two races, fusion of two acrocentric chromosomes in the $n = 5$ race reducing the basic number to $n = 4$.

Subsequently, Jones (1977) postulated a much more complex Robertsonian series to account for an initially baffling series of karyotypes in various species of *Tradescantia* sect. *Cymbispatha*. The clue was the realization that the "nombre fondamental" (NF) of all the species and races (i.e., the total number of chromosome arms) was a multiple of 7 (Table 14.6).

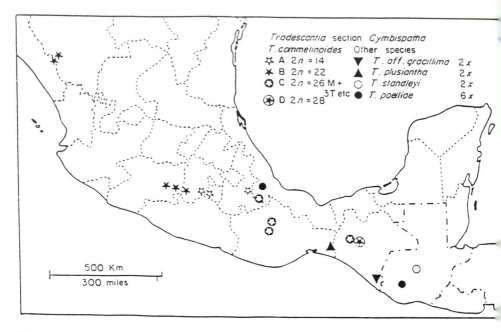

Figure 14.3. *Tradescantia* sect. *Cymbispatha.* Source Jones et al. (1981).

After this it was possible to see the relationship between all the observed karyotypes in terms of successive fusions and chromosome doubling. Figure 14.3 shows that species demonstrating early stages in the series still occur in Mesoamerica, whereas the true *Tradescantia commelinoides*, which demonstrates the later stages, is distributed from Oaxaca to Sinaloa. (Only specimens examined cytologically are plotted; *T. commelinoides* is actually a common element in pine-oak woodland at 2,500–2,900 m).

Figure 14.4 is a model based on Jones's hypothesis of karyotype evolution in *Tradescantia* sect. *Cymbispatha.* The cytological and morphological evidence are consistent and duly point to a tropical origin for the group and its subsequent migration northward into more temperate and seasonal habitats. Though not yet worked out in such detail, the pattern of morphology and karyotypes throughout the Tradescantiinae suggests that the hypothesis will be applicable to the whole of the subtribe.

The concentration of primitive tradescantioids in Oaxaca and the diversity of the Tradescantiinae north of Panama may seem to favor the possibility that the Tradescantiinae as a whole could have originated in North rather than South America. However, there are several lines of evidence pointing to the contrary. One is that the most *Tradescantia*-like of the "Tradescantia-precursors" (*Elasis hirsuta* [Kunth] D. Hunt) is endemic to Ecuador, not Mexico; another is a key section of*Tradescantia* that is endemic to South America and prominent in Southeastern Brazil, Uruguay, and northern Argentina, with several assumedly primitive

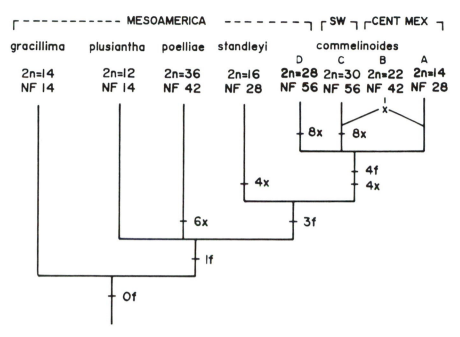

Figure 14.4. *Tradescantia* sect. *Cymbispatha*: chromosome evolution.

features morphologically and karyotypically. This group, *T.* sect. *Austrotradescantia* D. Hunt, which has bimodal complements of small chromosomes, cannot be readily explained in terms of recent dispersal or derivation from one of the other *Tradescantia* sections present in South America, and seems to suggest that an ancestral dichotomy occurred before the subtribe reached Central America.

APPENDIX

Checklist of Mexican Commelinaceae
___

Dichorisandra
D. hexandra (Aublet) Standley

Thyrsanthemum
T. floribundum (Martens & Galeotti) Pichon*
T. macrophyllum (Greenman) Pichon*
T. goldianum D. Hunt*

Gibasoides
G. laxiflora (C.B. Clarke) D. Hunt*

Tinantia
T. leiocalyx C.B. Clarke ex J.D. Smith
T. glabra (Standley & Steyermark) Rohweder
T. longepedunculata Standley & Steyermark
T. parviflora Rohweder
T. standleyi Steyermark
T. violacea Rohweder
T. erecta (Jacquin) Schlechtendal
T. macrophylla S. Watson*
T. pringlei (S. Watson) Rohweder (*Commelinantia pringlei* [S. Watson] Tharp)*

Matudanthus
M. nanus (Martens & Galeotti) D. Hunt (*Tradescantia nana* M. & G.)*

Weldenia
W. candida Schultes f.

Gibasis
G. oaxacana D. Hunt*
G. geniculata (Jacquin) Rohweder
G. pellucida (Martens & Galeotti) D. Hunt (*G. schiedeana* [Kunth] D. Hunt)*
G. pulchella (Kunth) Rafinesque
G. matudae D. Hunt*
G. triflora (Martens & Galeotti) D. Hunt
G. chihuahuensis (Standley) Rohweder*
G. consobrina D. Hunt*
G. karwinskyana (Schultes f.) Rohweder*
        subsp. *karwinskyana*
        subsp. *palmeri* D. Hunt
G. venustula (Kunth) D. Hunt*
        subsp. *venustula*
        subsp. *robusta* D. Hunt
        subsp. *peninsulae* D. Hunt (*G. heterophylla* [Brandegee] Reveal & Hess)
G. linearis (Bentham) Rohweder*
        subsp. *linearis*
        subsp. *rhodantha* (Torrey) D. Hunt

Appendix (cont.)

Tradescantia

T. zanonia (Linnaeus) Swartz (*Campelia zanonia* [L.] Rich.)
T. soconuscana Matuda (*Campelia standleyi* Steyermark)
T. schippii D. Hunt (*Campelia hirsuta* Standley)
T. huehueteca (Standley & Steyermark) D. Hunt (*Zebrina huehueteca* Standley & Steyermark)
T. zebrina Bosse (*Zebrina pendula* Schnizlein)
  var. *zebrina*
  var. *flocculosa* (Brueckner) D. Hunt
  var. *mollipila* D. Hunt
T. spathacea Swartz (*Rhoeo spathacea* [Swartz] Stearn)
T. guatemalensis C.B. Clarke ex J.D. Smith
T. commelinoides Schultes f.*
T. deficiens Brandegee
T. plusiantha Standley
T. poelliae D. Hunt
T. gracillima Standley
T. sillamontana Matuda*
T. rozynskii Matuda*
T. pinetorum Greene
T. wrightii Rose & Bush
T. gypsophila B.L. Turner*
T. cirrifera Martius (*T. tolimanensis* Matuda)*
T. maysillesii Matuda*
T. nuevoleonensis Matuda*
T. potosina D. Hunt*
T. monosperma Brandegee (incl. *T. stenophylla* Brandegee?)*
T. orchidophylla Rose & Hemsley ex Hooker*
T. mirandae Matuda*
T. brevifolia (Torrey) Rose (*Setcreasea brevifolia* [Torrey] Pilger)
T. buckleyi (I.M. Johnston) D. Hunt
T. pallida (Rose) D. Hunt (*Setcreasea pallida* Rose)*
T. hirta D. Hunt (*Setcreasea hirsuta* Markgraf)*
T. pygmaea D. Hunt (*Separotheca pumila* [Greene] Waterfall)*
T. llamasii Matuda*
T. masonii Matuda*
T. exaltata D. Hunt*
T. tepoxtlana Matuda*
T. crassifolia Cavanilles
T. peninsularis Brandegee*
T. burchii D. Hunt*
T. andrieuxii C.B. Clarke (*Setcreasea australis* Rose)*

Callisia

C. warszewicziana (Kunth & Bouche) D. Hunt (*Hadrodemas warszewicianum* (Kunth & Bouche) H.E. Moore)
C. laui (D. Hunt) D. Hunt*
C. micrantha (Torrey) D. Hunt (*Tradescantia micrantha* Torrey)
C. navicularis (Ortgies) D. Hunt (*Tradescantia navicularis* Ortgies)*
C. cordifolia (Swartz) Anderson & Woodson

Appendix (cont.)

C. filiformis (Martens & Galeotti) D. Hunt (*Leptorhoeo filiformis* [M. & G.] C.B. Clarke)

C. multiflora (Martens & Galeotti) Standley

C. monandra (Swartz) Schultes f.

C. gentlei Matuda

[var. *gentlei* is native to Belize]
var. *macdougallii* (Miranda) D. Hunt
var. *tehuantepecana* (Matuda) D. Hunt (ined.)

C. fragrans (Lindley) Woodson*

C. soconuscensis Matuda*

C. insignis C.B. Clarke*

C. repens Linnaeus

Tripogandra

T. angustifolia (Robinson) Woodson*

T. amplexans Handlos*

T. amplexicaulis (Klotzsch ex C.B. Clarke) Woodson*

T. sylvatica Handlos*

T. purpurascens (Schauer) Handlos

subsp. *purpurascens*
[subsp. *australis* Handlos is native to Bolivia and Argentina]

T. disgrega (Kunth) Woodson

T. serrulata (Vahl) Handlos (*T. cumanensis* misapplied)

T. montana Handlos

T. grandiflora (J.D. Smith) Woodson

T. palmeri (Rose) Woodson*

T. guerrerensis Matuda*

T. kruseana Matuda*

T. saxicola (Greenman) Woodson*

Murdannia

M. nudiflora (Linnaeus) Brenan

Commelina

C. erecta Linnaeus (incl. *C. elegans* Kunth)

C. diffusa Burmann f.

C. obliqua Vahl (*C. robusta* Kunth)

C. rufipes Seubert

var. *rufipes*
var. *glabrata* (D. Hunt) Faden & D. Hunt (*C. persicariifolia* misapplied)

C. tuberosa Linnaeus (divisible into well-marked varieties corresponding to *C. tuberosa* sensu stricto, *C. coelestis* Willd., *C. dianthifolia* Delile, and *C. elliptica* Kunth)

C. scabra Bentham*

C. standleyi Steyermark

C. texcocana Matuda (*C. pallida* misapplied)

C. leiocarpa Bentham (*Phaeosphaerion leiocarpum* [Bentham] Hasskarl ex C.B. Clarke)

---

*Endemic to Mexico.

REFERENCES

Brenan, J.P.M. 1966. The classification of Commelinaceae. J. Linn. Soc. Bot. 59:349–370.

Brückner, G. 1926. Beiträge zur Anatomie, Morphologie und Systematik der Commelinaceae. Bot. Jahrb. Syst. 61, Beibl. 137:1–70.

Clarke, C.B. 1881. Commelinaceae. *In* De Candolle A. & De Candolle, C. (ed.), Monographiae Phanerogamarum. Vol. 3. Paris: G. Masson. pp. 113–324.

Dahlgren, R.M.T., H.T. Clifford & P. Yeo. 1985. The families of the Monocotyledons. Berlin: Springer-Verlag.

Faden, R.B. & D.R. Hunt. 1987. Reunion of *Phaeopshaerion* and *Commelinopsis* with *Commelina* (Commelinaceae). Ann. Missouri Bot. Gard. 74:121–122.

———. 1991. The classification of Commelinaceae. Taxon 40(1):19–31.

Hunt, D.R. 1975. The reunion of *Setcreasea* and *Separotheca* with *Tradescantia*. Kew Bull. 30:443–458.

———. 1978. Three new genera in Commelinaceae. Kew Bull. 33:331–334.

———. 1980. Sections and series in *Tradescantia*. Kew Bull. 35:437–442.

———. 1986. A revision of *Gibasis* Rafin. Kew Bull. 41:107–129.

Jones, K. 1974. Chromosome evolution by Robertsonian translocation in *Gibasis* (Commelinaceae). Chromosoma (Berlin) 353–368.

———. 1977. The role of Roberstonian change in karyotype evolution in higher plants. *In* A. de la Chapplle & M. Sorsa (eds.), Chromosomes Today. Vol. 6. Amsterdam: Elsevier/North-Holland. pp. 121–129.

———, A. Kenton & D.R. Hunt. 1981. Contributions to the cytotaxonomy of the Commelinaceae: Chromosome evolution in *Tradescantia* section *Cymbispatha*. J. Linn. Soc. Bot. 83:157–188.

Matuda, E. 1956. Las Commelinaceas Mexicanas. Anales Inst. Biol. Univ. Nac. Méx. 26:303–432.

Meisner, C.F. 1842. 311. Commelinaceae. *In* Plantae vascularium genera. Leipzig: Weidmann. pp. 406–407.

Raven, P.H. & D.I. Axelrod. 1974. Angiosperm biogeography and past continental movements. Ann. Missouri Bot. Gard. 61:539–673.

Rohweder, O. 1956. Die Farinosae in der Vegetation von El Salvador. In Abh. Geb. Auslandsk. 61(C.18):(Commelinaceae: 98–178).

Woodson, R.E. 1942. Commentary on the North American genera of Commelinaceae. Ann. Missouri Bot. Gard. 31:138–151.

Chorology of Mexican Grasses

JESÚS VALDÉS REYNA
AND ISMAEL CABRAL CORDERO

A chorological study of Mexican grasses describing their distribution and diversity is presented. Based on herbarium studies and the literature, a two-way matrix was produced with number of species by states as unit areas. Maps showing the distribution patterns of selected subfamilies, tribes, and endemic species are provided. High diversity regions of subfamilies Pooideae, Chloridoideae, and Panicoideae are identified, and centers of endemism in the family in Mexico are pointed out.

Poaceae, variously estimated between 700 and 800 genera and 8,000 and 10,000 species, is the fourth largest vascular plant family. This ecologically important cosmopolitan family, which includes important cereals, is economically a major group. Grasslands comprise about one-fourth of the world's vegetation cover (Schantz, 1954), and in Mexico they comprise 10–12% of the national territory (Flores M. et al., 1971). Mexican grasses are composed of 183 genera and about 1,151 species, including introduced and cultivated ones (Beetle, 1983, 1987; Gould, 1979). Most studies pertaining to Mexican grasses have been floristic, and some have contributed to our understanding of the evolution and distribution of species in the country (Hernández X., 1959, 1964; Johnston, 1940; Miranda, 1960; Rzedowski, 1962, 1965, 1975, 1978; Sharp, 1953). This chapter presents a chorological study of the grasses of Mexico and the distribution patterns in selected subfamilies and tribes. This information has implications for programs aiming to raise the benefits obtained from better management of grasslands and rangelands.

   The methods used in this study are those discussed and elaborated by Clayton and Hepper (1974) and Valdés and Espinosa (1987). However, for the purposes of this study, the unit areas, which are the basic units of the inclusion of data, are equivalent to the states of the country. This method has facilitated easy access to information from herbarium specimens. The species were the basic sampling units. The data used were obtained from agrostological lists, both regional and national. The lists of Mexican Gramineae presented by Beetle (1977, 1987) were the primary source.

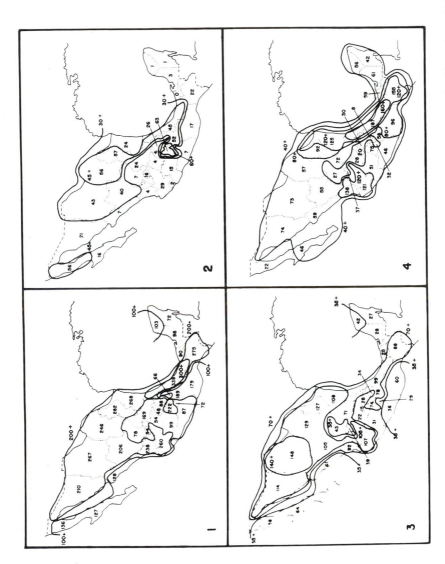

Figure 15.1. (Part one) Mexican grasses (species numbers) by state. 1. Distribution of family. 2. Subfamily Pooideae (*n* = 171). 3. Subfamily Chloridoideae (*n* = 370). 4. Subfamily Panicoideae (*n* = 378).

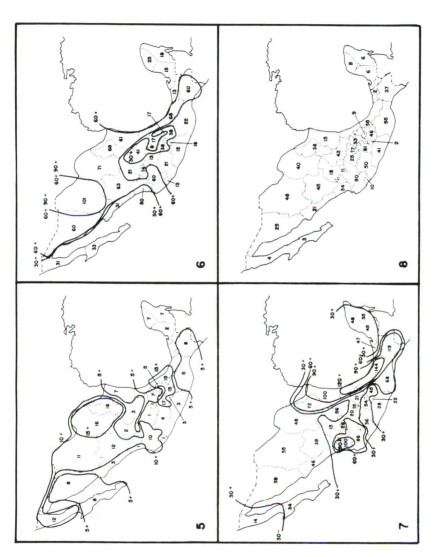

Figure 15.1. (Part two) Mexican grasses (species numbers) by state. 5. Tribe Poeae (*n* = 49). 6. Tribe Eragrostideae (*n* = 221). 7. Tribe Paniceae (*n* = 291). 8. Endemics.

Additions and nomenclatural changes were made based on ongoing studies (Valdés, unpublished). The phylogenetic arrangement of subfamilies, tribes, and genera follows Clayton and Renvoize (1986). Collections of Mexican grasses in several herbaria were consulted.

Several groups of species were excluded from this study: The 153 introduced or cultivated species; the bamboos (42 species), which are insufficiently known because of poor collecting; and 13 species that proved to be conspecific and seven species whose taxonomy was uncertain. The inclusion of such groups in the general analysis could lead to difficulties in interpretation and possible erroneous conclusions (Clayton & Hepper, 1974). The remaining 936 species were analyzed chorologically. A matrix of similarity, using Jaccard's coefficient (Sneath & Sokal, 1973), was produced to compare the unit areas in terms of species numbers.

Each species for each subfamily and tribe was analyzed, and the results are presented in maps. On these maps, isochores, defined by Good (1964) as lines that join numbers of species belonging to the same chorological order were traced that contain areas with a similar range of diversity. They are regions that present a high concentration of heterogeneity and species numbers. As a means of detecting the most important chorological elements of Mexican grasses, the representative data of each subfamily and tribe by unit area were analyzed following Valdés and Espinosa (1987).

The endemic factor was considered separately, following Clayton and Hepper (1974), who stressed its importance, as it represents the smallest chorological element. In this study species with highly restricted or regional distribution in Mexico were considered endemic. Species with wide distributions throughout the country were not considered endemic. A total of 272 species were found to be endemic to the country.

RESULTS AND DISCUSSION

The distribution pattern of 936 species are presented. Figure 15.1 gives the total number of species for each unit area. The presence of more than 48 species in all unit areas emphasizes the cosmopolitan nature of the family. The unit areas with the lowest number of species are Querétaro (48) and Guanajuato (54), which may be due to their small area. Moreover, these states have not been well collected. The results of this study generally support the observation by Cross (1980) that grasses are abundant in "open communities." Thus semiarid grasslands in the Central Plateau stretch as a belt from the Sierra Madre Occidental (including northeastern Sonora) toward northeastern Jalisco. The area includes portions of Chihuahua, Coahuila, Durango, Nuevo León and San Luis Potosí. Grasslands in alpine areas have a distribution restricted to high mountain areas in central and southern Mexico (Rzedowski, 1975, 1978).

Taxonomic Groups: Distribution of Selected Subfamilies

Subfamily Pooideae (Fig. 15.1) consists of C_3 species. It is characteristic of temperate climates of the Northern Hemisphere but is also present in South America (Hartley, 1973). The subfamily has the highest concentrations of species in the states of México (63), Nuevo León (57), Coahuila (56), Baja California Norte (56), and Puebla (52), with a total of 171 species for the country. The areas in which these species occur are those with some of the highest mountains and peaks and are also areas of highest species diversity. This finding suggests a relation between dry to subhumid temperate environments and the occurrence of these species. In the north and center of the country these species have adapted to temperate climates. Baja California Norte, with 41.1%, is the highest center of diversity for Pooideae.

Subfamily Chloridoideae (Fig. 15.1) is considered to be tropical and dry subtropical in origin, consisting mostly of C_4 species. It is best represented in northern Mexico where semiarid grasslands abound. The unit areas with the highest number of species are the states of Chihuahua (148), Coahuila (129), Nuevo León (127), Sonora (114), Tamaulipas (108), Durango (105), and Jalisco (107), with a total of 370 species for the entire country. Chloridoideae are most heterogeneous in the subtropical and arid regions of the north and center and bear an affinity with tropical environments or those that range from subtropical to arid. These results support the views of Cross (1980), Hartley (1950), and Hartley and Slater (1960), who suggested that this primarily tropical subfamily has secondarily radiated in temperate latitudes.

Subfamily Panicoideae (Fig 15.1) consists largely of C_4 plants (some species of *Panicum* are C_3) and is considered to be tropical in origin. The highest species-rich areas of this subfamily are in southern Mexico as well as along the Sierra Madre Oriental in the states of Tamaulipas and Nuevo León. The species distribution by unit area is Veracruz (187), Chiapas (155), Nayarit (138), Tamaulipas (125) and Jalisco (121), with a total of 378 for the country. Panicoideae and Chloridoideae are the two subfamilies that, at the species distribution level, occur in environments ranging from subtropical to arid. Panicoideae, however, is dominant in the south; in the north it is found in areas with tropical or subtropical climate and in aquatic habitats.

Distribution of Selected Tribes

Tribe Poeae (Fig. 15.1), which belongs to subfamily Pooideae, has two important centers of diversity. The first is in the north of the country, where it has 16 species each in the states of Coahuila and Nuevo León, and 12 each in Baja California Norte and Durango. Their occurrence in these areas may be considered range extensions of the species that are widely distributed in the western and southwestern United States (mainly species of *Bromus*, *Poa* and *Festuca*). The second area is in central Mexico in the states of México

Table 15.1. Genera with five or more species endemic to Mexico

Genus	No. of species
Stipa	10
Poa	7
Festuca	6
Bromus	5
Agrostis	9
Calamagrostis	6
Zeugites	5
Aristida	17
Muhlenbergia	52
Sporobolus	9
Bouteloua	12
Panicum	12
Paspalum	20
Axonopus	8
Andropogon	6

(17 species), Puebla (15) and Veracruz (15). It is found in the Trans-Mexican Neovolcanic Belt and in high mountain regions where endemism is high. This tribe, temperate in origin, is poorly represented in the south.

Tribe Eragrostideae (Fig. 15.1), like Chloridoideae, is most abundant in northern Mexico. Chihuahua, with 101 species representing 30% of its grass flora, may be considered the tribe's center of diversification in Mexico.

Tribe Paniceae (Fig. 15.1) is abundant throughout the country. The south has the highest concentration of species, mainly in the area of the Gulf of Mexico. It is common in Chiapas (119), Veracruz (144), Tamaulipas (100), and Sinaloa (100). Its distribution extends from tropical to subtropical climates, being most prolific in grassland communities dominated by soils with poor drainage and in habitats with high humidity levels.

Endemism

Centers of endemism are important, as endemics represent the smallest chorological units (Fig. 15.1). About 30%, or 272 species, are endemic to Mexico. Many of these elements may be paleoendemics or neoendemics, but in order for their nature to be known cytological and other studies are necessary. Genera with five or more species endemic to the country are listed in Table 15.1.

Chloridoideae has the highest number of endemics, with 73 species, of which 61 belong to tribe Eragrostideae. Fifty-two species of *Muhlenbergia* (Eragrostideae) are endemic. Panicoideae is third with 46 and Pooideae fourth with 43. The states with the highest number of endemic species are Jalisco (96), Mexico (81), Veracruz (55), and Oaxaca (55).

CONCLUSIONS

Poaceae (with 1,151 species, including the cultivated and introduced) is the third largest family in Mexico after Asteraceae and Fabaceae. Rzedowski (this volume) has suggested that Orchidaceae may in fact exceed grasses in number but the latter is more cosmopolitan.

The tropical Panicoideae is the richest subfamily followed by the subtropical Chloridoideae and the temperate Pooideae. These proportions illustrate the latitudinal species richness gradients in the distribution of grasses in general.

Tribe Poeae has two important centers of diversity: northern and central Mexico. The affinities of the northern species are with those of the grassland regions of western and southwestern United States. The high montane species of central Mexico show affinities with those of the vegetation of some areas of South America. Tribe Eragrostideae is most diverse in northern Mexico and in the central plateau, where semiarid grasslands abound. This region has characteristic soil conditions, where numerous grasses are adapted to live under special edaphic conditions, e.g., high alkalinity and excess of soluble salts, gypsum, and other substances. This tribe is most diverse here. The Paniceae is best represented in the humid and subhumid zones of central and southern Mexico.

Mexico has a total of 272 endemic species of grasses. The highest number of endemics are in Eragrostideae. The high level of endemism in the family (30%) points to the autochthonous origin of this flora, which is associated with the climatic conditions and the special soil conditions that restrict the distribution of these species (Rzedowski, 1975).

Acknowledgments

We thank the following colleagues of Universidad Autónoma Agraria Antonio Narro: Antonio Martínez H. for the maps, Miguel A. Carranza P. for bibliographic help, and Javier Espinosa A. for help with statistical analysis. T.P. Ramamoorthy, Marguerite Elliott and Fernando Chiang generously provided editorial assistance.

REFERENCES

Beetle, A.A. 1977. Noteworthy grasses from Mexico. V. Phytologia 37(4):317-407.
———. 1983. Las Gramíneas de México. 1. Comisión Técnico Consultiva para la determinación de los coeficientes de Agostadero. Mexico City: SARH.

————. 1987. Noteworthy grasses from Mexico. XIII. Phytologia 63(4):209–297.

Clayton, W.D. & F.N. Hepper. 1974. Computer aided chorology of west African grasses. Kew Bull. 29:213-234.

———— & S.A. Renvoize. 1986. Genera Graminum. Grasses of the world. Kew Bull. Additional series XIII:534 pp.

Cross, R.A. 1980. Distribution of subfamilies of Gramineae in the Old World. Kew Bull. 35(2):279-289.

Flores Mata, G., J. Jiménez López, X. Madrigal Sánchez, F. Moncayo Ruiz & F. Takaki Takaki. 1971. Memoria del mapa de tipos de vegetación de la Republica Mexicana. Mexico City: SRH.

Good, R. 1964. The Geography of the Flowering Plants. London: Longman.

Gould, F.W. 1979. A key to the genera of Mexican grasses. Tex. Agric. Exp. Sta. Texas A & M Univ. Syst. MP-1422.

Hartley, W. 1950. The global distribution of tribes of the Gramineae in relation to historical and environmental factors. Austral. J. Agric. Res. 1(4):355-373.

————. 1958. II. The tribe Paniceae. Austral. J. Bot. 6:343-357.

————. 1973. V. The subfamily Festucoideae. Austral. J. Bot. 21:201-234.

———— & C. Slater. 1960. III. The tribes of subfamily Eragrostoideae. Austral. J. Bot. 8:256-276.

Hernández X., E. 1959. Patrones de distribución de algunos zacates mexicanos. Chapingo 12(77,78):392-398.

————. 1964. Los pastos y pastizales. *In* E. Beltrán (ed.), Las Zonas Aridas del Centro y Noreste de México y el Aprovechamiento de sus Recursos. Mexico: IMERNAR. pp. 97-133.

Johnston, I.M. 1940. The floristic significance of shrubs common to north and south American deserts. J. Arnold Arb. 21:356-363.

Miranda, F. 1960. Posible significación del porcentaje de géneros bicontinentales en América tropical. Anales Inst. Biol. Univ. Nac. Méx. 30:117-150.

Rzedowski, J. 1962. Contribuciones al fitogeografía florística e histórica de México I. Algunas consideraciones acerca del elemento endémico en la flora mexicana. Bol. Soc. Bot. México. 27:52-65.

————. 1965. Relaciones geográficas y posibles orígenes de la flora de México. Bol. Soc. Bot. México. 29:121-177.

————. 1975. An ecological and phytogeographical analysis of the grasslands of Mexico. Taxon 24(1):67-80.

————. 1978. Vegetación de México. Mexico City: Limusa.

Schantz, H.L. 1954. The place of grasslands in the earth's cover of vegetation. Ecology 35:143-145.

Sharp, A.J. 1953. Notas acerca de la flora mexicana: distribución mundial de las familias de dicotiledoneas y el origen de la vegetación actual. Mem. Cong. Cient. Mex. 6:343-351.

Sneath, P.H.A. & R.R. Sokal. 1973. Numerical Taxonomy. The Principles and Practice of Numerical Classification. San Francisco, CA: W.H. Freeman.

Valdés Reyna, Jesus & J. Espinoza A. 1987. Corología de las gramíneas de Coahuila. Agraria 3(2):109-136.

16

The Genus *Quercus* in Mexico

KEVIN C. NIXON

There are about 135–150 species of *Quercus* in Mexico, compared with about 87 species in the United States and Canada. The highlands of central and eastern Mexico are a major center of diversity for the genus. Three sections of *Quercus* occur in Mexico, but delimitation of species groups within sections is difficult. However, within Mexican white oaks (section *Quercus*, with about 75 species in Mexico), two clearly defined species groups (subsections *Glaucoideae* and *Virentes*) can be identified. White oaks as a group have broad ecological tolerances and are relatively more diverse in xeric areas than are the red oaks. This point is particularly true of subsection *Glaucoideae*, which is concentrated in the drier parts of Mexico. Western North American leaf fossils from Tertiary deposits resemble some modern species of Mexican red oaks. Further studies of variation within the red oak group may provide insight into early diversification of the genus in North America and shed light on the relationships between the three North American sections of oak.

Quercus L. ("encino," "roble," oak) is one of the most important genera of woody plants in the Northern Hemisphere. The genus is widely distributed in the New World from southern Canada to Colombia, in South America. In the Old World, oaks are found from northern Europe and North Africa across the Mediterranean region, through the montane parts of southern Asia to Kamchatka, Korea, Japan, and south to southeast Asia and Malaysia. Although *Quercus* is widely thought of as a temperate genus, there are numerous species in the montane tropics and subtropics of southeast Asia and Latin America. In eastern Asia and Malesia, *Quercus* (as subgenus *Cyclobalanopsis*) is an important component of lowland tropical forests in many areas (Soepadmo, 1972). Although some lowland species occur in the American tropics, including Mexico, they are relatively few in number compared to the large number of montane species. In Mexico, *Quercus* species are concentrated in what is often termed the oak-pine forest, but they occur in a diverse array of habitats. Oaks exhibit a tremendous range of growth forms in Mexico, from low rhizomatous shrubs on dry slopes and high mountain peaks to massive forest trees with buttress roots in wet lowland forests.

Throughout the range of the genus, *Quercus* species are of exceptional value for high quality lumber and provide numerous other products including firewood, charcoal, cork, tannins, dyes, food for humans and livestock, ornamental shade trees, and habitat for wildlife and human recreation.

Of the six genera of Fagaceae native to the Western Hemisphere, only *Fagus* L. (beeches) and *Quercus* are known to occur naturally in Mexico. *Fagus*, with about 10 species distributed broadly in the temperate parts of the Northern Hemisphere, is represented by only one species in Mexico, *F. mexicana* Martinez. *Fagus mexicana* appears to be closely related to *F. grandifolia* Ehrh. of the eastern United States and is considered a variety of that species by some authors (Little, 1971). However, no phylogenetic studies of the genus *Fagus* have been undertaken to resolve the question of the relationships of these two species. *Fagus mexicana* appears to have a relictual distribution and is known only from four cloud forest localities in Puebla, Hidalgo, and Tamaulipas.

SPECIES CONCEPTS IN *QUERCUS*

Although the focus of this chapter is not taxonomic, species concepts greatly affect estimates of diversity for any group. Oaks are often cited as somehow unusual because of the great number of reported natural hybrids in the genus. Such hybrids are often reported between species that are morphologically and phylogenetically distant. It is clear that application of a "biological species concept" in *Quercus* would result in a nonsensical pattern of species, with low information content. Such a species concept would lump together polyphyletic elements, as well as elements that occur sympatrically in some areas without interbreeding but that interbreed freely in other areas. In order to avoid these problems, I have adopted a phylogenetic species concept similar to the concepts proposed by Rosen (1979) and Cracraft (1982). As I have applied this concept, species should be differentiable on the basis of stable attributes, regardless of whether morphological intergradation may occur in zones of contact (Nixon & Wheeler, 1990). The ability to interbreed is an ancestral feature that cannot be used to determine phylogenetic relationships (Rosen, 1979). The phylogenetic species concept used here is essentially similar to the morphological species concept used by most other "classical" workers in the genus. However, the theoretical basis for a phylogenetic species concept is explicit and related to phylogenetic information of the classification. Application of a phylogenetic species concept in general results in higher numbers of recognized species and fewer numbers of subspecies and varieties. This finding is consistent with classical of treatments in *Quercus*, where few infraspecific taxa have been recognized.

FOSSIL HISTORY

Megafossils assignable to *Quercus* are not known from Mexico. However, fossil *Quercus* leaves, as both impressions and compressions, are relatively common at numerous Tertiary localities in the western United States. In addition, other fagaceous remains occur at both western and eastern sites in the United States.

There remains some difficulty in distinguishing between fossil acorns of *Quercus* and *Lithocarpus*, which have a superficially similar cupule and fruit morphology. Earliest fossil acorns are from the Eocene of Oregon (Manchester, 1983). These acorns have cups with concentrically arranged scales, a character found only in extant Asian *Quercus* and *Lithocarpus* and that is absent from all North American Fagaceae. The Eocene fossils cannot be assigned unequivocally to *Quercus* at this time.

Fossil leaf compressions from the Oligocene of Texas clearly resemble modern species of red oak now found in Mexico, e.g., *Quercus sartorii* Liebm or *Q. acutifolia* Nee (Daghlian & Crepet, 1982). Additional fossils of staminate catkins and pollen associated with these leaves also suggest a red oak affinity (Crepet & Nixon, 1989b). These fossils are the earliest unequivocal remains of oak, although leaves that have been assigned to *Quercus* occur in older (Eocene) deposits. Occurring with *Quercus* in the Texas Oligocene deposits are well-preserved fossil infructescences, fruits, and leaves with characteristics of the extant fagaceous genus *Trigonobalanus* Forman. *Trigonobalanus* as defined by Forman (1964) and others (Lozano-C. et al., 1979) is paraphyletic and should be treated as three monotypic genera: *Trigonobalanus*, *Formanodendron* Nixon & Crepet, and *Colombobalanus* Nixon & Crepet. The three trigonobalanoid species, two from Southeast Asia and one from Colombia, South America, appear to be relicts of a widely distributed ancestral plexus from which modern *Quercus* emerged (Nixon & Crepet, 1989; Crepet & Nixon, 1989a,b).

Numerous fagaceous fossils assignable to subfamily Castaneoideae also occur in western North American Tertiary deposits, and the earliest unequivocal fossils of Fagaceae belong in this subfamily (Crepet & Nixon, 1989a). Subfamily Castaneoideae includes the modern genera *Lithocarpus* Blume (ca. 150–200 species, Asia; 1 species, western United States), *Castanopsis* (D. Don) Spach (ca. 200 species, Asia), *Chrysolepis* Helmq. (2 species, western United States), and *Castanea* L. (ca. 10 species, North America, Europe, and Asia). In tropical and subtropical Asia, numerous species of the genera *Lithocarpus* and *Castanopsis* are found. The absence of any modern trigonobalanoids or castaneoids in Mexico, in light of their broad distribution in North America during the Tertiary, is of great interest. The extinction of these elements in western North America (and, by implication, in Mexico) is correlated with the general drying and cooling trends of the late Tertiary. *Quercus*, on the other hand, survived this climatic deterioration, and is now dominant over large forested areas in both eastern North America and Mexico.

DISTRIBUTION AND DIVERSITY OF *QUERCUS* IN MEXICO

All of the major mountain ranges in Mexico are rich in species of *Quercus*, and oak is a major component in terms of biomass, particularly in the pine-oak forests, chaparral (matorral), encinal, and cloud forests (Rzedowski, 1978). In addition, the mountain ranges of the Chihuahuan Desert region and the altiplano to the south and west harbor numerous species of oak. The most comprehensive treatment of the oaks of Mexico was included in Trelease's monograph of the American oaks (Trelease, 1924). However, this treatment is not useful for determining numbers of species and distributions because Trelease recognized a far greater number of species than are recognized by more recent workers (McVaugh, 1974; Muller & McVaugh, 1972). Although a complete survey of the species of *Quercus* in Mexico is not possible at this time, some recent treatments, as well as work in progress by the author, allow a preliminary evaluation of species diversity in some parts of Mexico. The estimates provided here are intended only for broad comparisons and certainly will change slightly as our knowledge of the oaks of Mexico improves. These estimates are weakest for estimates of Central American species and strongest for the areas that are best known, e.g., where recent treatments are completed or in progress (Jalisco, Michoacán, Chihuahuan Desert, Coahuila, Nuevo León, and Tamaulipas).

BIOGEOGRAPHIC PATTERNS

On a worldwide basis, *Quercus* can be divided into two subgenera and five sections. Under this classification, three sections of *Quercus* are found in Mexico, section *Quercus* (the white oaks), section *Rubrae* (the red oaks), and section *Protobalanus* (the intermediate oaks). Sections *Rubrae* and *Protobalanus* are endemic to the New World, and section *Quercus* is widespread in Europe and Asia as well. The three sections that occur in Mexico are discussed in greater detail below.

Section *Rubrae*, the red oaks, although retaining several plesiomorphic (ancestral) features, are clearly a monophyletic assemblage and are defined by a derived pistillate floral morphology (a calyx "flange") that is absent in the rest of *Quercus*. My studies have concentrated on the white oaks of Mexico, and consequently estimates of the numbers of species of red oak are much less certain at this time. In general, the red oaks seem to be more restricted ecologically in their distribution than are the white oaks. This point is indicated by the relative paucity of red oak species in the drier parts of North America, where white oak species clearly predominate. Although this general pattern holds, there are particular species of red oak in Mexico that may occur in relatively xeric habitats.

Section *Protobalanus* is perhaps the most interesting group of North American oaks. It includes five species, four of which occur in Mexico. One species, *Q. cedrosensis* Muller, is apparently endemic to Mexico but comes within 10 km of the international border near Tijuana. This group is of great interest because of its uncertain phylogenetic affinities and restricted distribution. The five species are restricted to northwestern Mexico and the southwestern United States and tend to occur in relatively dry habitats.

Quercus tomentella Engelm. is the only North American oak entirely re-
stricted to islands. It is found on Isla Guadalupe off the coast of Baja
California Norte, as well as several of the channel islands off coastal
California.

Section *Quercus* (the white oaks) seems to have the broadest ecological
distribution of the three oak sections, ranging from massive trees of wet
tropical forests (e.g., *Q. insignis* M. & G.) to low shrubs of dry mountain
slopes (e.g., *Q. intricata* Trel.). Taxonomic studies of this group have also
been successful in identifying some potentially monophyletic subunits,
such as subsection *Glaucoideae* (Trelease) Camus and subsection *Virentes*
(Trelease) Camus. These subsections are important in Mexico and deserve
some discussion here. Both subsection *Glaucoideae* and subsection *Virentes*
have fused cotyledons, a character that is clearly apomorphic in *Quercus*
and suggests that two groups are closely related. Subsection *Virentes* also
is characterized by fused-stellate foliar trichomes, which are absent in
Glaucoideae and throughout most of *Quercus* with few, clearly independently derived exceptions, e.g., *Q. cornelius-mulleri* Nixon & Steele (Nixon
& Steele, 1981). Whether *Glaucoideae* is paraphyletic and ancestral to
Virentes cannot be determined without further work.

In some genera, e.g., *Fagus, Carpinus, Ostrya, Liquidambar, Acer,* and
Carya, species in eastern Mexico are considered to be conspecific with or
closely related to eastern North American species. However, this pattern
does not hold for *Quercus*. With few exceptions, there are no close
phylogenetic relationships between the endemic oak species of Mexico and
those of the eastern United States. A few species that are widespread in the
eastern United States, e.g., *Quercus muhlenbergii* Engelmann and *Q. sinuata*
Walt., have spotty distributions in the Sierra Madre Oriental.

SPECIES DIVERSITY THROUGHOUT MEXICO

Based on existing treatments and work in progress, a reasonable estimate
of oak species for the entire Western Hemisphere is about 200–225 species.
Approximately 135 species of *Quercus* occur wholly or partly in Mexico;
and with further study this estimate could increase to 150 or more species.
The 135 species of oak in Mexico contrasts with about 87 species of oak in
the United States and Canada. The ranges of 31 species of oak cross the
border between the United States and Mexico. Mexico also has more
endemic species (as defined by political boundaries), with 86 endemics
compared to 53 species found completely north of the border. In Central
America, there are about 30 species of oak, of which 19 also occur in Mexico.
The single species found in South America, *Q. humboldtii* Bonpland, is a red
oak that has now been discovered in the Darien of Panama as well.

The taxonomic sections within *Quercus* show a pattern of diversity in
North America that is similar to the genus as a whole. Mexico has 55 species
of red oak (with 41 species endemic), whereas the United States has 34
species of red oak (with fully 27 of them endemic). White oaks are

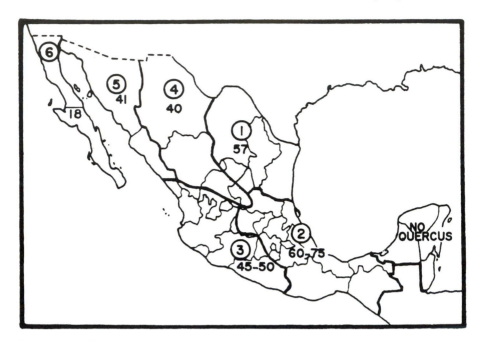

Figure 16.1. Diversity of the genus *Quercus* in Mexico on a regional basis. The six regions shown are defined and discussed in the text. Numbers indicate the approximate number of oak species in each region. Endemism is not indicated, so widespread species are counted in more than one region.

considerably more diverse in Mexico, with 76 species (44 endemic), whereas the United States has 49 species of white oak (28 of which are endemic). These numbers indicate that only about six red oak species have geographic ranges that cross the Unites States-Mexico border. This figure is in contrast to about 22 white oak species that have ranges that cross the border. This pattern reflects the scarcity of red oak species in the dry regions of northern Mexico and the southwestern United States. Perhaps of more interest is the fact that all 31 species of the white oak subsection *Glaucoideae* are found wholly or partly in Mexico, and only eight of these 31 are found also in the United States. The ranges of only two species of Glaucoideae extend from Mexico into central America, where oaks are generally found in relatively mesic areas.

REGIONAL DIVERSITY

The discussion of regional diversity provided here is not based on an a priori assessment of phytogeographic regions of Mexico. Our current knowledge of Mexican oaks is fragmentary, with some areas well studied and others not. The regions discussed here have been selected and delimited on the basis of the availability of data. This organization is not meant to reflect biogeographic areas in any strict sense. The six regions of Mexico discussed below are indicated, along with the total number of oak species for each, in Figure 16.1. Note that there is geographic overlap between northeastern Mexico, which has been defined on the basis of political boundaries, and the Chihuahuan Desert region, which is defined on the basis of phytogeographical considerations.

Partial lists of the better known species are included for each region. These lists are by no means exhaustive, nor should they be considered to be definitive for the species listed. Thus the number of species presented in the text and on the map (Fig. 16.1) do not match the number of species listed. Certain species have been omitted from the lists if there are insufficient data to apply a previously published name with confidence. Including doubtful names in such a list would only create additional problems, so names that have been applied inconsistently in the past have been avoided here wherever possible. Some widespread species complexes, such as those embodied by *Q. castanea* Nee and *Q. laeta* Liebm., are in great need of close study, and several species might be recognized within each of these complexes. Other problems exist with the correct application of particular names when types have been lost or are unavailable for study. Because of such problems, these lists must be taken only as rough indications of which species are present in each region.

Northeastern Mexico

Northeastern Mexico is defined arbitrarily here to include the states of Coahuila, Nuevo León, and Tamaulipas. A monographic treatment for the species found in these three states is near completion (Nixon & Muller, in prep.) and the number of species can be accurately estimated at this time. We recognize 57 species in this area, which includes the northern part of the Sierra Madre Oriental, the dry Tamaulipan lowlands, and the eastern part of the Chihuahuan Desert. White oaks account for 31 of the species, with red oaks making up the remaining 26 species. Of the 31 white oak species, 13 belong to subsection *Glaucoideae* and two to subsection *Virentes*. The greatest concentration of species occurs in the highlands of the Sierra Madre Oriental of Nuevo León and Tamaulipas. Comparison with the figures for the Chihuahuan Desert region, which covers much of Coahuila, shows a great increase in red oak species in the wetter mountains of the Sierra Madre Oriental in contrast to the desert ranges to the west. Large areas of limestone occur in the northern Sierra Madre Oriental and Chihuahuan Desert, and some species appear to be largely restricted to soils derived from these formations. More species may become known with intensive collecting of southern Tamaulipas, where diverse habitats may harbor endemics as well as unreported populations of species that are more common to the south in the Sierra Madre Oriental.

The partial species list for northeastern Mexico is as follows: *Quercus affinis* Scheidweiler, *Q. canbyi* Trel., *Q. castanea* Nee, *Q. hypoxantha* Trel., *Q.*

coccolobifolia Trel., *Q. convallata* Trel., *Q. crassifolia* H. & B., *Q. crassipes* H. & B., *Q. cupreata* Trel. & Mueller, *Q. deliquescens* Muller, *Q. depressa* H. & B., *Q. durifolia* von Seemen, *Q. eduardii* Trel., *Q. emoryi* Torrey, *Q. fulva* Liebm., *Q. fusiformis* Small, *Q. galeanensis* Mueller, *Q. gambellii* Nutt., *Q. germana* S. & C., *Q. ghiesbreghtii* Martens & Galeotti, *Q. gravesii* Sudworth, *Q. greggii* (A. DC.) Trel., *Q. grisea* Liebm., *Q. hypoleucoides* Camus, *Q. hypoxantha* Trelease, *Q. intricata* Trelease, *Q. invaginata* Trelease, *Q. laceyi* Small, *Q. laeta* Liebm., *Q. laxa* Liebm., *Q. martensiana* Trelease, *Q. mexicana* H. & B., *Q. microlepis* Trel. & Mueller, *Q. mohriana* Buckley in Rydberg, *Q. muhlenbergii* Engelm., *Q. obtusata* H. & B., *Q. oleoides* S. & C., *Q. opaca* Trelease, *Q. pinnativenulosa* Mueller, *Q. polymorpha* S. & C., *Q. potosina* Trelease, *Q. pringlei* von Seemen, *Q. pungens* Liebm., *Q. rugosa* Nee, *Q. rysophylla* Weatherby, *Q. saltillensis* Trelease, *Q. sartorii* Liebm., *Q. sideroxyla* H. & B., *Q. sillae* Trelease, *Q. sinuata* Walter, *Q. striatula* Trelease, *Q. tinkhamii* Muller, *Q. tuberculata* Liebm., *Q. vaseyana* Buckley.

Central and Southern Mexico

The area comprising central and southern Mexico is arbitrarily defined to include the highlands of south central and eastern Mexico and the entire states of Oaxaca and Chiapas. Because of incomplete information and a lack of separate data for the eastern part of Mexico, the states of Hidalgo and San Luis Potosí south of the Rio Verde are included in this region. No oaks are known from the Yucatan peninsula. The mountains of central and southern Mexico are probably the richest area of *Quercus* diversity in the Western Hemisphere. More work in this area is necessary before the diversity of oaks in this area can be accurately estimated. Based on the estimates for some adjacent areas, extrapolating provides an estimate of about 60-75 species for central and southern Mexico. As one travels south and east into Central America, species diversity in *Quercus* diminishes dramatically, as evidenced by the relatively low number (26 species) reported for Chiapas by Breedlove (1986).

The partial species list for central and southern Mexico is as follows: *Quercus affinis* Scheidweiler, *Q. brenesii* Trel., *Q. candicans* Nee, *Q. castanea* Nee, *Q. conspersa* Bentham, *Q. corrugata* Hooker, *Q. crassifolia* H. & B., *Q. crassipes* H. & B., *Q. crispipilis* Trel., *Q. elliptica* Nee, *Q. eugeniifolia* Liebm., *Q. excelsa* Liebm., *Q. gentryi* Muller, *Q. germana* S. & C., *Q. ghiesbreghtii* Martens & Galeotti, *Q. glabrescens* Bentham, *Q.glaucescens* H. & B., *Q. glaucoides* Martens & Galeotti, *Q. greggii* (A. DC.) Trel., *Q. insignis* M. & G., *Q. laeta* Liebm., *Q. laurina* H. & B., *Q. liebmannii* Oersted, *Q. magnoliifolia* Nee, *Q. mexicana* H. & B., *Q. microphylla* Nee, *Q. elliptica* Nee, *Q. obtusata* H. & B., *Q. oleoides* S. & C., *Q. opaca* Trel. (synonym: *Q. sebifera* Trel.), *Q. pacayana* Muller, *Q. paxtalensis* Muller, *Q. peduncularis* Nee, *Q. polymorpha* S. & C., *Q. repanda* H. & B., *Q. rysophylla* Weatherby, *Q. rugosa* Nee, *Q. salicifolia* Nee, *Q. sartorii* Liebm., *Q. sapotifolia* Liebm., *Q. scytophylla* Liebm., *Q. segoviensis* Liebm., *Q. splendens* Nee, *Q. toxicodendrifolia* Trel., *Q. xalapensis* H. & B.

Southwestern Mexico

Treatments of the oaks of Nueva Galicia (McVaugh, 1974), Michoacán (Gonzalez & Labat, 1987), and Jalisco (Villareal, 1986) provide a strong basis for an estimate of species diversity in southwestern Mexico. The region in general is drier than eastern and southern Mexico, which is reflected in the strong representation of white oaks and in particular members of subsection *Glaucoideae* (e.g., *Q. laeta* Liebm., *Q. splendens* Nee, *Q. glaucescens* H. & B.). In Jalisco, Villareal (1986) recognized 21 red oaks and 21 white oaks, 15 of which are members of subsection *Glaucoideae*. In Michoacán, Gonzalez and Labat (1987) recognized 15 red oak species and 15 white oak species, 10 of which are *Glaucoideae*. McVaugh (1974) recognized 21 red oaks and 23 white oaks, of which 17 are members of the *Glaucoideae*.

The partial species list for southwestern Mexico is as follows: *Quercus acutifolia* Nee, *Q. aristata* Hooker & Arnott, *Q. candicans* Nee, *Q. castanea* Nee, *Q. chihuahuensis* Trel., *Q. coccolobifolia* Trel., *Q. conspersa* Bentham, *Q. convallata* Trel., *Q. crassifolia* H. & B., *Q. crassipes* H. & B., *Q. diversifolia* Nee, *Q. eduardii* Trel., *Q. elliptica* Nee, *Q. excelsa* Liebm., *Q. fulva* Liebm., *Q. gentryi* Muller, *Q. glaucescens* H. & B., *Q. grisea* Liebm., *Q. insignis* M. & G., *Q. magnoliifolia* Nee, *Q. martinezii* Muller, *Q. obtusata* H. & B., *Q. peduncularis* Nee, *Q. planipocula* Trel., *Q. potosina* Trel., *Q. praeco* Trel., *Q. praineana* Trel., *Q. resinosa* Liebm., *Q. rugosa* Nee, *Q. salicifolia* Nee, *Q. scytophylla* Liebm., *Q. sideroxyla* H. & B., *Q. splendens* Nee, *Q. subspathulata* Trel., *Q. urbanii* Trel., *Q. uxoris* McVaugh, *Q. viminea* Trel.

Chihuahuan Desert Region

The Chihuahuan Desert region (CDR) as discussed here has been geographically defined by Henrickson and Straw (1976). The diversity of oak species in the CDR is high relative to other desert areas of North America. Muller (in press) recognized 40 species within the CDR. Of these 40 species, 28 were white oaks, 11 were red oaks, and 1 species (which barely enters the area) was a member of the *Protobalanus* group. Of the 28 species of white oak, 11 were members of subsection *Glaucoideae*, and one of subsection *Virentes*.

Within the CDR, oaks are generally restricted to mountains within the desert and particularly to moist canyons and northern slopes. The distributions of Chihuahuan Desert species therefore are often "island" distributions, with expanses of dry desert between relatively small areas of suitable oak habitat. Although most of the oaks of the CDR have wide distributions, some are narrow endemics, e.g., *Quercus carmenensis* Muller and *Q. deliquescens* Muller.

The partial species list for the CDR is as follows: *Quercus arizonica* Sargent, *Q. chihuahuensis* Trel., *Q. chrysolepis* Liebm. (extremely rare in Mexico), *Q. coccolobifolia* Trel., *Q. depressipes* Trel., *Q. durifolia* von Seemen,

Q. emoryi Torrey, *Q. gambellii* Nutt., *Q. gravesii* Sudworth, *Q. greggii* (A. DC.)
Trel., *Q. grisea* Liebm., *Q. hypoleucoides* Camus, *Q. hypoxantha* Trel., *Q.
intricata* Trel., *Q. invaginata* Trel., *Q. mexicana* H. & B., *Q. mohriana* Buckley
in Rydberg, *Q. oblongifolia* Torrey in Sitgreaves, *Q. potosina* Trel., *Q. pringlei*
von Seemen, *Q. saltillensis* Trel., *Q. sideroxyla* H. & B., *Q. striatula* Trel., *Q.
tinkhamii* Muller, *Q. toumeyi* Sargent, *Q. tuberculata* Liebm., *Q. turbinella*
Greene, *Q. vaseyana* Buckley.

Northwestern Mexico

The northwestern region of Mexico is arbitrarily and roughly defined as the
area bounded by Nueva Galicia (McVaugh, 1974) to the south and by the
Chihuahuan Desert region to the east. It includes the northern part of the
Sierra Madre Occidental, as well as the southern part of the Sonoran Desert.
I have not included Baja California in this area. In contrast to northeastern
Mexico, which has abundant limestone (see above), the Sierra Madre
Occidental is largely volcanic. There is no complete floristic treatment for
the oaks of this area, and the following estimates are rough. There are about
41 species of oak in this area, of which 23 are white oaks, 17 red oaks, and
1 species of *Protobalanus*. Subsection *Glaucoideae* is strongly represented
with about 17 species. When compared with northeastern Mexico (as
arbitrarily defined above) 19 species are found in common. As one travels
to the south in both mountain ranges, the number of species in common
increases, not surprisingly, as does the overall number of species.

The partial species list for northwestern Mexico is as follows: *Quercus
ajoensis* Muller (doubtfully entering Mexico), *Q. albocincta* Trel., *Q. arizonica*
Sargent, *Q. chihuahuensis* Trel., *Q. chrysolepis* Liebm., *Q. hypoxantha* Trel., *Q.
coccolobifolia* Trel., *Q. convallata* Trel., *Q. crassifolia* H. & B., *Q. depressipes* Trel.,
Q. durifolia von Seemen, *Q. emoryi* Torrey, *Q. fulva* Liebm., *Q. gambellii* Nutt.,
Q. gentryi Muller, *Q. gravesii* Sudworth, *Q. grisea* Liebm., *Q. hypoleucoides*
Camus, *Q. laeta* Liebm., *Q. laxa* Liebm., *Q. oblongifolia* Torrey in Sitgreaves,
Q. obtusata H. & B., *Q. pallescens* Trel., *Q. perpallida* Trel., *Q. planipocula* Trel.,
Q. potosina Trel., *Q. praeco* Trel., *Q. praineana* Trel., *Q. rugosa* Nee, *Q. scytophylla*
Liebm., *Q. sideroxyla* H. & B., *Q. splendens* Nee, *Q. striatula* Trel., *Q. subspathulata*
Trel., *Q. toumeyi* Sargent, *Q. transmontana* Trel., *Q. tuberculata* Liebm., *Q.
turbinella* Greene, *Q. urbanii* Trel., *Q. viminea* Trel.

Baja California

The peninsula of Baja California is of great interest biogeographically
because of its geographic isolation and desert climate. My estimate for the
number of oak species throughout the peninsula is about 18 species, of
which 9 species are white oak, 5 species are red oak, and 4 species are
Protobalanus. Four endemic oaks occur on the peninsula. The northern part

of the peninsula has a dry Mediterranean climate, with significant winter rains, whereas the climate in the southern part is more like that of north-western mainland Mexico. In the northern part of the peninsula, in the Sierra Juarez and Sierra San Pedro Martir, the oaks are mostly species that occur also in the adjacent area of the southwestern United States. There are five species of white oak (all found in the Unites States), four species of *Protobalanus* (one of which is endemic to Baja California), and four species of red oak (also with one endemic, *Q. peninsularis* Trel., which is only marginally distinct from *Q. emoryi* Torr.). Overall, of the 18 species of oak in Baja California, 11 are also found in the United States and six are found in mainland Mexico.

The southern half of the Baja California peninsula is somewhat separated from the northern half by desert areas, which are not generally suitable for oaks except on the highest peaks. In the mountains of the southern half, e.g., the Sierra La Laguna, oak species occur that are found also in the Sierra Madre Occidental of western Mexico. In addition, there are two endemics in the Cape region (*Quercus devia* Trel., a red oak, and *Q. brandegei* Goldm., a white oak of subsection *Virentes*).

The species list for the Baja California peninsula is as follows: *Quercus agrifolia* Nee, *Q. albocincta* Trel., *Q. arizonica* Sargent, *Q. berberidifolia* Liebm., *Q. brandegeei* Goldman, *Q. cedrosensis* Muller, *Q. chrysolepis* Liebm., *Q. cornelius-mulleri* Nixon & Steele, *Q. devia* Goldman, *Q. engelmannii* Greene, *Q. kelloggii* Newberry, *Q. oblongifolia* Torrey in Sitgreaves, *Q. palmeri* Engelmann, *Q. peninsularis* Trel., *Q. perpallida* Trel., *Q. tomentella* Engelmann, *Q. tuberculata* Liebm., *Q. turbinella* Greene.

CONCLUSIONS

The genus *Quercus* is one of the most important genera of woody plants in the Northern Hemisphere, and the center of diversity for the genus in the Western Hemisphere is montane Mexico. Central and southern Mexico and the Sierra Madre Oriental have the greatest concentrations of oak species. White oaks, and in particular subsection *Glaucoideae*, are more diverse than red oaks in the drier areas, but the two groups seem to be equally speciose in the wet areas.

The character of much of the Mexican vegetation is greatly influenced by oak species. Oaks provide a habitat for a large number of animals and are of great economic value. Although many oak species appear to be "resilient" in the sense that they are capable of stump sprouting after being cut, the long-term effects of continual cutting of forests is unknown. Undoubtedly, continual cutting will reduce genetic variation within species and select against species that cannot reproduce clonally. Because of the great diversity of plants and animals in the pine-oak forests, loss of these forests through the impact of man affects the diversity of the Mexican biota on a large scale.

REFERENCES

Breedlove, D. 1986. Listados Floristicos de Mexico IV. Flora de Chiapas. Mexico City: UNAM. Inst. Biol.

Cracraft, 1982. Species concepts and speciation analysis. In R.F. Johnston (ed.), Current Ornithology. New York: Plenum Press. pp. 159–187.

Crepet, W.L. & K.C. Nixon. 1989a. Earliest megafossil evidence of Fagaceae: phylogenetic and biogeographic implications. Amer. J. Bot. 76:842–855.

————— & —————. 1989b. Extinct transitional Fagaceae from the Oligocene and their phylogenetic significance. Amer. J. Bot. 76:1493–1509.

Daghlian, C.P. & W.L. Crepet. 1982. The evolutionary significance of Quercus sect. Erythrobalanus from the Oligocene of east Texas. Amer. J. Bot. 70:639–649.

Dorr, L.J. & K.C. Nixon. 1985. Typification of the oak (Quercus) taxa described by S.B. Buckley (1809–1884). Taxon 34:211–228.

Forman, L.L. 1964. Trigonobalanus, a new genus of Fagaceae, with notes on the classification of the family. Kew Bull. 17:381–396.

González, M.A B. & J. Labat. 1987. Los Encinos (Quercus) de Michoacán, México. Cuadernos de Estudios Michoacanos I. Mexico City: Centre d'etudes Mexicaines et Cetramericaines.

Henrickson, J. & R.M. Straw. 1976. A Gazetteer of the Chihuahuan Desert Region. A Supplement to the Chihuahuan Desert Flora. Los Angeles, CA: Published by the authors.

Little, E.L. 1971. Atlas of United States Trees. Vol. 1. Conifers and Important Hardwoods. U.S.D.A. Misc. Publ. 1146. Washington, DC: U.S. Government Printing Office.

Lozano-C., G., J. Hernández-Camacho & J.E. Henao-S. 1979. Hallazgo del género Trigonobalanus Forman, 1962 (Fagaceae) en el Neotropico. I. Caldasia 12:518–537.

Manchester, S.R. 1983. Eocene fruits, wood and leaves of the Fagaceae from the Clarno Formation of Oregon. Amer. J. Bot. 70:74 (Abstract).

McVaugh, R. 1974. Fagaceae. Flora Novo-Galiciana. Contr. Univ. Mich. Herb. 12:1–93.

Muller, C.H. In press. Quercus. In J. Henrickson & M. Johnston (eds.), Flora of the Chihuahuan Desert Region. In preparation.

Muller, C.H. & R. McVaugh. 1972. The oaks (Quercus) described by Nee (1801) and by Humboldt and Bonpland (1809), with comments on related species. Contr. Univ. Michigan Herb. 9:507–522.

Nixon, K.C. & W.L. Crepet. 1989. Trigonobalanus (Fagaceae): Taxonomic status and phylogenetic relationships. Amer. J. Bot. 76:828–841.

————— & K.P. Steele. 1981. A new species of Quercus (Fagaceae) from southern California. Madroño 28:210–219.

————— & Q.D. Wheeler. 1990. An amplification of the phylogenetic species concept. Cladistics 6:211–223.

Rosen, D.E. 1979. Fishes from the uplands and intermontane basins of Guatemala: Revisionary studies and comparative geography. Bull. Amer. Mus. Nat. Hist. 162:267–376.

Rzedowski, J. 1978. Vegetación de México. Mexico City: Limusa.

Soepadmo, E. 1972. Fagaceae. Flora Malesiana 7:265–403.

Trelease, W. 1924. The American Oaks. Mem. Natl. Acad. Sci. 20:1–255.

Villareal, L.M.G. 1986. Contribución al Conocimiento del Género Quercus (Fagaceae) en el estado de Jalisco. Colección Flora de Jalisco. Guadalajara, Jalisco: Inst. Bot. Univ. Guad.

17

Mexican Leguminosae: Phytogeography, Endemism, and Origins

MARIO SOUSA S. AND ALFONSO DELGADO S.

Aspects of the phytogeography, endemism, and origins of the second largest family of flowering plants in the Mexican flora, the Leguminosae, are discussed. The family is richest in numbers of species principally in the states of Chiapas and Oaxaca, but the nature of diversity varies in different parts of the country. Forty-nine genera displaying different phytogeographic and distribution patterns were studied. Brief diagnoses and estimates of species numbers for these genera are provided. Mexican Leguminosae were organized into six artificial groups based on the degree of endemism. Five phytogeographic patterns were discernible in the family. It is suggested that although Mexico is not the primary center of radiation of the family it is one of the major secondary centers, with 135 genera and 1,724 species. Many of these genera or even groups of genera present their principal distribution in territorial Mexico and therefore are considered Mexican. Their distributional ranges provide a natural phytogeographic province, which here is referred to as the Mexican Phytogeographic Legume Area. As many as 896 (51.9%) species are endemic to Mexico. A rich legume flora was already in existence in Mexico at the beginning of the Tertiary (65 Ma). Paleoendemic elements have been identified, and it is suggested that the sources of Mexican legumes are diverse and include in addition to South America: Europe and Asia (e.g., *Havardia*) and Africa (e.g., *Adenopodia*). Many genera (e.g., *Muellera*) are recent arrivals in Mexico. What accounts for the high number of legumes is both ancient and recent explosive speciations in different groups.

The Leguminosae are the second largest family of flowering plants in Mexico after the Compositae. The family is represented by 26 tribes, 135 genera, and 1,724 species. This richness and diversity presents a tremendous challenge. The taxonomy, systematics, phytogeography, evolution, speciation patterns, and nature of endemism are subjects of ongoing studies. This chapter reports on aspects of phytogeography, endemism at all levels, generic and species affinities, and age and fossil history where available. One of the primary difficulties is the continuing lack of informa-

459

Table 17.1. Group A: genera endemic to Mexico

Caesalpinioideae
  *Caesalpinieae
  Heteroflorum M. Sousa, inedit.
  Conzattia

Mimosoideae
  *Mimoseae
  Calliandropsis

Papilionoideae
  *Millettieae
  Hesperothamnus

---

*Tribe.

tion, both taxonomic and systematic, on this group of plants in Mexico. As more information becomes available, some of the conclusions arrived at from present data may be subject to change.

The family occurs in all parts of the country and in all possible habitats. It is more numerous in the tropical zones. The states of Oaxaca and Chiapas (Sousa, 1986) lead with the highest numbers of species due to the fact that several floristic provinces overlap within their boundaries. However, the nature of endemism in the family varies in different parts of the country. For instance, it is striking in the Rio Balsas Basin. This area may not be richer in species, but the diversity there represents some of the oldest as well as recent elements in the Mexican flora, e.g., *Lonchocarpus* (Sousa & Soto, 1989).

To interpret the phytogeography, origins, and endemism of the family in Mexico, 49 genera with different distributions and data on comparative morphology, cytology, fossil history, and in a few cases phytochemistry have been selected for this study. The selection of genera may seem arbitrary, but the aim is to give an overview of Mexican legumes and compare them to non-Mexican ones. Many genera that would have warranted inclusion in any discussion of Mexican Leguminosae are, as a consequence, not represented in this study. The genera analyzed here constitute 36% of the total of the family in Mexico. For each genus, details of their present (and, when possible, past) distributions are given along with estimates of species numbers. Taxonomic and phylogenetic problems are highlighted where relevant. Discussions regarding their time and place of origin are provided when data permitted such conclusions. An understanding of the phytogeography of these genera led to ideas regarding the origin of the family in Mexico, the sources of Mexican legumes, their possible time of arrival, and their speciation in the country. The tribes and genera treated in this chapter follow, with some modifications, the system proposed by Polhill (Polhill & Raven, 1981).

The genera have been arranged in six artificial groups (groups A–F) based on degrees of endemism in their taxa in the country. The first division has been made between endemic and nonendemic genera. The nonendemic group has been divided according to the percentages of transspecific and specific endemisms from the highest to the lowest in each genus.

TAXONOMY, PHYTOGEOGRAPHY, AND ENDEMISM IN THE MEXICAN LEGUMINOSAE: GROUPS A–F

Group A

Group A includes the genera endemic to Mexico (Table 17.1). The four genera (3% of the total) include two from the Caesalpinioideae, one from the Papilionoideae and from the Mimosoideae. This group is composed of ten species. The low generic endemism suggests that territorial Mexico may not be a natural phytogeographic province for this family.

Caesalpinieae

Conzattia ROSE. *Conzattia* is genus with three species. It principally occurs along the Pacific slopes from Los Cabos in Baja California Sur through Sinaloa to Chiapas, always in hot, dry climates.

Polhill and Vidal (1981) associated it with the group *Caesalpinia* despite its actinomorphic calyx. Miranda (1955) discussed its affinities within the Caesalpinieae and placed it near *Peltophorum*; he considered it to be a residual group that has retained some of the characters from the now extinct primary stock that gave rise to *Peltophorum*. A recent find, *Heteroflorum*, an as-yet undescribed genus, belongs to this group. *Conzattia coriacea*, a fossil species, was described from the Oligocene from the deposits in Florissant, Colorado, United States (MacGinitie, 1953).

Millettieae

Hesperothamnus BRANDEGEE. *Hesperothamnus* is a genus with five species. It is found mainly in western Mexico from Los Cabos in Baja California Sur to the Tehuantepec Isthmus in Oaxaca as well as in Hidalgo and the Tehuacan Desert. Species have restricted distributions.

Hesperothamnus is one of the most primitive representatives of the tribe in Mexico (Sousa & Peña de Sousa, 1981) with *Piscidia* as its closest relative, an affinity suggested by its nonprotein amino acids and amines (Evans et al., 1985). Geesink (1981, 1984a) placed this genus in the African-Asiatic *Millettia* sect. *Millettia* on the grounds that there is "no single morphological character different" from it; nonetheless, he did not propose Africa-Asia as its center of origin.

Table 17.2. Group B: nonendemic genera but with more than 75% of the species in Mexico

---

Subgroup I. Monotypic genera and no endemics
   Caesalpinioideae: *Poeppigia.* Mimosoideae: none. Papilionoideae: *Robinieae,
   Sphinctospermum, Olneya, Genistidium; *Phaseoleae, *Cymbosema*;
   *Amorpheae, *Apoplanesia*; *Aeschynomeneae, *Pachecoa*; *Thermopsideae,
   Pickeringia.

Subgroup II. Genera with species endemics
   Caesalpinioideae: *Cercidium.* Mimosoideae: *Schrankia, Leucaena,
   Lysiloma, Havardia.* Papilionoideae: *Millettieae, *Willardia*; *Robinieae,
   Peteria, Hybosema, Lennea, Gliricidia; *Phaseoleae, *Cologania, Ramirezella,
   Phaseolus*; *Amorpheae, *Eysenhardtia, Dalea, Marina*; *Aeschynomeneae
   (Ormocarpinae), *Diphysa, Nissolia*; *Brongniartieae, *Brongniartia.*

---

*Tribe.

Hesperothamnus is similar to the endemic Cuban genus *Behaimia* in having stipels and a similar type of inflorescence. The genus is ancient in Mexico; its species are narrowly distributed, suggesting they are paleoendemics.

Group B

Group B contains nonendemic genera, but with more than 75% of the species in Mexico (Table 17.2). They in turn are grouped into subgroup B I with monotypic genera and subgroup B II with genera of two or more species. Subgroup B II includes elements basically centered in Mexico that in some cases extend northward (marginally) into the southern United States or southward into Central America and even tenuously into South America, as is the case of *Phaseolus, Diphysa,* and *Nissolia.* As well, cases such as *Cercidium* and *Havardia* are included of which Mexico has a greater diversity but that have continental disjunctions, the first with South America and the second with Asia. The monotypic genera are treated within the context of their tribes and subtribes. For example, the Robinieae, which are centered in North America and the Caribbean, have three genera in Mexico and in the southwestern United States. On the other hand, the monotypic *Cymbosema* has its affinities with Diocleinae centered in South America. Subgroup B II is squarely centered in Mexico with 19 genera that have 370 species, of which 251 (67.8%) are endemic.

Robinieae

Sphinctospermum ROSE. *Sphinctospermum* is a monotypic genus, extending from the southwestern United States to the Isthmus of Tehuantepec. It is an annual with a marked tendency to reduction in number of parts (leaflets and inflorescences), which makes interpretation of its affinities difficult. Polhill and Sousa (1981) considered it a member of Robinieae, but Lavin (1987) included it in the Millettieae near *Tephrosia* subg. *Macronyx* based on similarities between it and *T. strigosa* and *T. tenuis* of India and Pakistan. Whether these similarities have a genetic basis is open to discussion: the seeds of *Sphinctospermum* accumulate canavanine, and those of *Tephrosia strigosa* do not (Lavin, 1987); the chromosome number of the former is $2n = 16$ and that of latter $2n = 22$ (Wood, 1949). At present little can be said about its phylogeny and origin. Both groups, Robinieae and Millettieae, are perennial, and the annual habit is a rarity; the latter habit in *Sphinctospermum* may be in part due to its evolutionary tendency toward simplification.

Aeschynomeneae

Pachecoa STANDLEY & STEYERMARK. *Pachecoa* is a monotypic genus from southern Mexico to Guatemala and Venezuela. Rudd (1981a) suggested that *Pachecoa prismatica* is a possible recent introduction in Venezuela.

Norman and Gunn (1982) considered its affinities to be with *Chapmania* (Florida, United States) and *Arthrocarpum* (Somalia, Africa and Socotra Island of the PDR of Yemen in the Indian Ocean), as the leaves underneath are red owing to tannin-concentrated stellate cells. They further considered the genus to be closer to *Arthrocarpum* because of shared flower characters, although the two are disjunct.

Gillett in 1966 (*fide* Norman & Gunn, 1982) noted that the distributions of these three genera indicate an ancient origin. It is possible that *Pachecoa*, a genus of Laurasian origin, has been present since the Tertiary.

Caesalpinioideae

Cercidium TULASNE. *Cercidium* is an American genus. The nine taxa in the genus are distributed as follows: two in Texas and northern Mexico, five in the Sonoran Desert of which *Cercidium praecox* is found in southern Mexico and southern Peru, and two in Argentina (Carter, 1974a). The Sonoran Desert, which has the most diversity, is not considered to be the center of origin of the genus (Carter, 1974a). The South American species are found in the Andes and extend to Patagonia (Burkart, 1952). *Cercidium australe* of Argentina is doubtfully distinct from *C. praecox* (Macbride, 1943). Hybridization is fairly common among the species of *Cercidium*. Carter considered *C. sonorae* a species of hybrid origin (*C. macrophyllum* x *C. praecox*). In areas where the putative parents overlap, *C. sonorae* is found. *Cercidium* is closely related to *Parkinsonia* and natural hybrids are known between *Cercidium praecox* and *Parkinsonia aculeata* (Carter, 1974b).

Polhill and Vidal (1981) considered *Cercidium* and *Parkinsonia* congeneric based on their relationship although the latter is better represented in

Africa with three endemics (Johnston, 1924; Polhill & Vidal, 1981). Irrespective of the taxonomic decisions, it is evident that the two are related and *Parkinsonia s.s.* is of African origin. *Cercidium*, which originated and developed in North America, probably reached South America via the Andes.

Ingeae

Havardia SMALL. *Havardia* is a Mexican, North American, Caribbean, and Asian genus of approximately 12 species. Nine species are centered in Mexico with extensions to the Caribbean and northern South America (Rico, pers. comm.) and three in Asia: one in Sri Lanka, another in Thailand, and the third in Vietnam (Nielsen, 1981).

Havardia has traditionally been included in *Pithecellobium s.l.* Small (1901), based on a combination of characters, rightly considered it a distinct taxon (Nielson, 1981; Rico, pers. comm.). The distribution of *Havardia* is Laurasian, and hence it may have had a continuous distribution during the Lower Tertiary. It may have reached the New World by way of eastern Asia or by way of the now extinct rich, warm flora of European mimosoids (see macrofossils of leaflets and pods in Unger, 1864, and Saporta, 1889), even though none of these fossils has been identified as belonging to *Havardia*.

Phaseoleae

Ramirezella ROSE. *Ramirezella* is a genus that ranges from Mexico to Nicaragua. It is composed of about seven species, one of which, *R. strobilophora*, occurs from Chihuahua to Nicaragua.

Generally, species of this genus are perennial vines that grow on the slopes and mountains along the Pacific coast. *Ramirezella nitida* is a disjunct species in the Sierra Madre Oriental. *Ramirezella* differentiated from an ancestral vignoid, probably of austral origin. The similarities between *Ramirezella* and South American species of *Vigna* are so strong that Lackey (1983) considered it part of *Vigna* subg. *Sigmoidotropis*. This transfer might be premature, as Lackey (1983) did not include the critical taxa in his study (Ochoterena-Booth, 1991).

Phaseolus L. *Phaseolus* is an American genus of approximately 40 species, mainly found in the mountains of Mexico. Literature on fossil *Phaseolites* is fairly common; those by Unger (1864) are noteworthy. From materials of leaf and fruit remains of European Tertiary, he described ten species similar to five genera in the tribe Phaseoleae: *Physolobium, Hardenbergia, Erythrina, Phaseolites,* and *Dolichites.* The leaflets of *Phaseolites oligantherus* are significant; according to Unger (1864) these leaflets are similar to those of *Phaseolus pauciflorus* of the present. Elsik (1968) described *Phaseoliidites standleyi,* from Paleocene pollen from Texas and compared them to those of *Phaseolus adenanthus* (= *Vigna adenantha*). However, the published photomicrograph shows few of the pollen characteristics, hence it is difficult to express an opinion on it.

Its natural range is from southern Quebec and Ontario in Canada to Florida and eastern Texas (*Phaseolus polystachios*). Several species of northwestern Mexico extend into the southwestern United States. In Mexico, most of the species inhabit the mountains and rarely the lowlands (principally on the Pacific slope), with some species extending to Central America (Delgado, 1985). In South America *Phaseolus* is distributed principally in the Andes, above all in the eastern section, where the climate is less extreme and more mesic (*P. augustii* complex). Nevertheless, it is important to mention the presence of *P. lunatus* populations in the lowlands as well as in the mountains and that of *P. mollis* in the Galapagos Islands. No native bean grows in Chile, and the genus reaches no farther south than northern Argentina. Domestication has altered the distribution of *Phaseolus* on the American continent. Delgado (1985) has discussed the distribution of the four cultivated species prior to and after domestication.

The concentration of *Phaseolus* in North America (Mexico), covering a great diversity of habitats, the presence of species with primitive attributes, and the ever less deniable separation of *Vigna* and *Macroptilium* (Delgado, in prep.) suggest a possible Laurasian origin for this genus. The genus is recent in South America, having reached there possibly after the formation of Central America.

Amorpheae

Dalea LUCANUS. *Dalea* is an American genus of approximately 161 species. It is present in North America, mainly in Mexico, with a secondary center of radiation in the Andean region from Colombia to northwestern Argentina and northern Chile.

Barneby (1977) showed that *Dalea* and allied genera (*Errazurizia*, *Psorothamnus* and *Marina*), had their origins during the Tertiary in a dry area north of the Neotropics. He referred to *D. filiciformis*, endemic to Mexico, as a primitive relict, similar in several aspects to *Psorothamnus* and *Marina* and that probably resembles this ancestral lineage. The chromosome number of this species, $n = 8$, could be an aneuploid originating from $x = 10$, the basic chromosome number of both *Marina* and *Psorothamnus*, the presumed ancestors. The basic chromosome number of *Dalea* ($x = 7$) may represent a further reduction. *Dalea* is recognized by the spiral hairs and the keel of the corolla, which is adnate to the stamens. The latter character is seen in *Marina*. Barneby considered *Marina* and *Psorothamnus* to be closer to each other than *Marina* and *Dalea*.

Marina LIEBMANN. *Marina* is an American genus, centered in Mexico. It consists of 38 species, almost all of which (81%) are endemic to Mexico. Three species extend into the southwestern United States and three to the mountains of Guatemala, one of which reaches Venezuela and Cuba. The presence of light sinuous lines on the leaves below and the uniovulate ovaries identify the genus immediately. *Marina,* which originated north of the Neotropics (Barneby, 1977), has been considered to be related to *Psorothamnus.*

Aeschynomeneae

Diphysa JACQUIN. *Diphysa* ia a genus extending from southern Arizona to Central America and reaching South America. It is made up of approximately 16 species, all occurring in Mexico. One species extends northward to southern Arizona and southward to Central America. Five species are reported for Nicaragua, one (*Diphysa americana*), for Panama (Sousa, 1990; Sousa & Antonio, in press), and another reaches Bolivia.

This genus was, until recently, included in the Robinieae (Polhill & Sousa, 1981), but Lavin (1987) rightly transferred it to Aeschynomeneae subtr. Ormocarpineae based on morphological and chemical data. *Ormocarpum,* a genus of the Old World tropics occurs in South Africa, Madagascar, southern Asia to the Philippines, northern Australia, and the Fiji Islands (Rudd, 1981b). Indeed, *Diphysa* may not be distinct from *Ormocarpum*: The fruits of *D. spinosa,* which do not develop vesicular valves, are flat and strongly nerved, and their loment-like articulations are conspicuous. It therefore is most likely an ancient genus with links to Asia and Africa.

Nissolia JACQUIN. *Nissolia* is a genus of 13 species centered in Mexico. It extends from southern Arizona and Texas to southern Argentina and Paraguay. All the taxa are represented in Mexico (Rudd, 1956a).

This genus is characterized by its samaroid fruit, with two to five fertile articulated segments and a sterile terminal winged segment. *Nissolia wislzenii* is the most primitive species (Rudd, 1956a), in which the terminal segment, larger than the fertile segments, is scarcely winged. It extends from southern Arizona through the Sierra Madre Occidental and the northern highlands to Hidalgo. *Nissolia fruticosa* is a variable and aggressive species with weedy tendencies that reaches South America.

Rudd (1956a, 1958) considered *Chaetocalyx,* whose center of diversity is in South America, as closely related to *Nissolia* but with differing habitat preferences: *Nissolia* is found in dry to clearly arid habitats, whereas *Chaetocalyx* is from more mesic areas.

Brongniartieae

Brongniartia KUNTH. *Brongniartia* is an American genus of 65 species, all of which are found in Mexico. One species extends into Texas (Dorado, pers.

comm.). Two South American species once considered *Brongniartia* differ in the manner their stamens are fused and in lacking bracteoles (Dorado, pers. comm.).

The genus is of phytogeographic interest. Although it has radiated into numerous taxa, it has not been able to colonize contiguous areas with hot-warm-dry climates, such as the central depression of Chiapas or the Yucatan Peninsula, which are similar to areas it inhabits. Its members are restricted almost entirely to the western slopes of Mexico, especially in Nayarit, Jalisco and Oaxaca, and are only marginally distributed in Chiapas. Thus it appears to be an ancient genus with a recent history of explosive speciation that has not had the time to expand in geographic areas of similar climates and even less time to explore new ecosystems.

Within the tribe in the Americas its other, more specialized relative is *Harpalyce*, from which it differs greatly. Its affinities also seem to be with two Australian genera that Crisp and Weston (1987) split into four. In a cladistic study of the tribe, these authors included *Brongniartia*, *Templetonia*, and a nearby group together in a clade because of the reduced inflorescence, placing *Harpalyce* as a sister group. They suggested that *Templetonia* is more closely related to *Brongniartia* and *Harpalyce* than to any of the Australian genera of the tribe.

Group C

Group C is composed of nonendemic genera (Table 17.3), but with 40-69% of their species in Mexico and with a high percentage of species endemism. This group consists of two subgroups: C I has genera with two species, of which Mexico has one. C II contains genera with three or more species, of which Mexico has two or more. *Prosopidastrum* and *Goldmania* of subgroup C I present wide disjunctions with South America. Most in subgroup C II belong to groups clearly centered in Mexico, with extensions with or without disjunctions. *Macroptilium* shows more affinities with South American groups, but its species are widely distributed (some associated with communities of savannahs); only one species is endemic to Mexico, hence its inclusion in this subgroup. In subgroup C II, 174 species are included, of which 101 (58%) are endemic.

Mimosoideae

Goldmania ROSE EX MICHELI. *Goldmania* is an American genus of two closely related species: *Goldmania foetida* from western Mexico and Honduras; and *Goldmania paraguarensis* from Paraguay and Argentina (Burkart, 1969a).

It is related to *Piptadenia* and *Parapiptadenia* groups centered in South America, and its advanced characteristics (reduction of leaflets and

Table 17.3. Group C: nonendemic genera but with endemic species representation and with 40–69% of the total species represented in Mexico

Subgroup I. Genera with two species, one in Mexico
Caesalpinioideae: none. Mimosoideae: *Prosopidastrum, Goldmania.* Papilionoideae: none.

Subgroup II. Genera with three or more species and two or more in Mexico
Caesalpinioideae: *Hoffmannseggia.* Mimosoideae: *Adenopodia, Desmanthus, Zapoteca.* Papilionoideae: *Sophoreae, Myrospermum, Ateleia, Styphnolobium*; *Millettieae, Piscidia, Lonchocarpus*; *Robinieae, Coursetia*; *Phaseoleae, Macroptilium, Oxyrhynchus*; *Psoraleeae, Hoita*; *Amorpheae, Errazurizia.*

*Tribe.

dehiscense of fruits along one of its sutures) indicate its possible origin from these groups.

Adenopodia PRESL. *Adenopodia* is a genus with seven species, three in Mexico and four in Africa. Species in Mexico occur on the west coast from Nayarit to Chiapas, with one extending to Nicaragua (Brenan, 1986). Brenan included three more species for South America, but Barneby (1986) correctly placed them in *Piptadenia*, commenting on their possible relation with *Mimosa* and *Piptadenia*. He further pointed out that within the group *Piptadenia*, "it appears to be a primitive circumtropical stock that may have given rise to *Mimosa* and *Adenopodia*," as suggested by Brenan (1955, 1963). Mexican *Adenopodia* species show a certain degree of differentiation compared to the African ones concerning the flower's smaller size and the sessile to subsessile ovary. *Adenopodia* is a relictual element in Mexican flora and may have arrived from Africa via a warmer (Paleogene) Europe to North America.

Sophoreae

Myrospermum JACQUIN. *Myrospermum* is an American genus of three species: two in Mexico with one extending into Central America, and the third in Colombia. The genus belongs to the "*Myroxylon* group" (Polhill, 1981a). The arboreal genera *Myrospermum* and *Myroxylon* are distinguished by the presence of gums or balsams in all their components. The leaves are distinctive with their translucent glands of lines or punctations.

Like *Myroxylon, Myrospermum* extends from Mexico to northern South America (Rudd, 1968). In Mexico *Myrospermum sousanum* is found in the moist canyons of northeastern Mexico (Nuevo León) in communities of xerophilous scrub, and it is disjunct from *M. frutescens*. The latter has a wider distribution in the Pacific Basin from Jalisco to northern Colombia.

The third species, *M. secundum*, is found in Colombia and Venezuela (Rudd, pers. comm.).

Delgado and Johnston (1984) noted that *Myrospermum sousanum* is remarkable not only for its unusual and disjunct distribution for the genus but also for its unusual morphology. The fruit is oblong and flat and has two narrow marginal wings, being less specialized than the basally winged fruits (samaroid) present in the other two species.

These observations and its present distribution suggest that the genus *Myrospermum* may have once had a wider distribution than now. Like other genera of the Northern Hemisphere, it has remained distinct since before the Cenozoic and has only recently reached South America.

Styphnolobium SCHOTT. *Styphnolobium* is a genus with approximately nine species: mainly North American with eight species and one in Southeast Asia. In North America, Mexico has five, the southern United States one, and Central America three, one extending to Colombia. These species (except one) are narrow endemics and occur in cold to hot areas and in dry to wet climates. The genus was reinstated by Yakovlev (1967), who segregated it from *Sophora*. His conclusions, based on its fruit morphology, its seeds, the presence of stipels, and its smooth wing petals, can be confirmed by its chromosome number of $2n = 28$ (Bernal & Martínez, 1989). In contrast, the chromosome number of *Sophora* is $2n = 18$.

The fleshy fruits suggest dispersal by birds and mammals. Bullock (pers. comm.) has observed birds feeding on these fruits on the Jalisco coast.

Within the group of *Sophora* and *Styphnolobium*, the latter possesses a series of primitive characters and may indeed be one of the predecessors of the group. The American species show less specialization than their Asiatic counterpart.

There are several fossil records of leaflets attributed to *Sophora*. The fossil fruit and the associated leaflets from sediments of the Claiborne formation leave no doubt that they are remains of *Styphnolobium*. These deposits are dated from the Middle Eocene from western Kentucky, Tennessee and northern Mississippi (Herendeen & Dilcher, 1986). This fact as well as the report by Crepet and Taylor (1985), suggests that the Leguminosae constituted an important element in the vegetation and flora of the Paleogene in the southeastern United States with a seasonal warm subtropical climate.

Millettieae

Piscidia L. *Piscidia* is an American genus with seven species, four of which are in Mexico (Rudd, 1969). Rudd observed that "it is probable that the genus originated on the old geological nucleus of Guatemala and southern Mexico. From there it spread northward to Sonora, or beyond, southward into South America, and eastward into the Antilles and southern Florida. Although the winged pods are readily dispersible by wind, the major eastward migration could have taken place during the Tertiary Period, when there was a land connection between Central America and the Greater Antilles." It may have taken place from the Late Cretaceous to the Paleocene. Macrofossil records identified as two species of *Piscidia* from the Lower Tertiary from Krakow, Poland (Unger, 1864) have not been verified.

Lonchocarpus s.s. KUNTH. The genus *Lonchocarpus s.s.* includes species with flowers in pairs on short peduncles and are mostly found in the American tropics. Mexico is the major center of diversity, with 74 species in 12 sections. In Central America it is numerically important in Nicaragua with 23 species (Sousa, in press) diminishing to 14 in Panama. In South America it reaches its greatest diversity in Colombia and Venezuela with approximately 15 species (Pittier, 1928; Sousa, pers. obs.) and diminishes in Brazil to eight species (Azevedo-Tozzi, 1989), including ser. *Pubiflori* Pittier, sect. *Lonchocarpus* and sect. *Densiflori* Benth. Interestingly, the Brazilian state of Bahia has only two wide-ranging species (Lewis, 1987), although a large part of the vegetation in the state, particularly in the west, is similar to the hot-dry types in Mexico, which are rich in species of *Lonchocarpus s.s.* The paucity of *Lonchocarpus* in Brazil and particularly Bahia may be due to historical factors.

Ongoing studies by Sousa have resulted in the modifications in Bentham's (1860) and Pittier's (1917) classifications, with radically differing concepts emerging. All 12 sections of *Lonchocarpus s.s.* are found in Mexico, two of which are endemic to the country: Sect. *Eriophylli* Benth. *p.p.*, with seven species, is almost endemic to the Rio Balsas Basin on the central western slope; and the other, a monotypic section (unnamed), is restricted to the Gulf slope in the Sierra Madre Oriental. Sect. *Densiflori* Benth. *p.p.* (*L. heptaphyllus* and related species) with about 18 species, of which eight occur in Mexico (two endemics), is the most primitive in the genus. Species of this section occur in the hot-humid to warm climates on the Gulf Coastal Plain and associated highlands.

There are several reports of macrofossils among the fossil remains of *Lonchocarpus*, of which some are clearly erroneous, as are those concerning the pods of "*L. anceps*" (Berry, 1924) from the Eocene of Tennessee, which more closely resembles *Gleditsia aquatica*. The latter possibility was suggested by Berry himself. The surrounding environment here was a *Liquidambar* forest, a vegetation unlikely to harbor *Lonchocarpus*. Another doubtful report is that of *Lonchocarpus novae-caesareae* from the Pleistocene of New Jersey (Hollick, 1896). On the other hand, *Lonchocarpus oregonensis* from the Upper Eocene of Oregon (Sanborn, 1935) is a good candidate. The vegetation was similar to present-day lower montane rain forests with a warm, humid climate. Sanborn pointed out that the fruits and associated leaflets are similar to those of *L. rugosus*, a species of the ser. *Obtusifolii* Benth. *p.p.*, which has 70% of its species in Mexico. The leaflets of *L. standleyi* from the same deposit bear an obtuse apex, but without the prominent nervations, and is thus similar to those of sect. *Eriophylli* Benth. *p.p.*, which is endemic to Mexico. *Lonchocarpus standleyi* is not related to *L. heptaphyllus* as was supposed by Sanborn. Remains from the Eocene of California (Sanborn et al., 1937) have not been easy to classify. Finds from the Miocene of Puerto Rico (Hollick, 1928) may be associated with *L. heptaphyllus*. Fossils from the Pliocene of the eastern Andes in Bolivia (Engelhardt, 1894) have been attributed to *Lonchocarpus*. The assignment, however, is questionable (Berry, 1917; Sousa, pers. obs.).

Lonchocarpus s.s. may have originated in North America during the Lower Tertiary and spread out from Mexico and northern Central America to the Greater Antilles in the east and to South America in the south from where it may have reached the Lesser Antilles. *Lonchocarpus sericeus* from the west coast of Africa may be a recent arrival there from South America by means of long-distance dispersal.

Robinieae

Coursetia A. CANDOLLE. *Coursetia* is an American genus of 39 species from the southern United States, Mexico, and South America. It is poorly represented in southern Central America and the Caribbean (Lavin, 1987, 1988).

Lavin (1988) reduced *Cracca* to *Coursetia* which is now represented in Mexico by three of the five sections and 18 species. Sect. *Madrense* is comprised of ten species, with seven in Mexico (six endemics) and extends southward to Costa Rica. It is represented in Nicaragua by four species (Lavin & Sousa, 1987; Sousa, 1986 [1987]). Lavin (1988) considered this section as the most primitive in the genus and suggested its place of origin in an area near its present distribution. He placed *Coursetia* with the "basal genera" *Lennea* and *Robinia* of Robinieae.

The other sections represented in Mexico are sect. *Coursetia* with three species (one endemic) and sect. *Craccoides* with eight (five endemic). Sect. *Coursetia* has a disjunct distribution and is found in northern Mexico and the Peruvian and Argentine Andes. Lavin suggested that this distribution was reached during the Quaternary, when the tropical deciduous forest had its greatest distribution. He argued that the ancestors of the tribe Robinieae diverged early as did the ancestors of *Robinia*, *Olneya*, *Peteria*, and *Genistidium*, which migrated northward, and *Coursetia*, which reached South America. His observations were based on the present distribution pattern of the tribe Robinieae, which is predominantly found in seasonally dry tropical forests of Mesoamerica and the Greater Antilles.

Phaseoleae

Macroptilium URBAN. *Macroptilium* is an American genus with about 16 species, all but one (*Macroptilium pedatum*) occurring in South America. Six of the South American species extend their distribution into Mexico.

Macroptilium pedatum is endemic to the Rio Balsas Depression and is closely related to the widely distributed *M. gibbosifolium*.

Generally, the genus is distinguished by the sessile flowers, which are aggregated in large inflorescences. The wings of the corolla are well developed, and the left one is modified to assume the position and function of the standard. Frequently, both chasmogamic and cleistogamic inflorescences are found in the members of this genus. Lackey (1983) suggested that *Macroptilium* may be related to *Phaseolus*, which is not supported by studies of floral morphology and biology (Delgado, in prep.). *Macroptilium* seems to be originally from South America and most likely arrived in the Northern Hemisphere during the Pliocene.

Oxyrhynchus BRANDEGEE. *Oxyrhynchus* is a genus with three species, two of which, *Oxyrhynchus volubilis* and *O. trinervius*, are distributed from Mexico to northwestern Colombia, with the former extending to the Bahamas and Cuba. The third species, *O. papuanus*, inhabits New Guinea (Verdcourt, 1978).

Oxyrhynchus is considered an intermediate between the subtribes Diocleinae and Phaseolineae (Lackey, 1981). Its seeds and those of *Dioclea* share certain characteristics that contribute to their capacity to float (Gunn & Dennis, 1976) and are dispersed by water currents. Its presence in the Yucatán Peninsula, the Bahamas, and Cuba is evidence for this dispersal form as are the reports of its occurrence in southeastern Florida and discoveries of it in the Gulf (Gunn & Dennis, 1976). Verdcourt (1978) attributed its occurrence in New Guinea to long-distance dispersal or perhaps to transportation by man, as its seeds are used in necklaces.

The genus represents an ancient tropical element in the Northern Hemisphere, where it probably arrived during the Tertiary. It may have reached the Bahamas recently. It is known from one locality in northwestern Colombia.

Amorpheae

Errazurizia PHILIPPI. *Errazurizia*, an American genus close to *Psorothamnus*, has only four species. Three are present in Baja California and nearby islands, Sonora, and Arizona; and the fourth occurs in the arid zones of Chile. This vicariant (amphitropical) distribution between Baja California and Chile suggests a wider distribution in the past and a great age for the group (Barneby, 1977).

Group D

Group D comprises nonendemic genera (Table 17.4), but with supraspecific endemic taxa or with more than 60% of the species represented in Mexico and the total species of the genus with less than 35% representation in the

Table 17.4. Group D: nonendemic genera, but with supraspecific endemic elements or these taxa with more than 60% species represented in Mexico but with less than 35% of the total species represented in Mexico

Caesalpinioideae
Caesalpinia ("*Brasilettia,*" "*Poincianella*" in part), *Cassia* (ser. *Heterospermae*), *Senna* (from sect. *Chamaefistula*: ser. *Skinneranae*, ser. *Confertae*, ser. *Brachycarpae*, ser. *Tharpia*, ser. *Armatae*, ser. *Galeottianae*; sect. *Astroites*; sect. *Peiranisia*: ser. *Deserticolae*, ser. *Isandrae*, ser. *Excelsae*[1]), *Chamaecrista* (from sect. *Chamaecrista*, ser. *Greggianae*), *Bauhinia* (five groups with distinct pollination syndromes, taxonomic recognition awaits studies[2]).

Mimosoideae
Mimosa (sect. *Habbasia*, ser. *Malacophyllae*, ser. *Sinaloensis*, ser. *Monancistrae*, ser. *Argillotrophae*; sect. *Mimosa*: ser. *Xantiae*), *Acacia* ("*Acaciella,*" "*Myrmecodendrum*"), *Calliandra* (ser. *Racemosae*).

Papilionoideae
Sophora (sect. *Calia*), *Tephrosia* (3 informal groups[3]), *Indigofera* (ser. *Microcarpae*, ser. *Leptosepalae*, ser. *Thibaudianae*, ser. *Disperinae*, ser. *Platycarpae*[4]), *Desmodium* (Mexico is one of the centers of diversity[5]), *Erythrina* (sect. *Breviflorae*, sect. *Leptorhizae*, sect. *Olivianae*[6]), *Clitoria* (sect. *Mexicana*), *Aeschynomene* (sect. *Americanae*[7]), *Astragalus* (sect. *Strigulosi*, sect. *Scalares*, sect. *Scytocarpi*: subsect. *Antonini*, sect. *Quinqueflori*, sect. *Greggiani*, sect. *Micranthi*, sect. *Hypoleuci*, sect. *Diphaci*, sect. *Scutanei*[8]), *Harpalyce* (sect. *Harpalyce*), *Lupinus* (it is probable that for its size in Mexico, it forms discrete units, but we lack information).

[1]Irwin & Barneby (1981–1982); [2]R. Torres (pers. comm.); [3]Wood (1949); [4]Rydberg (1923); [5]Ohashi et al. (1981); [6]Krukoff (1982); [7]Rudd (1955); [8]Barneby (1964).

country. Included here are pantropical elements or those with wide distribution in temperate-cold climates but that have taxonomically isolated taxa native to Mexico. In several cases they are single species, but their differentiation merits supraspecific ranks. Examples are *Senna villosa* which constitutes *S.* sect. *Astroites*, or *Erythrina olivia, E.* sect. *Olivianae*. In group D 19 genera account for 788 species, of which 430 (54.5%) are endemic. It should be emphasized that 19 genera, or 14% of the genera of the family in Mexico, represent 45.7% of all the Mexican legumes.

Caesalpinieae

***Caesalpinia* L.** *Caesalpinia* is a pantropical genus of approximately 150 species, with probably 45 in Mexico (31 endemics) (Contreras, pers. comm.). The included species are heterogeneous. Recognition of several genera here may be justified. Britton and Rose (1930) recognized the following groups in Mexico: (1) *Guilandina* with two species, no endemics; (2)

Poincianella with 29 species, 21 endemics; (3) *Brasilettia* with eight species, six endemics; (4) *Russellodendron* (including *Nicarago*) with two species, one endemic; (5) *Libidibia* with two species, one endemic; (6) *Caesalpinia* with one species (endemic); (7) *Poinciana* (including *Erythrostemon*) with one species, doubtfully native.

1. *Guilandina* is a group whose efficient means of dispersal defines its current distribution pattern. These species are dispersed by ocean currents in almost all the tropical seas. There are two ways to deal with this pantropical group: either to divide the group as Britton and Rose (1930) did, recognizing 17 species for North America and the Caribbean, or reduce it to fewer species with wide distributions (Sousa, pers. obs.).

2. *Poincianella*, well represented in Mexico and South America is just beginning to be understood. The core of the group in Mexico centers on *Caesalpinia hintonii, C. caladenia, C. exostemma* and *C. mexicana*. Several species in hot-dry areas of the Pacific slope are in an active state of speciation. Some of these caesalpinias may be considered ancient in North America, whereas others, e.g., *C. gaumeri*, related to *C. pyramidalis* complex of Bahia (Brazil), may be a more recent acquisition from South America. *Hoffmannseggia*, maintained by Eifert (1970), may be placed in this group. It is, however, defined by a combination of weak characters, and its status is uncertain. Studies by Simpson (in prep.), may lead to the definition of its taxonomic status.

3. *Brasilettia* (including *Guaymasia*), which is defined by the glandular-lacerate lower calyx lobe and indehiscent fruits, is centered in Mexico; its members are unarmed. *Caesalpinia mollis*, found in the Caribbean and the Yucatan Peninsula, extends into northern South America. *Brasilettia* and *Peltophorum* have been placed together in the past, but their relationship has been questioned (Polhill and Vidal, 1981). Taxonomically, *Brasilettia* is an isolated relict in Mexican legumes. Its distribution presently extends into the Caribbean and northern South America.

4. *Russellodendron* (including *Nicarago*) with two species in Mexico; one is endemic to the western coast, and the other extends to the Caribbean.

5. *Libidibia*, with perhaps six species, extends from Mexico to South America. Two species occur in Mexico: *Caesalpinia coriaria*, which ranges from Mexico through Central America to northern South America and the Caribbean, is closely related to *C. paraguariensis* of South America; and *Caesalpinia sclerocarpa*, endemic to Mexico (coastal plains of Sinaloa to Oaxaca), is doubtfully distinct from the variable Brazilian *C. ferrea*.

6. *Caesalpinia* is well represented in the Caribbean. One species that occurs in Mexico (*C. sessilifolia*) is endemic to the Chihuahuan Desert.

7. *Poinciana* (including *Erythrostemon*) is South American. It is represented in Mexico by two species: *Caesalpinia pulcherrima* (probably native) and *C. gilliesii* introduced by man.

The oldest fossil pollen reports on *Caesalpinia* and related genera are from the Lower Eocene of Assam in northeastern India (Mueller, 1981) in the foothills of the Himalayas, which may have served as a corridor for floristic elements from west to east and vice versa (Ramamoorthy, pers. comm.). The impressive variability of the samples led Mueller to suggest that India may have been the cradle of this group of caesalpinioids that developed the pollen characteristic of *Caesalpinia*.

Cassieae

Cassia L. *Cassia* is a pantropical genus of 30 species (Irwin & Barneby, 1981), including 11 in America, 10 in Africa (Lock, 1988) and nine in south-southeastern Asia. *Cassia* is represented in Mexico by four species of which one is endemic. The Mexican *Cassia* fall into two groups: the nonendemic species, which are clearly related to South American ones, and the endemic *Cassia hintonii*, forming the monotypic ser. *Heterospermae* (Irwin & Barneby, 1982), distinguished by the aseptate pod, the arrangement of the seeds in the fruit, and the hilum of the seed, among other characters. Its relationship with other Cassias and its possible geographic link are not clear. It is obvious that *C. hintonii* is a well defined paleoendemic found on the coastal plain of western Mexico (Jalisco to Guerrero) and from the upper basin of the Rio Balsas in the states of Mexico and Morelos.

Mimoseae

Mimosa L. *Mimosa* is a tropical American genus with approximately 450 species, minimally represented in Africa and Asia. Mexico harbors 102 species, of which 59% are endemic (Grether, 1978, 1984, pers. comm.).

Mimosa, defined by fruit and stamen characters, is a complex genus that Britton and Rose (1928) divided into many smaller genera and series. Presently, its taxonomy is the subject of study by Grether and by Barneby. Its generic delimitation from *Schrankia* is not yet clear. *Mimosa*, with one of its major centers of diversification in north-northwestern Mexico, may have originated in Brazil (Elias, 1974; Grether & Chehaibar, pers. comm.).

Mexican members of the genus belong to two sections: *Habbasia* with 87 species and *Mimosa* with 15 (Grether, 1978). *Habbasia* is the more primitive of the two, as there are twice as many stamens in the section as the number of corolla lobes. This section includes the ser. *Distachyae* of Britton and Rose (1928) with approximately 30 species, 11 of which are found in Mexico, six being endemic. This series corresponds to Bentham's (1875) ser. *Leptostachyae* (including *Spirocarpae* Britton & Rose), considered the most primitive because of its spicate inflorescence. The well-preserved inflorescences and pollen grains of the fossil *Eomimosoidea plumosa* described by Crepet and Dilcher (1977) identified it with ser. *Leptostachyae* (Sousa & Grether, pers. obs.). These remains are known from the Middle Eocene of Tennessee to the Oligocene of Texas (Daghlian et al., 1980). Of the other series of sect. *Habbasia*, ser. *Rubicaules* Bentham, including *Malacophyllae* and *Leucaenoides* of Britton and Rose (Grether, pers. comm.), has an interesting geographic distribution. It is represented by nine (five endemic) species in Mexico, six species in South America, all the African species, and the two species of *Mimosa* from India. The African *M. onilahensis* seems more closely related to the Mexican *M. malacophylla* and *M. leucanoides* than to any South American ones (Grether, pers. comm.).

Mimosa ser. *Xantiae* and ser. *Sensitivae* are of phylogenetic interest (Chehaibar, 1988). The former has five species (four endemic) in Mexico, one of which extends to Guanacaste, Costa Rica. All five are found on the Pacific slope, ranging from the Los Cabos region of Baja California Sur to the Isthmus of Tehuantepec, where they are highly concentrated, and into Central America, suggesting the Isthmus of Tehuantepec as the center of diversification for this series (Chehaibar, 1988). It is intimately related to ser. *Sensitivae*, with *M. acapulcensis* as the link between the two series. The two are often difficult to distinguish from one another. Ser. *Sensitivae* is considered the culmination of the sect. *Mimosa*, on the basis of the reduction in its leaflets and its shrubby nature.

Acacieae

Acacia MILLER. *Acacia* is a pantropical genus with 1,250 species, 900 of which are exclusively Australian, the rest being from Africa and America. In the neotropics there are around 200 species. In Mexico it is composed of 85 species of which 46 are endemic to the country (Rico, 1984).

The genus includes numerous heterogenous elements but has been treated as one for historical rather than biological reasons. The unifying characters have been the numerous free stamens and the eglandular anthers, although there are numerous exceptions to the latter character. Within the neotropics, the South American and Mesoamerican species represent different phyletic lines. The Mexican group is notable for its high degree of endemism (54%, Rico, pers. comm.) and the number of taxa (Table 17.5) it shares with the United States (88.9%), Central America (87.5%), the Antilles (22%), and South America (10.7%).

Guinet and Vassal (1978) divided *Acacia* into three subgenera: *Aculeiferum*, *Heterophyllum*, and *Gummiferae* (*Acacia*). *Heterophyllum* is almost endemic to Australia, and the other two are pantropical and are represented in Mexico: *Acacia* (*Gummiferae sensu* Guinet & Vassal, 1978) with 26 species and *Aculeiferum* with 56 (Rico, pers. comm.). The species of the latter fall into two phyletic lines: group *Senegalia* (recognized as genus by Britton &

Table 17.5. Genus *Acacia* in the Americas

Region	No. of species	No. common to Mexico	% of common species
United States	18	16	88.9
Mexico	85	85	100.0
Central America	24	21	87.5
South America	103	11	10.7
Antilles	32	7	22.0

Rose, 1928) and the sect. *Filicinae* or genus *Acaciella* of Britton and Rose (1928). Rico considered *Senegalia* with its wider distribution (Africa, Australia, and America) to be the oldest group of the acacias. On the other hand, it is difficult to maintain *Acaciella* within *Acacia*. It is nonaculeate and lacks glands on the petioles. Its pollen is a polyad of eight instead of 16. The type of pore seems to have had an independent origin (Rico, pers. comm). *Acaciella* is represented in Mexico by 20 species, of which 13 are endemic. The group is also represented in the Caribbean and South America. There exists a fossil record for *Acaciella* preserved in amber from Simojovel, Chiapas, from the end of the Oligocene to the beginning of the Miocene (Miranda, 1963).

In the subgenus *Acacia, Myrmecodendron*, given generic status by Britton and Rose (1928) is distinguished by advanced characters related to the mirmecofilia: partly hollow stipules harboring the ants with which it shares a close symbiotic relationship (Janzen, 1966), and nutritionally rich modified leaflet tips (Beltian bodies) for feeding them; complex inflorescence and pollen; and the great specialization of fruit and arillate seeds. Such specialization suggests that it is a natural group. *Myrmecodendron* is made up of 13 species (Ebinger & Seigler, 1987; Janzen, 1974) of which 10 are found in humid, warm areas of Mexico with the remaining three endemic to Central America.

Finally, Rico (pers. comm.) considered *Acacia* to be a genus from Gondwanaland, probably centered in Australia, and *Acaciella* a Laurasian element that had its origin in Mexico.

Ingeae

Calliandra BENTHAM. *Calliandra* is predominantly a tropical American genus with about 120 species. It has minor representation in eastern Africa, Madagascar, and India (Hernández, 1984; pers. comm.).

The taxonomy of this genus is the subject of study by Hernández in Mexico and Central America and Forero (1984) in South America. Of the five series in *Calliandra*, Hernández (1986 [1987], 1987, 1989) segregated the ser. *Laetevirentes*, a monophyletic group, to *Zapoteca*, leaving Bentham's

(1875) four series within it. Three of them are found in Mexico. The ser. *Nitidae* is the most primitive of the genus, with three centers: Mexico with 15 species (eight endemics), the dry zone of Bahia in Brazil, and Africa where it is represented by two species. The Mexican ser. *Racemosae* is composed of 12 species (nine endemics) with two extensions, one into Panama and another into Surinam. The third ser. *Macrophyllae* is well represented in the Andes, and in Mexico it forms a complex of nine species (four endemics); this series is closely related to ser. *Nitidae*, and *C. laevis* of western Mexico provides a link between the series (Hernández, pers. comm.).

Millettieae

Tephrosia PERSOON. *Tephrosia* is a pantropical genus of 400-450 species, with a great diversity in Africa. In America, Wood's study (1949) of American taxa is being continued by Téllez (1979, 1985, 1986), particularly for the neotropical area.

In the Americas, the genus *Tephrosia* is made up of two subgenera: *Barbistylia* with bearded style (approximately 180 species) and *Tephrosia* with glabrous style (approximately 220 species). The former is represented in Mexico by an estimated 47 species (37 endemics), in the United States by eight and in South America by two: *T. sinapou* and *T. nitens* with weedy tendencies. The latter is poorly represented in Mexico. Its four species are widely distributed (South America and the Caribbean). The subgenus is better represented in South America. In Africa and southern Asia both groups are well represented. *Barbistylia* species of India are closely related to the Mexican and northern central American species, e.g., *T. belizensis*. Likewise, the African species of this group are related to those such as *T. rhodantha* of Mexico (Téllez, pers. comm.).

The Mexican species have an interesting phytogeographic history. They may have reached Mexico from different areas at different times: from Africa via Asia or from warmer Europe during the Lower Tertiary (Paleogene) or more recently from South America. On the other hand, some of the South American taxa may have had their origins from Mexican ones in recent times.

Phaseoleae

Clitoria L. *Clitoria* is a tropical genus of 59 species: 47 of them native to the neotropics and subtropics, 11 native to the paleotropics, mainly in Africa and Southeast Asia, and one in the temperate regions of North America and Southeast Asia. Fantz (1977) divided *Clitoria* into three subgenera: (1) *Bractearea* (29 species, concentrated in South America, considered primitive), (2) *Clitoria* (five species, inhabiting Africa and Southeast Asia), (3) *Neurocarpum* (24 species, with three centers of diversification: Mexico, South America, and Asia).

The South American subgenus *Bractearea* is represented in Mexico by *C. glaberrima* and may be a recent arrival in the country dating from the Pliocene-Pleistocene.

In the subgenus *Neurocarpum*, sect. *Mexicana* is especially noteworthy. It has five species, of which three are endemic, and the other two extend into Central America. This section is closely related to *Clitoria mariana*, which suggests a possible Laurasian origin. On the other hand, the affinities of sect. *Neurocarpum* (represented by *C. guianensis* and *C. falcata*) are to South American and Antillian species. It is possible that *Clitoria* in Mexico owes its origin to ancestral elements from both hemispheres.

A significant taxonomic change concerning this genus was the reduction of monotypic *Barbieria* to *Clitoria* by Lackey (1981); *Barbieria pinnata*, usually found in small and isolated populations in moist forests, extends from Guerrero in Mexico to the Greater Antilles and South America.

Galegeae

Astragalus L. *Astragalus* is the most numerous genus of the family and one of the largest of the angiosperms. It has approximately 2,000 species, of which 88 are found in Mexico. It occurs in Europe, Asia, Africa, and the Americas. The three areas of high diversity are Southeast Asia, western North America, and western South America.

Polhill and Raven (1981) suggested that the explosive speciation occurred in the genus during the Tertiary with the opening up of dry regions. The genus, however, is well represented in temperate and mesic areas.

Barneby (1964) postulated that seven groups of the genus reached North America from Eurasia during the middle of the Tertiary. He further suggested that these invasions may have taken place at different times, the evidence of which comes from the presence of groups considered primitive in the Mexican mountains but whose affinities are to Old World groups. He commented further on the significance of these Mexican groups, as groups with advanced features are found further north.

Corroborating evidence for the multiple invasion idea came from the studies of Spellenberg (1976). His studies suggested that the chromosome numbers of Mexican species of *Astragalus* sect. *Strigulosi* are $n = 15$. According to him, the ancestors of these species may have been tetraploids ($2n = 32$), and aneuploidy may have contributed to the origin of groups with $2n = 30$. These tetraploids perhaps came from the Old World where the euploids with the basic number $x = 8$ are common. In addition to species with $n = 15$, others with numbers $n = 11, 12, 13$, and 14 are common. This chromosomal diversity and the dry and mesic climates may have contributed to speciation in the genus. Aneuploidy could have been another evolutionary force in the genus leading to the generation of numerous species, many of them endemic.

It is of interest that Burkart (1939) reported *Astragalus* fossils of doubtful identification from the Eocene from Patagonia, Argentina.

Brongniartieae

Harpalyce MOC. & SESSÉ EX A. CANDOLLE. *Harpalyce* has 20 species forming three geographically isolated sections (Arroyo, 1976): (1) sect. *Harpalyce* with six species is essentially Mexican with one extending to Nicaragua; (2) sect. *Brasilianae* with seven species, all of southeastern Brazil; and (3) sect. *Cubensis* with seven Cuban species. Borhidi and Muñiz (1977), however, recognized 17 species in Cuba.

The nature of endemism in the genus is noteworthy. Species are often restricted to specific soil types (Arroyo, 1976). Arroyo discounted the idea of long-distance dispersal to explain disjunct distributions and argued that the present disjunctions in the genus were caused by the fragmentation of a continuous distribution during the dry Pleistocene by the Post-Pleistocene wet period. She considered the sect. *Brasilianae* to be the most primitive, based on the morphology of the calyx in bud; nonetheless, the calyx in the section seems to be more specialized (its apex is asymmetrical owing to the vexillar gibbous lip). On the other hand, the Mexican sect. *Harpalyce*, "derived from the above," possesses an apiculated calyx without a gibbous lip. The Cuban sect. *Cubensis*, which is distinguished by the elongated calyx as well as its possible pollination by birds (in contrast to pollination by bees in the other sections), appears to be the most advanced. The great differences of *Harpalyce* from the other members of the tribe *Brongniartieae* led Arroyo to suggest that it may be an ancient group whose immediate ancestors have since become extinct.

Group E

Group E contains nonendemic genera (Table 17.6) but with less than 35% Mexican species, without endemic supraspecific taxa, but with specific endemism. This group includes tribes and genera whose major development is outside Mexico but that have been able to colonize the country. Species are not strongly differentiated and can be taxonomically correlated with others outside Mexico. It includes 33 genera (24.6% of the total) containing 319 species, of which 93 (29.15%) are endemic.

Mimoseae

Prosopis L. *Prosopis* is a genus with 44 species, nine of which are in Mexico, of which three (33%) are endemic. Burkart (1940, 1976) considered *Prosopis* to be a primitive genus within Mimosoideae, which probably originated in tropical Africa, where *P. africana*, the least specialized species in the genus, is found. In western Asia and ranging from Arabia to India, a prickly group of the sect. *Prosopis* is well developed in arid zones. The American species are made up of two parallel groups, also of arid climates: sect. *Strombocarpa* (with stipular spines) and sect. *Algarobia* (with cauline spines). A third, the

Table 17.6. Group E: with nonendemic genera, without supraspecific endemic taxa, but with specific endemism with less than 35% of the total species

Caesalpinioideae: none.

Mimosoideae: *Prosopis, Inga, Albizia, Pithecellobium, Zygia, Cojoba.*

Papilionoideae: *Swartzieae, Swartzia; *Sophoreae, Ormosia, Dussia;
*Dalbergieae, Andira, Machaerium, Dalbergia, Platymiscium, Pterocarpus;
*Robinieae, Sesbania; *Phaseoleae, Mucuna, Cleobulia, Canavalia, Galactia,
Centrosema, Vigna, Rhynchosia, Eriosema; *Psoraleeae, Orbexilum;
*Amorpheae, Amorpha; *Aeschynomeneae, Amicia, Zornia, Stylosanthes;
*Loteae, Lotus; *Vicieae, Vicia, Lathyrus; *Trifolieae, Trifolium; *Crotalarieae,
Crotalaria.

*Tribe.

small sect. *Monilicarpa* with one species, *P. argentina*, is also found in the Americas. The origin of *Prosopis* and *Prosopidastrum*, a related genus, may have occurred at the end of the Mesozoic and beginning of the Tertiary when the ancestors of these groups spread from Africa to the east and west, not by long distance but by endozoic dispersal of their seeds in fleshy fruits by mammals and birds. Rzedowski (1988) observed that "moreover, one cannot completely discard the possibility that despite its Gondwanian origin and the current concentration of the diversity of *Prosopis* in Argentina, these plants may have arrived in North America at the end of the Cretaceous or during the Eocene by way of the Laurasian meridional and later may have migrated southward." The current distribution of the genus in America is notably disjunct with a Mexican-Texan and an Argentine-Paraguayan-Chilean center, which, according to Burkart (1976), "does not exclude the possibility of a common ancient desert flora between both areas that was later divided with both centers remaining separate. The fact that in both centers endemic species exist indicates their antiquity and suggests that long-distance dispersal did play only a secondary role, if any." To conclude, Burkart affirmed that "several facts indicate that in America the main center of origin is Argentina, and Mexico is a secondary one. Or perhaps they are of the same age, but factors such as stronger competition with boreal floristic elements, prevented a major species development in the northern area of the genus." The paucity of its species in the north may indeed be due to the severe competition with elements there. It should be mentioned here that there is a fossil report (*Prosopis linearifolia*) for the Oligocene of Florissant, Colorado (MacGinitie, 1953).

Prosopis juliflora is continuously distributed from Mexico to northern South America and the Antilles. It is a recent distribution pattern of an aggressive species that invades marginal coastal areas and lacustrine habitats, whose dispersal was probably favored by large herbivores until about 12,000 years ago (Janzen & Martin, 1982).

Ingeae

Inga MILLER. *Inga* is a tropical American genus with approximately 300 species. Poncy (1985) recognized three centers of diversity: Guiana-Amazonia, Mexico-Central America (including northern Colombia and Venezuela), and Bahia (east-southeastern Brazil); each region is defined by its own set of species, and between them they share a small number of species (five to nine). The genus is the most diverse legume in the Amazonia (89 species) (Ducke, 1949) and may have originated there (León, 1966). However, Poncy (1985) argued that the ancestral elements of *Inga* may have originated in the ancient shields contiguous to the Amazonian basin, which is relatively young, formed by the accumulated sediments during the Miocene-Pliocene.

Inga is a genus of mesic to hydrophytic habitats, certainly due to the heavy requirements of moisture of the seeds during germination. Although the seed covering has been reinterpreted as endocarpic tissue (Gunn, 1984), there is evidence of the presence of a thin testa in some species (Rico, pers. comm.). Viviparity is known in the genus. The embryos remain protected inside the fruit until it begins to decompose. This characteristic, along with a probably short viability, has restricted the genus to humid climates, both hot and temperate (there are species in Soconusco, Chiapas, that grow at 2,200 m); in conditions of dry seasonal climate, it is confined to stream and lake banks. Thus in Mexico *Inga* is well represented on the Gulf slope and the area of Soconusco on the Pacific slope in the tropical and montane rain forests.

In Mexico *Inga* is represented by sect. *Bourgonia* (four species, no endemics); sect. *Diadema* (three species, no endemics); sect. *Pseudinga* (24 species, eight endemics) and sect. *Inga* (five species, one endemic). Fourteen species are distributed in a more or less continuous form from Mexico to South America. The species *I. affinis*, *I. semialata*, *Inga thibaudiana* and *I. vera* are widespread.

Sophoreae

Ormosia JACKSON. *Ormosia* is a tropical genus of about 100 species, with two major centers: southern and Southeastern Asia and South America, extending to Mexico (Rudd, 1965). It is represented in Mexico by five species (Rudd, 1981a; Rudd & Wendt, 1988 [1989]). Four of the Mexican species are allied to groups centered in Brazil and one, *O. panamensis*, to an Asiatic group formed by 17 species. Yakovlev (1971) raised this group to genus *Fedorovia*, which is distinguished from its American relatives by the septate fruit.

The species of *Ormosia s.l.* in Mexico are made up of two phyletic lines: one from South America and one from Asia. The former includes two groups: (1) one with *O. macrocalyx* which has a continuous distribution from Brazil and *O. isthmensis* from Venezuela, and (2) another with two endemic species, one from Oaxaca related to *O. isthmensis* and the other from the moist Atlantic coasts of Mexico and Belize, belonging to an Amazonian group. The first species group of *O. macrocalyx* and *O. isthmensis* may be recent in the country, and the other of two endemics may be more ancient but have retained their South American links. On the other hand, *O. panamensis* is a relict of an ancient flora dating back to warmer North America, which was connected to floristic elements of Asia directly or by way of Europe.

Dalbergieae

Pterocarpus JACQUIN. *Pterocarpus* is a pantropical genus of more than 20 species, perhaps 25 to 30 (Lewis, 1987), with three centers of diversity: tropical America with six species (perhaps more than 10), Africa with 11 species, and the Indo-Malayan region and the Pacific with five species (Rojo, 1972).

Except for *Pterocarpus officinalis*, whose pods are water-dispersed, the species in the genus are wind-dispersed. The propagules of *P. officinalis* are frequently found on the beaches of Quintana Roo; they have not been able to establish themselves on Mexican soil probably because of the lack of surface rivers in the Peninsula. The fruits of the three Mexican species are winged. In two of these species the ovary and fruit are sessile, and in *P. acapulcensis* they are stipitate. In contrast, in the African (except *P. santalinoides* with bicontinental distribution in Africa and America) and Indo-Malayan species the ovary and fruit are stipitate (Rojo, 1972), and in all the South American species they are sessile. *Pterocarpus acapulcensis* has a disjunct distribution along the Pacific coastal plain from Guerrero to Chiapas and appears again in Panama and northern South America. Like many of the elements of "Los llanos" in Venezuela, it may have arrived there from Mexico.

The species in Mexico represent two phyletic lines: one with South American affinities, represented by *Pterocarpus rohrii*, which has a continuous distribution from Brazil, and the endemic *P. orbiculatus*. *Pterocarpus acapulcensis* represents the other line with its affinities to Afro-Indo-Malayan species.

Phaseoleae

Cleobulia MARTIUS EX BENTHAM. *Cleobulia* is an American genus of the subtribe Diocleinae with a main center of diversity in Brazil (Maxwell, 1977). Of its four species, three occur in Brazil; the fourth, *Cleobulia crassistyla*, is disjunct in Guerrero, Mexico (Maxwell, 1982). The last differs from the rest in its shrubby habitat, in having a corolla whose wings are as long as the petals of the keel, and in having a gynoecium with an indurate swelling on the dorsal, distal end of the ovary.

Cleobulia crassistyla is a relict in Mexico. The genus must have had a wider distribution before the union of the two hemispheres, i.e., at the beginning or middle of the Tertiary. Berry (1923) reported fossil leaflets similar to those of the genus *Dioclea* for the Miocene in Oaxaca, although a reexamination by Delgado of the illustration of the fossil left Berry's conclusion questionable.

Vigna SAVI. *Vigna* is a cosmopolitan genus of approximately 150 species with three important centers of diversity: Africa, Asia, and South America. In Mexico it is represented by 15 species, three of which are endemic. American species were grouped by Maréchal et al. (1978) into two subgenera, *Lasiospron* and *Sigmoidotropis*. These subgenera are probably not natural units.

The species of *Lasiospron* are morphologically similar to the vignas of the Old World and are characterized by peltate stipules, yellow flowers whose keels form spirals, and triporate pollen. Two of the species have strophiolate seeds. The species inhabit flood plains and are usually found in rice paddies. One of them has recently been introduced into Africa by man (Delgado, in prep.).

The species of *Vigna* subgenus *Sigmoidotropis* are characterized by stipules without appendices and inflorescences with developed swollen nodes in extrafloral nectaries. The keel of the corolla shows different configurations: sigmoid, erect, or spiral. Most Mexican species belong to subg. *Sigmoidotropis*.

Fossil pollen of *Vigna* has been reported from the Paleocene of Texas (Elsik, 1968), which suggests that *Vigna*, which is of austral origin, was part of the Pre-Tertiary flora in Mexico. These ancestral elements could well have given rise to the segregate genus *Ramirezella*.

Aeschynomeneae

Amicia KUNTH. *Amicia* is an American genus with about seven species. *Amicia*, along with *Weberbauerella*, *Zornia*, and *Poiretia*, forms the subtribe Poiretiinae (Rudd, 1981a). This subtribe is an unnatural assemblage and is distributed mainly in the Southern Hemisphere. *Amicia* is represented in Mexico by *A. zygomeris*, which occurs in the montane rain forests of the Pacific and Gulf coasts.

The only chromosome count (Rudd, 1981a) for *Amicia* suggests that it is a polyploid, $2n = 38$ in contrast to $2n = 20$, which characterizes other genera in the subtribe. The present distribution of the subtribe and its polyploid nature suggest that *Amicia zygomeris* is a recent introduction to the Northern Hemisphere, spreading from South America with a notable disjunction in Central America.

Loteae

Lotus L. A cosmopolitan genus with around 100 species, *Lotus* is centered in the Mediterranean region with extensions in Africa, Australia, and the mountains of South America. Polhill (1981b) has reported that there exists a group of fairly old species spreading from the Himalayas to California.

Isely (1981) divided the North American species into four groups ("subgenera") and identified *Hosackia* as a unique group because its stipules are not reduced to glandular protuberances. The four groups are represented in Mexico. Endemism is low. Three species of the group *"Microlotus"* are remarkable, as their chromosome number is $n = 6$, an aneuploid originating from the basic number of the genus, $x = 7$.

Lotus is an element from Laurasian floras that reaches northwestern Mexico in large numbers. Extant collections suggest that it may be actively speciating in the mountains. The Mexican taxa of this genus are the subject of study by Corral.

Trifolieae

Trifolium L. A cosmopolitan genus of approximately 240 species, *Trifolium* has important centers of diversity in Eurasia and western North America with secondary centers in Africa and South America.

Zohary and Heller (1984) documented an exchange of species between the two areas from the Tertiary. They reported cases of species of North American origin that have moved to Asia and that have diversified in Europe. Saporta (1889) described *Trifolium protocalyx* (= *T. squamosum*) from a fossil calyx from the French Eocene and *T. palaeogaeum* based on a fossil leaflet. The origin of *Trifolium* is from Laurasian elements. Zohary and Heller (1984) reported that in the Americas 65% of the species are found in the Rocky Mountains and the West Coast of the United States, and only 11 species inhabit South America. They divided *Trifolium* into eight sections, two of which are represented in Mexico. The sect. *Lotoidea*, occurring in Eurasia, Africa, and America, is heterogeneous and includes species considered primitive. They recognized 24 species in the American sect. *Involucrarium*. In Mexico both these sections are represented, the former by five and the latter by six species. The Mexican representatives are range extensions of those from the United States, except *T. wigginsii*, of sect. *Involucrarium*, which is endemic to Baja California.

Crotalarieae

Crotalaria L. *Crotalaria* is a pantropical genus of about 650 species, of which 500 are found in Africa where it may have originated (Polhill, 1982). The genus is well represented in southern Asia and the Americas. *Crotalaria* is represented in Mexico by three sections (Soto & Moreno, pers. comm.). *Chrysocalycinae* (six species, two endemics) is the least specialized in its flowers (Bisby & Polhill, 1973). Sect. *Crotalaria* is represented by three species (no endemics). Sect. *Calycinae* (ten species, six endemics) displays the most advanced characteristics in habit, flowers, and the reduction of the leaf to a single leaflet (Windler, 1970, 1971). The species of this section are adapted to warm mountainous climates in the pine-oak forests of Mexico. It is interesting that the center of diversity of *Calycinae* (70 species) is Asia (India), but its range extends, even if tenuously, to Australia, Africa, and America (Polhill, 1982); the Mexican species characteristically have smaller flowers that distinguish them from those of South America.

Group F

Group F comprises of genera with no endemic taxa (Table 17.7). This group is divided into two subgroups: F I with two to five species in Mexico and F II with only one species per genus in the country; subgroup F I with 18 genera and 44 species, and subgroup F II with 34 genera. Thus 26.12% of all the genera in Mexico contain only 2.05% of the species.

Subgroup F I is the most marginal in Mexico, mainly when the genera are speciose, as are *Piptadenia* with 20 (centered in South America) and *Dialium* with 40 (centered in Africa). *Hoita* seems to be the exception, considering that 66.66% of its species are located in Mexico, but none is endemic.

Subgroup F II appears marginal in some cases, although its species may be ancient in the country, as is the case of *Hymenaea courbaril*, which is known to exist in Chiapas from the Oligocene-Miocene and in the south-eastern United States from the Upper Paleocene; another example is *Entada*, known from fruits of the Lower Tertiary of Europe (Unger, 1864). *Peltophorum* (related to *Conzattia* and *Heteroflorum*) has been discovered in Veracruz and is a relict of the legume flora. It is represented in the Greater Antilles by two other species; its distribution illustrates the Mexican Phytogeographic Legume Area (see below). Monotypic genera are also found in this subgroup; in seven cases such groups are centered in Mexico (see group B, subgroup B I, with the exception of *Cymbosema*).

Phaseoleae

Pachyrhizus A. CANDOLLE. A neotropical genus (subtr. Diocleinae), *Pachyrhizus* has five species that range from central Mexico to the Andean region of Peru and Bolivia (Sorensen, 1988). Three species are mainly Mexican-Central American. One of them, *Pachyrhizus panamensis*, reaches northern and northwestern South America. *Pachyrhizus tuberosus* and *P. ahipa* are cultivated; the latter is known only in cultivation, and the former's natural distribution is difficult to ascertain as it is widely cultivated (Sorensen, 1988). The present distribution of *Pachyrhizus* has thus been

Table 17.7. Group F: genera with no endemic taxa

Subgroup I: Genera with 2–3(–5) species in Mexico
Caesalpinioideae: *Haematoxylum, Cynometra.* Mimosoideae: *Piptadenia, Entadopsis, Neptunia, Abarema, Enterolobium, Chloroleucon.* Papilionoideae: *Gliricidia, Dioclea, Pachyrhizus, Calopogonium, Teramnus, Oxyrhynchus, Pediomelum, Hoita, Psorothamnus, Chaetocalyx.*

Subgroup II: Genera with one species in Mexico
Caesalpinioideae: *Gleditsia, Peltophorum, Schizolobium, Parkinsonia, Poeppigia, Dialium, Cercis, Peltogyne, Hymenaea.* Mimosoideae: *Entada, Entadopsis.* Papilionoideae: *Acosmium, Myroxylon, Vatairea, Muellera, Robinia, Olneya, Genistidium, Sphinctospermum, Cymbosema, Amphicarpaea, Barbieria, Strophostyles, Psoralidium, Orbexilum, Psoralidium, Rupertia, Apoplanesia, Psorothamnus, Poiretia, Pachecoa, Oxytropis, Glycyrrhiza, Pickeringia.*

---

considerably altered by man. It is possible that the genus had its origins in the Mexico-Central America area and spread to South America during the Pliocene-Pleistocene implying that *Pachyrhizus* or an ancestor of it may have reached Mexico by way of Laurasia. Hence its origin is contrary to that proposed for genera such as *Dioclea* and close relatives, which came from the south.

Amorpheae

Psorothamnus RYDBERG. *Psorothamnus* is a North American genus that is found in desert zones of northwestern Mexico, the southeastern United States and the states of Nevada, southern Utah, and southeastern California.

It is recognized by its petals, which are inserted on the hypanthium. The ovary in one species is pluriovulate. Barneby (1977) used this evidence to suggest that Amorpheae may have been derived from a pluriovulate ancestor. Information about its origin and phylogenetic relations are discussed under the genera *Dalea* and *Marina*.

Caesalpinieae

Gleditsia L. *Gleditsia* is a genus with 12 species (Polhill & Vidal, 1981). Mexico shares its only species with the eastern and northeastern United States.

Along with *Gymnocladus, Gleditsia* is considered one of the most archaic genera of the family. Fossils exist, and they are attributed to the Late Cretaceous. Its floral structure is simple and atypical of the family. *Gleditsia* occurs in the temperate-cold zones of North America (two species) and

Asia (ten species), and it has one species, *G. amorphoides*, in northern Argentina. Raven and Polhill (1981) explained this disjunction by suggesting that *Gleditsia* was perhaps present in Africa, where it became extinct, but it may have arrived in South America during the Lower Tertiary. *Gleditsia triacanthos* has recently been collected in the Sierra Madre Oriental in Mexico (Briones, 1988 [1989]).

Cercideae

Cercis L. *Cercis* is a genus with eight species that are mainly found in the Northern Hemisphere in temperate forests. Five are centered in Asia (China), one in southern Europe, and two in North America.

Wunderlin et al. (1981) grouped it along with clearly tropical genera (*Bauhinia*, *Griffonia*, and *Adenolobus*) in the subtribe Cercidinae. In this setting, however, it stands apart as a Laurasian element, and Yakovlev (1972) considered that the genus should be the only member in the tribe.

The genus *Cercis* is ancient and is the only one in the subfamily whose chromosome number is $n = 7$, the same as the basic number ($x = 7$) of the caesalpinioids (Wunderlin et al., 1981).

Cercis occidentalis is found in the eastern United States, and *C. canadensis* extends from Canada to the state of Puebla, Mexico. It occurs in the wet deciduous *Liquidambar* forests. Hopkins (1942) proposed *C. canadensis* as the ancestor of *C. occidentalis* and related it to the Asiatic species. *Cercis* has been reported in fossil beds of the Early Tertiary (Eocene) in Europe (Unger, 1864; Saporta, 1889; see Burkart, 1952).

Detarieae

Peltogyne VOGEL. A neotropical genus of 23 species, *Peltogyne* is distributed in Mexico, Panama, South America (Colombia, Guiana, Bolivia, southeastern Brazil), and Trinidad (Freitas da Silva, 1976). Its greatest diversity is in central Amazonia, the same as with its related genera *Hymenaea* and *Cynometra*. In Mexico, only a few populations are known of *Peltogyne mexicana* in the medium-sized evergreen seasonal forests near Chilpancingo, Guerrero (Contreras, pers. comm.). Freitas da Silva (1976) reported *P. mexicana* also in Panama and Colombia and considered its affinities to be with *P. gracilipes* of Brazil.

The disjunct distribution of *Peltogyne mexicana* is the result of a wider distribution in the past. The present Mexican populations represent a relictual and endangered element dating back to the pre-Cenozoic times.

Hymenaea L. *Hymenaea* is a genus with 17 species, mainly South American (one in Mexico). Langenheim and Lee (1974) reduced the monotypic African *Trachylobium* to *Hymenaea*, thus extending its distribution to eastern Africa.

The variable American *Hymenaea courbaril* has a wide distribution and is found in Mexico. Lee and Langenheim (1975) recognized five varieties, of which the typical element has the widest distribution, from northern South America to Mexico and the Caribbean. Distributional evidence suggests that this taxon may have reached Mexico after the Central American bridge was formed, but proving the contrary, the fossilized resin of this species in the form of amber dating back to the Late Oligocene to the Early Miocene has been found in Simojovel, Chiapas, Mexico, Colombia and Brazil (Langenheim, 1966). This period is approximately 27 Ma or a little more than 22 Ma before the Central American bridge was formed. Its arrival in Mexico must have been earlier, because in order to form fossils sufficiently abundant populations are necessary; and, in fact, Berry (1930), and Herendeen and Dilcher (1986) reported macrofossils from the Late Paleocene to the Middle Eocene from Kentucky and Tennessee, opening the time span to 45 Ma.

It has been proposed that marine currents may have brought the species to Mexico. Fruits have been found on the western coast of Mexico in Oaxaca (Barra de Colotepec), which one might conclude are from the local flora and not from South America, as sea currents from the Southern Hemisphere are absent here. Despite intense collections in beach deposits by Sousa and collaborators, it has not been found in the Gulf or the Atlantic coast of Puerto Morelos (Quintana Roo), Playa de Balsapote (Veracruz), or in the Yucatán Peninsula (Gunn et al., 1984). Gunn and Dennis (1976), however, reported a collection of a fruit with viable seeds on a Massachusetts beach, and there is another such report from the coasts of Ireland (Blake, 1825, in Gunn & Dennis, 1976).

Hymenaea courbaril var. *courbaril* forms important populations in the tropical deciduous to evergreen seasonal forests, fundamentally in riparian habitats on the Pacific coastal plains from Nayarit to Chiapas, but it is scarce on the Gulf coast. During the Oligocene-Miocene, the Simojovel group, according to Langenheim et al. (1967), was formed in a hot climate by deposits of sediment from the ancestral Sierra Madre in superficial marine conditions. There were occasional fluctuations of the coastline whose nearby vegetation included mangrove swamps and freshwater plants such as *Pachira aquatica*, as well as mesophilous elements originating in the mountains, vegetation types that are poor in *Hymenaea* now. It is, however, possible that the Oligocene-Miocene resin deposits may have been transferred from drier areas of the southern slopes of the Sierra, as happens today with the Grijalva River, which flows from the high basin through the dry central depression of Chiapas.

If *Hymenaea* reached Mexico by long distance dispersal, it reached the Pacific coast of Mexico through the Atlantic oceanic currents, as Central America did not exist then to form a barrier. These seeds would have had to beach together repeatedly to form dense populations, as the species is self-incompatible (Janzen, 1983). The number of seeds arriving in earlier times would have had to have been more numerous than now.

Millettieae

Muellera L. F. Mostly South American, *Muellera* is a genus of about four species. *Muellera frutescens*, which extends to Mexico, is found in coastal habitats of Veracruz and Tabasco. Burkart (1969b) recognized only two species. Its taxonomic distinction from *Lonchocarpus* has always been unclear, even though it has been based fundamentally on the moniliform pods with a thick pericarp, notably that of *M. frutescens* (Bentham, 1860). Burkart (1969b) cited characters such as apiculate or acute and generally pilose anthers to emphasize *Muellera*'s differences from *Lonchocarpus*, even though there are species of *Lonchocarpus* related to *L. sericeus* with pilose anthers. Geesink (1984b) reduced *Muellera* to *Lonchocarpus* sect. *Punctati* based on these weak characters and, during the same year (1984a), proposed to conserve *Lonchocarpus* over *Muellera*. The above taxonomic conclusions are erroneous, as they have been derived from studies based on wrong features when the characters should have been the vegetative buds and the inflorescences (Sousa, in press).

Muellera frutescens is dispersed by water currents, and it is common to find its fruits on the Veracruz and Tabasco coasts. *Muellera* may recently have arrived in Mexico from the lower Amazon Basin, probably progressively, as it has a continuous distribution up to Belize, with an interruption in the Yucatan Peninsula.

Phaseoleae

Amphicarpaea NUTTALL. The genus *Amphicarpaea* has three species: *Amphicarpaea africana* (Africa), *A. edgeworthii* (Asia), and *A. bracteata* (United States and eastern Mexico). Turner and Fearing (1964) suggested that the genus originated in the temperate forests of Africa from an ancestor similar to the present-day *Amphicarpaea africana*. They further suggested that its affinities are to the Old World *Shuteria* as well as to the American *Cologania*. The East Asian and North American species are found in mesophilous forests, and they are probably relictual elements of the Arcto-Tertiary flora (Turner & Fearing, 1964).

In Mexico, *Amphicarpaea bracteata* is found in an isolated form in secondary vegetation of *Liquidambar* forests and as a crop-field weed in Puebla and Veracruz states.

ENDEMISM, PHYTOGEOGRAPHIC PATTERNS, AND MEXICAN PHYTOGEOGRAPHIC LEGUME AREA

Tables 17.8 and 17.9 summarize and present the generic and species numbers in Mexican Leguminosae and point out their percentages in the world legume flora along with their endemic status in the country. Generic endemism in the family in Mexico is low, in sharp contrast to other regions of high endemism, such as tropical Africa and South America (Table 17.10). If generic endemism were considered an index for the origin and status of a plant family in a given area, the low endemism of Leguminosae in Mexico suggests that the family did not originate in Mexico and that it is not well represented there. However, the high species endemism of the family in the country (51.9%) and a comparison of generic diversity of the family with other regions of the world suggest that Mexico is an important world center of this economically important family. The diversity and degree of endemism of many leguminous groups rise (the latter to as much as 75% of the species or more) if their distribution is considered in a more natural floristic province which is here termed the "Mexican Phytogeographic Legume Area" (MPLA) to include the south-southwestern United States, northern limits of the Chihuahuan Desert Region, parts of Edward's Plateau, the south Texas plains and prairies, marshes of the Gulf of Mexico (Texas), Mexico, Central America as far south as Guanacaste in Costa Rica, and the Greater Antilles (Fig. 17.1). This area is somewhat similar to the Mega-Mexico 1 floristic province proposed by Rzedowski (this volume). The distribution of many legume groups suggests that MPLA is a natural phytogeographic province.

Five phytogeographic patterns are discernible in the Mexican Leguminosae.

1. Genera endemic to territorial Mexico (Table 17.1) but pertaining to tribes with related taxa, whose distributions may or may not be disjunct: Caesalpinieae—*Conzattia* (*Peltophorum*, of wide distribution, but possibly centered in Asia), *Heteroflorum* M. Sousa ined. (*Conzattia* in Mexico). Mimoseae—*Calliandropsis* (*Dichrostachys* group, especially with *Desmanthus*, Mexican centered and *Neptunia*, wide spread) (Hernández H. & Guinet, 1990). Millettieae—*Hesperothamnus* (*Piscidia* in Mexico and *Behaimia* from the Greater Antilles).

2. Genera endemic or mostly restricted to the MPLA (Table 17.11, group 1). They are (1) genera essentially of Mexican tribes with distributions principally without disjunctions (Table 17.11, first two tribes: Amorpheae and Robinieae); (2) groups of related genera, deserving higher taxonomic (subtribal) ranks with wider distributions resulting in relationships with groups and/or disjunctions from distant areas, predominantly Asia. They are in the tribe Caesalpinieae, part of the *Peltophorum* group (*Conzattia, Heteroflorum* and *Peltophorum*, the last with disjunctions); in the tribe Sophoreae, part of the *Myroxylon* group (*Myroxylon* and *Myrospermum*) and part of the *Sophora* group (*Styphnolobium, Cladrastis*), both with disjunctions and probably *Pickeringia* as suggested by Goldblatt (1981); part of the tribe Millettieae (*Hesperothamnus, Piscidia, Behaimia* in one group and *Lonchocarpus, Willardia, and Derris* in other, the last with disjunctions); in the tribe Aeschynomeneae, part of the subtribe Ormocarpinae (*Diphysa, Belairia, Pictetia*, and the distant *Ormocarpum*) and part of the subtribe Stylosanthineae (*Pachecoa, Chapmania*, and the distant *Arthrocarpum*);

Table 17.8. Genera of Mexican Leguminosae

		Estimated number			
Genera	Species in Mexico	Species endemic to Mexico	% of endemic species	Species worldwide	Mexican % of total species
CAESALPINIOIDEAE (20 genera)					
Caesalpinieae (11/47)					
Gleditsia	01	0	0	12	8.3
Heteroflorum	01	01	100	01	100
Peltophorum	01	0	0	7–9	14.2–11.1
Conzattia	03	03	100	03	100
Schizolobium	01	0	0	01–02	100–50
Caesalpinia	>45	>31	68	150	30
Hoffmannseggia	>19	>09	47	≈33	57
Haematoxylum	02	0	0	03	66.6
Parkinsonia	01	0	0	02	50
Cercidium	06	01	16	08	75
Poeppigia	01	0	0	01	100
Total	81	44	54		
Cassieae (4/20)					
Dialium	01	0	0	40	2.5
Cassia	04	01	25	30	13.5
Senna	63	16	25	240	26
Chamaecrista	20	02	10	250	8
Total	88	20	23		
Cercideae (5/2)					
Cercis	01	0	0	06	16.6
Bauhinia	30	19	63	250	12
Total	31	19	61		
Detarieae (3/55)					
Cynometra	02	0	0	70	2.8
Peltogyne	01	0	0	23	4.3
Hymenaea	01	0	0	17	5.9
Total	04	0	0		
TOTAL SPECIES	204	83	40		
MIMOSOIDEAE (26 genera)					
Mimoseae (13/39)					
Entada	01	0	0	20	5
Prosopis	09	03	33	44	20.4
Prosopidastrum	01	01	100	02	50
Goldmania	01	01	100	02	50
Piptadenia	03	0	0	15	20
Adenopodia	03	02	66.6	07	43
Entadopsis	01	0	0	02	50
Mimosa	102	60	59	450	23
Schrankia	04	03	75	05	80
Leucaena	11	07	63	12	91.6
Desmanthus	15	08	53.3	23	65.21

Table 17.8. (cont.)

	Estimated number				
Genera	Species in Mexico	Species endemic to Mexico	% of endemic species	Species worldwide	Mexican % of total species
Neptunia	03	0	0	12	25
Calliandropsis	01	01	100	01	100
Total	**155**	**86**	**58**		
Acacieae 1/(2)					
Acacia	85	46	54	1250	7
Total	**85**	**46**	**54**		
Ingeae (21/12)					
Inga	36	09	25	300	12
Abarema	2	0	0	40	5
Albizia	11	03	31	150	7.3
Lysiloma	08	03	11	09	89
Enterolobium	02	0	0	05	40
Calliandra	36	21	58	120	30
Zapoteca	09	06	66	18	50
Pithecellobium	≈14	03	21	30	46.6
Chloroleucon	02	0	0	10	20
Havardia	09	06	66	12	75
Zygia	08	03	37	53	15
Cojoba	09	01	11	24	37
Total	**146**	**55**	**37.7**		
TOTAL SPECIES	**386**	**187**	**48.4**		
Papilionoideae (89 genera)					
Swartzieae (1/11)					
Swartzia	05	01	20	135	3.7
Total	**05**	**01**	**20**		
Sophoreae (49/8)					
Acosmium	01	0	0	16	6.2
Myrospermum	02	01	50	03	66.6
Myroxylon	01	0	0	02	50
Ateleia	09	07	77	17	59
Ormosia	05	01	20	100	05
Dussia	02	01	50	10	20
Sophora	05	01	20	45	11
Styphnolobium	05	04	80	9	55
Total	**30**	**15**	**50**		
Dalbergieae (6/19)					
Andira	02	01	50	20	10
Vatairea	01	0	0	7	14
Machaerium	12	01	8	150–200	8–6
Dalbergia	10	03	30	100	10
Platymiscium	04	02	50	≈25	16
Pterocarpus	03	01	33	≈20	20
Total	**32**	**08**	**25**		

Table 17.8. (cont.)

Genera	Species in Mexico	Species endemic to Mexico	% of endemic species	Species worldwide	Mexican % of total species
			Estimated number		
Millettieae (7/41)					
Hesperothamnus	05	05	100	05	100
Piscidia	04	01	25	07	57
Derris	02	02	100	40	5
Lonchocarpus	74	45	61	115	64
Willardia	04	02	50	05	80
Muellera	01	0	0	04	20
Tephrosia	47	37	79	400–450	12–10
Total	**137**	**103**	**75**		
Robinieae (10/20)					
Sphinctospermum	01	0	0	01	100
Robinia	01	0	0	04	20
Olneya	01	0	0	01	100
Peteria	03	2	66	04	75
Genistidium	01	0	0	01	100
Gliricidia	02	0	0	02	100
Hybosema	02	1	50	02	100
Coursetia	18	12	66.6	39	46
Lennea	05	01	20	05	100
Sesbania	05	02	40	50	12
Total	**39**	**18**	**46**		
Indigofereae (1/4)					
Indigofera	25	10	40	700	3.6
Total	**25**	**10**	**40**		
Desmodieae (1/27)					
Desmodium	≈100	≈50	50	300	33
Total	**100**	**50**	**50**		
Phaseoleae (22/85)					
Erythrina	28	17	60	108	26
Mucuna	05	02	40	100	05
Dioclea	04	0	0	30	13.3
Cymbosema	01	0	0	01	100
Cleobulia	01	01	100	04	20
Canavalia	15	05	33	54	27.7
Pachyrhizus	02	0	0	05	40
Galactia	14	04	28	50	28
Calopogonium	04	0	0	08	50
Teramnus	03	0	0	08	37
Cologania	10	07	70	10	100
Amphicarpaea	01	0	0	03	33.3
Centrosema	08	01	12.5	45	17.7
Clitoria	09	03	33.3	59	15.2
Vigna	15	03	20	150	10
Ramirezella	07	06	85.7	07	100

Table 17.8. (cont.)

Genera	Estimated number				
	Species in Mexico	Species endemic to Mexico	% of endemic species	Species worldwide	Mexican % of total species
Oxyrhynchus	02	0	0	03	66.6
Strophostyles	01	0	0	03	33.3
Macroptilium	08	01	12.5	16	50
Phaseolus	34	18	52.9	37	87
Rhynchosia	26	09	34.6	193	13.5
Eriosema	10	07	70	112	09
Total	**207**	**84**	**40.5**		
Psoraleeae (5/9)					
Pediomelum	04	0	0	22	18.2
Orbexilum	01	01	100	08	12.5
Psoralidium	01	0	0	03	33.3
Hoita	02	0	0	03	66.66
Rupertia	01	0	0	03	33.3
Total	**09**	**01**	**11.11**		
Amorpheae (7/7)					
Apoplanesia	01	0	0	01	100
Eysenhardtia	11	07	64	11	100
Amorpha	02	01	50	15	13.3
Errazurizia	02	02	100	04	50
Psorothamnus	05	0	0	09	55.5
Marina	38	31	81	38	100
Dalea	113	74	65	161	70
Total	**172**	**115**	**67**		
Aeschynomeneae (9/26)					
Diphysa	16	08	50	16	100
Nissolia	13	10	77	13	100
Chaetocalyx	02	0	0	11	18.2
Aeschynomene	29	14	48	150	19.3
Amicia	01	01	100	07	14
Poiretia	01	0	0	06	16.6
Zornia	08	04	50	75	10.6
Pachecoa	01	0	0	01	100
Stylosanthes	08	02	25	50	16
Total	**79**	**39**	**49**		
Galegeae (3/20)					
Astragalus	88	55	62	2,000	4.4
Oxytropis	01	0	0	300	0.3
Glycyrrhiza	01	0	0	20	0.5
Total	**90**	**55**	**61**		
Loteae (1/4)					
Lotus	26	06	23	70–100	37–26
Total	**26**	**06**	**23**		
Vicieae (2/5)					
Vicia	06	01	16	140	4.3

Table 17.8. (cont.)

		Estimated number			
Genera	Species in Mexico	Species endemic to Mexico	% of endemic species	Species worldwide	Mexican % of total species
Lathyrus	08	02	25	150	5.3
Total	**14**	**03**	**21**		
Trifolieae (1/7)					
Trifolium	11	01	9.1	239	4.6
Total	**11**	**01**	**9.1**		
Brongniartieae (2/8)					
Brongniartia	65	64	98	65	100
Harpalyce	06	06	100	20	30
Total	**71**	**70**	**98**		
Crotalarieae (1/16)					
Crotalaria	21	08	38	550	3.8
Total	**21**	**08**	**38**		
Thermopsideae (1/6)					
Pickeringia	01	0	0	01	100
Total	**01**	**0**	**0**		
Genisteae (1/20)					
Lupinus	≈65	≈39	60	200	32.5
Total	**65**	**39**	**60**		
TOTAL SPECIES	**1,134**	**626**	**55.2**		

in the tribe Phaseoleae, part of the subtribe Glycininae (*Cologania, Amphicarpaea, Herpyza*, and the distant *Shuteria*) and part of the subtribe Phaseolinae; in the tribe Brongniartieae (*Harpalyce* with a disjunction, *Brongniartia*, and the distant *Templetonia* group); a great part of the tribe Psoraleeae (*Pediomelum, Orbexilum, Psoralidium, Hoita*, and *Rupertia*), which is endemic to North America (Grimes, 1990), *Cullen* (Asiatic and African) and *Bituminaria* (Mediterranean, European and African); they are all distant within the group of genera with a flower-pedicel not subtended by a cupulum (Stirton, 1981).

3. Genera that had their center of origin in Mexico; or, more likely, Mexico represents a small part of such a center and today is a refuge of ancient lineages from the Tertiary under the following criteria: (1) genera with rich diversity (more than 75% of their species) in territorial Mexico including groups A and B and group 1 (Tables 17.1, 17.2, 17.11), except for the monotypic *Cymbosema*; and (2) speciose genera with a greater concentration of primitive traits that have secondary disjunctions (Table 17.12, group 2). These disjuncts are primarily found in the Greater Antilles, which may have been part of Mexico at the end of the Cretaceous-Early Tertiary when it, along with the Proto-Caribbean Arc, broke off from Mexico (Oaxaca or Proto-Yucatan)

Table 17.9. Mexican Leguminosae

| Leguminosae | Genera in Mexico | Genera endemic to Mexico | Estimated number | | | |
			% Endemic genera	Species In Mexico	Species endemic in Mexico	Percent of endemic species
Caesalpinioideae	20	2	10	204	83	40
Mimosoideae	26	1	3.8	386	187	48.4
Papilionoideae	89	1	1.1	1,134	626	55.2
Total	**135**	**4**	**3**	**1,724**	**896**	**51.9**

Table 17.10. Comparison of numbers of endemic genera by subfamilies

	No. of Endemic Genera			
	Caesalpinioideae	Mimosoideae	Papilionoideae	Total
Tropical Africa	54	8	34	96
South America	27	11	97	135
Mexico	2	1	1	4

Data from Brenan (1965); Polhill & Raven (1981).

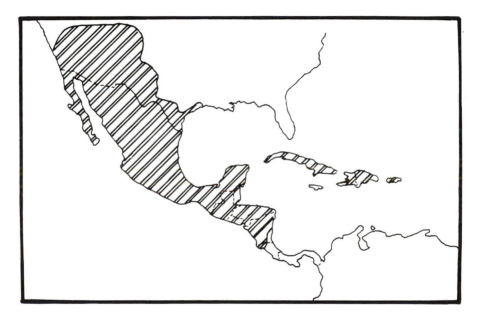

Figure 17.1. Mexican phytogeographical legume area.

to form the present-day Cuba, Hispaniola, and Puerto Rico (Khudoley & Meyerhoff, 1971; Pindell & Dewey, 1982). The legume endemism in these islands corresponds in great part to the same groups as in Mexico (Liogier, 1985; Sauget & Liogier, 1951).

Disjunctions with Asia suggest ancient distribution patterns (Paleogene). The sudden drop in temperature at the end of the Eocene may have broken up fairly continuous past distributions of many of these elements (except *Oxyrhynchus*) resulting in the present-day disjunctions. The relics of the Paleocene-Eocene flora, isolated in various areas today, may have been closer together in the past. Mexico and south-southeastern Asia with Paleocene-Eocene relics

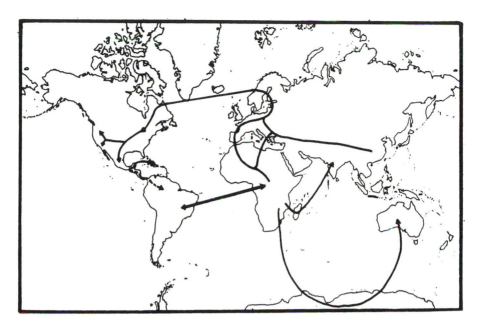

Figures 17.2-4. Suggested past migration routes for floristic elements during various ages. See text for details. **Figure 17.2**. Upper Cretaceous–Paleocene– Eocene. Broken line indicates meager migration; unbroken light line, substantial migration; continuous thick line, substantial and significant migration.

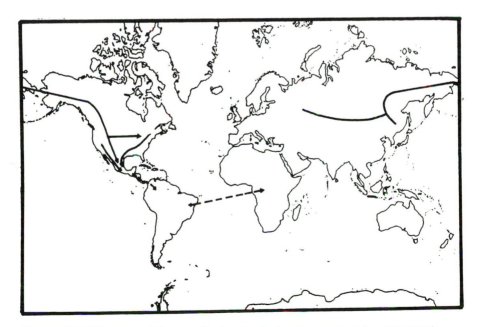

Figure 17.3. Oligocene–Miocene. Broken line indicates meager migration; continuous line, considerable migration.

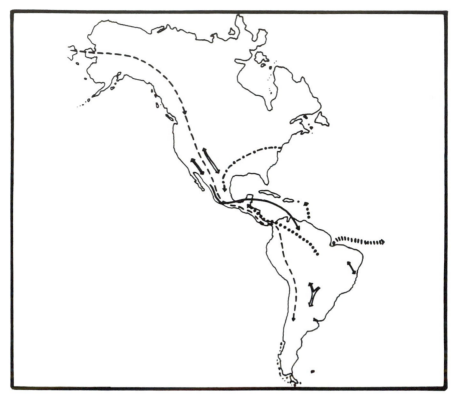

Figure 17.4. Pliocene–Pleistocene–Recent.   Broken line indicates temperate forests; continuous double line, desert; continuous line, tropical seasonal forests; solid line interrupted by closed circles, deciduous forests; closed circles, tropical rain forests; vertical lines, long-distance.

were probably connected by Europe and parts of the United States during the Paleogene. Part of the ancestral elements of the South American legume flora may have arrived there via Mexico after the Central American bridge was formed, although in some cases (*Coursetia, Harpalyce,* and *Zapoteca*) it appears to have occurred when Central America was a group of volcanic islands (stepping stone dispersal) and there had already been enough time for differentiation and speciation of taxa at transspecific levels.

Thus this group is mostly composed of the Mexican generic element, made up of 41 genera (31% of the total), with 508 species (29.6%) including 342 endemics (38.5%).

4. Genera of worldwide distributions but that have formed discrete groups in Mexico (group D, Table 17.4) made up of 19 genera (14.4% of the total), containing 788 species (45.7% of the total) and 430 endemics (48% of the total). The genera in this group are not numerically impressive, but they have speciated prolifically in Mexico. Included here are lineages from the Paleogene and more recent ones;

Table 17.11. Group 1: phytogeographic Mexican tribes and genera without disjunctions; related groups, and their distribution

---

*Amorpheae (North America): *Apoplanesia, Eysenhardtia (Amorpha* from North America), *Psorothamnus, Dalea, Marina.*

*Robinieae (North America and Greater Antilles): *Sphinctospermum, Olneya, Peteria, Genistidium, Hybosema* (Corynellanae from the Greater Antilles[a]), *Lennea, Gliricidia.*

*Millettieae: *Hesperothamnus (Behaimia* from the Greater Antilles and *Piscidia), Piscidia (Hesperothamnus* from Mexico), *Lonchocarpus (Derris* from Mexico and *Derris s. s.* and *Paraderris* from Asia), *Willardia (Lonchocarpus* from Mesoamerica).

*Phaseoleae: *Pachyrhizus* (Diocleinae from South America), *Cologania (Herpyza* from the Greater Antilles, in lesser degree *Amphicarpaea* of North America and *Shuteria* from Asia), *Ramirezella (Vigna* from South America), *Phaseolus* (Africa).

*Psoraleeae: *Psoralidium, Hoita* (North America).

*Aeschynomeneae: *Diphysa (Ormocarpum* from Asia), *Nissolia (Chaetocalyx* from South America), *Pachecoa (Arthrocarpum* from East Africa and Indian Ocean *Chapmania* from North America).

*Brongniartieae: *Brongniartia (Templetonia* from Australia).

*Caesalpinieae: *Conzattia (Peltophorum,* of wide distribution, but possibly centered in Asia), *Heteroflorum* Sousa ined. *(Conzattia* from Mexico), *Poeppigia* (without known relatives).

*Ingeae: *Lysiloma (Albizia* from Asia).

*Mimoseae: *Schrankia (Mimosa diplotricha* from tropical America), *Leucaena (Schleinitzia* from the Pacific), *Calliandropsis (Desmanthus* centered in Mexico), *Neptunia* in Tropical America extending through Africa, Asia and Australia).

---

*Tribe.
[a]M. Lavin, pers. comm.

sometimes both are present in the same genus (e.g., *Cassia, Caesalpinia, Mimosa, Ormosia,* and *Erythrina*).

Together, the Mexican and "Mexicanized" groups total 61 genera (45.2% of the total) accounting for 1,300 species (75.4% of the total) and 775 endemics (86.5% of the total).

5. The group of genera with marginal presence, most of which invaded Mexico from neotropical regions after the union of the Americas, colonizing fundamentally hot-humid to hot-dry areas. A somewhat

Table 17.12. Group 2: Mexican genera with secondary disjunctions and the general areas where they occur

Genera	Location
Caesalpinioideae	
*Caesalpinieae	
Cercidium	South America
Mimosoideae	
*Ingeae	
Havardia	Asia
Zapoteca	Greater Antilles, South America
*Mimoseae	
Desmanthus	South America
Papilionoideae	
*Amorpheae	
Errazurizia	South America
*Brongniartieae	
Harpalyce	Greater Antilles, South America
*Phaseoleae	
Oxyrhynchus	New Guinea
*Robinieae	
Coursetia	South America
*Sophoreae	
Myrospermum	South America
Ateleia	Greater Antilles, South America
Styphnolobium	Southeast Asia

*Tribe.

smaller group may have arrived here earlier (see Migration Patterns, below). This group is formed in part by elements of groups E (Table 17.6) and F (Table 17.7) and *Macroptilium*.

MIGRATION PATTERNS

The Leguminosae are an ancient group in North America dating back at least to the Paleocene, i.e., 60 Ma, even though fossils to suggest so are scarce (Table 17.13) and, when available, are difficult to interpret. The group was well diversified by the Eocene (Berry, 1930; Crepet & Dilcher, 1977; Crepet & Taylor, 1985, 1986; Daghlian et al., 1980; Herendeen & Dilcher, 1986; Sanborn, 1935; Sanborn et al., 1937). Most of them are elements that are found today in hot dry seasonal climates, forming part of the lower to medium-sized tropical deciduous forests, as well as in more humid and clearly warm to cold habitats. This Paleo-Eocenic flora is made up of paleoendemic elements (see groups 1 and 2, Tables 17.11 and 17.12) as well as pantropical and northern temperate genera that have developed evolutionary lines in North America (group D, Table 17.4). Consequently, the flora of the Mexican legumes begins with these paleoendemics and has been enriched by colonizations by groups of different origins at varying intervals (Figs. 17.2–17.4) from Europe (Paleo-Eocene), Asia (Paleocene-Miocene), and South America (Late Cretaceous, Miocene, but mostly Pliocene to present). The routes of entry from the north were open not only for groups of Mediterranean and Asiatic origin but also for lines that radiated from South America (e.g., *Peltogyne, Cleobulia*), Africa (e.g., *Parkinsonia s.l., Adenopodia, Calliandra, Indigofera*), and Australia (*Acacia, Brongniartia*). The existence of noticeable disjunctions at the tribal, generic, and infrageneric levels, among the different zones and Mexico, support the connections to the Paleogene. At the same time, groups and species were dispersed from different Mexican habitats, enriching the flora of other regions of the American continent (North America, Greater Antilles, and South America). Temperate- and cold-adapted groups of Euroasiatic origin (e.g., *Trifolium, Astragalus, Vicia, Lupinus*) arrived in South America through the mountain chains after the hemispheres were joined. Using the same route, although at a quantitatively lesser level, elements adapted to mild conditions originating in South America reached Mexico (e.g., *Amicia*). The groups with distributions in tropical regions in South America were able to reach hot-humid regions of Mexico and Central American countries after the Central American bridge was formed. This group comprises in Mexico a good part of the legumes' marginal generic element, without notable disjunctions in its current distribution. The groups that have clearly adapted to dry-hot tropical climates represent a large number of disjunctions that in part include the western slopes of Mexico to Bahia (Brazil) or in other cases "Los Llanos" of Venezuela and Colombia. With regard to the elements of desert climates, the disjunctions are even wider, from the Chihuahuan and Sonoran Deserts to northern Argentina. It is interesting to point out these disjunctions at the level of one species or species pairs (vicariants?), mostly doubtfully different, that have established themselves up to recently, because of the increase in humidity during the postglacial periods. Here the course of migration has been in both directions, although there is a dominance from north to south; it is noteworthy that large parts of the flora, particularly the arboreal, of "Los Llanos" from northern South America are formed by elements of the "selvas caducifolias" of western Mexico.

The rich legume flora of Mexico is the result of explosive speciation in a few genera, principally at infrageneric levels, and it represents a major secondary center of diversification of the Leguminosae in the world.

ACKNOWLEDGMENTS

We are pleased to dedicate this chapter to Velva E. Rudd in recognition of her valuable contribution to the understanding of Mexican Leguminosae.

Table 17.13. Paleocene-Miocene macrofossil and microfossil records (in North America* and Europe**) of Legumes related to present-day Mexican genera

Age	Fossil records
Miocene	*Desmanthus** (Graham, 1976), *Albizia**
Oligocene	*Conzattia**, *Prosopis**, *Acacia** [*Acaciella*], *Vicia**
Eocene-Paleocene	*Swartzia***, *Styphnolobium**, *Dalbergia***, *Piscidia***, *Lonchocarpus**, *Robinia**, *Erythrina***, *Dioclea** [?], *Canavalia**[?] Berry, 1930), *Vigna** [probably proto-*Ramirezella*], *Phaseolus***, *Amorpha***, *Glycyrrhiza***, *Trifolium***
	*Entada***, *Mimosa**, *Acacia***, *Inga**, *Pithecellobium s.l.**
	*Gleditsia**, *Caesalpinia**, *Parkinsonia** (Berry, 1930), *Cassia** [*Senna*], *Cercis**, *Bauhinia***, *Hymenaea**

Note: For Europe only the Paleocene and Eocene are considered. Only the oldest record is given for each genus.

We are indebted to colleagues and students who shared valuable information, both published and unpublished, with us. There are numerous unpublished dissertations covering several genera in Mexican Leguminosae that their authors generously permitted us to consult. Our valuable resource, in addition to these works, has been the excellent holding of Mexican legumes in the National Herbarium of Mexico (MEXU). We are grateful to the following friends and collaborators who helped in various ways: Gloria Andrade (*Lysiloma*), Rupert Barneby (*Abarema, Albizia, Chloroleucon*), Teresita Chehaibar (*Schrankia, Mimosa*), José L. Contreras (*Caesalpinia, Peltogyne*), Oscar Dorado (*Brongniartia*), Teresa Germán (*Galactia*), Rosaura Grether (*Mimosa*), James Grimes (Psoraleeae), Hector Hernández (*Calliandra, Zapoteca*), Matt Lavin (*Coursetia*), Aaron Liston (*Astragalus*), Melissa Luckow (*Desmanthus*), René Moreno (*Crotalaria*), Lourdes Rico (*Acacia, Inga, Cojoba, Pithecellobium, Zygia, Havardia*), Velva Rudd (*Dalbergia, Myrospermum*), Carmen Soto (*Crotalaria*), Oswaldo Téllez (*Tephrosia*), Rafael Torres (*Bauhinia*), and Sergio Zárate (*Leucaena*). Comments by Colin Hughes, Gwilym Lewis, and Lourdes Rico resulted in changes and corrections. We are equally grateful to M. Sousa P. for his skillful typing of the Spanish text. Marguerite Elliott translated the original text into English and typed it. The authors express their appreciation and thanks to Robert Bye, Marguerite Elliott, and T.P. Ramamoorthy for their help in the preparation of this chapter.

REFERENCES

Arroyo, M.T.K. 1976. The systematics of the legume genus *Harpalyce* (Leguminosae: Lotoideae). Mem. New York Bot. Gard 26(4):1–80.

Azevedo-Tozzi, A.M.G. de. 1989. Estudos taxonômicos dos gêneros *Lonchocarpus* Kunth e *Deguelia* Aubl. no Brasil. Ph.D. dissertation, Universidade Estadual de Campinas, Brasil.

Barneby, R.C. 1964. Atlas of North American *Astragalus*. Mem. New York Bot. Gard. 13(1–2):1–1188.

——— 1977. Daleae imagines. Mem. New York Bot. Gard. 27:1–891.

——— 1986. A contribution to the taxonomy of *Piptadenia* (Mimosaceae) in South America. Brittonia 38:222–229.

Bentham, G. 1860. Synopsis of Dalbergieae, a tribe of Leguminosae. J. Linn. Soc., Bot. 4(Suppl.):1–128.

——— 1875. *Mimosa. In* Revision of the Suborder Mimoseae. Trans. Linn. Soc. London. 30:335–664.

Bernal G., M.C. & M.P. Martínez A. 1989. Determinación del cariotipo de algunas especies del género *Sophora* L. *s. l.* (Familia: Leguminosae). Thesis, UNAM (Zaragoza).

Berry, E.W. 1917. Fossil plants from Bolivia and their bearing upon the age of uplift of the Eastern Andes. Proc. U.S. Natl. Mus. 54:151.

———. 1923. Miocene Plants from Southern Mexico. Proc. U.S. Natl. Mus. 62:1–27.

———. 1924. Middle and Upper Eocene Floras of Southeastern North America. U.S. Geol. Surv. Prof. Paper No. 92. pp. 171–172.

———. 1930. Revision of the Lower Eocene Wilcox flora of the southeastern states. U.S. Geol. Surv. Prof. Paper No. 156.

Bisby, F.A. & R.M. Polhill. 1973. The role of taximetrics in Angiosperm taxonomy, II. Parallel taximetric and orthodox studies in *Crotalaria* L. New Phytol. 72:727–742.

Borhidi, A. & O. Muñiz. 1977. Adiciones al conocimiento de la flora cubana. 1. El género *Harpalyce* DC. en Cuba. Cien. Biol. 1:131–139.

Brenan, J.P.M. 1955. Notes on Mimosoideae: I. Kew Bull. 10:161–192.

———. 1963. Notes on Mimosoideae: VIII. A further note on *Piptadenia* Benth. Kew Bull. 17:227–228.

———. 1965. The geographical relationships of the genera of Leguminosae in Tropical Africa. Webbia 19(2):545–578.

———. 1986. The Genus *Adenopodia* (Leguminosae). Kew Bull. 41:73–90.

Briones V., O.L. 1988 [1989]. Nuevo registro para México de *Gleditsia* (Leguminosae). Bol. Soc. Bot. México 48:143–144.

Britton, N.L. & J.N. Rose. 1928. *Myrmecodendron, Acaciella, Senegalia, Mimosa, Neomimosa, Mimosopsis, Acanthopteron, Haitimimosa, Pteromimosa, Lomoplis*. North Amer. Flora 23:1–124.

——— & ———. 1930. *Libidibia, Nicarago, Russellodendron, Brasilettia, Poinciana, Caesalpinia, Erythrostemon, Poincianella, Guilandina*. North Amer. Flora 23:318–341.

Burkart, A. 1939. Estudios sistemáticos sobre las Leguminosas Hedisareas de la República Argentina y regiones adyacentes. Darwiniana 3:170–175.

———. 1940. Materiales para una monografía del género *Prosopis* (Leguminosae). Darwiniana 4:57–128.

———. 1952. Las Leguminosas Argentinas Silvestres y Cultivadas, 2nd ed. Buenos Aires: Acme Agency.

————. 1969a. 4. El género americano bicéntrico "*Goldmania*" Rose ex Micheli. *In* Leguminosas nuevas o críticas, VII. Darwiniana 15:506–513.

————. 1969b. 9. "*Muellera fluvialis*" (Lindm.) Burk., nov. comb.; del Chaco y notas sobre "*Muellera*." *In* Leguminosas Nuevas o Críticas. VII. Darwiniana 15:535–542.

————. 1976. A monograph of the genus *Prosopis* (Leguminosae subfam. Mimosoideae). J. Arnold Arb. 57:219–249, 450–525.

Carter, A.M. 1974a. The genus *Cercidium* (Leguminosae: Caesalpinioideae) in the Sonoran Desert of Mexico and the United States. Proc. Calif. Acad. Sci. 60(2):17–57.

————. 1974b. Evidence for the Hybrid origin of *Cercidium sonorae* (Leguminosae: Caesalpinioideae) of Northwestern Mexico. Madroño 22:266–272.

Chehaibar N., M.T. 1988. Estudio taxonómico de la serie *Xantiae* y especies afines del género *Mimosa* (Leguminosae). M.S. Thesis, UNAM.

Crepet, W.L. & D.L. Dilcher. 1977. Investigations of Angiosperms from the Eocene of North America: a mimosoid inflorescence. Amer. J. Bot. 64:714–725.

———— & D.W. Taylor. 1985. The diversification of the Leguminosae: first fossil evidence of the Mimosoideae and Papilionoideae. Science 228:1087–1089.

———— & ————. 1986. Primitive Mimosoid flowers from the Paleocene-Eocene and their systematic and evolutionary implications. Amer. J. Bot. 73:548–563.

Crisp, M.D. & P.H. Weston. 1987. Cladistics and legume systematics with an analysis of the Bossiaeeae, Brongniartieae and Mirbelieae. *In* C.H. Stirton (ed.), Advances in Legume Systematics. Part 3. Kew: Royal Botanic Gardens. pp. 65–130.

Daghlian, C.P., W.L. Crepet & T. Delevoryas. 1980. Investigations of Tertiary angiosperms: a new flora including *Eomimosoidea plumosa* from the Oligocene of Eastern Texas. Amer. J. Bot. 67:309–320.

Delgado S., A. 1985. Systematics of the genus *Phaseolus* in Mexico and Central America. Ph.D. dissertation. Univ. of Texas, Austin.

———— & M.C. Johnston. 1984. A new species of *Myrospermum* (Leguminosae: Papilionoideae) from Northeastern Mexico. Syst. Bot. 9:356–358.

Ducke, A. 1949. As leguminosas da Amazônia Brasileira. *In* Notas sôbre a Flora Neotrópica II. Bol. Téc. Inst. Agro. Norte 18:1–248.

Ebinger, J.E. & D.S. Seigler. 1987. A new species of Ant-Acacia (Fabaceae) from Mexico. Southw. Naturalist 32:245–249.

Eifert, I.J. 1970. *Caesalpinia, Hoffmannseggia*. *In* D.S. Correll & M.C. Johnston (eds.), Manual of the Vascular Plants of Texas. Renner: Texas Research Foundation.

Elias, T.S. 1974. The genera of Mimosoideae (Leguminosae) in the Southeastern States. J. Arnold Arb. 55:67–118.

Elsik, W.C. 1968. Palynology of a Paleocene Rockdale Lignite, Milam County, Texas. II. Morphology and taxonomy. Pollen & Spores 10:599–664.

Engelhardt, H. 1894. I. Ueber neuer fossile Pflantzenreste vom Cerro de Potosí. Sitz. Naturev. Gesell. Isis in Dresden, 1:7.

Evans, S.V., L.E. Fellows & E.A. Bell. 1985. Distribution and systematic significance of basic non-Protein Amino acids and Amines in the Tephrosieae. Biochemical Systematics & Ecology 13:271–302.

Fantz, P. 1977. A monograph of the genus *Clitoria* (Leguminosae: Glycineae). Ph.D. dissertation. Univ. of Florida, Gainesville.

Forero, E. 1984. Revision of *Calliandra*: a multidisciplinary approach. Bull. IGSM 12:14–15.

Freitas da Silva, M. 1976. Revisao taxonômica do gênero *Peltogyne* Vog. (Leguminosae-Caesalpinioideae). Acta Amazônica [Supl]. 6(1):5–61.

Gambill, W.G. 1953. The Leguminosae of Illinois. Ill. Biol. Monographs 22(4):1–117.

Geesink, R. 1981. Tribe 6. Tephrosieae (Benth.) Hutch.(1964). *In* R.M. Polhill & P.H. Raven (eds.), Advances in Legume Systematics. Part 1. Kew: Royal Botanic Gardens. pp. 245–260.

———. 1984a. Three proposals for conservation of generic names in the tribe Millettieae. Taxon 33:742–744.

———. 1984b. Scala Millettiearum. Leiden Bot. Ser. 8:1–131.

Goldblatt, P. 1981. Cytology and the phylogeny of the Leguminosae. *In* R.M. Polhill & P.H. Raven (eds.), Advances in Legume Systematics. Part 2. Kew: Royal Botanic Gardens. pp. 427–463.

Graham, A. 1976. Late Cenozoic evolution of tropical lowland vegetation in Veracruz, Mexico. Evolution 29:723–735.

Grether, R. 1978. A general review of the genus *Mimosa* L. (Leguminosae) in Mexico. Bull. IGSM 6:45–50.

———. 1984. Notes on the genus *Mimosa* in Mesoamerica. Bull. IGSM 12:43–48.

Grimes, J.W. 1990. A revision of the New World species of Psoraleeae (Leguminosae; Papilionoideae). Mem. New York Bot. Gard. 61:1–114.

Guinet, Ph. & J. Vassal. 1978. Hypothesis on the differentiation of the major groups in the genus *Acacia* (Leguminosae). Kew Bull. 32:509–527.

Gunn, C.R. 1984. Fruits and seeds of genera in the subfamily Mimosoideae (Fabaceae). U.S. Dept. Agr. Agr. Ser. Tech. Bull. (1681):1–194.

———, J.M. Andrews & P.J. Paradine. 1982 [1984]. Stranded seeds and fruits from the Yucatan Peninsula. An. Inst. Biol. Univ. Nal. Autón. México Ser. Botánica 47–53:21–60.

——— & J.V. Dennis. 1976. World Guide to Tropical Drift Seeds and Fruits. Quadrangle/New York Times Book, New York. 240 pp.

Herendeen, P.S. & D.L. Dilcher. 1986. Fossil Leguminosae from W. Kentucky and Tennessee. Second Int. Legume Conference, "Biology of the Leguminosae." Missouri Bot. Gard. Poster Session.

Hernández M., H.M. 1984. Contribution to the systematics of *Calliandra*, with Particular Reference to its Infrageneric Relationships. Bull. IGSM 12:16–18.

———. 1986 [1987]. *Zapoteca*: a new genus of Neotropical Mimosoideae. Ann. Missouri Bot. Gard. 73:755–763.

———. 1987. Systematics of the genus *Zapoteca* (Leguminosae: Mimosoideae). Ph.D. dissertation, St. Louis Univ., St. Louis.

———. 1989. Systematics of *Zapoteca* (Leguminosae). Ann. Missouri Bot. Gard. 76(3):781–862.

——— & Ph. Guinet. 1990. Calliandropsis: a new genus of Leguminosae: Mimosoideae from Mexico. Kew Bull. 45(4):609–620.

Hollick, A. 1896. New species of leguminous pods from the Yellow Gravel at Bridgeton, N.J. Bull. Torrey Bot. Club 23:46–66.

———. 1928. Paleobotany of Puerto Rico. New York Acad. Sci. Scient. Surv. Puerto Rico 7:207.

Hopkins, M. 1942. *Cercis* in North America. Rhodora 44:193–211.

Irwin, H.S. & R.C. Barneby, 1981. Tribe 2. Cassieae Brown (1822). *In* R.M. Polhill & P.H. Raven (eds.), Advances in Legume Systematics. Part 1. Kew: Royal Botanic Gardens. pp. 97–106.

——— & ———. 1982. The American Cassiinae, a synoptical revision of Leguminosae tribe Cassieae, subtribe Cassiinae in the New World. Mem. New York Bot. Gard. 35(1–2):1–918.

Isely, D. 1981. Leguminosae of the United States. III. Subfamily Papilionoid-eae: tribes Sophoreae, Podalyrieae, Loteae. Mem. New York Bot. Gard. 25(3):1–264.

Janzen, D.H. 1966. Coevolution of mutualism between ants and acacias in Central America. Evolution 20:249–275.

———. 1974. Swollen-thorn Acacias of Central America. Smithsonian Contr. Bot. 13:1–131.

———. 1983. *Hymenaea courbaril* (Guapinol, Stinking Toe). *In* D.H. Janzen (ed.), Costa Rican Natural History. Species Accounts. Chicago, IL: Univ. of Chicago Press. pp. 253–256.

——— & P.S. Martin. 1982. Neotropical anachronisms: the fruits the Gomphotheres ate. Science 215:19–27.

Johnston, I.M. 1924. *Parkinsonia* and *Cercidium*. Contr. Gray Herb. 70:61–68.

Kearney, T.H. & R.H. Peebles. 1960. Arizona Flora. Berkeley, CA: Univ. of Calif. Press.

Khudoley, K.M. & A.A. Meyerhoff. 1971. Paleogeography and geologic history of the Greater Antilles. Mem. Geol. Soc. Amer. 129:1–199.

Krukoff, B.A. 1982. Notes on the species of *Erythrina*. XVIII. *In Erythrina* Symposium IV. Allertonia 3(1):121–138.

Lackey, J.A. 1981. Tribe 10. Phaseoleae DC. (1825). *In* R. M. Polhill & P.H. Raven (eds.), Advances in Legume Systematics. Part 1. Kew: Royal Botanic Gardens. pp. 301–327.

———. 1983. A review of generic concepts in American Phaseolinae (Fabaceae: Faboideae). Iselya 2(2):21–64.

Langenheim, J.H. 1966. Botanical source of amber from Chiapas, Mexico. Ciencia (México) 24:201–211.

———, B.L. Hackner & A. Bartlett. 1967. Mangrove pollen at the depositional site of Oligo-Miocene Amber from Chiapas, Mexico. Bot. Mus. Leafl. 21:289–324.

——— & Y.T. Lee. 1974. Reinstatement of the genus *Hymenaea* (Leguminosae: Caesalpinioideae) in Africa. Brittonia 26:3–20.

Lavin, M. 1987. A cladistic analysis of the tribe Robinieae (Papilionoideae, Leguminosae). *In* C.H. Stirton (ed.), Advances in Legume Systematics. Part 3. Kew: Royal Botanic Gardens. pp. 31–64.

———. 1988. Systematics of *Coursetia* (Leguminosae–Papilionoideae). Syst. Bot. Monog. 21:1–167.

——— & M. Sousa, S. 1987. The madrensis group of *Coursetia* (Leguminosae: Robinieae). Syst. Bot. 12:106–115.

Lee, Y.T. & J.H. Langenheim. 1975. A systematic revision of the genus *Hymenaea* (Leguminosae: Caesalpinioideae; Detarieae). Univ. Calif. Publ. Bot. 69:1–109.

León, J. 1966. Central American and West Indian Species of *Inga* (Leguminosae). Ann. Missouri Bot. Gard. 53:265–359.

Lewis, G.P. 1987. Legumes of Bahia. Kew: Royal Botanic Gardens.

Liogier, A.H. 1985. Familia 101: Leguminosae. *In* La Flora de La Española. Univ. Central Este 56 (ser. Cient. 22) 3:11–300.

Lock, J.M. 1988. *Cassia* sens. lat. (Leguminosae–Caesalpinioideae) in Africa. Kew Bull. 43:333–342.

Macbride, J.F. 1943. *Cercidium*. *In* J. Macbride (ed.), Flora of Peru. Field Mus. Nat. Hist., Bot. Ser. 13:187, 188.

MacGinitie, H.D. 1953. Fossil plants of the Florissant Beds, Colorado. Publ. Carnegie Inst. Wash. 599:198.

Maréchal, R., J.M. Mascherpa & F. Stainier. 1978. Etude taxonomique d'un groupe complexe d'espèces des genres *Phaseolus* et *Vigna* (Papilionaceae) sur la base de données morphologiques et polliniques, traitées par l'analyse informatique. Boissiera 28:1–273.

Maxwell, R.H. 1977. A résumé of the genus *Cleobulia* (Leguminosae) and its relations to the genus *Dioclea*. Phytologia 38:51–65.

———. 1982. A disjunct new species of *Cleobulia* (Leguminosae) from Mexico. Phytologia 51:361–368.

Miranda, F. 1955. Ensayo de evaluación de las relaciones entre los géneros *Conzattia*, *Peltophorum* y *Cercidium*. Bol. Soc. Bot. México 18:7–10.

———. 1963. Two plants from the Amber of the Simojovel, Chiapas, Mexico area. J. Paleontol. 37:611–614.

Mueller, J. 1981. Fossil pollen records of extant Angiosperms. Bot. Rev. 47:1–142.

Nielsen, I. 1981. Tribe 5. Ingeae Benth. (l865). *In* R. M. Polhill & P.H. Raven (eds.), Advances in Legume Systematics. Part 1. Kew: Royal Botanic Gardens. pp. 173–190.

Norman, E.M. & C.R. Gunn. 1982. *Pachecoa prismatica* (Fabaceae): taxonomy and phylogeny. Brittonia 37:78–84.

Ochoterena-Booth, H. 1991. Revisión Taxonómica de Génera *Ramirezella* Rose (Fabaceae, Papilionoideae). Thesis, UNAM.

Ohashi, H., R.M. Polhill & B.G. Schubert. 1981. Tribe 9. Desmodieae (Benth.) Hutch. (l964). *In* R.M. Polhill & P.H. Raven (eds.), Advances in Legume Systematics. Part 1. Kew: Royal Botanic Gardens. pp. 292–300.

Pindell, J. & J.F. Dewey. 1982. Permo-Triassic reconstruction of western Pangea and the evolution of the Gulf of Mexico/Caribbean region. Tectonics 1:179–211.

Pittier, H. 1917. The Middle American species of *Lonchocarpus*. Contr. U.S. Natl. Herb. 20:37–93.

———. 1928. Contribuciones a la dendrología de Venezuela. Arboles y arbustos del orden de las Leguminosas. III. Papilionaceae. Trab. Mus. Com. Venezuela (Bol. Minist. R.R.E.E. no. 4–7) 4:179–259.

Polhill, R.M. 1981a. Tribe 2. Sophoreae Sprengel (1818). *In* R.M. Polhill & P.H. Raven (eds.), Advances in Legume Systematics. Part 1. Kew: Royal Botanic Gardens. pp. 213–230.

———. 1981b. Tribe 19 Loteae DC. (1825). *In* R.M. Polhill & P.H. Raven (eds.), Advances in Legume Systematics. Part 1. Kew. Royal Botanic Gardens. pp. 371–375.

———. 1982. *Crotalaria* in Africa and Madagascar. Rotterdam: A.A. Balkema.

——— & P.H. Raven. 1981. Advances in Legume Systematics. Part 1. Kew: Royal Botanic Gardens.

———, P.H. Raven & C.H. Stirton. 1981. Evolution and systematics of the Leguminosae. *In* R.M. Polhill & P.H. Raven (eds.), Advances in Legume Systematics. Part 1. Kew: Royal Botanic Gardens. pp. 1–26.

——— & M. Sousa, S. 1981. Tribe 7. Robinieae (Benth.) Hutch. (l964). *In* R.M. Polhill & P.H. Raven (eds.), Advances in Legume Systematics. Part 1. Kew: Royal Botanic Gardens. pp. 283–288.

——— & J.E. Vidal. 1981. Tribe 1 Caesalpinieae. *In* R.M. Polhill & P.H. Raven (eds.), Advances in Legume Systematics. Part 1. Kew: Royal Botanic Gardens. pp. 81–95.

Poncy, O. 1985. Le genre *Inga* (Légumineuses, Mimosoideae) en Guyane Française. Mém. Mus. Natl. Hist. Nat., Sér. B, Bot. 31:1–124.

Raven, P.H. & R.M. Polhill, 1981. Biogeography of the Leguminosae. *In* R.M. Polhill & P.H. Raven (eds.), Advances in Legume Systematics. Part 1. Kew: Royal Botanic Gardens. pp. 27–34.

Rico A., L. 1980. El género Acacia (Leguminosae) en Oaxaca. Thesis, UNAM.

————. 1984. The Genus *Acacia* in Mexico. Bull. IGSM. 12:50–59.

Rojo, P.J. 1972. *Pterocarpus*. Phanerogamarum Monographieae 5:1–119.

Rudd, V.E. 1955. The American species of *Aeschynomene*. Contr. U.S. Natl. Herb. 32(1):1–172.

————. 1956. A revision of the genus *Nissolia*. Contr. U.S. Natl. Herb. 32(2):173–205.

————. 1958. A revision of the genus *Chaetocalyx*. Contr. U.S. Natl. Herb. 32(3):207–243.

————. 1965. The American species of *Ormosia* (Leguminosae). Contr. U.S. Natl. Herb. 32(5):279–384.

————. 1968. Leguminosae of Mexico-Faboideae. I. Sophoreae and Podalyrieae. Rhodora 70:492–532.

————. 1969. A synopsis of the genus *Piscidia* (Leguminosae). Phytologia 18:473–499.

————. 1981a. *Ormosia* (Leguminosae) in Mexico, including a new species from Oaxaca. Bol. Soc. Bot. México 41:153–159.

————. 1981b. Tribe 14. Aeschynomeneae (Benth.) Hutch. (1964). *In* R.M. Polhill & P.H. Raven (eds.), Advances in Legume Systematics. Part 1. Kew: Royal Botanic Gardens. pp. 347–354.

———— & T. Wendt. 1988 [1989]. Una adición al género *Ormosia* (Leguminosae) en México: *O. panamensis*. Bol. Soc. Bot. México 48:155–158.

Rydberg, A. 1923. Tribe 7. Indigofereae. North Amer. Fl. 24:137–153.

Rzedowski, J. 1988. Análisis de la distribución geográfica del Complejo *Prosopis* (Leguminosae, Mimosoideae) en Norteamérica. Acta Bot. Mex. 3:7–19.

Sanborn, E.I. 1935. The Comstock flora of West Central Oregon. Publ. Carnegie Inst. Wash. (465):21–22.

————, S.S. Potbury & H.D. MacGinitie. 1937. Eocene flora of Western America. Publ. Carnegie Inst. Wash. (465):70.

Saporta, M.G. de. 1889. Flora fossile d'Aix-en-Provence. Ann. Sci. Nat. Ser 7. Bot. 10:1–192.

Sauget, J.S. [Hno. León] & A.H. Liogier [Hno. Alain]. 1951. Leguminosas. *In* Flora de Cuba. Contr. Ocas. Mus. Hist. Nat. Col. La Salle No. 10. 2:224–367.

Small, J.K. 1901. The Mimosaceae of the southeastern United States. Bull. New York Bot. Gard. 2:89–101.

Sorensen, M. 1988. A taxonomic revision of the genus *Pachyrhizus* (Fabaceae-Phaseoleae). Nord. J. Bot. 8:167–192.

Sousa S., M. 1986. Fabaceae. *In* D.E. Breedlove, Listados Florísticos de México IV. Flora de Chiapas. Mexico City: UNAM. Inst. Biol. pp. 90–112.

————. 1986 [1987]. Adiciones a las Leguminosas de la Flora de Nicaragua. Ann. Missouri Bot. Gard. 73:722–737.

————. 1990. Adiciones a las Papilionadas de la Flora de Nicaragua y una nueva combinación para Oaxaca, México. Ann. Missouri Bot. Gard. 77:573–577.

————. *Lonchocarpus. In* D.W. Stevens (ed.), Flora de Nicaragua. Missouri Bot. Gard. In press.

————. *Muellera. In* D. W.Stevens (ed), Flora de Nicaragua. Missouri Bot. Gard. In press.

———— & R. Antonio O. *Diphysa. In* D.W. Stevens (ed.), Flora de Nicaragua. Missouri Bot. Gard. In press.

———— & M. Peña de Sousa, 1981. New World Lonchocarpinae. *In* R.M. Polhill & P.H. Raven (eds.), Advances in Legume Systematics. Part 1. Kew: Royal Botanic Gardens. pp. 261–281.

———— & J.C. Soto. (1987) 1989. Nuevos taxa de *Lonchocarpus* (Leguminosae) de las cuencas baja y media del Río Balsas, México. An. Inst. Biol. Univ. Nal. Autón. México, Ser. Bot. 58:69–86.

Spellenberg, R. 1976. Chromosome numbers and their cytotaxonomic significance for North American *Astragalus* (Fabaceae). Taxon 25:463–476.

Stirton, C.H., 1981. Tribe II. Psoraleeae (Benth.) Rydb. 1919. *In* R. M. Polhill & P.H. Raven, Advances in Legume Systematics. Part 1. Kew: Royal Botanic Gardens. pp. 337–343.

Téllez V., O., 1979. *Tephrosia woodii* (Leguminosae) una nueva especie del estado de Oaxaca, México. Bol. Soc. Bot. México 38:77–82.

———. 1985. Two new species and a new combination in *Tephrosia* (Leguminosae). Iselya 2:101–107.

———. 1986. El género *Tephrosia* (Leguminosae) en Oaxaca. Thesis. UNAM.

Turner, B.L. & O.S. Fearing. 1964. A taxonomic study of the genus *Amphicarpaea* (Leguminosae). Southw. Naturalist 9:207–218.

Unger, F. 1864. Sylloge Plantarum Fossilium. Wien: pp. 26–27.

Verdcourt, B. 1978. A new combination in *Oxyrhynchus* (Leguminosae-Phaseoleae). Kew Bull. 32:779–780.

Windler, D.R. 1970. Systematic studies in *Crotalaria sagittalis* L. and related species in North America (Leguminosae). Ph.D. dissertation. Univ. of North Carolina at Chapel Hill.

———. 1971. New North American Unifoliate *Crotalaria* Taxa (Leguminosae). Phytologia 21:257–266.

Wood, C.E. 1949. The American barbistyled species of *Tephrosia* (Leguminosae). Contr. Gray Herb. 170:193–384.

Wunderlin, R.P., K. Larsen & S.S. Larsen. 1981. Tribe 3. *Cercideae* Bronn (1822). *In* R.M. Polhill & P.H. Raven (eds.), Advances in Legume Systematics. Part 1. Kew: Royal Botanic Gardens. pp. 107–116.

Yakovlev, G.P. 1967. [Systematic and geographical studies of the genus *Sophora* L. and related genera.] Trud. Leningr. Khim. Farm. Inst. 21:42–62 [in Russian].

——— 1971. [A contribution to the revision of the genus *Ormosia* Jacks. I. The genera *Ruddia* Yakovl. and *Fedorovia* Yakovl. (Leguminosae).] Bot. Zhurn. USSR. 56:652–658 [in Russian].

——— 1972. [Contributions to the system of the order Fabales.] Bot. Zhurn. USSR. 57:585–595. [in Russian].

Zohary, M. & D. Heller, 1984. The genus *Trifolium*. Tel Aviv: Israel Academy of Sciences and Humanities.

18

Mexican Lamiaceae: Diversity, Distribution, Endemism, and Evolution

T.P. RAMAMOORTHY AND MARGUERITE ELLIOTT

Mexican Lamiaceae are comprised of 27 genera belonging to six tribes, and about 512 species. Of the tribes, the most diverse is Mentheae with 11 genera; Salvieae with two genera is the most speciose. Affinities of Mexican Lamiaceae are to elements of Laurasian (including North American) and Gondwanan (including South American) origins, although phytogeographically the Mexican genera may be organized into groups with links to Californian, Laurasian-Mediterranean, North American, and Gondwanan-South American ones. It is significant that the Laurasian-Mediterranean group is the most speciose in which considerable transspecific evolution has also taken place. Mexican Lamiaceae are predominantly montane; they are richly represented in the desert and arid vegetation types but poorly so in the tropical lowlands of Mexico. Generic endemism is low (0.03%), whereas species endemism is over 77%. The extreme local occurrences of numerous species in several genera suggest rapid, local evolution through fragmentation. The combination of pollination biology, a favorable small population system, a complex geological history, the breakup of an ancient flora, and the evolution of deserts and mountains has led to the extant diversity in the family in Mexico. The role of chromosomal diversity in the evolution of the various groups is significant. In some groups, e.g., *Salvia*, the morphological differentiation parallels chemical differentiation.

Distribution patterns and endemism of the family in Mexico suggest that radiation and multiplication of species have primarily occurred in the North American group in the desert and arid vegetation types (Baja California and the northern States of Chihuahua, Coahuila, Nuevo León, and Tamaulipas), whereas speciation in the Laurasian elements has occurred principally in the mountains (Sierra Madre Occidental, Sierra de Oaxaca and Chiapas and the Transmexican Volcanic Belt in the central states).

The distribution of the family indicates that several groups have evolved independently in various parts of the world, suggesting great antiquity for the family: it may date back to the Cretaceous. The evolution and differentiation in Mexican Lamiaceae suggest that at least some of the elements in Mexican Lamiaceae with links to Laurasia were already present in the country during the Cretaceous, whereas those with South American connections may be younger.

Lamiaceae are composed of approximately 224 genera and 5,600 species (Mabberley, 1987). Its members are primarily herbaceous. Small trees in *Hyptidendron* (Harley, 1988), and large "arboreal" shrubs in *Salvia* (Ramamoorthy, pers. obs.) are rare. In addition to the usually bilabiate calyx and corolla, the intrusive placenta that divides the maturing ovary into four nutlets, and the gynobasic style, the most distinguishing feature of the family is the terpene-containing glands found on the epidermal cells. The oils from these glands lend an aromatic odor to the plant, and as a consequence members of the family in different parts of the world including Mexico are used in folk medicine and traditional foods. The data now available provide an empirical base to their use in traditional medicines (Gonzáles et al., 1989). Outstanding examples of such use in Mexico are various species of *Salvia* (Esquivel et al., 1989) and *Agastache mexicana* subsp. *xolocotziana* (Bye et al., 1987). Members of "*Satureja*" and *Cunila* are among others that are extensively used.

The members of the family are found in the tropics, subtropics, and temperate parts of the world. Diverse taxonomic groups in the family have developed in isolation in various parts, some examples of which are *Eriope* and *Hyptis* in South America (Epling, 1935, 1936a,b, 1937, 1942; Harley, 1976, 1988), *Monarda* (McClintock & Epling, 1942) and *Monardella* (Epling, 1925) in North America, *Sideritis* in southwest Asia (Hedge, 1986), and *Pogostemon* in Indo-Malaysia. Several well-developed genera are also endemic to these areas: *Eriope* and *Hypenia* to South America, *Monarda* and *Monardella* to North America (including northwestern Mexico), and several genera to southwest Asia (Hedge, 1986). Such distribution patterns and endemism in the family suggest that it is of fair antiquity and may date back to the Cretaceous (Hedge, 1986; Harley, 1988). Fossil data to support this conclusion are scarce. The only fossil report pertaining to this family (Emboden, 1965) suggests that the family is less than 11 million years old. Cronquist (1988) indicated that the family may have originated during the Upper Miocene.

The areas of high diversity in Lamiaceae include the region extending from the Mediterranean through Central Asia, the Americas, the Pacific islands, tropical Africa, and China. For diversity in number and kind, the first-mentioned area—which combines the Mediterranean and the Irano-Tauranean floristic provinces of Takhtajan (1969, 1986) as well as southwest Asia whose Lamiaceae was analyzed by Hedge (1986)—is the most noteworthy. Hedge remarked on the diversity of the family in southwest Asia, which harbors 66 genera (about 30% of the world total) and over 1,100 species (about one-fifth of the world total), writing that "the protean range of the global variation" seen here is comparable to those elsewhere. The diversity at both generic (the number of endemic genera may exceed 22 [Takhtajan, 1986]) and species levels in the Mediterranean-Central Asia

Table 18.1. Mexican Lamiaceae: genera and species numbers

Genus	Total No. of species in genus	No. of species in Mexico	No. of endemics in Mexico
Acanthomintha	3	1	0
Agastache	30	14	13
Asterohyptis	3	3	2
Catoferia	4	3	1
Chaunostoma	1	1	0
Cunila	15	8	6
Hedeoma	38	20	17
Hesperozygis	8	3	2
Hyptis	300	32	22
Lepechinia	30	7	3
Marsypianthes	5	1	0
Monarda	12	6	6
Monardella	19	7	1
Neoeplingia	1	1	1
Ocimum	150	2	0
Physostegia	12	1	0
Pogogyne	5	3	1
Poliomintha	4	2	2
Prunella	7	1	0
Salvia	900	312	270
Satureja	150	6	4
Scutellaria	300	37	24
Stachys	300	24	16
Teucrium	100	2	0
Tetraclea	2	1	?
Trichostema	17	8	2
TOTALS		512	393

Numbers for species provided in this and following tables are mostly based on Epling's contributions and unpublished data and are subject to revision.

region may be considerably higher, suggesting the possibility that the progenitors of the present-day Lamiaceae may have originated in these parts during the early to mid-Cretaceous (140–100 Ma) when the land masses of the earth were closer to each other, facilitating movement of taxa to and from North America directly, to South America through Africa, to Asia directly, and to Australia through Asia.

In Mexico the family is richly represented by over 512 species and 27 genera that belong to six tribes (Table 18.1). The generic composition is a mixture of Laurasian, North American, South American, and Gondwanan elements. A high number of species, however, belong to genera of Laurasian origin. Introduced and naturalized genera are excluded from this study. Prominent among them are *Coleus* (now considered to be *Solenostemon*, usually cultivated in many parts of the world for its variegated leaves, *C. blumei* is known in cultivation and is an escape in many parts of Mexico); *Leonotis* (tropical African, *L. nepetifolia*, with eight to ten spine-tipped calyx lobes, is a weed in the country); *Leonurus* (temperate European, *L. sibiricus*, with palmately three-lobed leaves and spine-tipped calyx lobes is weedy in Mexico); *Marrubium* (European-Mediterranean, *M. vulgare*, usually cultivated for medicinal purposes, is an escape and is found in many parts of the country [rugose veined leaves and the ten claw-like calyx teeth distinguish it]); *Mentha* (temperate, Old World, cultivated extensively for flavorings, many hybrids are known); *Origanum* (Eurasian, *O. vulgare*, the common European oregano and *O. majorana* [marjoram] are known in cultivation); and *Rosmarinus* (*R. officinalis*, the common rosemary, shrubby with revolute leaf margins, is extensively cultivated).

GENERAL SURVEY OF MEXICAN LAMIACEAE

Distribution

The family, which shows a predilection for montane areas, is, however, found in almost all vegetation types in Mexico. It occurs from lowland tropical conditions to deserts and alpine areas in the country. In the tropical lowlands, members of Ocimieae such as *Catoferia*, *Hyptis*, *Marsypianthes*, and *Ocimum* are typical. Generally, species of the latter three genera tend to have wider distribution. Several species of these genera, excepting those of *Catoferia*, tend to be weedy. *Catoferia spicata* extends from lowland Veracruz to Central America through Chiapas and occurs locally. In the cloud forests of Mexico several genera are found: *Catoferia* (*C. chiapensis*), *Chaunostoma* (*C. mecistandrum*—endemic to Chiapas and Guatemala and distinguished by its cauliflory and flaring calyx and arching stamens), *Cunila*, *Salvia*, "*Satureja*," *Stachys*, and *Scutellaria*. Numerous species of these genera are restricted in their occurrence. Vegetation types composed of pines, oaks, *Abies*, or conifer elements support a large number of species of the family. *Asterohyptis*, *Hyptis*, *Lepechinia*, *Salvia*, "*Satureja*," *Scutellaria*, and *Trichostema*, among others, are found in these communities. In fact, species of *Salvia* are numerous in these places, and many are highly restricted as well: *S. dichlamys*, for example, in Central Mexico is found mostly in oak forests. Several species of *S.* sect. *Sigmoideae* and *Lavanduloideae* among others are restricted to pine forests. Several genera are represented in the deserts and arid lands of Mexico. Among the more notable are *Hedeoma*, *Hesperozygis*, *Neoeplingia*, *Poliomintha*, *Salvia*, and *Tetraclea*. Monotypic *Neoeplingia* (*N. leucophylloides*) is presently known only from its type locality and is restricted to the Barranca de Tolantongo, an interesting

Table 18.2. Mexican Lamiaceae: tribes and genera

Ajugeae:	*Tetraclea, Teucrium, Trichostema*
Lamieae:	*Physostegia, Prunella, Scutellaria, Stachys*
Mentheae:	*Acanthomintha, Chaunostoma, Cunila, Hedeoma, Hesperozygis, Lepechinia, Monardella, Neoeplingia, Pogogyne, Poliomintha, Satureja*
Nepeteae:	*Agastache*
Ocimieae:	*Asterohyptis, Catoferia, Hyptis, Marsypianthes, Ocimum*
Salvieae:	*Monarda, Salvia*

Nomenclature after Cantino & Sanders (1986).

area in the state of Hidalgo where numerous endemics are found. *Salvia hidalgensis* with yellow flowers is also known only from this area. *Poliomintha incana* may be seen in the sand dunes of Samalayuca in the state of Chihuahua. In the arid lands of the country, numerous species of *Salvia* are encountered. Highly restricted among them are *Salvia anastomosans, S. semiatrata, S. aspera, S. candicans, S. fruticulosa,* and several members of sect. *Scorodonia.* Species of *Hedeoma* and *"Satureja"* show similar restricted distributions. *Trichostema purpusii* with its showy flowers is known only from the Tehuacan Valley. In the deciduous forests of Mexico are found species of *Agastache, Catoferia, Hyptis, Ocimum,* and *Salvia.* Notably, members of the family are not found in aquatic conditions, although *Salvia concolor,* with its hollow internodes, tolerates water well. It is often found along streams in Central Mexico. *Salvia clinopodioides* may actually be paludose. In addition to the pine forests, this species has been found in wet areas near the lake in Valle de Bravo in Central Mexico. Species of *Stachys* are montane and often found in moist or wet areas.

No epiphytes, saprophytes, or parasites are known in the family.

DIVERSITY AND TAXONOMY

Among the more notable students who have contributed to our understanding of Mexico's diverse Lamiaceae are Bentham (1832–36, 1848, 1876), Epling (see references below), and Fernald (1900). The 27 genera in Lamiaceae belong to six tribes (Table 18.2): Ajugeae (3 genera), Lamieae (4), Mentheae (11), Nepeteae (2), Ocimeae (5), and Salvieae (2). In terms of genera, Mentheae is the largest of the tribes and is the most diverse in Mexico. *Lepechinia* of this tribe has its center of diversity in South America whereas *Hedeoma* and *Poliomintha* have radiated extensively in the arid northern Mexico and southern United States and are clearly Mexican-North American, although the former is represented in South America by three species. *Hedeoma*'s amphitropical distribution is by no means unique: Harley (1983) has discussed it for *Hyptis alata,* Sousa and Delgado (this volume) discussed it for Leguminosae. Mexican Ocimeae includes ele-

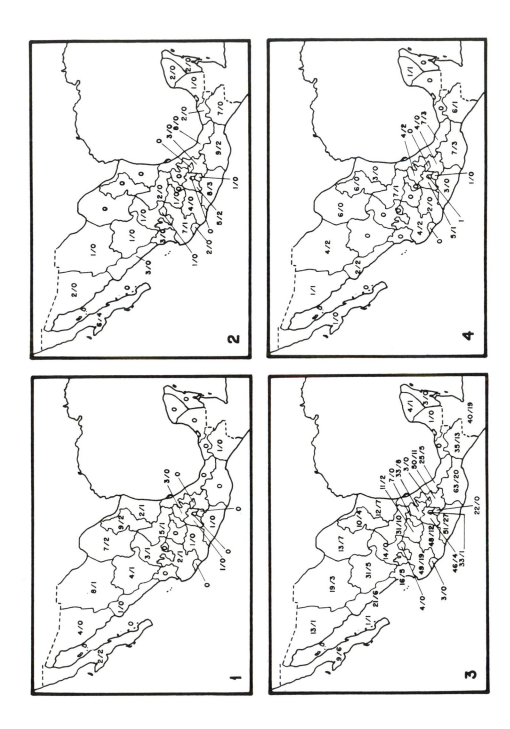

ments (*Ocimum*) that are obviously Gondwanan. *Catoferia*, which is American, is most closely related to *Orthosiphon* of Asia and Africa (Ramamoorthy, 1986). *Hyptis* with over 250 species (Epling, 1942) has its center of diversity in the highlands of Brazil, where it is represented by narrow endemics. The center of diversity of Salvieae seems to be Laurasian, and the tribe is represented in Mexico by *Salvia*. Lamieae is represented in the country by the widespread *Stachys* and the cosmopolitan *Scutellaria*.

The 27 genera may be divided, based on the number of their species in the country, into three groups: large (30–300 species), medium (10–30), and small (1–10). The large genera together constitute about 78% of all the Lamiaceae in Mexico; *Salvia* alone contributes 64% of the total. The medium-sized genera comprise about 13% of the total, and the remaining 9% are represented by 18 genera of the small group. Figure 18.1 provides information regarding species numbers in selected genera in various states of the country.

The genera in the large group are distinctive. *Hyptis* (the lower lip of the corolla is saccate and eventually deflexed, and at first enveloping the stamens) with over 250 species is primarily South American and its center of diversity lies in Brazil. Epling (1938b, 1949) recognized 27 sections in the genus. Studies by Harley of the Royal Botanic Gardens, Kew, suggested a need for a taxonomic reevaluation of the genus (Harley, 1985, 1986), and he has already recognized two segregates: *Hypenia* and *Hyptiodendron* (Harley, 1988). Eight of the 27 sections, recognized by Epling (1949), comprising 32 species, are represented in Mexico (Table 18.3). One section and 22 species are endemic to the country. Betraying its tropical origins, the genus in Mexico is found principally in the tropical parts in the south and along the two coasts. Three of the eight sections in Mexico (*H.* sect. *Rhytidea, Umbellatae,* and *Laniflorae*) are noteworthy. *Rhytidea* is endemic and is found in montane areas in western Mexico: *H. rhytidea* occurs from the western central state of Michoacán to Sinaloa through Nayarit in the west and to Durango through Jalisco; *H. psedolantana* is locally endemic in the state of Guerrero. Of the three species in the *Umbellatae,* two (*H. subtilis* and *H. iodantha*) are found in central Mexico, and the third (*H. tafallae*) is found in Peru and Bolivia. The Mexican species are montane and are found at higher elevations in deciduous forests. *Hyptis* sect. *Laniflorae* is nearly endemic (five of six species) to the country. The species of this section are mostly found along the western coast of Mexico including Baja California. The five nonendemic sections include species endemic to Mexico (Table 18.3) but have a more tropical distribution in southern Mexico. These sections are *Minthidium, Mesophaeria* subsect. *Pectinaria, Polydesmia* subsect. *Vulgaris, Cephalohyptis* (subsections *marrubiastrae* and *Hyptis*), and *Pusillae*.

Figure 18.1. Species numbers/endemics of selected Lamiaceae in the states of Mexico. 1. *Hedeoma*. 2. *Hyptis*. 3. *Salvia*. 4. *Scutellaria*. Data based on Epling (see references) and T.P. Ramamoorthy (unpublished data). These numbers are tentative and subject to change with further studies.

Table 18.3. *Hyptis* in Mexico

Section	Total No. of species in section	No. of species in Mexico	No. of species endemic to Mexico
Rhytidea	2	2	2
Umbellatae	3	2	2
Laniflorae	6	6	5
Minthidium	11	3	2
Mesophaeria			
subsec. *Pectinaria*	14	6	5
Polydesmia			
subsec. *Vulgaris*	11	2	1
Cephalohyptis			
subsec. *Marrubiastrae*	25	4	1
subsec. *genui*	20	6	4
Pusillae	4	1	
TOTAL	96	32	22

Unlike the Laurasian genera *Salvia* and *Scutellaria*, which are mostly montane and nontropical, *Hyptis* is predominantly found in the tropics in southern Mexico, but it includes species that are secondarily adapted to nontropical conditions (*H. rhytidea*).

Of the other large genera in the country, *Salvia* is the most remarkable as it is the most numerous. Its two stamens, one-celled anther cells, and nonfertile connective that is modified into the rudder, help distinguish it. Worldwide, it is represented by over 900 species (Standley & Williams, 1973) organized into four subgenera (Bentham, 1876): *Salvia, Calosphace, Leonia,* and *Sclarea*. The American Salvias excluding the Californian Audibertias, particularly those found in Mexico, and Central and South America, mostly belong to subg. *Calosphace* and are about 500 species strong in 104 sections (Epling, 1939b, 1940, 1941, 1944, 1947, 1951, 1960; Epling & Jativa, 1962, 1963, 1964, 1965; Epling & Mathias, 1957; Ramamoorthy, 1984a,b,c, 1985; Standley & Williams, 1973), making it one of the larger vascular plant genera in the New World. The major centers of diversity of the genus *Salvia* are the Mediterranean (including southern Europe) extending into West Asia, southern Africa and the mountains of Mexico and South America. The diversity of the genus found in the central highlands of Mexico may be one of the highest known for any genus. Ramamoorthy (1984a) estimated that in Mexico there may be as many as 275 species with an astonishing 88% species endemism. Subsequent studies suggested that this number may have to be revised to 300 species, and additional sections may be needed to reflect a more natural taxonomy of Mexican Salvias. In addition to the subgenus *Calosphace, Salvia* (?)subg. *Audibertia* with 12 species in Mexico (northwestern Mexico, Baja Califor-

Table 18.4. *Salvia* Subgenus *Calosphace* in Mexico: nonendemic sections

Section	Total No. of species in section	No. of species in Mexico	No. of species encemic to Mexico
Microphace	6	6	3
Tomentellae	19	10	8
Caducae	2	2	1
Lavanduloideae	12	12	11
Uliginosae	29	22	21
Micranthae	11	3	2
Blakea	5	4	3
Glareosae	4	4	3
Potiles	1	1	0
Mitratae	2	2	1
Subrotundae	3	1	0
Incarnatae	2	2	1
Chariantha	2	1	1
Membranacea	13	13	8
Flocculosae	18	11	9
Scorodonia	12	11	11
Uricae	2	2	0
Rudes	12	1	1
Farinaceae	10	6	5
Dusenostachys	9	7	7
Donnelsmithia	4	3	2
Polystachyae	14	14	12
Carneae	9	5	3
Angulatae	44	24	20
Maxonia	7	3	3
Briquetia	5	3	3
Fulgentes	6	6	5
Nobilis	12	3	3
Erythrostachys	4	3	3
Cardinales	8	8	4
Tubiflorae	12	2	2
Purpureae	8	5	4
Curtiflorae	9	3	3
Secundae	10	1	1

nia) is a major element in the Mexican flora. The genus *Salvia* in Mexico is currently estimated at 312 species.

Some comments concerning its variation and evolution are pertinent here, as *Salvia* may be the most speciose in the country. The genus in Mexico has radiated into all possible habitats, including tropical lowlands of the Yucatan Peninsula where *Salvia fernaldii*, endemic to that part of Mexico, is found. The vegetation types where the genus occurs predominantly in

Table 18.5. *Salvia* Subgenus *Calosphace* in Mexico: endemic sections

Section	No. of species
Axillares	1
Gentryana	1
Hintoniana	2
Lanatae	1
Nelsonia	1
Longipes	1
Pedicellata	1
Purpusiana	1
Cucullatae	1
Fernaldia	1
Sigmoideae	14
Standleyana	1
Bracteata	1
Iodophyllae	1
Pennelia	1
Atratae	1
Silvicolae	1
Conzattiana	1
Sphacelioides	1
Brandegeia	3
Pruinosae	1
Palmerostachys	1
Peninsularis	1
Iodantheae	3
Albolanatae	1

Mexico include *Abies*, pine, pine-oak, oak, cloud and deciduous forests, deserts, and arid areas. Species are also known in alpine conditions (*S. unicostata* in Tamaulipas). About 92% of Mexican Salvias are perennials, and the perennating organs are diverse: they include stolons, thick root stocks, and tubers. Rarely, the stem is seen rooting along nodes (*S. hamulus*); it may be terete to four-multiangled with the internodes sometimes hollow (*S. concolor*). Sometimes the plant is scapose (*S. scaposa*). The leaves are narrow, filiform (*S. filifolia*), to spectacularly large and foliose (*S. gravida*). They are membranous to chartaceous (*S. hintonii*). The flowers are never solitary; they are in pairs or in clusters and arranged in verticils which in turn are usually arranged in paniculate racemes. The bracts are subulate to occasionally foliose and are mostly early deciduous. In several sections (sects. *Membranacea* and *Potiles*, among others) they are persistent. The calyx is fairly uniform with most variation being in its size and vesture. However, the variation in the corolla is impressive and suggests the underlying variation in the staminal length, the length and orientation of the rudder and the connective. The last-mentioned characters play a significant role in the pollination biology and the maintenance of variation in the genus in sympatric species, as is discussed below.

Epling (1939b) used a combination of characters to delimit the sections in the genus; these character sets vary significantly from each other and suggest transspecific evolution of a higher order in the genus in the New World. Of the 104 sections recognized by Epling, 59 occur in Mexico (Table 18.4), and as many as 25 are endemic to the country (Table 18.5). Among the latter, the most significant is the monotypic *Salvia* sect. *Axillares* (*S. axillaris*), which has retained all the anther cells; the connectives separated from each other are all fertile. Epling (1938a) remarked that, on the basis of this character, this species may be related to the Californian Audibertias. Another monotypic section that shares characters with the Audibertias is sect. *Cucullatae* (*S. clinopodioides*). *Salvia* sect. *Conzattiana* is of interest: *S. aspera* (yellow flowers, bee-pollinated) and *S. oaxacana* and *S. pexa* (red flowers, hummingbird-pollinated) are all restricted to the Tehuacan Valley. Sect. *Skeptostachys* (*S. gravida*), with its large calyx and huge corolla, is restricted to the Coalcoman areas in Michoacán. Sect. *Fernaldia* is known only from the lowlands in Quintana Roo. *Salvia axillaris* and several species of sections *Tomentellae*, *Scorodonia*, and *Atratae* are highly local and restricted to the arid parts or the deserts. Several are paleoendemics and may represent some of the oldest surviving elements of the ancestral elements of the present-day Mexican flora. On the other hand, species of the sect. *Lavanduloideae* are restricted to the montane and geologically younger areas of Mexico, suggesting that they may represent some of the younger elements of the Mexican flora.

Many of these sections, which were established by Epling on morphological grounds, seem to be chemically distinct. Studies by Rodriguez-Hahn and collaborators (Esquivel et al., 1985, 1986a,b, 1987; Galicia et al., 1988; Hernández M. et al., 1987; Rodriguez-Hahn et al., 1986) suggested that many sections are characterized by unique diterpenes: sects. *Erythrostachys* and *Tomentellae* are distinct in their abietane-type diterpenes (Ramamoorthy et al., 1988). The diterpenes of *S.* sect. *Fulgentes* and *S.* sect. *Scorodonia* provide further examples. The variations in the clerodane-type diterpenes seen in various populations of *S. melissodora/S. keerlii* highlight the importance of the local populations as evolutionary units in this group (Ramamoorthy et al., in prep.).

Ongoing investigations by Rodriguez-Hahn and collaborators (unpublished data) suggest that novel diterpenes await discovery in many species of Mexican Salvias. The data from phytochemical studies in association with taxonomic understanding of the genus suggest that subgenus *Calosphace* may have differentiated from groups ancestral to Old World Salvias or even prior to the differentiation of groups in Salvias (Ramamoorthy et al., 1988).

Chromosomal diversity is equally noteworthy. The genus is polybasic with different chromosome base numbers in different parts of the world (Palomino et al., 1986). The numbers in species of Mexican *Calosphace* are generally $2n = 22$ but "younger" species such as *Salvia lavanduloides* are

Table 18.6. *Scutellaria* in Mexico

Section	Total No. of species in section	No. of species in Mexico	No. of species endemic to Mexico
Cardinales	11	2	2
Crassipedes	1	1	1
Galericulata	17	2	1
Mixtae	5	1	0
Resinosae	11	9	5
Speciosae	11	4	4
Pallidiflorae	2	2	1
Spinosae	1	1	1
Uliginosae	21	12	9
TOTAL	80	34	27

polyploids ($2n = 66$). Annual species such as *S. hispanica* possess a low number ($2n = 12$). "B" chromosomes and satellites are also known in many species. Chromosomal diversity in number and morphology have played significant roles in the evolution of the genus in Mexico.

This diversity in Mexican Salvias is maintained through an elegant mechanism in areas of sympatry. As many as five to seven species with numerous flowers at anthesis may be seen growing in sympatry (Ramamoorthy, pers. obs.). Usually the pollinating vector does not discriminate against any of these species. Bees generally visit all the blue-flowered species and hummingbirds all the red-flowered ones (Dieringer, 1991). Yet hybrids seem rare. The corolla lengths along with staminal and connective lengths which may be termed "functional stamen length" (Dieringer, pers. comm.) in species in sympatry always differ from one another, which seems to contribute to the maintenance of genetic distinctness of species of the associated species. The pollinating agent visits neighboring flowers before moving to an adjacent plant. The differing "functional stamen lengths" probably ensure deposition of the pollen on different parts of the vector, thus achieving pollination for the plant and population. Such inbreeding reinforces genetic isolation of these species leading to higher homozygosity over time, but rare incidences of hybridization should result in 100% heterozygous F1 populations, which would enhance variation and evolution in the group.

The uniformity of the morphology of the flower structure leads to the suggestion that *Salvia* subg. *Calosphace* is monophyletic (Ramamoorthy & Lorence, 1987). Personal observations (T.P.R.) suggest that hybridization is rare. The multiplication of taxa in the genus in Mexico seems mostly to have been due to fragmentation of the ancestral populations (Epling, 1938a, 1939a; Ramamoorthy & Lorence, 1987) brought about by geological and other factors such as orogeny and evolution of deserts and climatic changes. The isolation and differentiation of these populations are reinforced through self-pollination and evolution of chromosomal diversity in numbers as well as morphology.

Scutellaria (including *Salazaria*), defined by its scutellum of the upper calyx lobe, is a cosmopolitan genus represented by 114 species (Epling, 1942; Hiriart, 1984) in the Americas. Epling (1942) recognized 18 sections in the American species, of which nine sections with about 36 species are found in Mexico (Table 18.6). Most of these sections, however, are well represented outside the country. Two sections and 24 species are endemic to Mexico. The centers of diversity of the genus in Mexico lie mostly in the northern arid and semiarid zones and the pine and pine-oak forests in the north and central zones. In the tropical south they are noticeably fewer. Sections *Spinosae* and *Crassipedes* are monotypic and endemic to Mexico. The ditypic sect. *Pallidiflorae* is Mexican and Central American: *Scutellaria pallidiflora* is endemic to Jalisco, and *S. orichalcea* has a southern distribution ranging from Chiapas to Costa Rica. The sections that seem to have speciated in Mexico are *Resinosae* and *Uliginosae*; the former is found in the arid northern and eastern parts of the country whereas the latter is richly represented in the pine and pine-oak forests of the central and southern zones. Eight of ten species in sect. *Resinosae* are restricted to Mexico. The only "tropical" section in Mexico is *Cardinales*, and it is represented by three species, of which the best known is the beautiful *S. mociniana*.

Among the medium-sized genera are *Stachys* (24 species), *Hedeoma* (20 species), *Agastache* (14 species), and *Cunila* (10 species). Together they account for 68 species in Mexico and belong to three tribes: Lamieae, Mentheae, and Nepeteae.

Stachys, a large genus of 300 species (Mabberley, 1987), principally occurs in temperate parts of the world. Its center of diversity seems to lie from Europe to Turkey. It is distinguished by a turbinate calyx terminating in equal to subequal lobes and a ring of hairs on the inside of the corolla tube. In the Americas the genus is represented by over 81 species (Epling, 1934b; Ramamoorthy, 1985; Rzedowski & Rzedowski, 1988). Mexican species are poorly known taxonomically and are usually found in the montane areas. Epling divided the genus rather artificially into ten groups, seven of which are represented in Mexico (Table 18.7). Of the 24 species in the country, 16 are endemic. These numbers are arbitrary and will change with the improved understanding of the taxonomy of the genus.

Hedeoma, a medium-sized Mexican-north Mexican genus of 38 species (Epling & Stewart, 1939; Irving, 1980), is easily distinguished by its gibbous calyx and two stamens. It is represented in Mexico by over 20 species, of which 17 are endemic (Table 18.8). Irving (1980) divided the genus into four subgenera, three of which (subg. *Ciliatum*, *Poliominthoides* and *Saturejoides*) occur in Mexico. The fourth (subg. *Hedeoma* with three species) has an amphitropical distribution in the Americas. Harley (1983) has discussed a similar distribution in *Hyptis alata*. Subg. *Saturejoides* is the largest with 26 species, of which 15 (12 endemics) occur in Mexico. Subg. *Ciliatum* and *Poliominthoides* are composed of five (two endemic to the country) and four (three endemic to Mexico) species, respectively. The many subspecies and

Table 18.7. *Stachys* in Mexico

Group I (3 species)	
Stachys inclusa	E
S. keerlii	E
S. nepetifolia	E
Group II (2 species)	
S. aristata	E
S. bigelovii	
Group III (4 species)	
S. eriantha	
Group IV (6 species)	
S. agraria	
S. parviflora	
S. grahamii	E
S. radicans	
Group V	
S. globosa	E
S. exilis	E
S. collina	E
Group VI (24 species)	
S. bullata	E
S. pringlei	E
S. flaccida	E
S. tenerrima	E
S. mexicana	E
S. vulcanica	E
S. costaricensis	
Group VII (7 species)	
S. coccinea	
S. lineata	
S. lindenii	
Group VIII (9 species)	
No representation	
Group IX (1 species)	
No representation	
Group X (14 species)	
No representation	

Groups after Epling (1936).
E, endemic to Mexico.

varieties recognized by Irving suggest that the genus is probably in an active state of evolution.

Agastache is a genus of about 30 species and occurs from Central Asia to North America through East Asia. It is comprised of 13 species in Mexico, which form a natural section, *A.* sect. *Brittonastrum* (Lint & Epling, 1945; Sanders, 1987). Its distribution in Mexico is from the Sierra Madre Occidental (northwestern Mexico) through Durango to the Trans-Mexican Volca-

Table 18.8. *Hedeoma* in Mexico

H. subgenus *Ciliatum*
H. ciliolatum (E), *H. pusillum* (E), *H. pilosum, H. todsenii, H. apiculatum.*

H. subgenus *Poliominthoides*
H. palmeri (E), *H. patrinum* (E), *H. montanum* (E), *H. molle*

H. subgenus *Saturejoides*
Sect. *Alpine. H. polygalaefolium, H. bellum* (E), *H. jucundum* (E), *H. piperitum* (E), *H. patens* (E). *H. floribundum* (E); Sect. Saturejoides. *H. pulcherrima, H. drummondii, H. reverchonii, H. multiflorum, H. diffusum, H. nanum, H. microphyllum* (E), *H. medium, H. costatum, H. johnstonii* (E), *H. plicatum, H. tenuiflorum* (E), *H. martirenze* (E), *H. quercetorum* (E), *H. tenuipes* (E), *H. hyssopifolium, H. oblongifolium, H. dentatum* (E), *H. acinoides, H. hispidum*

H. subgenus *Hedeoma*
H. pulegioides, H. crenatum, H. mandonianum

E, endemic to Mexico.

nic Belt in the central parts. Sanders divided the section into five series and recognized many varieties, which indicates that the genus is in an active state of evolution. Geographic isolation, hybridization and pollination biology have been the principal forces for the multiplication of taxa in the genus in Mexico (Bye et al., 1987; Sanders, 1987).

Cunila is an American genus of nearly 20 species. Its distribution is wide-ranging, from the state of New York (United States) to northern Argentina. Mexico with ten species has the largest number of species of any area. It is characterized by a tubular 10- to 13-nerved, five-dentate calyx, a sub-bilabiate corolla, and two stamens; it seems to be related to *Hedeoma* (Irving, 1980), which is easily distinguished from *Cunila* by its gibbous calyx. In Mexico the species of this genus is primarily found in the pine-oak forests and occasionally in the cloud forests. Revisionary studies by Garcia P. (1989, in prep.) suggested that although some species have wider distributions in the country (*C. lythrifolia*) many others occur locally. Flowers of some of the species are fragrant (*C. ramamoorthiana*) and are visited by bees.

Neoeplingia (*N. leucophylloides*), similar to *Hedeoma* and *Poliomintha* but differing in its nongibbous calyx and four stamens, is endemic to the deserts of the Barranca de Tolantongo, Hidalgo (Ramamoorthy et al., 1982). It is related to the *Hedeoma* and *Poliomintha* group of genera and may be a paleoendemic. The peculiar *Chaunostoma* (*C. mecistandrum* with its cauliflory, said to be related to *Lepechinia* (Epling, 1948), is endemic to the cloud forests of Chiapas and Guatemala. *Marsypianthes*, similar to *Hyptis* but distinguished from it by its broadly campanulate calyx, is a tropical American genus with five species; *M. chamaedrys* represents its northern limit in southern Mexico.

Monardella is a Californian (western United States) genus of 25 species (Epling, 1925, 1939c; Hardham, 1966) adapted to arid conditions. The almost regular corolla with subequal lobes, the four stamens, and the dense arrangement of flowers in heads distinguish this genus. Epling divided the genus into two subgenera: *Macrantheae* with one section and *Pycnantheae* with four sections. The eight Mexican species are mostly from Baja California and represent both subgenera. The two endemic species are *M. lagunensis* and *M. thymifolia*, both restricted to Baja California, the latter to the Cedros Islands. *Monardella arizonica* is the only mainland species found in Mexico in the state of Sonora. *Physostegia* (identified by its nonbilabiate, faintly nerved, inflated calyx), composed of 12 species, is North American; at its southern distributional ranges it is represented by *P. correlii* and *P. virginiana* ssp. *proemorsa* in Mexico. *Acanthomintha*, characterized by the broad bracts armed with needle-like spines, is a western United States genus of three species and is represented in Mexico in Baja California by *A. ilicifolia* (Wiggins, 1980). *Prunella* is a north temperate genus of three or four species. The calyx, which is dorsally flattened, readily distinguishes the genus from other Mexican Lamiaceae. *Prunella vulgaris* is a widespread weed in the higher elevations in the country.

Among the remaining 12 genera, two clearly of Gondwanan links are *Catoferia* and *Ocimum*. *Catoferia*, which is distinguished by the exserted stamens, is closely related to the African/Asian *Orthosiphon* (Ramamoorthy, 1986) and is composed of four species. In Mexico it is represented by three species, of which *C. martinezii* is endemic to the state of Guerrero in western Mexico. *Catoferia chiapensis* and *C. spicata* are southern in their distributions. *Ocimum*, on the other hand, is African-Asian with about 150 species: it has no endemic representation in Mexico, and its two species (*O. micranthum* and *O. americanum*) are predominantly found in the tropical lowlands.

Asterohyptis is a Mexican-Central American genus of three species (Epling, 1933). Its affinities are not clear, and it may belong in Ocimeae. Two of its three species are endemic to Mexico: *A. seemanii* is restricted to northwestern Mexico, whereas the other, *A. stellulata*, is fairly widespread in the country. *Asterohyptis mociniana* extends into Guatemala. It is likely that the genus, easily distinguished by its stellately spreading calyx teeth, may have had a tropical origin.

Poliomintha, with a tubular symmetrical calyx whose teeth reflex to close the orifice and the annulate corolla tube, is comprised of five species all of which are represented in northern Mexico. The exception is *Poliomintha conjunctrix*, which is endemic to Baja California (Irving, 1972). *Hesperozygis*, related to *Hedeoma*, is represented by *H. marifolia* in the arid parts of northern and central Mexico (Valle de Mesquital in the state of Hidalgo). The three genera *Hedeoma*, *Hesperozygis*, and *Poliomintha* are closely related and according to Irving could represent one genus. *Teucrium*, recognized by its irregular corolla, reduced upper lip, and elaborately enlarged lower one, is a Mediterranean but widespread genus of over 100 species; it is represented in Mexico by four nonendemic elements (McClintock & Epling, 1946). The four species belong in two sections: *Stachybotris* (*T. vesicarium* and *T. canadense*) and *Teucris* (*T. cubense* subsp. *laevigatum*, subsp. *depressum*, *chamaedrifolium*, and *T. glandulosum*) are tropical and subtropical. The recently described *T. proctorii* (Williams, 1972) from Guatemala may be found in Chiapas.

Lepechinia, Monarda, Pogogyne, "*Satureja*," and *Trichostema* are represented in the country by three or more species. Transspecific evolution is particularly notable in *Trichostema*. All these genera have locally radiated: both specific and transspecific endemisms are known. Three species of the Californian *Pogogyne* (distinguished by its dense heads and round to spatulate leaves) are found in northwestern Mexico; one is endemic to Baja California.

Monarda, with a tubular five-cleft calyx and an elongated strongly bilabiate corolla (lips linear or oblong, somewhat equal), is a Mexican-North American genus of 17 species (McClintock & Epling, 1942; Scora, 1965). McClintock and Epling recognized two subgenera in *Monarda*—*Monarda* and *Cheilyctis*—both of which are represented in Mexico. Subgenus *Monarda* seems to have a northeastern distribution in Mexico: *M. pringlei* occurs in Coahuila and Nuevo León, *M. bartletti* in Tamaulipas, and *M. fistulosa* to Coahuila. In the subgenus *Cheilyctis*, excepting *M. citriodora*, which is found in Coahuila and Tamaulipas, all are western and southern: *M. austromontana* (Chihuahua to Michoacán), *M. mexicana* (Durango), and *M. punctata* subsp. *occidentalis* (Chihuahua).

Lepechinia, as defined by Briquet (1896–97) and Epling (1948), is primarily a South American genus with 31 species but has an interesting distribution. Its distributional range includes Mexico, the western United States, Hawaii, and the Reunion Islands. It is recognized by its nongaleate upper lip of corolla, the accrescent calyx, and the exserted stamens. It is represented in Mexico by six species, of which three are endemic. *Lepechinia nelsonii* and *L. glomerata* are central Mexican, the former endemic to Temascaltepec and the latter to Guerrero. *Lepechinia mexicana* (*Sphacele mexicana*) is endemic to the arid parts of Hidalgo and the Tehuacan Valley in central Mexico. *Lepechinia hastata* of Baja California is found in the Hawaiian islands. *Lepechinia schiedeana* and *L. caulescens* (including *L. spicata*) are Mexican and Central American and have wider distributions in the country. It should be noted that the genus most closely related to *Lepechinia, Chaunostoma*, is endemic to the mesic mountains of Chiapas and Guatemala.

Satureja is a large Mediterranean genus of 150 species that presently includes the American material, which falls into about 64 species (Epling, 1927; Epling & Jativa, 1964, 1966; Henrickson, 1981; McVaugh & Schmidt, 1967). Studies by Doroszenko in Edinburgh (Harley, pers. comm.) suggested that the American material does not belong in *Satureja* and that several new genera may be recognized to accommodate these species. In Mexico *Satureja*, as it is currently interpreted, is represented by over eight species. The endemics of the country are *S. macrostema, S. mexicana, S. oaxacana, S. jaliscana,* and *S. maderensis,* most of which belong to *S.* sect. *Gardoquia. Satureja macrostema,* although widespread, is restricted to Mexico. *Satureja jaliscana* (McVaugh & Schmidt, 1967) is restricted to the state of Jalisco. *Satureja brownei* is a widespread, weedy species in Mexico and Central America.

Trichostema, with its almost equally five-lobed corolla (lobes declined) and exserted stamens, is a Mexican-North American genus of 19 species (Henrickson, 1982; Lewis, 1945; Lewis & Rzedowski, 1978). Lewis divided the genus into five sections, four of which are represented in Mexico. Sections *Chromocephalum* (*T. lanatum* and *T. parishii*) and *Rhodanthum* (*T. purpusii*) are endemic to Mexico, the former to northwestern Mexico (including Baja California) and the latter to the Tehuacan Valley. The two other sections represented in Mexico are *Orthopodium* (two species) and *Paniculatum* (two species).

Tetraclea is a North American genus of three or four species. In Mexico, where it has a northern distribution, it occurs principally in arid areas in the states of Chihuahua, Coahuila, Tamaulipas, Durango, Zacatecas, and San Luis Potosí. Of the two species in the country, *T. coulteri* has a wider distribution; *T. subinclusa* is restricted to Coahuila. Among the Mexican Lamiaceae, it may be distinguished by its nearly actinomorphic corolla and exserted stamens. Gray (1853), who described the genus, included it in Verbenaceae; but Bentham (1876) and Briquet (1896–97), both students of Lamiaceae, included it in Lamiaceae. Opinions differ among authors as to where it belongs: Correl and Johnston (1970) and Henrickson (pers. comm.) follow Gray and include it in Verbenaceae. Among the Lamiaceae it seems to be related to *Trichostema* but shares several pollen characteristics with *Clerodendron* (Kim, pers. comm.) in Verbenaceae. The genus is of systematic interest as it seems to be intermediate between the two families.

POLLINATION BIOLOGY

Information on the pollination biology of many genera in the family in Mexico is scarce. Personal observations in the field, floral morphology and corolla color generally suggest that the flower visitors include bees, butterflies, hummingbirds and moths(?). Although the members of the family attract a diversity of visitors to their flowers, the pollinators are mainly bees and hummingbirds. For example, the primary dichotomy in Mexican Salvias is between bee- and bird-pollinated species. The diversity in the family is in many ways linked to the diversity of pollinators, even though no specificity between pollinators and species in the family has been established. The general nature and consequences of pollination biology in Salvias has been discussed under the section on *Salvia.* Species of *Hyptis,* which is predominantly bee-pollinated, have a specialized anther-exploding mechanism (Burkart, 1937; several reports by Harley, particularly 1976) similar to that found in *Eriope* (Harley, 1971). In *Catoferia* the stamens are generally exserted, and in *C. martinezii* they are extraordinarily so (Ramamoorthy, 1986). It would be interesting to find out what pollinates the members of this genus. Their floral morphology suggests hawkmoths as possible visitors, but it has not been documented. It will also be interesting to see what pollinates the cauliflorous *Chaunostoma mecistendrum*, whose stamens are arching. For most other genera the pollinators are either bees (e.g., *Hedeoma, Lepechinia, Neoeplingia, Salvia, Trichostema*) or hummingbirds (*Agastache, Salvia, "Satureja"*). The flowers of *Cunila* (*C. ramamoorthiana*) (García-P., 1989) are fragrant, and the principle pollinators of the species of *Cunila* including the above are species of *Bombus* (Garcia P. pers. comm.).

CYTOLOGY

The chromosomal diversity in the Mexican members of this family is significant. In terms of chromosome numbers, aneuploidy and polyploidy have played an important role. Moreover, evolution in chromosomal morphology seems to have contributed to speciation in the family in Mexico.

Different ploidy levels are known in several genera, among which are *Hedeoma, Hesperozygis,* and *Poliomintha* (Irving, 1976), *Salvia* (Epling et al., 1962; Mercado et al., 1989; Palomino et al., 1986) and *Trichostema* (Lewis, 1945, 1960; Lewis & Rzedowski, 1973). Irving (1976), who discusses the chromosomal diversity in a group of related genera (*Hedeoma, Hesperozygis* and *Poliomintha*), concluded that in the evolution of *Hedeoma*, which is dibasic (x = 18, 22), both ploidy levels have influenced the course of evolution. He suggested that the predominantly Mexican (five of of its six species are endemic to Mexico) *H.* sect. *Alpineae* (x = 22) may have had a hybrid origin ($2n$ = 18 X $2n$ = 72) followed by aneuploid reduction. Incidentally, the species of this section in Mexico, which are mostly found in montane areas in the Sierra Madre Occidental, may be relatively young and may have secondarily adapted to their habitats. *Hedeoma* may be an epibiotic and its species in this section all neoendemics. The genera related to *Hedeoma, Hesperozygis,* and *Poliomintha* are both polyploids. Knowledge of the cytology of *Trichostema* is fairly complete; ploidy levels have had a major role in the evolution and differentiation of the taxa. Lewis (1945, 1960) and Lewis and Rzedowski (1973) reported that the perennial species of Mexican *Trichostema* (*T. arizonicum, T. lanatum, T. parishii,* and *T. purpusii*) are characterized by a haploid number of n = 10. In the annual species (sect. *Orthopodium*: *T. lanceolatum, T. micrathum,* and *T. austromontanum*) the haploid number is n = 14.

There is no numerical diversity ($2n$ = 22) in *Agastache* (Sanders, 1987), and evolution and differentiation here seem to have been mostly at the diploid level. In Mexican Salvias (including the Audibertias) multiplication of taxa seems to have been greatly influenced by variation in the number of chromosomes as well as chromosome morphology (Epling et al.,

1962; Palomino et al., 1986; Mercado et al., 1989). *Salvia* is polybasic (Palomino et al., 1986). The base number in *Salvia* subg. *Calosphace* seems to be $n = 11$, but both polyploidy and aneuploidy are known. In the Audibertias the base number is higher ($n = 15$), and aneuploidy is prevalent in the group. The perennial *Salvia lavanduloides*, which is widespread through the Trans-Mexican Volcanic Belt, is a polyploid with $2n = 66$. In the annual species *Salvia hispanica*, on the other hand, the number is $2n = 12$. The variation in the morphology of the chromosomes is notable. There are metacentric, submetacentric and telocentric chromosomes in the genus. "B" chromosomes are known and satellites are fairly common. Aneuploidy is known in the related genus *Poliomintha* ($2n = 26, 36$) (Irving, 1973).

EVOLUTION

The family, as was noted earlier, probably dates back to the Cretaceous. It is possible that the progenitors of today's Mexican Lamiaceae have been present in these areas since then and conceivably were part of the Madro-Tertiary flora. The differentiation and multiplication of taxa in the family seem to be intimately tied to the breakup of this flora, the evolution of the deserts, vulcanism, and processes of orogeny. The general morphological evidence of usual uniformity of flower structure, the rarity of hybrids, and the extreme local nature of many species in several genera in the family in Mexico (*Salvia, Scutellaria, Hedeoma*, and *Agastache*, among others) suggest that evolution in Mexican Lamiaceae has been generally through fragmentation of ancestral populations (Bye et al., 1987; Irving, 1972,1980; Sanders, 1987; Ramamoorthy, 1984c; Ramamoorthy & Lorence, 1987; Ramamoorthy et al., 1988). Their breakup and differentiation may have followed the breakup of the Madro-Tertiary flora into woodlands of which pines and oaks were primary constituents (Axelrod, 1950). Several genera of the present-day Mexican Lamiaceae are mostly found in the *Abies*, pine, pine-oak, and oak forests or their derivatives in Mexico. The "woodlands," isolated by deserts and valleys, served as islands for these differentiating populations. Additionally, the breakup of the flora may have led to the extinction of many of the tree species, allowing radiation of the herbaceous elements.

The nature of the pollination biology that usually results in selfing of the plant or population in many species (see discussion in section pertaining to *Salvia*) may have led to the establishment of genetically isolated populations that over time would differentiate into distinct species. Aneuploidy, polyploidy, and variation in chromosomal morphology as in species of *Salvia* have also played a crucial role in the evolution of the family in Mexico (Epling et al., 1962; Irving, 1976, 1980; Mercado et al., 1989; Palomino et al., 1986).

Transspecific evolution in Mexican Lamiaceae is significant. *Salvia* subg. *Calosphace* is composed of over 104 sections in the Americas. In

Mexico over 59 of these sections are represented, of which at least 25 are endemic to the country. Many of these endemic sections are monotypic. Among the nonendemic sections, many may be considered Mexican because of their major representation in Mexico (*Salvia* sect. *Lavanduloideae, S.* sect. *Scorodonia,* among others). Transspecific differentiation is notable in *Hyptis* (eight sections, one endemic), *Scutellaria* (nine sections, two endemics) *Stachys* (seven of ten groups recognized by Epling are represented in Mexico, of which two are endemic), and *Hedeoma* (four subgenera recognized by Irving are all represented in the country), among others. Interestingly, this supraspecific differentiation is most prominent in the montane groups but not in the tropical lowland groups in Mexico.

ENDEMISM IN MEXICAN LAMIACEAE

The percentage of species endemism in Mexican vascular plant flora is estimated at between 55 and 60% (Ramamoorthy & Lorence, 1987; Rzedowski, this volume). This may be considered high for a continental flora but Mexico's flora is in many ways similar to an island flora. The high species endemism is in part due to the rugged topography and the complex geological history of the country. The countless ravines, gorges, and arid areas ("deserts") that crisscross the mountainous landscape of the country dissect it into "islands," providing for geographical and genetic isolation of plant and animal populations. Most genera of Mexican Lamiaceae are montane, and the extremely local incidence of numerous species in the family as elsewhere suggest the pivotal role that geological history in combination with climatic changes has played in the evolution of these species.

About 75% of the Mexican Lamiaceae are endemic to the country. Generic endemism is low, however. Only one genus, *Neoeplingia,* is endemic to Mexico. Another, *Chaunostoma,* may be considered semiendemic as it is presently known only from the mesic mountains of Chiapas and Guatemala. The former occurs locally in the arid ravines in the Barranca de Tolantongo, Hidalgo and is isolated from *Hedeoma* and *Poliomintha,* its generic allies. It may be considered to be a paleoendemic. On the other hand, *Chaunostoma* may be recent as its closest generic ally *Lepechinia* (Epling, 1948) occurs in the same area. Generally, species of Lamiaceae in the montane cloud forests and in the Trans-Mexican Volcanic Belt and other geologically young areas are neoendemics. Examples include the following sections of *Salvia: Lavanduloideae* and *Sigmoideae.* The polyploid nature of *Salvia lavanduloides* (Palomino et al., 1987) supports this view as well. On the other hand, those species that are found in the geologically old deserts of Mexico may be relictual. Examples are species of the following sections of *Salvia: Tomentellae, Conzattiana,* and *Scorodonia.*

The endemism in the family (especially *Salvia*) is high along the cordilleras of the Sierra Madre Occidental and Oriental, the Trans-Mexican

Volcanic Belt and the mountains of Chiapas and Oaxaca. The arid and desertic parts of northern Mexico harbor many endemics in *Hedeoma* and allied genera. The highest concentration of endemism is seen in the states of Oaxaca, Puebla, Guerrero, Michoacán, and Jalisco (Fig. 18.1).

PHYTOGEOGRAPHY

A comparison of Mexican Lamiaceae with those of California (Munz, 1960), Texas (Correl & Johnston, 1970), and South America (Epling, 1933, 1934a, 1935, 1936) suggests the existence of four distinct groups in Mexican Lamiaceae: genera of Californian, North American, Gondwanan-South American, and Laurasian affinities. The Californian group has its primary center of distribution in northwestern Mexico (including Baja California) and includes *Acanthomintha, Monarda, Monardella, Pogogyne, Salvia* subg. *Audibertia,* and *Trichostema.* Some of these genera are monotypic, and some are represented in Mexico by endemic taxa. The North American connection is provided by the following genera: *Agastache, Cunila, Hedeoma, Hesperozygis, Poliomintha, Physostegia,* and *Tetraclea.* The generic status of *Hesperozygis* seems questionable (Irving, 1980). Most of these genera are represented in Mexico by endemic taxa, at both specific and transspecific levels. Three of them, (*Agastache, Cunila,* and *Hedeoma*) have radiated extensively. The Gondwanan-South American connection is seen in *Asterohyptis, Catoferia, Chaunostoma, Hyptis, Lepechinia, Ocimum,* and *Marsypianthes.* Except *Ocimum* and *Marsypianthes,* the other genera have endemic taxa in the country. The remaining group of genera represent the Laurasian connections: *Salvia, Satureja, Scutellaria, Stachys,* and *Teucrium.* It is significant that *Salvia, Scutellaria,* and *Stachys* represent more than 65% of the species of the family in the country. These genera have either had a long history in Mexico or a higher speciation rate than the others. The transspecific differentiations in these genera, especially *Salvia,* suggest greater antiquity for it.

It was noted earlier that Mentheae, with 11 genera in Mexico, is the most diverse. Seven of them (*Acanthomintha, Cunila, Hedeoma, Monardella, Neoeplingia, Pogogyne,* and *Poliomintha*) are Californian-North American and New World species of "*Satureja*" generically distinct from the Mediterranean *Satureja.* Five of them (*Cunila, Hedeoma, Neoeplingia, Poliomintha,* and *Hesperozygis*) are closely related to each other and in turn are allied to *Satureja,* suggesting the Mediterranean as the source area for the Mexican Mentheae. The progenitors of the group in North America may have diverged early into the distinct Californian and North American groups in the Mentheae, and they may have contributed to the Mentheae in South America.

Tribes Ajugeae, Lamieae, Nepeteae, and Salvieae all seem to have their sources from Laurasian elements and the Ocimeae of Mexico from Gondwanan-South American ones. Thus Mexican Lamiaceae is derived from both Laurasian (including Mediterranean) and Gondwanan (including South American) elements. South American genera of Laurasian affinities represented in Mexico (e.g., *Salvia, Scutellaria*) may have arrived there in a manner similar to that of Mentheae from Mexico. These elements are primarily cordilleran, and the mountains of Mexico may have served as corridors. It seems reasonable to suppose that the arrival of these genera in South America postdates the evolution of the mountains in Mexico. The age of Mexican Ocimeae is, on the other hand, intriguing. They may date back to the Cretaceous when an island arc may have connected North and South America (Coney, 1982). Ramamoorthy (1986) has suggested this possibility for *Catoferia*. The extant diversity at transspecific and specific levels in Mexican Lamiaceae, however, is due primarily to the explosive nature of speciation in Mexico. A similar scenario may be seen with several other groups of plants in Mexico (e.g., legumes of Mexico, Sousa & Delgado, this volume).

USES AND CONSERVATION NEEDS

Many members of Mexican Lamiaceae are used in folk medicines and traditional foods the world over. In Mexico such outstanding uses involve members of *Agastache* (Bye et al., 1987), *Cunila* spp., *Salvia* spp., and *"Satureja"* spp. among others. The members of *Agastache* that are used are commonly called "toronjil." Incidentally, it must be mentioned here that one of the recently described taxa was from among the cultivated members of this genus (Bye et al., 1987). *Cunila* is locally known as "Poleo." *Salvia* spp. is generally called "mirto" and species of *Satureja* "té del monte." Usually an infusion of tea is made with the leaves and is said to be effective against various stomach ailments.

The studies by Rodriguez-Hahn and her collaborators (Esquivel et al., 1985, 86a,b; Rodriguez-Hahn et al, 1986) are significant in this respect. Fruticulin, an abietane diterpene from *S. fruticulosa* (Rodriguez-Hahn et al., 1986), which has been isolated and reported by this group, is structurally similar to known cancer-suppressant organic compounds. The diterpenes from related species such as *S. goldmanii* and *S. anastomosans* may be equally promising. Their use awaits further research and study.

The greatest threat to the threatened and endangered species in the family in Mexico is from loss of habitats. The varied needs of human populations in Mexico, as elsewhere, constantly encroach on habitats where many of the endemic species of the family grow. The lumber operation in the pine forests is one example. Many of the species can be rescued by introducing them into cultivation. Habitat preservation alone in the long run will save many of the threatened species. The preservation of these habitats in the tropics is probably one of the most serious scientific and socioeconomic challenges humans are likely to face as they may be required to redefine social needs and material comforts not only in the developing parts of the world but in the developed parts as well.

Acknowledgments

We express our appreciation to all those plant collectors and field biologists who must of necessity remain anonymous and whose efforts have led to our understanding of diversity of this group in Mexico. We are grateful to Dr. Raymond Harley of Royal Botanical Gardens, Kew, whose constructive review benefited the chapter.

REFERENCES

Axelrod, D.I. 1950. Evolution of the Madrotertiary Geoflora. Bot. Rev. (Lancaster) 24:433–509.

Bentham, G. 1832–36. Labiatarum Genera et Species. London: James Ridgeway & Sons.

———. 1848. Labiatae. *In* A. P. de Candolle (ed.), Prodromus 12:27–603.

———. 1876. Labiatae. *In* G. Bentham & J. Hooker (eds.), Genera Plantarum. Vo. 2. London: W. Pamplin, L. Reeve & Co., Williams & Norgate. pp. 1160–1223.

Briquet, A. 1896–97. Labiatae *In* A. Engler & K. Prantl (eds.), Pflanzenfam. 4(3):183–375.

Burkart, A. 1937. El mecanismo floral de la labiada *Hyptis mutabilis*. Darwiniana 3:425–427.

Bye, R., E. Linares, T.P. Ramamoorthy, F. Garcia, O. Collera, G. Palomino & V. Corona. 1987. *Agastache mexicana* subsp. *xolocotziana* (Lamiaceae), a new taxon from among the Mexican medicinal plants. Phytologia. 62(3):157–168.

Cantino, P.D. & R.W. Sanders. 1986. Subfamilial classification of Labiatae. Syst. Bot. 11:163–185.

Coney, 1982. Biogeography of Middle America. Ann. Missouri Bot. Gard. 1–10.

Correl, D.S. & M.C. Johnston. 1970. Manual of the Vascular Plants of Texas. Renner, TX: Texas Research Foundation.

Cronquist, A. 1988. The evolution and classification of flowering plants. Bronx, NY: New York Botanical Garden.

Dieringer, G., T.P. Ramamoorthy and P. Tenorio. 1991. Floral visitors in Mexican Salvias. Acta Bot. Mex. 13:75–83.

Emboden, W.A. 1965. Pollen morphology of the genus *Salvia* sect. *Audibertia*. Pollen et Spores 6:527–536.

Epling, C.C. 1925. Monograph of the genus *Monardella*. Ann. Missouri Bot. Gard. 12:1–106.

———. 1933. *Asterohyptis*: a newly proposed genus of Mexico and Central America. Bull. Torrey Bot. Club 60:17–21.

———. 1934a. Synopsis of the genus *Hyptis* in North America. Repert. Spec. Nov. Regni Veg. 34:73–130.

———. 1934b. Preliminary revision of American *Stachys*. Repert. Spec. Nov. Regni Veg. Beih. 80:1–75.

———. 1935. Synopsis of South American Labiatae (1). Repert. Spec. Nov. Regni Beih. 85:1–96.

———. 1936a. Synopsis of South American Labiatae (2). Repert. Spec. Nov. Regni Beih. 85:97–192.

————. 1936b. Synopsis of South American Labiatae (3). Repert. Spec. Nov. Regni Beih. 85:193–288.

————. 1937. Synopsis of South American Labiatae (4). Repert. Spec. Nov. Regni Beih. 85:289–341.

————. 1938a. The Californian Salvias. Ann. Missouri Bot. Gard. 25:95–188.

————. 1938b. Scylla, Charybdis and Darwin. Amer. Naturalist 72:547–561.

————. 1939a. A note on the occurrence of *Salvia* in the New World. Madroño 5:34–37.

————. 1939b. A revision of *Salvia* subgenus *Calosphace*. Repert. Spec. Nov. Regni Veg. Beih. 110:1–383.

————. 1939c. A new species of *Monardella* . Wash. Acad. Sc. 29:489.

————. 1940. Supplementary Notes on American Labiatae I. Bull. Torrey Bot. Club 67:509–537.

————. 1941. Supplementary Notes on American Labiatae II. Bull. Torrey Bot. Club 68:553–568.

————. 1942. A revision of American *Scutellaria*. Univ. Calif. Publ. Bot. 20:1–147.

————. 1944. Supplementary Notes on American Labiatae III. Bull. Torrey Bot. Club 71:487–497.

————. 1947. Supplementary Notes on American Labiatae IV. Bull. Torrey Bot. Club 74:512–518.

————. 1948. A synopsis of the tribe Lepechenieae (Labiatae). Brittonia 6: 352–364.

————. 1949. Revisión del género *Hyptis* (Labiatae). Revista Mus. La Plata, Secc. Bot. 7:153–497.

————. 1951. Supplementary Notes on American Labiatae V. Brittonia 7:129–142.

————. 1960. Supplementary Notes on American Labiatae VII. Brittonia 12: 140–150.

———— & C. Jativa. 1962. Supplementary Notes on American Labiatae VIII. Brittonia 15:366–376.

———— & ————. 1963. Supplementary Notes on American Labiatae IX. Brittonia 18:255–265.

———— & ————. 1964. Revisión del género *Satureja* en America del Sur. Brittonia 16:393–416.

———— & ————. 1965. Supplementary Notes on American Labiatae X. Brittonia 20:295–312.

———— & ————. 1966. A descriptive key to the species of *Satureja* indigenous to North America. Brittonia 18:244–248.

————, H. Lewis & P.H. Raven. 1962. Chromosomes of *Salvia* section *Audibertia*. Aliso 5:217– 221.

———— & M.E. Mathias. 1957. Supplementary Notes on American Labiatae VI. Brittonia 8:297–313.

———— & W.S. Stewart. 1939. A revision of *Hedeoma*. Repert. Spec. Nov. Regni Veg. Beih. 115:1–49.

Esquivel, B., A. Méndez, A. Ortega, M.S. García, A. Toscano & L. Rodriguez-Hahn. 1985b. Two new clerodane type diterpenoids from *Salvia keerlii*. Phytochemistry 24:1769.

————, J. Cárdenas, T.P. Ramamoorthy & L. Rodriguez-Hahn. 1986a. Clerodane diterpenoids of *Salvia lineata*. Phytochemistry. 25:2381.

————, M. Hernández, J. Cárdenas, T.P. Ramamoorthy & L. Rodriguez-Hahn. 1986b. Semiatrin, a neoclerodane diterpenoid from *Salvia semiatrata*. Phytochemistry 25:1484.

————, L. Rodriguez-Hahn & T.P. Ramamoorthy. 1987. The diterpenoid constituents of *Salvia fulgens* Cav. and *Salvia microphylla* Kunth (Labiatae). J. Nat. Prod. 50:738–740.

————, N. Martínez, J. Cárdenas, T.P. Ramamoorthy & L. Rodriguez-Hahn. 1989. The pimarane-type diterpenoids from *Salvia microphylla* var. *neurepea*. Planta Medica 55:62–63.

Fernald, M.A. 1900. Synopsis of the Mexican and Central American species of *Salvia*. Proc. Amer. Acad. Arts 35:489–573.

Galicia M., A., B. Esquivel, A.A. Sánchez, J. Cárdenas, L. Rodriguez-Hahn & T.P. Ramamoorthy. 1988. Abietane diterpenoids from *Salvia pubescens*. Phytochemistry 27:217.

García-Peña, M.R. 1989. A new species of *Cunila* (Lamiaceae) from southwestern Mexico. Kew Bull. 44(4):727–730.

Gonzáles, A.G., T. Abad, I. Jiménez, A. Ravelo, J. Luís, Z. Aguir, L. Andres, M. Plasencia, J. Herrera & L. Moujir. 1989. A first study of antibacterial activity of diterpenes isolated from some *Salvia* spp. (Lamiaceae). Biochem. Syst. Ecol. 17:293–296.

Gray, A. 1853. Characters of *Tetraclea*, a new genus of Verbenaceae. Amer. J. Sci. Arts (Ser. 2) 16:97–98.

Hardham, C.B. 1966. Three diploid species of *Monardella villosa* complex. Leafl. W. Bot. 10:237–256.

Harley, R.M. 1971. An explosive pollination mechanism in *Eriope crassipes*, a Brazilian labiate. Biol. J. Linn. Soc. 3(2):159–164.

————. 1976. A review of *Eriope* and *Eriopidion* (Labiatae). Hooker's Icones Plantarum 8, 3:1–107.

————. 1983. *Hyptis alata*, amphitropically disjunct in the Americas. Notes on New World Labiatae V. Kew Bull. 38(1):47–52.

————. 1985. New Taxa in *Hyptis* sect. *Polydesmia* Benth. from Bahia, Brazil. Notes on New World Labiatae VI. Kew Bull. 40(3):609–625.

————. 1986. New species of *Hyptis* (Labiatae) from South America. Notes on New World Labiatae VIII. Kew Bull. 41(1):141–150.

————. 1988. Evolution and distribution of *Eriope* (Labiatae), and its relatives in Brazil. *In* W.R. Heyer & P.E. Vanzolini (eds.), Proceedings of a Workshop on Neotropical Distribution Patterns. Rio de Janeiro: Academia Brasileira de Ciencias.

————. 1988. Revision of generic limits in *Hytpis* Jacq. (Labiatae) and its allies. Bot. J. Linn. Soc. 98:87–95.

Hedge, I. 1986. Labiatae of Southwest Asia: diversity, distribution and endemism. Proc. Royal Soc. Edinburgh 89B:23–35.

Henrickson, J. 1981. A new species of *Satureja* (Lamiaceae) from Chihuahuan desert region. Brittonia 33(2):211–213.

————. 1982. On the recognition of *Trichostema mexicanum* (Lamiaceae). Madroño 29(2):104–108.

Hernández M., B. Esquivel, J. Cárdenas, L. Rodriguez-Hahn & T.P. Ramamoorthy. 1987. Diterpenoid abietane quinones from *Salvia regla*. Phytochemistry 26(12):3297–3299.

Hiriart V., P. 1984. Una nueva especie de *Scutellaria* (Labiatae). Bol. Soc. Bot. Méx. 46:43–46.

Irving, R.S. 1972. A revision of the genus *Poliomintha* (Labiatae) Sida 5(1):8–22.

————. 1976. Chromosome numbers of *Hedeoma* (Labiatae) and related genera. Syst. Bot. 1:46–56.

————. 1980. The systematics of *Hedeoma* (Labiatae). Sida 8(3):218–295.

Lewis, H. 1945. A revision of the genus *Trichostema*. Brittonia 5:276–303.

————. 1960. Chromosome numbers and phylogeny in *Trichostema*. Brittonia 12:93–97.

————— & J. Rzedowski. 1978. The genus *Trichostema* (Labiatae) in Mexico. Madroño 25:151–154.

Lint, H. & C.C. Epling. 1945. A revision of *Agastache*. Midl. Naturalist 33:207–230.

Mabberley, D.J. 1987. The Plant Book. A Portable Dictionary of Flowering Plants. Cambridge: Cambridge Univ. Press.

McClintock, E. & C.C. Epling. 1942. A review of the genus *Monarda* (Labiatae). Univ. Calif. Publ. Bot. 20:147–194.

————— & —————. 1946. A revision of *Teucrium* in the New World, with observations on its variation, geographical distribution and history. Brittonia 5:491–510.

McVaugh, R. & R. Schmidt. 1967. Novelties in *Satureja* sect. *Gardoquia* (Labiatae). Brittonia 19:261–267.

Mercado, P., T.P. Ramamoorthy & G. Palomino. 1989. Karyotypic analysis of five species of *Salvia* (Lamiaceae). Cytologia 54:605–608.

Munz, P. 1960. A Californian Flora. Berkeley, CA: Univ. of California Press.

Palomino, G., P. Mercado & T.P. Ramamoorthy. 1986. Chromosomes of *Salvia* subgenus *Calosphace*, a preliminary report. Cytologia 51:381–386.

Ramamoorthy, T.P. 1984a. Notes on *Salvia* (Lamiaceae) in Mexico with three new species. J. Arnold Arb. 65:135–143.

—————. 1984b. A new species of *Salvia* (Lamiaceae) from Mexico. Brittonia 36:297–299

—————. 1984c. A new species of *Salvia* (Lamiaceae) from the Sierra de Los Tuxtlas, Mexico. Pl. Syst. and Evolution 141:141–143.

—————. 1985. *Salvia. In* J. Rzedowski & G. Rzedowski (eds.), Flora Fanerogamica del Valle de México. Mexico City: Inst. Ecol.

—————. 1986. A revision of Catoferia (Lamiaceae). Kew Bull. 41(2):299–305.

—————, P. Hiriart & F. Medrano. 1982. *Neoeplingia* Ramamoorthy, Hiriart & Medrano (Labiatae), Un nuevo género de Hidalgo, México. Bol. Soc. Bot. Méx. 43:61–65.

—————, B. Esquivel, A.A. Sánchez & L. Rodriguez-Hahn. 1988. Phytogeographical significance of the occurrence of abietane type diterpenoids in *Salvia* sect. *Erythrostachys* (Lamiaceae). Taxon 37(4):908–912.

————— & D.H. Lorence. 1987. Species vicariance in the Mexican Flora and a description of a new species of *Salvia* (Lamiaceae). Adansonia 2:167–175.

Rodriguez-Hahn, L., B. Esquivel, C. Sánchez, J. Cárdenas, L. Estebanez, M.S. García, R.A. Toscano, & T.P. Ramamoorthy. 1986. New highly oxidized diterpene quinones from *Salvia fruticulosa* (Labiatae). Tetrahedron Let. 27:5459–5462.

Rzedowski, J. & G.C. Rzedowski. 1988. Tres nuevas especies de *Stachys* (Labiatae) de México. Acta Bot. Méx. 3:1–5.

Sanders, R. 1987. Taxonomy of *Agastache* sect. *Brittonastrum* (Lamiaceae-Nepeteae). Syst. Bot. Mono. 15:1–92.

Scora, R.W. 1965. New taxa in the genus *Monarda* (Labiatae). Madroño 18(4):119–122.

Shreve, F. & I. Wiggins. 1964. Vegetation and Flora of the Sonoran Desert. Vol. 2. Stanford, CA: Stanford Univ. Press.

Standley, P. & L. Williams. 1973. Labiatae. Flora of Guatemala. Fieldiana, Bot. 24(9):237–317.

Takhtajan, A. 1969. Flowering Plants of the World. Origins and Dispersal. Washington, DC: Smithsonian Institution.

—————. 1986. Floristic Regions of the World. Berkeley, CA: Univ. of California Press.

Wiggins, I. 1980. Flora of Baja California. Stanford, CA: Stanford Univ. Press.

Williams, L.O. 1972. Teucrium proctorii. Fieldiana, Bot. 34:114.

19

Mexican Acanthaceae: Diversity and Distribution

THOMAS F. DANIEL

About 360 species of Acanthaceae belonging to 46 genera representing each of the four subfamilies occur in Mexico. Eight genera and nearly 50% of the species of Mexican Acanthaceae are endemic to the country. The family appears to have had a Gondwanan origin and to have undergone a secondary radiation in Mexico. Although Acanthaceae are found in all of the major vegetation types of Mexico, they are best represented in regions of tropical dry forest. *Justicia* and *Ruellia*, the largest genera in the family, are each represented in Mexico by more than 60 species. Nearly all of the ecological life forms and floral types encountered in the family are represented among Mexican Acanthaceae. This chapter summarizes information on the distribution, reproductive biology, and chromosome numbers of Mexican Acanthaceae. A list of the genera of Acanthaceae in Mexico, their classification in two relevant systems, and a summary of the literature are provided.

The Acanthaceae are a predominantly tropical family comprising about 4,350 species in some 350 genera (Mabberley, 1987) with centers of distribution in Indo-Malaysia, Africa (including Madagascar), South America, and Mexico-Central America. Based on Mabberley's estimate of the number of species in the family, the Acanthaceae are the eleventh largest family of flowering plants and the eighth largest family of Magnoliopsida (i.e., dicotyledonous plants). About 12 genera, including the two largest, *Justicia* (350–600 species) and *Ruellia* (ca. 250 species), are pantropical. Of the remaining genera, about 40% are confined to the Western Hemisphere (Long, 1970).

The family is included by Cronquist (1988) in the Scrophulariales, where it differs from the Scrophulariaceae primarily by the lack of seminal endosperm, the specialized funiculus, and the explosively dehiscent capsules. Many genera of the Acanthaceae can be further distinguished from the Scrophulariaceae by their cystoliths in the epidermis of vegetative organs.

There have been two major monographs of the family, a complete taxonomic treatment by Nees von Esenbeck in 1847 that included descriptions of all species known to the author and a generic account by Lindau in 1895 that included an extensive infrafamilial classification. Lindau treated the family in a broad sense to include four subfamilies that can be characterized by the following key.

Fruit drupaceous .... Mendoncioideae
Fruit capsular
   Vines .... Thunbergioideae
   Herbs, shrubs, or trees
      Retinacula hook-shaped; cystoliths often present .... Acanthoideae
      Retinacula papilliform or lacking; cystoliths
      never present .... Nelsonioideae

Considerable attention has been given to suprageneric classification of the Acanthaceae. Extensive studies of morphology, palynology, embryology, and leaf anatomy have addressed the circumscriptions and taxonomic relationships of the four subfamilies identified above. There is little doubt that the Acanthoideae, the largest subfamily, is a monophyletic taxon. It is

Table 19.1. Genera of Mexican Acanthaceae (including naturalized and excluding cultivated taxa)

Genus	No. of species in Mexico	Worldwide distribution[a]
Anisacanthus	9	NA, CA, SA
Aphanosperma	1	EN
Aphelandra	12	NA, CA, SA
Barleria	1	NA, CA, SA, OW
Blechum	1	NA, CA, CB, SA
Bravaisia	3	NA, CA, CB, SA
Buceragenia	4	NA, CA
Carlowrightia	23	NA, CA
Chaetothylax	2	CA, SA
Chalarothyrsus	1	EN
Chileranthemum	3	EN
Dicliptera	15	NA, CA, CB, SA, OW
Dyschoriste	22	NA, CA, CB, SA, OW
Elytraria	4	NA, CA, CB, SA, OW
Geissomeria	1	NA, SA
Gypsacanthus	1	EN
Habracanthus	2	NA, CA, SA
Hansteinia	4	NA, CA, SA
Henrya	2	NA, CA
Holographis	15	EN
Hoverdenia	1	EN
Hygrophila	1	NA, CA, CB, SA, OW
Hypoestes	1	OW
Justicia	78	NA, CA, CB, SA, OW

Table 19.1. (cont.)

Genus	No. of species in Mexico	Worldwide distribution[a]
Lophostachys	3	CA, SA
Louteridium	8	NA, CA
Mendoncia	2	CA, SA
Mexacanthus	1	EN
Mirandea	4	EN
Nelsonia	1	NA, CA, CB, SA, OW
Neohallia	1	CA
Odontonema	9	NA, CA, CB, SA
Pachystachys	1	NA, SA
Poikilacanthus	3	NA, CA, SA
Pseuderanthemum	10	NA, CA, SA, OW
Razisea	1	CA, SA
Ruellia	65	NA, CA, CB, SA, OW
Schaueria	1	NA, SA
Siphonoglossa	5	NA, CA, CB, SA
Spathacanthus	3	NA, CA
Stenandrium	9	NA, CA, CB, SA
Teliostachya	1	NA, CA, CB, SA, OW
Tetramerium	21	NA, CA, SA
Thunbergia	1	OW
Tribliocalyx	1	NA, CA
Yeatesia	2	NA

[a]Excluding regions where taxa have become naturalized or are cultivated only. EN, endemic to Mexico; NA, North America; CB, Caribbean region; CA, Central America; SA, South America; OW, Old World.

readily definable by the presence of retinacula (specialized funiculi) on which the seeds are borne. The relationships of the other subfamilies are problematic. Cronquist (1981) and others have treated the Mendoncioideae as a separate family—Mendonciaceae—in the Scrophulariales with unknown phylogenetic affinities. The Nelsonioideae and Thunbergioideae appear to represent intermediary links between the Scrophulariaceae and the Acanthaceae. Both of the latter subfamilies have likewise been treated as distinct families at one time or another (e.g., Bremekamp, 1965; Sreemadhavan, 1977).

In contrast to the attention accorded to higher taxa in the family, generic monographs of Acanthaceae are few in number. In fact, plants of Acanthaceae offer few characters for distinguishing genera. Minor traits of the anthers and pollen morphology have been much utilized, resulting in a proliferation of small genera. Among the 11 largest families of flowering plants, only the Gramineae have a lower species/genera ratio (7,950:737) than the Acanthaceae (based on numerical estimates in Mabberley, 1987). The number of monotypic genera in the family (about 120, according to

Long, 1970) with restricted distributions is particularly large. It is doubtful that monophylesis can be established for many genera of Acanthaceae as they are traditionally defined. As a result, generic circumscriptions are often inadequate or problematic. Ultimately, taxonomic problems at the generic level obscure understanding of the phylogenetic and geographic patterns in the family.

A significant portion of my research efforts during the past decade has been directed toward providing a taxonomic treatment of this family for Mexico while monographing the genera that occur there. Monographs utilizing various systematic techniques (often including chromosome number analysis, pollination biology, and hybridization experiments) have been completed for *Mexacanthus* (Daniel, 1981), *Mirandea* (Daniel, 1982a), *Carlowrightia* (Daniel, 1983a), *Holographis* (Daniel, 1983b), *Stenandrium* (Daniel, 1985a), *Tetramerium* (Daniel, 1986a), *Aphanosperma* (Daniel, 1988a), *Bravaisia* (Daniel, 1988b), and *Henrya* (Daniel, 1990). These monographs and other recent studies of Mexican Acanthaceae by various authors are summarized in the Appendix. Regional floristic accounts of the family have been completed for Baja California (Wiggins, 1980), the Sonoran Desert (Leonard & Wiggins, 1964), and the Valley of Mexico (Lott et al., 1985). Floristic treatments are under way in west-central Mexico for McVaugh's Flora Novo-Galiciana, north-central Mexico for Henrickson and Johnston's Flora of the Chihuahuan Desert, and Chiapas for Breedlove's Flora of Chiapas.

TAXA IN MEXICO

Mexican Acanthaceae have received little taxonomic attention since Standley's treatment for his "Trees and Shrubs of Mexico" in 1926. He included 78 species in 17 genera in that account. A conservative estimate of the numbers of genera and species of the family in Mexico based on the studies cited above and others reveals the presence of about 360 species in 46 genera (Table 19.1), representing 13% of the genera of Acanthaceae and eight percent of the estimated number of species. Each of the four traditionally recognized subfamilies is represented by either native (Acanthoideae, Mendoncioideae, Nelsonioideae) or naturalized (Thunbergioideae) taxa. In addition to the genera listed in Table 19.1, several nonnative genera (e.g., *Acanthus* L., *Sanchezia* Ruiz & Pavon) are cultivated for ornament in Mexico. The overdescription of genera in the family is well represented in Mexico where at least 12 unispecific genera have been described and only six are presently recognized. The largest genera of Mexican Acanthaceae are *Justicia* (ca. 78 species), *Ruellia* (ca. 65 species), *Carlowrightia* (23 species), *Dyschoriste* (22 species), and *Tetramerium* (21 species)

Classifications of Mexican Acanthaceae are presented in Tables 19.2 and 19.3. Table 19.2 utilizes the traditional classification scheme of Lindau (1895). Several genera unknown to him have been appropriately placed.

Table 19.2. Classification of Mexican Acanthaceae according to the system of Lindau (1895)

Sfam. Acanthoideae
  Tr. Aphelandreae: *Aphelandra, Geissomeria, Holographis, Stenandrium*
  Tr. Asystasieae: *Chalarothyrsus, Spathacanthus*
  Tr. Barlerieae: *Barleria, Lophostachys, Teliostachya*
  Tr. Graptophylleae: *Anisacanthus, Aphanosperma, Carlowrightia, Gypsacanthus, Mexacanthus, Mirandea, Pachystachys, Tribliocalyx*
  Tr. Hygrophileae: *Hygrophila*
  Tr. Isoglosseae:
    Str. Porphyrocominae: *Yeatesia, Poikilacanthus*
    Str. Isoglossinae: *Habracanthus, Hansteinia*
  Tr. Justicieae: *Chaetothylax, Justicia, Neohallia*
  Tr. Louteridieae: *Louterdium*
  Tr. Odontonemeae:
    Str. Diclipterinae: *Dicliptera, Henrya, Hypoestes, Tetramerium*
    Str. Odontoneminae: *Chileranthemum, Hoverdenia, Odontonema, Razisea, Schaueria, Siphonoglossa*
  Tr. Petalidieae: *Blechum*
  Tr. Pseuderanthemeae: *Buceragenia, Pseuderanthemum*
  Tr. Ruellieae: *Ruellia*
  Tr. Strobilantheae: *Dyschoriste*
  Tr. Trichanthereae: *Bravaisia*

Sfam. Mendoncioideae-*Mendoncia*

Sfam. Nelsonioideae—*Elytraria, Nelsonia*

Sfam. Thunbergioideae—*Thunbergia*

Sfam., Subfamily; Tr., Tribe; Str., Subtribe.

The classification shown in Table 19.3 is based on the system proposed by Bremekamp (1965). Unfortunately, he did not list the genera comprising his suprageneric taxa, nor did he deal with several taxa treated by Lindau (1895), e.g., subtribe Porphyrocominae. Despite these problems, Bremekamp's system is considerably better than that of Lindau in many respects. Additional refinement is especially needed in his Justicieae. For example, several genera of his subtribe Odontoneminae (e.g., *Hoverdenia, Siphonoglossa*) certainly belong in the Justiciinae.

All morphological life forms represented in the family occur in Mexico. They include acaulescent herbs, e.g., *Elytraria bromoides* Oerst., *Stenandrium dulce* (Cav.) Nees; caulescent herbs, e.g., *Dicliptera resupinata* (Vahl) Juss., *Carlowrightia hapalocarpa* Robs. & Greenm.; shrubs, e.g., *Justicia linearis* Robs. & Greenm., *Holographis virgata* (Harv. ex Bentham. & Hook.) T. Daniel; trees, e.g., *Louteridium donnell-smithii* Wats., *Bravaisia integerrima* (Spreng.) Standl.; and vines, e.g., *Mendoncia retusa* Turrill, *Thunbergia alata*

Table 19.3. Classification of Mexican Acanthaceae according to system of Bremekamp (1965)

Fam. ACANTHACEAE
Sfam. Acanthoideae
  Tr. Aphelandreae: *Aphelandra, Geissomeria, Holographis, Stenandrium.*
Sfam. Ruellioideae
  Tr. Justicieae:
    Str. Justiciinae: *Chaetothylax, Justicia, Neohallia.*
    Str. Odontoneminae: *Anisacanthus, Aphanosperma, Buceragenia, Carlowrightia, Chalarothyrsus, Chileranthemum, Dicliptera, Gypsacanthus, Henrya, Hoverdenia, Hypoestes, Mexacanthus, Mirandea, Odontonema, Pachystachys, Pseuderanthemum, Razisea, Schaueria, Siphonoglossa, Spathacanthus, Tetramerium, Tribliocalyx, Yeatesia*
    Str. Rhytiglossinae: *Habracanthus, Hansteinia.*
    "Genera incertae sedis": *Poikilacanthus*
  Tr. Lepidagathideae:
    Str. Lepidagathidinae: *Teliostachya*
  Tr. Louteridieae: *Louteridium*
  Tr. Ruellieae:
    Str. Barleriinae: *Barleria, Lophostachys*
    Str. Blechinae: *Blechum*
    Str. Hygrophilinae: *Hygrophila*
    Str. Petalidiinae: *Dyschoriste*
    Str. Ruelliinae: *Ruellia*
  Tr. Trichanthereae: *Bravaisia*

Fam. MENDONCIACEAE: *Mendoncia*

Fam. SCROPHULARIACEAE
Sfam. Rhinanthoideae
  Tr. Nelsonieae: *Elytraria, Nelsonia*

Fam. THUNBERGIACEAE: *Thunbergia*

Fam., family; Sfam., subfamily; Tr., Tribe; Str., Subtribe

Bojer ex Sims. The diversity of ecological life forms includes hygrophiles, e.g., *Justicia americana* (L.) Vahl; xeromorphs, e.g., *Holographis ilicifolia* T. Brandegee; epiphytes, e.g., *Louteridium parayi* Miranda; and mangroves, e.g., *Bravaisia berlandieriana* (Nees) T. Daniel.

PHYTOGEOGRAPHY AND ENDEMISM

The distribution of the family is suggestive of some antiquity in its age; however, pollen attributable to Acanthaceae does not appear until the Lower Miocene (Muller, 1970). In 1964 Chandler (cited in Raven & Axelrod, 1974) referred fruits from the Eocene of England to *Acanthus*. Stearn (1971) argued that phytogeographical evidence (e.g., disjunctions in genera such as *Oplonia* between tropical America and Madagascar) indicates a Cretaceous origin for the Acanthaceae. Present concentrations of species in Africa, Madagascar, India, and South America certainly suggest a Gondwanan origin. Interestingly, however, the family is relatively poorly represented in Australia (only 57 native and naturalized species) (Barker, 1986) and is entirely absent from New Zealand. Both of these regions are believed to have been part of the southern continent Gondwanaland, which began to break up during the Cretaceous. In order to reconcile the suggested antiquity of the family with its more recent appearance in the fossil record, Van Steenis (1972) proposed that either the record is incomplete or that the pollen of early Acanthaceae cannot be recognized as such.

The Acanthaceae of Mexico and Central America have strong affinities with their South American counterparts. Thirty genera of Acanthaceae occurring in Mexico also occur in South America (28 of them occur naturally in both regions) and at least 20 species are common to the two regions. By contrast, only 13 genera occur naturally in Mexico and the West Indies and only 11 in Mexico and the Old World. It appears likely that Mexican Acanthaceae are derived primarily from South American stock. The contention by Raven and Axelrod (1974) that "moderately large genera such as *Carlowrightia* in North America seem to provide evidence of an early Laurasian element" in the acanthaceous flora there is unfounded. *Carlowrightia* appears to have diversified, particularly in the Chihuahuan Desert region, relatively recently (Daniel, 1983a), and its affinities are clearly with South American taxa. Most of the species of Acanthaceae in the United States, the majority of which occur near the Mexican border, represent mere extensions of Mexican taxa.

Approximately one-half of the species of Acanthaceae indigenous to Mexico are endemic there. Eight genera are endemic to the country (Table 19.1), and several others are either nearly endemic to Mexico or have their major concentration of species there (e.g., *Anisacanthus, Bravaisia, Carlowrightia, Henrya, Louteridium, Siphonoglossa,* and *Tetramerium*). Most of the endemic genera are known from few collections in a limited geographic area and have close taxonomic affinities to larger, more widely distributed genera.

The relatively high percentage of endemism among Mexican Acanthaceae is likely a secondary phenomenon resulting from radiation into or diversification within the dry regions of southern North America (see discussion in Gentry, 1982). For example, the largest genus endemic to Mexico, *Holographis,* is related to two widespread neotropical genera, *Aphelandra* and *Stenandrium.* All species of *Holographis* occur in regions of tropical dry forest or desert scrub. The high percentage of species of Acanthaceae endemic to the Chihuahuan Desert region (19/36 species, or 53%) of north-central Mexico further underscores the affinity of the family for dry regions in the country.

Mexico encompasses regions of both North and Central America, which meet at the Isthmus of Tehuantepec. Among genera native to Mexico, 32

Table 19.4. Genera and species of Acanthaceae in ten regions of the New World

Region	Genera	Species
Sonoran Desert Region[1]	11	25
Chihuahuan Desert Region[2]	12	36
Valle de México[3]	8	11
Nueva Galicia Region[4]	23	105
Chiapas[5]	31	103
Guatemala[6]	32	127
Nicaragua[7]	34	73
Costa Rica[8]	39	119
Panama[9]	34	109
Colombia[10]	36	326

[1]Daniel (unpublished data), [2]Daniel (1985b), [3]Lott et al (1985), [4]Daniel (unpublished data), [5]Daniel (1986b), [6]Gibson (1974), [7]Durkee (1988), [8]Durkee (1988), [9]Durkee (1988), [10]Leonard (1951–58).

occur on both sides of the Isthmus, six are restricted to North America and six occur only in Central America. There is a rather sharp discontinuity in species distributions at the Isthmus, across which usually only the widely distributed or weedy species spread.

Acanthaceae occur in all ten of the major vegetation types recognized by Rzedowski (1978) for Mexico, with especially high concentrations of species in regions of tropical dry forest. Four species are found universally in tropical deciduous and subdecidous forests throughout Mexico: _Carlowrightia arizonica_ A. Gray, _Elytraria imbricata_ (Vahl) Persoon, _Henrya insularis_ Nees ex Benth., and _Tetramerium nervosum_ Nees. These four species are weedy and have spread into artificially created habitats, e.g, roadsides and fence rows in other vegetational zones. In terms of numbers of individuals, they are undoubtedly the most abundant species of Acanthaceae in Mexico. The number of nonweedy species occurring in Mexican dry forests is considerable and includes both narrow endemics (e.g., _Holographis_ spp.) and species found wherever dry forests are encountered (e.g., _Tetramerium tenuissimum_ Rose). Other species of Acanthaceae are restricted to regions of tropical perennial forest (e.g., _Neohallia borrerae_ Hemsley, _Poikilacanthus macranthus_ Lindau), thorn forest (e.g., _Aphanosperma sinaloensis_ T. Daniel), xerophytic scrub (e.g., _Mirandea grisea_ Rzedowski), and pine-oak forest (e.g., _Dyschoriste rubiginosa_ Ramam. & Wassh.). Many genera (e.g., _Aphelandra, Justicia, Louteridium, Mirandea, Pseuderanthemum_, and _Ruellia_) possess species occurring in both dry (_A. lineariloba_ Leonard, _J. candicans_ [Nees] Benson, _L. koelzii_ Miranda & McVaugh, _M. huastecensis_ T. Daniel, _P. praecox_ [Benth.] Leonard, and _R. foetida_ Willd.) and moist (_A. aurantiaca_ [Scheidw.] Lindl., _J. aurea_ Schlecht., _L. parayi_ Miranda, _M. sylvatica_

Acosta C., *P. verapazense* J.D. Smith, and *R. longituba* D. Gibson) regions. Other genera are entirely restricted to either dry (e.g., *Holographis*) or moist (e.g., *Lophostachys, Spathacanthus*) regions of Mexico. The family is conspicuously absent from the coniferous forests at high elevations. In fact, species of Acanthaceae are rarely encountered above 2,000 m elevation in Mexico. Exceptions include *Tetramerium nervosum* and *Carlowrightia parviflora* (Buckley) Wassh., the latter being recorded from as high as 2,800 m.

Compilations of species have been completed for five regions within the country and five regions to the south (Table 19.4). Comparisons among these regions generally reveal a gradual increase in total numbers of genera and species in a southward direction, underscoring the tropical nature of the family. Within Mexico it is not surprising that the acanthaceous flora of Chiapas, which lies to the east of the Isthmus of Tehuantepec, is more similar to those of the Central American republics than to those in regions of Mexico located west of the Isthmus. The states of Veracruz and Oaxaca, which include most of the Isthmus, are enriched by Acanthaceae from regions to the east and west. Unfortunately, there are no recent compilations of taxa from either state.

REPRODUCTIVE BIOLOGY

Plants of Mexican Acanthaceae usually produce flowers continuously for several months. Flowers of many species are ephemeral, lasting only a portion of one day. Flowers of Mexican Acanthaceae exhibit all major colors except black. Most have colored nectar guides and produce some quantity of nectar from the large disc at the base of the ovary. Nectar and pollen are the only known floral rewards. Despite floral rewards and floral mechanisms (e.g., heterostyled floral morphs in *Odontonema*) that might promote outcrossing by insect-mediated pollination, self-pollination can be an effective means of seed production (see data on *Tetramerium nervosum* in Daniel, 1986a). Experimental studies of 51 species in 16 genera of Mexican Acanthaceae reveal most to be self-compatible and at least partially autogamous (Daniel, unpublished data). Autogamy is accomplished in several taxa by modifications of floral structures (e.g., cleistogamy in *Ruellia* species). A developing capsule is normally evident within a few days after pollination and fertilization. Seeds can be projected up to five meters from the parent plant by the explosive dehiscence of the capsules. For all Mexican genera studied, germination of the seeds can take place immediately after expulsion from the capsule, and sexual maturation can occur within a single growing season.

Projective expulsion of the seeds, autogamy, and a rapid life cycle are common reproductive features of *Carlowrightia arizonica, Elytraria imbricata, Henrya insularis,* and *Tetramerium nervosum* and doubtless help to explain their weedy and invasive natures as noted above.

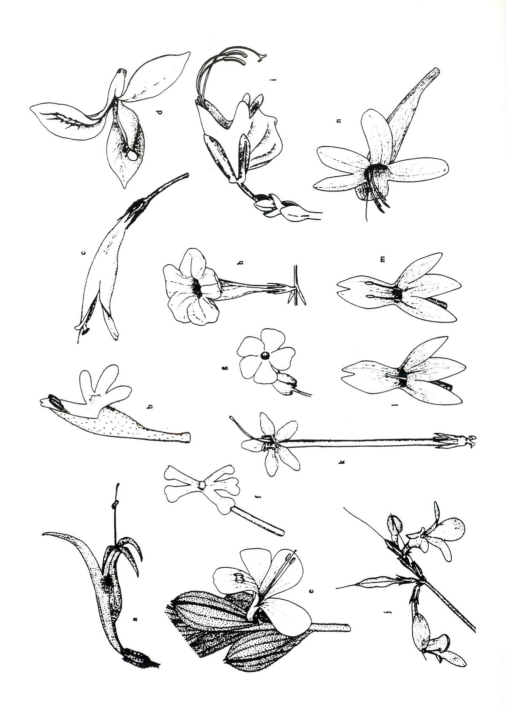

A diverse array of floral forms is evident among Mexican Acanthaceae (Fig. 19.1). Floral visitors observed or reported include bees (e.g., *Bravaisia*), flies (e.g., *Carlowrightia*), bats (e.g., *Louteridium*), butterflies (e.g., *Pseuderanthemum*), hawkmoths (e.g., *Siphonoglossa*), and hummingbirds (e.g., *Anisacanthus*). Differences in floral form relating to different pollinators are commonly used for distinguishing Mexican Acanthaceae at the generic, e.g., *Anisacanthus* vs. *Carlowrightia* (Daniel, 1982b, 1983a), and subgeneric, e.g., sections *Tetramerium* and *Torreyella* of *Tetramerium* (Daniel, 1986a), levels. Knowledge of floral adaptations to particular types of pollinator is also useful for assessing similarities due to convergence. For example, *Anisacanthus abditus* T. Brandegee was described in *Anisacanthus* because of its large, red, funnelform corollas. Studies of this species (Daniel, 1986a) ultimately revealed it to possess all of the diagnostic characters of the genus *Tetramerium*. The similarity of floral form between this species and species of *Anisacanthus* was found to be the result of similar floral visitors (Daniel, 1986a) rather than systematic affinity.

CHROMOSOME NUMBERS

Chromosome numbers have been determined for more than 80 species of Mexican Acanthaceae (Daniel et al., 1984, 1990; Hilsenbeck, 1983). Fifteen haploid chromosome complements, varying from $n = 9$ in the naturalized species *Thunbergia alata* to $n = 40$ in five native species of *Dicliptera*, are known for Mexican taxa. This finding compares with 32 haploid complements, varying from $n = 8$ to $n = 68$, reported for the family worldwide. Both polyploidy and aneuploidy are evident among genera occurring in Mexico. The genera of Mexican Acanthaceae for which the chromosome number of more than one species is known can be divided into three groups: those with a single constant number in Mexico (*Anisacanthus, Carlowrightia, Dicliptera, Elytraria, Henrya, Mirandea, Ruellia,* and *Tetramerium*), those exhibiting polyploidy (i.e., euploidy) in Mexico (*Dyschoriste, Holographis*), and those with more than one number (other than euploids) in Mexico (*Justicia, Siphonoglossa*).

In a study of chromosome numbers of ten species in the largely Mexican genus *Siphonoglossa*, Hilsenbeck (1983) found a dichotomy in numbers that correlates with morphological and chemical differences among the species. Three of the four species with five calyx lobes comprising *Siphonoglossa* sect. *Pentaloba* have chromosome numbers based on $x = 14$ (i.e., $n = 14$, $n=28$), whereas counts of all species with four calyx lobes comprising the

---

Figure 19.1. Floral diversity of Mexican Acanthaceae. A. *Anisacanthus thurberi.* B. *Mirandea sylvatica.* C. *Dicliptera sciadephora.* D. *Carlowrightia arizonica.* E. *Henrya insularis.* F. *Elytraria macrophylla.* G. *Thunbergia alata.* H. *Ruellia* sp. I. *Louteridium parayi.* J. *Pseuderanthemum* sp. K. *P. cuspidatum.* L. *Chileranthemum* sp. ("pin"). M. *Chileranthemum* sp. ("thrum"). N. *Ruellia* sp.

nominate section are based on $x = 11$ (i.e., $n = 11$, $n = 22$). Based in part of this chromosomal evidence, Hilsenbeck (1990) transferred the section *Pentaloba* to *Justicia*, most of whose species in Mexico exhibit a meiotic complement of $n = 14$.

The different chromosome patterns of the two largest and most widely distributed genera of Acanthaceae are well represented in Mexico. Both *Justicia* ($n = 11, 12, 14, 28$ in Mexico) and *Ruellia* ($n = 17$ in Mexico) are polymorphic and have been divided into numerous segregate genera at one time or another. Daniel et al. (1984) speculated that the apparent constancy of chromosome numbers in diverse species of *Ruellia* may be reflective of a monophyletic, though morphologically diverse, taxon, whereas the heterogeneity of chromosome numbers prevalent in *Justicia* s.l. may be indicative of a polyphyletic assemblage.

Based on the numbers now known for Mexican Acanthaceae, chromosome number correlations with habit or distribution are not evident. Likewise, few correlations between chromosome numbers and suprageneric taxa are apparent. Two probable correlations were discussed by Daniel et al. (1984, 1990), however.

1. Aphelandreae in Mexico include three morphologically similar genera: *Aphelandra*, *Holographis*, and *Stenandrium*. *Aphelandra* appears to have a base number of $x = 14$, whereas all counts in *Holographis* and *Stenandrium* are multiples of 13. Daniel et al. (1984) have suggested that *Holographis* represents a specialized lineage of some part of *Aphelandra* or its ancestor that underwent an aneuploid reduction from 14 to 13. Available chromosome numbers in this tribe also suggest the possibility of a closer relationship between *Holographis* and *Stenandrium* (both $x = 13$) than between either genus and *Aphelandra*.

2. Ten genera of Mexican Justicieae subtribe Odontoneminae (*Anisacanthus*, *Aphanosperma*, *Carlowrightia*, *Chalarothyrsus*, *Gypsacanthus*, *Henrya*, *Mexacanthus*, *Mirandea*, *Tetramerium*, *Yeatesia*) have a chromosome number of $n = 18$. They differ from Bremekamp's (1965) definition of subtribe Odontoneminae by having an androecium of only two stamens (no staminodes). They probably represent an assemblage worthy of suprageneric recognition. *Dicliptera*, another Mexican member of subtribe Odontoneminae lacking staminodes, has a chromosome number of $n = 40$ in the New World, a pantropical distribution, and a suite of morphological features that clearly distinguish it from the assemblage of predominantly Mexican taxa with $n = 18$. Limited chromosome counts of Mexican *Pseuderanthemum* and *Odontonema*, both of which have two stamens and two staminodes and therefore more accurately fit Bremekamp's definition of Odontoneminae, have a meiotic complement of $n = 21$ (Daniel et al. 1990). Chromosome counts of Mexican Odontoneminae suggest minor changes or refinements in the composition of this subtribe.

FUTURE STUDIES

A significant body of data on pollination biology and chromosome numbers is accumulating for Mexican Acanthaceae. More chromosome numbers are now known for taxa occurring in Mexico than for taxa of Acanthaceae in any other country except India. Although chromosome numbers of Acanthaceae have been little studied in the past, their importance in assessing systematic relationships has become obvious. A taxonomic account of the family in Mexico is nearing completion. Additional studies of *Justicia* and its relatives are especially needed to resolve generic limits in the Justicieae. Knowledge of Acanthaceae in the three states (Guerrero, Oaxaca, and Veracruz) with perhaps the greatest number of species is still incomplete. Collections from each of these states continue to add taxa and distributional records to the known flora of Acanthaceae in Mexico. Monographic studies at the generic level are particularly desirable so accurate phytogeographic patterns and evolutionary relationships can be determined for Mexican Acanthaceae.

Acknowledgments

I am grateful for the assistance of Drs. T.P. Ramamoorthy, F. Chiang, A. Delgado S., and J. Rzedowski during my studies in Mexico. Financial support has been received from the U.S. National Science Foundation, the American Philosophical Society, The University of Michigan, Arizona State University, and the California Academy of Sciences.

APPENDIX

<u>Selected recent literature on Mexican Acanthaceae</u>

Anisacanthus
Daniel T.F. 1982. *Anisacanthus andersonii* (Acanthaceae), a new species from north-western Mexico. Bull. Torrey Bot. Club. 109:148–151.
———— & J. Henrickson. 1982. On the recognition of *Anisacanthus junceus* (Acanthaceae). Brittonia 34:177–180.
Henrickson, J. 1986. *Anisacanthus quadrifidus* sensu lato (Acanthaceae). Sida 11:286–299.
———— & E. Lott. 1988. New combinations in Chihuahuan Desert *Anisacanthus* (Acanthaceae). Brittonia 34:170–176.

Aphanosperma
Daniel, T.F. 1988. *Aphanosperma*, a new genus of Acanthaceae from Mexico with unusual diaspores. Amer. J. Bot. 75:545–550.

Aphelandra
Wasshausen, DC. 1975. The genus *Aphelandra* (Acanthaceae). Smithsonian Contr. Bot. 18:1–157.

Blechum
See Ramamoorthy (1988) under *Ruellia*.

Bravaisia
Daniel, T.F. 1988. A systematic study of *Bravaisia* (Acanthaceae). Proc. California Acad. Sci. 45:111–131.

Carlowrightia
Daniel, T.F. 1983. *Carlowrightia* (Acanthaceae). Flora Neotropica 34:1–116.
————. 1988. Taxonomic, nomenclatural, and reproductive notes on *Carlowrightia* (Acanthaceae). Brittonia 40:245–255.

Chaetothylax
Acosta C., S. 1989. *Chaetothylax rzedowskii* (Acanthaceae), una especie nueva de Chiapas, México. Acta Bot. Méx. 5:5–11.

Dyschoriste
Daniel, T.F. 1990. New and reconsidered Mexican Acanthaceae. IV. Proc. California Acad. Sci. 46:279–287.
Ramamoorthy, T.P. & D.C. Wasshausen. 1985. A new name in *Dyschoriste* (Acanthaceae). Brittonia 37:358–359.

Elytraria
Fryxell, P.A. & S.D. Koch. 1987. New or noteworthy species of flowering plants from the Sierra Madre Sur of Guerrero and Michoacán, Mexico. Aliso 11:539–561.

Gypsacanthus
Lott, E., V. Jaramillo & J. Rzedowski. 1986. Un género nuevo de la parte meridional de México: *Gypsacanthus* (Acanthaceae, Justicieae, Odontoneminae). Bol. Soc. Bot. México 46:29–35.

Appendix (cont.)

Henrya

Daniel, T.F. 1990. Systematics of *Henrya* (Acanthaceae). Contr. Univ. Michigan Herb. 17:99–131.

See Daniel (1984) under *Pseuderanthemum*.

Holographis

Daniel, T.F. 1983. Systematics of *Holographis* (Acanthaceae). J. Arnold Arb. 64:129–160
———. 1988. Three new species of *Holographis* (Acanthaceae) from Mexico. Proc. Calif. Acad. Sci. 46:73–81.

See Daniel (1984) under *Pseuderanthemum* and Daniel (1986) under *Mirandea*.

Justicia

Daniel, T.F. 1980. The genus *Justicia* (Acanthaceae) in the Chihuahuan Desert. Contr. Univ. Michigan Herb. 14:61–67.
———. 1990. New and reconsidered Mexican Acanthaceae. III. *Justicia*. Contr. Univ. Michigan Herb. 17:133–137.

Henrickson, J. & P. Hiriart. 1988. New species and transfers into *Justicia* (Acanthaceae). Aliso 12:45–58.

Hilsenbeck, R.A. 1990. Systematics of *Justicia* sect. *Pentaloba* (Acanthaceae). Pl. Syst. Evol. 169:219–235.

Wasshausen, D.C. 1981. New species of *Justicia* (Acanthaceae). Phytologia 49:65–68.
See Daniel (1990) under *Dyschoriste*.

Lophostachys

Acosta C., S. 1985. Algunas especies interesantes de la familia Acanthaceae en México. Phytologia 57:249–260.

Louteridium

Richardson, A. 1972. Revision of *Louteridium* (Acanthaceae). Tulane Stud. Zool. and Bot. 17:63–76.

See Daniel (1984) under *Pseuderanthemum*.

Mexacanthus

Daniel, T.F. 1982. *Mexacanthus*, a new genus of Acanthaceae from western Mexico. Syst. Bot. 6:288–293.

Mirandea

Daniel, T.F. 1982. The genus *Mirandea* (Acanthaceae). Contr. Univ. Michigan Herb. 15:171–175.
———. 1986. New and reconsidered Mexican Acanthaceae. II. Southw. Naturalist 31:169–175.

See Acosta C. (1985) under *Lophostachys*.

Odontonema

Baum, V. 1982. A revision of the genus *Odontonema* (Acanthaceae). M.S. Thesis, Univ. of Maryland. College Park.

See Daniel (1986) under *Mirandea*.

Pachystachys

Wasshausen, D.C. 1986. The systematics of the genus *Pachystachys* (Acanthaceae). Proc. Biol. Soc. Wash. 99:160–185.

Appendix (cont.)

Poikilacanthus
Ramamoorthy, T.P. 1989. *Poikilacanthus capitatus*: a new combination in Mexican Acanthaceae. Syst. Bot. 14:150–151.

Pseuderanthemum
Daniel, T.F. 1984. New and reconsidered Mexican Acanthaceae. Madroño 31:86–92.

Ruellia
Daniel, T.F. 1990. New, reconsidered, and little-known Mexican species of *Ruellia* (Acanthaceae). Contr. Univ. Michigan Herb. 17:139–162.
Ramamoorthy, T.P. 1988. A new species of *Ruellia* (Acanthaceae) from western Mexico. Ann. Missouri Bot. Gard. 75:1664–1665.
——— & Y. Hornelas U. 1988. A new name and a new species in Mexican *Ruellia* (Acanthaceae). Pl. Syst. Evol. 159:161–163.
——— & D.H. Lorence. 1987. Species vicariance in the Mexican flora and a description of a new species of *Salvia* (Lamiaceae). Bull. Mus. Nation. Hist. Nat., B, Adansonia 9:167–175.

Schaueria
Hilsenbeck, J. & D.L. Marshall. 1983. *Schaueria calycobractea* (Acanthaceae), a new species from Veracruz, Mexico. Brittonia 35:362–366.
See Daniel (1990) under *Dyschoriste*.

Siphonoglossa
Henrickson, J. & R.A. Hilsenbeck. 1979. New taxa and combinations in *Siphonoglossa* (Acanthaceae). Brittonia 31:373–378.
Hilsenbeck, R.A. 1983. Systematic studies of the genus *Siphonoglossa sensu lato* (Acanthaceae). Ph.D. dissertation, Univ. of Texas, Austin.
———. 1989a. Generic affinities and typification of eleven species excluded from *Siphonoglossa* Oerst. (Acanthaceae). Phytologia 67:227–234.
———. 1989b. A new species of *Siphonoglossa* (Acanthaceae) and some infrageneric transfers. Madroño 36:198–207.
———. 1990. Pollen morphology and systematics of *Siphonoglossa sensu lato* (Acanthaceae). Amer. J. Bot. 77:27–40.

Stenandrium
Daniel, T.F. 1985. A revision of *Stenandrium* (Acanthaceae) in Mexico and adjacent regions. Ann. Missouri Bot. Gard. 71:1028–1043.

Tetramerium
Daniel, T.F. 1986. Systematics of *Tetramerium* (Acanthaceae). Syst. Bot. Monogr. 12:1–134.

Tribliocalyx
Gibson, D.N. 1974. *Tribliocalyx*. Fieldiana, Bot. 24(10):459–461.

Yeatesia
Hilsenbeck, R.A. 1989. Taxonomy of *Yeatesia* (Acanthaceae). Syst. Bot. 14:427–438.

REFERENCES

Barker, R. 1986. Revision of Australian Acanthaceae. J. Adelaide Bot. Gard. 9:1–292.

Bremekamp, C.E.B. 1965. Delimitation and subdivision of the Acanthaceae. Bull. Bot. Surv. India 7:21–30.

Cronquist, A. 1981. An Integrated System of Classification of Flowering Plants. New York, NY: Columbia Univ. Press.

———. 1988. The Evolution and Classification of Flowering Plants. 2nd. ed. Bronx, NY: New York Botanical Garden.

Daniel, T.F. 1981. *Mexacanthus*, a new genus of Acanthaceae from western Mexico. Syst. Bot. 6:288–293.

———. 1982a. The genus *Mirandea* (Acanthaceae). Contr. Univ. Michigan Herb. 15:171–175.

———. 1982b. Contrasting reproductive strategies in two desert Acanthaceae. Abstracts, Botanical Society of America, Misc. Ser. Pub. No. 162:90.

———. 1983a. *Carlowrightia* (Acanthaceae). Flora Neotropica 34:1–116.

———. 1983b. Systematics of *Holographis* (Acanthaceae). J. Arnold Arb. 64:129–160.

———. 1985a. A revision of *Stenandrium* (Acanthaceae) in Mexico and adjacent regions. Ann. Missouri Bot. Gard. 71:1028–1043.

———. 1985b. Taxonomy and phytogeography of the Chihuahuan Desert Acanthaceae. Amer. J. Bot. 72:948–949.

———. 1986a. Systematics of *Tetramerium* (Acanthaceae). Syst. Bot. Monogr. 12:1–134.

———. 1986b. Acanthaceae. *In* D. Breedlove (ed.), Flora de Chiapas pp. 27–30. Listados Floristicos de México IV. Mexico City: UNAM. Inst. Biol.

———. 1988a. *Aphanosperma*, a new genus of Acanthaceae from Mexico with unusual diaspores. Amer. J. Bot. 75:543–548.

———. 1988b. A systematic study of *Bravaisia* (Acanthaceae). Proc. California Acad. Sci. 45:111–131.

———. 1990. Systematics of *Henrya* (Acanthaceae). Contr. Univ. Michigan Herb. 17:99–131.

———, B.D. Parfitt & M.A. Baker. 1984. Chromosome numbers and their systematic implications in some North American Acanthaceae. Syst. Bot. 9:346–355.

———, T.I. Chuang, & M.A. Baker. 1990. Chromosome numbers of American Acanthaceae. Syst. Bot. 15:13–25.

Durkee, L.H. 1988. A checklist of Acanthaceae in Costa Rica, Nicaragua, and Panama. Acanthus 3:3–4.

Gentry, A. 1982. Phytogeographic patterns as evidence for a Chocó refuge. *In* G. Prance (ed.), Biological Diversification in the Tropics. New York, NY: Columbia Univ. Press.

Gibson, D.N. 1974. Acanthaceae. *In* P. Standley & L. Williams (eds.), Flora of Guatemala. Fieldiana, Bot. 24(10):328–461.

Hilsenbeck, R. 1983. Systematic studies of the genus *Siphonoglossa sensu lato* (Acanthaceae). Ph.D. dissertation, Univ. of Texas, Austin.

———. 1990. Systematics of *Justicia* sect. *Pentaloba* (Acanthaceae). Pl. Syst. Evol. 169:219–235.

Leonard, E.C. 1951–58. The Acanthaceae of Colombia, I–III. Contr. U.S. Natl. Herb. 31:1–781.

Leonard, E.C. & I. Wiggins. 1964. Acanthaceae. *In* F. Shreve & I. Wiggins (eds.), Vegetation and Flora of the Sonoran Desert. Vol. 2. Stanford, CA: Stanford Univ. Press. pp. 1375–1392.

Lindau, G. 1895. Acanthaceae. *In* A. Engler & K. Prantl (eds.), Die Natürlichen Pflanzenfamilien. Vol. IV (3b). Leipzig: Wilhem Engelmann. pp. 274–354.

Long, R.W. 1970. The genera of Acanthaceae in the southeastern United States. J. Arnold Arb. 51:257–309.

Lott, E.J., J.A.A. Malfavon & T.P. Ramamoorthy. 1985. Acanthaceae. *In* J. Rzedowski & G. Rzedowski (eds.), Flora Fanerogámica del Valle de México. Vol. 2. Mexico City: Inst. Pol. Nac./Inst. Ecol. pp. 376–384.

Mabberley, D.J. 1987. The Plant-Book. Cambridge: Cambridge Univ. Press.

Muller, J. 1970. Palynological evidence on early differentiation of angiosperms. Biol. Rev. 45:417–450.

Nees von Esenbeck, C.G. 1847. Acanthaceae. *In* A.P. de Candolle (ed.), Prodromus Systematis Naturalis Regni Vegetabilis. Vol. 11. Paris: Treutell & Wurtz. pp. 46–519.

Raven, P.H. & D.I. Axelrod, 1974. Angiosperm biogeography and past continental movements. Ann. Missouri Bot. Gard. 61:539–673.

Rzedowski, J. 1978. Vegetación de México. Mexico City: Limusa.

Sreemadhavan, C.P. 1977. Diagnosis of some new taxa and some new combinations in Bignoniales. Phytologia 37:412–416.

Standley, P.C. 1926. Trees and shrubs of Mexico (Bignoniaceae—Asteraceae). Contr. U.S. Natl. Herb. 23:1313–1721.

Stearn, W.T. 1971. A survey of the tropical genera *Oplonia* and *Psilanthele* (Acanthaceae). Bull. British Mus. (Nat. Hist.) Bot. 4:261–323.

Van Steenis, C.G. 1972. *Nothofagus*, key genus to plant geography. *In* D.H. Valentine (ed.), Taxonomy, Phytogeography and Evolution. London: Academic Press.

Wiggins, I. 1980. Flora of Baja California. Stanford, CA: Stanford Univ. Press.

20

Biogeography, Diversity, and Endangered or Threatened Status of Mexican Asteraceae

BILLIE L. TURNER AND GUY L. NESOM

Mexico is estimated to contain over 2,700 species of Asteraceae distributed among 323 genera. Of these species, about 2,600 species are believed to be native. The species are categorized as wide-ranging (922 species), regional (953 species), or local (848 species) in distribution. From among the latter it is calculated that 663 species and 23 genera are endangered or threatened; representing about 24% of the species and 7% of the genera. As regards species diversity, the tribes Heliantheae and Eupatorieae are especially well developed in Mexico, followed in order by the Astereae, Helenieae, and Senecioneae. South-central Mexico—mainly the Sierra Madre del Sur (with about 386 endemic species) and the Trans-Volcanic Belt (with about 370 endemic species)—is the primary region of diversity of this prolific family. With the exception of the tribe Astereae, which appears to have strong affinities with temperate North America, most of the genera in the major tribes in Mexico appear to have developed autochthonously. Most of the tribe Helenieae is believed to be a natural phyletic grouping that has largely developed in the desert regions of northwestern Mexico, much as suggested by Rzedowski in his earlier biogeographic studies. Based on the number of species and genera of the Asteraceae native to North America, it is estimated that the number of described species and genera of this family for the world is approximately 25,500 and 2,230, respectively. It is suggested that yet undescribed species and genera will bring these totals to about 28,000 and 2,500. Mexico, with less than 1% of the world's vegetated land surface, is estimated to contain over 10% of the world's species. Along with South America and Africa, it must rank as one of the centers of diversity for the family.

The Asteraceae, by many accounts the largest family of flowering plants, reportedly contains approximately 32,000 described species (i.e., formal names) distributed among approximately 1,500 genera (Hendrych, 1985). Despite this apparent diversity, the family is remarkably uniform in terms of its floral morphology: All of the species are uniformly gamopetalous, synantherous, epigynous, two-carpellate with a single ovule in an unilocular ovary, the stylar shaft having two, well-defined branches. Indeed, the

Table 20.1. Tabulation of the number of taxa of the Asteraceae (arranged by tribes) largely restricted to one of the 11 provinces

Province[a]	No. of taxa, by Tribe[b]													Total
	1	2	3	4	5	6	7	8	9	10	11	12	13	Total
BCP	1	17	42	7	67	34	1	6	—	—	4	4	4	187
NPS	2	21	9	1	22	35	—	0	—	—	3	5	1	99
SMC	4	38	48	1	123	37	—	31	—	—	6	6	6	300
CCPR	4	19	43	1	48	53	—	5	—	—	4	5	5	187
SMO	3	48	27	1	29	14	—	21	—	—	5	3	—	151
GCP	3	22	23	1	37	5	—	5	—	—	—	1	—	97
CPM	1	24	6	—	46	12	1	9	—	—	—	—	1	100
TVB	19	64	24	14	150	16	1	56	—	—	14	16	3	377
SMS	17	80	16	1	179	34	—	24	—	—	2	3	—	356
SMCH	10	57	10	—	48	4	—	11	—	—	2	2	1	145
YPM	3	5	2	1	15	—	—	0	—	—	—	1	—	27
Total	67	395	250	28	764	244	3	168	0	0	40	46	21	2,026

[a]**BCP**, Baja California Peninsula Morphotectonic Province (MP); **NPS,** Northwestern Plains and Sierras MP; **SMC,** Sierra Madre Occidental MP; **CCPR,** Chihuahua-Cohuila Plateaus and Ranges MP; **SMO,** Sierra Madre Oriental MP; **GCP,** Gulf Coast Plain MP; **CPM,** Central Plateau MP; **TVB,** Trans-Mexican Volcanic Belt MP; **SMS,** Sierra Madre del Sur MP; **SMCH,** Sierra Madre de Chiapas MP; **YPM,** Yucatan Platform MP. These MPs are recognized by Ferrusquia-Villafranca (this volume) for Mexico.
[b]Widely distributed and introduced taxa are not tabulated (cf. Appendix). 1. Vernonieae (including Liabeae). 2. Eupatorieae. 3. Astereae. 4. Inuleae. 5. Heliantheae. 6. Helenieae (including Tageteae). 7. Anthemideae. 8. Senecioneae. 9. Arctoteae. 10. Calenduleae. 11. Cardueae. 12. Mutiseae. 13. Lactuceae.

family is so homogeneous in its floral morphology that there is not a single species known which might link the Asteraceae to yet another family. In fact, nearly all of the species are readily accommodated within the 12–18 tribes that have been conventionally proposed for the family. Therefore there has been considerably controversy as to the age and origin of the group (Bremer, 1987; Turner, 1977a).

The Asteraceae is distributed throughout the world; and except for the lowland tropical regions, it usually accounts for 10–15% of the species in any large floristic or biogeographic region. It is especially diverse in mountainous areas of the tropics, subtropics, and warmer temperate regions. Thus it is not surprising to find that Mexico, with less than 1% of the world's vegetated terrestrial regions, houses some 23% of the family's genera and about 12% of its species (Bentham, 1873). Rzedowski (1972), citing Reko (1946), estimated that about 13% of the Mexican genera of flowering plants belong to the Asteraceae. The family probably also contains about 15% of its flowering species, assuming that the native flora of Mexico is approximately 21,600 species (Rzedowski, this volume).

MEXICAN ASTERACEAE

As indicated in Table 20.1, we counted (as of September 1990) slightly more than 2,700 species of Asteraceae in Mexico; only 80 of them represent introduced weeds or escaped, cultivated taxa. Most of the genera are strongly centered in Mexico; indeed, we calculated that 40 of the genera and 1,707 species are *confined* to Mexico. No doubt these figures are underestimates, for new genera and species in this family are routinely described from regions newly opened to automobile travel. This considerable diversity is undoubtedly due to its mostly subtropical climates and well-developed mountainous habitats of long standing with their adjacent desert regions. For example, China, with over five times the area of Mexico, reportedly has only 2,027 asteroid species in 167 genera (Hu, 1958). This figures includes a large assortment of introduced species, many of which are weeds from elsewhere (e.g., the tribe Heliantheae with 24 genera and 46 species, most of them introduced). Mexico, in contrast, contains about 2,700 species of Asteraceae in considerably more than 300 genera, relatively few of which are introduced.

As noted above, Mexico contains at least 323 genera of Asteraceae. This number compares favorably with the estimates of Rzedowski (1978), who counted 297 native genera, and to those of Cabrera-Rodriguez and Villaseñor (1987), who accounted for 340 genera. Although Rzedowski did not estimate the number of species of the family occurring in Mexico, he did call to the fore those regions of Mexico where the Asteraceae seemed best represented, as judged by the percent of asteraceous genera occurring within a given broad region. Thus he showed that the generic diversity of the family (i.e., percent of genera of the floras concerned) is seemingly highest in those regions with montane and desert habitats, ranging from a high of nearly 20% in the central plateau regions of northern Mexico to a low of 4.4% in the lowland regions of Gulf Coastal Mexico (near Tuxtepec, Oaxaca).

Our summaries in Table 20.1 are based upon the tabulations in the Appendix. These figures, in turn, are based upon evaluations of these taxa by the present authors in their preparation of a systematic treatment of the Asteraceae for Mexico.

In our systematic studies (Appendix), we have attempted to ascertain the geographical distribution and likely "endangered" status of the various taxa occurring in Mexico. We have categorized the various species as either wide-ranging, regional, or local in distribution. Thus of the 2,723 species estimated to be native to Mexico, about 1,707 are believed to be restricted to Mexico; of them, 922 species are categorized as wide-ranging (i.e., occurs over a broad region of Mexico, usually in five or more of the larger states); 953 are regional in distribution (occurring in relatively few states, usually one to five, although within its region perhaps rather abundant or common); and 848 are thought to be local in distribution (known from only one or a few localities in a restricted region of Mexico, although they may be, on occasion, somewhat common locally). From among the latter we have calculated the number of species we consider to be likely threatened or endangered. Our estimate of 663 "endangered or threatened" is surely an underestimate, as many of the species listed as regionally distributed might represent relatively small, spotty populations that are precarious. At least many of the records upon which our listings are founded often represent older collections by important collectors near the turn of the century, most notably Palmer, Pringle, and others of their ilk.

Davis et al. (1986) have estimated that of the world's 250,000 species of vascular plants 2% are endangered and perhaps 10–20% are rare or threatened. With regard to the Asteraceae of Mexico, this figure seems to be an underestimate, as we calculated that 31% have ranges localized enough to be considered threatened, and more than two-thirds of these species are clearly endangered.

BIOGEOGRAPHY AND EVOLUTION OF MEXICAN ASTERACEAE

The family Asteraceae is believed to have had its ancestral origin in the more montane regions of northern South America, as was first suggested by Bentham (1873). This view has also been ascribed to by Turner (1977a) and is supported by comparative studies on chloroplast DNA (Jansen & Palmer, 1988). Nevertheless, as regards the numbers of species and genera and tribal development, the montane regions of Mexico also figure as one of the major centers of diversity of the family.

Rzedowski (1972) noted that among the Mexican Asteraceae the tribe Heliantheae is richest in number of genera and species, followed by the tribes Helenieae, Eupatorieae, and Astereae. Actually, as regards the number of species, the tribes Eupatorieae and Astereae are more richly developed in Mexico than is the Helenieae (Appendix). In terms of generic diversity, however, the Helenieae is more richly developed than the Eupatorieae, at least as conceived by us. The tribe Heliantheae, with about 914 native species in approximately 112 genera, is far and away the largest and most diverse tribe to be found in Mexico. It is followed by the Eupatorieae with about 483 species in 27 genera, the Astereae with about 387 species in 40 genera, the Helenieae with 222 species in 43 genera, and the tribe Senecioneae with about 200 species in 13 genera.

Those genera with the largest number of species are also found in the larger tribes: *Verbesina* (Heliantheae) 119 species, *Ageratina* (Eupatorieae) 113 species, *Erigeron* (Asteraceae) 90 species, *Senecio sensu lato* (Senecioneae) 154 species.

The concentration of tribal groupings, as determined from species number and diversity, within the Asteraceae on given continents (e.g., Heliantheae and Eupatorieae to mostly tropical and subtropical regions of North and South America; Mutiseae to mostly South America; Inuleae to mostly Eurasia, Africa, and Australia) prompted Turner (1977a) to suggest that continental drift was important in the distributional patterns concerned, although this idea is countered by the arguments of Raven and Axelrod (1974), who viewed the family as of relatively more recent origin, largely because it appears to lack a fossil record much below the Miocene-Oligocene boundary. In any case, the Mexican Asteraceae, as regards species diversity, is clearly dominated by the tribe Heliantheae and this tribe must have had an early development in Mexico, perhaps as a largely montane element within the Madro-Tertiary flora (Axelrod, 1958) with secondary differentiation (as the tribe Helenieae) in the desert regions of northern Mexico. Indeed, Rzedowski (1972) thought the tribe Helenieae to be a monophyletic element, largely because of its restriction to such regions. However, the tribe Helenieae as classically delimited is believed by many workers to be polyphyletic (Robinson, 1981; Turner &Powell, 1977); in short, the Helenieae has been thought to have arisen from a number of independent phyletic lines within the Heliantheae, including perhaps a few from near, or in, yet other tribes. Chloroplast DNA restriction site studies (K.-J. Kim et al., in prep.), however, suggest that Rzedowski's surmise may be correct, and we have therefore retained this tribe in the present account. The tribes Helenieae and Astereae have their centers of diversity in the more temperate regions of northern Mexico. The greatest portion of Mexican taxa, however, particularly in the large tribes Eupatorieae, Heliantheae, and Senecioneae, appear to have developed largely *in situ* within the Sierra Madre del Sur and Trans-Mexican Neovolcanic Belt of south-central Mexico (Table 20.1; Fig. 20.1). The Sierra Madre Occidental also has been a major center of evolutionary radiation of these tribes as well as of the whole family. The Isthmus of Tehuantepec appears to form a rather sharp division between the rich diversity of the Asteaceae to the north and the relatively poor diversity to the south. The Chiapan region (Fig. 20.1) does not have an especially rich asteraceous flora, and most of the taxa there have phyletic ties to Central and South America.

ASTERACEAE ON A WORLD BASIS: AN ESTIMATE

Apart from the Mexican species, it is interesting to speculate as to the number of asteraceous genera and species to be found in all of North America. To this end we have drawn upon the data of Kartesz and Bell (1980) who estimated that about 2,690 species of Asteraceae in approximately 346 genera occur in the mainland United States and Canada. Because there is a natural, or fairly sharp, break in the species that range into the United States from Mexico, and vice versa, we can calculate, after making allowances for those weedy species common to both and those that partially extend into one or the other's boundaries, that approximately 5,000 species are involved; if we add an estimated additional 500 species from Central America (Guatemala to Panama), the figure for North America approximates 5,500 species in perhaps 500 genera.

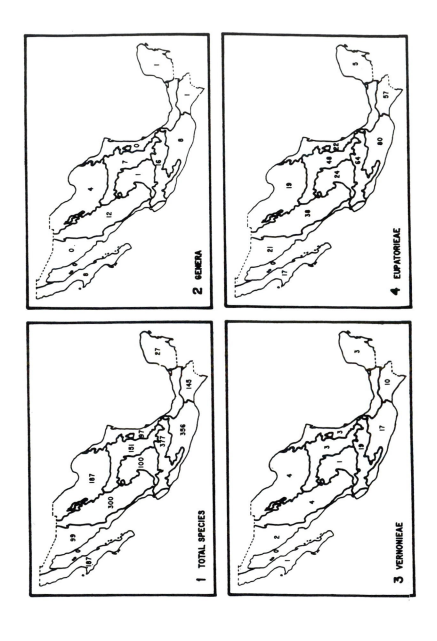

1 TOTAL SPECIES

2 GENERA

3 VERNONIEAE

4 EUPATORIEAE

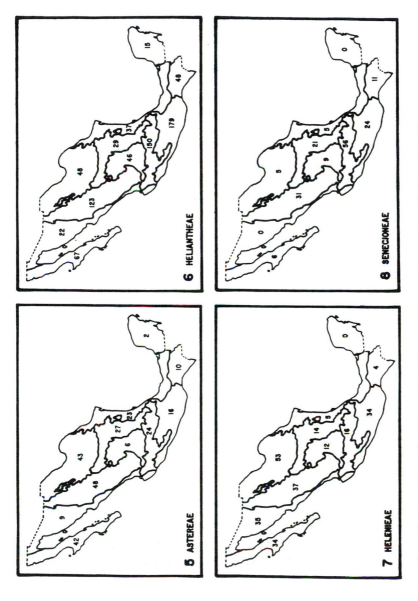

Figure 20.1. 1. Number of species largely confined to each of the morphotectonic provinces indicated in **Figure 1.1.** 2. Number of genera endemic to each of the morphotectonic provinces. 3-8. Numbers of species of the six largest tribes of Asteraceae as related to those provinces in which they principally occur (wide-ranging and introduced species not included).

Table 20.2. Currently recognized species of Asteraceae in the major geographic regions of the world

Continent	No. of species	No. of genera
North America	5,500	500
South America	6,000	430
Africa (including Madagascar)	6,500	600
Eurasia	3,500	300
Australia	3,500	300
Misc. islands, etc.	500	100
Total	25,500	2,230

For the world as a whole we can project similar figures from our knowledge of the North American Asteraceae, assuming that similar diversities hold for yet other continents, albeit among different tribal elements, as seems to be the case (Bentham, 1873; Hu, 1958; Komarov, 1959–64). Our calculations of the currently recognized species of Asteraceae in the major geographic regions of the world are shown in Table 20.2.

The above figures are surely underestimates, for new species and genera of Asteraceae are routinely described throughout the world, especially in poorly explored tropical montane regions. For example, Gibbs Russell et al. (1987) estimated that approximately 2,260 species of Asteraceae occur in South Africa alone; the two largest genera, *Senecio* and *Helichrysum*, possess 310 and 245 species, respectively. Thus our estimate for African taxa in the above accounting may be on the low side. We venture that the family will ultimately be found to possess about 28,000 species (based on a 10% increase) distributed among 2,500 genera (*if* current trends in generic sculpturing and splintering continue).

Mexico itself will almost certainly be found to house approximately 3,000 of these species, or some 10% of the world's diversity in the family. Especially distressing is the realization that, of the new species that are likely to be discovered, nearly all will represent endangered or threatened taxa. Furthermore, we estimate that 600–800 species, or about one-fourth of the Mexican Asteraceae, will be eliminated in the near future (within 50 years) if rampant population growth with its concomitant agronomic pressure continues, as we suspect it will.

APPENDIX

Asteraceae of Mexico

Genus	Species[a]								
	1	2	3	4	5	6	7	8	9
Achillea	1	0	1	0	1	0	0	0	N
Achyrachaena	1	0	1	0	1	0	0	0	N
Achyrocline	4	0	4	2	2	0	2	2	N
*Achyropappus**	1	0	1	1	0	1	0	0	N
Acmella	4	0	4	1	3	0	0	0	N
Acourtia	56	0	56	50	11	22	23	20	N
Adenocaulon	1	0	1	0	0	1	0	0	N
*Adenopappus**	1	0	1	1	0	1	0	0	N
*Adenothamnus**	1	0	1	1	0	0	1	0	N
Ageratella	1	0	1	1	0	1	0	0	N
Ageratina	113	1	112	94	30	43	40	30	N
Ageratum	12	0	12	8	3	6	3	2	N
Agoseris	1	1	0	0	1	0	0	0	N
Aldama	1	0	1	0	1	0	0	0	N
Alloispermum	9	0	9	8	4	3	2	2	N
Alomia	4	0	4	4	0	1	3	3	N
*Alvordia**	4	0	4	4	0	1	3	3	N
*Amauria**	3	0	3	3	0	2	1	1	N
Amblyolepis	1	0	1	0	0	1	0	0	N
Amblyopappus	1	0	1	0	1	0	0	0	N
Ambrosia	24	0	24	13	5	5	3	2	N
Anaphalis	1	1	0	0	1	0	0	0	N
Ancistrocarphus	1	0	1	0	1	0	0	0	N
Anisocoma	1	0	1	0	1	0	0	0	N
Antennaria	2	0	2	0	2	0	0	0	N
Anthemis	1	0	1	0	1	0	0	0	N
Aphanostephus	3	0	3	0	3	0	0	0	N
Archibaccharis	25	0	25	14	7	10	8	5	N
*Arnicastrum**	2	0	2	2	0	0	2	2	N
Artemisia	8	1	7	2	6	2	0	0	N
Aster	20	2	18	7	15	4	1	0	N
Astranthium	10	0	10	9	3	6	1	1	N
Atrichoseris	1	0	1	0	1	0	0	0	N
*Axiniphyllum**	5	0	5	5	0	1	4	4	N
Baccharis	38	0	38	15	28	8	2	2	N
*Baeriopsis**	1	0	1	1	0	0	1	1	Y
Bahia	6	0	6	2	3	3	0	0	N
Baileya	3	0	3	0	0	3	0	0	N
Baltimora	2	1	1	0	2	0	0	0	N
Bartlettia	1	0	1	0	0	1	0	0	N
Bebbia	2	0	2	1	1	1	0	0	N
Bellis	1	1	0	0	1	0	0	0	N
Berlandiera	1	0	1	0	1	0	0	0	N

Appendix (cont.)

Genus	Species								
	1	2	3	4	5	6	7	8	9
Bidens	56	2	54	30	30	15	11	11	N
Blumea	1	1	0	0	1	0	0	0	N
*Boeberoides**	1	0	1	1	0	1	0	0	N
Borrichia	2	1	1	0	2	0	0	0	N
Brachyactis	1	0	1	0	1	0	0	0	N
Brickellia	76	0	76	60	15	40	21	19	N
Calea	7	0	7	3	3	1	3	3	N
Calycadenia	1	0	1	0	0	1	0	0	N
Calycoseris	1	0	1	0	1	0	0	0	N
Calyptocarpus	2	1	1	0	2	0	0	0	N
Carduus	1	1	0	0	1	0	0	0	N
Carminatia	3	0	3	1	3	1	0	0	N
Carphochaete	7	0	7	6	2	2	3	2	N
Carthamus	1	1	0	0	1	0	0	0	N
Centaurea	7	5	2	0	7	0	0	0	N
Chaenactis	8	0	8	1	1	7	0	0	N
Chaetopappa	9	0	9	3	4	3	2	2	N
*Chaetymenia**	1	0	1	1	0	1	0	0	N
Chamomilla	2	2	0	0	2	0	0	0	N
Chaptalia	11	0	11	6	5	4	2	1	N
*Chromolepis**	1	0	1	1	0	1	0	1	Y
Chrysactinia	5	0	5	4	1	3	1	1	N
Chrysanthellum	9	1	8	8	2	1	6	6	N
Chrysanthemum	3	3	0	0	3	0	0	0	N
Chrysothamnus	2	0	2	0	2	0	0	0	N
Cichorium	1	1	0	0	1	0	0	0	N
Cirsium	54	1	53	33	19	19	16	13	N
Clappia	1	0	1	0	0	1	0	0	N
Clibadium	1	0	1	0	1	0	0	0	N
Conyza	6	0	6	0	6	0	0	0	N
Coreocarpus	12	0	12	11	0	6	6	4	N
Coreopsis	17	0	17	14	3	4	10	6	N
Corethrogyne	1	0	1	0	1	0	0	0	N
*Correllia**	1	0	1	1	0	0	1	1	Y
Cosmos	28	0	28	20	6	5	17	7	N
Cotula	2	1	1	1	1	1	0	0	N
*Coulterella**	1	0	1	1	0	0	1	1	Y
Crepis	1	0	1	0	1	0	0	0	N
Croptilon	1	0	1	0	0	1	0	0	N
Cymophora	2	0	2	2	1	1	0	0	N
Cynara	1	1	0	0	1	0	0	0	N
Dahlia	25	0	25	22	5	2	18	5	N
Decachaeta	7	0	7	6	5	1	1	1	N
Delilia	1	0	1	0	1	0	0	0	N
Desmanthodium	6	0	6	5	2	3	1	1	N
Dichaetophora	1	0	1	0	0	1	0	0	N

Appendix (cont.)

Genus	Species								
	1	2	3	4	5	6	7	8	9
Dicoria	2	0	2	1	1	0	1	1	N
Dicranocarpus	1	0	1	0	1	0	0	0	N
*Digitacalia**	5	0	5	5	1	1	3	3	N
Dugaldia	2	0	2	1	0	1	1	1	Y
*Dugesia**	1	0	1	1	0	1	0	0	N
*Dyscritothamnus**	2	0	2	2	0	1	1	1	N
Dyssodia	28	0	28	15	12	11	5	4	N
Eclipta	1	1	0	0	1	0	0	0	N
Egletes	2	0	2	0	2	0	0	0	N
Elephantopus	4	0	4	0	4	0	0	0	N
Emilia	2	2	0	0	2	0	0	0	N
Encelia	12	0	12	7	4	5	3	3	N
Engelmannia	1	1	0	0	1	0	0	0	N
Enhydra	1	1	0	0	1	0	0	0	N
Epaltes	1	0	1	0	1	0	0	0	N
Erechtites	2	2	0	0	2	0	0	0	N
Ericameria	10	0	10	2	4	4	2	2	N
Erigeron	90	2	88	51	21	28	41	30	N
Eriophyllum	6	0	6	2	0	4	2	1	N
*Eryngiophyllum**	2	0	2	2	0	0	2	2	Y
Espejoa	1	0	1	0	0	1	0	0	N
Eupatoriastrum	5	0	5	3	1	3	1	1	N
Eupatorium									
Eup-Bartlettina	15	0	15	8	2	7	6	5	N
Eup-Campuloclinium	1	1	0	0	1	0	0	0	N
Eup-Chromolaena	19	0	19	12	9	6	4	4	N
Eup-Conoclinium	2	0	2	0	2	0	0	0	N
Eup-Critonia	15	0	15	3	10	2	3	1	N
Eup-Fleischmanniopsis	2	0	2	1	1	0	1	1	N
Eup-Fleischmannia	20	0	20	6	12	5	3	1	N
Eup-Hebeclinium	1	0	1	0	1	0	0	0	N
Eup-Koanophyllon	15	0	15	11	5	3	7	3	N
Eup-Kyrsteniopsis	9	0	9	5	0	6	4	2	N
Eup-Neomirandea	2	0	2	0	1	1	0	0	N
Eup-Peteravenia	3	0	3	1	1	2	0	0	N
*Euphrosyne**	1	0	1	1	0	1	0	0	N
*Eutetras**	2	0	2	2	0	1	1	0	N
Euthamia	1	0	1	0	1	0	0	0	N
Evax	1	0	1	0	1	0	0	0	N
*Faxonia**	1	0	1	1	0	0	1	1	Y
Filago	4	1	3	0	2	2	0	0	N
Flaveria	14	1	13	12	2	6	6	4	N
Florestina	8	0	8	6	2	4	2	2	N
Flourensia	13	0	13	11	1	4	8	6	N
Flyriella	4	0	4	3	1	0	3	1	N
Gaillardia	11	1	10	7	3	3	5	4	N

Appendix (cont.)

Genus	Species								
	1	2	3	4	5	6	7	8	9
Galeana	1	0	1	0	1	0	0	0	N
Galinsoga	7	0	7	5	2	2	3	3	N
Gamochaeta	8	1	7	2	5	2	1	1	N
Gazania	1	1	0	0	1	0	0	0	N
*Geissolepis**	1	0	1	1	0	0	1	1	Y
Geraea	2	0	2	0	2	0	0	0	N
Gnaphaliothamnus	9	0	9	9	1	0	8	6	N
Gnaphalium	45	4	41	30	26	12	7	2	N
Gochnatia	6	0	6	5	1	3	2	0	N
Goldmanella	1	0	1	0	0	1	0	0	N
*Greenmaniella**	1	0	1	1	0	0	1	1	Y
Grindelia	24	1	23	16	6	11	7	4	N
Guardiola	9	0	9	8	2	4	3	2	N
Gutierrezia	13	0	13	5	8	3	2	0	N
Gymnocoronis	1	0	1	0	1	0	0	0	N
*Gymnolaena**	3	0	3	3	0	1	2	2	N
Gy...nosperma	1	0	1	0	1	0	0	0	N
Haploesthes	3	0	3	2	0	1	2	2	N
Hazardia	10	0	10	6	0	2	8	4	N
Helenium	11	0	11	8	6	4	1	1	N
Helianthella	4	0	4	3	1	0	3	1	N
Helianthus	9	0	9	1	7	2	0	0	N
Heliopsis	8	0	8	7	3	3	2	2	N
Hemizonia	12	0	12	1	9	2	1	0	N
*Henricksonia**	1	0	1	1	0	0	1	1	Y
Heterosperma	2	0	2	1	1	1	0	0	N
Heterotheca	18	0	18	8	7	6	5	2	N
Hidalgoa	3	0	3	2	2	0	1	1	N
Hieracium	15	0	15	10	7	5	3	3	N
*Hofmeisteria**	7	0	7	7	2	1	4	0	N
Hulsea	1	0	1	2	0	1	2	2	N
*Hybridella**	1	0	1	0	1	0	0	0	N
*Hydrodyssodia**	1	0	1	1	0	0	1	1	Y
*Hydropectis**	1	0	1	1	0	0	1	1	Y
Hymenoclea	3	0	3	2	2	0	1	1	N
Hymenopappus	6	0	6	2	3	1	2	1	N
Hymenothrix	4	0	4	2	0	3	1	1	N
Hymenoxys	6	0	6	2	2	2	2	1	N
Hypochaeris	1	1	0	0	1	0	0	0	N
*Iostephane**	4	0	4	4	2	1	1	1	N
Isocarpha	2	1	1	0	2	0	0	0	N
Isocoma	7	0	7	3	4	3	0	0	N
Iva	6	1	5	1	5	0	1	1	N
Jaegeria	7	0	7	6	2	4	1	1	N
*Jaliscoa**	3	0	3	3	0	2	1	1	N
Jaumea	1	0	1	0	1	0	0	0	N

Appendix (cont.)

Genus	Species								
	1	2	3	4	5	6	7	8	9
Jungia	2	0	2	1	0	1	1	0	N
Krigia	1	0	1	0	1	0	0	0	N
Lactuca	4	2	2	0	4	0	0	0	N
Laennecia	10	0	10	0	6	2	2	1	N
Lagascea	8	0	8	5	3	5	0	0	N
Lapsana	1	1	0	0	1	0	0	0	N
Lasianthaea	19	0	19	15	5	7	7	6	N
Lasthenia	4	0	4	0	0	4	0	0	N
Layia	3	0	3	1	2	1	0	0	N
Leibnitzia	2	0	2	1	1	0	1	0	N
Lepidospartum	1	0	1	0	1	0	0	0	N
Lessingia	1	0	1	0	0	1	0	0	N
*Leucactinea**	1	0	1	1	0	1	0	0	N
Liabum	1	0	1	0	1	0	0	0	N
Liatris	1	0	1	0	1	0	0	0	N
Lindheimera	1	0	1	0	1	0	0	0	N
*Loxothysanus**	2	0	2	2	0	0	2	1	N
Lygodesmia	2	0	2	0	1	1	0	0	N
Machaeranthera	31	0	31	14	12	10	9	4	N
Macvaughiella	1	0	1	0	0	1	0	0	N
Madia	4	1	3	1	4	0	0	0	N
Malacothrix	4	0	4	3	0	2	2	0	N
Malperia	1	0	1	0	0	1	0	0	N
*Marshalljohnstonia**	1	0	1	1	0	0	1	1	Y
Melampodium	36	0	36	24	15	14	7	5	N
Melanthera	3	1	2	0	3	0	0	0	N
*Mexerion**	2	0	2	2	0	1	1	1	N
*Mexianthus**	1	0	1	1	0	0	1	1	Y
Micropus	1	0	1	0	1	0	0	0	N
Microseris	4	0	4	0	3	1	0	0	N
Microspermum	7	0	7	7	2	0	5	5	N
Mikania	16	0	16	3	9	4	3	3	N
Milleria	1	0	1	0	1	0	0	0	N
Monoptilon	2	0	2	0	2	0	0	0	N
Montanoa	20	0	20	13	8	11	1	1	N
Neurolaena	8	0	8	6	1	2	5	5	N
Nicolletia	3	0	3	2	0	3	0	0	N
*Olivaea**	2	0	2	2	0	1	1	2	Y
Onoseris	1	0	1	0	1	0	0	0	N
Osbertia	1	0	1	0	1	0	0	0	N
Oteiza	2	0	2	1	0	1	1	1	N
Otopappus	13	0	13	8	5	5	3	2	N
Oxylobus	4	0	4	1	2	1	1	0	N
*Oxypappus**	1	0	1	1	0	1	0	0	N
Palafoxia	7	0	7	2	1	5	1	1	N
Parthenice	1	0	1	0	1	0	0	0	N

Appendix (cont.)

Genus	Species								
	1	2	3	4	5	6	7	8	9
Parthenium	11	0	11	7	4	4	3	2	N
Pectis	30	0	30	15	5	20	5	3	N
*Pelucha**	1	0	1	1	0	1	0	0	N
Pentachaeta	1	0	1	0	0	1	0	0	N
Pericome	1	0	1	0	1	0	0	0	N
Perityle	35	0	35	20	5	20	10	5	N
Perymenium	35	0	35	30	6	16	13	13	N
Peucephyllum	1	0	1	0	1	0	0	0	N
Philactis	3	0	3	2	1	1	1	1	N
Picris	1	1	0	0	1	0	0	0	N
Pinaropappus	10	0	10	8	1	5	4	2	N
*Pippenalia**	1	0	1	0	0	1	0	0	N
Piptocarpha	1	0	1	0	1	0	0	0	N
Piqueria	6	0	6	5	1	3	2	2	N
Pityopsis	1	0	1	0	1	0	0	0	N
*Plagiolophus**	1	0	1	1	0	0	1	1	Y
*Plateilema**	1	0	1	1	0	0	1	1	Y
Pleurocoronis	3	0	3	2	1	1	1	0	N
Pluchea	9	1	8	2	6	1	2	0	N
Podachaenium	2	0	2	0	1	1	0	0	N
Porophyllum	20	0	20	16	6	3	11	6	N
Prenanthella	1	0	1	0	1	0	0	0	N
Psacalium	39	0	39	36	6	25	8	2	N
Psathyrotes	3	0	3	1	0	2	1	1	N
Pseudoclappia	2	0	2	0	0	1	1	1	N
Pseudogynoxys	2	0	2	0	2	0	0	0	N
Psilocarphus	3	0	3	0	3	0	0	0	N
Psilostrophe	3	0	3	0	3	0	0	0	N
Pterocaulon	1	0	1	0	1	0	0	0	N
Pyrrhopappus	1	0	1	0	1	0	0	0	N
Rafinesquia	2	0	2	0	2	0	0	0	N
Ratibida	6	0	6	2	2	2	2	2	N
Rensonia	1	0	1	0	1	0	0	0	N
Rojasianthe	1	0	1	0	0	0	1	1	N
Rudbeckia	1	0	1	0	1	0	0	0	N
Rumfordia	6	0	6	4	2	1	3	3	N
Sabazia	14	0	14	12	2	1	11	7	N
Salmea	5	0	5	2	1	2	2	2	N
Sanvitalia	5	0	5	2	4	0	1	1	N
Sartwellia	3	0	3	2	0	2	1	1	N
Schistocarpha	7	0	7	4	3	2	2	1	N
Schkuhria	3	0	3	1	2	1	0	0	N
Sclerocarpus	7	0	7	4	2	4	1	1	N
*Selloa**	1	0	1	1	0	1	0	0	N
Senecio									
Sen-Annui	8	1	7	0	3	4	1	0	N

Appendix (cont.)

Genus	Species								
	1	2	3	4	5	6	7	8	9
Sen-Aurei	17	0	17	14	3	10	4	4	N
Sen-Barkleyanthus	1	0	1	1	1	0	0	0	N
Sen-Fruticosa	7	0	7	5	5	2	0	0	N
Sen-Lugentes	15	0	15	15	2	7	6	3	N
Sen-Mulgedifolia	11	0	11	8	3	5	3	2	N
Sen-Multinervi	1	0	1	0	1	0	0	0	N
Sen-Nelsonianthus	2	0	2	0	1	0	1	1	N
Sen-Pentacalia	4	0	4	0	2	2	0	0	N
Sen-Pittocaulon*	4	0	4	4	0	1	3	3	Y
Sen-Roldana	51	0	51	43	6	10	35	33	N
Sen-Suffruticosa	6	0	6	4	1	2	3	3	N
Sen-Telanthophora	11	0	11	9	0	6	5	4	N
Sen-Triangularis	16	0	16	14	0	7	9	7	N
Sigesbeckia	6	0	6	4	2	1	3	3	N
Silybum	1	1	0	0	1	0	0	0	N
Simsia	12	0	12	8	8	2	2	2	N
Sinclairia	22	0	22	14	10	6	6	6	N
Smallanthus	4	0	4	2	1	2	1	1	N
Solidago	21	0	21	7	13	4	4	3	N
Soliva	1	1	0	0	1	0	0	0	N
Sonchus	3	3	0	0	3	0	0	0	N
Sparganophorus	1	1	0	0	0	0	0	0	N
Sphaeromeria	1	0	1	1	0	0	1	1	N
Spilanthes	1	1	0	0	1	0	0	0	N
Spiracantha	1	1	0	0	1	0	0	0	N
Squamopappus	1	0	1	0	0	1	0	0	N
Stenocarpha*	1	0	1	1	0	1	0	0	N
Stenotus	1	0	1	1	0	0	1	0	N
Stephanodoria*	1	0	1	1	0	0	1	1	Y
Stephanomeria	7	0	7	3	2	2	3	3	N
Stevia	82	0	82	61	21	35	26	24	N
Steviopsis	8	0	8	8	3	3	2	2	N
Strotheria*	1	0	1	1	0	0	1	1	Y
Stuessya*	3	0	3	3	0	1	2	2	N
Stylocline	3	0	3	0	2	1	0	0	N
Synedrella	1	1	0	0	1	0	0	0	N
Tagetes	22	2	20	11	13	4	5	5	N
Tanacetum	2	1	1	0	1	0	1	1	N
Taraxacum	1	1	0	0	1	0	0	0	N
Tetrachyron	7	0	7	6	0	1	6	4	N
Tetradymia	1	0	1	0	0	1	0	0	N
Tetragonotheca	1	0	1	0	0	1	0	0	N
Thelesperma	8	0	8	5	3	1	4	3	N
Tithonia	11	0	11	5	5	3	3	2	N
Townsendia	2	0	2	0	2	0	0	0	N
Tragopogon	1	1	0	0	1	0	0	0	N

Appendix (cont.)

					Species				
Genus	1	2	3	4	5	6	7	8	9
Trichocoronis	3	0	3	1	1	1	1	1	N
*Trichocoryne**	1	0	1	1	0	0	1	1	Y
Trichoptilium	1	0	1	0	0	1	0	0	N
Trichospira	1	1	0	0	1	0	0	0	N
Tridax	17	0	17	15	7	6	4	4	N
Trigonospermum	4	0	4	3	1	2	1	1	N
Trixis	18	0	18	14	5	9	4	3	N
*Urbinella**	1	0	1	1	0	0	1	1	Y
Varilla	2	0	2	1	0	2	0	0	N
Venegazia	1	0	1	0	0	1	0	0	N
Verbesina	119	0	119	95	17	36	66	61	N
Vernonia									
Vern-Eremosis	20	1	19	16	6	10	4	2	N
Vern-Leiboldia	9	0	9	8	0	1	8	7	N
Vern-Polyanthes	2	0	2	1	1	1	0	0	N
Vern-Scorpioides	9	1	8	2	4	3	2	2	N
Vern-Vernonia	18	0	18	12	7	7	4	4	N
Vigethia	1	0	1	1	0	0	1	1	Y
Viguiera	75	0	75	68	5	35	35	35	N
Villanova	1	0	1	1	0	1	0	0	N
Wedelia	19	0	19	14	3	7	9	7	N
Werneria	1	0	1	0	1	0	0	0	N
Wyethia	1	0	1	0	0	1	0	0	N
Xanthisma	1	0	1	0	1	0	0	0	N
Xanthium	2	1	1	0	2	0	0	0	N
Xanthocephalum	6	0	6	4	3	2	1	1	N
Xylorhiza	2	0	2	1	0	1	1	0	N
Xylothamia	9	0	9	8	1	5	3	2	N
Youngia	1	1	0	0	1	0	0	0	N
Zaluzania	10	0	10	10	3	5	2	2	N
Zexmenia	8	0	8	6	3	2	3	2	N
Zinnia	10	0	19	14	9	9	1	1	N

[a]1. Total species. 2. Alien species. 3. Native species. 4. Species restricted to Mexico. 5. Species of wide distribution. 6. Species of regional distribution. 7. Species of local distribution. 8. Endangered species. 9. Endangered genera.
Asterisks denote genera endemic to Mexico. N, not endangered; Y, endangered.

REFERENCES

Axelrod, D.L. 1958. Evolution of the Madro-tertiary geoflora. Bot. Rev. 24:433–509.
Bentham, G. 1873. Classification, history and geographical distribution of Compositae. J. Linnean Soc. 13:335–577.
Bremer, K. 1987. Tribal interrelationships of the Asteraceae. Cladistics 3:210–253

Cabrera-Rodriguez, L. & J.L. Villaseñor. 1987. Revisión bibliographica sobre el conocimiento de la familia compositae en México. Biotica 12:131–147.

Davis, S.D. et al. 1986. Plants in Danger: What Do We Know? 461 pp., Bernan-Unipub.

Gibbs Russell, G.E., W. G. Wellman, E. Retief, K.L. Immelman, G. Germishuizen, B.J. Pienaar, M. Van Wyk & A. Nicholas. 1987. List of species of Southern African Plants. Mem. Bot. Surv. S. Afr. 56 [ed. 2, pt. 2].

Hendrych, R. 1985. Quantitative Übersicht rezenter Cormobionten. Preslia Praha 57:359–370.

Hu, S. 1958. Statistics of Compositae in relation to the flora of China. J. Arnold Arb. 39:347–371; 405–419.

Jansen, R.K. & J.D. Palmer. 1988. Phylogenetic implications of chloroplast DNA restriction site variation in the Mutisieae (Asteraceae). Amer. J. Bot. 75:753–766.

Kartesz, J.T. & C.R. Bell. 1980. A summary of the taxa in the vascular flora of the United States, Canada and Greenland. Amer. J. Bot. 67:1495–1500.

Komarov, V.L. (ed.), 1959–1964. Flora U.R.S.S. Acad. Sci. (Leningrad) vol. 25–30.

Raven, P.H. & D.I. Axelrod. 1974. Angiosperm biogeography and past continental movements. Ann. Missouri Bot. Gard. 61:539–673.

Reko, B.P. 1946. Los generos fanerogamicos mexicanos. Bol. Soc. Bot. Mex. 4:19–45.

Robinson, H. 1981. A revision of the tribal and subtribal limits of the Heliantheae (Asteraceae). Smithsonian Contr. Bot. 51:1–102.

Rzedowski, J. 1972. Contribuciones a la fitogeografia floristica e historica de México. Ciencia (Mexico) 27:123–132.

———. 1978. Claves para la identificación de los generos de la familia Compositae en México. Acta Cienc. Potosina 7:1–145.

Turner, B.L. 1977a. Fossil history and geography. *In* V.H. Heywood, J.B. Harborne & B.L. Turner (eds.), Biology and Chemistry of the Compositae Vol. I. London: Academic Press.

———. 1977b. Summary of the biology of the Compositae. *In* V.H. Heywood, J.B. Harborne and B.L. Turner (eds.), Biology and Chemistry of the Compositae. Vol. II. London: Academic Press.

——— & Powell. 1977. Helenieae-systematic review. *In* V.H. Heywood, J.B. Harborne & B.L. Turner (eds.), Biology and Chemistry of the Compositae. Vol. II. London. Academic Press.

21

Diversity of Mexican Aquatic Vascular Plant Flora

ANTONIO LOT, ALEJANDRO NOVELO,
AND PEDRO RAMÍREZ-GARCIA

The considerable diversity of aquatic habitats in Mexico supports a variety of communities, some of which constitute the aquatic flora of Mexico. The diversity of this flora, with representatives from 84 families, is composed of 258 genera and 747 species. It includes ferns, gymnosperms, and angiosperms. The last, which includes 38 dicotyledonous families, is the most diverse. The species are grouped according to the medium in which they are found.

The complex mosaic of ecosystems, which range from the coral reefs that shelter a notable diversity of sea grasses, through the coastal lakes with large extensions of land covered by mangroves, to the freshwater communities, which are widely distributed into swamps, flood plains, rivers, lakes, ponds, and springs in both warm, low-lying and cold, high-altitude areas, has played an important role in the development of aquatic communities in Mexico.

This chapter reviews the diversity of Mexico's aquatic vascular plant flora. It cites the characteristic families in each of the aquatic habitats, their species richness, and their contribution to the world's aquatic flora. It should be stressed that there remain gaps in the information on the biodiversity of large areas of the country. We update extant data and suggest new ways to understand better the aquatic plant resources of Mexico.

The aquatic vascular flora of Mexico is remarkably diverse, represented by 84 families, 258 genera, and 747 species. It includes ferns, gymnosperms, and angiosperms. Among these, the most diverse are dicotyledonous plants, represented by over 38 families. This flora is equally diverse in its life forms with trees, shrubs, and herbs. Many of these elements show habitat-related adaptations, and species in the same genus show considerable variation in morphology (e.g., *Ludwigia sedoides* and *L. octovalvis*).

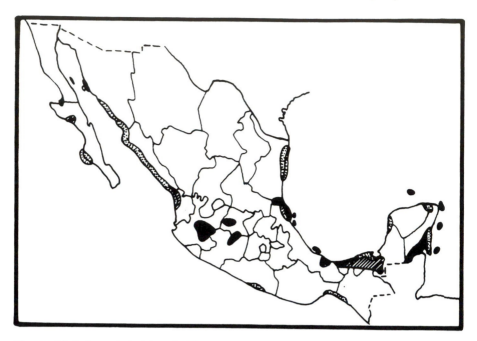

Figure 21.1. Aquatic habitats in Mexico. Darkened areas, marine; stippled areas, saline; striped areas, freshwater.

This flora is equally interesting from an evolutionary point of view. The dogma in evolutionary biology is that genetic differentiation is a result of restricted gene flow between populations. Populations in aquatic habitats are comparable to island populations as intervening habitats are uninhabitable, which suggests the possibility that aquatic plant populations may experience a higher evolutionary rate than their counterparts on land. Mexico's aquatic flora provides rich material for study by evolutionary biologists.

The aquatic habitats of Mexico are diverse, and in several states they cover a great deal of surface area. Figure 21.1 illustrates three types of aquatic habitat in the country. The marine habitats are best represented by the expanses that include the coasts of Quintana Roo, Yucatán, Campeche, Veracruz, and Tamaulipas in warm seas and the coasts of Sinaloa and Baja California in colder waters. The vascular aquatic plants, generally known as sea grasses, grow in these waters and reproduce under totally submerged conditions. The saltwater habitats, dominated by woody communities known commonly as mangroves, have their northern limits in Mexico. The areas principally covered by this type of vegetation are found in Quintana Roo, in the north and east of the Yucatan Peninsula, and in a large part of the coasts of Campeche, Tabasco, Veracruz, and Tamaulipas in the Gulf of Mexico. On the Pacific coast they are found in the states of Chiapas, Oaxaca, Guerrero, and the salt marshes of the Sonora and Sinaloa coasts. Other small extensions are found in northern and southern Baja California. The freshwater habitats cover equally large areas of Mexico, especially in the states of Campeche, Tabasco, and southern Veracruz. Among the most wide-ranging habitats are rivers, swamps, and tree-covered flood plains. Other important areas are found near the coasts in Quintana Roo, Tamaulipas, and Nayarit. The Mexican Altiplano has important lakes such as Chapala, Patzcuaro, Cuitzeo, and Zirahuen.

In contrast to the marine and saltwater habitats, it is difficult to state where freshwater habitats begin or end, as the areas between water and land are not clearly defined. Such transitional zones are called ecotones (Sculthorpe, 1967). Classifying plants that grow in these ecotones as aquatics is problematical because of their morphological and physiological adaptations. For example, within the group of true aquatic plants, several species reproduce sexually during the dry season when the soil is completely dry, whereas during the wet season they reproduce vegetatively. On the other hand, many weeds or tolerant species colonize the ecotone during the dry period, remaining partially or temporarily flooded during the wet season. The above observations suggest the inherent difficulties in defining an "aquatic plant." The concept as it is presently understood is applicable to many species belonging to a great number of families.

Based on the experience of approximately 20 years of collecting and studying aquatic plants in Mexico and parts of Central America, it is proposed here that these plants be grouped into the following categories.

1. Strictly aquatic. Such plants complete their entire life cycle totally submerged, partially emerging from the water or floating on the surface. Most do not survive being outside water even for short periods of time.
2. Subaquatic. These plants complete most of their life cycle at the water's edge. They tolerate dry soil temporarily, during which time they reproduce.
3. Tolerant. These plants complete most of their life cycle in a dry environment; nonetheless, during a short period of time coinciding with the rainy season, they are temporarily and partially flooded. This category includes a great number of weeds that have wide ranges of tolerance, not only of humid conditions but also of nutrients and types of substrate.

This chapter discusses aquatic plants of Mexico belonging to the first two categories.

AQUATIC PLANT GROUPS

Tables 21.1–21.6 present families that grow in diverse aquatic habitats. They provide generic and species numbers, and their ratio with the world total.

Table 21.1. Marine angiosperms

Family	Genera			Species		
	Worldwide (No.)	Mexico (No.)	%	Worldwide (No.)	Mexico (No.)	%
Cymodoceaceae	5	2	40	14	3	21.4
Hydrocharitaceae	3	2	66.6	11	3	27.2
Posidoniaceae	1	0	0	3	0	0
Zosteraceae	3	2	66.6	18	3	16.6
Total	12	6	50	46	9	19.5

Source: After Hartog (1970).

Marine Angiosperms

Table 21.1 presents an overview of the marine angiosperms, known world-wide as "sea grasses" or locally (Gulf of Mexico and the Caribbean) as "ceibadales" (Lot, 1971); the term "sea grass" does not denote the family Gramineae. Their gross morphologies suggest grasses.

Seventy-five percent of the families of these plants are represented in the shallow areas of Mexican seas. The dominant families in the warm waters of the Mexican Caribbean and the Atlantic are Hydrocharitaceae (*Thalassia* and *Halophila*) and Cymodoceaceae (*Halodule* and *Syringodium*) (Lanza & Tovilla, 1986; Lot, 1971, 1977; Novelo, 1976, 1978). Whereas Hydrocharitaceae includes both saline and fresh water plants, others in Table 21.1 consist exclusively of marine taxa. Along the temperate coasts of the Gulf of California the families Zosteraceae (*Zostera*) (Dawson, 1951; Felger & Moser, 1973; Felger et al., 1980; Phillips & Backman, 1983) and Cymodoceaceae (*Halodule*) (McMillan & Phillips, 1979) are found. Popula-tions of *Halodule* on the west coast are disjunct from those in the Gulf of Mexico and deserve further study. Zosteraceae is found on the western coast of the Baja CaliforniaPeninsula; these species belong predominantly to *Phyllospadix* (Hartog, 1970).

The diverse sea grass genera discussed above show both neotropical and neartic affinities. They represent almost 20% of all species, and 50% of the genera at a worldwide level.

Saltwater Angiosperms

The geographic locality of Mexico, characterized by the transition from tropical to subtropical zones, permits a study of latitudinal variations of the vegetation with relation to habitat gradients, principally climatic, of which the northern limit of mangroves in the Gulf of Mexico is a wide transitional

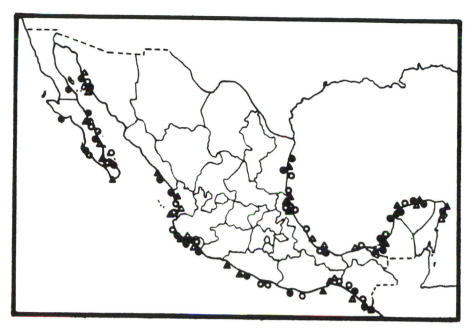

Figure 21.2. Distribution of mangroves in Mexico. Closed circles, *Rhizophora mangle*; open circles, *Laguncularia racemosa*; closed triangles, *Conocarpus erectus*; open triangles, *Avicennia germinans*; closed squares, *Rhizophora harrisonnii*.

zone (Lot et al., 1975). On the Pacific slope, the mangroves reach their northern distributional limit on the coasts of Baja California and Sonora (Fig. 21.2).

Three families—Avicenniaceae (*Avicennia*), Combretaceae (*Conocarpus* and *Laguncularia*), and Rhizophoraceae (*Rhizophora*)—characterize Mexico's mangroves (Lot & Novelo, 1990), as they do mangroves of other Latin American countries. These genera are trees and in general have a marked distribution pattern in the areas where they grow. Species of *Rhizophora* colonize new areas and grow in the deeper waters. *Conocarpus* is found on the shallower side of the mangrove in drier, less salty areas (Lot et al., 1975; Novelo, 1978). The intermediate part of the mangrove is commonly inhabited by *Laguncularia* and *Avicennia*, which may grow together or displace one another depending on the type of substratum, flooding rate, and concentration of salinity.

These plants are facultative halophytes (Bowman, 1917; Dawes, 1986; McMillan, 1974; Walsh, 1974), having a wide range of tolerance to salinity; they tolerate marine waters, as in the case of the communities around the Lagartos River, on the border of Quintana Roo and Yucatan, as well as several protected bays in the Gulf of California. On the other hand, species of *Rhizophora* can be found as one more element of a riparian vegetation in fresh waters in the San Pedro River in Tabasco state, about 200 km from the coast (Lundell, 1942).

Table 21.2. Principal woody mangroves

Family	Genera			Species		
	Worldwide (No.)	Mexico (No.)	%	Worldwide (No.)	Mexico (No.)	%
Avicenniaceae	1	1	100	8	1	12.5
Combretaceae	3	2	66.6	4	2	50
Palmae	1	0	0	1	0	0
Rhizophoraceae	4	1	25	17	2	11.8
Sonneratiaceae	1	0	0	5	0	0
Total	10	4	40	35	5	14.3

Source: After Tomlinson (1982).

Mangroves in saline conditions are known to be a limiting factor in harboring a large number of species. This community is relatively poor in herbaceous species, even more so if the soil is totally or temporarily under water. Among herbaceous species usually associated with the mangrove and that tolerate a certain amount of flooding are *Acrostichum danaeifolium, Batis maritima, Borrichia frutescens, Distichlis spicata, Fimbristylis* spp., *Sesuvium portulacastrum, S. maritimum, Spartina alterniflora,* and several others (López-P., 1982; Lot et al., 1975; Menéndez, 1976; Novelo, 1978; Rico-Gray, 1979; Vázquez-Y., 1971).

The diversity of woody species in mangroves is richer and better represented in Southeast Asia. Mexico has a little more that 14% of the total species (Table 21.2).

Freshwater Angiosperms

In the freshwater communities, there is a clear predominance of herbaceous species that, in some regions, form pure colonies covering large areas (Lot & Novelo, 1978, 1988). In this aquatic habitat not only genera and species but whole families have evolved and adapted.

As mentioned above, the aquatic plants may be divided into two large groups: strictly aquatic plants encompassing families, genera, and species that live completely associated with different aquatic habitats and plants that have evolved and adapted to living in water, although they belong to fundamentally terrestrial families.

In angiosperms, a great diversity is found among the strictly aquatic families. Table 21.3 provides an overview of the monocotyledonous families—the number of genera, species, and their percentages—at world and national levels. Many of these families are related. Fourteen of the 19

Table 21.3. Strictly aquatic monocots

Family (n = 19)	Genera			Species		
	Worldwide (No.)	Mexico (No.)	%	Worldwide (No.)	Mexico (No.)	%
Alismataceae*	11	2	18.1	≈100	17	17
Aponogetonaceae*	1	0	0	≈45	0	0
Butomaceae*	1	0	0	1	0	0
Cymodoceaceae*	5	2	40	14	3	21.4
Hydrocharitaceae*	16	6	37.5	118	7	5.9
Juncaginaceae*	3	1	33.3	16	2	12.5
Lemnaceae	5	4	80	35	15	42.8
Liliaceae*	1	1	100	1	1	100
Limnocharitaceae*	3	2	66.6	12	3	25
Mayacaceae	1	1	100	≈10	1	10
Najadaceae*	1	1	100	50	3	6
Pontederiaceae	8	5	62.5	≈32	12	37.5
Posidoniaceae*	1	0	0	3	0	0
Potamogetonaceae*	3	2	66.6	≈108	10	9.2
Sparganiaceae	1	1	100	≈19	2	10.5
Scheuchzeriaceae*	1	0	0	1	0	0
Typhaceae	1	1	100	≈10	2	20
Zannichelliaceae*	4	1	25	10	1	10
Zosteraceae*	3	2	66.6	18	3	16.6
Total	70	32	45.7	≈603	82	13.6

Asterisks identify families that belong to order Helobiae.

families belong to the same order—the Helobiae—and ten families of this order are to be found in Mexico. All the sea grasses are monocots.

Families notable for their diversity in Mexico are the Alismataceae, Lemnaceae, Pontederiaceae, and Potamogetonaceae with ten or more species. At the generic level, Hydrocharitaceae (six genera), Pontederiaceae (five), and Lemnaceae (four) are the most important. About 13% of monocotyledonous species, 50% of their genera, and 80% of the families are represented in Mexico.

The dicotyledonous aquatic plant families are a diverse group and belong to different orders (Table 21.4). Among them are monogeneric and monospecific families with restricted distributions in other parts of the world; they include Tetrachondraceae, Trapaceae, and Trapellaceae, and, from North America, Hippuridaceae.

Podostemaceae, rich in species, is the only family that grows adhering to rocky substrates in clear rivers with continuous flow. The Mexican Podostemaceae have not been well studied. It is likely that several undescribed species will be found in the rivers close to the coasts of the

Table 21.4. Strictly aquatic dicots

Family	Genera			Species		
(n = 13)	Worldwide (No.)	Mexico (No.)	%	Worldwide (No.)	Mexico (No.)	%
Cabombaceae	2	2	100	8	3	37.5
Ceratophyllaceae	1	1	100	2	2	100
Elatinaceae	2	2	100	32	3	9.3
Hippuridaceae	1	0	0	1	0	0
Hydrostachyaceae	1	0	0	22	0	0
Menyanthaceae	5	1	20	39	2	5.1
Nelumbonaceae	1	1	100	2	1	50
Nymphaeaceae	6	2	33.3	65	9	13.8
Podostemaceae	≈46	4	8.6	≈260	9	3.5
Sphenocleaceae	1	1	100	2	1	50
Tetrachondraceae	1	0	0	1	0	0
Trapaceae	1	0	0	3	0	0
Trapellaceae	1	0	0	2	0	0
Total	≈69	14	20.2	≈439	30	6.8

states of Nayarit, Jalisco, Colima, Michoacán, Guerrero, Oaxaca and Chiapas on the Pacific side and Tamaulipas, Veracruz, and southern Tabasco in the Gulf of Mexico. It is fairly common to find a genus with one or several species endemic to the same river. However, genera such as *Tristicha* and *Podostemum* have a worldwide distribution. Podostemaceae and Nymphaeaceae, with nine species each, are the most species rich in Mexico. Several are endemic (Novelo & Lot, 1984; Royen, 1951, 1953, 1954). On a worldwide basis almost 7% of the dicotyledonous species, 20% of the genera, and a little more than 60% of the families are represented in Mexico.

The species richness (monocots and dicots together) of each state provides interesting information (Fig. 21.3). The states with the greatest species numbers, in order of importance, are Veracruz (51 species), Chiapas (42), México (41), Michoacán (39), Jalisco (36), Tamaulipas (34), Tabasco (33), Federal District (32), and Campeche (30) (Fig. 21.1). The states with least diversity are Tlaxcala (six species), Colima (six), Zacatecas (seven), Guanajuato (nine), Aguascalientes (nine), and Querétaro, Nuevo León, and Sinaloa (11 species each) (Lot et al., 1986). These states, which are poorly studied regions, are either small area-wise or include arid and semiarid zones.

In general, there is a great need in Mexico for a program on a wide scale that would allow for greater exploration, especially in those states whose biological diversity is poorly known: principally Tabasco, Campeche, Chiapas, Oaxaca, Guerrero, Jalisco, Michoacán, Nayarit, Tamaulipas, Chi-

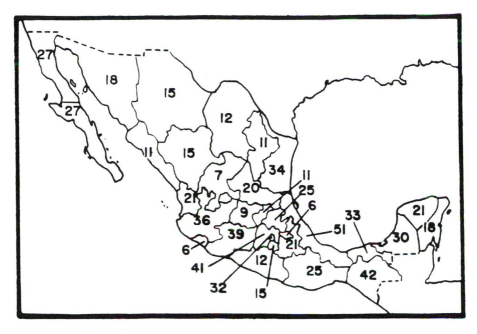

Figure 21.3. Taxa of strictly aquatic angiosperm species recorded per state.

huahua, Durango, San Luis Potosí, Hidalgo, and Puebla. It should be mentioned that the states not mentioned here have not necessarily been well collected. The incomplete knowledge of the biodiversity of these states underscores the importance of their study.

Freshwater Pteridophytes and Gymnosperms

Table 21.5 presents a list of the aquatic Pteridophytes and Gymnosperms. Among the Pteridophytes, there are several families (known as the heterosporic ferns) with marked morphological and physiological modifications (including reproductive structures) for the aquatic habitat. The families included in this group are Azollaceae, Salviniaceae, and Marsileaceae. The members of the first two float on the surface of water and sometimes have been considered aquatic weeds, as they tend to cover large areas of their habitat (Anonymous, 1976; Sculthorpe, 1967).

The family Polypodiaceae, although basically terrestrial, includes several genera adapted to living in water, e.g., *Blechnum* and *Acrostichum*. The families Parkeriaceae and Isoetaceae are strictly aquatic. A large number of the Equisetaceae are found along river and lake banks, although several species live in areas that are almost dry year-around.

Among the Gymnosperms, *Taxodium*, commonly known as "ahuehuete" or "sabino," occurs along river and spring banks in temperate areas at altitudes almost always over 500 m. The basal trunk is usually submerged

Table 21.5. Aquatic ferns and gymnosperms

Family	Genera			Species		
	Worldwide (No.)	Mexico (No.)	%	Worldwide (No.)	Mexico (No.)	%
Ferns and fern allies (n = 7)						
Azollaceae	1	1	100	6	3	50
Equisetaceae	1	1	100	29	8	27.5
Isoetaceae	1	1	100	≈75	4	5.3
Marsileaceae	3	1	33.3	72	4	5.5
Parkeriaceae	1	1	100	6	2	33.3
Polypodiaceae	2	2	100	10	3	30
Salviniaceae	1	1	100	≈12	2	16.6
Subtotal	10	8	80	≈210	26	12.4
Gymnosperms						
Taxodiaceae	1	1	100	3	1	33.3
Total (n = 8)	11	9	81.8	≈213	27	12.7

Table 21.6. Flowering plant families with representatives in Mexico's aquatic flora

Family	Genus (No.)	Species (No.)
Acanthaceae	5	17
Amaranthaceae	2	7
Amaryllidaceae	3	7
Anacardiaceae	1	1
Annonaceae	1	1
Aquifoliaceae	1	1
Araceae	3	5
Asclepiadaceae	1	1
Bombacaceae	1	1
Boraginaceae	3	4
Callitrichaceae	1	3
Cannaceae	1	1
Capparidaceae	1	7
Caprifoliaceae	4	5
Caryophyllaceae	7	25
Compositae	59	131
Convolvulaceae	2	3
Crassulaceae	1	2
Cruciferae	2	8
Cyperaceae	7	69
Droseraceae	1	1
Ebenaceae	1	2

Table 21.6. (cont.)

Family	Genus (No.)	Species (No.)
Elatinaceae	2	3
Eriocaulaceae	2	7
Euphorbiaceae	2	5
Graminae	23	57
Guttiferae	2	4
Haloragaceae	2	5
Hydrophyllaceae	1	3
Iridaceae	1	6
Juncaceae	2	33
Leguminosae	11	26
Lentibulariaceae	1	15
Lobeliaceae	1	4
Lythraceae	4	6
Marantaceae	1	1
Onagraceae	4	20
Orchidaceae	3	10
Palmae	4	6
Polygonaceae	2	18
Portulacaceae	1	1
Ranunculaceae	1	10
Rubiaceae	1	1
Salicaceae	1	20
Scrophulariaceae	7	15
Solanaceae	3	5
Umbelliferae	7	15
Verbenaceae	1	1
Xyridaceae	1	4
Total (n = 49)	199	603

for a greater part of the year. The vegetation formed by what is termed "bosque de galería" (Rzedowski, 1978) or semievergreen riparian forest (Lot & Novelo, 1990).

About 13% of the species, a little more than 80% of the genera, and 100% of the families are represented in the country.

Terrestrial Families with Some Aquatic Representatives

Table 21.6 summarizes the list of known families in Mexico that have one genus or species growing in distinct aquatic ecosystems. The list excludes those families already mentioned in previous tables. The species in Table 21.6 are freshwater aquatic plants whose habit may be arboreal, shrubby, or herbaceous.

These aquatic plants belong to several genera, or in extreme cases to only one species, and may show habitat-related adaptations. Many of their relatives are found in terrestrial habitats.

Among the more important families with arboreal species are Acanthaceae (*Bravaisia integerrima*), Anacardiaceae (*Metopium brownei*), Annonaceae (*Annona glabra*), Bombacaceae (*Pachira aquatica*), Ebenaceae (*Diospyros* spp.), Guttiferae (*Calophyllum brasiliense*), Leguminosae (*Andira galeottiana, Haematoxylum campechianum, Pithecellobium* spp., *Inga* spp.), Palmae (*Acoelorrhaphe wrightii, Bactris* spp.), and Salicaceae (*Salix* spp.).

Families with shrubby species are Acanthaceae (*Bravaisia tubiflora, B. berlanderiana*), Convolvulaceae (*Ipomoea fistulosa*), Leguminosae (*Mimosa pigra*), and Rubiaceae (*Cephalanthus occidentalis*).

The rest of the species in Table 21.6 correspond to herbs. There are several genera within the herbs that have developed important morphological and physiological modifications, including the ability to live as free-floating plants. The most important of these genera and species are: *Pistia stratiotes* (Araceae), *Phyllanthus fluitans* (Euphorbiaceae), *Utricularia inflata, U. foliosa, U. gibba, U. hydrocarpa* (Lentibulariaceae), *Neptunia oleracea* (Leguminosae), and *Ludwigia helminthorrhiza* (Onagraceae).

There are herbaceous species as well that are completely submerged, and a few of them still depend on insects or on the wind for reproduction. Among the more common are *Callitriche* spp. (Callitrichaceae), *Eriocaulon salzmanii* (Eriocaulaceae), *Elatine* spp. (Elatinaceae), *Myriophyllum* spp. (Haloragaceae), *Juncus repens* (Juncaceae), *Utricularia* spp. (Lentibulariaceae), *Didiplis diandra* (Lythraceae), *Ludwigia palustris* (Onagraceae), *Ranunculus* spp. (Ranunculaceae), and *Benjaminia reflexa* (Scrophulariaceae).

The rest of the species are frequently found rooted at different depths on the banks of rivers, lakes, lagoons, marshes, swamps, and temporary pools; some form colonies of a considerable size. The more important ones include *Hymenocallis* spp. and *Crinum* spp. (Amaryllidaceae); *Canna glauca* (Cannaceae); *Cladium jamaicense* known locally as "Sibal"; *Cyperus giganteus, Eleocharis* spp., *Scirpus* spp. locally called "tulares" and "chuspatales" (Cyperaceae); *Phragmites australis*, called "carrizal" (Gramineae); *Juncus* spp. (Juncaceae); and *Thalia geniculata*, commonly known as "popal" (Maranthaceae) (Lot & Novelo, 1988; Novelo, 1978; Orozco & Lot, 1976).

CONCLUSIONS

The diverse species of the aquatic flora of Mexico discussed in this chapter represent approximately 50% of the aquatic genera and 13% of the aquatic species of the world. These figures, detailed in the tables according to life forms, are as highly significant as those of other natural groups with regard to their ecological affinities.

The following premises, which make these numbers pertinent, should be borne in mind. (1) Generally, aquatic plants have a wide distribution in the world and in some cases are cosmopolitan. Although initital considerations may not suggest it, this natural group is useful as an indicator of biodiversity of a region or a country. (2) The partial knowledge of these plants in Mexico and the poorly explored various species-rich aquatic systems of the country suggest that its aquatic flora may be even richer. These considerations are significant for families with aquatic and subaquatic species that have not been considered in depth in this study. New records and reports of species for various parts of Mexico and the country in general support this conclusion. Of interest are reports relating to *Phyllanthus fluitans* (Euphorbiaceae) (Lot et al., 1980), *Didiplis diandra* (Lythraceae) (Novelo, 1981b), *Myriophyllum quitense* (Haloragaceae) (Novelo, 1984), *Luziola subintegra* and *L. spruciana* (Gramineae) (Novelo, 1986), *Pontederia rotundifolia* (Pontederiaceae) (Novelo, 1981a), *Heteranthera spicata* (Pontedariaceae), *Hydrilla verticillata* (Hydrocharitaceae) (Novelo & Martínez, 1989); *Nymphaea amazonum* (Nymphaeaceae) (Ramírez-G. & Novelo, 1989a); *Spirodela intermedia* (Ramírez-G., 1989b).

This preliminary synthesis can be a useful point of departure to continue and reorganize floristic, taxonomic, and phytogeographic studies with the collaboration of a large number of students and interested botanists in integrating the inventory of the aquatic plant resources that are part of the flora of Mexico.

REFERENCES

Anonymous 1976. Making Aquatic Weeds Useful: Some Perspectives for Developing Countries. Report of an Ad Hoc Panel of the Advisory Committee on Technology for International Development, Commission on International Relations. Washington, D.C: National Academy of Sciences.

Bowman, H.M. 1917. Ecology and physiology of the red mangrove. Proc. Am. Phil. Soc. 61:589-672.

Dawes, C.J. 1986. Botánica Marina. Mexico City: Limusa.

Dawson, E.Y. 1951. A further study of upwelling and associated vegetation along Pacific Baja California, Mexico. J. Mar. Res. 10:39–58.

Felger, R. & M.B. Moser. 1973. Eelgrass (*Zostera marina* L.) in the Gulf of California: discovery of its nutritional value by the Seri Indians. Science 181:355–356.

———, E.W. Moser & M.B. Moser. 1980. Seagrasses in Seri Indian culture. *In* R.C. Phillips & C.P. McRoy (eds.), Handbook of Seagrass Biology: An Ecosystem Perspective. New York, NY: Garland Press. pp. 260–276.

Hartog, C. den. 1970. The Sea-Grasses of the World. Amsterdam: North Holland.

Lanza, E.G. de la & H.C. Tovilla. 1986. Una revisión sobre taxonomía y distribución de pastos marinos. Univ. Ciencia 3(6):17–38.

López-Portillo, G. 1982. Ecología de manglares y otras comunidades de halófitas en la costa de la laguna de Mecoacán, Tabasco. Thesis, UNAM.

Lot H., A. 1971. Estudios sobre fanerógamas marinas en las cercanias de Veracruz, Ver. An. Inst. Biol. UNAM. Ser. Bot. 42(1):1–48.

———. 1977. General status of research on Sea grasses ecosystems in Mexico. *In* C.P. McRoy & C. Hellferich (eds.), Seagrass Ecosystems. A Scientific Perspective. New York, NY: Marcel Dekker. pp. 233–245.

————— & A. Novelo. 1978. Laguna de Tecocomulco, Hidalgo. Guías botánicas de excursiones en México. Mexico City: Sociedad Botánica de México.

————— & —————. 1988. Vegetación y flora acuática del Lago de Pátzcuaro; Michoacán, México. Southw. Naturalist 33(2):167–175.

————— & —————. 1990. Forested wetlands of Mexico. *In* A.E. Lugo, M.M. Brinson & S. Brown (eds.), Forested wetlands of the world. 15:287–298. Ecosystems of the world. Amsterdam: Elsevier.

—————, C. Vazquez-Yanes & F. Menéndez. 1975. Physiognomic and floristic changes near the northern limit of mangroves in the Gulf Coast of Mexico. *In* G.E. Walsh, S.C. Snedaker & H.J. Teas (eds.), International Symposium on Biology and Management of Mangroves. 1:52–61. Univ. of Florida, Gainesville: Institute of Food and Agricultural Science.

—————, A. Novelo & P. Cowan. 1980. Hallazgo en México de una euforbiaceae acuática originaria de Sudamérica. Bol. Soc. Bot. Méx. 39:87–93.

—————, ————— & P. Ramírez-García. 1986. Angiospermas Acuáticas Mexicanas 1. Vol. V. Listados Florísticos de México. Mexico City: UNAM, Inst. Biol.

Lundell, C.L. 1942. Flora of eastern Tabasco and adjacent Mexican areas. Contrib. Univ. Mich. Herbarium 8:1–74.

McMillan, C. 1974. Salt tolerance of mangroves and submerged aquatic plants. *In* R.J. Reimold & W.H. Queen (eds.), Ecology of Halophytes. New York: Academic Press. pp. 379–390.

————— & R.C. Phillips. 1979. *Halodule wrightii* Aschers in the Sea of Cortes, Mexico. Aquatic Bot. 6:393–396.

Menéndez L., F. 1976. Los manglares de la Laguna de Sontecomapan, Los Tuxtlas, Veracruz, estudio florístico-ecológico. Thesis, UNAM.

Novelo R., A. 1976. Observaciones ecológicas de las poblaciones de *Thalassia testudinum* Koenig (Hydrocharitaceae marina) en una zona arrecifal de Veracruz. Thesis, UNAM.

—————. 1978. La vegetación de la Estación Biologica del Morro de la Mancha, Veracruz. Biotica 3(1):9–23.

—————. 1981a. Nuevo registro para México de *Pontederia rotundifolia* L. f. Bol. Soc. Bot. Méx. 41:161.

—————. 1981b. *Didiplis diandra* (Lythraceae) in Southeastern Mexico. Sida 9(2):182.

—————. 1984. Registros nuevos de plantas acuáticas mexicanas I: *Myriophyllum quitense* HBK. (Haloragaceae). Bol. Soc. Bot. Méx. 45:147–149.

—————. 1986. Registros nuevos de plantas acuáticas mexicanas II. *Luziola subintegra* Swallen y *L. sprucena* Benth. ex Doell. (Gramineae). Bol. Soc. Bot. Méx. 46:90–91.

————— & A. Lot. 1984. Esclarecimiento taxonómico de *Nymphaea gracilis* Zucc., planta acuática endémica de México. Bol. Soc. Bot. Méx. 45:85–95.

————— & M. Martínez. 1989. *Hydrilla verticillata* (Hydrocharitaceae), problemática maleza acuática de reciente introducción en México. An. Inst. Biol. UNAM, Ser. Bot. 58:97–102.

Orozco S., A & A. Lot. 1976. La vegetación de las zonas inundables del sureste de Veracruz. Biotica 1(1):1–44.

Phillips, R.C. & T.W. Backman. 1983. Phenology and reproduction biology of eelgrass (*Zostera marina*) at Bahia Kino, Sea of Cortes, Mexico. Aquatic Bot. 17:185–199.

Ramírez-García, P. & A. Novelo. 1989a. *Nymphaea amazonum* (Nymphaeaceae) en México; clave de las especies del subgénero *Hydrochallis* en el país. An. Inst. Biol. UNAM, Ser. Bot. 58:87–91.

————— & —————. 1989b. Nota sobre *Spirodela intermedia* (Lemnaceae) en México y Costa Rica. An. Inst. UNAM, Ser. Bot. 59(1):103–105.

Rico-Gray, V. 1979. El manglar de la Laguna de la Mancha, Veracruz, estructura y productividad neta. Thesis, UNAM.

Royen, P. van. 1951. The Podostemaceae of the New World. Part I. Meded. Bot. Mus. Herb. Rijks Univ. Utrecht 107:1–151.

———. 1953. The Podostemaceae of the New World. Part II. Acta Bot. Neerl. 2(1):1–21.

———. 1954. The Podostemaceae of the New World. Part III. Acta Bot. Neerl. 3(2):215–263.

Rzedowski, J. 1978. Vegetación de México. Mexico City: Limusa.

Sculthorpe, C.D. 1967. The biology of aquatic vascular plants. London: Edward Arnold.

Tomlinson, P.B. 1982. *In* C.R. Metcalfe (ed.). Anatomy of the monocotyledons. VII. Helobiae. Oxford: Clarendon Press.

Vázquez-Yanes, C. 1971. La vegetación de la Laguna de Mandinga, Veracruz. An. Inst. Biol. UNAM, Ser. Bot. 42:49–94.

Walsh, G.E. 1974. Mangroves: a review. *In* R.J. Reimold & W. H. Queen (eds.), Ecology of Halophytes. New York: Academic Press.

IV

PHYTOGEOGRAPHY OF SELECTED VEGETATION TYPES IN MEXICO

22

Composition, Floristic Affinities, and Origins of the Canopy Tree Flora of the Mexican Atlantic Slope Rain Forests

TOM WENDT

Important factors in the floristic history of the Mexican rain forests include the (1) subtropical location and climate; (2) geological and topographic diversity; (3) continued connection from the Cretaceous onward to the North American continent; (4) Eocene or earlier immigration routes for megathermal taxa from South America, Eurasia, and Africa via Laurasia; (5) Neogene immigration routes for megathermal taxa from South America; (6) limited nature in space and time of the immigration routes mentioned; (7) Neogene and Pleistocene climatic cycles; (8) marine climate and low relief of insular southern Mesoamerica through much of the Cenozoic; and (9) large extensions of karst terrain in northern Mesoamerica. Combined phytogeographical and paleobotanical evidence suggests that the modern flora is composed of (1) an ancient element, descended from lineages present in North America by Eocene times; and (2) a recent element composed of descendents of Neogene arrivals from South America. Both elements are large and diverse in the modern flora. The ancient element includes lineages that arrived from South America and others that arrived via Laurasian connections, the latter including representatives of both Laurasian and Gondwanan families. The idea that the modern rain forest flora of Mexico is almost entirely South American in origin is not supported by the data presented here. The past and continuing floristic interrelationships of tropical wet, tropical dry, and cloud forest floras of Mexico are emphasized.

A preliminary list is presented of the species of trees that reach at least 18 m in height in the rain forests of the Gulf of Mexico and Caribbean slopes of Mexico, as well as data on Mexican and worldwide distributions of these species. The list includes 64 families, 230 genera, and 452 species or intraspecific taxa. The flora is clearly less rich than many equatorial lowland floras but overall is about four times as rich as the richest temperate forest of equal area in the United States. The largest numbers of species are found in Leguminosae, Moraceae, Lauraceae, Sapotaceae, and Rubiaceae; the richest genera are *Ficus*, *Inga*, *Nectandra*, *Lonchocarpus*, *Pouteria*, and *Coccoloba*. The high diversity of *Lonchocarpus* in the rain forests of Mexico and northern Central America contrasts with the relative unimportance of this genus in other neotropical rain forests.

Of the species included, 9.6% are considered to be endemic to Mexico, 30.9% to Mexico and northern Central America, and 55.0% to Mexico and Central America as a whole; 25.1% reach Amazonia and 20.6% the Greater Antilles. The direct relationship with the Antilles is not strong but is more notable in the Yucatan Peninsula.

Mexican endemics are concentrated in three areas of high precipitation: the Los Tuxtlas area of Veracruz, the Tuxtepec area of northern Oaxaca, and the Crescent area of extreme southeastern Veracruz (Uxpanapa) and adjacent Oaxaca (Chimalapa) to southern Tabasco and northern Chiapas. Two important broader areas of endemism are recognized for the Mexico-northern Central America area: (1) an area of wet although seasonal evergreen forests distributed from southern Veracruz (including the three endemic areas cited above) through the wetter part of the Lacandon forests to the Izabal area of Guatemala and Belize; and (2) the drier forests of the Yucatan Peninsula, in the broad sense, to the north of area 1, including parts of the Lacandon forests.

The comparatively wet "Gulf" forests of Veracruz and neighboring states are far richer in canopy tree species than the drier "Yucatan" forests of Campeche, Quintana Roo, and Yucatan. Most of the major families are likewise more diverse in the Gulf area, but the palms (Arecaceae) and Sapindaceae are richer in the Yucatan area.

Seven basic distributional patterns of species within the Mexican rain forest area are recognized: widespread, widespread Gulf, southern Gulf, high precipitation, widespread Yucatan Peninsula, restricted Yucatan Peninsula, and Lacandon. It is likely that floristic interchange between Atlantic and Pacific slope floras has occurred within Mexico via the Isthmus of Tehuantepec and perhaps other routes.

Based on the floristically and historically unique nature of Mexican rain forests, the importance of the conservation of these forests and their flora is clear. Five areas are suggested as especially important: the three areas of high precipitation, the Lacandon area, and the southern Yucatan Peninsula.

The Spanish Civil War had an important effect on the study of Mexican rain forests. Faustino Miranda, already settled on a life of studying marine algae in Spain, was forced to leave his homeland and arrived in Mexico City in 1939. Finding himself too far from the coast to efficiently pursue marine botany, he turned his talents to the study of tropical trees and vegetation. Although he started with the drier vegetation types, he soon discovered the exuberance of Mexico's rain forests. Since that time, these rain forests have been not only an object of continual scientific study but also the testing ground on which many of today's leading Mexican botanists have proved themselves. Miranda had planned a tree flora of the Mexican rain forests, but his untimely death in 1964 robbed us of what surely would have been a magnificent and insightful work. To this day, a complete list of the trees of the Mexican rain forests is lacking, although Pennington and Sarukhán's (1968) guide to the common tropical trees of Mexico has admirably filled a part of this need. An objective of this chapter, therefore, is to provide a preliminary list of the canopy trees of the rain forests of the Gulf of Mexico and Caribbean slopes of Mexico.

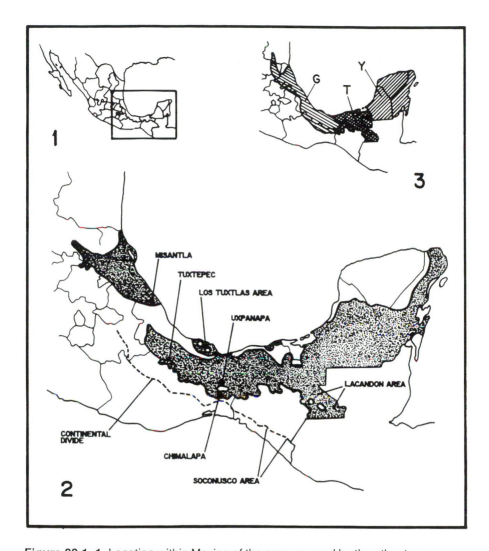

Figure 22.1. 1. Location within Mexico of the area covered by the other two maps in this Figure as well as those in Figures 22.4 and 22.5. **2.** Generalized distribution of evergreen and semievergreen rain forest on the Atlantic slope of Mexico, and location of localities mentioned in text. Distribution of rain forest based on Rzedowski (1978) with modifications based on Breedlove (1981), Flores et al. (1971), López (1980), Olmsted et al. (1983), Puig (1976), Téllez & Sousa (1982), and the author's observations. Limits are approximate, and many small inclusions of other types of vegetation (lowland oak forest, savanna, etc.) are not shown. **3.** Generalized floristic provinces of the Atlantic slope of southeastern Mexico, based on Rzedowski (1978). Rzedowski recognizes a "Peninsula of Yucatan" province (Y), and a "Coast of the Gulf of Mexico" province (T + G). For the species distribution patterns used in the present work, the areas within the states of Tabasco, Chiapas, and extreme western Campeche (T) are considered to be a zone of overlap between the Gulf (G) and Yucatan (Y) distributions.

A second objective is to use the data of the list to investigate the phytogeographic relationships of the rain forest canopy tree flora. These results are then combined with paleobotanical data to discuss the origins of the Mexican rain forest flora.

MODERN FLORA

Methodology

A list was compiled of the species of trees that meet the following criteria: (1) occur in lowland (below 1,000 m elevation) evergreen and semievergreen rain forest on the Atlantic slope of Mexico (Gulf of Mexico and Caribbean drainages); and (2) include individuals that have been recorded to reach at least 18 m (ca. 60 feet) in height in Mexico, Belize, or Guatemala. The objective is a list of those species that at least occasionally form part of the canopy tree flora of these rain forests. Although I consider that a minimum height of 20 m should include these species, I have used an 18 m (60 feet) cutoff and have included height data from neighboring countries because of the incomplete nature of our knowledge and collections of Mexican rain forest trees. Only species documented to occur in the Atlantic rain forests of Mexico, based on herbarium specimens, monographs, or other reliable reports, are included. All rain forests of the Pacific slope of Mexico are excluded from the study.

"Tree" is defined broadly to include all plants that reach the prescribed minimum height and that are at least sometimes self-supporting at that height. "Strangler" hemiepiphytes, which eventually become self-supporting trees (e.g., *Ficus*) are included, as are monocotyledons.

The definition of rain forest employed corresponds to that of Rzedowski (1978) for "bosque tropical perennifolio" (tropical evergreen forest) (Fig. 22.1), which combines the "selva alta perennifolia" (high evergreen forest) and the "selva alta o mediana subperennifolia" (high or medium semievergreen forest) of Miranda and Hernández (1963). As with any floristic work that is delimited ecologically, decisions on the inclusion of certain species are difficult and debatable. Excluded in this study are species that occur in other lowland communities, e.g., oak, pine, mangrove, or subdeciduous forests, but that do not occur in rain forest as here defined, or that occur in rain forest only in ecotones. Likewise, species of cloud forest and other montane vegetation that occur in lowland rain forest only in ecotones are excluded. For these reasons, the list includes no species of *Quercus, Liquidambar, Ulmus, Pinus*, etc. Also excluded are highland species that occur in lowland forests only as scattered individuals in areas near large montane populations of these species and that do not appear to form self-sustaining lowland populations. Essentially temperate species that descend into some rain forest areas only along rivers (e.g., *Platanus mexicana* Moric.) are excluded, whereas the essentially lowland riparian species *Populus mexicana*[1] is included. The list thus excludes some species that are found in the rain forests under some conditions but attempts to include those that are well adapted to and self-sustaining in the lowland rain forest—the "core" species.

The exclusion of some temperate elements can be criticized on the grounds that these elements have played an important role in the evolution of Mexican rain forests (Sarukhán, 1968; Toledo, 1982) and that to eliminate them gives a biased view of the floristic composition of these forests. Nevertheless, in my experience most of these species occur either in separate, discrete lowland vegetation types (e.g., the lowland savanna-like forests of *Quercus oleoides* Schldl. & Cham. and of *Pinus caribaea* Morelet) or in ecotonal areas, as noted above. Where such species appear to be fully integrated into normal rain forest communities (e.g., *Clethra* sp.), they are included, although surely a few species of this type have been missed. Because one objective of the present work is to investigate the floristic affinities of the tropical component of the Mexican rain forests, the exclusion of clearly temperate species diminishes a source of potential error.

The angiosperm families used are those recognized by Cronquist (1981). Intraspecific taxa are used where appropriate, and in these cases all information on distribution, ecology, and so on refers to these taxa and not to the species as a whole.

The distribution range of each species within Mexican Atlantic rain forest is given in a generalized fashion. The two basic areas used (Fig. 22.1, Map 3) correspond to two floristic divisions recognized by Rzedowski (1978): the "Costa del Golfo de México" (here the "Gulf" distribution) and the "Península de Yucatán" (here the "Yucatan" distribution). The principal difference between the present work and that of Rzedowski is the recognition here of the area that includes Tabasco, northern and eastern Chiapas, and extreme western Campeche as a zone of floristic overlap between the two divisions, instead of as part of the Gulf distribution pattern.

Total distribution range is given for all species except those that have not yet been identified (listed as "sp.") and those that are poorly understood taxonomically. In the case of cultivated species, only those for which there is good evidence of native status in Mexican rain forests are included; for those that have been cultivated to such an extent that their original total distribution outside Mexico is unclear, an estimate of their total native distribution is given, but these taxa are excluded from the phytogeographic analysis.

Many species included apparently reach canopy tree size only rarely. Therefore an indication is given of those that commonly exceed 20 m in height in order to identify the subset of species that commonly reach canopy size.

---

[1]Authors for species are included in the text only for species not included in the Appendix.

The ecological amplitude of Mexican rain forest species, especially with respect to the ability of these species to exist in drier types of lowland vegetation, has been emphasized by Toledo (1982) as an important factor in the evolution of the Mexican rain forests. Data on the common occurrence of the species in these drier lowland forests is thus also provided, although this evaluation is clearly subjective. Species that occur in dry regions only in riparian or inundated areas (e.g., *Ficus insipida, Bravaisia integerrima*) are considered not to occur ecologically in dry forests.

Data on distribution and ecology in Mexico, and on height in Mexico, Belize, and Guatemala, are based on: (1) a partial review of the herbarium collections at CHAPA, MEXU, TEX/LL, and US (herbarium acronyms throughout this chapter follow Holmgren et al., 1990); (2) data in relevant monographs and in *Flora of Guatemala* and *Flora de Veracruz*; (3) the author's personal observations; (4) consultation with specialists (see acknowledgments); and (5) the judicious use of other published sources, especially: Breedlove (1981); Gómez (1966); Ibarra & Sinaca (1987); León & Gómez (1970); López (1980); Lott (1985); Lundell (1942a); Meave (1983); Miranda (1952-53, 1958, 1961); Pennington & Sarukhán (1968); Pérez & Sarukhán (1970); Puig (1976); Rzedowski (1963, 1978); Rzedowski & McVaugh (1966); Sánchez (1987); Sousa (1968); Sousa & Cabrera (1983); Téllez & Sousa (1982); Vázquez (1989); and Vera (1988). Data on total distribution are based on sources 1, 2, and 4 above, and the use of other floras, including *Flora of Panama.*

The list is preliminary. Neither the taxonomic work nor the nomenclatural verification was exhaustive, and many areas of Mexican rain forest remain poorly known. Nevertheless, I believe that the list is sufficiently accurate and complete to provide a meaningful picture of the size, floristic composition, and geographic affinities of the flora.

Species Richness

The list of species (Appendix) includes 64 families, 230 genera, and 452 species or intraspecific taxa, of which 322 frequently exceed 20 m in height (Table 22.1). Only one species (*Trichilia moschata*) is represented by more than one intraspecific taxon. Therefore for ease of discussion "species" is understood to include species and intraspecific taxa.

The dicotyledonous families account for 61 families, 221 genera, and 439 species. Gymnosperms are represented by a single species of Podocarpaceae, and the monocots by 12 species in two families, Agavaceae and Arecaceae.

The present list does not consider the much smaller area of rain forests on the Pacific slope of Mexico, which includes at least those of the Soconusco region of southern Chiapas (Fig. 22.1, Map 2) (Rzedowski, 1978) and probably an area of the Sierra Madre del Sur of Oaxaca (Pennington & Saruhkán, 1968). These areas are excluded because: (1) detailed floristic-vegetational studies are lacking for them; (2) these forests, at least in the

Table 22.1. Synthesis of species data

Species distribution	Total No. of species	No. of species commonly > 20 m	Species adapted to drier forests	
			No.	%
Gulf	272	200	35	12.9
Yucatan	39	24	19	48.7
Gulf/Yucatan (widespread)	113	75	75	66.4
Only Tabasco and/or Chiapas	28	23	1	3.6
Total	452	322	130	28.8

For definitions of species distributions, see explanation in the Appendix and Fig. 22.1, map 3.

Soconusco, are reported to reach 1,400 m in elevation and interdigitate in a complex way with cloud forests (Miranda, 1952–53), making it difficult to distinguish consistently rain forest species from cloud forest species based on herbarium material; and (3) I have little field experience in this area. A preliminary list that included rain forest species only from below 800 m elevation in the Soconusco contained fewer than ten species not found on the Atlantic slope, the most notable of these species being *Terminalia oblonga* (R. & P.) Steudel. However, the Soconusco rain forests include many other species above 800 m (Miranda, 1952–53).

Equivalent reliable data for large, more equatorial areas in the Neotropics are scarce. It is perhaps more instructive to compare the present data with those of the *Tree Flora of Malaya*. The estimate for the total flora of trees that reach three feet in girth—roughly equivalent to the present criteria—for West Malaysia and Singapore (Symington, 1943; Whitmore, 1972) includes about 1,830 species in 490 genera and 95 families, including gymnosperms but excluding monocots. The total area included is roughly three-fifths that of the area originally covered by rain forest in Mexico and lies close to the equator. Even taking into account that the Malay flora includes a number of montane and other species, the relatively impoverished nature of the Mexican rain forest canopy tree flora (one-fourth of that found in the smaller Malayan area) is clear. Perhaps even more dramatic is the comparison of the 452 Mexican species with the 375 species of canopy trees reported by Poore (1968) for just 23.04 hectares (ha) (transects within a 1.04 km² area) of lowland rain forest in West Malaysia. Neotropical equatorial forests may be richer in canopy tree species than the Malayan ones (Gentry, 1988a). On the other hand, all of North America north of Mexico contains only about 220 native species at least 18 m tall (almost one-third of which are gymnosperms) in about 37 families and 70 genera (based on data in Elias, 1980). The greatest number of North American species in any area equal to that occupied by Mexican rain forests is found in a swath of the southeastern-east central United States, where perhaps slightly over one-

Table 22.2. Families with ten or more species in the canopy tree flora

Family	Total No. of genera/species	No. of species with individuals commonly > 20 m
Leguminosae	35/84	63
Papilionoideae	17/36	27
Mimosoideae	8/33	25
Caesalpinioideae	10/15	11
Moraceae	8/30	27
Lauraceae	5/30	22
Sapotaceae	5/23	18
Rubiaceae	18/23	16
Euphorbiaceae	12/17	14
Sapindaceae	8/14	9
Flacourtiaceae	7/13	4
Myrtaceae	5/12	2
Tiliaceae	6/11	11
Meliaceae	4/11	8
Polygonaceae	2/11	8
Arecaceae	7/11	8

half of the species occur. These data suggest that at the level of regional tree floras in equal areas, the subtropical Mexican rain forest flora is about four times as rich as the richest temperate U.S. flora and perhaps one-fourth to one-fifth as rich as equatorial Amazonian floras.

At a more local level, about 290 of the species listed occur within the approximately 2,000 km^2 of the Uxpanapa rain forests of southern Veracruz (Wendt, unpub.). About 150 of the species are reported for the approximately 600 ha of rain forest in the Los Tuxtlas Tropical Biological Station in Veracruz (Ibarra & Sinaca, 1987), of which about 130 have been observed to reach 20 m in height at that site (G. Ibarra, pers. comm.)

Plot inventories of 1 ha or more for canopy trees in Mexican rain forest are few. In the Uxpanapa rain forests, our studies of plots of five contiguous hectares reveal 24–41 species per hectare of trees at least 28.5 cm diameter breast high (d.b.h.), and 60–80 species per 5 ha plot. The highest diversity occurs in karst forest at the wettest site, which also presents the greatest substrate heterogeneity (Vázquez, 1989; Vera, 1988; Wendt, unpub.). Meave (1983), in his study of 1 ha of relatively homogeneous rain forest in the Lacandon area of eastern Chiapas, reported 44 tree species over 20 m in height. Meave's four 0.25-ha subplots were not contiguous, being separated by up to several kilometers, and thus his numbers are not immediately comparable with the Uxpanapa data. Toledo (1982), summarizing data from a series of 2,000 m^2 inventories of woody vegetation 3.3 cm or more and 6.6 cm or more d.b.h. from rain forests along a transect from

Table 22.3. Genera with five or more species in the canopy tree flora

Genus	No. of species	No. of species with individuals commonly > 20 m
Ficus	19	19
Inga	15	11
Nectandra	12	10
Lonchocarpus	11	8
Pouteria	10	10
Coccoloba	10	7
Ocotea	9	7
Sideroxylon	8	6
Zanthoxylum	7	6
Cordia	7	5
Casearia	7	1
Eugenia	7	0
Licaria	6	2
Sloanea	5	5
Albizia	5	5
Cupania	5	3

southern San Luis Potosí to eastern Chiapas, concluded that the Lacandon forests present the highest local richness in Mexican rain forests; however, no data were available from the Uxpanapa area at the time, and it remains uncertain where the highest local richness in canopy tree species occurs in Mexican rain forest.

On the other hand, it is clear that Mexican rain forests are less rich in canopy tree species at the local level than many Amazonian forests. Gentry (1988a) reported 49-81 species of trees 30 cm or more d.b.h. per hectare for six internally homogeneous rain forest plots in the Peruvian Amazon. The same author (1988b) presented data from a large series of 1,000 m² rain forest plots in which all species of plants 2.5 cm or more d.b.h. were inventoried. The 108–109 species reported for a plot at Los Tuxtlas, Veracruz, contrast with figures that exceed 250 species for some South American rain forests.

Floristic Composition

The best represented families are listed in Table 22.2. Leguminosae is the largest family at the species level, with 84 species, or 18.6% of the total. Within this family, the subfamilies Papilionoideae (36 species) and Mimosoideae (33 species) are each larger than any other family within the flora, whereas Caesalpinioideae (15 species) is exceeded only by Moraceae (30), Lauraceae (30), Sapotaceae (23), Rubiaceae (23), and Euphorbiaceae

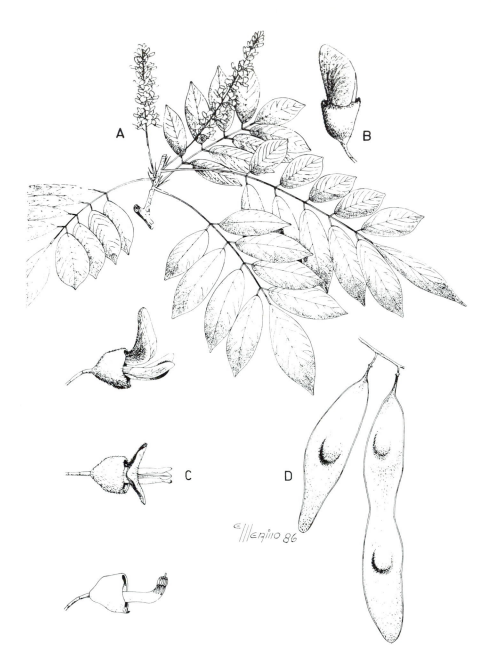

Figures 22.2 and 22.3 (above and right). Genera that are more diverse in Mexican rain forests than in South American rain forests. **Figure 22.2** *Lonchocarpus* (*L. verrucosus*), Leguminosae. A, flowering branch; B, flower bud; C, flower and its parts; D, fruit. **Figure 22.3**. *Sideroxylon* (*S. portoricense* subsp. *minutiflorum*), Sapotaceae. A, flowering branch; B, flower and its parts; C, fruit; D, seed. Illustrations by Eduardo Merino.

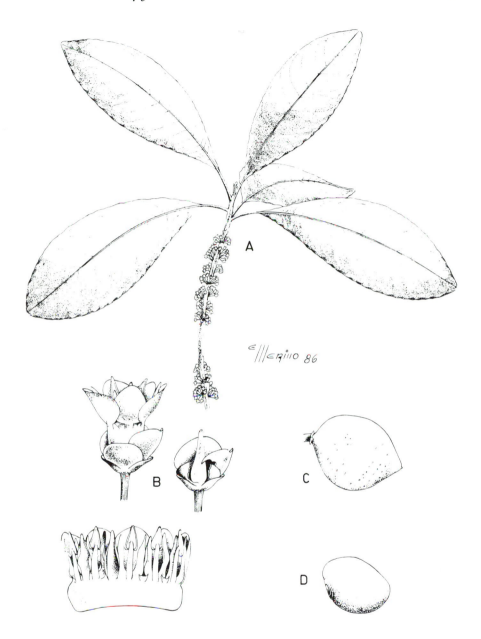

(17). The legumes have been found by Gentry (1988b) to be the most species-rich family in almost all of his neotropical small plots sampling woody vegetation 2.5 cm or more or 10.0 cm or more d.b.h. All of the other families listed, except Tiliaceae and Polygonaceae, are generally well developed throughout the neotropical mainland rain forests; specifically, Moraceae, Lauraceae, Sapotaceae, Rubiaceae, Euphorbiaceae, Meliaceae, and Arecaceae are cited by Gentry (1988b) as usually being among the ten most species-rich families on local plots. The relative richness in our tree flora of Polygonaceae is due to the diversity in the genus *Coccoloba*, discussed below. The family Tiliaceae is well developed in Mexican rain forests, and most of the genera included in the present list either do not reach South America (*Berrya*) or are poorly developed there (*Heliocarpus*, *Mortoniodendron*, *Trichospermum*). Two of the tiliaceous genera (*Berrya*, *Trichospermum*) are also present in the Southeast Asia region. In Amazonia, the family is not notably more diverse than in Mexico. Thus, to refer to this family in the New World as an "Amazonian-centered Gondwanan family" (Gentry, 1982a) seems to be an over-simplification, as several of the neotropical lines may well have arrived via Laurasian connections (see Origins below), if indeed the family is ultimately Gondwanan in origin. It is notable that over half of the species of Tiliaceae included are character-istic of secondary forests.

The best represented genera are listed in Table 22.3. It is of interest that the largest genus, *Ficus*, consists mostly of hemiepiphytic ("strangler") species in Mexico. Most of the genera listed are well developed throughout neotropical rain forests. *Coccoloba* is most diverse in circum-Caribbean areas but nevertheless presents numerous species in Brazil (Howard, 1961). On the other hand, the diversity of *Lonchocarpus* (*sensu stricto*, see Sousa & Delgado, this volume) in the Mexican rain forest flora contrasts sharply with the relative unimportance of this genus in South American rain forest. Sousa and Delgado (this volume) note that, although Mexico boasts a total of 76 species of *Lonchocarpus* (in habitats ranging from very dry to rain forest) and Nicaragua has 23, Panama has but 13, Colombia and Venezuela about 15 between them, and Brazil 8. They also note that the genus probably originated in the general area of Mexico and northern Central America. The diversity of *Lonchocarpus* is thus a unique character-istic of the rain forests of this region (Fig. 22.2). In addition, *Sideroxylon* (Fig. 22.3), in the broad sense as recognized by Pennington (1990) to include *Bumelia*, *Dipholis*, and *Mastichodendron*, is poorly developed in the Amazon region, whereas it is diverse in Mexico, Central America, and especially the Antilles.

Species Distributions in Mexico

Table 22.1 summarizes the data on the distribution patterns of the species in Mexico. Most species (272, or 60.2%) present a "Gulf" distribution (Fig. 22.1, Map 3); all of these species occur in the rain forests of Veracruz or Oaxaca and often in bordering states, but not in the Yucatan Peninsula (except extreme southwestern Campeche in many cases). A small group (39 species, or 8.6%) is restricted to the Yucatan Peninsula in Mexico, not reaching Veracruz or Oaxaca, whereas a considerably larger group com-prises species widespread in Mexican Atlantic rain forest (113 species, or 25.0%). A few species (28 or 6.2%) are known in the area only from Chiapas or Tabasco. Thus the Gulf rain forests (excluding Chiapas and Tabasco)

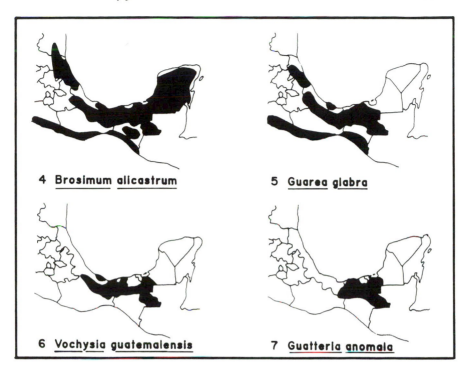

4 **Brosimum alicastrum**

5 **Guarea glabra**

6 **Vochysia guatemalensis**

7 **Guatteria anomala**

Figure 22.4. Generalized distributions of representative species; see text for explanation of patterns. **4.** *Brosimum alicastrum* var. *alicastrum*, pattern 1. **5.** *Guarea glabra (sensu lato)*, pattern 2. **6.** *Vochysia guatemalensis*, pattern 3. **7.** *Guatteria anomala*, pattern 3. *Sources*: all maps, Pennington & Sarukhán (1968) and author's data base; map 4, Berg (1972); map5, *Guarea* Pennington (1981).

include 385 canopy tree species, of which 70.6% are not found in the rain forests of the Yucatan Peninsula, whereas the Yucatan rain forests (excluding Chiapas and Tabasco) include 152 species, of which only 25.7% do not reach Veracruz or Oaxaca. The Chiapas-Tabasco rain forests, containing most of the species of both the Gulf and Yucatan elements plus a small unique element, are probably at least as rich in total species as the rest of the Gulf forests.

As a rough estimate (Table 22.1), about 29% of the total number of species are adapted to drier forest types in Mexico. However, the estimates for the different distribution types are markedly dissimilar. Over 65% of the widespread species and almost half of the Yucatan species also occur frequently in drier forests, whereas the percentages are low for the Gulf species and those known only from Tabasco and Chiapas. This point is not surprising: When the Atlantic rain forests of Mexico are divided into evergreen and semievergreen types (e.g., Pennington & Sarukhán, 1968), the evergreen (wetter) types are restricted to the southern Gulf and Tabasco-Chiapas areas. Toledo (1982) recognized a basic dichotomy in dominant trees of Mexican rain forests: (1) species usually with a Gulf

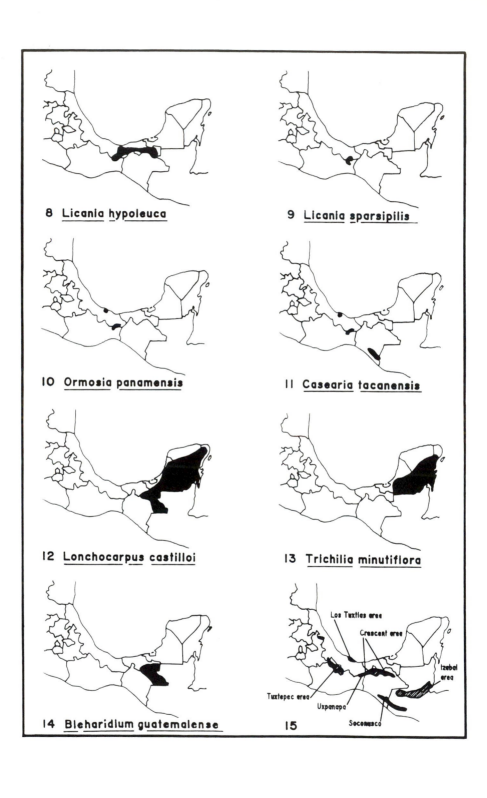

8 Licania hypoleuca

9 Licania sparsipilis

10 Ormosia panamensis

11 Casearia tacanensis

12 Lonchocarpus castilloi

13 Trichilia minutiflora

14 Bleharidium guatemalense

15

Los Tuxtles area

Crescent area

Izabal area

Tuxtepec area

Uxpanapa

Socorusco

distribution (as here defined) and a relative intolerance to dry conditions; and (2) species widespread in Mexican rain forests and well adapted to drier conditions. These two groups stand out clearly in the present data, as does a third group of species restricted to the Yucatan Peninsula in which almost half of the species are also well adapted to dry conditions.

The use of an a priori classification of distributions within Mexico limits the analyses possible, but I consider that the present data base is incomplete and that the presentation of more detailed distributions would be misleading in many cases. A detailed analysis of geographic distribution patterns of Mexican rain forest trees thus awaits further study. Nevertheless, based on recurring patterns noted during the compilation of the present list, I believe that it is useful to recognize seven basic distributional patterns for nonendemic species (the endemic species are considered separately below). The patterns of nonendemics (Figs. 22.4 and 22.5), based primarily on distribution only within the rain forests under study except as noted, are as follows.

1. *Widespread distribution* (Fig. 22.4, map 4; *Brosimum alicastrum*): species occurring in both Veracruz-Oaxaca and the Yucatan Peninsula, often also on the Pacific slope (e.g., *Bursera simaruba, Cedrela odorata, Dendropanax arboreus, Protium copal*).
2. *Widespread Gulf distribution* (Fig. 22.4, map 5; *Guarea glabra*): species occurring from eastern Chiapas to at least the Misantla region of central Veracruz (see Fig. 22.1, map 2 for localities mentioned in the text), not in the Yucatan Peninsula (excluding extreme southwestern Campeche), often also on the Pacific slope (e.g., *Apeiba tibourbou, Sideroxylon persimile, Licaria capitata, Pithecellobium arboreum*).
3 *Southern Gulf distribution* (Fig. 22.4, maps 6 and 7): as in pattern 2, but not reaching as far north as the Misantla region of Veracruz, often not on the Pacific slope of Mexico, e.g., *Dialium guianense, Guarea grandifolia, Poulsenia armata, Terminalia amazonia, Vochysia guatemalensis* (Fig. 22.4, map 6). Included here are more restricted species found only from eastern Chiapas to the Uxpanapa-Chimalapa area of extreme southeastern Veracruz and adjacent Oaxaca, e.g., *Coutaportla*

Figure 22.5. Generalized distributions of representative species; see text for explanation of patterns. **8.** *Licania hypoleuca*, pattern 4. **9.** *Licania sparsipilis*, pattern 4. **10.** *Ormosia panamensis*, pattern 4. **11.** *Casearia tacanensis*, pattern 4. **12.** *Lonchocarpus castilloi*, pattern 5. **13.** *Trichilia minutiflora*, pattern 6. **14.** *Blepharidium guatemalense*, pattern 7. **15.** Distribution of areas of high precipitation in southeastern Mexico and northern Central America. The lined areas represent regions with over 3000 mm. average annual precipitation; the "three areas of high precipitation" mentioned in the text are the Crescent area, the Los Tuxtlas area, and the Tuxtepec area. *Sources*: maps 8–14, author's data base; map 11, Sleumer (1980); maps 12 & 14, Pennington & Sarukhán (1968); map 13, Pennington (1981); map 15, based on Wendt (1989) and García (1970).

guatemalensis, Guatteria anomala (Fig. 22.4, map 7), *Lennea modesta, Ormosia macrocalyx.*

4. *High precipitation distribution* (Fig. 22.5, maps 8–11): species restricted in Mexico to one or more of the three lowland areas of high precipitation (Fig. 22.5, map 15): (1) "Crescent area" (Wendt, 1989) of extreme southeastern Veracruz (Uxpanapa), adjacent Oaxaca (Chimalapa), southern Tabasco, and northern Chiapas; (2) Los Tuxtlas area of southern Veracruz; and (3) Tuxtepec area of northern Oaxaca. Crescent area species sometimes extend slightly northward into the adjacent lowlands of southern Veracruz and northern Tabasco (Fig. 22.5, map 8). Examples: Crescent area only (either Uxpanapa-Chimalapa only, or entire area): *Casearia arborea, Licania hypoleuca* (Fig.22.5, map 8), *Inga alba, Licaria sparsipilis* (Fig. 22.5, map 9); Crescent area plus Los Tuxtlas area: *Cordia megalantha, Ormosia panamensis* (Fig. 22.5, map 10), *Sloanea petenensis, Virola guatemalensis.* An interesting variant involves species known from the Atlantic rain forests only in or near one or more of the three mentioned areas but also occurring on the Pacific slope of Mexico, where they are either restricted to the Soconusco (e.g, *Casearia tacanensis* [Los Tuxtlas, Uxpanapa] [Fig. 22.5, map 11], *Micropholis melinoniana* [Uxpanapa], *Trichilia moschata* subsp. *matudae* [Los Tuxtlas, Uxpanapa]) or widespread (e.g., *Nectandra* sp. "globosa" [Tuxtepec, Los Tuxtlas], *Willardia schiedeana* [Los Tuxtlas, Uxpanapa]). (Note: "Pacific slope" is considered to include the eastern slope of the Sierra Madre of Chiapas, just across the continental divide from the Soconusco area [Fig. 22.1, map 2]. Although technically part of the Gulf of Mexico drainage, it is separated from the Gulf rain forests by extensive arid and montane areas.)

5. *Widespread Yucatan Peninsula distribution* (Fig. 22.5, map 12; *Lonchocarpus castilloi*): species occurring more or less throughout the rain forest area of the Yucatan Peninsula (Campeche, Yucatan, Quintana Roo) and to the Lacandon area of eastern Chiapas; either restricted to this area in Mexico (e.g., *Alseis yucatanensis, Bucida buceras, Caesalpinia yucatanensis, Lonchocarpus castilloi*) or also occurring on the Pacific slope (e.g., *Acacia dolichostachya, Vitex gaumeri*).

6. *Restricted Yucatan Peninsula distribution* (Fig. 22.5, map 13; *Trichilia minutiflora*): species occurring in the Yucatan Peninsula but not extending to Tabasco or Chiapas; either restricted to this area in Mexico(e.g., *Antirhea lucida, Celtis trinervia, Thouinia canescens, Trichilia minutiflora*) or also occurring on the Pacific slope (e.g., *Licaria campechiana*).

7. *Lacandon distribution* (Fig. 22.5, map 14; *Blepharidium guatemalense*): species restricted in Mexico to the Lacandon area of eastern Chiapas and frequently nearby parts of northern Chiapas and Tabasco (e.g., *Luehea seemannii, Orthion subsessile, Pourouma guianense, Sebastiana tuerckheimiana*).

With almost all of these patterns, the species may also occur on the Pacific slope of Mexico (as exemplified by Fig. 22.4, maps 4 and 5, and Fig. 22.5, map 11); in some cases, the two areas of distribution are quite distant in Mexico (e.g., pattern 6). This relationship has been noted previously for the disjunct rain forest floras of northern Chiapas and the Soconusco by Miranda (1952–53), and in a more general way by Rzedowski (1978). The two slopes are today separated by large montane areas except in part of the Isthmus of Tehuantepec, where at present low, dry hills separate the Gulf forests from the arid Pacific slope. Whereas in many cases the occurrence on both slopes could be due to past northward migrations of rain forest species along both slopes from common areas in Central America— perhaps Nicaragua, where the Caribbean and Pacific lowlands are not separated by high mountains—it seems probable that direct interchange across the Isthmus of Tehuantepec has occurred during more humid periods. This hypothesis is supported by: (1) the existence on both slopes of taxa endemic to Mexico, e.g., the genus *Recchia* (see Fig. 22.9, below) (Wendt & Lott, 1985); and (2) the above-mentioned variant of pattern 4 in which the same taxa occur on the Pacific slope (Fig. 22.1, map 11). In the latter case, because the three Gulf high precipitation areas mentioned (Fig. 22.1, map 15) are located in the general area of the Isthmus and the taxa are not known to occur elsewhere on the Atlantic slope, migration across the Isthmus, probably south to north, is a likely explanation of the pattern. It furthermore suggests the possibility of the arrival of southern elements into Mexican Atlantic rain forests by at least two routes: (1) direct northward migration through the Caribbean lowlands; and (2) Pacific slope migration north through the Soconusco and across the Isthmus of Tehuantepec into the isthmian rain forests. Given the discontinuous nature of the present distribution of potential rain forest habitats on the Pacific slope of Central America, the latter route would probably have been viable only under more humid conditions than exist at present, at least for species not also adapted to drier climates.

These proposed patterns and hypotheses are based on the data available regarding distribution, which are incomplete. Furthermore, much-needed collecting in Mexican rain forests will provide a test for the present speculations.

Total Species Distributions in the New World

The total distributions of the species in Mexico and Central and South America are summarized in Table 22.4 and the occurrence of the species in the Greater Antilles in Table 22.5. One species, *Sabal mexicana*, occurs in native form in southern Texas. Otherwise, all species native to North America north of Mexico are restricted to the southern subtropical portion of Florida; because all such Floridian species also occur in the Greater Antilles, Florida is here included in the Greater Antilles area, as noted in the

Appendix. The total number of species (427) considered excludes some widely cultivated and other problem species, as previously noted.

The combined data of Tables 22.4 and 22.5 indicate that 41 species (9.6% of the total) are endemic to Mexico, 132 species (30.9%) are restricted to Mexico and northern Central America (hereafter "northern Mesoamerica"), and 235 species (55.0%) are restricted to Mexico and Central America ("Mesoamerica"); 107 species (25.1%) reach Amazonia and 88 (20.6%) the Greater Antilles. Species reaching Amazonia are here considered to be those that reach Peru, Brazil, or further south; although this classification is a clear simplification, it seems sufficiently precise for the present purposes. These data indicate that the flora is clearly Mesoamerican, with over half of the species not found outside this area. Rzedowski (1978) has previously noted that most species of Mexican tropical lowland vegetation are restricted to Mesoamerica, an observation borne out here for rain forest canopy tree species. Gentry (1982a) estimated endemism of 42% for all rain forest canopy tree and liana species of the Mesoamerican region.

The above data also indicate a clearly defined northern Mesoamerican element within the flora. Gentry (1982b) stated that a number of southern Central American species are to be expected in the Chocó area of Colombia and Ecuador; thus the present estimate for Mesoamerican endemism may be a bit high, but it should not affect substantially the figures for northern Mesoamerica. Indeed, the present estimate for northern Mesoamerican endemism is probably conservative, as a number of species reach their southern limits of distribution in northern Nicaragua and are thus not counted as northern Mesoamerican endemics because of the arbitrary use of the Nicaragua-Honduras border to separate northern and southern Mesoamerica. A more natural dividing line would fall in central Nicaragua, as discussed below.

The equivalent data for the different groups based on Mexican distribution recognized in Tables 22.4 and 22.5 (e.g., Gulf and Yucatan) reveal notable differences.

1. *Widespread species* (Gulf + Yucatan) are in general also widespread in the New World. Only 13 species (12.4%) are restricted to northern Mesoamerica and 38 (36.2%) to Mesoamerica in general, whereas 31.4% reach Amazonia and 41.9% reach the Greater Antilles.
2. *Gulf species* include almost all Mexican endemics (39 species, 15.2%); 80 species (31.1%) are restricted to northern Mesoamerica and 151 (58.8%) to Mesoamerica. A good percentage reach Amazonia (26.5%), but only 12.8% reach the Greater Antilles.
3. *Yucatan species* include one Mexican endemic, 21 species (53.8%) endemic to northern Mesoamerica, and 26 species (66.7%) endemic to Mesoamerica; only 5.1% reach Amazonia, whereas 25.6% reach the Greater Antilles.
4. *Tabasco-Chiapas species* not found elsewhere in Mexico include one Mexican endemic, 18 species (69.2%) endemic to northern Mesoamerica, and 20 species (76.9%) restricted to Mesoamerica;

Table 22.4. New World distributions (excluding Greater Antilles) of Mexican Atlantic rain forest canopy tree species

Distribution in Mexico	No. of species considered	No. of species reaching southern extreme of New World continental distribution				
		Mexico	Guatemala, Belize, Honduras, El Salvador	Nicaragua, Costa Rica, Panama	Colombia, Venezuela, Guyanas, Ecuador	Brazil, Peru, or further south
Gulf	257	39 (15.2%)	44 (17.1%)	77 (30.0%)	29 (11.3%)	68 (26.5%)
Yucatan	39	1 (2.6%)	24 (61.5%)	6 (15.4%)	6 (15.4%)	2 (5.1%)
Gulf + Yucatan (widespread)	105	0	19 (18.1%)	36 (34.3%)	17 (16.2%)	33 (31.4%)
Only Tabasco and/or Chiapas	26	1 (3.8%)	17 (65.4%)	2 (7.7%)	2 (7.7%)	4 (15.4%)
Total	427	41 (9.6%)	104 (24.4%)	121 (28.3%)	54 (12.6%)	107 (25.1%)

For definitions of Mexican and New World distribution types, see Appendix.

Table 22.5. Presence in Greater Antilles of Mexican Atlantic rain forest canopy tree species

Distribution in Mexico	No. of species reaching Greater Antilles/No. of species within phytogeographical groups from Table 22.1					
	Mexico	Guatemala, Belize, Honduras, El Salvador	Nicaragua Costa Rica, Panama	Colombia, Venezuela, Guyanas, Ecuador	Brazil, Peru, or further south	Total
Gulf	0/39	3/44 (6.8%)	6/77 (7.8%)	4/29 (13.8%)	20/68 (29.4%)	33/257 (12.8%)
Yucatan	0/1	4/24 (16.7%)	1/6 (16.7%)	4/6 (66.7%)	1/2 (50.0%)	10/39 (25.6%)
Gulf + Yucatan (widespread)	0/0	6/19 (31.6%)	11/36 (30.6%)	10/17 (58.8%)	17/33 (51.5%)	44/105 (41.9%)
Only Tabasco and/or Chiapas	0/1	0/17	0/2	0/2 (25.0%)	1/4 (3.8%)	1/26
Total	0/41 (0%)	13/104 (12.5%)	18/121 (14.9%)	18/54 (33.3%)	39/107 (36.4%)	88/427 (20.6%)

For explanations of Mexican and New World distribution types and definition of Greater Antilles, see Appendix.

TABLE 22.6. Groups recently monographed or revised for New World: all relevant monographs in *Flora Neotropica* and other works: 1969 to present

Acanthaceae: *Bravaisia* (Daniel, 1988)
Actinidiaceae: *Saurauia* (Hunter, 1966; Soejarto, 1980, 1984)
Annonaceae: *Cymbopetalum* (N. Murray, pers. comm.); *Rollinia* (P.J.M. Maas, pers. comm.); *Sapranthus* (G. Schatz, pers. comm.)
Apocynaceae: *Tabernaemontana, Stemmadenia* (Allorge, 1985)
Arecaceae: *Gaussia* (Quero & Read, 1986); *Sabal* (Zona, 1990); *Thrinax* (Read, 1975)
Bombacaceae: *Quararibea* (W. Alverson, pers. comm.)
Boraginaceae: *Ehretia* (Miller, 1989)
Cecropiaceae: *Coussapoa* (Akkermans & Berg, 1982)
Chrysobalanaceae (Prance, 1972)
Euphorbiaceae: *Margaritaria* (Webster, 1979)
Flacourtiaceae (Sleumer, 1980; Wendt, 1988)
Hernandiaceae (Kubitzki, 1969; A. Espejo, pers. comm.)
Lacistemataceae (Sleumer, 1980)
Lauraceae: *Licaria* (Kurz, 1983; van der Werff, 1988); *Nectandra* (J. Rohwer, pers. comm.)
Leguminosae: *Acosmium, Ateleia, Dussia, Myroxylon, Ormosia* (Rudd, 1972); *Cassia, Senna* (Irwin & Barneby, 1982); *Hymenaea* (Lee & Langenheim, 1974); *Lysiloma* (Thompson, 1980); *Pterocarpus* (Rojo, 1972); *Swartzia* (Cowan, 1968)
Lythraceae: *Ginorea* (S. Graham, pers. comm.)
Malvaceae: *Hampea, Robinsonella* (Fryxell, 1969, 1973, 1988)
Melastomataceae: *Bellucia* (Renner, 1989); *Mouriri* (Morley, 1976, 1989)
Meliaceae (Pennington, 1981)
Moraceae: *Brosimum, Pseudolmedia* (Berg, 1972)
Myrtaceae: *Pimenta* (Landrum, 1986)
Olacaceae (Sleumer, 1984)
Rhamnaceae: *Colubrina, Krugiodendron* (Johnston, 1971, pers. comm.)
Rubiaceae: *Hamelia* (Elias, 1976)
Rutaceae: *Esenbeckia* (Kaastra, 1982)
Salicaceae: *Populus* (Eckenwalder, 1977)
Sapindaceae: *Thouinia* (Votava, 1973)
Sapotaceae (Pennington, 1990)
Turneraceae: *Erblichia* (Arbo, 1979)
Ulmaceae: *Ampelocera* (Todzia, 1989)

species in common with either Amazonia or the Greater Antilles are few.

The Mexican canopy tree flora does not appear to present strong direct links to the Antillean flora at the species level. No species are found only in Mexico and the Greater Antilles. Within the group of species that are not endemic to Mexico but are restricted to northern Mesoamerica in their continental distributions, only 12.5% occur in the Greater Antilles. Both within the total flora and within each subgroup based on Mexican distribution (Table 22.5), there is in general a positive correlation between more widespread continental distribution and higher percentage of occurrence in the Greater Antilles. These data indicate that for the most part the floristic relation between Mexico and the Antilles is probably indirect. Nevertheless, among species restricted to northern Mesoamerica in their continental distribution (excluding Mexican endemics), the 43 species that occur in the Peninsula of Yucatan (Yucatan and widespread groups) have a much higher percentage also occurring in the Greater Antilles (10 species, 23.2%) than do the 61 species not present in the peninsula (3 species, 4.9%). Thus such direct floristic relationship with the Greater Antilles as exists is strongest in the Yucatan rain forest flora. This relationship has been noted previously for the overall flora of the Yucatan Peninsula (in the broad sense) by several authors (e.g., Miranda, 1958; Rzedowski, 1978; Standley, 1930). The present data support the conclusion of Miranda (1958), who noted that although the Peninsular-Antillean floristic links are clearly closer than for other parts of Mexico they are not strong.

A recurrent problem in studies that involve large-scale species distributions is that the same species may receive different names in different parts of its range, owing to the lack of monographic studies for the entire area of the group involved. This problem can lead to overestimates of local and regional endemism. To investigate this possibility, a subset of 126 species included in monographs in which all of the New World species were studied (Table 22.6) was analyzed. Included are species found in all *Flora Neotropica* monographs (1968 to the present) and in most other monographs (some unpublished) from 1969 to the present, with updated distributions from the present data base (Appendix). These 126 species—well over one-fourth of the full set of 427—reveal the following pattern: 12 species (9.5%) endemic to Mexico (three of the 12 were described after the monographer's work by other authors); 37 (29.4%) restricted to northern Mesoamerica; 61 (48.4%) restricted to Mesoamerica; 36 (28.6%) reaching Amazonia; 28 (22.2%) reaching the Greater Antilles. These figures are roughly comparable with the figures for the total flora, the most noticeable differences being an increase in Amazonian species and a decrease in species restricted to Mexico and Central America as a whole. The relative figures for Mexican endemics and northern Mesoamerican endemics are similar between the two data sets. Thus although the full data set may underestimate somewhat the number of species in common with South America, the figures for local and regional endemism seem relatively reliable.

Familial Patterns

Most of the large families listed in Table 22.2 roughly parallel the trends already noted for the tree flora in general. Thus, except as noted, these families present the following phytogeographic features.

1. They are better represented in the Gulf area than in the Yucatan area (excluding Chiapas and Tabasco), with only 10–45% of their species

Figures 22.6 and 22.7. Species endemic to Mexico. **Figure 22.6.** *Eschweilera mexicana*, the only known native species of Lecythidaceae in Mexico. **Figure 22.7.** *Elaeagia uxpanapensis* (Rubiaceae). Photographs by the author.

Figures 22.8-22.11 (above and below). Endemic and disjunct genera. **Figure 22.8** (left). Undescribed species of *Styphnolobium* (Leguminosae) endemic to Mexico and known only from the Los Tuxtlas and Uxpanapa regions of Veracruz. The genus *Styphnolobium* is restricted to Mesoamerica, southeastern United States, and Southeast Asia. It probably represents an ancient Laurasian element in the Mexican rain forest flora. **Figure 22.9** (right). *Recchia simplicifolia* (Simaroubaceae), a Mexican endemic. *Recchia*, a genus of three species, is also apparently endemic to Mexico.

Figure 22.10 (left). *Blomia prisca* (Sapindaceae). The monotypic genus *Blomia* is endemic to northern Mesoamerica, where it occurs in southeastern Mexico and eastern Guatemala. **Figure 22.11** (right). *Chiangiodendron mexicanum* (Flacourtiaceae) on karst terrain, Uxpanapa, Veracruz. This monotypic Mexican endemic genus is closely related only to genera of the Southeast Asia-Australia region. It seems clearly to represent an ancient Laurasian element in the Mexican karst rain forest flora. Photographs by the author.

present in the Yucatan area and less than 20% restricted to that area in Mexico. *Exceptions*: Sapindaceae (86% of species in Yucatan area, 29% restricted to that area in Mexico); Arecaceae (73%/36%); also the caesalpinioid subfamily of legumes (53%/27%).

2. About 15–47% of their species are endemic to northern Mesoamerica. *Exceptions*: Arecaceae (55%), Polygonaceae (60%).
3. About 20% or more of their species occur in South America. *Exceptions*: Arecaceae (9%), Polygonaceae (10%).
4. Fewer than 30% of their species are present in the Greater Antilles. *Exceptions*: Myrtaceae (55%), Meliaceae (50%), Sapindaceae (36%).

The palms (Arecaceae) and Sapindaceae are relatively much better developed in the Yucatan Peninsula than are the other large families, as far as their canopy tree elements are concerned. The palms are highly endemic at the regional level, with eight of 11 species restricted to Mesoamerica (none of the eight occurring south of Nicaragua) and no species reaching the Amazon basin; indeed, six of the seven genera (all except *Orbignya*) are circum-Caribbean in distribution and do not reach the Amazon basin. The Sapindaceae display a relatively high affinity to the Antillean flora (36% at the specific level) and include three genera (of eight) that do not reach South America: *Blomia* (see Fig. 22.10), *Exothea*, and *Thouinia*. Other notable features are the strong Antillean relationship of the Myrtaceae, a family well developed in that area, and the regionally endemic nature of the Polygonaceae.

Endemism

At present, 41 (9.6%) of the 427 species considered are thought to be endemic to Mexico (Figs. 22.6–22.9 and 22.11; see also 22.15 below). These species are found in 19 families, the best represented of which are Leguminosae (13), Lauraceae (4), Malvaceae (3), Myrtaceae (3), and Rubiaceae (3).

The ranges within Mexico of these endemics are given in Table 22.7. The three areas of high precipitation (Fig. 22.5, map 15) previously defined are of great importance in understanding the distribution patterns of the endemic species. All but two of these species (*Acacia dolichostachya*, *Populus mexicana*) occur in one or more of these three areas, and 23 (56.1%) are restricted to one or more (Table 22.7). Thus the endemics are strongly concentrated in southern Veracruz and the immediately adjoining parts of other states, including northern Chiapas. Only *Acacia dolichostachya* occurs in the Yucatan Peninsula in the strict sense (Yucatan, Quintana Roo, Campeche), to which it is confined. Some of the endemic species are relatively widely distributed, reaching northern Veracruz (Table 22.7), but none is restricted to that area. Five species also occur on the Pacific slope in Mexico. The Tuxtepec area and parts of the

Table 22.7. Distributions of Mexican endemics

Restricted to the three areas of high precipitation

Crescent area only (* = only in Uxp. and/or Chimalapa): *Bauhinia* sp. nov.,*
Chiangiodendron mexicanum,* *Elaeagia uxpanapensis*, *Eschweilera
mexicana*,* *Eugenia uxpanapensis*,* *Hernandia* sp. nov.,* *Inga* sp. nov. 1,*
Inga sp. nov. 2,* *Lonchocarpus* sp. & var. nov. 1,* *Nectandra* sp. nov., *Ocotea*
sp. nov.,* *Recchia simplicifolia*, *Robinsonella samaricarpa*, *Sterculia* sp.
nov.,* *Zygia conzattii* var. nov.*

Tuxtepec area only: *Styphnolobium* sp. nov. 2.

More than one of areas: *Erythrina tuxtlana* (all 3), *Gaussia gomez-pompae*
(Tuxt., Crescent), *Licaria velutina* (LT., Crescent), *Ocotea uxpanapana* (LT.,
Crescent—Uxp.), *Robinsonella brevituba* (Tuxt., Crescent), *Rondeletia galeottii*
(LT., Crescent), *Styphnolobium* sp. nov. 1 (LT., Crescent).

**More widespread on Gulf slope, not occurring on Pacific slope or in Yucatan
Peninsula**

Occurring in southern Veracruz and often adjacent states but not reaching the
Misantla area of northern Veracruz: *Andira galeottiana*, *Coussarea mexicana*,
Dalbergia glomerata, *Robinsonella mirandae*, *Sapium bourgeaui*, *Tapirira*
sp. nov.

As in B-4, but also in northern Veracruz (at least to the Misantla area):
Cymbopetalum baillonii, *Eugenia colipensis*, *Eugenia mexicana*, *Ficus
lapathifolia*, *Senna multijuga* subsp. *doylei*, *Scheelea liebmannii*.

Occurring on both Gulf of Mexico and Pacific slopes, not in Yucatan Peninsula

Known on Gulf slope only from one or more of the three areas of high
precipitation: *Capparis mollicella*, *Coccoloba matudae*.

More widespread on Gulf slope: *Albizia purpusii*, *Ginorea nudiflora*, *Populus
mexicana* subsp. *mexicana*.

Restricted to Yucatan Peninsula (Yucatan, Quintana Roo, Campeche)

Acacia dolichostachya.

The "three areas of high precipitation" are defined in the text and in Fig. 22.5, map 15; see also
Fig. 22.1, map 2 for localities mentioned.
LT., Los Tuxtlas area; Tuxt., Tuxtepec area; Uxp., Uxpanapa.

crescent area remain poorly known botanically, and both areas can be
expected to yield more novelties.

About 35 species are endemic to the Yucatan Peninsula in the broad
sense, which includes Yucatan, Quintana Roo, Campeche, Tabasco, eastern Chiapas, Belize, and eastern Guatemala (some species extending to
adjacent northwestern Honduras). This number includes most of the
species of the Yucatan and Tabasco-Chiapas groups that are restricted to
northern Mesoamerica (including two Mexican endemics, one of which
is restricted to the eastern part of the Crescent area in Chiapas); but it
excludes those that reach the Pacific slope of Mexico or of northern
Central America. The total number of species endemic to this area is of

course higher owing to endemism in Guatemala and Belize, excluded from the present study.

On a broader scale, 132 (30.9%) of the species in the present study are endemic to northern Mesoamerica. This larger inclusive group breaks down as follows.

1. *Gulf and Tabasco-Chiapas species*: 98 species, of which 40 are endemic to Mexico; as a group, not well adapted to dry forest types
2. *Yucatan and widespread species*: 34 species, one endemic to Mexico; many species adapted to drier forest types

I have shown elsewhere (Wendt, 1989) that the Crescent area, although clearly floristically related to the rest of the southern Veracuz area, displays strong floristic ties to the wet Izabal area of Guatemala (Fig. 22.5, map 15) in the southern part of the Yucatan Peninsula. Combining this information with that of the preceding paragraph, I believe that, instead of recognizing the southern Veracruz area and the Yucatan Peninsula in the broad sense as the two important centers of endemism in our area, the following areas are more meaningful.

1. *An area of endemism of species of wet forest.* It includes the southern Veracruz area (which includes adjacent parts of Oaxaca and the crescent area to southern Tabasco and northern Chiapas), the wettest portions of the Lacandon rain forests of eastern Chiapas, and the Izabal area of eastern Guatemala and Belize (Fig. 22.5, map 15). This total area corresponds fairly closely to most of Brown's (1982, 1987) hypothesized Guatemalan paleoecological forest refuge area and is an area of endemisms for other organisms such as butterflies (Brown, 1982, 1987) and birds (Haffer, 1987). It includes a number of smaller, discrete centers of endemism, e.g., the three Mexican areas of high precipitation mentioned above, and parts of eastern Guatemala and southern Belize(Lundell, 1942b; Miranda, 1958; Toledo, 1982). I have discussed aspects of the phytogeographic history of this area elsewhere (Wendt, 1989).
2. *A less rich and more diffuse area of endemism of species of more seasonal (semievergreen) rain forests.* It includes the Yucatan Peninsula of Mexico, the drier types of forest in eastern Tabasco and the Lacandon area, and northern parts of Belize and the Petén of Guatemala— essentially, the area of forest dominated by *Manilkara zapota*, as defined by Lundell (1934). It corresponds to an area of plant species endemism recognized by many authors including Standley (1930), Lundell (1937), Miranda (1958), Rzedowski (1978), and Durán & Olmsted (1987), based on all life forms and vegetation types. This area merges southward in southern Petén and southern Belize into the first, wetter area of endemism.

These two broad areas correspond closely to the northern and Yucatan endemic areas, respectively, of the Atlantic Versant region of Central America recognized by Savage (1982) based on herpetofaunal evidence. It is of note that in the cited studies of butterflies, birds, and herpetofauna no areas of significant endemism are recognized on the Atlantic slope between the Izabal area (including adjacent Honduras) to the north and extreme southeastern Nicaragua or Costa Rica to the south. The same situation seems to be true for rain forest canopy trees as well, with a distinct gap in important areas of endemism from northwestern Honduras to extreme southeastern Nicaragua or Costa Rica, although my data for the intervening areas are incomplete.

The figures presented here for canopy tree endemism are somewhat higher than might be expected based on the literature. The 9.6% endemism in Mexico is nearly double Rzedowski's estimate (this volume) of 5% for the Mexican rain forest flora as a whole. It is nevertheless fairly low and in no way compares with the 40–70% Mexican endemism Rzedowski cites for several other vegetation types, nor with the figures cited by Gentry (1982a) for rain forest canopy tree endemism in other neotropical areas: e.g., Amazonia 80%, Coastal Brazil 67%, Antilles 72%. On the other hand, the Mexico–northern Central America area as a whole appears to represent a significant area of endemism for rain forest flora on a continental scale, especially considering its small size. The fact that almost one-third of the Mexican species are restricted to this area is indicative, although of course the total endemism may be higher or lower depending on the composition of the other rain forests of northern Central America.

Many authors have commented on some aspect of the distinct nature of the rain forest flora of northern Central America and Mexico. Standley (1930; Standley & Record, 1936) concluded that the Yucatan Peninsula (in the broad sense) represents a distinct floral district within the American tropics. Lundell (1942b) emphasized that the rain forests of southern Belize contain "relict floral elements of relatively great age," combining endemic taxa with species and genera of disjunct distributions (often with respect to South America), within a context of mostly widespread neotropical species. Miranda (1952–53) stated that the relationship between the Chiapan and Guatemalan-Belizian floras is marked, and that many species are restricted to that total area or to that area plus Tabasco and the Yucatan Peninsula; many rain forest species were among his cited examples. The same author, noting that many South American species reach their northern extreme in Belize and Guatemala without reaching Mexico, also noted the significant endemism of the Maya Mountains area of Belize (Miranda, 1952–53, 1958). Rzedowski (1978), discussing the Caribbean floristic region, which includes tropical lowland areas of Mexico, Central America, northern South America, the Antilles, and southern Florida, noted that subregions (above provinces) should probably be recognized, one of which would include Mexico and an unspecified portion of northern Central

America. He defined this subregion by the occurrence of a number of holarctic genera (e.g., *Salix, Quercus, Populus*) among the predominant neotropical ones. Gómez (1982) recognized a "Yucatán-Petén" flora on the Atlantic slope of Central America from central Nicaragua northward to Mexico, distinguishing it from the "Amazonian" flora from southern Nicaragua southward. Gentry (1982a,b) likewise noted that many Amazonian rain forest families do not extend north of southeastern Nicaragua, and that there is a general northward decrease in South American elements northward in Central America.

ORIGINS OF THE FLORA

On a world map of rain forest distribution (e.g., Richards, 1952), the Mexican rain forests appear as the northern extension of the vast rain forests of South America (although discontinuous across the northern Andes), but this has not always been the case. During the early Eocene, when megathermal vegetation reached its maximum global areal extent, Mexico was part of Laurasia and without direct connections to South America, and quite possibly it might have had many more genera in common with Europe than with South America. Tectonic and climatic changes from the Upper Cretaceous onward have caused shifting possibilities for biotic exchange between Mexico and other parts of the world, and the total distribution of megathermal vegetation has varied dramatically from the Eocene maximum to the minima of the Pleistocene glacial periods. Unless the present rain forests of Mexico are composed entirely of South American post-Pleistocene invaders from the south—and the data presented below indicate that they are not—one should expect the complexities of tectonic and climatic history and of shifting biotic connections to be reflected in a phytogeographically complex composition of the present Mexican rain forests. Furthermore, it cannot be assumed that the taxa of these present rain forests are necessarily all descendents of phyletic lines that have always inhabited rain forests.

By the time angiosperms first appeared in the known fossil record in the Lower Cretaceous (Crane, 1987; Taylor & Hickey, 1990; Upchurch & Wolfe, 1987), Mexico had become part of the North American Plate (Donnelly, 1989; Gose, 1985; Smith, 1985) and thus was part of the Laurasian supercontinent. Large parts of Mexico were submerged during the Upper Jurassic and even more during the Lower Cretaceous, but Upper Cretaceous uplift generally concomitant with that of the western United States led to most of Mexico being emergent from Paleocene onward. A complex geological history has led to large areas of diverse land habitats being present throughout the Tertiary, with continuous land connections to the western United States, although significant portions of southeastern Mexico were covered by seas at least intermittently until the Miocene or later (de Cserna, 1989; Ferrusquía-V., this volume).

Angiosperms probably originated in West Gondwanaland (Cronquist, 1988; Raven & Axelrod, 1974) but appear to have spread rather quickly throughout both the Gondwanan and Laurasian areas, before the origin and initial diversification of most modern families. Raven and Axelrod (1974) thus identified "Gondwanan" and "Laurasian" families based on their proposed places of origin. Angiosperms were in North America shortly after their initial appearance in the fossil record (Crabtree, 1987), and it is thus possible that some elements in the present Mexican rain forests may be descendents of Laurasian angiosperm lines that have been in North America since shortly after the origin of the flowering plants. However, factors such as climatic and tectonic events, which may have led to enrichments and extinctions within this flora, must be considered further.

Paleotemperature curves from oxygen isotope studies of benthic and planktonic foraminifera (Savin & Douglas, 1985; Shackleton, 1984; see also Graham, 1987) and climatic inferences from paleobotanical studies (e.g., Wolfe, 1985; Wolfe & Upchurch, 1987) combine to present a congruent picture of Upper Cretaceous and Cenozoic climates. In general terms, pre-Upper Eocene climates showed long-term stability and were warmer than those of later periods, especially at middle and high latitudes, reaching a thermal maximum during the early Eocene; latitudinal decrease in temperature was not nearly as precipitous as at present. The Upper Eocene marked a period of dramatic climatic deterioration that continued through most of the Oligocene, and the worldwide equable conditions of earlier times were never regained.

Along with sporadic long distance dispersal of species into North America, which has surely occurred, continuous or stepping-stone land routes were clearly important to the subsequent enrichment of the original Lower Cretaceous North American angiosperm flora. These routes, which also could have served in the spread of North American elements to other areas, are discussed below.

In the following discussion, I follow the definitions of Wolfe (1979, 1985) for the classification of climates based on annual average temperature: megathermal, over 20°C, including tropical (>25°C) and paratropical (20–25°C); mesothermal, 13–20°C; and microthermal, under 13°C. The same terms are applied directly to taxa or lineages to indicate their adaptive modes with reference to temperature.

Laurasian Connections

Since the breakup of Pangea, North America and Eurasia have maintained tectonic contact almost constantly through Beringian or North Atlantic connections (or both), although these connections have not always been emergent. The resulting long-term, if sporadic, biotic interchange has led to a certain continued coherence of the Laurasian biota, notable today in such phenomema as the temperate eastern North American–eastern Asian plant disjunctions (Boufford & Spongberg, 1983) and the strongly pan-Laurasian distribution of many angiosperm families and genera (Raven & Axelrod, 1974). The Beringian connection was apparently present and emergent from the Mesozoic onward, except for some marine transgressions especially during the late Cenozoic (McKenna, 1983; Tiffney, 1985b). North Atlantic land connections probably included separate northern and southern areas, the southern area extending through the present British Isles to southern Greenland. The emergent land of this latter connection was interrupted by a narrow sea during most of the Cretaceous but apparently was more or less continuous during parts of the Paleocene and Lower Eocene (Hamilton, 1983; McKenna, 1983; Parrish, 1987; Tiffney, 1985b).

Major epicontinental seas affected the pattern of emergent land and thus of biotic interchange within Laurasia during the Mesozoic and early Tertiary. In North America, a continuous north–south seaway completely separated the eastern and western parts of the continent during the Middle Cretaceous. Reconnection first occurred in the Canadian area during the mid-Upper Cretaceous, but direct east–west dry land contact across the present United States did not occur until the end of the Cretaceous, concomitant with the beginning of the uplift of the southern Rocky Mountains (Hamilton, 1983; Tiffney, 1985b). In Eurasia, similar continuous north–south seaways completely separated Europe from eastern Asia several times, including during parts of the Cretaceous and again during the Eocene. In addition, southwestern Europe was separated from northern and eastern Europe during parts of the Cretaceous and Paleogene (Barron, 1987; Hamilton, 1983; Mai, 1989; McKenna, 1983; Tiffney, 1985b). Many interruptions of the Laurasian "land mass" thus existed at different times.

Wolfe and Upchurch (1987; Upchurch & Wolfe, 1987) suggested that during the Upper Cretaceous megathermal climates extended to lat. 45–50° N, and Wolfe (1985) showed that they reached lat. 65° N (to 70° N in coastal areas) during the latest Paleocene–Lower Eocene thermal maximum. Wolfe and Upchurch suggested a subhumid, nonseasonal climate with open-canopy woodland for most megathermal areas during the Upper Cretaceous. Multistratal megathermal closed canopy forests (megathermal rain forest) may have been absent or restricted in equatorial areas at this time, although these authors acknowledged the scarcity of Upper Cretaceous paleobotanical data from these low-latitude regions. Parrish (1987), on the other hand, maintained that zonal climatic circulation had already become established by this time and that the equatorial areas should therefore have been relatively humid. At any rate, a global increase in precipitation at the Cretaceous–Paleocene boundary led to either the origin or rapid expansion of megathermal rain forests (Upchurch & Wolfe, 1987; Wolfe, 1987; Wolfe & Upchurch, 1987). During the latest Paleocene–early Eocene thermal maximum, tropical rain forest reached as far as 50° N and S, and paratropical rain forest to 60–65° (70° in coastal areas) (Wolfe, 1985, 1987; see also Collinson & Hooker, 1987; Mai, 1989; Romero, 1986). Seasonality of precipitation in megathermal areas may have begun to develop locally by the Upper Paleocene (Wing, 1987) and was well developed in parts of Eurasia and eastern and west-central North America by Middle Eocene times (Dilcher, 1973; Leopold & MacGinitie, 1972; Upchurch & Wolfe, 1987; Wing, 1987; Wolfe, 1975, 1985, 1987).

Thus the North Atlantic intercontinental connection, at 45–50° N paleolatitude (Tiffney, 1985b), presents a plausible route for terrestrial interchange of biota, including megathermal plant lineages, during periods of the Upper Cretaceous, Paleocene, and Lower to Middle Eocene, as suggested by McKenna (1983), Tiffney (1985b), and Wolfe (1985). These same authors suggested that the northern position of the Beringian connection, which was at higher latitudes (ca. 75° N) during the early Tertiary than at present, would have prevented effective exchange of megathermal evergreen angiosperms by this route, although McKenna (1983) and Tiffney (1985b) noted the possibility of a more southern but narrow Aleutian land bridge. Wolfe (1985) suggested that the vegetation of Beringia during the Eocene thermal maximum was mostly mesothermal forest of deciduous aspect due to seasonality of light rather than of temperature. In general, the North Atlantic connection was probably more important for the exchange of megathermal elements during the early Tertiary.

Abundant plant mega- and microfossil evidence indicates that considerable North American–Eurasian floristic interchange in megathermal and other lineages did indeed occur during parts of the Late Cretaceous and especially early Tertiary. This interchange led to the emergence of a recognizable tropical Laurasian or "boreotropical" flora by the Paleocene (Batten, 1984; Crane, 1987; Wolfe, 1975), although megafossil evidence probably overestimates the similarity of Paleocene floras between the two continents (Wing, 1987; Wolfe, 1985). Although floristic regionalization and impeded exchange are apparent, they are often related to epicontinental sea barriers. For instance, during much of the Upper Cretaceous and Early Tertiary, Europe and eastern North America appear to have been more similar floristically than eastern and western North America (Batten, 1984; Crane, 1987; Tiffney, 1985b; Wolfe, 1975). After the retreat of the central North American seaway, cross-continental interchange during the Paleocene may still have been impeded by dry areas of central North America caused by increased continentality and rain shadow effects of the emerging Rocky Mountains (Tiffney, 1985b). Nevertheless, some east–west exchange appears to have occurred from the Upper Cretaceous onward (Batten, 1984; Tiffney, 1985a,b), and Leopold and MacGinitie (1972) suggested that it may have reached a maximum during the Middle Eocene.

Similar patterns are indicated by fossil mammal evidence, which suggests interchange between North America and both Asia and Europe during periods from the Upper Cretaceous onward (Gingerich, 1985; McKenna, 1975; Webb, 1985). Particularly notable is the strong similarity in mammal faunas of western Europe and North America during the EarlyEocene, a similarity that decreases sharply after that time (McKenna, 1975; Webb, 1985).

Laurasian connections could account for the arrival in North America of Laurasian elements of Eurasian origin. They would also provide a route for the arrival of Gondwanan elements via an Africa–Europe–North America route. Africa was close to Eurasia into the early Tertiary and may have been directly connected at times as late as the Paleocene. Numerous smaller blocks existed between Africa and Eurasia, some of them accreting onto southern Eurasia during the Upper Cretaceous and Cenozoic, and India collided with southern Eurasia prior to Mid-Eocene (Bally et al., 1989; Hamilton, 1983; McKenna, 1983; Parrish, 1987; Raven & Axelrod, 1974). Thus Raven and Axelrod (1974) suggested that much of the interchange between Laurasia and Gondwanaland from the origin of angiosperms through the Paleocene probably occurred via the western Eurasia–Africa area, although Parrish (1987) suggested that continuous pre-Oligocene land connections may have been limited to a brief period during the Upper Cretaceous.

The explanation of present generic disjunctions due to Eocene floristic connections is plausible, as many extant genera were present by that time, whereas a much higher percentage of Paleocene and earlier fossil plants cannot be assigned to modern genera (Hickey, 1977; Raven & Axelrod, 1974; Wing, 1987). Post-Eocene floristic connections via North Atlantic or Beringian routes would have involved mostly microthermal elements and are thus not important in the present context.

South American Connections

South America and Africa were continuous or separated by short oceanic gaps through the middle part of the Upper Cretaceous, forming West Gondwanaland (Bally et al., 1989; Raven & Axelrod, 1974), although continuous land connection may have been severed as early as late Lower Cretaceous (Parrish, 1987). Meanwhile, the North and South American Plates were separated by 800 km and more by the middle part of the Lower Cretaceous (Bally et al., 1989; Donnelly, 1985) when the first angiosperms appeared. The geological history of Central America and the Caribbean is thus critical to understanding possible interchange of angiosperm elements between the two continents, but it is one of the most complex and controversial areas for plate tectonic reconstruction. Nevertheless, Rosen (1985), summarizing various recent geological models, concluded that although they sometimes differ dramatically in mechanism, they are similar as to their biogeographical implications for biotic exchange. I here follow the apparently balanced scenario provided by Donnelly (1989, which see for useful maps).

During Upper Jurassic times, three major plates—the North and South American and the Farallon of the eastern Pacific—were diverging relative to each other. The future northern or nuclear Central America (southern Guatemala to northern Nicaragua) existed as the separate Chortis block, apparently previously joined to western Mexico. This block was moving southward off western Mexico as the Farallon Plate moved in the same direction. During the later Lower Cretaceous, the Farallon Plate started converging northeastward. This generally correlated with and perhaps caused a major "flood basalt" episode behind a new northeastwardly-advancing proto-Antillean arc, initiating the Caribbean Plate between Mexico and northwestern South America with its leading edge far to the southwest of its present Antillean position. This plate, "pushed" by the Farallon Plate, moved northeastward relative to North and South America into the present Caribbean area to fill the void left by the separating North and South American Plates. Island arc and other biotically significant stepping stone land connections between North and South America may have existed during this or later periods, as the Caribbean was geologically an active area; Parrish (1987), using geological data, argued for intercontinental connections through the Antilles area during part of the Upper Cretaceous and perhaps during the Eocene. During the middle part of the Upper Cretaceous the basalt event ended as the North and South American Plates stopped diverging. The pressure from all three major plates now started a period of major deformation of the Caribbean Plate, and the North and South American Plates began converging by Middle Eocene. Because the Farallon Plate was still pushing northeastward, by Upper Cretaceous a new island arc was formed on its leading edge as it was subducted under the Caribbean Plate. The southern part of this arc would eventually form southern Central America. Near the end of the Cretaceous, when the eastward-moving Chortis block sutured onto the Maya block to form present northern Central America, the southern Central American arc probably extended southward from that area, not southeastward as at present, and thus was far removed from northwestern South America. Slowly the south end of this arc swung eastward on the leading edge of the Farallon Plate (which separated into the Cocos and Nazca Plates during the mid-Cenozoic), and an emergent Central American connection to South America was not formed until latest Miocene or Pliocene, perhaps as recently as 3 Ma (Savin & Douglas, 1985). As in the case of the Caribbean in general, the patterns of emergent land in Central America during different parts of the Tertiary are not clear, although emergent lands did exist at times; major changes in sea level occurred during mid to late Cenozoic times (Savin & Douglas, 1985).

Biotic exchange between North and South America could have occurred via the relatively direct proto-Antillean arc during the Cretaceous, via Antillean connections during the Tertiary, or via southern Central America during the late Tertiary. The relative importance of each route depends in large part on the presence and arrangement of emergent land, which is poorly understood for the earlier periods. Gradually increasing Neogene biotic exchange between the continents, culminating in the dramatic "great American biotic interchange" as the Central American isthmus was completed during the Pliocene, is well documented (Stehli & Webb, 1985). Earlier exchange appears to have been on a far lesser scale, but fossil vertebrate evidence strongly indicates a period of interchange during parts of the Upper Cretaceous and Paleocene, followed by relative isolation until the Neogene (Estes & Baéz, 1985; Gingerich, 1985; Pascual et al., 1985; Webb, 1985—but compare to Cifelli, 1985). South-to-north migration may have been more common than the reverse. Such early interchange has been used to explain present distribution patterns in diverse groups of animals (e.g., Bussing, 1985; Rosen, 1976; Vanzolini & Heyer, 1985) and plants (Gentry, 1982a; Taylor, 1988).

Unlike the Laurasian intercontinental migration routes mentioned above, possible connections to South America would have been in areas of megathermal climate throughout much of the history of the area (Wolfe, 1985). Thus they could have been effective routes for the exchange of megathermal plant taxa at almost any time that land connections existed, except during short periods such as the Pleistocene glacial maxima.

Hypotheses

Based on the foregoing analysis, it may be hypothesized that the following floristic elements might have played a part in the present composition of the Mexican rain forest canopy tree flora.

1. Laurasian arrivals
   a. Laurasian lineages, either originally North American groups or Eurasian taxa that arrived via Cretaceous–Eocene Laurasian connections
   b. Gondwanan lineages that arrived via an Africa–Europe–North America route during the Cretaceous–Eocene
2. South American arrivals
   a. Gondwanan lineages that arrived during the Cretaceous–Paleocene(–Eocene), probably via proto-Antillean or Antillean routes
   b. Gondwanan lineages that arrived recently, during the Neogene, via southern Central America

The existence of other elements, e.g., Laurasian taxa that arrived via a Eurasia–Africa–South America–North America route, is possible, although in that particular case the early isolation of South America from Africa indicates that this route would probably be of minor importance. On the other hand, reinvasion events, such as the introduction of Laurasian-arrival elements into South America during the Cretaceous or Paleogene, and their subsequent reinvasion of Mexico during the Neogene, must be considered definite possibilities that complicate the picture. Additionally, sporadic arrivals via long distance dispersal have surely occurred.

Essentially all recent authors who have opined on the origins of the Mexican rain forest flora agree that elements of South American origin are an important component of the flora (e.g., Gentry, 1982a; Raven & Axelrod, 1974; Rzedowski, 1978, this volume). They differed, however, as to the relative importance of Laurasian-arrival elements. Gentry (1982a) argued strongly that the canopy tree flora of neotropical (including Mexican) rain forests does not include a significant Laurasian-arrival component, especially with regard to the locally more common species. He minimized the contribution of Laurasian families and attributed ancient North American Gondwanan elements to Late Cretaceous arrivals from South America via terrestrial connections or stepping stone migration. Raven and Axelrod (1974) also stated that the tropical lowlands of Central America and Mexico are populated by a largely South American flora; and they likewise argued that some South American Gondwanan taxa have been in North America since the Upper Cretaceous or Tertiary, but they attributed it to long-distance dispersal. On the other hand, Rzedowski (1978, this volume), following Sharp (1966), argued that a significant number of Mexican rain forest genera, including many of the "Gondwanan" families, are probably derived from taxa that arrived via Laurasian connections, a stance also taken by van der Hammen and Cleef (1983) with respect to the submontane flora of northwestern South America. All of the above-mentioned authors agreed that Laurasian taxa are clearly important in montane neotropical areas.

Thus a central point of contention with respect to the origins of the Mexican rain forest canopy tree flora concerns the relative importance of the Laurasian-arrival element. As noted previously, these taxa would have been in North America by the Eocene, and then either survived in southern North America until the present or invaded South America, reinvading tropical North America recently. Data supporting the significant contribution of this element are presented next, and important factors in its persistence are discussed.

Floristic Relationship of Eocene North American Floras to Modern Mexican Rain Forest

A revision of publications on the Eocene floras of North America (Table 22.8) reveals that at least 51 genera (22%) of the modern Mexican rain forest canopy tree flora were reported from North America during that period. Older paleofloristic works are notorious for their taxonomic misidentifications. For instance, Dilcher (1971) estimated that over 60% of Berry's familial or generic assignments for southeastern Unites States Eocene macrofossils are incorrect. Therefore in the present study only works published after 1960 are included. Although some fossil genera are surely misidentified, the inclusion of almost one-fourth of the present rain forest genera in the list strongly indicates that the extant rain forest generic flora was well represented in the North American Eocene forests. Equally notable is the disproportionately strong representation of the more species-rich genera of the modern Mexican rain forests in the fossil floras; the 22% of the genera include 36% of the modern species and nine of the 12 largest genera (Table 22.8). MacGinitie (1969, 1974; Leopold & MacGinitie, 1972) has previously emphasized the floristic similarity of Eocene floras of the Rocky Mountain area with the extant tropical flora of Mesoamerica.

Table 22.8 was intended to include only genera reported from Eocene floras from Mexico or northward; due to the scarcity of recent publications on Mexican Eocene floras, all of the reports are from further north. The single recently studied Mexican Eocene flora consulted (Martínez-Hernández et al., 1980), from extreme northeastern Mexico along the Texas border, is a palynological study that includes mostly form-genera along with a few extant genera such as the presently Asian *Platycarya*, which is common in many North American floras of the Early and Middle Eocene (Leopold & MacGinitie, 1972). Notable in this Mexican flora is the diversity of Sapotaceae, an important family in the modern Mexican rain forest flora. The only other recently studied Eocene flora from northern Latin America

Table 22.8. Mexican rain forest canopy tree genera reported from Eocene paleofloras of North America north of Mexico

Acacia*
Alchornea
Allophyllus
Apeiba
Astronium
Bauhinia
Beilschmiedia*
Bernoullia
Bursera
Caesalpinia*
CASEARIA
Cassia*
Cedrela*
Celtis* (L)
Chrysophyllum*
Clethra* (L)
CORDIA* (L)
Dalbergia*
Dendropanax
Diospyros*
Erythrina
EUGENIA

Table 22.8. (cont.)

---

*FICUS**
Homalium
*Ilex** (L)
LONCHOCARPUS
Luehea
Mabea
*NECTANDRA**
Ochroma
*OCOTEA**
*Oreopanax**
*Persea**
*Podocarpus**
*Populus** (L)
*POUTERIA (*incl.*Lucuma)*
*Sabal**
*Saurauia** (L)
*Salix** (L)
*Sapindus**
*Sapium**
*Sterculia**
Styphnolobium
*Talauma** (L)
Tapirira
Thouinia
*Terminalia**
*Trema** (L)
Trichilia
*Turpinia** (L)
*ZANTHOXYLUM** (incl. *Fagara)*

The list is based on post-1960 publications (Dilcher, 1971; Frederiksen, 1980; Gregory, 1971; Hergert, 1961; Hickey, 1977; Leopold & MacGinitie, 1972; MacGinitie, 1969, 1974; Manchester, 1977; Sousa & Delgado, this volume [as to *Lonchocarpus*]; Taylor, 1989; Tidwell et al., 1981; Wing, 1987; Wolfe, 1972).
Genera in all capital letters are those among the 16 most species-rich in the present flora (Table 22.3). Asterisk (*) indicates genera reported for the Old World portion of Laurasia during the Eocene (Chandler, 1964; Hsü, 1983; Mai, 1981; Tanai, 1972). (L) denotes genera of Laurasian families according to Gentry (1982a). Determinations questioned in the publications ("?," "cf.") are excluded.

is from Panama (Graham, 1985); all of the tree genera reported in this flora presently occur in Central America, with about three-fourths in Mexico (Table 22.9). Mexican rain forest canopy tree genera in this flora but not included in Table 22.8 are *Coccoloba* and *Mortoniodendron*, the latter of which may be equivalent to *Tiliaepollenites* of the Upper Eocene of Mississippi (Graham, 1985).

Persistence of Elements of the Eocene North American Tropical Flora in Mesomerica

Worldwide Late Eocene cooling led to a gradual shrinking equatorward of the megathermal forests (Leopold & MacGinitie, 1972; Wolfe, 1985), and at the end of the Eocene a major climatic deterioration occurred (Graham, 1987; Parrish, 1987; Savin & Douglas, 1985; Wolfe, 1985, 1987). Wolfe hypothesized that during the Oligocene and Lower Miocene evergreen vegetation was restricted to below 35° latitude, and that in the New World tropical climates occurred only in northern South America and paratropical climates only as far north as Cuba and southern Mexico, finding their northern limit somewhat to the south of their present boundaries. A temporary warming during the late Lower to Middle Miocene led to the expansion of megathermal vegetation to somewhat higher latitudes but far short of the early Eocene extremes (Wolfe, 1985). This period was followed by an overall cooling to approximately present temperature levels at low and middle latitudes during the late Middle Miocene, followed by a generally cooler Pliocene (Graham, 1987, 1989a; Savin & Douglas, 1985;

Table 22.9. Mexican rain forest canopy tree genera reported from Eocene through Pliocene paleofloras of northern Latin America (Central America, Mexico, Caribbean)

Acacia OM,P
Alchornea OM,P
Allophyllus P
Bernoullia OM,P
Bravaisia P
Bursera OM,P
Calyptranthes OM
CASEARIA E,OM,P
Cedrela P
Ceiba P
Celtis OM,P
(*Chrysophyllum*) OM
COCCOLOBA E,P
CUPANIA OM,P
Cymbopetalum P
(*Dendropanax*) OM
Erythrina P
EUGENIA/Myrcia E,OM,P
Guarea OM,P
Hampea/Hibiscus OM,P
Hymenaea OM
Hura OM
Ilex E,OM,P
Iresine P
Matayba OM,P
Mortoniodendron E,OM,P

Table 22.9. (cont.)

Oreopanax OM
Podocarpus OM,P
Populus P
Posoqueria P
POUTERIA OM
Protium P
Pseudobombax OM,P
Psychotria OM
Salix OM
Sapium OM
SIDEROXYLON OM
Symphonia P
Tapirira OM
Terminalia/Combretum E,OM,P
Tetrorchidium OM
(*ZANTHOXYLUM*) OM

This list is based on post-1960 publications (derived from summarized lists in Graham, 1988, 1991, this volume; Langenheim et al., 1967; Rzedowski & Palacios, 1977).
E, Eocene (Panama); OM, Oligocene through Lower Miocene (Mexico, Panama, Costa Rica, Puerto Rico); P, Pliocene (Panama, Mexico). Capitalized genera as in Table 22.8. Genera reported only from Puerto Rico are included in parentheses. Determinations questioned in the publications ("?," "cf.") are excluded.

Wolfe, 1985). Pleistocene glacial cycles probably included the coolest low-latitude climates of the Cenozoic and thus probably caused the most extreme reduction or fragmentation of the neotropical megathermal forests. Could elements of the Eocene megathermal forests have survived in Mexico or Mesoamerica in general through these adverse post-Eocene periods?

Table 22.9 lists the canopy tree genera of Mexican rain forests that are reported for Tertiary floras of northern Latin America (Mexico, Central America, Caribbean) in post-1960 studies. Studies in this area are still few, and almost all work has been done on fossil pollen, leading to probable underrepresentation of key groups such as Lauraceae and *Ficus* (Graham, 1975; Muller, 1981). Nevertheless, the list includes 42 genera, all of which were present in the area during post-Eocene times; 22 of these genera (52%) are reported for the North American Eocene floras (Table 22.8). The works cited reported a total of 50 genera that at present include rain forest canopy tree species; 45 of these genera occur in the modern flora of Mexico, Belize-Guatemala, or both (those in Table 22.9 plus *Laetia*, *Crudia*, and *Jacaranda*). Only three are not present in the modern Central American flora (*Catostemma* in South America; *Pleodendron* in the West Indies; *Glycydendron* in South America); the first two of these genera are reported only from the Puerto Rican fossil flora. These data clearly show a continued similarity between the post-Eocene floras of Mesoamerica and the present Mexican rain forest flora; they also support the hypothesis of the persistence of many of the North American Eocene genera in Mesoamerica after their disappearance from more northern floras following the climatic deterioration of the Late Eocene and Lower Oligocene.

Clearly, the Pliocene localities might include South American genera that arrived after a continuous terrestrial connection was established to South America, depending on the exact timing of that event. Eliminating genera reported only from these later paleofloras, and combining the data for North America, Mexico, and Central America from Eocene, Oligocene, and Lower Miocene paleofloras, it can be seen that of the present 230 genera of Mexican rain forest canopy trees: (1) at least 64 (28%) were present; (2) at least 12 of the largest 16 genera, including 11 of the largest 12, were already present; and (3) conspicuously absent are genera that include some of the most abundant canopy tree species at present, e.g., *Dialium, Brosimum, Vochysia, Pseudolmedia,* and *Calophyllum.*

The two Pliocene floras include 28 genera of present Mexican rain forest canopy tree genera—Mexico 21 (Graham, 1988), Panama 19 (Graham, 1991)—of which 50% are reported for North American Eocene floras (compare Tables 22.8 and 22.9), 61% for Eocene through Lower Miocene floras of northern Latin America (Table 22.9), and 75% for pre-Pliocene floras of either area. These data suggest that many genera of the Eocene floras persisted through the Pliocene in both northern and southern parts of Mesoamerica, although other genera seem clearly to have arrived from South America during the Neogene, e.g., *Symphonia* (Germeraad et al., 1968).

Critical, then, is the survival of these lineages during the subsequent, more extreme Pleistocene climates. Pleistocene glacial cycles led to the equatorward displacement of most plant distributional ranges in midlatitudes (Davis, 1983), and the effects of these cycles are documented for neotropical areas, where cool or dry cycles alternated with wetter, warmer periods such as the present Holocene interglacial. Van der Hammen (1982) demonstrated that at least 20 cooler periods occurred during the Pleistocene in the northern Andes, and similar cycles appear to have occurred in the South American equatorial lowlands (Haffer, 1987). Pollen evidence and other data from the central highlands of Mexico clearly show repeated periods of cooler or drier climate (or both) during that epoch (Bradbury, 1989; Heine, 1984; Sears & Clisby, 1955; Straka & Ohngemach, 1989; White & Valastro, 1984). Toledo(1982) extrapolated such information to the lowlands of southeastern Mexico for the last 40,000 years—from just before the last glacial maximum to the present—and concluded that average annual temperatures were at times at least 5°C cooler than at present. The only Pleistocene pollen record for lowland northern Mesoamerica that includes data from before the Holocene interglacial (i.e., before ca. 10,000 BP) is one from the Petén of Guatemala studied by Leyden (1984), who documented a significantly cooler, drier climate for that site during late Glacial times (13,000–10,000 BP). Late Pleistocene data for Panama (Bartlett & Barghoorn, 1973) indicated temperatures of about 2.5°C lower than those at present at around 35,500 BP and 11,300–9,600 BP, but some rain forest genera were nevertheless present during the former period and many during the latter.

Both Toledo (1982) and Wendt (1989) argued that present floristic evidence indicates rain forest taxa have a history much longer than 10,000 years in northern Mesoamerica, based on endemic species and genera of strictly rain forest habitats, and thus that elements of rain forest did indeed survive adverse Pleistocene climates in the region. The data of the present study, which indicate 9.6% Mexican and 30.9% northern Mesoamerican endemism, support this assertion. Although little is yet known of speciation rates in tropical trees, it seems improbable that these levels of endemism could be attained during 10,000 years in these long-generation organisms.

Generic distributions also support a relative antiquity in Mexico of many of the taxa of the rain forest flora. Particularly notable are genera endemic to Mexico and restricted to rain forest habitats. The only canopy tree genus in this category is *Chiangiodendron* (Fig. 22.11), although *Blepharidium*, *Blomia* (Fig. 22.10), and *Oecopetalum* are endemic to northern Mesoamerica (the latter also occurring in cloud forest). However, at least two other woody genera are not only endemic to Mexican rain forest but display marked adaptations that are strongly correlated with rain forest habitats (Richards, 1952; Soderstrom, 1981), indicating that their invasion of rain forest habitats is probably not a recent occurrence: *Olmeca*, a bamboo with fleshy fruits (Soderstrom, 1981), and a new genus of Annonaceae with flagelliflory (Schatz & Wendt, unpub.). These genera are restricted to one or more of the areas of high precipitation previously mentioned.

At a broader regional scale, the occurrence of only 38% of the Mexican canopy tree species in South America (Table 22.4) indicates that the northern Latin American (Mesoamerica, Antilles) rain forest flora is not entirely a recent derivative of the South American flora. At the generic level, genera from the Appendix endemic to Mesoamerica include *Beaucarnea*, *Orthion*, *Rehdera*, *Robinsonella*, *Sapranthus*, and *Wimmeria*, whereas other genera are endemic to the Antilles (including Florida in some cases) and Mexico (*Ginorea*), northern Mesoamerica (*Gaussia*, *Metopium*, *Thrinax*), or Mesoamerica in general (*Chione*, *Exothea*, *Lysiloma*, *Thouinia*). Furthermore, a number of neotropical genera occurring in South America are clearly more diverse in northern Latin America, e.g., *Amyris*, *Bravaisia*, *Bucida*, *Exostemma*, *Hampea*, *Lonchocarpus* (one South American species disjunct to West Africa), *Mortoniodendron*, *Rondeletia*, *Roystonea*, *Sabal*, *Simarouba*, *Stemmadenia*, and *Zuelania*. Additional examples come from genera occurring in the Old World that are restricted to or centered in northern Latin America in the New World, discussed below.

Paleobotanical and present distributional information thus combine to indicate that the flora of the North American Eocene forests likely has contributed to the present flora of the Mexican rain forests. Further support for this hypothesis comes from a consideration of the origins of those Eocene forests.

Origins of the North American Eocene Megathermal Flora

It seems clear that North American Eocene megathermal elements have survived to become incorporated into the present Mexican rain forest canopy tree flora. This fact, however, does not answer a central question: What was the ultimate origin of these elements, South American-arrival or Laurasian-arrival?

Gentry (1982a, modified from Raven & Axelrod, 1974) provided lists of Gondwanan, Laurasian, and "unassigned" angiosperm families in the neotropical flora based on their hypothesized area of origin. In the following discussion, as a convenience, I refer families to these three groups following Gentry's assignments, even though these assignments are in many cases highly speculative and based more on present distribution than on firm data on the area of origin. This is especially true in the case of the "Gondwanan" families, as "tropical" and "Gondwanan" are nearly synonymous owing to the present restricted nature of Laurasian tropical forests. Furthermore, even if a given family were Gondwanan in origin, early dispersal and diversification could produce Gondwanan and Laurasian elements within the family, a point emphasized by Raven and Axelrod (1974) but apparently not considered by Gentry (1982a) in his consideration of the origins of the neotropical flora based on family-level analysis. For instance, Wolfe (1975) and Krutzsch (1989), using fossil evidence, argued that at least some lines of the "Gondwanan" family Bombacaceae are Laurasian in origin.

Based on Gentry's lists, ten (20%) of the 50 Eocene angiosperm genera (i.e., excluding *Podocarpus*) in Table 22.8 belong to Laurasian families, 39 (78%) are Gondwanan, and one genus (*Zanthoxylum*) is unassigned. Although the overall representation of Laurasian families in these Eocene floras was much higher than 20% owing to the presence of many Laurasian genera that are at present restricted to temperate vegetation types (numerous genera of Betulaceae, Cornaceae, Hamamelidaceae, Juglandaceae, Platanaceae, Symplocaceae, Ulmaceae, and others), the strong representation of putative Gondwanan families is clear, as noted by numerous authors (e.g., Gentry, 1982a; Leopold & MacGinitie, 1972; Raven & Axelrod, 1974; Taylor, 1988).

In the present Mexican rain forest canopy tree flora, 14 Laurasian families (21.9%) contain 26 genera (11.4%) and 36 species (8.0%). As noted earlier, all or most of these few taxa can be assumed to be Laurasian arrivals, although Gentry (1982a) noted that the case of *Cordia* is not clear. If these species are the only ones whose presence is due to Laurasian connections,

Gentry (1982a) is correct in stating that South American elements outnum-
ber Laurasian arrivals by better than 10:1. However, evidence discussed
below suggests that a significant number of genera of the 45 Gondwanan
(70.3%) and five unassigned (7.8%) families in the present flora may also be
Laurasian arrivals.

The origins of the Gondwanan elements of the megathermal forests of
the North American Eocene have been much discussed (e.g., Gentry, 1982a;
Leopold & MacGinitie, 1972; Raven & Axelrod, 1974; Taylor, 1988; Tiffney,
1985a; Wolfe, 1975), and opinions range from an almost entirely South
American-arrival (Gentry, 1982a) to Laurasian-arrival (Wolfe, 1975) source.
In one study aimed specifically at this problem, Taylor (1988) analyzed well
studied fossil evidence of taxa of 13 Gondwanan families found in the
southeastern North American Eocene. He concluded that in 11 cases an
arrival from South America into North America is most likely, as in the
cases of some woody genera of Annonaceae, Bombacaceae, Myrtaceae,
Euphorbiaceae, Malpighiaceae, and Sapindaceae. However, he also argued that two lineages, including the *Ocotea/Cinnamomum* group, show
strong evidence of exchange by Laurasian routes; he concludes that bidirectional exchange did occur between North and South America during the
Late Cretaceous to Early Tertiary , with northward flow probably greater
than the reverse. Nevertheless, it should be noted that the taxa represented
by these Laurasian links did not necessarily migrate into North America
from Laurasia and thence to South America; the reverse process could have
occurred, with origin in North or South America. Nevertheless, Taylor's
data supported the hypothesis that biotic exchange of Gondwanan taxa
occurred via both the South American and Laurasian connections.

The fact that 59% of the Mexican rain forest genera reported for North
American Eocene floras also occur in Eocene floras of Eurasia (Table 22.8)
supports the hypothesis of Laurasian biotic connections as well. The list of
such Eocene pan-Laurasian genera includes all ten genera of the Laurasian
familes, genera that are at present often best developed in temperate (*Ilex*,
Populus, *Salix*) or tropical montane (*Clethra*, *Saurauia*) areas in the New
World. However, also included are 20 genera of Gondwanan families,
including many best developed in modern tropical lowland forests, including evergreen and seasonal types, e.g., *Ficus*, *Nectandra*, *Ocotea*, *Sabal* (Zona,
1990), *Sterculia*. Directionality, as in Taylor's (1988) study, is not clear in
most cases.

Some clarification of the latter point, as well as further support for
Eocene Laurasian floristic links, can be found in present distributional
patterns. For temperate elements, an East Asian–North American distribution is routinely interpreted as a relict of a formerly more widespread
Laurasian distribution involving North Atlantic or Beringian connections
(Tiffney, 1985a,b). A somewhat similar pattern for tropical taxa, the
amphipacific tropical distribution, involves taxa known from only tropical
eastern Asia and tropical America (usually including South America)
(Thorne, 1972; van Steenis, 1962). This pattern has been interpreted as
involving relicts of: (1) a Laurasian distribution (Eocene or earlier) with

invasion into South America and extinction in intervening areas due to post-Eocene cooling; or (2) a former Asia–Africa–America distribution, probably from a Gondwanan origin, with subsequent extinction in Africa due to Neogene and Quaternary aridity (Raven & Axelrod, 1974; Thorne, 1972). Thorne (1972) suggested that the amphipacific pattern probably is the product of both processes plus some long distance-dispersal, whereas Raven and Axelrod (1974) argued that in Gondwanan families it most likely results largely from the second process, a stance concordant with Gentry's (1982a) overall views. Van Steenis (1962) and Thorne (1972) recorded 20 genera of Mexican rain forest canopy trees in non-Laurasian families that present this pattern, and several more are included below. It should be noted that in the case of Laurasian tropical distributions, a given taxon did not necessarily exist in eastern Asia during the Paleocene and early Eocene, even in the case of taxa at present endemic to that area. Due to Asian epicontinental seas, it may have been restricted to Europe and North America at the time, with subsequent dispersal to southeastern Asia after the drying of the central Asian seaway during the Eocene (Tiffney, 1985a).

That an Eocene or earlier Laurasian distribution is indicated for at least some amphipacific tropical taxa in Gondwanan families is strongly supported by genera in which the total present distribution is clearly centered in southeastern Asia and northern Latin America, with few or no species in South America. In this case, extinction in both Africa and South America would be hypothesized to explain this distribution as a relict of a former Gondwanan distribution. However, unlike the case of Africa, the extinction of numerous tropical lines in South America would not be easy to explain (Raven & Axelrod, 1974), and thus a relict Laurasian distribution seems indicated. For the Mexican rain forest canopy tree flora, the following taxa exhibit this distribution: *Aphananthe*, Mexico, Central America, Asia to Australia; Flacourtiaceae subtribe Hydnocarpinae, Mexico (*Chiangiodendron*), Southeast Asia to Australia; *Styphnolobium* (Fig. 22.8), Mexico, Central America, southeastern United States, Southeast Asia; *Trichospermum*, Mexico, Central America, northwestern South America, Antilles, Malaysia to western Pacific; *Federovia* segregate genus of *Ormosia* (Yakovlev, 1971), Mexico and Central America (*Ormosia panamensis*), eastern and southeastern Asia. Only one of these taxa (*Aphananthe*) belongs to a Laurasian family. The others, if the families to which they belong are indeed correctly regarded as Gondwanan in origin, may represent lineages of these families that invaded Laurasian areas early from Africa, India, or South America. The present distribution of sister groups in each case could help to identify the route of interchange, but phylogenetic studies are lacking. In the case of the subtribe Hydnocarpinae, the evidence is congruent with an early Africa-Eurasia divergence, as the other subtribe of the tribe Pangieae is African (Gilg, 1925), with subsequent spread to North America via a Laurasian route; the Mexican *Chiangiodendron* does not appear to be most primitive genus in the tribe (Wendt, 1988). In this case a spread from Eurasia to North America via Laurasian connections is indicated.

The present distributions of other Old World–northern Latin American groups also support the hypothesis of the arrival of many tropical taxa of Gondwanan families via a European–North Atlantic route from African or Eurasian origins. They include *Antirhea* (Mexico, Central America, Antilles, Asia, and Australia to Madagascar); *Berrya* (Mexico, Antilles, Asia, and Polynesia to eastern Africa); *Bourreria* (Mexico, Central America, Antilles, southern United States, northern South America [few species], apparently also Africa) (see Miller, 1989); *Ehretia* (Mexico, Central America, Antilles, southern United States, Asia to Africa); *Erblichia* (Mexico, Central America, Madagascar); and *Sideroxylon* (Mexico, Central America, Antilles, southern United States, South America [poorly represented], eastern Asia to Africa). Only *Bourreria* and *Ehretia* belong to a "Laurasian" family (Boraginaceae). For the other genera, presence in Africa and absence in South America indicates that, if they are indeed ultimately Gondwanan in origin, the most likely dispersal route was Africa–Europe–North America. Total distribution of *Sideroxylon* and related genera (Aubreville, 1973; Pennington, pers. comm.) strongly suggests an Old World origin and subsequent spread to North America. *Lonchocarpus*, essentially endemic to the New World and strongly centered in northern Mesoamerica (see above), probably represents the same pattern, as the most closely related genus is probably the essentially Southeast Asian to Australian *Derris* (Sousa, pers. comm.). The suite of related genera is strongly centered in the Old World, including both Africa and Southeast Asia (Geesink, 1981; Sousa & Peña de Sousa, 1981), suggesting an Old World origin of the group with subsequent immigration into North America. Rzedowski (this volume) suggests, based on the present and fossil distribution of the genus *Bursera* and related genera, that it likewise is descended from African ancestors and arrived in the New World via Laurasian connections.

By analogy, these data suggest that the more generalized amphipacific tropical distribution (as well as pantropical distributions) of taxa of Gondwanan families in some cases also represents a relict of an ancient Laurasian distribution involving dispersal from Eurasia to North America, with present diversity in South America resulting from prior immigration from North America, as suggested by Taylor (1988), Thorne (1972), and van der Hammen and Cleef (1983). In Taylor's (1988) previously discussed study, examples from four amphipacific tropical Gondwanan taxa are included, with the conclusion that two represent a southern and two a Laurasian route of migration. Berg (1983) hypothesized that the two New World sections of *Ficus*, a genus apparently of Old World origin, as well as *Trophis* and *Chlorophora*, entered the New World via Laurasian connections and subsequently spread to South America. For taxa such as these that are well developed in subtropical, seasonal climates, the alternate hypothesis involving extinction in Africa during Neogene and Quaternary dry periods seems less likely due to their ecological amplitudes. For this reason,

Miranda's (1960) finding that neotropical megathermal subtropical arboreal floras show a stronger generic relationship to Asia and a weaker one to Africa, compared to neotropical equatorial floras, may likewise be a reflection of former Laurasian connections in the New World subtropical flora. Miranda's results should be rechecked using a more recent, fuller data base.

In conclusion, a significant portion of the canopy tree flora of the Mexican rain forests appears to consist of taxa derived from ancient Eocene North American tropical forests, and Laurasian-arrival taxa in putative Gondwanan as well as Laurasian families seem to be an important part of that ancient element. Sousa and Delgado's (this volume) synthesis for Mexican legumes is strongly in accord with this interpretation with respect to rain forest genera in that "Gondwanan" family. Using the largest 16 genera in the present work (Table 22.3) as an example, a Laurasian-arrival element seems clear. The previous discussion indicates that *Ficus*, *Lonchocarpus*, *Sideroxylon*, and *Ocotea*, all in Gondwanan families, seem to represent Laurasian connections, with a Eurasia-to-North America direction of migration probable for at least the first three. *Cordia* belongs to the Laurasian family Boraginaceae, although its history is unclear (Gentry, 1982a). Present plus fossil distribution of the tribe Zanthoxyleae (Gregor, 1989) of the unassigned family Rutaceae strongly suggests a former Laurasian distribution and origin of *Zanthoxylum*. *Sloanea* (Thorne, 1972; van Steenis, 1962) and *Pouteria* (Pennington, pers. comm.) have essentially amphipacific tropical distributions. *Cordia*, *Ficus*, *Nectandra*, *Ocotea*, and *Zanthoxylum* are reported for the Eocene of both North America and Eurasia (Table 22.8). Eocene or earlier immigration of some but probably not all of these genera from Eurasia via Laurasian connections seems likely. On the other hand, strong arguments for early immigration from South America of at least the following large genera can be made based on the combination of probable South American origin, lack of Eurasian Eocene fossils, and present high diversity in Mexico, denoting antiquity in that area: *Eugenia* (Taylor, 1988) present in North and Mesoamerica during the Eocene and later (Tables 22.8 and 22.9); *Inga* (Sousa & Delgado, this volume); and *Licaria* (Kurz, 1983).

Based on the foregoing considerations as applied to the list of genera in the Appendix, I estimate that at least one-fourth of the species of canopy trees in extant Mexican rain forests are descendents of Laurasian-arrival lineages of both Laurasian and Gondwanan families. However, a relatively exact quantification of this element is not possible at present for at least two reasons. First and foremost, phylogenetic studies of individual taxa at the species-complex to suprageneric level are necessary to be able to carry this study beyond its present preliminary stage, and such studies are essentially lacking for neotropical forest tree groups. Second, the composition of the Eocene and earlier floras of South America, although obviously relevant, is not clear. Many recent studies (e.g., Romero, 1986) appear to be based largely on old identifications of the macrofossils cited.

Even so, it is clear that diverse families are represented. Fossil pollen data (Muller, 1981; Romero, 1986) show that, by Eocene times, many rain forest canopy tree families were present in South America, including Annonaceae, Apocynaceae, Arecaceae, Bombacaceae, Euphorbiaceae (including perhaps *Alchornea*), Leguminosae (Caesalpinioideae), Lythraceae, Malphigiaceae, Malvaceae, Myrtaceae, Olacaceae, Proteaceae, Sapindaceae, Tiliaceae (perhaps *Trichospermum*), and others. Also present already were families such as the South American Caryocaraceae, which may never have reached Mexico, and Ulmaceae (perhaps *Ampelocera*), an apparently Laurasian family that may have entered South America from North America during the Cretaceous. Other families present by Eocene times include Anacardiaceae (including *Astronium*), Elaeocarpaceae, Lauraceae, and Rutaceae (Raven & Axelrod, 1974; Romero, 1986). Further information on Eocene floras of South America, especially from the northern half, will help to elucidate the history of exchange of elements with the Laurasian rain forests to the north.

Vegetational Interrelationships and the Persistence of Taxa

The preceding discussion emphasizes the immigration and persistence of individual taxa rather than the history of an integrated rain forest flora. Although the geoflora concept has been much used by authors such as Chaney (1947) and Axelrod (1958, 1975) to explain the evolution of modern American floras, many studies indicate that present floras can be understood only as the amalgam of different taxa with differing histories (Davis, 1983; Wolfe, 1972, 1979). Numerous authors have noted that Eocene plant communities of North America almost invariably include genera that at present grow under widely divergent ecological conditions and are not sympatric (Dilcher, 1973; MacGinitie, 1974; Tiffney, 1985b; Wing, 1987; Wolfe, 1972). For instance, paleocommunities that apparently represent tropical or paratropical rain forest, based on physiognomic analysis and other data, include genera at present restricted to temperate forests, as previously noted. Shifts in the adaptive modes of many lineages apparently occurred as microthermal and seasonal megathermal climates increased in importance in the Northern Hemisphere from the Eocene onward.

The elements that have persisted in Mexico since the Eocene have done so under radically changing environmental conditions, to which each taxon has adapted in individualistic fashion. Thus most of the components of the Eocene forests have become adapted principally to either montane (e.g., *Symplocos, Liquidambar, Tilia, Quercus*, Juglandaceae in general), or lowland tropical environments; nevertheless, many genera occur in both (e.g., *Beilschmiedia, Cedrela, Ilex, Oreopanax, Persea*), and the lowland genera that do not are mostly found in both dry and wet tropical forests. Thus Graham (1976) noted that almost all potential rain forest genera in the

Mexican Paraje Solo fossil pollen flora (Pliocene) (Graham, 1989a) also occur at present in other vegetation types, making vegetational reconstruction difficult. This ecological amplitude may be the result of adaptation to the strong climatic changes in temperature and rainfall these lineages have survived since the Eocene. Leopold and MacGinitie (1972) and MacGinitie (1974) suggested that North American Eocene lineages that have persisted to the present in Mesoamerica are those that were well adapted to the seasonal subhumid climates that became more common during the Middle Eocene of the Rocky Mountain area. Lineages not adapted to strong seasonality have since died out in North America, although many of the latter persist in southeastern Asia.

A preliminary analysis of the 230 present rain forest canopy tree genera indicates that approximately 37% have other, non-rain-forest species in Mexican dry tropical vegetation and 20% in Mexican cloud forest, with a combined total of 48% (including overlap). Of the 12 largest genera (Table 22.3), all except *Pouteria* include other, non-rain-forest species of dry tropical or cloud forest habitats in Mexico (some of the rain forest species of *Pouteria* also occur in drier forests). About 59% of the total genera have at least one species (which may also occur in rain forest) in Mexican dry tropical vegetation. These facts are strong indications that broad ecological amplitude of lineages has been important in the history of the Mexican flora and that the floristic histories of these vegetation types are intertwined. The floristic relationship between Mexican tropical rain forest and tropical dry forests seems particularly strong. For instance, rain forest and semideciduous forests in Mexico share many codominant species (e.g., *Aphananthe monoica*, *Brosimum alicastrum*, *Bursera simaruba*, *Calophyllum brasiliense*, *Cedrela odorata*, *Bernoullia flammea*, *Manilkara zapota*) (Rzedowski, 1978), the so-called "tolerant" species of Toledo (1982). The Mexican endemic genus *Recchia* includes three species, two in tropical deciduous woodlands and one in rain forest (Wendt & Lott, 1985). MacGinitie (1974) pointed out the generic similarity of Middle Eocene floras of the western United States to present tropical semideciduous forests of southwestern Mexico as well as, to a lesser extent, to neighboring cloud forests. However, all of the fossil genera he listed as typical of semideciduous forests also occur in Mexican rain forests, as do most of the cloud forest genera. This statement is not to suggest that MacGinitie erred; rather, it suggests the more basic interrelationship of the floristic histories of Mexican tropical wet and dry forests and cloud forest. The interplay through time of the climatic effects of Mexico's subtropical position between two oceans with its complex geological history, causing a complex mosaic of environments and thus allowing diverse vegetation types to coexist in mutual proximity, has obviously been important in this regard.

Adaptation to subhumid, cool, or strongly seasonal conditions has not only been an important factor in the persistence of lineages of the Mexican lowlands, but because of Mexico's subtropical position it continues to be critical in the modern floristic composition. The subtropical zones or "trade wind tropics" between about 12° and 25° latitude—from Nicaragua northward in the northern Neotropics—have in general more extreme mean temperatures during both the warmest and the coolest month, and a more strongly seasonal rainfall regime compared to equatorial areas (Granger, 1987; Holdridge et al., 1971; Strahler & Strahler, 1978). Limited precipitation data for the Atlantic lowlands of Mesoamerica (Cardoso, 1979; Coen, 1983; Mosiño and García, 1974; Portig, 1976) indicate a much more complex situation, but in general a smaller percentage of the annual rainfall occurs during the dry season in northern parts compared to the more southern areas, although strongly seasonal climates occur in some Atlantic slope parts of Panama. García (1973) has shown that all Mexican rain forests areas have strong seasonality of precipitation, even those with average annual precipitations of over 5,000 mm, and thus the floristic similarity of dry and wet lowland forest types in Mexico is not surprising. As to winter temperatures in the Atlantic lowlands of Mesoamerica, the average temperature of the coolest month is clearly lower in northern Mesoamerica (see data in Cardoso, 1979; Mosiño & García, 1974; Portig, 1976). Perhaps even more important are the effects of winter outbreaks of cool modified continental polar air ("nortes"), which generally do not have strong effects south of central Nicaragua. These nortes can drop temperatures to well below 10° C in the tropical lowlands of northern Mesoamerica, whereas to the south temperatures rarely if ever fall below 15° C (Cardoso, 1979; Mosiño & García, 1974; Portig, 1976).

A considerable number of South American rain forest genera and families occur no further north in Central America than Costa Rica or southeastern Nicaragua (Raven & Axelrod, 1974; Gentry, 1982a,b; W.D. Stevens, pers. comm.). Although the relatively short duration of the present more benign Holocene climates in northern Mesoamerica may be a causal factor in this pattern, it seems more likely that the different nature of the present climate of the humid lowlands to the north of this area is the determining factor in most cases, as suggested by Raven and Axelrod (1974). Immediately to the north of this area is the southern limit of the nortes and the drier, more seasonal lowland climate in the "bulge" of Central America (data in Portig, 1976), signs of the generally more seasonal nature of northern Mesoamerica already mentioned. Conversely, it can be argued that Eocene lineages that have persisted in Mexico will form a progressively smaller part of the flora southward due not only to historical factors of migration routes but also to their general adaptation to subtropical conditions because of their history.

Based on endemisms, it appears that many Mexican rain forest species have survived recent adverse climatic periods by remaining in Mexico or at least in northern Mesoamerica. Some probably continued as widespread species because of their ecological amplitudes, as hypothesized by Toledo (1982); for instance, almost 30% of the present species are also adapted to drier vegetation types (Table 22.1). It has been suggested that others survived the most adverse periods in restricted areas, or refuges (Graham,

1982; Toledo, 1982; Wendt, 1989). However, this does not necessarily imply that small areas of rain forest vegetation survived intact through these adverse periods ("vegetational refuges") (Wendt, 1989), as suggested by early workers on the refuge theory (e.g., see definition in Haffer, 1982). Refuges in northern Mesoamerica are probably better characterized as areas where environmental and biotic parameters were propitious for the survival of one or more rain forest species whose distributions had become markedly reduced owing to regional climatic deterioration, without neces-sarily implying the preservation of rain forest per se ("floristic refuges") (Wendt, 1989). Areas emphasized as probable refuge areas, e.g., the Crescent area (Wendt, 1989), are those where many species apparently survived these adverse periods. Under this concept, rain forest as a vegetation type may or may not have a continuous history from the Eocene to the present in northern Mesoamerica, and its floristic composition could have varied considerably depending on the dynamics of persistence and immigration, but many elements of the modern Mexican rain forests are nevertheless ancient in the area. This point is in keeping with the dynamic and individualistic nature of the history of the Mexican flora discussed above.

Graham (1989a,b) suggested that the insular nature of southern Mesoamerica during most of the Cenozoic was also important in the survival of many taxa. His paleobotanical studies indicate that paleofloras of northern Mesoamerica (e.g., Mexico) were under cooler climatic condi-tions and included more apparently temperate elements than coeval paleofloras of southern Mesoamerica (Costa Rica–Panama). Graham attributed it not only to differences of latitude but also to the greater continentality and proximity of highlands in Mexico, in contrast to the low-relief archipelago nature and resulting maritime climate of southern Mesoamerica. Thus some Mesoamerican elements probably survived adverse periods in these southern insular areas. It seems likely that some species, including some Mesoamerican endemics, have repeatedly become extinct in northern Mesoamerica during adverse periods only to reinvade subsequently from southern Central American refuges. Gentry (1982a) suggested a similar scenario for putative Laurasian groups such as Sabiaceae with present centers of diversity in southern Central America.

Recent Southern Arrivals

For reasons discussed previously, most of the foregoing discussion has concerned the origin and persistence of the "ancient" element in the Mexican rain forest flora. Nevertheless, it is clear that a relatively recent influx of southern taxa has dramatically altered these rain forests, not only floristically but vegetationally. For instance, many present Mexican rain forests are dominated by *Dialium guianense* or *Terminalia amazonia* (Rzedowski, 1978), both of which also occur in Amazonia and are probably recent arrivals to northern Central America. These recent arrivals do not form a homogeneous group, however, and two quite distinct phenomena are involved: (1) the arrival in Mesoamerica of taxa from South America during the Neogene and later, mostly via the Panamanian land bridge; and (2) the subsequent history of the taxon within Mesoamerica, including the possibility of repeated extinction in Mexico and reintroduction from Central America especially during the Pleistocene glacial cycles. Climatic cycles could of course also lead to repeated extinction in Mesoamerica and subsequent reintroduction (or not) from South America. An added twist is the possibility of Laurasian-arrival lineages having reached South America from North America during the early Tertiary, followed by diversification there and subsequent reinvasion by new lines into Mesoamerica.

It does not seem possible based on present data to decide if, for instance, *Dialium guianense* has invaded Mexico for the first time during the present interglacial period, or was present in Mexico during many interglacials and became locally extinct during glacial maxima. The presence of this same species in South America does not necessarily imply a recent (late Pleistocene or Holocene) immigration into Mesoamerica. *Symphonia globulifera*, like *Dialium guianense* the only neotropical species in its genus, is similarly widespread but was clearly in Mesoamerica during the Pliocene (Table 22.9). It is notable that many genera of probable recent (Neogene) arrival from South America include species endemic to Mesoamerica (e.g., *Cecropia*, *Couepia*, *Pseudolmedia*, *Vatairea*, *Virola*, *Vochysia*), including many with northern Mesoamerican endemics (*e.g., Coussapoa, Eschweilera, Hernandia, Licania, Pera*), indicating that many of these genera probably are not very recent, Holocene arrivals to the area.

Importance of Karst

In addition to the factors previously mentioned, a feature that distinguishes the rain forest areas of northern Mesoamerica from those further to the south is the dominance of limestone substrates forming karst terrain in large portions of the area (Figs. 22.11 and 22.13, below). Karst, also common in the Greater Antilles, is rare in the humid lowlands of southern Central America and South America; indeed, although it is a dominant feature of a considerable proportion of Laurasian rain forest areas in general, it is uncommon in Gondwanan ones (Middleton & Waltham, 1986; Sweeting, 1972). Tropical karst areas differ from other rain forest areas in their usually shallower and less acidic soils with more active calcium and magnesium cycling, and their different pattern of microtopography, including abundant limestone outcrops (Crowther, 1982, 1987; Furley & Newey, 1979). The shallower soils and rock outcrops also probably accentuate seasonal climatic drought. The flora of karst rain forest areas can be strikingly different from that of other substrates (Crowther, 1982; Richards, 1952; Wendt, unpub.) The historical continuity, age, and extension of present and former karst in Mexico are not clear. Although the creation and erosion of karst landscapes in tropical areas can be rapid, rates vary enormously according to regional and local factors, and tropical karst can be many millions of years old (Jennings, 1985).

Karst and variations in seasonality may be factors that help to explain how Mexican rain forest niches have been partitioned between the ancient Mexican elements and the more recently arrived southern ones. Apparent recent South American arrivals are conspicuous in the flora, but these species are strongly concentrated in the less seasonal areas with deep soils. *Dialium guianense* and *Terminalia amazonia* are typical of such areas, and many of the recent South American arrivals discussed in the previous section are more or less confined to these habitats, which are most widespread in the Gulf (including overlap) area. On the other hand, of 43 identified tree species (>28.5 cm d.b.h.) rooted directly in karst outcrops of a 5-ha plot in southern Veracruz and less common or lacking on nearby deep soils (Wendt, unpub.), 9.3% are endemic to Mexico, 40.9% are endemic to northern Mesoamerica, 68.2% are endemic to Mesoamerica, and only 9.1% occur in Amazonia. This outcrop flora is noticeably rich in genera endemic to or centered in northern Latin America and genera of probable Laurasian-arrival lineages, e.g., *Blomia, Bursera, Chiangiodendron, Chione, Cordia, Exostema, Lonchocarpus, Ficus, Mortoniodendron, Robinsonella, Sideroxylon, Stemmadenia, Wimmeria,* and *Zanthoxylum*. The ancient elements and the endemic elements in the Mexican rain forest flora seem to be especially well represented in these karst forests, perhaps in part due to their accentuated seasonality of soil moisture.

Conclusions

The following factors seem particularly important for understanding the origins and history of the canopy tree flora of the present Mexican rain forests.

1. Subtropical location and resultant strongly seasonal climate.
2. Geological and topographic diversity throughout the Cenozoic, which, combined with the first factor and the location between two oceans, has allowed diverse types of vegetation to coexist in proximity.
3. Inclusion of Mexico in the North American continent from the Jurassic–Lower Cretaceous onward, with terrestrial connections to western North America continual from at least Upper Cretaceous.
4. Eocene or earlier immigration of megathermal taxa from Eurasia and from Africa via Eurasia, especially via North Atlantic land routes, and from South America via proto-Antillean or Antillean routes, during a period of great worldwide extension of megathermal climates.
5. Miocene and later (especially Pliocene to present) immigration of South American megathermal elements into Mesoamerica via the Panamanian land bridge.

6. Relatively limited nature, in both time and space, of the migration routes in items 4 and 5, above.
7. Neogene and Pleistocene climatic cycles leading to local extinction and often subsequent reinvasions in all or part of southeastern Mexico or Mesoamerica.
8. Ocean-ameliorated nature of the climate of southern Mesoamerica through much or all of the Cenozoic, allowing the persistence of more "sensitive" taxa during periods of adverse climate.
9. Large extensions, at least in the recent past, of karst terrain in northern Mesoamerica.

These factors have led to moderately high endemism in the canopy tree flora of northern Mesoamerica and the presence of at least the following floristic elements.

1. Ancient arrivals
   a. Laurasian-arrival elements, descended from lineages that immigrated during the Eocene or earlier and that subsequently survived in Mesoamerica or invaded South America and later reinvaded Mesoamerica. They include lineages in both "Laurasian" and "Gondwanan" families, the latter probably mostly following an Africa–Eurasia–North America route, and comprise at least 25% of the flora.
   b. South American-arrival elements, descended from lineages that immigrated via proto-Antillean or Antillean routes during or probably mostly well before the Eocene.
2. Recent arrivals from South America
   a. Neogene arrivals that survived through the Pleistocene climatic cycles in at least southern Mesoamerica.
   b. More recent arrivals, some arriving in Mesoamerica perhaps as recently as during the Holocene, although not necessarily for the first time.

Persistence of lineages through adverse climatic conditions has been due to great adaptive amplitude, restriction to floristic refuge areas within northern Mesoamerica, or survival in less extreme climates in southern Mesoamerica. Even during the present, relatively benign conditions of the Holocene, the climatic conditions of northern Mesoamerica are probably too extreme for many South American rain forest taxa.

Graham (1975, 1977, 1982) is surely correct in his assessment that the Mexican rain forest, *in its present form*, is a recent phenomenon. Rain forests may have existed through much or all of the Cenozoic in Mexico, but they have at times been limited in extent and were floristically different from present ones. Although modern Mexican rain forests include many lineages that date to the ancient widespread Eocene rain forests of North America (including Mexico), the modern flora is due to the interaction of these ancient lineages with recent immigrants from South America. The

Figures 22.12-15 (above and below). The unique Mexican rain forests and their destruction. **Figure 22.12** (left). Large extensions of undisturbed rain forest such as this one in the southern part of the Uxpanapa region of Veracruz are now rare and are threatened in Mexico. **Figure 22.13** (right). Rain forest on karst terrain, Uxpanapa, Veracruz.

Figure 22.14 (left). Destruction of Mexican rain forest, Uxpanapa, Veracruz. **Figure 22.15** (right). Two Mexican endemics—*Albizia purpusii* (Leguminosae, left) and *Ocotea uxpanapana* (Lauraceae)—as remnants of a once-luxuriant riparian forest, Uxpanapa, Veracruz. Photographs by the author.

ancient elements and endemic taxa may be better represented at present in more "marginal" environments (e.g., more seasonal or karst areas), as suggested by Gentry (pers. com.), and show strong floristic relationships to drier or cooler Mexican vegetation types.

The extent of the applicability of these ideas to other life forms in Mexican rain forest is not clear, since different life forms can have significantly different phytogeographic histories within a vegetation type (Gentry, 1982a).

CONSERVATION

The data presented here substantially strengthen the already strong arguments for the conservation of Mexican rain forests, as these forests are shown to be unique floristically and historically (Figs. 22.12–22.15). Although the conservation of the genetic resources represented by these tree species must be pursued along many lines, including *ex situ* living collections and germplasm collections, the preservation of the natural ecosystems *in situ* is obviously preferable when feasible. The information in this chapter provides some bases for establishing priorities for conservation among the remaining areas of lowland rain forest of the Atlantic slope.

It seems obvious that biological reserves in the area should be placed in both the Gulf and Yucatan floristic areas, include high diversity areas, and protect as many of the endemic species as possible. At present, two large Biosphere Reserves exist in the area: (1) Sian Ka'an, on the Caribbean coast of central Quintana Roo, which includes some areas of semievergreen forest; and (2) Montes Azules, in the western Lacandon area of Chiapas, which includes diverse types of rain forest. The small (700 ha) Los Tuxtlas Tropical Biological Station of UNAM protects a small remnant of rain forest in this part of Veracruz. Based on the above criteria, the following areas seem the most critical for conservation within the rain forest area of Mexico (order is west to east and does not infer priority).

1. *Tuxtepec area*: center for endemism; much of the critical area is on karst substrate, which makes it difficult to develop for other purposes, thus providing fewer potential conflicts during the establishment of reserves.
2. *Los Tuxtlas area*: center for endemism; forest is rapidly disappearing and extremely threatened; any enlargement of the Los Tuxtlas station or establishment of other viable reserves soon is critical.
3. *Crescent area*: extraordinary center for endemism, also high diversity. The Uxpanapa-Chimalapa area seems especially important owing to higher demonstrated endemism and the presence of a broad, essentially undisturbed transect from lowland rain forest to montane cloud forest. Important are both the karst forests of northern Uxpanapa, which present the logistical advantages mentioned for Tuxtepec, and

the floristically different hill forests on deep soils in southern Uxpanapa and Chimalapa, which are rapidly disappearing and extremely threatened.

4. *Lacandon area*: part of the Yucatan Peninsula endemic area, especially as to wet forest species; also overlap of Gulf and Yucatan floras and high diversity. The forests of this area have been enormously reduced in areal extent since 1980. Enlargement of the Montes Azules reserve or establishment of other reserves is important.

5. *Southern Yucatan Peninsula*: part of the Yucatan Peninsula endemic area, especially as to semievergreen forest species. The proposed reserve in the Calakmul area, in southeastern Campeche, presents the logistical advantage of light population pressures due largely to the lack of surface water. The establishment of a reserve in the wetter, extreme southern portion of Quintana Roo is also important.

It is clear that the establishment of reserves in any part of the Mexican rain forests is important, and thus the above suggestions are not meant in any way to deter work in other areas, especially as they are based on only a single set of floristic criteria using canopy tree species. In the end, the establishment of functioning, viable reserves that do not lead to insurmountable local conflicts is very difficult, and any successes anywhere in Mexico are vital.

ACKNOWLEDGMENTS

This chapter would not have been possible without the enormous amount of work that has been done by collectors and floristic botanists in Mexico; articles by many of these workers are cited in the bibliography. I am greatly indebted also to the specialists listed below, who freely shared their unpublished data on critical families and areas; as imperfect as the present list may be, it would have been far more so without their help. I must single out Mario Sousa Sánchez for his continual help on the Leguminosae; he also read and commented on a draft of the chapter. Alwyn Gentry reviewed a draft of this manuscript and offered many useful suggestions and criticisms, as well as distributional information on selected taxa. I am indebted to Amy Pool of the Flora de Nicaragua project (directed by W.D. Stevens, MO) for information on the occurrence of selected species in Nicaragua. Alan Graham and Beryl Simpson reviewed the manuscript and offered valuable suggestions. Alan Graham and Scott Wing graciously made available unpublished paleobotanical data. Eduardo Merino, of the Centro de Estudios de Desarrollo Rural, Colegio de Postgraduados, drew Figures 22.2 and 22.3. The Centro de Botánica of the Colegio de Postgraduados, Chapingo, has continually supported these studies; financial support for our field work has come from the Centro de Botánica, from the Consejo

Nacional de Ciencia y Tecnología (CONACyT) through a grant for Flora Mesoamericana, and more recently from C.L. Lundell. The Plant Resources Center of the University of Texas at Austin and its director, Dr. Billie Lee Turner, have graciously allowed me space and use of facilities during a part of the preparation of this manuscript. Lastly, I thank the many people with whom I have had the great fortune to work in the field in the rain forests of Mexico, including Fernando Chiang C., Kathleen Collins, Heriberto Hernández G., Mario Ishiki I., Emily Lott, Salomón Maya J., Agustín Montero, Isidro Navarrete S., Mario Vázquez T., Patricia Vera C., and Agustín Villalobos C.

The following specialists provided critical information on their groups or areas of expertise: W. Alverson (WIS: *Quararibea*), R. Barneby & J. Grimes (NY: *Abarema*), F. Chiang (MEXU: Rutaceae), A. Espejo (UAMIZ: Hernandiaceae), P. Fryxell (U.S. Dept. Agriculture, College Station, Texas: Malvaceae), A. Gentry (MO: Bignoniaceae), S. Graham (KE: Lythraceae), W. Hahn (US: Aquifoliaceae), L. Hernández S. (LL/TEX: Agavaceae), G. Ibarra (MEXU & CHAPA: *Ficus*, also trees of Los Tuxtlas Tropical Biological Station, Veracruz), H. Iltis (WIS: Capparaceae), M.C. Johnston (LL/TEX: Rhamnaceae), L.R. Landrum (ASU: *Psidium*), D. Lorence (PTBG: Rubiaceae), P.J.M. Maas (U: *Rollinia*), E. Martínez (MEXU: Trees of Lacandon rain forests, Chiapas), W. Meijer (KY: Tiliaceae), N. Murray (NY: *Cymbopetalum*), T.D. Pennington (K: Sapotaceae), H. Quero (MEXU: Arecaceae), L. Rico (K: *Abarema, Acacia, Albizia* [in part], *Cojoba, Zygia*), R.L. Robbins (KSTC: Sapindaceae), J. Rohwer (HBG: *Nectandra*), G. Schatz (MO: Annonaceae), M. Sousa (MEXU: Leguminosae), W.T. Stearn (BM: Oleaceae), W. Thomas (NY: *Simarouba,*), P. Vera (CHAPA: *Calatola*), H. van der Werff (MO: Lauraceae).

APPENDIX—CANOPY TREES OF THE TROPICAL RAIN FORESTS OF THE MEXICAN ATLANTIC SLOPE: PRELIMINARY LIST

Key to Table

Families and Species

Families are listed in alphabetical order. Angiosperm families are according to Cronquist (1981), except for the more common broad circumscription of the Leguminosae, which follows Sousa and Delgado (this volume); in the latter case, the species are listed by subfamily. In parentheses following each family name are the number of genera/number of species and intraspecific taxa included. Criteria for inclusion are found in the text. Author abbreviations follow the list of the Royal Botanic Gardens, Kew (1980), with the following exceptions: "R. & P." instead of the cumbersome "Ruíz Lopez & Pavón"; "Vellozo" instead of the unusual "Vell. Conc."; "H.B.K." instead of the unacceptable "Kunth" when referring to species published in *Nova Genera et Species Plantarum*. Short-form author citation is used (see articles 46.2 and 46.3 of the 1988 *International Code of Botanical Nomenclature*). Synonymy is included only in a few cases of commonly used synonyms where information is not readily available elsewhere. Necessary new names or combinations not already published are indicated by "sp.," "subsp.," or "var." followed immediately by a synonym in parentheses. In the case of unpublished new species, the authority is included in brackets (see acknowledgments); in the case of unidentified species ("sp." without synonymy), the known distribution in Mexican rain forest is included in brackets using the following abbreviations:

Chim.: Chimalapa area of eastern Oaxaca
L.T.: Los Tuxtlas-Catemaco area of Veracruz
n. Chis.: northern Chiapas
Uxp.: Uxpanapa area of extreme southeastern Veracruz

Distributional and Other Information (Columns 1–5)

Column 1. Generalized range in Atlantic slope rain forest of Mexico (note: distribution refers only to range within this type of vegetation and on the Atlantic slope), using the following code.

G: "Gulf" distribution, occurring in at least one of the states of Vercruz, Oaxaca, Puebla, Hidalgo, and San Luis Potosí; often also in Tabasco, Chiapas, and/or extreme southwestern Campeche; not in rest of Campeche, Yucatan, or Quintana Roo
Y: "Yucatan" distribution, occurring in at least one of the states of Yucatan, Quintana Roo, and Campeche; often also in Tabasco and/ or Chiapas; not in Veracruz, Oaxaca, or further to northwest
G,Y: Combination of the above two distributions
T: Recorded only from Tabasco and/or Chiapas

Column 2. Southern limit of continental distribution of the taxon in the New World, using the following code.

* *: Mexico
* 1: Belize, Guatemala, El Salvador, or Honduras
* 2: Nicaragua, Costa Rica, or Panama
* 3: Colombia, Ecuador, Venezuela, or the Guyanas (see also discussion of column 3)
* 4: Peru, Brazil, or further south

Information is not included for unidentified species or those that are very poorly understood. Information in parentheses refers to taxa which have been cultivated to such an extent that the native range is unclear; the data given are a "best guess" but are excluded from the phytogeographical analysis.

Column 3. Occurrence in the Greater Antilles (+, present; -, absent); "Greater Antilles" is defined as a broad phytogeographical area from Puerto Rico northwestward to Cuba, southern Florida, and the Bahamas. Taxa that occur in the Lesser Antilles but not in the Greater Antilles are included in column 2 as occurring in northern South America; those that occur in both the Greater and Lesser Antilles are included with the Greater Antilles in column 3. For parenthetical and lacking information, see discussion of column 2.

Column 4. Taxa in which individuals often reach 20 m in height in Mexico (+) *versus* those that only rarely do so (-).

Column 5. Taxa that frequently also occur in much drier lowland forests in Mexico, such as tropical semideciduous and deciduous forests (+), *versus* those that rarely or never do so (-).

APPENDIX

Canopy trees of the tropical rain forests of the Mexican Atlantic slope: Preliminary list

Taxa	1[a]	2	3	4	5
ACANTHACEAE (1/1)					
Bravaisia integerrima (Sprengel) Standley	G	2	-	+	-
ACTINIDIACEAE (1/1)					
Saurauia yasicae Loes.	G	3	-	-	-
AGAVACEAE (1/1)					
Beaucarnea pliabilis (Baker) Rose (syn. *B. ameliae* Lundell)	Y	1	-	-	+
AMARANTHACEAE (1/1)					
Iresine arbuscula Uline & W. Bray	G	2	-	-	-
ANACARDIACEAE (5/7)					
Astronium graveolens Jacq.	G,Y	4	-	+	+
Metopium brownei Urban	G,Y	1	+	+	+
Mosquitoxylum jamaicense Krug & Urban	G	3	+	+	-
Spondias mombin L.	G,Y	(4)	(+)	+	+
Spondias radlkoferi J.D. Smith	G	3	-	+	-
Tapirira mexicana Marchand	G	2	-	+	-
Tapirira sp. nov. [Wendt]	G	*	-	+	-
ANNONACEAE (7/9)					
Annona liebmanniana Baillon (syn. *A. scleroderma* Saff.)	G	2	-	-	-
Cymbopetalum baillonii R.E. Fries	G	*	-	+	-
C. mayanum Lundell	T	1	-	+	-
Guatteria anomala R.E. Fries	G	1	-	+	-
Oxandra maya Miranda	T	1	-	+	-
Rollinia membranacea Triana & Planchon (syn. *R. rensoniana* Standley)	G	3	-	-	-
R. mucosa (Jacq.) Baillon (syn. *R. jimenezii* Saff.)	G	(4)	(+)	+	-
Sapranthus violaceus (Dunal) Saff.	G	2	-	-	+
Xylopia frutescens Aublet	G,Y	4	-	-	-
APOCYNACEAE (4/5)					
Aspidosperma cruentum Woodson	G,Y	4	-	+	-
A. megalocarpon Muell. Arg.	G,Y	3	-	+	+
Plumeria rubra L.	G,Y	3	-	-	+

Appendix (cont.)

Taxa	1	2	3	4	5
Stemmadenia donnell-smithii					
(Rose) Woodson	G,Y	2	-	+	-
Tabernaemontana arborea Rose	G	3	-	+	-
AQUIFOLIACEAE (1/3)					
Ilex belizensis Lundell	G	1	-	+	-
I. guianensis (Aublet) Kuntze	T	4	+	-	-
I. tectonica Hahn	G	2	-	+	-
ARALIACEAE (3/3)					
Dendropanax arboreus (L.)					
Decne. & Planchon	G,Y	4	+	+	+
Didymopanax morototonii (Aublet)					
Decne. & Planchon	G	4	+	+	-
Oreopanax peltatus Regel	G	1	-	-	-
ARECACEAE (7/11)					
Acrocomia mexicana C. Martius	G,Y	1	-	-	+
Gaussia gomez-pompae (Quero) Quero	G	*	-	-	-
G. maya (Cook) Quero & Read	Y	1	-	+	-
Orbignya cohune (C. Martius) Standley	Y	2	-	+	-
Roystonea dunlapiana P. Allen	G,Y	2	-	+	-
Sabal mexicana C. Martius	G,Y	1	-	+	+
S. mauritiiformis (Karsten)					
Griseb. & H.A. Wendl.	G,Y	3	-	+	-
S. yapa Becc.	Y	1	+	+	+
Scheelea liebmannii Becc.	G	*	-	+	-
S. lundellii Bartlett	T	1	-	+	-
Thrinax radiata Schultes & Schultes f.	Y	1	+	-	+
BIGNONIACEAE (1/4)					
Tabebuia chrysantha (Jacq.) Nicholson	G,Y	4	-	+	+
T. donnell-smithii Rose					
(syn. *T. millsii* (Miranda) A. Gentry)	G	3	-	+	+
T. guayacan (Seemann) Hemsley	G	4	-	+	-
T. rosea (Bertol.) DC.	G	3	-	+	+
BOMBACACEAE (6/8)					
Bernoullia flammea Oliver	G	2	-	+	+
Ceiba pentandra (L.) Gaertner	G,Y	(4)	(+)	+	+
Ochroma pyramidale (Lam.) Urban	G	4	+	+	-
Pachira aquatica Aublet	G,Y	4	-	+	-
Pseudobombax ellipticum					
(H.B.K.) Dugand	G,Y	2	-	+	+
Quararibea funebris (Llave) Vischer	G,Y	2	-	+	-
Q. yunckeri Standley	G	1	-	-	-
Quararibea sp. nov. [Alverson]	T	1	-	-	-

Appendix (cont.)

Taxa	1	2	3	4	5
BORAGINACEAE (3/9)					
Bourreria oxyphylla Standley	Y	3	-	+	-
Cordia alliodora (R. & P.) Oken	G,Y	4	+	+	+
C. bicolor A.DC.	G	4	-	+	-
C. dodecandra A.DC.	G,Y	1	+	-	+
C. gerascanthus L.	G,Y	3	+	+	+
C. megalantha S.F. Blake	G	2	-	+	-
C. prunifolia I.M. Johnston	G	2	-	-	-
C. stellifera I.M. Johnston	G	2	-	+	-
Ehretia tinifolia L.	G,Y	1	+	-	+
BURSERACEAE (2/2)					
Bursera simaruba (L.) Sarg.	G,Y	3	+	+	+
Protium copal (Schldl. & Cham.) Engl.	G,Y	1	-	+	+
CAPPARACEAE (2/4)					
Capparis discolor J.D. Smith					
(syn. *C. tuerckheimii* J.D. Smith)	G	2	-	+	-
C. mollicella Standley	G	*	-	+	+
C. quiriguensis Standley	G	1	-	-	-
Crateva tapia L.	G,Y	4	+	+	+
CECROPIACEAE (3/5)					
Cecropia obtusifolia Bertol.	G,Y	3	-	+	+
C. peltata L.	G,Y	3	+	-	+
Coussapoa oligocephala J.D. Smith	Y	1	-	+	-
C. purpusii Standley	G	1	-	+	+
Pourouma guianensis Aublet	T	4	-	+	-
CELASTRACEAE (3/5)					
Maytenus schippii Lundell	G,Y	2	-	-	-
Maytenus sp. [Chim.]	G			+	-
Maytenus sp. [Uxp., Chim.]	G			+	-
Wimmeria bartlettii Lundell	G	1	-	+	-
Zinowiewia integerrima (Turcz.) Turcz.	G	2	-	+	-
CHRYSOBALANACEAE (3/6)					
Couepia polyandra (H.B.K.) Rose	G	2	-	+	+
Hirtella americana L.	Y	3	+	+	-
H. triandra Sw. subsp. *media*					
(Standley) Prance	G	2	-	+	-
Licania hypoleuca Benth. var. *hypoleuca*	G	4	-	+	-
L. platypus (Hemsley) Fritsch	G	3	-	+	-
L. sparsipilis S.F. Blake	G	2	-	+	-
CLETHRACEAE (1/1)					
Clethra sp. [Uxp., L.T.]	G			+	-

Appendix (cont.)

Taxa	1	2	3	4	5
CLUSIACEAE (4/7)					
Calophyllum brasiliense Cambess.					
var. *rekoi* (Standley) Standley	G,Y	2	-	+	+
Clusia belizensis Standley	T	2	-	-	-
C. flava Jacq.	G,Y	2	+	-	+
C. rosea Jacq.	G	3	+	-	-
Garcinia intermedia (Pittier) Hammel	G	4	-	-	-
G. macrophylla C. Martius	G,Y	4	-	-	-
Symphonia globulifera L.f.	G	4	+	+	-
COMBRETACEAE (2/2)					
Bucida buceras L.	Y	3	+	+	-
Terminalia amazonia (J. Gmel.) Exell	G	4	-	+	-
EBENACEAE (1/4)					
Diospyros campechiana Lundell	G,Y	1	-	-	-
D. digyna Jacq.	G	(2)	(+)	+	+
D. verae-crucis (Standley) Standley	G,Y	2	-	-	+
D. yatesiana Standley	Y	1	-	+	-
ELAEOCARPACEAE (1/5)					
Sloanea meianthera J.D. Smith	G	3	-	+	-
S. medusula Schumann & Pittier	G	3	-	+	-
S. petenensis Standley & Steyerm.	G	3	-	+	-
S. terniflora (DC.) Standley	G	4	-	+	-
S. tuerckheimii J.D. Smith	G	4	-	+	-
EUPHORBIACEAE (12/17)					
Alchornea latifolia Sw.	G,Y	4	+	+	-
Croton draco Schldl. & Cham.					
subsp. *draco*	G,Y	1	-	+	+
C. schiedeanus Schldl.	G	4	-	-	-
Drypetes brownii Standley	T	1	-	+	-
D. lateriflora (Sw.) Krug & Urban	G,Y	1	+	-	+
Hura polyandra Baillon	G,Y	2	-	+	+
Hyeronima oblonga (Tul.) Muell. Arg.	G	4	-	+	-
Mabea excelsa Standley & Steyerm.	G	2	-	+	-
Margaritaria nobilis L.f.	G,Y	4	+	-	+
Omphalea oleifera Hemsley	G	2	-	+	-
Pera barbellata Standley	G	1	-	+	-
Sapium bourgeaui Croizat	G	*	-	+	-
S. lateriflorum Hemsley	G	2	-	+	+
S. nitidum (Monach.) Lundell	G	1	-	+	-
S. oligoneuron Schumann & Pittier	Y	2	-	+	-
Sebastiana tuerckheimiana					
(Pax & K. Hoffm.) Lundell					
(syn. *S. longicuspis* Standley)	T	1	-	+	-
Tetrorchidium rotundatum Standley	G	2	-	+	-

Appendix (cont.)

Taxa	1	2	3	4	5
FLACOURTIACEAE (7/13)					
Casearia arborea (Rich.) Urban	G	4	+	-	-
C. arguta H.B.K.	G	3	-	-	+
C. bartlettii Lundell	T	1	-	-	-
C. commersoniana Cambess.	G	4	-	-	-
C. corymbosa H.B.K.	G,Y	3	-	-	+
C. sylvestris Sw. var. *sylvestris*	G	4	-	-	+
C. tacanensis Lundell	G	2	-	+	-
Chiangiodendron mexicanum T. Wendt	G	*	-	-	-
Homalium racemosum Jacq.	G	4	+	+	-
Lunania mexicana Brandegee	G	2	-	-	-
Pleuranthodendron lindenii					
(Turcz.) Sleumer	G	4	-	+	-
Xylosma chloranthum J.D. Smith	G	2	-	-	-
Zuelania guidonia (Sw.)					
Britton & Millsp.	G,Y	3	+	+	+
HERNANDIACEAE (1/3)					
Hernandia sonora L.	G	2	+	+	-
H. stenura Standley	G	3	-	+	-
Hernandia sp. nov. [Espejo]	G	*	-	+	-
HIPPOCRATEACEAE (1/1)					
Hemiangium excelsum					
(H.B.K.) A.C. Smith	Y	2	-	-	+
ICACINACEAE (3/3)					
Calatola sp. nov. [Vera]	G	1	-	+	-
Mappia racemosa Jacq.	G	2	+	-	-
Oecopetalum mexicanum					
Greenman & C. Thompson	G	1	-	-	-
LACISTEMATACEAE (1/1)					
Lacistema aggregatum (Bergius) Rusby	G	4	+	-	+
LAURACEAE (5/30)					
Beilschmiedia anay					
(S.F. Blake) Kosterm.	G	3	-	+	-
Licaria campechiana					
(Standley) Kosterm.	Y	1	-	+	-
L. capitata (Cham. & Schldl.) Kosterm.	G	1	-	+	-
L. misantlae (Brandegee) Kosterm.	G	3	-	-	-
L. peckii (I.M. Johnston) Kosterm.	G,Y	1	-	-	-
L. triandra (Sw.) Kosterm.	G	4	+	-	+
L. velutina van der Werff	G	*	-	-	-
Nectandra ambigens (S.F. Blake) Allen	G	1	-	+	-
N. cissiflora Nees	G	4	-	+	-
N. colorata Lundell	G	1	-	-	-

Appendix (cont.)

Taxa	1	2	3	4	5
N. cuspidata Nees	G	4	-	+	-
N. lundellii Allen	G	1	-	+	-
N. membranacea (Sw.) Griseb.	G	4	+	+	-
N. nitida Mez (syn. *N. perdubia* Lundell)	G	2	-	+	+
N. reticulata (R. & P.) Mez	G	4	-	+	-
N. salicifolia (H.B.K.) Nees	G,Y	2	-	-	+
N. turbacensis (H.B.K.) Nees	G	4	+	+	+
Nectandra sp. (*N. globosa* auth. in part, not (Aublet) Mez)	G	4	+	+	+
Nectandra sp. nov. [Rohwer]	T	*	-	+	-
Ocotea cernua (Nees) Mez	G	4	-	-	-
O. dendrodaphne Mez	G	2	-	-	-
O. effusa (Meissner) Hemsley	G	1	-	+	-
O. rubriflora Mez	G	2	-	+	-
O. uxpanapana T. Wendt & van der Werff	G	*	-	+	-
Ocotea sp. (syn. *Nectandra sinuata* Mez)	G	2	-	+	+
Ocotea sp. nov. [van der Werff]	G	*	-	+	-
Ocotea sp. [Chim.]	G			+	-
Ocotea sp. [Uxp.]	G			+	-
Persea longipes (Schldl.) Meissner	G	1	-	+	-
P. schiedeana Nees	G	3	-	+	-
LECYTHIDACEAE (1/1)					
Eschweilera mexicana T. Wendt, S. Mori & Prance	G	*	-	+	-
LEGUMINOSAE (35/84)					
Caesalpinioideae (10/15)					
Bauhinia sp. nov. [Wendt]	G	*	-	+	-
Caesalpinia gaumeri Greenman	Y	1	-	+	+
C. mollis (H.B.K.) Sprengel	Y	3	+	+	-
C. yucatanensis Greenman	Y	1	-	-	+
Caesalpinia sp.	G			+	-
Cassia grandis L.f.	G,Y	(4)	(+)	+	-
C. moschata H.B.K.	G	4	-	+	-
Cynometra retusa Britton & Rose	G	2	-	+	-
Dialium guianense (Aublet) Sandw.	G	4	-	+	-
Hymenaea courbaril L. var. *courbaril*	G,Y	4	+	+	+
Peltophorum dubium (Sprengel) Taubert	G	4	-	-	-
Schizolobium parahyba (Vellozo) S.F. Blake	G,Y	4	-	+	-
Senna spectabilis (DC.) H. Irwin & Barneby var. *spectabilis*	G,Y	(4)	(-)	-	-

Appendix (cont.)

Taxa	1	2	3	4	5
S. multijuga (Rich.) H. Irwin & Barneby subsp. *doylei* (Britton & Rose) H. Irwin & Barneby	G	*	-	-	-
Swartzia cubensis (Britton & P. Wilson) Standley var. *cubensis*	Y	2	+	+	+
Mimosoideae (8/33)					
Abarema sp. (syn. *Pithecellobium idiopodum* (S.F. Blake)Pittier, *P. halogenes* Standley)	G	2	-	+	-
Acacia dolichostachya S.F. Blake	Y	*	-	-	+
A. gentlei Lundell	Y	1	-	-	-
A. glomerosa Benth.	G,Y	4	-	+	+
A. usumacintensis Lundell	G,Y	1	-	+	-
Albizia adinocephala (J.D. Smith) Rec.	T	3	-	+	+
A. guachapele (H.B.K.) Dugand	T	4	-	+	-
A. niopoides (Benth.) Burkart (syn. *A. caribaea* (Urban) Britton & Rose)	G,Y	4	-	+	+
A. purpusii Britton & Rose	G	*	-	+	-
Albizia sp. (syn. *Pithecellobium leucocalyx* (Britton & Rose) Standley)	T	1	-	+	-
Cojoba arborea (L.) Britton & Rose (syn. *Pithecellobium arboreum* (L.) Urban)	G	2	+	+	-
C. donnell-smithii Britton & Rose (syn. *Pithecellobium donnell-smithii* (Britton & Rose) Standley)	G,Y	2	-	-	-
Cojoba sp. nov. [Rico]	G	2	-	+	-
Enterolobium cyclocarpum (Jacq.) Griseb.	G,Y	4	-	+	+
E. schomburgkii Benth.	G	4	-	+	-
Inga alba (Sw.) Willd.	G	4	-	+	-
I. belizensis Standley	T	1	-	+	-
I. brevipedicellata Harms	G	2	-	+	-
I. fagifolia (L.) Benth.	G	4	+	+	+
I. paterno Harms	G	2	-	+	-
I. punctata Willd.	G	4	-	+	-
I. quaternata Poeppig	G	4	-	-	-
I. sapindoides Willd.	G	3	-	-	-
I. tenuipedunculata León	G	2	-	+	-
I. thibaudiana DC.	T	4	-	+	-
I. vera Willd. subsp. *spuria* (Willd.) León	G,Y	4	-	-	+
Inga sp. nov. 1 [Sousa]	G	*	-	+	-

Appendix (cont.)

Taxa	1	2	3	4	5
Inga sp. nov. 2 [Sousa]	G	*	-	+	-
Inga sp. nov. 3 [Sousa]	G	2	-	-	-
Inga sp. nov. 4 [Sousa]	G	1	-	+	-
Lysiloma auritum (Schldl.) Benth.	G	2	-	+	+
L. latisiliquum (L.) Benth.	Y	1	+	+	+
Zygia conzattii (Standley) Britton & Rose var. nov.[Rico]	G	*	-	-	-
Papilionoideae (17/36)					
Acosmium panamense (Benth.) Yakovlev	G,Y	3	-	+	+
Andira galeottiana Standley	G	*	-	+	-
A. inermis (Wright) DC.	G	4	+	+	+
Ateleia pterocarpa D. Dietr.	G	1	-	-	+
Dalbergia glomerata Hemsley	G	*	-	+	-
D. tucurensis J.D. Smith	G	2	-	+	-
Diphysa americana (Miller) M. Sousa	G	2	-	-	+
D. carthagenensis Jacq.	Y	4	-	-	+
Dussia mexicana (Standley) Harms	G	2	-	+	-
Erythrina folkersii Krukoff & Mold.	G	2	-	-	-
E. tuxtlana Krukoff & Barneby	G	*	-	-	-
Lennea modesta (Standley & Steyerm.) Standley & Steyerm.	G	1	-	+	-
Lonchocarpus atropurpureus Benth.	G	3	-	-	-
L. castilloi Standley	Y	1	-	+	-
L. cruentus Lundell	G	1	-	+	-
L. guatemalensis Benth.	G,Y	2	-	+	+
L. hondurensis Benth.	G,Y	1	-	-	-
L. lasiotropis F.J. Herm.	G	2	-	+	-
L. robustus Pittier	G	1	-	-	-
L. verrucosus M. Sousa	G	2	-	+	-
Lonchocarpus sp. & var. nov. 1 [Sousa]	G	*	-	+	-
Lonchocarpus sp. nov. 2 [Sousa]	G	1	-	+	-
Lonchocarpus sp. nov. 3 [Sousa]	G	1	-	+	-
Myroxylon balsamum (L.) Harms var. *pereirae* (Royle) Harms	G,Y	2	-	+	-
Ormosia isthmensis Standley	G	3	-	+	-
O. macrocalyx Ducke	G	4	-	+	-
O. panamensis Benth.	G	2	-	+	-
O. schippii Standley & Steyerm.	Y	1	-	+	-
Piscidia piscipula (L.) Sarg.	G,Y	3	+	-	+
Platymiscium pinnatum (Jacq.) Dugand	G	3	-	+	-
P. yucatanum Standley	Y	1	-	+	+
Pterocarpus rohrii M. Vahl	G,Y	4	-	+	-

Appendix (cont.)

Taxa	1	2	3	4	5
Styphnolobium sp. nov. 1 [Sousa]	G	*	-	+	-
Styphnolobium sp. nov. 2 [Sousa]	G	*	-	+	-
Vatairea lundellii (Standley) Killip	G,Y	2	-	+	-
Willardia schiedeana (Schldl.) F.J. Herm.	G	2	-	+	+
LYTHRACEAE (1/1)					
Ginoria nudiflora (Hemsley) Koehne	G	*	-	+	+
MAGNOLIACEAE (1/1)					
Talauma mexicana (DC.) G. Don	G	1	-	+	-
MALPIGHIACEAE (3/3)					
Bunchosia lindeniana Adr. Juss.	G	2	-	-	-
Byrsonima cotinifolia H.B.K.	G			+	-
Malpighia wendtii W.R. Anderson	G	1	-	-	-
MALVACEAE (2/4)					
Hampea stipitata S. Watson	G	1	-	-	-
Robinsonella brevituba Fryxell	G	*	-	-	-
R. mirandae Gómez Pompa	G	*	-	+	-
R. samaricarpa Fryx.	G	*	-	+	-
MELASTOMATACEAE (3/6)					
Bellucia grossularioides (L.) Triana	G	4	-	-	-
Miconia argentea (Sw.) DC.	G	2	-	-	-
M. elata (Sw.) DC.	G	4	+	-	-
M. trinervia (Sw.) Loudon	G	4	+	+	-
Mouriri gleasoniana Standley var. *gleasoniana*	G	2	-	-	-
M. myrtilloides (Sw.) Poiret subsp. *parvifolia* (Benth.) Morley	G	4	-	-	-
MELIACEAE (4/11)					
Cedrela odorata L.	G,Y	4	+	+	+
Guarea glabra M. Vahl	G	4	+	+	+
G. grandifolia DC.	G	4	-	+	-
Guarea sp. (Uxp.)	G			+	-
Swietenia macrophylla King	G,Y	4	-	+	-
Trichilia hirta L.	G,Y	4	+	-	+
T. martiana C.DC.	G	4	-	+	-
T. minutiflora Standley	Y	1	-	-	-
T. moschata Sw. subsp. *matudae* (Lundell) Penn.	G	1	-	+	-
T. moschata Sw. subsp. *moschata*	G,Y	3	+	+	-
T. pallida Sw.	G	4	+	-	-

Appendix (cont.)

Taxa	1	2	3	4	5
MORACEAE (8/30)					
Brosimum alicastrum Sw.					
var. *alicastrum*	G,Y	2	+	+	+
B. guianense (Aublet) Huber	G	4	-	+	-
B. lactescens (S. Moore) C.C. Berg	G	4	-	+	-
Castilla elastica Sessé var. *elastica*	G,Y	(2)	(-)	+	+
Chlorophora tinctoria (L.) Gaudich.	G,Y	4	+	-	+
Clarisia biflora R. & P.	G	4	-	+	-
Ficus aurea Nutt.	G	1	+	+	-
F. colubrinae Standley	G	2	-	+	-
F. cookii Standley	G	1	-	+	+
F. cotinifolia H.B.K.	G,Y	2	-	+	+
F. insipida Willd.	G	4	-	+	-
F. isophlebia Standley	G	2	-	+	-
F. jimenezii Standley	G	2	-	+	-
F. lapathifolia (Liebm.) Miq.	G	*	-	+	-
F. maxima Miller	G,Y	4	+	+	+
F. morazaniana W. Burger	G,Y	2	-	+	+
F. obtusifolia H.B.K.	G,Y	4	-	+	+
F. paraensis (Miq.) Miq.	G	4	-	+	-
F. perforata L.	G	3	+	+	-
F. pertusa L.f.	G,Y	4	+	+	+
F. petenensis Lundell	G	1	-	+	-
F. tecolutensis (Liebm.) Miq.	G,Y	1	-	+	-
F. trigonata L.	Y	4	+	+	+
F. turrialbana W. Burger	G	2	-	+	-
F. yoponensis Desv.	G	3	-	+	-
Poulsenia armata (Miq.) Standley	G	4	-	+	-
Pseudolmedia oxyphyllaria J.D. Smith	G,Y	2	-	+	-
Ps. spuria (Sw.) Griseb.	G	2	+	+	-
Trophis mexicana (Liebm.) Bureau	G	2	-	-	-
Trophis racemosa (L.) Urban subsp.					
ramon (Schldl. & Cham.) W. Burger	G,Y	2	-	-	+
MYRISTICACEAE (2/2)					
Compsoneura sprucei (A.DC.) Warb.	G	4	-	-	-
Virola guatemalensis (Hemsley) Warb.	G	2	-	+	-
MYRTACEAE (5/12)					
Calyptranthes pallens Griseb.					
var. *pallens*	G,Y	1	+	-	+
Eugenia acapulcensis Steudel	G,Y	3	+	-	+
E. aeruginea DC.	G	1	+	-	-
E. choapamensis Standley	G	1	-	-	-
E. colipensis O. Berg	G	*	-	-	-

Appendix (cont.)

Taxa	1	2	3	4	5
E. mexicana Steudel	G	*	-	-	-
E. tikalana Lundell	Y	1	-	-	-
E. uxpanapensis P. Sánchez &					
L.M. Ortega	G	*	-	-	-
Myrcia splendens (Sw.) DC.	G	4	+	-	-
Pimenta dioica (L.) Merr.	G,Y	2	+	+	-
Psidium friedricksthalianum					
(O. Berg) Niedenzu	G	(3)	(-)	-	-
Ps. sartorianum (O. Berg) Niedenzu	G,Y	3	+	+	+
OLACACEAE (1/1)					
Heisteria media S.F. Blake	G	2	-	+	-
OLEACEAE (1/1)					
Chionanthus domingensis Lam.	G	2	+	+	-
PODOCARPACEAE (1/1)					
Podocarpus guatemalensis Standley	G	2	-	+	-
POLYGONACEAE (2/11)					
Coccoloba acapulcensis Standley	Y	2	-	-	+
C. barbadensis Jacq.	G,Y	1	-	+	+
C. belizensis Standley	G	2	-	+	-
C. hondurensis Lundell	G	1	-	+	-
C. matudae Lundell	G	*	-	-	-
C. montana Standley	G	1	-	+	+
C. spicata Lundell	Y	1	-	-	+
C. tenuis Lundell	T	1	-	+	-
C. tuerckheimii J.D. Smith	G	3	-	+	-
Coccoloba sp. [n. Chis.]	T			+	-
Ruprechtia pallida Standley	G	2	-	+	+
PROTEACEAE (1/1)					
Roupala montana Aublet	G	4	-	+	-
RHAMNACEAE (2/2)					
Colubrina arborescens (Miller) Sarg.	G,Y	2	+	-	+
Krugiodendron ferreum					
(M. Vahl) Urban	G,Y	3	+	+	+
RHIZOPHORACEAE (1/1)					
Cassipourea guianensis Aublet	G,Y	4	+	-	-
RUBIACEAE (18/23)					
Alseis yucatanensis Standley	Y	1	-	+	+

Appendix (cont.)

Taxa	1	2	3	4	5
Antirhea lucida (Sw.) Benth. & J.D. Hook.	Y	1	+	-	-
Blepharidium guatemalense Standley (syn. *B. mexicanum* Standley)	Y	1	-	+	-
Calycophyllum candidissimum (M. Vahl) DC.	G,Y	3	+	+	+
Chione chiapasensis Standley	G	1	-	+	-
Ch. guatemalensis Standley	T	1	-	+	-
Ch. mexicana Standley	G	2	-	+	-
Coussarea mexicana Standley	G	*	-	-	-
Coutaportla guatemalensis (Standley) Lorence	G	1	-	-	-
Elaeagia uxpanapensis Lorence	G	*	-	+	-
Exostema mexicanum A. Gray	G,Y	2	-	+	+
Genipa americana L.	G	4	+	+	+
Guettarda combsii Urban	G,Y	2	+	+	+
G. macrosperma J.D. Smith	G,Y	2	-	-	+
Hamelia calycosa J.D. Smith	G	4	-	-	-
Morinda panamensis Seemann	G	3	-	+	-
Posoqueria latifolia (Rudge) Roemer & Schultes	G	4	-	+	-
Psychotria chiapensis Standley	G	4	-	-	-
Ps. simiarum Standley	G	2	-	+	-
Randia matudae Lorence & Dwyer	G	2	-	+	-
Rondeletia galeottii Standley	G	*	-	-	-
Simira salvadorensis (Standley) Steyerm.	G,Y	1	-	+	+
Simira sp. (syn. *Sickingia lancifolia* Lundell)	T	1	-	+	-
RUTACEAE (3/9)					
Amyris elemifera L.	G,Y	2	+	-	+
Esenbeckia pentaphylla (Macfad.) Griseb. subsp. *belizensis* (Lundell) Kaastra	G	1	-	+	-
Zanthoxylum belizense Lundell	G	2	-	+	-
Z. caribaeum Lam.	G,Y	4	+	-	+
Z. kellermanii P. Wilson	G	2	-	+	-
Z. mayanum Standley	T	2	-	+	-
Z. microcarpum Griseb.	G	4	-	+	-
Z. panamense P. Wilson	G	2	-	+	-
Z. procerum J.D. Smith	G,Y	2	-	+	-
SALICACEAE (2/2)					
Salix humboldtiana Willd.	G,Y	4	-	-	-
Populus mexicana Wesm. subsp. *mexicana*	G	*	-	+	-

Appendix (cont.)

Taxa	1	2	3	4	5
SAPINDACEAE (8/14)					
Allophylus psilospermus Radlk.	G,Y	3	-	-	-
Blomia prisca (Standley) Lundell	G,Y	1	-	+	-
Cupania belizensis Standley	Y	1	-	-	-
C. cubensis M. Gómez & Molinet (syn. *C. macrophylla* A. Rich.)	G	2	+	-	-
C. dentata DC.	G,Y	2	-	+	+
C. glabra Sw.	G,Y	2	+	+	+
C. mayana (Lundell) Lundell	T	1	-	+	-
Exothea diphylla (Standley) Lundell	G,Y	1	-	+	+
E. paniculata (A.L. Juss.) Radlk.	G,Y	2	+	+	+
Matayba apetala (Macfad.) Radlk.	G,Y	2	+	+	-
Sapindus saponaria L.	G,Y	4	+	-	+
Talisia floresii Standley	Y	1	-	+	-
T. oliviformis (H.B.K.) Radlk.	Y	3	-	+	+
Thouinia canescens Radlk. var. (syn. *T. paucidentata* Radlk.)	Y	1	-	-	+
SAPOTACEAE (5/23)					
Chrysophyllum mexicanum Brandegee	G,Y	2	-	-	-
C. venezuelanense (Pierre) Penn.	G	4	-	+	-
Manilkara chicle (Pittier) Gilly	G	3	-	+	-
M. zapota (L.) P. Royen	G,Y	(2)	(-)	+	+
Micropholis melinoniana Pierre	G	4	-	+	-
Pouteria belizensis (Standley) Cronq.	T	1	-	+	-
P. campechiana (H.B.K.) Baehni	G,Y	(2)	(-)	+	+
P. durlandii (Standley) Baehni subsp. *durlandii*	G	4	-	+	-
P. glomerata (Miq.) Radlk. subsp. *glomerata*	G,Y	(4)	(-)	+	-
P. reticulata (Engl.) Eyma subsp. *reticulata*	G,Y	4	-	+	-
P. sapota (Jacq.) H.E. Moore & Stearn	G	(2)	(-)	+	-
P. torta (C. Martius) Radlk. subsp. *tuberculata* (Sleumer) Penn.	G	4	-	+	-
P. viridis (Pittier) Cronq.	T	(2)	(-)	+	-
Pouteria sp. [Uxp.]	G			+	-
Pouteria sp. [Uxp.]	G			+	-
Sideroxylon capiri (A.DC.) Pittier subsp. *tempisque* (Pittier) Penn.	G	3	-	+	+
S. contrerasii (Lundell) Penn.	G	2	-	-	-
S. eucoriaceum (Lundell) Penn.	G	1	-	+	-
S. foetidissimum Jacq. subsp. *gaumeri* (Pittier) Penn.	Y	1	-	+	+
S. persimile (Hemsley) Penn. subsp. *persimile*	G	3	-	+	+

Appendix (cont.)

Taxa	1	2	3	4	5
S. portoricense Urban					
subsp. *minutiflorum* (Pittier) Penn.	G	2	-	+	+
S. salicifolium (L.) Lam.	G,Y	1	+	-	+
S. stevensonii (Standley) Penn.	T	1	-	+	-
SIMAROUBACEAE (2/2)					
Recchia simplicifolia T. Wendt & Lott	G	*	-	-	-
Simarouba glauca DC.	G,Y	2	+	+	+
SOLANACEAE (1/1)					
Cestrum racemosum R. & P.	G	4	-	-	-
STAPHYLEACEAE (2/2)					
Huertea sp. [Chim.]	G			+	-
Turpinia occidentalis (Sw.) G. Don					
subsp. *breviflora* Croat	G	3	+	+	-
STERCULIACEAE (2/4)					
Guazuma ulmifolia Lam.	G,Y	4	+	+	+
Sterculia apetala (Jacq.) Karsten	G	4	-	+	-
S. mexicana R. Br.	G	2	-	+	-
Sterculia sp. nov. [Wendt & E. Taylor]	G	*	-	+	-
THEACEAE (3/3)					
Freziera grisebachii Krug & Urban	G	4	+	+	-
Laplacea grandis Brandegee	G	2	-	+	-
Ternstroemia tepezapote					
Schldl. & Cham.	G	2	-	+	-
TILIACEAE (6/11)					
Apeiba tibourbou Aublet	G	4	-	+	+
Berrya cubensis (Griseb.) M. Gómez					
(syn. *Carpodiptera ameliae* Lundell)	G	1	+	+	-
Heliocarpus appendiculatus Turcz.	G	2	-	+	-
H. donnell-smithii Rose	G,Y	2	-	+	+
Luehea seemannii Triana & Planchon	T	3	-	+	-
L. speciosa Willd.	G,Y	4	+	+	+
Mortoniodendron palaciosii Miranda	G	1	-	+	-
M. ruizii Miranda	G	1	-	+	-
M. vestitum Lundell	G	1	-	+	-
Trichospermum galeottii					
(Turcz.) Kosterm.	G	4	-	+	-
T. grewiifolium (A. Rich.) Kosterm.	G,Y	2	+	+	-
TURNERACEAE (1/1)					
Erblichia odorata Seemann var. *odorata*	G	2	-	+	-

Appendix (cont.)

Taxa	1	2	3	4	5
ULMACEAE (4/4)					
Ampelocera hottlei (Standley) Standley	G,Y	2	-	+	-
Aphananthe monoica (Hemsley)					
J. Leroy	G	2	-	+	+
Celtis trinervia Lam.	Y	3	+	-	-
Trema micranthum (L.) Blume	G,Y	4	+	-	+
VERBENACEAE (4/5)					
Citharexylum affine D. Don	G	1	-	+	+
C. caudatum L.	G	4	+	-	-
Lippia myriocephala Schldl. & Cham.	G,Y	2	-	-	-
Rehdera penninervia Standley & Mold.	T	1	-	+	-
Vitex gaumeri Greenman	Y	2	-	+	+
VIOLACEAE (1/2)					
Orthion malpighiifolium (Standley)					
Standley & Steyerm.	G	2	-	-	-
O. subsessile (Standley)					
Steyerm. & Standley	T	1	-	-	-
VOCHYSIACEAE (1/1)					
Vochysia guatemalensis J.D. Smith	G	2	-	+	-

ªFor explanation of codes, see introduction to list.

REFERENCES

Allorge, L. 1985. Monographie des Apocynacées-Tabernaemontanoïdées Americaines. Mem. Mus. Natl. Hist. Nat., Sér. B, Bot. 30:1–216.

Akkermans, R.W.A.P., & C.C. Berg. 1982. New species and combinations in *Coussapoa* (Cecropiaceae), and keys to its species. Proc. Kon. Ned. Akad. Wetensch. C, 85:441–471.

Arbo, M.M. 1979. Revisión del género *Erblichia* (Turneraceae). Adansonia, ser. 2, 18:459–482.

Aubreville, A. 1973. Géophylétique florale des Sapotacées. Adansonia, ser. 2, 13:255–271.

Axelrod, D.I. 1958. Evolution of the Madro-Tertiary geoflora. Bot. Rev. (Lancaster) 24:433–509.

———. 1975. Evolution and biogeography of Madrean-Tethyan sclerophyll vegetation. Ann. Missouri Bot. Gard. 62:280–334.

Bally, A.W., C.R. Scotese, & M.I. Ross 1989. North America; plate-tectonic setting and tectonic elements. *In* A.W. Bally & A.R. Palmer (eds.), The Geology of North America. Vol. A. The Geology of North America—An Overview. Boulder, CO: Geological Society of America. pp. 1–15.

Barron, E.J. 1987. Global Cretaceous paleogeography—International Geologic Correlation Program project 191. Palaeogeogr. Palaeoclimatol. Palaeoecol. 59:207–214.

Bartlett, A.S. & E.S. Barghoorn. 1973. Phytogeographic history of the Isthmus of Panama during the past 12,000 years (a history of vegetation, climate, and sea-level change). *In* A. Graham (ed.), Vegetation and Vegetational History of Northern Latin America. Amsterdam: Elsevier. pp. 203–299.

Batten, D.J. 1984. Palynology, climate and the development of Late Cretaceous floral provinces in the Northern Hemisphere; a review. *In* P. Brenchley (ed.), Fossils and Climate. Chichester: John Wiley & Sons. pp. 127–164.

Berg, C.C. 1972. Olmedieae, Brosimeae (Moraceae). Fl. Neotropica 7:1–229.

———. 1983. Dispersal and distribution in the Urticales—an outline. *In* K. Kubitzki (ed.), Dispersal and Distribution. An International Symposium. Sonderb. Naturwiss. Ver. Vol. 7. Hamburg: Verlag Paul Parey. pp. 219–229.

Boufford, D.E. & S.A. Spongberg. 1983. Eastern Asian-eastern North American phytogeographic relationships—a history from the time of Linnaeus to the twentieth century. Ann. Missouri Bot. Gard. 70:423–439.

Bradbury, J.P. 1989. Late Quaternary lacustrine paleoenvironments in the Cuenca de México. Quat. Sci. Rev. 8:75–100.

Breedlove, D.E. 1981. Flora of Chiapas. Part 1. Introduction to the Flora of Chiapas. San Francisco: California Academy of Sciences.

Brown, Jr., K.S. 1982. Paleoecology and regional patterns of evolution in Neotropical forest butterflies. *In* G.T. Prance (ed.), Biological Diversification in the Tropics. New York: Columbia Univ. Press. pp. 255–308.

———. 1987. Biogeography and evolution of neotropical butterflies. *In* T.C. Whitmore & G.T. Prance (eds.), Biogeography and Quaternary History in Tropical America. Oxford: Clarendon Press. pp. 66–104.

Bussing, W.A. 1985. Patterns of distribution of the Central American ichthyofauna. *In* F.G. Stehli & S.D. Webb (eds.), The Great American Biotic Interchange. New York: Plenum Press. pp. 453–473.

Cardoso C., M.D. 1979. El Clima de Chiapas y Tabasco. Mexico City: UNAM. Inst. Geogr.

Chandler, M.E.J. 1964. The Lower Tertiary Floras of Southern England. IV. A Summary and Survey of Findings in the Light of Recent Botanical Observations. London: British Museum (Natural History).

Chaney, R.W. 1947. Tertiary centers and migration routes. Ecol. Monogr. 17:139–148.

Cifelli, R.L. 1985. South American ungulate evolution and extinction. *In* F.G. Stehli & S.D. Webb (eds.), The Great American Biotic Interchange. New York: Plenum Press. pp. 249–266.

Coen, E. 1983. Climate. *In* D.H. Janzen (ed.), Costa Rican Natural History. Chicago: Univ. Chicago Press. pp. 35–46.

Collinson, M.E. & J.J. Hooker. 1987. Vegetational and mammalian faunal changes in the Early Tertiary of southern England. *In* E.M. Friis, W.G. Chaloner & P.R. Crane (eds.), The Origins of Angiosperms and their Biological Consequences. Cambridge: Cambridge Univ. Press. pp. 259–304.

Cowan, R.S. 1968. *Swartzia* (Leguminosae, Caesalpinioideae, Swartzieae). Fl. Neotropica 1:1–228.

Crabtree, D.R. 1987. Angiosperms of the northern Rocky Mountains: Albian to Campanian (Cretaceous) megafossil floras. Ann. Missouri Bot. Gard. 74:707–747.

Crane, P.R. 1987. Vegetational consequences of angiosperm diversification. *In* E.M. Friis, W.G. Chaloner & P.R. Crane (eds.), The Origins of Angiosperms and their Biological Consequences. Cambridge: Cambridge Univ. Press. pp. 107–144.

Cronquist, A. 1981. An Integrated System of Classification of Flowering Plants. New York: Columbia Univ. Press.

————. 1988. The Evolution and Classification of Flowering Plants. 2nd ed. Bronx: New York Botanic Garden.

Crowther, J. 1982. Ecological observations in a tropical karst terrain, West Malaysia. I. Variations in topography, soils and vegetation. J. Biogeogr. 9:65–78.

————. 1987. Ecological observations in tropical karst terrain, West Malaysia. II. Rainfall interception, litter fall and nutrient cycling. J. Biogeography 14:145–155.

Daniel, T.F. 1988. A systematic study of *Bravaisia* DC. (Acanthaceae). Proc. Calif. Acad. Sci. 45:111–132.

Davis, M.B. 1983. Quaternary history of deciduous forests of eastern North America and Europe. Ann. Missouri Bot. Gard. 70:550–563.

De Cserna, Z. 1989. An outline of the geology of Mexico. *In* A.W. Bally & A.R. Palmer (eds.), The Geology of North America. Vol. A. The Geology of North America— An Overview. Boulder, CO: Geological Society of America. pp. 233–264.

Dilcher, D.L. 1971. A revision of the Eocene flora of southeastern North America. Palaeobotanist 20:7–18.

————. 1973. A paleoclimatic interpretation of the Eocene floras of southeastern North America. *In* A. Graham (ed.), Vegetation and Vegetational History of Northern Latin America. Amsterdam: Elsevier. pp. 39–59.

Donnelly, T.W. 1985. Mesozoic and Cenozoic plate evolution of the Caribbean region. *In* F.G. Stehli & S.D. Webb (eds.), The Great American Biotic Interchange. New York: Plenum Press. pp. 89–121.

————. 1989. Geologic history of the Caribbean and Central America. *In* A.W. Bally & A.R. Palmer (eds.), The Geology of North America. Vol. A. The Geology of North America—An Overview. Boulder, CO: Geological Society of America. Pp. 299–321.

Durán G., R. & I. Olmsted. 1987. Listado Floristico de la Reserva Sian Ka'an. Puerto Morelos, Quintana Roo: Amigos de Sian Ka'an.

Eckenwalder, J.E. 1977. North American cottonwoods (*Populus*, Salicaceae) of sections *Abaso* and *Aigeros*. J. Arnold Arbor. 58:193–208.

Elias, T.S. 1976. A monograph of the genus *Hamelia* (Rubiaceae). Mem. New York Bot. Gard. 26:81–144.

————. 1980. The Complete Trees of North America. New York: Van Nostrand Reinhold (Reprinted 1987. New York: Gramercy).

Estes, R. & A. Báez. 1985. Herpetofaunas of North and South America during the Late Cretaceous and Cenozoic: evidence for interchange? *In* F.G. Stehli & S.D. Webb (eds.), The Great American Biotic Interchange. New York: Plenum Press. pp. 139–197.

Flores M., G., J. Jiménez L., X. Madrigal S., F. Moncayo R. & F. Takaki T. 1971. Mapa y Descripción de los Tipos de Vegetación de la República Mexicana. Mexico City: Dirección de Agrología, Secretaría de Recursos Hidráulicos.

Frederiksen, N.O. 1980. Sporomorphs from the Jackson Group (Upper Eocene) and Adjacent Strata of Mississippi and Western Alabama. U.S. Geol. Surv., Prof. Pap. 1084.

Fryxell, P.A. 1969. The genus *Hampea* (Malvaceae). Brittonia 21:359–396.

————. 1973. A revision of *Robinsonella* Rose & E.G. Baker (Malvaceae). Gentes Herb. 11:1–26.

————. 1988. Malvaceae of Mexico. Syst. Bot. Monographs 25:1–522.

Furley, P.A. & W.W. Newey. 1979. Variations in plant communities with topography over tropical limestone soils. J. Biogeogr. 6:1–15.

García, E. 1970. Los climas del estado de Veracruz (según el sistema de clasificación climática de Köppen modificado por la autora). Anales Inst. Biol. Univ. Nac. México, Ser. Bot. 41:3–42.

———. 1973. Modificaciones al Sistema de Clasificación Climática de Köppen (para adaptarlo a las condiciones de la República Mexicana). 2nd ed. Mexico City: UNAM. Inst. Geogr.

Geesink, R. 1981. Tribe 6. Tephrosieae (Benth.) Hutch. (1964). *In* R.M.Polhill & P.H. Raven (eds.), Advances in Legume Systematics Part 1. Kew: Royal Botanic Gardens. pp. 245–260.

Gentry, A.H. 1982a. Neotropical floristic diversity: phytogeographical connections between Central and South America, Pleistocene climatic fluctuations, or an accident of the Andean orogeny? Ann. Missouri Bot. Gard. 69:557–593.

———. 1982b. Phytogeographic patterns as evidence for a Chocó refuge. *In* G.T. Prance (ed.), Biological Diversification in the Tropics. New York: Columbia Univ. Press. pp. 112–136.

———. 1988a. Tree species richness of upper Amazonian forests. Proc. Natl. Acad. Sci. USA 85:156–159.

———. 1988b. Changes in plant community diversity and floristic composition on environmental and geographical gradients. Ann. Missouri Bot. Gard. 75:1–34.

Germeraad, J.H., C.A. Hopping, & J. Muller. 1968. Palynology of Tertiary sediments from tropical areas. Rev. Palaeobot. Palynol. 6:189–348.

Gilg, E. 1925. Flacourtiaceae. *In* A. Engler & K. Prantl (eds.), Die Natürlichen Pflanzenfamilien. 2nd ed. Vol. 21. Leipzig: Wilhelm Engelmann. pp. 377–457.

Gingerich, P.D. 1985. South American mammals in the Paleocene of North America. *In* F.G. Stehli & S.D. Webb (eds.), The Great American Biotic Interchange. New York: Plenum Press. pp. 123–137.

Gómez P., A. 1966. Estudios Botánicos en la Región de Misantla, Veracruz. Mexico City: IMERNAR.

Gómez P., L.D. 1982. The origin of the pteridophyte flora of Central America. Ann. Missouri Bot. Gard. 69:548–556.

Gose, W.A. 1985. Caribbean tectonics from a paleomagnetic perspective. *In* F.G. Stehli & S.D. Webb (eds.), The Great American Biotic Interchange. New York: Plenum Press. pp. 285–301.

Graham, A. 1975. Late Cenozoic evolution of tropical lowland vegetation in Veracruz, Mexico. Evolution 29:723–735.

———. 1976. Studies in neotropical paleobotany. II. The Miocene communities of Veracruz, Mexico. Ann. Missouri Bot. Gard. 63:787–842.

———. 1977. The tropical rain forest near its northern limits in Veracruz, Mexico: recent and ephemeral? Bol. Soc. Bot. México 36:13–19.

———. 1982. Diversification beyond the Amazon basin. *In* G.T. Prance (ed.), Biological Diversification in the Tropics. New York: Columbia Univ. Press. pp. 78–90.

———. 1985. Studies in neotropical paleobotany. IV. The Eocene communities of Panama. Ann. Missouri Bot. Gard. 72:504–534.

———. 1987. Tropical American Tertiary floras and paleoenvironments: Mexico, Costa Rica, and Panama. Amer. J. Bot. 74:1519–1531.

———. 1988. Some aspects of Tertiary vegetational history in the Gulf/Caribbean region. Trans. 11th Caribbean Geol. Conference Barbados 3:1–18.

———. 1989a. Paleofloristic and paleoclimatic changes in the Tertiary of northern Latin America. Rev. Paleobot. Palynol. 60:283–293.

———. 1989b. Late Tertiary paleoaltitudes and vegetational zonation in Mexico and Central America. Acta Bot. Neerl. 38:417–424.

————. 1991. Studies in neotropical paleobotany. X. The Pliocene communities of Panama—composition, numerical representation, and paleocummunity/ paleoenvironmental reconstructions. Ann. Missouri Bot. Gard. 78:465–475.

Granger, O.E. 1987. Precipitation distribution. *In* J.E. Oliver & R.W. Fairbridge (eds.), The Encyclopedia of Climatology. New York: Van Nostrand Reinhold. pp. 690–697.

Gregor, H.J. 1989. Aspects of the fossil record and phylogeny of the family Rutaceae (Zanthoxyleae, Toddalioideae). Pl. Syst. Evol. 162:251–265.

Gregory, J. 1971. An ancient *Acacia* wood from Oregon. Palaeobotanist 20:19–21, pl. 1.

Haffer, J. 1982. General aspects of the refuge theory. *In* G.T. Prance (ed.), Biological Diversification in the Tropics. New York: Columbia Univ. Press. pp.6–24.

————. 1987. Biogeography of neotropical birds. *In* T.C. Whitmore & G.T. Prance (eds.), Biogeography and Quaternary History in Tropical America. Oxford: Clarendon Press. pp. 105–150.

Hamilton, W. 1983. Cretaceous and Cenozoic history of the northern continents. Ann. Missouri Bot. Gard. 70:440–453.

Heine, K. 1984. Comment of "Pleistocene glaciation of Volcano Ajusco, central Mexico, and comparison with the standard Mexican glacial sequence" by Sidney E. White and Salvatore Valastro, Jr. Quaternary Res. 22:242–246.

Hergert, H.L. 1961. Plant fossils in the Clarno formation, Oregon. Ore-Bin 23:55–62.

Hickey, L.J. 1977. Stratigraphy and Paleobotany of the Golden Valley Formation (Early Tertiary) of Western North Dakota. Geol. Soc. Amer. Mem. 150:1–181.

Holdridge, L.R., W.C. Grenke, W.H. Hatheway, T. Liang & J.A. Tosi, Jr., 1971. Forest Environments in Tropical Life Zones. A Pilot Study. Oxford: Pergamon Press.

Holmgren, P.K., N.H. Holmgren, & L.C. Barnett. 1990. Index Herbarium. Part I: The Herbaria of the World. 8th Ed. (Regnum Veg. 120). Bronx: New York Bot. Gard.

Howard, R.A. 1961. Studies in the genus *Coccoloba*, X. New species and a summary of distribution in South America. J. Arnold Arbor. 62:87–95.

Hsü, J. 1983. Late Cretaceous and Cenozoic vegetation in China, emphasizing their connections with North America. Ann. Missouri Bot. Gard. 70:490–508.

Hunter, G.E. 1966. Revision of Mexican and Central American *Saurauia* (Dilleniaceae). Ann. Missouri Bot. Gard. 53:47–89.

Ibarra M., G. & S. Sinaca C. 1987. Listados Florísticos de México VII. Estación de Biología Tropical Los Tuxtlas, Veracruz. Mexico City: UNAM. Inst. Biol.

Irwin, H.S. & R.C. Barneby. 1982. The American Cassiinae—A synoptical revision of Leguminosae tribe Cassieae subtribe Cassiinae in the New World. Mem. New York Bot. Gard. 35:1–918.

Jennings, J.N. 1985. Karst Geomorphology. Oxford: Basil Blackwell.

Johnston, M.C. 1971. Revision of *Colubrina* (Rhamnaceae). Brittonia 23:2–53.

Kaastra, R.C. 1982. Pilocarpinae (Rutaceae). Fl. Neotropica 33:1–198.

Krutzsch, W. 1989. Paleogeography and historical phytogeography (paleochorology) in the Neophyticum. Pl. Syst. Evol. 162:5–61.

Kubitzki, K. 1969. Monographie der Hernandiaceen. Bot. Jahrb. Syst. 89:78–209.

Kurz, H. 1983. Fortpflanzungsbiologie einiger Gattungen Neotropischer Lauraceen und Revision der Gattung *Licaria* (Lauraceae). Ph.D. dissertation, Univ. of Hamburg.

Landrum, L.R. 1986. *Campomanesia, Pimenta, Blepharocalyx, Legrandia, Acca, Myrrhinium*, and *Luma* (Myrtaceae). Fl. Neotropica 45:1–179.

Langenheim, J.H., B.L. Hackner & A. Bartlett. 1967. Mangrove pollen at the depositional site of Oligo-Miocene amber from Chiapas, Mexico. Bot. Mus. Leafl. 21:289–324.

Lee, Y.-T. & J.H. Langenheim. 1974. Systematics of the genus *Hymenaea* L. (Leguminosae, Caesalpinioideae, Detarieae). Univ. Calif. Publ. Bot. 69:1–109.

León C., J.M. & A. Gómez P. 1970. La vegetación del sureste de Veracruz. Publ. Esp. Inst. Nal. Invest. For. México 5:13–48. (Reprinted 1982.)

Leopold, E.B. & H.D. MacGinitie. 1972. Development and affinities of Tertiary floras in the Rocky Mountains. *In* A. Graham (ed.), Floristics and Paleofloristics of Asia and Eastern North America. Amsterdam: Elsevier. pp. 147–200.

Leyden, B.W. 1984. Guatemalan forest synthesis after Pleistocene aridity. Proc. Natl. Acad. Sci. USA 81:4856–4859.

López M., R. 1980. Tipos de Vegetación y Su Distribución en el Estado de Tabasco y Norte de Chiapas. Cuadernos Universitarios Ser. Agronomía No. 1. Chapingo, Mexico: Universidad Autónoma Chapingo.

Lott, E.J. 1985. Listados Florísticos de México III. La Estación de Biología Chamela, Jalisco. Mexico City: UNAM. Inst. Biol.

Lundell, C.L. 1934. Preliminary sketch of the phytogeography of the Yucatan Peninsula. Carnegie Inst. Washington Publ. 436:255–321.

————. 1937. The Vegetation of Petén. Carnegie Inst. Wash. Publ. 478:1–244.

————. 1942a. Flora of eastern Tabasco and adjacent Mexican areas. Contr. Univ. Michigan Herb. 8:5–74.

————. 1942b. The vegetation and natural resources of British Honduras. Chron. Bot. 7:169–171.

MacGinitie, H.D. 1969. The Eocene Green River flora of northwestern Colorado and northeastern Utah. Univ. Calif. Publ. Geol. Sci. 83:1–140.

————. 1974. An early Middle Eocene flora from the Yellowstone-Absaroka volcanic province, northwestern Wind River basin, Wyoming. Univ. Calif. Publ. Geol. Sci. 108:1–103.

Mai, D.H. 1981. Entwicklung und klimatische Differenzierung der Laubwaldflora Mitteleuropas im Tertiär. Flora 171:525–582.

————. 1989. Development and regional differentiation of the European vegetation during the Tertiary. Pl. Syst. Evol. 162:79–91.

Manchester, S.R. 1977. Wood of *Tapirira* (Anacardiaceae) from the Paleogene Clarno Formation of Oregon. Rev. Palaeobot. Palynol. 23:119–127.

Martínez-Hernández, E., H. Hernández-Campos & M. Sánchez-López. 1980. Palinología del Eoceno en el noreste de México. UNAM. Inst. Geol. Rev. 4:155–166.

McKenna, M.C. 1975. Fossil mammals and Early Eocene North Atlantic land continuity. Ann. Missouri Bot. Gard. 62:335–353.

————. 1983. Holarctic landmass rearrangement, cosmic events, and Cenozoic terrestrial organisms. Ann. Missouri Bot. Gard. 70:459–489.

Meave del C., J.A. 1983. Estructura y Composición de la Selva Alta Perennifolia en los Alrededores de Bonampak, Chiapas. Thesis, UNAM.

Middleton, J. & T. Waltham. 1986. The Underground Atlas: A Gazetteer of the World's Cave Regions. New York: St. Martin's Press.

Miller, J.S. 1989. Revision of the New World species of *Ehretia* (Boraginaceae). Ann. Missouri Bot. Gard. 76:1050–1076.

Miranda, F. 1952–53. La Vegetación de Chiapas. Chiapas, Mexico: Ediciones del Gobierno del Estado, Tuxtla Gutiérrez. Vol. 1 & 2.

————. 1958. Estudios acerca de la vegetación. *In* E. Beltran (ed.), Los Recursos Naturales del Sureste y Su Aprovechamiento. Vol. 2. Mexico City: IMERNAR. pp. 213–271.

————. 1960. Posible significación del porcentaje de géneros bicontinentales en América tropical. Anales Inst. Biol. Univ. Nac. México 30:117–150.

————. 1961. Tres estudios botánicos en la Selva Lacandona, Chiapas, México. Bol. Soc. Bot. México 26:133–176.

———— & E. Hernández X. 1963. Los tipos de vegetación de México y su clasificación. Bol. Soc. Bot. México 28:29–178.

Morley, T. 1976. Memecyleae (Melastomataceae). Fl. Neotropica 15:1–296.

————. 1989. New species and other taxonomic matters in the New World Memecyleae (Melastomataceae). Ann. Missouri Bot. Gard. 76:430–443.

Mosiño A., P.A. & E. García. 1974. The climate of Mexico. In R.A. Bryson & F.K. Hare (eds.), World Survey of Climatology. Vol. 11. Climates of North America. Amsterdam: Elsevier. pp. 345–404.

Muller, J. 1981. Fossil pollen records of extant angiosperms. Bot. Rev. 47:1–142.

Olmsted, I.C., A. López O., & R. Durán G. 1983. Vegetación de Sian Ka'an: Reporte Preliminar. In C.I.Q.Ro. (compiler), Sian Ka'an. Puerto Morelos, Q.R: Centro de Investigaciones de Quintana Roo. pp. 63–84.

Parrish, J.T. 1987. Global palaeogeography and palaeoclimate of the Late Cretaceous and Early Tertiary. In E.M. Friis, W.G. Chaloner, & P.R. Crane (eds.), The Origins of Angiosperms and their Biological Consequences. Cambridge: Cambridge Univ. Press. pp. 51–73.

Pascual, R., M.G. Vucetich, G.J. Scillato-Yané & M. Bond. 1985. Main pathways of mammalian diversification in South America. In F.G. Stehli & S.D. Webb (eds.), The Great American Biotic Interchange. New York: Plenum Press. pp. 219–247.

Pennington, T.D. 1981. Meliaceae. Fl. Neotropica 28:1–472.

————. 1990. Sapotaceae. Fl. Neotropica 52:1–771.

———— & J. Sarukhán. 1968. Arboles Tropicales de México. Mexico City: Instituto Nacional de Investigaciones Forestales/F.A.O.

Pérez J., L.A., & J. Sarukhán K. 1970. La vegetación de la región de Pichucalco, Chis. Publ. Esp. Inst. Nal. Invest. For. México 5:49–123. (Reprinted 1982.)

Poore, M.E.D. 1968. Studies in Malaysian rain forest I. The forest on Triassic sediments in Jengka Forest Reserve. J. Ecol. 56:143–196.

Portig, W.H. 1976. The climate of Central America. Pp. 405–478, In W. Schwerdtfeger (ed.), World Survey of Climatology, Vol. 12. Climates of Central and South America. Elsevier, Amsterdam.

Prance, G.T. 1972. Chrysobalanaceae. Fl. Neotropica 9:1–410.

Puig, H. 1976. Végétation de la Huasteca, Mexique. Mission Archeologique et Ethnologique Française au Mexique, Mexico City. (Etudes Mésoaméricaines, Vol. 5).

Quero, H.J. & R.W. Read. 1986. A revision of the palm genus *Gaussia*. Syst. Bot. 11:145–154.

Raven, P.H. & D.I. Axelrod. 1974. Angiosperm biogeography and past continental movements. Ann. Missouri Bot. Gard. 61:539–673.

Read, R.W. 1975. The genus *Thrinax* (Palmae: Coryphoideae). Smithsonian Contr. Bot. 19:1–98.

Renner, S.S. 1989. Systematic studies in the Melastomataceae: *Bellucia, Loreya*, and *Macairea*. Mem. New York Bot. Gard. 50:1–112.

Richards, R.W. 1952. The Tropical Rain Forest. An Ecological Study. Cambridge: Cambridge Univ. Press.

Rojo, J.P. 1972. *Pterocarpus* (Leguminosae—Papilionaceae) Revised for the World. (Phanerogamarum Monographiae 5.) Lehre, West Germany: Cramer.

Romero, E.J. 1986. Paleogene phytogeography and climatology of South America. Ann. Missouri Bot. Gard. 73:449–461.

Rosen, D.E. 1976. A vicariance model of Caribbean biogeography. Syst. Zool. 24:431–464.

————. 1985. Geological hierarchies and biogeographic congruence in the Caribbean. Ann. Missouri Bot. Gard. 72:636–659.

Royal Botanic Gardens, Kew. 1980. Draft Index of Author Abbreviations. London: Her Majesty's Stationery Office.

Rudd, V.E. 1972. Leguminosae-Faboideae-Sophoreae. North American Flora Ser. II, 7:1–36.

Rzedowski, J. 1963. El extremo boreal del bosque tropical siempre verde en Norteamérica continental. Vegetatio 11:173–198.

————. 1978. Vegetación de México. Mexico City: Limusa.

———— & R. McVaugh. 1966. La vegetación de Nueva Galicia. Contr. Univ. Michigan Herb. 9:1–123.

———— & R. Palacios C. 1977. El bosque de *Engelhardtia (Oreomunnea) mexicana* en la región de la Chinantla (Oaxaca, México)—Una reliquia del Cenozoico. Bol. Soc. Bot. México 36:93–123.

Sánchez M., V. 1987. Estudio Fitosociológico de una Selva Alta Perennifolia en una Zona de Uxpanapa, Oaxaca. Thesis, Universidad Autónoma Chapingo, Chapingo, Mexico.

Sarukhán K., J. 1968. Análisis sinecológico de las selvas de *Terminalia amazonia* en la planicie costera del Golfo de México. M.S. thesis, Colegio de Postgraduados, Chapingo, Mexico.

Savage, J.M. 1982. The enigma of the Central American Herpetofauna: dispersals or vicariance? Ann. Missouri Bot. Gard. 69:464–547.

Savin, S.M. & R.G Douglas. 1985. Sea level, climate, and the Central American land bridge. *In* F.G. Stehli & S.D. Webb (eds.), The Great American Biotic Interchange. New York: Plenum Press. pp. 303–324.

Sears, P.B. & K.H. Clisby. 1955. Palynology in southern North America. Part IV. Pleistocene climate in Mexico. Bull. Geol. Soc. Amer. 66:521–530.

Shackleton, N.J. 1984. Oxygen Isotope Evidence for Cenozoic Climatic Change. *In* P.J. Brenchley (ed.), Fossils and Climate. Chichester (W. Sussex)/New York: John Wiley & Sons. pp. 27–34.

Sharp, A.J. 1966. Some aspects of Mexican phytogeography. Ciencia (Mexico) 24:229–232.

Sleumer, H.O. 1980. Flacourtiaceae. Fl. Neotropica 22:1–499.

————. 1984. Olacaceae. Fl. Neotropica 38:1–159.

Smith, D.L. 1985. Caribbean plate relative motions. *In* F.G. Stehli & S.D. Webb (eds.), The Great American Biotic Interchange. New York: Plenum Press. pp.17–48.

Soderstrom, T.R. 1981. *Olmeca*, a new genus of Mexican bamboos with fleshy fruits. Amer. J. Bot. 68:1361–1373.

Soejarto, D.D. 1980. Revision of South American *Saurauia* (Actinidiaceae). Fieldiana Bot. n.s. 2:1–141.

————. 1984. Actinidiaceae. Flora de Veracruz 35:1–25.

Sousa S., M. 1968. Ecología de las leguminosas de Los Tuxtlas, Veracruz. Anales Inst. Biol. Univ. Nac. México, Ser. Bot. 39:121–161.

———— & E.F. Cabrera C. 1983. Listados Florísticos de México II. Flora de Quintana Roo. Mexico City: UNAM, Inst. Biol.

———— & M. Peña de Sousa. 1981. New World Lonchocarpinae. *In* R.M. Polhill & P.H. Raven (eds.), Advances in Legume Systematics Part 1. Kew: Royal Botanic Gardens. pp. 261–281.

Standley, P.C., 1930. Flora of Yucatan. Field Mus. Nat. Hist., Bot. Ser. 3:157–492.

———— & S.J. Record. 1936. The forests and flora of British Honduras. Field Mus. Nat. Hist., Bot. Ser. 12:1–432.

Stehli, F.G. & S.D. Webb (eds.). 1985. The Great American Biotic Interchange. New York: Plenum Press.

Strahler, A.N. & A.H. Strahler. 1978. Modern Physical Geography. New York: John Wiley & Sons.

Straka, H. & D. Ohngemach. 1989. Late Quaternary vegetation history of the Mexican highland. Pl. Syst. Evol. 162:115–132.

Sweeting, M.M. 1972. Karst Landforms. London: Macmillan.

Symington, C.F. 1943. Foresters' Manual of Dipterocarps. Malayan Forest Records No. 16. Reprinted (with plates) 1974 by Penerbit Universiti Malaya, Kuala Lumpur.

Tanai, T. 1972. Tertiary history of vegetation in Japan. *In* A. Graham (ed.), Floristics and Paleofloristics of Asia and Eastern North America. Amsterdam: Elsevier. pp. 235–255.

Taylor, D.W. 1988. Paleobiogeographic relationships of the Paleogene flora from the southeastern U.S.A.: implications for West Gondwanaland affinities. Palaeogeogr. Palaeoclimatol. Palaeoecol. 66:265–275.

———. 1989. Eocene floral evidence of Lauraceae: corroboration of the North American megafossil record. Amer. J. Bot. 75:948–957.

——— & L.J. Hickey. 1990. An Aptian plant with attached leaves and flowers: implications for angiosperm origin. Science 247:702–704.

Téllez V., O. & M. Sousa S. 1982. Imágenes de la Flora Quintanarroense. Centro de Investigaciones de Quintana Roo, Puerto Morelos, Q. R., Mexico.

Thompson, R.L. 1980. A Revision of *Lysiloma* (Leguminosae). Ph.D. dissertation, Southern Illinois Univ. at Carbondale.

Thorne, R.F. 1972. Major disjunctions in the geographic ranges of seed plants. Quart. Rev. Biol. 47:365–411.

Tidwell, W.D., S.R. Ash & L.R. Parker. 1981. Cretaceous and Tertiary floras of the San Juan Basin. *In* S.G. Lucas, J.K. Jigby, Jr. & B.S. Kues (eds.), Advances in San Juan Basin Paleontology. Albuquerque: Univ. New Mexico Press. pp. 307–332.

Tiffney, B.H. 1985a. Perspectives on the origin of the floristic similarity between eastern Asia and eastern North America. J. Arnold Arbor. 66:73–94.

———. 1985b. The Eocene North Atlantic land bridge: its importance in Tertiary and modern phytogeography of the Northern Hemisphere. J. Arnold Arbor. 66:243–273.

Todzia, C. 1989. A revision of *Ampelocera* (Ulmaceae). Ann. Missouri Bot. Gard. 76:1087–1102.

Toledo, V.M. 1982. Pleistocene changes of vegetation in tropical Mexico. *In* G.T. Prance (ed.), Biological Diversification in the Tropics. New York: Columbia Univ. Press. pp. 93–111.

Upchurch, G.R. Jr. & J.A. Wolfe. 1987. Mid-Cretaceous to Early Tertiary vegetation and climate: evidence from fossil leaves and woods. *In* E.M. Friis, W.G. Chaloner, & P.R. Crane (eds.), The Origins of Angiosperms and their Biological Consequences. Cambridge: Cambridge Univ. Press. pp.75–105.

Van der Hammen, T. 1982. Paleoecology of tropical South America. *In* G. Prance (ed.), Biological Diversity in the Tropics. New York: Columbia Univ. Press. pp. 60–66.

——— & A.M. Cleef. 1983. *Trigonobalanus* and the tropical amphi-Pacific element in the North Andean forest. J. Biogeogr. 10:437–440.

Van der Werff, H. 1988. Eight new species and one new combination of neotropical Lauraceae. Ann. Missouri Bot. Gard. 75:402–419.

Van Steenis, C.G.G.J. 1962. The land bridge theory in botany, with particular reference to tropical plants. Blumea 11:235–372.

Vanzolini, P.E. & W.R. Heyer. 1985. The American herpetofauna and the interchange. *In* F.G. Stehli & S.D. Webb (eds.), The Great American Biotic Interchange. New York. Plenum Press. pp. 475–487.

Vázquez T., S.M. 1989. Riqueza de Plantas Vasculares y la Diversidad de Especies
Arboreas del Dosel Superior en 5 Ha. de Selva Tropical Cálido-Húmeda en la Zona
de Uxpanapa, Ver. M.S. thesis, Colegio de Postgraduados, Chapingo, Mexico.

Vera C., M.P. 1988. Diversidad de Arboles en una Selva Alta Perennifolia de Santa
María Chimalapa, Oaxaca. B.S. thesis, Escuela Nacional de Estudios Profesionales
Iztacala, UNAM.

Votava, F.V. 1973. A Taxonomic Revision of the Genus *Thouinia* (Sapindaceae). Ph.
D. dissertation, Columbia Univ.

Webb, S.D. 1985. Main pathways of mammalian diversification in North America.
In F.G. Stehli & S.D. Webb (eds.), The Great American Biotic Interchange. New
York: Plenum Press. pp. 201–217.

Webster, G.L. 1979. A revision of *Margaritaria* (Euphorbiaceae). J. Arnold Arbor.
60:403–444.

Wendt, T. 1988. *Chiangiodendron* (Flacourtiaceae: Pangieae), a new genus from
southeastern Mexico representing a new tribe for the New World flora. Syst. Bot.
13:435–441.

———. 1989. Las selvas de Uxpanapa, Veracruz-Oaxaca, México: evidencia de
refugios florísticos cenozoicos. Anales Inst. Biol. Univ. Nac. México, Ser. Bot.
58:29–54.

——— & E.J. Lott. 1985. A new simple-leaved species of *Recchia* (Simaroubaceae)
from southeastern Mexico. Brittonia 37:219–225.

White, S.E. & S. Valastro, Jr. 1984. Pleistocene glaciation of Volcano Ajusco, central
Mexico, and comparison with the standard Mexican glacial sequence. Quaternary Res. 21:21–35.

Whitmore, T.C. (ed.). 1972. Tree Flora of Malaya, Vol. 1. Malayan Forest Records
No. 26. Kuala Lumpur: Longman Malaysia Sdn. Berhad.

Wing, S.L. 1987. Eocene and Oligocene floras and vegetation of the Rocky Mountains. Ann. Missouri Bot. Gard. 74:748–784.

Wolfe, J.A. 1972. An interpretation of Alaskan Tertiary floras. *In* A. Graham (ed.),
Floristics and Paleofloristics of Asia and Eastern North America. Amsterdam:
Elsevier. pp. 201–233.

———. 1975. Some aspects of plant geography of the Northern Hemisphere during
the Late Cretaceous and Tertiary. Ann. Missouri Bot. Gard. 62:264–279.

———. 1979. Temperature Parameters of Humid to Mesic Forests of Eastern Asia
and Relation to Forests of Other Regions of the Northern Hemisphere and
Australasia. U.S. Geol. Surv. Prof. Paper 1106.

———. 1985. Distribution of major vegetational types during the Tertiary. *In* E.T.
Sundquist & W.S. Broecker (eds.), The Carbon Cycle and Atmospheric CO_2:
Natural Variation Archean to Present. Monogr. 32. Washington, D.C: Amer.
Geophys. Union Geophys. pp. 357–375.

———. 1987. An overview of the origins of the modern vegetation and flora of the
northern Rocky Mountains. Ann. Missouri Bot. Gard. 74:785–803.

——— & G.R. Upchurch, Jr. 1987. North American nonmarine climates and
vegetation during the Late Cretaceous. Palaeogeogr. Palaeoclimatol. Palaeoecol.
61:33–77.

Yakovlev, G.P. 1971. A contribution to the revision of the genus *Ormosia* Jacks. I. The
genera *Ruddia* Yakovl. and *Fedorovia* Yakovl. (Leguminosae). Bot. Zurn. (Mos-
cow & Leningrad) 56:652–658.

Zona, S. 1990. A monograph of *Sabal* (Arecaceae: Coryphoideae). Aliso 12:583–666.

23

Phytogeography and History of the Alpine-Subalpine Flora of Northeastern Mexico

J. ANDREW MCDONALD

The present-day distribution of alpine-subalpine vegetation in Mexico has been greatly influenced by events during glacial maxima. The literature on paleopollen profiles, mammalian macrofossils, packrat middens, and extinct glaciers suggests that timberlines descended about 1,000 m during the Wisconsin glacial in both southern and northern regions of Mexico. If such were to have occurred, two closely situated subregions of alpine vegetation would have developed in northeastern Mexico, contrasting with the present-day, highly restricted distribution of seven disjunct refugia. The southern subregion would have centered around the present-day timberline refugium of Sierra Peña Nevada, Tamaulipas, and the northern subregion would have extended from the south of Cerro Potosí to Sierra la Viga, Nuevo León, linking in a continuous distribution the five most northerly alpine refugia of northeastern Mexico. Varying floristic similarities between peaks provide supporting empirical evidence for the hypothesized 1,000 m displacement of timberline during the Wisconsin glacial. Using the Sorensen index of species similarity, the flora of Peña Nevada is less similar to other alpine-subalpine refugia (53–58%, x = 55.2%) than the remainder are between themselves (55–76%, x = 66.7%), suggesting a greater degree of isolation of the former. Additionally, Peña Nevada sequesters more narrow endemics (nine species) than do the other six refugia taken as a whole (eight species). Extrapolated for the whole of Mexico, a 1,000 m depression of timberline vegetation along altitudinal gradients would have also resulted in an extensive corridor of alpine vegetation in the Sierra Madre Occidental, a broken but somewhat continuous alpine corridor across the Tran-Mexican Volcanic Belt, and an isolated band of alpine vegetation traversing the Sierra Madre del Sur. This newly proposed scenario for the past distribution of alpine vegetation provides a different perspective on the possible migrational routes followed by montane elements of Mexico during colder times.

Climatic vacillations since the Pliocene have affected the expansion and recession of various vegetation zones in northern Mexico, including desert (Lanner & van Devender, 1981; van Devender, 1977; van Devender &

Spaulding, 1979; Wells 1977, 1979), Madro-Tertiary sclerophyll forest (Axelrod, 1975; Lanner & van Devender, 1981; van Devender, 1977; van Devender & Spaulding, 1979; Wells 1977, 1979), temperate forest (Martin & Harrell, 1957; Sarukhán, 1968), tropical forest (Rzedowski, 1978; Sarukhán, 1968; Toledo, 1982), coniferous forest and alpine vegetation (Beaman, 1959; Beaman & Andresen, 1966; Crum 1951; Delgadillo, 1971). The full complement of these vegetation types are presently centered in northeastern Mexico, where the patchy distributions of these varied biotic provinces form a complex mosaic in the dissected terrain and abrupt elevational gradients (0–3700 m) of the northern Sierra Madre Oriental (Muller, 1937, 1939, 1947).

The recurrent episodes of glaciation during the Quaternary (van Devender, 1978; Wells, 1977) undoubtedly made possible the distributional expansion of floristic elements inhabiting the high montane zones of the region. Forests dominated by *Abies, Picea, Pseudotsuga,* and *Pinus* were probably more widespread than their contemporary, disjunct distributions in elevations above 2800 m from Monterrey (Muller, 1937, 1939; Patterson, 1988; Rojas, 1965) to southern Tamaulipas (Martin & Harrell, 1957; Rzedowski, 1978). Although these boreal elements have been present in northern Mexico since the upper Cretaceous (Rzedowski, 1978), and their expansion into the southern Sierra Madre Oriental occurred at least by the Miocene (Graham, 1973), the maximal extension of these genera probably occurred during glacial peaks of the Quaternary. Cooler regional climates allowed coniferous genera to descend altitudinal gradients and migrate further south, reuniting disjunct populations in a manner reminiscent of that which occurred with similar vegetation in the southwestern United States (Martin & Mehringer, 1965). *Picea,* for example, which presently does not extend south of the Tropic of Cancer (Patterson, 1988), has been reported as far south as the Valley of Mexico for the Pleistocene (Clisby & Sears, 1955) and in southern Veracruz for the mid-Pliocene (Graham, this volume).

Associated with the uppermost elevational limits of coniferous forest, alpine vegetation of Mexico probably experienced a concomitant waxing and waning with the ascending and descending of arborescent vegetation. Owing, however, to the absence of fossil data (the well-drained and windswept sites above timberline are not conducive to fossil formation), the history of this vegetation type can only be inferred from its present distribution and floristic compositions or from indirect evidence of past migrations of associated vegetation types.

Alpine vegetation is now known in seven localities in northeastern Mexico (Fig. 23.1), including Cerro Potosí, Nuevo León (Beaman & Andresen,1966; Muller, 1939), high montane ridges east of Saltillo, and ranges north of Miquihuana, including Sierra Borrado and Sierra Peña Nevada (McDonald, 1990). Species compositions of these refugia are assessed here to provide perspectives on the extent to which glacial events may have affected northeastern Mexico. They suggest a vegetation of

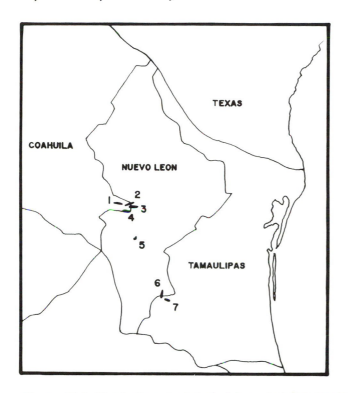

Figure 23.1. Distribution of relictual alpine-subalpine vegetation in northeastern Mexico. 1. Sierra la Viga. 2. Sierra Portrero de Abrego. 3. Sierra Coahuilon. 4. Sierra la Marta. 5. Cerro Potosí. 6. Sierra Peña Nevada. 7. Sierra Borrado.

autochthonous origin, with relatively closer biogeographical relations with the Rocky Mountains than with the neotropical alpine flora of the Trans-Mexican Volcanic Belt (TNB).

DELIMITATION OF THE ALPINE-SUBALPINE VEGETATION IN NORTHEASTERN MEXICO

The timberline vegetation of northeastern Mexico, much like alpine vegetation around the world, resists precise definition owing to several unavoidable ambiguities imposed by nature. Lower distributional limits of alpine species do not necessarily coincide with the upward distribution of tree species (Billings, 1978; Lloyd & Mitchell, 1973). Irregular timberlines often interdigitate and intergrade between forest and alpine vegetation, or alpine microcosms can interrupt dense forests, depending on exposures, microclimatic effects, soils, fires, and other environmental influences (Beaman, 1962; Billings, 1974; Major & Taylor, 1977; Marr, 1977; Stebbins, 1982). It is not always clear when high montane trees become dwarfed enough to be considered subalpine shrubs, e.g., wind-timber, krummholtz (Lloyd & Mitchell 1973). Wardle (1971) arbitrarily used a 2 m limit, whereas Beaman and Andresen (1966) accepted up to 5 m. Difficulties also arise when alpine zones are bordered by deserts, chaparral, steppe, or other semiarid types of vegetation that lack trees, in which case an observable "timberline" is lacking (Bell & Johnston, 1980; Billings, 1978; Hollerman, 1973; Lloyd & Mitchell, 1973; Major & Taylor, 1977).

The definition of alpine and subalpine vegetation in northeastern Mexico presents similar difficulties. For example, Riskind and Patterson (1975) questioned Beaman and Andresen's (1966) definition of *Pinus culminicola* forest as subalpine vegetation, as this species commonly occurs as a subdominant component of montane chaparral (e.g., Mediterranean scrub, Madro-Tertiary sclerophyll vegetation) as low as 2,900 m. This interpretation could be regarded as too restrictive, as the literature on North American alpine zones in the Rocky Mountains and the Sierra Nevada of California is replete with examples of timberline shrubs occurring as trees in marginal forests. Some examples include *Pinus albicaulis* (Chabot & Billings, 1972; Major & Taylor, 1977), *Pinus aristata* (Schaack, 1983), *Picea engelmannii* (Hutchins, 1974; Marr, 1977; Schaack,1983; Wardle, 1968), and *Abies lasiocarpa* (Hutchins, 1974; Marr, 1977). Nevertheless, the lower reaches of *Pinus culminicola* forest on Cerro Potosí (ca. 3500 m) often become tall and dense enough to shade out most of the alpine or subalpine elements.

Another difficulty in delimiting alpine regions of northeastern Mexico is encountered on dry, south-facing slopes of the higher ridges east of Saltillo, and to a limited extent the southeastern slopes of Peña Nevada, where alpine or subalpine zones are bordered on their lower reaches by montane chaparral. A treeless ecotone blurs the boundary between chaparral and alpine vegetation.

As the vegetational boundaries of most of the varying alpine-subalpine floras in northeastern Mexico have yet to be characterized, the working definitions of these zones here follow those described for Cerro Potosí by Beaman and Andresen (1966). Alpine zones include treeless vegetations composed of hemicryptophytes and chamaephytes above 3,600 m in elevation. Subalpine zones occur above 3,450 m and include associations dominated by erect forbs and grasses (in the presence or absence of dwarfed wind-timber [*Pinus hartwegii* or *P. culminicola*]), the latter of which when present make up less than 50% of the ground cover.

Subalpine or alpine vegetation in northeastern Mexico is found in three principal centers (Fig. 23.1). The northernmost center begins 36 km east of Saltillo in a series of ranges running from east to west in the Sierra Madre Oriental. The ranges with sufficient elevation (3,450–3,700 m) to support alpine-subalpine vegetation include Sierra Coahuilón, Sierra La Marta, Sierra Potrero de Abrego, and Sierra La Viga. Cerro Potosí, which is centrally located, occurs as a singular peak 38 km to the south of Sierra La Marta. The southern center of subalpine contacts in northeastern Mexico is located 125 km south of Cerro Potosí, including Sierra Borrado and Sierra Peña Nevada, whose peaks reach to 3,420 m and 3,650 m, respectively. Figure 23.1 summarizes localities explored and collected, giving approximate elevations of each based on the latest topographic DETENAL maps of Mexico, with latitude and longitude of the highest point of each range.

CLIMATE

Little climatic information for these areas of Mexico is available. The Mexican Secretaria de Programación y Presupuesto (SPP) classifies the climates of Peña Nevada and Cerro Potosí as C(E)W1)x', indicating that mean annual temperatures range from -3° to 18°C, and that precipitation exceeds more than 10.2 dm per year, with most of the precipitation occurring during the summer months. Sierra La Marta, Sierra Coahuilón, and Sierra La Viga are classified as C(E)W1, which is identical in regard to temperature but has precipitation that varies from only 5.0 to 10.2 dm per year. Data on temperature extremes (perhaps a more relevant parameter) are not available. Personal observations and interviews with native people reveal that during late fall, winter, and early spring frosts occur almost every night on the high ridges, and the summits often remain snow-covered. Because winter months coincide with the region's dry season, snow levels rarely reach significant depths. More important are the extreme temperatures that can occur during the growing season of the summer months. Sierra Coahuilón and Sierra La Marta receive daily afternoon precipitation during the summer, usually in the form of hail or sleet, which commonly covers the ground on the upper ridges to about 1 cm. The ice often persists until the following morning, lowering the ground temperatures to near freezing on most days. Local residents also report that strong summer freezes occur every 5–10 years. In July 1985, a strong frost reached as low as 2,900 m in the Ejido La Marcela at the eastern base of Sierra Peña Nevada, killing the year's crops of corn and barley.

In addition to low temperatures, frequent fires also apparently play a major role in the maintenance of these alpine-subalpine refugia. Almost every peak visited showed signs of fire, and several sierras (Sierra Borrado, Sierra Coahuilón, Sierra La Marta, Sierra Peña Nevada) have recently experienced extensive burns. Although man is the cause of some of the recent burns, the highly combustible montane chaparral that often borders the alpine-subalpine vegetation from below is easily ignitable by natural causes, as evidenced by occasional charred patches of vegetation after electrical storms. Fire is recognized as an important factor in the maintenance of chaparral vegetation (Axelrod, 1975), and rapid regeneration of vegetation after a burn in this region is observed. On high ridges, however, where the regenerative growth rate of trees and shrubs is retarded, alpine vegetation apparently has sufficient time to establish itself. Similar pyrogenic effects have been reported for other timberline habitats of both temperate and tropical origin (Cox, 1933; Hedberg, 1964; Janzen, 1973).

GEOLOGIC HISTORY

The geologic history of the high sierras of northeastern Mexico is complex and in need of investigation. Padilla (1982) theorized that the series of 32 anticlines comprising the "curvature of Monterrey" and adjacent regions were produced during the late Paleocene–early Eocene by a westward movement of the southern United States to a relative eastward movement of northern Mexico during the Laramide orogeny. The subsequent uplift of the Sierra Madre Oriental is generally thought to have occurred during the Pliocene (Beaman & Andresen, 1966), though a critical explanation of orogenic processes that led to the uplift is lacking (Ferrusquía-Villafranca, this volume). Because the alpine flora of the TNB has a probable Miocene–Pliocene origin (White, 1956), and the origins of the Rocky Mountains and Sierra Nevada alpine floras are generally thought to be of Pliocene and Pleistocene age (Billings, 1988; Moore, 1965), it is probable that the alpine-subalpine zones of northeastern Mexico are of roughly comparable age. Exposed rock in all of the study sites is composed of Cretaceous limestone (Ferrusquía-Villafranca, this volume). Soil depths are invariably thin (1–5 cm) and, in the more xeric or rocky stations such as the southern slopes of Sierra La Marta and Sierra La Viga, almost absent. These edaphic factors contrast markedly with the igneous substrates of alpine zones in the TNB.

PLANT COMMUNITIES

Beaman and Andresen (1966) recognized three types of alpine-subalpine vegetation on Cerro Potosí: alpine meadow dominated by shade-intolerant, prostrate dicots, subalpine meadow dominated by erect or cespitose forbs, and subalpine scrubland dominated by *Pinus culminicola*. The additional alpine-subalpine regions reported in northeastern Mexico (McDonald, 1990) deviate little from the characterizations by Beaman and Andresen (1966), but dominant species may vary among sites. In general, the dicot-dominated (up to 85% cover) aspect of the northeastern alpine enclaves is more similar to alpine zones of the Rocky Mountains than to the tussock grass-dominated alpine zones of the Mexican volcanic belt (Beaman, 1965; Narave, 1985) and Volcán Tacaná of Chiapas (Miranda, 1952-53).

Two additional alpine zones dominated by shade intolerant, prostrate forbs can be recognized in the region as a whole. Scattered and highly localized patches of this vegetation type occur on Sierra Peña Nevada in open glades where microcosmic conditions prevent the growth of trees (e.g., southern and southeastern exposures at summit or in mountain saddles). Of the 16 most frequent alpine meadow species on Cerro Potosí (Beaman & Andresen, 1966), 13 are prominent on Peña Nevada, along with the codominant, narrow endemics, *Erigeron onofrenis* Nesom and *Machaeranthera odysseus* Nesom. The presence of the following species distinguishes the timberline vegetation of Peña Nevada from Cerro Potosí: *Argemone subalpina* McDonald, *Alchemilla procumbens* Rose, *Commelina tuberosa* L., *Erigeron wellsii* Nesom, *Hedyotis wrightii* (A. Gray) Fosberg, *Piptochaetium virescens* (H.B.K.) Parod., *Salvia unicostata* Fern., *Solanum macropilosum* Correll, *Sedum papilicaulum* Nesom (ined.), *Senecio bellidifolium* H.B.K., and *Valeriana sorbifolia* H.B.K. var. *sorbifolia*. It is noteworthy that the dominant subalpine shrub of Cerro Potosí, *Pinus culminicola*, is absent from Peña Nevada.

Another zone of alpine vegetation is found on the southeastern flank of Sierra La Marta, though this steep, well-drained, exposed slope engenders a more xeric environment with a relatively depauperate vegetation. Distances between plants are observably greater than on other sites, and grasses are more prominent. *Juniperus monticola* Martinez, *Pinus culminicola* Andresen & Beaman, *Ceanothus buxifolius* Willd. ex Schult., and *C. greggii* A. Gray codominate as scattered, scandent shrubs generally not exceeding 10 cm in height.

The northern sierras (e.g., Sierra Coahuilón, Sierra La Marta, Sierra La Viga) support subalpine vegetation consisting of erect forbs, usually in the presence of *Pinus culminicola* on their upper, southern ridges. The northern slopes of these ranges are dominated by dense coniferous forest of *Pseudotsuga menziesii* (Mirb.) Franco, *Picea mexicana* M. Martinez, *Abies lasiocarpa* Hook., and *Pinus hartwegii* Benth. up to their summits. On the southern slopes, Madro-Tertiary sclerophyll scrub dominated by *Arctostaphylos pungens*, H.B.K., *Ceanothus buxifolius* Willd. ex Schult., and *Quercus greggii* (A. DC.) Trel. reaches close to the summit. Sandwiched between the coniferous forest and chaparral on the uppermost, south-exposed ridges (3,600–3,700 m) occur well-developed, herb-dominated strips of subalpine vegetation that are usually interrupted by patches of *Pinus culminicola* or scattered krummholtz individuals of *P. hartwegii*. In this curious arrangement, the alpine-subalpine vegetation paradoxically has a true timberline above, and a pseudotimberline with montane chaparral below, the clear and abrupt transitions between which are often observed within a 100 m transect. Dominant species of the relatively xeric alpine-subalpine habitat of Sierra La Marta include *Arenaria lanuginosa* Rohrb., *Arracacia schneideri* Mathias & Constance, *Brachypodium mexicana* (Roem. ex Schult.) Link., *Brickellia coahuilensis* (A. Gray) Harcombe & Beaman, *Castilleja scorzonerifolia* H.B.K., *Campanula rotundifolia* L., *Euphorbia beamanii* M.C. Johnston, *Lupinus cacumunis* Standl., *Paxistima myrsinites* Raf., *Penstemon leonensis* Straw, *Sedum chrysicaulum* McDonald, *Senecio carnerensis* Greenm., *S. coahuilensis* Greenm., and *Trisetum spicatum* (L.) Richter. All of the latter species are common on Cerro Potosí.

Sierra Coahuilón, situated between Sierra La Viga and Sierra La Marta but no further than 10 km from either, has a unique, rich and humid subalpine meadow on the southeastern slope. Several common species that occur here but not on other subalpine zones of northeastern Mexico include *Ageratina campylocladia* (B.L. Rob.) B. Turner, *Hymenopappus hintoniorum* B. Turner, *Nama whalenii* Bacon (ined.), *Oenothera tetraptera* Cav., and *Salvia sp. nov.* In contrast, the adjacent Sierra La Viga has a depauperate timberline vegetation, with alpine-subalpine elements persisting only marginally on the ridges and in rocky crevices of the highest southern exposures. Species encountered on the peak represent the most common elements of northeastern Mexico's alpine-subalpine flora, e.g., *Castilleja scorzonerifolia* H.B.K., *Penstemon leonensis* Straw, and *Sedum chrysicaulum* McDonald.

DISTRIBUTION OF ALPINE VEGETATION DURING GLACIALS IN MEXICO

Although the present-day distributions of alpine-subalpine floras in north-eastern Mexico (Figs. 23.1 and 23.4, below) are reduced and insular, the vegetation must have been a more prominent floristic component in the region during glacial cycles of the Quaternary, the last of which (Wisconsin) tapered off abruptly about 12,000–10,000 years BP (Martin & Mehringer, 1965; van Devender & Spaulding, 1979; Wells, 1979). Indirect geologic evidence suggests that montane vegetations in northeastern Mexico and adjacent regions descended about 1000–1300 m below their present-day positions during glacial cycles. This approximation is based in part on the low elevations at which moraines formed during the Quaternary in the adjacent southern Rocky Mountains, e.g., 3,000 m on San Francisco Peak, Arizona (Sharp, 1941); 3,580 m on Wheeler Peak, and 3,675 m on Sierra Blanca, New Mexico (Antevs, 1954) and the TNB. Richmond (1965) suggested the average depression of Pleistocene ice caps in the Rocky Mountains was about 1,200 m due to the estimated lowering of mean summer maximum temperatures by about 8°C. Moore (1965), in turn, extrapolated that vast extensions of alpine-subalpine vegetation covered considerable portions of New Mexico and northern Mexico. Although the argument can be made that the effects of glaciation were stronger on U.S. peaks than they were in northeastern Mexico owing to the more northerly latitude of the former, similar and simultaneous changes in snow line have also been reported much farther to the south in the TNB. White (1956) reported moraines at 3,10–3,500 m on Ixtaccihuatl (lat. 19 10' N), suggesting the lowering of snow lines by about 1,100–1,600 m during the Wisconsin (Weichselian) glaciation. Similar and contemporaneous movements of glaciers are reported for the Ajusco (White & Valastro, 1984), Malinche, and Pico de Orizaba (Heine, 1984; Hollin & Schilling, 1980) volcanoes.

Paleobiological evidence of glacial effects on northeastern Mexico is scanty, though studies of adjacent regions and varied sources, including fossilized packrat (*Neotoma*) middens, mammalian macrofossil remains (Gilmore, 1947), and fossil pollen profiles, generally support the purported vegetational changes extrapolated from glacial evidence. Pollen profiles of Pleistocene age in North America indicate a marked descent of vegetations along altitudinal gradients. Though pollen studies of northeastern Mexico are too few to be of much relevance here (Bryant & Riskind, 1980; Meyer, 1973), a review of sites studied in the southern Rocky Mountains (Martin & Mehringer, 1965) demonstrates that *Picea-Pinus* forest descended to 1,900 m, and subalpine spruce forest to 2,300 m, during the peak of the Wisconsin glacial (70,000–12,000 years BP). This finding suggests about a 1,000 m lowering of boreal vegetation, with subalpine-alpine zones hypothetically covering about one-fourth of Arizona and New Mexico (Martin & Mehringer, 1965). Glacial peaks had similar effects on timberline vegetation in the TNB. Straka and Ohngemach (1989) noted that alpine grasslands presently dominate the volcano Malinche, Tlaxcala, at around 4,000 m, whereas their palaeopalynological evidence suggests its occurrence at about 3,000 m 18,000 years BP..

Fossilized *Neotoma* middens provide further corroborative evidence for cooler climates and related vegetational changes in both northern Mexico and the southwestern United States during the Quaternary. The contributions of van Devender (1977, 1978), van Devender and collaborators (van Devender & Spaulding, 1979; van Devender et al., 1977), and Wells (1966, 1976, 1977, 1979) agree upon a marked decline in temperatures during glaciations and a lowering of pinyon pine forests along altitudinal gradients from 800 to 1,200 m throughout the southwestern United States and northern Mexico. These authors also indicate that vegetation zones above and below the favored habitats of *Neotoma*, including deserts and taiga, ascended and descended along with the packrat habitats.

In support of pollen and midden evidence, discoveries of boreal rodent fossils and other mammal groups (Findley, 1953; Stock, 1943) outside their present ranges suggest the distributional expansion of high-montane biotas. Van Devender et al. (1977) proposed that montane vegetation of the Guadalupe Mountains, New Mexico, descended from 1,075 to 1,230 m, based in part on the presence of fossilized bones of the yellow-bellied marmot (*Marmota flaviventris*) (Stearns, 1942), the bushy-tailed packrat (*Neotoma cinerea*) (Harris, 1970), and the masked shrew (*Sorex cinereus*) (Logan & Black, 1977). The southernmost fossil locality of *Marmota* has been reported about 10 km from Peña Nevada in the San Josecito caves, Nuevo León (Cushings, 1945), a considerable southern range extension for a genus whose extant, southern distributional limits are in southern New Mexico (Antevs, 1954; Stearns, 1942). Indeed, the altitude of the San Josecito caves (2,265 m) is about 1,000 m lower than the altitudes at which present-day marmots occur. If, as Brown (1971) has suggested, present-day distributions of poorly colonizing boreal mammals in the Great Basin are basically relictual and do not generally represent an equilibrium between rates of colonization and extinction, the *Marmota* fossils in northeastern Mexico could suggest the possibility of a cool corridor between the Rocky Mountains and northeastern Mexico.

Based on these geologic and paleobiological data, it can be conservatively estimated that alpine-subalpine zones of northeastern Mexico descended in altitude about 1,000 m during the peak of the Wisconsin glacial. Figures 23.2 and 23.3 depict (respectively) the hypothetical distribution of alpine-subalpine vegetation during the glacial for Mexico in general and in more detail for northeastern Mexico. The distributional limits of timberline

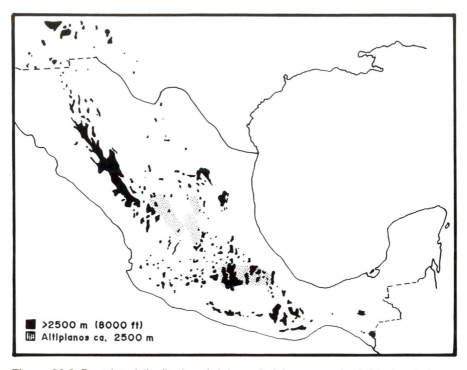

Figure 23.2. Postulated distribution of alpine-subalpine vegetation in Mexico during the Wisconsin glacial maxima. The darkened areas represent regions approximately 1,000 m below the present-day, altitudinal limits of timberline (3,500 m). The stippled regions indicate highland plateaus at the hypothetical, transitional timberline limits, across which subalpine elements also may have migrated.

vegetation during Wisconsin time are thought to follow near the 2,500 m contours, about 1,000 m below their current 3,500 m lower elevational limits. Alpine-subalpine floras presumably comprised about 5% of Mexico's land surface during the Wisconsin glacial and were relatively widespread. Most of the habitats were insular and disappeared during the interglacials. The TNB formed an east-west alpine archipelago along the Nineteenth parallel, and the Sierra Madre Occidental formed a north-south corridor through which alpine species of northern origin probably migrated. Paradoxically, the western corridor covered the largest region during glacial episodes; but because the Sierra Madre Occidental's altitudinal extremes are below the postulated timberline limits for interglacial periods (3,000–3,300 m), the corridor disappears completely during the present, with the exception of a single, highly restricted subalpine refugium, Cerro Mohinora of southwestern Chihuahua (Correll, 1960; McDonald & Martinez, in prep; Webb & Baker, 1984;).

Alpine vegetation in northeastern Mexico appears to have remained isolated during both glacial and interglacial periods. Being situated closer

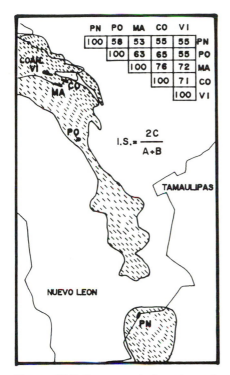

Figure 23.3. Sorensen index of species similarity between alpine-subalpine vegetation in northeastern Mexico. PN, Peña Nevada; PO, Cerro Potosí; MA, Sierra La Marta; CO, Sierra Coahuilón; VI, Sierra La Viga. Indices of species similarity are calculated by the formula I.S. = 2C ÷ (A + B), where A and B are the number of species present at two respective sites, and C is the number of species that occur at both sites. Shaded regions depict hypothetical distribution of alpine-subalpine vegetation (>2,500 m) during glacial episodes of the Quaternary. Darkened regions depict present-day, interglacial refugia.

to the hypothetical western alpine corridor than to any other present-day Mexican or U.S. alpine center, it seems possible that more propagules could have reached the region via the western sierra corridor during glacials than from the Rocky Mountains or the TNB due to long-distance dispersal. In addition, the hypothetical distribution also suggests that northeastern Mexico contained two isolated subregions. If the timberline failed to descend below 2,300 m, a southern and northern subregion of alpine vegetation would have persisted (Figs. 23.2 and 23.3), as the Cerro Potosí alpine refugium coalesces with the present-day northernmost centers of alpine-subalpine zones (Sierra Coahuilón, Sierra La Marta, Sierra La Viga), and the Sierra Peña Nevada and Sierra Borrado refugia merge and remain isolated in the south. These hypothetical distributions can be tested through known species compositions and distributions for the alpine-subalpine regions of northeastern Mexico (McDonald, 1990).

Table 23.1. Alpine-subalpine species endemic to a single peak of northeastern Mexico

Peña Nevada
Argemone subalpina McDonald, *Calochortus marcellae* Nesom, *Erigeron onofrensis* Nesom, *E. wellsii* Nesom, *Machaeranthera odysseus* Nesom, *Potentilla* sp. nov. (Ertter, in prep.), *Schoenocaulon* sp. nov. (Frame, in prep.), *Sedum clausenii* Nesom (ined.)., *Sisyrinchium* sp. nov., *Thelesperma graminiformis* (Sherff) Melchert (comb. in prep.)

Cerro Potosí
Dugaldia pinetorum (Standl.) Bierner, *Poa mulleri* Swallen, *Scutellaria potosina* Brandeg., *Senecio hintoniorum* B. Turner, *Thelesperma mulleri* (Sherff) Melchert

Sierra Coahuilón
Hymenoxys hintoniorum B. Turner, *Nama whalenii* (Bacon, ined.), *Salvia* sp. nov. McDonald

PHYTOGEOGRAPHY

If northeastern Mexico persists as an insular alpine center during both glacial and interglacial cycles (Fig. 23.2), an endemic flora of autochthonous origin would be expected in the region. Indeed, Beaman and Andresen (1966) calculated that 42% of the flora of Cerro Potosí is endemic to the Sierra Madre Oriental and 15% (12/81 species) to the peak itself. Further floristic studies of the region (McDonald, 1990) indicated that at least 35% of the timberline species from northeastern Mexico are endemic (Appendix). Twelve percent of the alpine-subalpine flora of Peña Nevada is endemic to this peak alone, accounting for about 8% of the endemic timberline elements of the region (Table 23.1). This degree of endemism is much higher than the rates of endemism recorded for the isolated mountains of the southwestern United States. For example, only 2% of the alpine species of both San Franciso Peak, Arizona (Schaak, 1983) and the White Mountains of California (Lloyd & Mitchell, 1973) are narrow endemics.

The overall floristic similarities between these Mexican peaks further point to the insular nature of this alpine-subalpine vegetation. Based on a species checklist for this region (McDonald, 1990) and alpine floras published for the White Mountains of southern New Mexico (Hutchins, 1974) and Cofre de Perote, Veracruz (Narave, 1985), indices of species similarity (Sorensen, 1948) were calculated by the formula

$$IS = 2C \div (A + B)$$

where A and B are the numbers of species present at two respective sites, and C is the number of species shared between the two sites. Floristic similarities between the northeastern alpine refugia are high (53–76%) (Fig.

Table 23.2. 22 Species common to northeastern Mexico and the White Mountains, New Mexico (132 species total)

Achillea millefolium L.
Elymus trachycaulum (Link.) Malte
Androsace septentrionalis L.
Aquilegia elegantula Greene
Blepharoneuron tricholepis (Torr.) Nash
Campanula rotundifolia L.
Commelina tuberosa L.
Draba helleriana Greene
Epilobium angustifolium L.
Erysimum capitatum (Dougl.) Greene
Fragaria aff. *californica* Newberry
F. speciosa Dougl.
Gentianella amarella (L.) Borner
Kohleria pyramidata (Lam.) Beauv.
Linum lewisii Pursh
Paxistima myrsinites Raf.
Phleum alpinum L.
Phacelia heterophylla Pursh
Populus tremuloides Michx.
Ranunculus praemorsus H.B.K. ex DC.
Taraxacum officinale Weber
Urtica spirealis Blume

23.3). A near equal index of similarity is found between this region as a whole and the alpine-subalpine flora of the White Mountains, New Mexico (15%; A = 170, B = 132, C = 22) (Table 23.2; Fig. 23.4), and that of Cofre de Perote, Veracruz (14%; A = 170, B = 52, C = 15) (Table 23.3, Fig. 23.4). This can be attributed, in part, to the depauperate alpine flora of Cofre de Perote (ca. 52 species), compared to the rich flora of the White Mountains (ca. 132 species), as the Sorensen index is dependent on the total number of species comprising the respective floras. Considered in absolute terms, there are 35% more shared species between northeastern Mexico and the White Mountains (C = 22) (Table 23.2) than between northeastern Mexico and Cofre de Perote (C = 15) (Table 23.3). In either case, the peaks representing the Rocky Mountains and the TNB share relatively few species with the alpine refugia of northeastern Mexico.

Indices of generic similarities show a stronger affinity between northeastern Mexico and the Rocky Mountains (66%; A = 116, B = 115, C = 77) (Fig. 23.4) than between northeastern Mexico and Cofre de Perote (40%; A = 116, B = 38, C = 31) (Fig. 23.4) despite the fact that the White Mountains are almost twice the distance to northeastern Mexico than is Cofre de Perote, Veracruz (960 km and 525 km, respectively). Possible explanations for the closer phytogeographic affinities to the north include the following. (1) The glacially induced alpine corridor found in the Sierra Madre Occi-

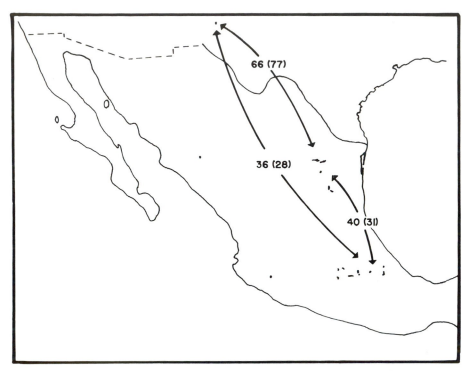

Figure 23.4. Sorensen index of generic similarity between alpine-subalpine vegetation of northeastern Mexico, White Mountains, New Mexico, and Cofre de Perote, Veracruz. Indices of generic similarity are calculated by the formula I.S. = 2C ÷ (A + B), where A and B are the number of genera at two respective sites, and C is the number of genera that occur at both sites. Numbers in parentheses denote number of genera shared by two floras.

dental more readily facilitated migration of northern alpine migrants from the southwestern United States to northeastern Mexico. (2) Climate and soils of the alpine habitats in northeastern Mexico are more similar to those of the Rocky Mountains than those of the TNB. (3) Northeastern Mexico is situated south of the Rocky Mountains, receiving the early autumn dissemination of Rocky Mountain alpine propagules by means of northerly winds and southward bird migrations.

In regard to the hypothetical distribution of alpine vegetation during the Wisconsin glacial, the contention that two subregions of alpine vegetation persisted within northeastern Mexico is also supported by floristic data. The fact that Peña Nevada (southern subregion) exhibits a slightly lower floristic similarity (53–58%; x = 55.2) to all adjacent refugia than the remainder of the refugia do between themselves (59–71%; x = 66.7) lends some support to the possibility of Peña Nevada's persistent isolation during the last glacial. Stronger evidence of this possibility are found in the distribution of narrow endemics and vicariads of the region. Between the

Table 23.3. 15 Species common to northeastern Mexico and Cofre de Perote, Veracruz (50 species total)

Alchemilla procumbens Rose
Arenaria lycopodioides Willd. ex Schlecht.
A. oresbia Greenm.
Bidens triplinervia H.B.K.
Brachypodium mexicanum (Roem. & Schult.) Link
Erysimum capitatum (Dougl.) Greene
Festuca rosei Piper
Geranium seemanii Peyr.
Juniperus monticola Martinez
Phacelia platycarpa (Cav.) Spreng.
Physalis orizabae Dunal
Ranunculus praemorsus H.B.K.
Ribes microphylla H.B.K.
Stellaria cuspidata Willd.
Trisetum spicatum (L.) Richter

putative northern and southern subregions of alpine-subalpine vegetation in northeastern Mexico there are several examples of vicariads, including the species pairs *Erigeron palmeri* A. Gray and *E. hintoniorum* Nesom, *Senecio bellidifolia* H.B.K. and *S. hintoniorum* B. Turner, *Thelesperma mulleri* (Sherff) Melchert (comb. ined.) and *T. graminiformis* (Sherff) Melchert (comb. ined.), and *Sedum papillicaulum* Nesom (ined.) and *S. clausenii* Nesom (ined.). These vicariads suggest ample isolation in time for the process of allopatric speciation to have occurred. However, there are no apparent vicariads found between the peaks that comprise the northern subregion alone (e.g., Cerro Potosí, Sierra Coahuilón, Sierra La Marta, Sierra La Viga), owing presumably to their probable recent floristic fusion during the Wisconsin glacial. In addition, the considerably higher number of species endemic (nine species) (Table 23.1) to the southern subregion (Sierra Peña Nevada), in comparison to few (none to five species) endemic to a single peak in the northern subregion (Table 23.1), further supports the likelihood that Peña Nevada remained isolated from the more northerly alpine refugia during the last glacial. These narrow endemics represent a significant 13% of the alpine-subalpine flora of northeastern Mexico.

CONSERVATION

The alpine regions of Mexico represent one of the richest centers, per unit area, of narrow endemics. In northeastern Mexico alone, the seven known alpine refugia sequester populations of 59 endemic species which inhabit regions totaling no more than about 6 km². About one-third of these species have been known to science for only a few years; and apart from their distributions, almost no information of these rare species is available (e.g., their reproductive biology, substrate specificity, development, natural history). Their habitats are more fragile than most, given their short growing seasons and slow regenerative processes after disturbance. Nevertheless, all of these habitats are grazed, and we have no information on what effect this practice has on the vegetation. A second source of disturbance also comes directly from the hand of man. Because high mountain peaks are favored localities for the placement of microwave stations, human presence, destruction, and continued disturbance are rife. A conservation ethic on the part of the developers has been wholly lacking in the past. Because Mexican alpine habitats are so restricted in distribution and so fragile, they could and should be recognized as endangered habitats with "untouchable" status. Conservation initiatives to save just a few square kilometers of alpine vegetations throughout Mexico could easily protect at least 100 rare plant species from extinction.

CONCLUSIONS

Several tentative conclusions can be drawn from the above discussion. The alpine-subalpine flora of Mexico has undergone significant distributional changes during the Quaternary. Today's highly restricted Mexican alpine flora contrasts considerably with its probable widespread distribution during times as recent as 12,000 years BP. The alpine-subalpine flora of northeastern Mexico, once thought to be endemic to a single peak, Cerro Potosí, is relatively widespread over about seven disjunct and narrowly insular refugia. This regional flora is largely autochthonous at the species level but shares closer floristic relation (generic level) with the Rocky Mountains than it does with the southern alpine floras of the TNB. The high proportion of endemics in the alpine flora of northeastern Mexico can be attributed to the region's isolation during both glacial and interglacial episodes of the Quaternary. Distributions of species vicariads and indices of species similarity between the present-day alpine refugia of the region support the probability of an approximate 1,000 m altitudinal descent of this vegetation type during glacials of the Quaternary. When extrapolated for the whole of Mexico, this conclusion provides new and testable perspectives on the possible modes of migration affecting the evolution of the alpine-subalpine floras of Mexico.

A<small>CKNOWLEDGMENTS</small>

The author gratefully acknowledges CONACyT (Mexico), the World Nature Association, and INIREB (1977–1988), Xalapa, Veracruz, for providing funds to undertake several summers of botanical exploration. José

Gómez A. assisted in field work, and Francisco González Medrano provided administrative support. David Riskind kindly made available a crucial bibliography on northeastern Mexican biogeography, and Guy Nesom, Marshall Johnston, Beryl Simpson, and Billie Turner made helpful comments on initial drafts of the manuscript.

APPENDIX

Species endemic to the alpine-subalpine flora of northeastern Mexico

BORAGINACEAE
Hackelia leonotis I.M. Johnston
Onosmodium dodrantale I.M. Johnston

COMPOSITAE
Astranthium beamanii De Jong
Brickellia nesomii B. Turner
B. coahuilensis (A. Gray) Harcombe & Beaman
B. hintoniorum B. Turner
Chaetopappa parryi A. Gray
Cirsium novoleonense G. Nesom (ined.)
Dugaldia pinetorum (Standl.) Bierner
Erigeron hintoniorum Nesom
E. onofrensis Nesom
E. potosinus Standl.
E. wellsii Nesom
Gnaphalium hintoniorum Nesom
Grindelia inuloides Willd.
Hymenoxys ursina Standl.
Hymenopappus hintoniorum B. Turner
H. hintoniorum B. Turner
Machaeranthera odysseus Nesom
Senecio carnerensis Greenm.
S. coahuilensis Greenm.
S. hintoniorum B. Turner
S. lorantifolius Greenm.
S. madrensis A. Gray
Thelesperma graminiformis (Sherff) B. Turner

CRASSULACEAE
Sedum chrysicaulum McDonald
S. clausenii Nesom (ined.)
S. papillicaulum Nesom (ined.)

CRUCIFERAE
Thlaspi mexicanum Standl.

EUPHORBIACEAE
Euphorbia beamanii M.C. Johnston

GERANIACEAE
Geranium crenatifolium H. Moore

GRAMINAE
Muhlenbergia sp. nov.
Poa mulleri Swallen

Appendix (cont.)

HYDROPHYLLACEAE
Nama whalenii Bacon (ined.)

IRIDACEAE
Sisyrinchium schaffneri Wats.
Sisyrinchium sp. nov.

LAMIACEAE
Agastache palmeri (B. Rob.) Standl. var. *leonensis* R. Sanders
Salvia sp. nov.
Scutellaria potosina Brandeg.

LEGUMINOSAE
Astragalus purpusii M.E. Jones
Trifolium schneideri Standl. Lupinus cacuminis Standl.

LILIACEAE
Calochortus marcellae Nesom
Schoenocaulon sp. nov. (Frame, in prep.)

PAPAVERACEAE
Argemone subalpina McDonald

PINACEAE
Pinus culminicola Andresen & Beaman

RANUNCULACEAE
Delphinium valens Standl.

ROSACEAE
Potentilla leonina Standl.
Potentilla sp. nov. (Ertter, in prep.)

SAXIFRAGACEAE
Ribes neglectum Rose

SCROPHULARIACEAE
Castilleja bella Standl.
Penstemon leonensis Straw

SOLANACEAE
Solanum macropilosum Correll

UMBELLIFERAE
Arracacia schneideri Mathias & Constance
Tauschia hintoniorum Constance & Affolter
T. madrensis Coult. & Rose

VIOLACEAE
Viola galeanaensis M.S. Baker

REFERENCES

Antevs, E. 1954. Climate of New Mexico during the last glacio-pluvial. J. Geol. 62:182–191.

Axelrod, D.A. 1975. Evolution and biogeography of Madrean-Tethyan sclerophyll vegetation. Ann. Missouri Bot. Gard. 62:280–334.

Beaman, J.H. 1959. The alpine flora of Mexico and Central America. Yearbook Amer. Phil. Soc. 266–268.

———. 1962. The timberlines of Iztaccihuatl and Popocatepetl, Mexico. Ecology 43:377–385.

———. 1965. A preliminary ecological study of the alpine flora of Popocatepetl and Iztaccihuatl. Bol. Soc. Bot. México. 29:63–75.

——— & J.W. Andresen. 1966. The vegetation, floristics, and phytogeography of the summit of Cerro Potosí, Mexico. Amer. Midl. Naturalist 75:1–33.

Bell, K.L. & R.E. Johnston. 1980. Alpine flora of the Wassuk range, Mineral County, Nevada. Madroño 27:25–35.

Billings, W.D. 1974. Adaptations and origins of alpine plants. Arctic Alpine Res. 6:129–142.

———. 1978. Alpine biogeography across the Great Basin. Great Basin Nat. Mem. 2:105–117.

———. 1988. Alpine vegetation. *In* M.G. Barbour & W.D. Billings (eds.), North American Terrestrial Vegetation. Cambridge: Cambridge Univ. Press.

Brown, J.H. 1971. Mammals on mountaintops: nonequilibrium insular biogeography. Amer. Naturalist 105:467–468.

Bryant, V.M. & D.H. Riskind. 1980. The paleoenvironmental record for northeastern Mexico: a review of pollen evidence. *In* J.F. Epstein, T.R. Hestor & C. Graves (eds.), Papers on the Pre-history of Northeastern Mexico and Adjacent Texas. Special Report No. 9. San Antonio: Univ. of Texas, Center for Archeological Research.

Chabot, B.F. & Billings. 1972. Origins and ecology of the Sierran alpine flora and vegetation. Ecol. Monogr. 42:163–199.

Clisby, K.H. & P.B. Sears. 1955. Palynology in southern North America. III. Microfossil profiles under Mexico City correlated with sedimentary profiles. Bull. Geol. Soc. Amer. 66:511–520.

Correll, D.S. 1960. A mule-train trip to Sierra Mohinora Chihuahua. Amer. Fern J. 50:66–78.

Cox, C.V. 1933. Alpine plant succession on James Peak, Colorado. Ecol. Monogr. 3:301–372.

Crum, H.A. 1951. The Appalachian-Ozarkian element in the moss flora of Mexico with a check-list of all known Mexican mosses. Ph.D. dissertation, Univ. of Michigan, Ann Arbor.

Cushings, J.E. 1945. Quaternary rodents and lagomorphs of San Josecito cave, Nuevo Leon, Mexico. J. Mammal. 26:182–185.

Delgadillo, C. 1971. Phytogeographic studies on alpine mosses of Mexico. The Bryologist 74:331–346.

Findley, J.S. 1953. Pleistocene Soricidae from San Josecito cave, Nuevo Leon, Mexico. Univ. Kansas Publ. Mus. Nat. Hist. 5:635–639.

Gilmore, R.M. 1947. Report on a collection of mammal bones from archeological cave sites in Coahuila, Mexico. J. Mammal. 28:147–165.

Graham, A. 1973. History of the arborescent temperate element in the Northern Latin American Biota. *In* A. Graham (ed.), Vegetation and Vegetational History of Northern Latin America. Amsterdam: Elsevier.

Harris, T.A. 1970. Tex. J. Sc. 22:3–27

Hedberg, O. 1964. Features of Afro-alpine plant ecology. Acta Phytogeogr. Suec. 49:1–144.

Heine, K. 1984. The classical late Weichselian climatic fluctuations in Mexico. *In* N.A. Morner & W. Karlen (eds.), Climatic Changes on a Yearly to Millenial Basis. Dordrecht: Reidel Publishing.

Hollerman, P. 1973. Some reflections on the nature of high mountains, with special reference to the western United States. Arctic Alpine Res. 5:149–160.

Hollin, J.T. & D.H. Schilling. 1980. Late Wisconsin-Weichselian mountain glaciers and small ice caps. *In* G.H. Denton & T.J. Hughes (eds.), The Last Great Ice Sheets. New York: John Wiley & Sons.

Hutchins, C.R. 1974. A Flora of the White Mountain Area, Southern Lincoln and Northern Otero Counties, New Mexico. Albuquerque, New Mexico: Publ. privately by author.

Janzen, D.H. 1973. Rate of regeneration after a tropical high elevation fire. Biotropica 5:117–122.

Lanner, R.M. & T.R. van Devender. 1981. Late Pleistocene piñon pines in the Chihuahuan Desert. Quaternary Res. 15:278–290.

Lloyd, R.M. & R.S. Mitchell. 1973. A Flora of the White Mountains, California and Nevada. Berkeley, CA: Univ. of California Press.

Logan, L.E. & L.L. Black. 1977. The Quaternary fauna of Upper Sloth Cave, Guadalupe Mountains National Park, Texas. *In* H.H. Genoways & R.J. Baker (eds.), Biological Investigations in the Guadalupe Mountains Texas, Proc. & Trans. Ser. No. 4. Washington D.C: U.S. Govt. & Printing Office.

Major, J. & D.W. Taylor. 1977. Alpine. *In* M.G. Barbour & J.Major (eds.), Terrestrial Vegetation of California. New York: Wiley Interscience.

Martin, P.C. & B.E. Harrell. 1957. The Pleistocene history of temperate biotas in Mexico and Eastern United States. Ecology 38:486–480.

Martin, P.C. & P.J. Mehringer. 1965. Pleistocene pollen analysis and bio-geography of the Southwest. *In* H.E. Wright & W.H. Osburn (eds.), The Quaternary of the U.S. Princeton, NJ: Princeton Univ. Press.

Marr, J.W. 1977. The development and movement of the tree islands near the upper limit of tree growth in the southern Rocky Mountains. Ecology 58:1159–1164.

McDonald, J.A. 1990. The alpine-subalpine flora of northeastern Mexico. Sida 14:21–28.

Meyer, E.R. 1973. Late Quaternary paleoecology of the Cuatro Cienegas Basin, Coahuila, Mexico. Ecology 54:982–995.

Miranda, F. 1952–53. La Vegetación de Chiapas. Chiapas, México: Ediciones del Gobierno del Estado, Tuxtla Gutierrez. Vols. 1 & 2.

Moore, T. 1965. Origin and disjunction of the alpine flora of San Francisco Mountain, Arizona. Ecology 46:860–864.

Muller, C.H. 1937. Plants as indicators of climate in Northeast Mexico. Amer. Midl. Naturalist 18:986–1000.

———. 1939. Relations of the vegetation and climatic types of Nuevo León, Mexico. Amer. Midl. Nat. 21:687–729.

———. 1947. Vegetation and climate of Coahuila, Mexico. Madroño 9:33–57.

Narave, H. 1985. La vegetación del Cofre de Perote. Biotica 10:35–64.

Padilla, R.J. 1982. Geologic Evolution of the Sierra Madren Oriental Between Linares, Concepcion del Oro, Saltillo and Monterrey, Mexico. Ph.D. Dissertation, The Univ. of Texas, Austin.

Patterson, T.F. 1988. A new species of *Picea* (Pinaceae) from Nuevo Leon, Mexico. Sida 13:131–135.

Richmond, G.M. 1965. Glaciation of the Rocky Mountains in the Quaternary of the U.S. Princeton, NJ: Princeton Univ. Press.

Riskind, D.H. & T.F. Patterson. 1975. Distributional and ecological notes on *Pinus culminicola*. Madroño 23:159–161.

Rojas, P. 1965. Generalidades Sobre la Vegetación del Estado de Nuevo León y Datos Acerca de su Flora. Thesis. UNAM.

Rzedowski, J. 1978. Vegetación de México. Mexico City: Limusa.

Sarukhán, K.J. 1968. Analisis Sinecologico de las Selvas de *Terminalia amazonia* en la Planicie Costera del Golfo de México. M.S. thesis, Colegio de Posgrado, Esc. Nac. de Agricultura, Chapingo, Mexico.

Schaack, C.G. 1983. The alpine vascular flora of Arizona. Madroño 30:79–88.

Sharp, R.P. 1941. Ancient glaciers of the San Francisco Peaks. Plateau 14:28–32

Sorensen, T. 1948. A method of establishing groups of equal amplitude in plant sociology, based on similarity of species content. Det. Kong. Danske Vidensk. Selsk. Biol. Skr. (Copenh) 3:1–34.

Stearns, C.E. 1942. A fossil marmot from New Mexico and its climatic significance. Amer. J. Sci. 240:867–878.

Stebbins, G.L. 1982. Floristic affinities of the high Sierra Nevada. Madroño 29:189–199.

Stock, C. 1943. The cave of San Josecito, Mexico. *In* Engineering and Science Monthly. No. 361. Pasadena: California Institute of Technology.

Straka, H. & D. Ohngemach. 1989. Late Quaternary vegetation history of the Mexican highlands. Pl. Syst. Evol. 162:115–132.

Toledo, V.M. 1982. Pleistocene changes of vegetation in tropical Mexico. *In* G. Prance (ed.), Biological Diversification in the Tropics. New York: Columbia Univ. Press.

Van Devender, T.R. 1977. Holocene woodlands in the Southwestern Deserts. Science 198:189–192.

———. 1978. Glaciopluvial woodlands in the Bolson de Mapimi, Durango and Coahuila, Mexico. Amer. Quaternary Assoc. Abstr. 5:234.

——— & W.G. Spaulding. 1979. Development of vegetation and climate in the Southwestern United States. Science 204:701–710.

———, P.S. Martin, A.M. Phillips & W.C. Spaulding. 1977. Late Pleistocene biotic communities from the Guadalupe Mountains, Culberson County, Texas. *In* R.H. Wauer and D.H. Riskind (eds.), Transactions of the Symposium on the Biological Resources of the Chihuahuan Desert Region, U.S. and Mexico. Ser. 3. U.S. Dept. of the Interior, National Park Service. pp. 107–114.

Wardle, P. 1968. Engelmann spruce (*Picea engelmanii* Engel.) at its upper limits on the Front Range, Colorado. Ecology 49:483–495.

———. 1971. A comparison of alpine timberlines in New Zealand and North America. New Zealand J. Bot. 3:113–135.

Webb, R.G. & R. Baker. 1984. Terrestrial vertebrates of the Cerro Mohinora region. Southwest. Naturalist 29:243–246.

Wells, P.V. 1966. Late Pleistocene vegetation and degree of pluvial climatic change in the Chihuahuan Desert. Science 153:970–975.

———. 1976. Macrofossil analysis of wood rat (*Neotoma*) middens as a key to the quaternary vegetational history of arid America. Quat. Res. 6:223–248.

————. 1977. Post glacial origin of the present Chihuahuan desert less than 11,500 years ago. *In* R.H. Wauer and D.H. Riskind (eds.), Transactions of the Symposium on the Biological Resources of the Chihuahuan Desert Region, U.S.–Mexico. Ser. 3. U.S. Dept. of the Interior, National Park Service.

————. 1979. An equable glaciopluvial in the West: Pleniglacial evidence of increased precipitation on a gradient from the Great Basin to the Sonoran and Chihuahuan basins. Quaternary Res. 12:311–325.

White, S.E. 1956. Probable substages of glaciation on Iztaccihuatl, Mexico. J. Geol. 64:289–295.

———— & S. Valastro. 1984. Pleistocene glaciation of Ajusco, Central Mexico, and comparison with the standard Mexican glacial sequence. J. Quaternary Res. 22:240–246.

V
PLANT DIVERSITY AND HUMANS

24

The Role of Humans in the Diversification of Plants in Mexico

ROBERT BYE

Mexico is one of the culturally and botanically richest countries in the New World with 8 million people speaking one of 54 native languages and with two major floristic kingdoms represented by ten vegetation types and 30,000 species. Ethnobotanical richness is reflected by the utilization of over 5,000 vascular plants, various rich and dynamic folk taxonomies, and the detailed cultural perception and management of vegetal resources. Human enterprises resulting in erosion, livestock grazing, and agriculture alter the Mexican landscape and threaten its botanical diversity. The gathering, incipient management, and cultivation of wild, weedy and domesticated plants produce positive as well as negative effects. Other responses to human activities are the changes in floristic composition, the hybridization of wild and domesticated species, the persistence of certain taxa in archaeological sites, and the rapid evolution of some plant genera in secondary vegetation.

The coevolutionary relationship between humans and other organisms has developed in various directions over a long period of time. Coevolution is "an evolutionary process in which the establishment of a symbiotic relationship between organisms, increasing the fitness of all involved" (Rindos, 1984) and may consist of natural (genetic) and cultural (nongenetic) selection (Durham, 1978). "Gene-culture coevolution" is generally recognized, but there is little consensus on the details (Grant, 1985). Cultural evolution, the progressive development and accumulation of cultural heritage, has its own momentum that is different from that of genetic evolution (Grant, 1985).

In Mexico, the mutual dependence between humans and plants has developed intricately over time and space. Mexico's cultural richness, combined with its floristic wealth, has diversified plant–human interactions. The derivation of an ethnobotanical diversity from the correlation of ethnic and biological richness is a useful exercise but may be less practical than considering the richness of plant species employed by Mexicans. The recognition and utility of plants as registered in ethnobotanical inventories approximate richness. Archaeological and historical records demonstrate the dynamics of vegetal taxa utilized throughout human history. The intensification of certain processes by humans, such as erosion and selective elimination, has a negative impact upon plants. However, other actions tend to favor biological diversity.

The human influence on plant evolution has many facets (Baker, 1972). The wild–domesticated continuum in association with human actions illustrates the role of people in plant diversification. Other examples of mankind's imprint on biological diversity is seen in hybridization, persistence of useful plants in archaeological sites, incorporation of new taxa in local flora, and accelerated speciation in secondary vegetation.

With the growth of the Mexican human population, it is important to consider its impact on diversification of the flora. Conservation policies for the maintenance of biological diversity must recognize the cumulative effects of human influence on the evolution and ecology of plants as well as seek a balance between natural and anthropogenic communities.

CULTURAL AND FLORISTIC RICHNESS

Richness refers to the variety or number of discrete elements within a given hierarchy. For the following discussion, I use languages (rather than the higher categories of language families and branches or the ethnic group) because each language facilitates the cultural inheritance of information. The background for floristic richness is summarized in terms of floristic kingdoms, regions and provinces as well as by vegetation types and number of species. More detail is given to cultural richness rather than floristic richness because the latter is covered in greater detail in other chapters in this book (e.g., Rzedowski, this volume).

The integrity of Mexican cultures may be indexed by the number of native speakers. Language is important in the formulation and communication of indigenous concepts, perceptions, and actions. Since the Spanish Conquest, not only have native peoples suffered drastic population decline but also a decrease in the number of their languages. Only 54 languages of the 120 spoken at the time of the Conquest are currently in use in Mexico (Martínez, 1986). Nonetheless, about 8 million individuals (or 7.5% of the 1980 Mexican population) speak an indigenous language, and the absolute number of native speakers has been on the increase since 1950 (Valdés & Menéndez, 1987). Four patterns of numerical change in indigenous speakers since 1930 have been identified (Olivera et al., 1982). One pattern is the increase of people who practice the languages of Amuzgo, Chol, Mazatec, Tlapanec, Tzeltal, and Tzotzil; however, these ethnic groups are small in population number. The second pattern consists of speakers whose number has remained constant; they are exemplified by the Chinantec, Huastec, Huichol, Mixe, and Mixtec. The third case demonstrates the decline in the number of native speakers. Unfortunately, this group includes the larger native populations such as those of the Maya, Chontal, Mayo, Nahuatl, Otomi, Mazahua, Tarahumara, and Zapotec. The fourth group includes such ethnic groups as the Chatino, Popoluca, Tojolabal, Totonaca, and Zoque in which the change in the number of native speakers has been variable over the years.

The most important groups of native speakers are located in nine geographic regions. The five major areas are (1) the central-east region in the high valleys (2,000 m) of Mexico, Toluca and Mezquital; (2) the Huastec region of the warm, subhumid Sierra Madre Oriental; (3) southern region in the highly dissected southern Sierra Madre Oriental, Sierra Madre del Sur, and Trans-Mexican Neovolcanic Belt with warm, subhumid to dry climate; (4) southeastern region in the warm, humid mountains of Chiapas; and (5) the Yucatan Peninsula on the warm, subhumid flats. Four minor regions include (6) the northwestern arid coastal plains; (7) the Sierra Tarahumara of northwestern Mexico; (8) the West-Central Sierra Madre Occidental; and (9) the lake zone of west-central Trans-Mexican Neovolcanic Belt.

Mexico's vegetal richness is as complex as its cultural richness. Two major floristic kingdoms converge in Mexico and their subordinate regions and provinces interdigitate along the length of the country (Rzedowski, 1978). The Holartic kingdom consists of two regions: the North American Pacific region in Baja California with two provinces and the Mesoamerican Mountain region of the higher elevations from Chihuahua to Chiapas with four provinces. The Neotropic kingdom contains two regions: (1) the Mexican Desert region with five provinces from sea level to the Central Plateau; and (2) the Caribbean region along both Atlantic and Pacific coasts and penetrating the adjacent lower inland drainages. In addition, ten basic vegetation types have been classified (Rzedowski, 1978), and the flora is estimated to contain over 30,000 species (Rzedowski & Equihua, 1987), 21,600 of which are vascular plants (Rzedowski, this volume).

The relationships between biological and cultural richness are difficult to determine. An attempt is made by considering ethnobotanical diversity.

ETHNOBOTANICAL DIVERSITY

Ethnobotany is an interdisciplinary study for which there is no uniformly accepted definition. The basic elements are plants and humans at different levels of organization (e.g., from individuals through populations and communities to ecosystems). For the purpose of this chapter, I assume that ethnobotany is the study of the biological, ecological and cultural bases of plant–human interactions and relationships over evolutionary time and sociogeographic space.

Diversity is an equally challenging term to define. Most scholars agree that the principal component is the variety or richness of species. The variance of this richness must be included along with the richness in order to calculate the index of diversity. This variance is a weighted measure (Pielou, 1975) and can be appraised by abundance of individuals, biomass, and energy, among others. The many forms of diversity (Magurran, 1988) suggest that the concept is developing vigorously. Ethnobotanical diversity may be expressed as the sum of the products of taxonomic richness and weighted cultural value. Although attempts have been made to quantify cultural significance of plants (Turner, 1988), such an index has limitations. The diversity of plant–human interactions may be explored conceptually by the correlations of botanical richness and cultural richness with a complex ethnobotanical interaction. Mexico is an appropriate place to examine the plant–human richness relationships because of the great degree of diversification of cultures and biota in comparison to the rest of the Americas and the world. An estimation of Mexico's baseline plant–human richness relative to that of Latin American has been made by Toledo (1987) and Caballero (1987). In terms of absolute number of indigenous individuals, Mexico ranks first with over 8 million people. On the other hand, it has 54 ethnic groups, fewer than the countries of Brazil, Colombia, and Peru. In terms of number of vascular plant species, Mexico stands second, after Brazil. About one-fifth of Mexico's indigenous societies has recorded ethnobotanical information.

The "man–agave symbiosis" (Gentry, 1982) is an appropriate Mexican example to examine a relationship between biological and cultural richness. The genus *Agave* has its natural range from the southern United States to South America. The major concentration of taxa is found in Mexico, which has 136 species according to Gentry's (1982) taxonomic treatment that reduced the plethora of named, unnatural species. Magueys, as the various taxa of *Agave* are called by Mexicans, have been an important food, beverage, and fiber source for humans since prehistoric times. These drought-tolerant plants concentrate liquid, carbohydrates, and fiber, which can be exploited by people living in environments where sustenance is seasonal or unreliable. Based upon the relative occurrence of plant remains recovered from archaeological sites in northeastern and south-central Mexico, magueys have been consistently used throughout time with a maximum representation about 4,000 years ago (Callen, 1965; Smith, 1967). With such prolonged contact and dependence upon magueys, one might assume that the greatest ethnobotanical diversification would be found where the highest *Agave* species richness occurs or the greatest cultural richness.

The production of pulque, the alcoholic beverage produced by fermenting the liquid extracted from maturing magueys, is one of the most advanced coevolutionary relationship between *Agave* and humans. The propagation of the maguey plants, the timely collection of the sweet juice (or "aguamiel"), and the batch fermentation technology with its associated microorganisms depend upon direct human intervention. On the other hand, the people of central Mexico relied upon this uncontaminated beverage to alleviate thirst and hunger during stressful periods of the year.

The antiquity of the critical importance of *Agave* is confirmed by the fact that many Aztec ceremonies included pulque along with human sacrifices and that a major deity was the pulque goddess, Mayahuel (Gonçalves, 1956). The human dependence upon fermented *Agave* juice was broken only in the early twentieth century. After the Mexican Revolution, the economic and social importance of pulque in urban areas declined owing to the abandonment of the maguey-producing haciendas and to the greater availability of drinking water and industrialized beverages (Zorrilla, 1988).

The dimensions of the *Agave*–human diversity are still present in rural areas where older people rely upon pulque as daily beverage providing a dietary complement of calories, vitamin B complex and protein (Ulloa et al., 1987). The farmers manipulate the maguey populations by conscious selection of different races and by transplanting vegetative propagules as part of traditional agroecosystems (Ruvalcaba, 1983). The selective transplant and the hybridization between different species brought into close proximity (Gentry, 1982) has resulted in the diversification of folk cultivars (Marino, 1966), which are readily recognized by "pulqueros" but defy scientific classification (Gentry, 1982). The maintenance of maguey populations is dependent upon humans because the central bud (which contains the inflorescence) is destroyed in the collection process and the basal offshoots are too dense and close to the mother plant to guarantee survival and dispersal. The maguey's sweet water, which rapidly decomposes, is transformed to nutritionally enriched pulque through traditional manipulation of a microecosystem involving ten species of bacteria and eight taxa of yeasts (Ulloa et al., 1987). Human behavior has adapted to pulque production owing to the requirements of maintaining permanent row cultivation of magueys as an integral part of the agroecosystem and of daily harvesting and fermenting routines.

The antiquity of the pulque process is unknown even though many cultures of central Mexico have their own myths. Analyses of these legends suggest the time of origin to be about 900 AD (Lobato, 1884; Zorrilla, 1988), but undoubtedly it is much older. Today mestizos as well as Indians, such as the Huastec, Matlatzinc, Mazahua, Nahua, Otomi, Purepecha, and Zapotec, maintain this intensive *Agave*–human symbiosis (Ulloa et al., 1987; Zorrilla, 1988). Given the broad distribution of the 136 species of *Agave* in Mexico, one would expect that this ethnobotanical process would be (1) confined to a related group of species or (2) correlated with centers of species richness. According to Gentry (1982), only five species distributed in four sections are the major pulque producers (Table 24.1); consequently there is no taxonomic restriction. These pulque magueys originated and are cultivated in the southern Central Plateau (Gentry, 1982), which is relatively poor in *Agave* species and outside of the three regions of highest species concentration: the Sierra Madre Occidental, central-southern Sierra Madre Oriental and southern Tehuacan Valley of Puebla (and adjacent northeastern Oaxaca) (Reichenbacher, 1985). Hence this plant–human interaction is not correlated with taxonomic affinity and assumed

Table 24.1. Species of *Agave* that produce pulque

Section	No. of species per section	Species used for pulque
Americanae	6	Agave americana
Crenatae	6	Agave hookeri
Hiemiflorae	12	Agave atrovirens
Salmianae	4	Agave salmiana
		Agave mapisaga
Total species	28	5

Source: Gentry (1982).

phylogeny nor with centers of species richness. Another expected association would be the concurrence of the centers of both *Agave* and cultural richness. The state of Oaxaca has the greatest number of ethnic groups as well as the largest population of native speakers (Valdés & Menéndez, 1987). Nonetheless, the nearest *Agave* richness center is located to the northeast of the populated areas. Also, the pulque process is recorded for one ethnic group in the region. The relationship between biological and cultural diversity is unclear and requires detailed analysis.

ETHNOBOTANICAL RICHNESS

Given that ethnobotanical diversity may not be reliably determined due to uncertainties in cause-and-effect relationships between biological and cultural factors, the richness or variety component of plant–human interaction may be the most practical manner to evaluate the human influence (or potential influence) on the diversity in Mexican flora. The utilization and perception of plants is based upon direct contact between *Homo sapiens* and plants. Ethnobotanical studies generate inventories of biological species with utilitarian and nomenclatural information. Despite the relatively healthy state of ethnobotanical studies of ethnic groups in Mexico described by Toledo (1987) and Caballero (1987), Mexico still lacks fundamental national ethnobotanical inventories. Understanding the naming and significance of plants is the first step in understanding the cultural perception of a culture's vegetal world.

The historical development of registers of useful Mexican plants and their names began with the de la Cruz and Badianus (1964) manuscript on medicinal plants. Early colonial compilations by Spaniards include the "Florentine Codex" by Sahagún (Estrada, 1989) and the "Natural History of Plants of New Spain" (Hernández, 1959). Explorations during the later colonial period and early national era of Mexico yielded information that was scattered in various reports. Systematic compilation began during the late 1800s by both national institutions, e.g., Instituto Médico Nacional (Fernández, 1961) and foreign agencies, e.g., United States National Museum (Rose 1899; Standley, 1920–26). During the 1920's and 1930's Maximino Martínez assembled historical and contemporary information on names (Martínez, 1979), medicinal plants (Martínez, 1969) and useful plants (Martínez, 1959). Contemporary inventories based upon literature and herbarium specimens are being computerized at such institutions as the Universidad Nacional Autónoma de México (Caballero et al., 1985) and the former Instituto Nacional de Investigaciones sobre Recursos Bióticos (whose installations are now part of Instituto de Ecología) (Gómez-Pompa & Nevling, 1988). Even though these data bases are not accessible in published form, partial listings are available for medicinal plants of Veracruz (del Amo, 1979) and Yucatan (Mendieta & del Amo, 1981) and for the ethnoflora of Yucatan (Sosa et al., 1985).

Despite constant activity in the field, herbarium, laboratories, and libraries, Mexico has published listings for only selected useful plants. A national inventory of medicinal plants (Díaz, 1976) based upon 22 publications yielded 2,196 vascular plants. A recent analysis of other publications and theses (Bye et al., 1990) added 1,156 species, bringing the total Mexican medicinal flora to 3,352. Contemporary inventories of other plant use categories are not available at this time. Nonetheless, Caballero (1987) estimated that Mexico has about 5,000 useful angiosperms, roughly 23% of its flora. Human interaction (from destructive exploitation to intensive management) with one-fourth of the Mexican vascular plants has an impact on the natural biological diversity of the Mexican flora.

The recognition of plants is reflected in the nomenclature and the classification applied by people. Although not considered in detail here, the reader is cautioned to consider the variability of knowledge and competence within human populations. It is important to recognize two extreme conditions when using ethnoscientific data (Werner & Fenton, 1970): the theoretical intersection of common elements that are shared by all members of a given culture and the theoretical union of all elements representing the collective competence of a culture. One of the major contributions to ethnoscience is based upon the studies of the Tzeltal Maya of Chiapas, Mexico (Berlin et al., 1974). There is a strong positive correlation of the increase of simple lexemes for folk generic names as one advances along the cultural significance continuum of useless–significant–protected–cultivated plants. More interestingly, polytypic folk genera (i.e., consisting of primary lexemes that are further subdivided) increase directly with the greater cultural significance of the plants. In the case of the Tzeltal Maya, 73 generics (15% of the total folk genera of 471) include 237 specific taxa; the remaining 85% are monotypic (i.e., they are terminal primary lexemes with no other categories). As a result, plants with more significant utility have a greater tendency to be lexically diversified. Another manner of considering cultural recognition of biological diversity is to examine the correspondence between folk genera and biological species. As one may expect, overdifferentiation (when two or more folk generics match a single botanical species) occurs for plants with greater cultural significance. The Tzeltal Maya have two or more generic classes for each of eight biological species that are of edible or medicinal importance. A similar pattern of overdifferentiation is found with folk species. The cultural practices of naming and classifying plants demonstrate greater diversification when the plants are of greater significance. Each overdifferentiation is based upon genetic differences within a biological species. Folk taxonomy promises to be a powerful tool in the detection and evaluation of biological diversity.

The importance of cultural perception is also apparent at the plant community level. The primary and secondary tropical forests of southeastern San Luis Potosí and adjacent Veracruz appear to be "natural" to the "institutionally trained" botanist; that is, the forest lacks directional alteration by humans. However, Alcorn (1984) has shown that the Huastec Indians constantly manage the "te'lom" or "group of trees" in which certain plants are removed, others are protected, and certain ones are favored.

The comparison between the folk taxonomy of hunters and gatherers and that of small-scale agriculturists reveals another aspect of the relationship between recognition of plants and their exploitation by humans. Brown (1985) found that agrarian societies (including the Amuzgo, Huastec, and Tzeltal of Mexico) have a larger folk taxa inventory than that of the foragers. This phenomenon is attributed to the diversification of ecotypes resulting from agricultural activities and the support of a wider range of plants and animals. Because of the susceptibility to periodic crop failure and the need to support a concentrated sedentary population, the farmers require an expanded traditional knowledge in order to survive by exploiting "wild" resources during periods of cultivated food shortage.

The diversification of plants correlated with exploitation and modification of the local environment is exemplified by a study of prickly pear cacti (*Opuntia* spp.) in the El Bajío region of central Mexico (Colunga G-M. et al., 1986). The lands that are not suited to extensive agriculture are covered with thorn scrub and cacti vegetation. Subsistence agriculturalists have modified this marginal land to create "huamil" (or dry, rocky, honeycomb cultivated plots) surrounded by "nopalera" (managed stands of cacti). The farmers of the region obtain edible fruits ("tuna" and "joconostle") and stem pads ("nopal") from 15 species of *Opuntia* that are wild, tolerated in the field, encouraged in the "nopalera," and cultivated (possibly as domesticates) in the gardens. There is a positive correlation between the species with a number of named and morphologically distinct forms and those that are encouraged and cultivated (Bye et al., 1989). The wild plants are maintained in order to produce sexually genetic variants; the desirable forms are then massively multiplied by vegetative propagation. It is interesting to note that these overdifferentiated biological taxa produce various distinctive forms that are sold in the local markets.

HUMAN INTERACTION WITH PLANTS: CHANGES OVER TIME

The diversity of the plants associated with humans changes over time. The archaeological record of plant remains in coprolites suggests major modifications in the consumption pattern as the inhabitants of arid zones of Mexico shifted from hunting and gathering to agriculture. The graphs from the excavations of Tamaulipas and Tehuacán show the change of importance in various foods over 8,000 years (Callen, 1967, 1973). Domesticated species (e.g., maize, beans, squash, chile, cassava, and amaranth) generally increased in the diet, whereas others decreased. Of those plants that had reduced dietary importance, some plants (e.g., maguey, prickly pear, and black sapote) drastically diminished in their significance even though they are occasionally consumed today, are others (e.g., pochote, mesquite, and *Setaria* grain) have been totally abandoned in contemporary meals.

A comparison of useful plants in early post-Conquest documents with those employed today reflects similar degeneration of the plant diversity. Early Colonial Period chroniclers (de la Cruz & Badianus, 1964; Hernández; 1959, Sahagún, 1963) reported 150 types of "quelites," or edible herbs. Today only 15 species are commonly consumed in central Mexico. Of the more than 3,000 plants recorded by Francisco Hernández from central Mexico 50 years after the Conquest, only half have been identified (Valdés & Flores, 1985). The remainder defy classification because the common names are no longer used and the descriptions are inadequate. Such lack of cultural recognition implies loss of ethnobotanical knowledge and may signal the extinction of certain species, at least at a local level.

HUMAN ACTIVITIES: NEGATIVE IMPACT

Human activities often destroy the natural vegetation and may selectively eliminate species. These plant–human interactions are also the consequences of perception—or in extreme cases the lack of perception of the consequences. Many times these perceptions form part of the culture that has coevolved with the changing environment and its biological components. It has been suggested that the Judeo-Christian tradition is one of the factors contributing to the Western world's current environmental crisis (Moncrief, 1970) and may be contradictory to indigenous cosmovision (Hernández X., this volume). Ethnobotanical studies can identify and explain indigenous traditions and perceptions that may provide alternatives to conservation of natural and cultural resources. Only erosion, urbanization, grazing, and agriculture are considered briefly here.

Erosion caused by changing land use patterns (e.g., deforestation, agriculture, overgrazing, overcollection, urbanization) is one of the most visible repercussions of human activities. According to a survey by the Mexican Agricultural Department (Secretaría de Agricultura y Recursos Hidráulicos, or SARH) (Anonymous, 1983), only 19% of Mexico was uneroded. Light erosion was reported for 23% of the territory, and the greatest area (32% or 63,168,834 hectare [ha]) was moderately eroded. Severe and total erosion was recorded for 18% and 8% of the national territory, respectively. When using government data, the reader should be aware of their incongruence. For example, the National Atlas of Physical Environment (Instituto Nacional de Estadística, Geografía e Informática, INEGI, 1980) reported about 41% of the vegetation on the national level had been altered (rather than 81%).

Human population growth increases the demand for land through urbanization. The Mexico City metropolitan area (27,091 ha) in the southern Valley of Mexico has increased by 431% (679 ha to 3,607 ha, or 2.5% to 13.3%) between 1959 and 1977 (Benítez et al., 1987). The urbanization process has destroyed 992 ha (29%) of oak forest, 800 ha (29.2%) of dry "palo loco" scrub, and 775 ha (13.8%) of agricultural land.

Mexican history is deeply ingrained in agricultural reform. The technological and socioeconomic changes of Mexican agriculture have altered the status of plant diversity as well. The increase of agrarian pressure can be indexed by various data. The allotment of land to the farmers has been a continuing interest since the end of the Mexican Revolution (INEGI & Instituto Nacional de Antropología e Historia [INAH], 1986]). Of the almost 103 million ha released over the 85-year period, about 24% of the land was distributed during the period 1965–70. Pasture land included 55,219,499 ha (53%), dry land agriculture 12,471,657 ha (12%), and forested areas 11,203,301 ha (11%).

Major livestock has increased by 4.7 times since the early twentieth century (INEGE & INAH, 1986). Using the "cow unit coefficient" of 2.41 ha to support one head of cattle, Mexico's pasture land use increased from 24,606,548 ha in 1902 to 116,004,496 ha in 1981. This calculation of required area for livestock represents 59% of the national territory. Because much less of the country is grazed, it appears that the domestic grazing animals are more concentrated on the land (in addition to managed feed lots) and hence represent a tremendous force in the alteration of the natural vegetation.

A few regional studies document the history and impact of livestock. In the Río Bravo (or Río Grande) drainage of New Mexico (which was once part of Mexico), the number of domestic animals increased with Mexican expansion but grew exponentially after the Mexican-American War and then declined at the turn of the twentieth century. The vegetation changes associated with this grazing pattern included the disappearance of cool season grasses and other plants upon which many Indians subsisted (Bohrer, 1975, 1978) and the invasion of mesquite as a woody weed (York & Dick-Peddie, 1969). During the twentieth century the most dramatic livestock increase occurred in Mexican cattle, which has had a serious impact on tropical forests (Toledo, 1991). Minor livestock such as sheep and goats has not increased as significantly. Because they tend to be found in drier or higher zones, their effect in recent times (and possibly in the past) probably has not been as dramatic as that of cattle.

The cultivation of the national territory is critical for feeding the Mexican population. As of 1983 there were 20,808,462 ha tilled, which represents an increase of 3.5 times the area cultivated in 1940 and accounts for 10.5% of the national area. Dry land agriculture increased by a factor of 2.75, whereas irrigated fields increased by 7.56. Irrigated land accounted for 23% of the agricultural land in 1983, the remainder being dry land. Cultivation practices, pesticides, fertilizer, and changing soil factors (pH, salinity, water saturation, and percolation) cause changes in the local flora and in the anthropogenic and natural communities.

HUMAN ACTIVITIES: EVOLUTIONARY INTERACTION

Human interactions with plants have tended to span an evolutionary continuum and involve special activities in which human behavior (conscious or not of expected results) affects the survival and reproduction of the plant population and modifies its natural genetic composition and ecological behavior. With emphasis on the plant, three evolutionary phases (wild, weed, and domesticate) are generally recognized even though they do not form mutually exclusive categories. The path of a species evolving along this continuum is not necessarily unidirectional but may be reversible under certain circumstances. Wild plant survive and reproduce naturally without the necessity of human intervention. Weeds, on the other hand, survive in habitats perturbed periodically by human activities, although their reproduction cycle is also successfully completed without humans. They have evolved in man-made habitats from natural colonizers, hybrid derivatives of wild and domesticated species, and feral plants originating from abandoned domesticates that reestablished seed dispersal (de Wet & Harlan, 1975). Subdivisions of weeds can be distinguished by their habitat preference (e.g, agrestals in cultivated fields, ruderals along vias of travel and near dwellings). Finally, domesticated plants' survival and reproduction cycles depend genetically and ecologically upon direct human intervention. Distinctive adaptation syndromes are products of various avenues that domestication processes have taken (Harlan, 1975; Hawkes, 1983; Schwanitz, 1967). This continuum is based upon a coevolutionary process in which plants undergo morphological and autecological changes (which eventually may be genetically maintained) and humans modify their behavior. Because of this process, the plants are undergoing unnaturally rapid change and most likely are not fixed end-products. Hence this continuum reflects a dynamic component of biological diversity.

Human activities exert variable selection pressures that propel plants along this continuum; they fall into three major categories that are not restricted to any particular phase of the continuum. *Gathering* products from plants that grow naturally or are concentrated in response to habitat modification may favor or diminish the population and possibly alter its genetic composition. *Incipient management* intensifies human actions whereby certain plants may be tolerated (e.g., selected plants are allowed to survive and reproduce rather than be eliminated), encouraged (e.g., people's efforts promote increased dispersal and distribution of sexual or vegetative propagules within the site or to new site), and protected (e.g., plants have an advantage gained by removal of competitors, special supports, exclusion of predators, and other special care). The third interaction is *cultivation* where people's special actions modify environmental conditions to promote optimal production and reproduction. As the intensification increases, human behavior becomes increasingly oriented toward modifying the population dynamics of the plant as well as its local environment and less directed toward opportune benefit from a vegetal product. These general actions are similar, but not exactly equivalent, to Rindos' (1984) evolutionary stages of domestication (incidental domestication, specialized domestication, and agricultural domestication) where there is a shift of importance from the human behavior that directly affects the evolution of the plant to the agroecology that mediates the direction and force of selective processes.

Wild Plants

The contribution of wild plants to human economy is important but difficult to quantify (Prescott-Allen & Prescott-Allen, 1986). In Mexico, about 50% of the plants are used as food and medicine. The Nahua and Mixtec Indians of the tropical deciduous forest of Puebla and Guerrero obtain over 50% (about 42 species) of their nondomesticated food species from wild plants (Vázquez, 1986; Viveros & Casas, 1985). The Nahua populations of the pine-oak forests of eastern state of México gather 25% (28 species) of medicinal herbs from the wild plants (Bye & Linares, 1983). Material for construction, handicrafts, and fuel are derived almost exclusively from wild plants extracted from primary and, now more frequently, secondary vegetation.

Gatherers recognize and select plants with desirable characteristics from the larger pool of natural variation. Medicinal herb collectors know different forms of sweet marigold (*Tagetes lucida*) based upon distinctive flower morphology, color, and odor. Traditional healers of the Sierra Norte de Puebla distinguish the chemical races of jimsonweed (*Datura stramonium*), when they collect material for making their psychotropic preparations so as to not select a plant that will cause a fatal reaction (Knab, 1982). Inhabitants of the dry tropical forest of coastal Jalisco prefer to collect the branching "finger-form" tubercles of the edible native yam (*Dioscorea nelsonii*) rather than the unbranched "candle-form" of the same species (Cervantes & Bye, in prep.).

Human subsistence activities may promote the abundance of wild plants; and in some cases the cessation of such practices triggers their disappearance. Today, sandfood (*Pholisma sonorae*), a root parasite on desert shrubs, is considered an endangered species, whereas a century ago it was an abundant wild food of the Sand Papago Indians of northern Sonora. The excavation to extract the buried edible stems promoted its propagation by permitting the seeds to fall into holes and come into contact with host roots (Nabhan, 1985). Normally, the seeds would be released on the sandy surface and would not come into direct contact with the subterranean roots. Hence the Indians had inadvertently dispersed the seeds to favorable germination sites while digging and promoted an increase in the plants. The collection of wild onions (*Allium* spp.) by the Tarahumara Indians of Chihuahua increases local populations because of the loosening of the normally compact soil, the reduction of competitive perennials, and dispersal of the daughter bulbs upon cleaning the central bulb (Bye, 1985).

The overcollection of wild plants, especially for commercial exploitation, can bring local populations to near extinction. In the pine forests of central Mexico wild hyssop (*Agastache mexicana*) and valerian (*Valeriana ceratophylla*) are rare owing to the demand of these medicinal herbs (Bye et al., 1987a; García, 1981). A similar situation occurs in the Sierra Tarahumara where medicinal roots of lovage (*Ligusticum porteri*) and "matarique" (*Psacalium decompositum*) have been overcollected in order to satisfy national and international markets (Bye et al., 1987b). In the tropical deciduous forest, local populations of cancer bark (*Hippocratea excelsa*) have disappeared (Barreto et al., in prep.).

Once one plant species becomes rare or expensive because of overcollecting, other species then become vulnerable. A medicinal plant complex consists of different species and genera that share vernacular names, curative uses, and biological properties (Linares & Bye, 1987). The dominant species is preferred because of its efficacy and is marketed beyond its natural phytogeographic range, whereas the other species are sold only within their natural extension. In the case of the "matarique" complex, *Psacalium decompositum* is the dominant species. Recently, the short supply due to declining natural populations in Chihuahua prompted an increase in demand for *P. peltatum* from México and Morelos, whose local populations, in turn, have begun to decline. The identification of these useful plant complexes may be a practical tool for predicting which species will be threatened with overexploitation in the future.

World attention was attracted to the problems of wild collected plants with the discovery of a steroid precursor in *Dioscorea*, the wild yam of the humid tropics of southeastern Mexico. The rapid exploitation prompted formation of the Mexican *Dioscorea* Commission to supervise the collection, initiate cultivation, and control extraction and transformation of diosgenin (Applezweig, 1977). Even though Mexico initially supplied the international market with an inexpensive source (compared to animal and synthetic sources of the period), the wild tubers declined in importance as these alternative sources became less expensive and more reliable in the world market (Applezweig, 1977). This situation may have saved many wild yam populations from extinction but it did not serve as a model for rational exploitation of wild resources. Similar problems exist with wild plants that produce raw materials used for industrial purposes, e.g., candelilla (*Euphorbia antisyphilitica*), guayule (*Parthenium argentum*), chicozopote (*Manilkara zapota*), and jojoba (*Simmondsia chinensis*).

Various forms of incipient management are applied to wild plants. Spontaneous shrubs of bird-pepper (*Capsicum annuum*) in northern Mexico are carefully guarded to ensure a good harvest of piquant red fruits. Humans unconsciously encourage the wild chile by spreading the seeds (made suitable for germination by chemical treatment in the digestive tract) in fecal matter, a packaged dispersal unit with fertilizer (Nabhan, 1985). It is interesting to note that this chile bush grows naturally in association with a protective sheltering tree (Nabhan, 1987); such nurse plant strategy is taken advantage of by people who protect the chile in natural communities. Useful plants of the "te'lom" or Huastec tropical forest of San Luis Potosí are favored by the removal of nearby unwanted vegetation (Alcorn, 1984). Transplanting wild plants into special habitats is one form of encouraging plants. "Bakánawi," a ritual psychotropic plant (*Scirpus* sp.) of Chihuahua, is moved from streamsides and set in irrigation ditches by the mestizos, who exchange it with the Tarahumara Indians (Bye, 1979a). In central Mexico, rooted plants, stem cuttings, or bulbs of such ornamental and medicinal plants as baby salvia (*Salvia microphylla*), shrubby stonecrop (*Sedum oxypetalum*), Mexican elderberry (*Sambucus mexicana*), and tiger flower (*Tigridia pavoni*) are planted around houses, fields, and cemeteries. Natural variation in Mexican trees that produce edible fruit or seeds such as cherry (*Prunus serotina* subsp. *capuli*), crabapple (*Crataegus mexicana*), "nanche" (*Byrsonima crassifolia*), and "guaje" (*Leucaena* spp.) have been selectively promoted through transplanting, discriminant elimination and grafting. As a result, several semidomesticated forms are found throughout Mexico.

Weeds

Weeds occur in all forms and defy standard definition (Harlan & de Wet, 1965). The weed flora of Mexico contains a high percentage of native species. Human tolerance of certain weeds with desirable characteristics has diversified the forms with species.

The weed flora in Mexico contains mostly native species even though it is not homogeneous. Agrestals are usually indigenous species that have evolved in traditional agroecosystems, whereas exotic species are found where mechanized farming is practiced (Rzedowski, 1978). In urban areas such as Mexico City, only 32% of the 564 weeds of streets and waste areas are introduced plants (Rapoport et al., 1983). Even though a high percent of the Mexican weeds is native, the tremendous ecological variation does not permit a common weed flora of Mexico (Rzedowski, 1978). Fewer than 4% of the species from sites in San Luis Potosí and the Valley of Mexico are shared with those from sites in Yucatán.

Weeds of major agronomic importance in Mexico (Cárdenas & Coulston, 1967) include 93 species, 56% of which are native. The most important families for all species as well as for native species (based upon order of importance using the Simpson diversity index) are Asteraceae, Poaceae, Amaranthaceae, and Fabaceae.

The special nature of native weeds is reflected in the fact that Mexico has not been an important contributor of major weeds to the world. A list of major plant pests of the world (Holm, 1969) does not contain any Mexican species. Nonetheless, such native species as *Galinsoga parviflora, Solanum rostratum,* and *Datura stramonium* have cosmopolitan distribution.

In addition to creating the ecological habitat that favors the evolution of weeds, humans also exploit them for medicine and food. About 45% of the medicinal plants gathered from the eastern part of the state of México and sold in the Mexico City market are collected weeds (Bye & Linares, 1983). In the tropical deciduous forest of Puebla and Guerrero, about 29% of the nondomesticated edible species are weeds (Vázquez, 1986; Viveros & Casas, 1985). Farmers readily recognize different stabilized forms of weeds in which one is favored over the other. This infraspecific diversification of useful weeds is found in such plants as erect sand-tomato with sweet fruits (versus prostrate herbs with tasteless berries; *Jaltomata procumbens*; Davis & Bye, 1982), nightshade with edible leaves (versus nonedible foliage; *Solanum nigrescens*), small versus large flowered edible "alache" (*Anoda cristata*),green pokeweed with edible leaves (versus red poke with soapy tasting foliage; *Phytolacca icosandra*) and nonrank palatable goosefoot (versus fetid, nonedible form; *Chenopodium berlandieri*).

Incipient interaction between people and weeds is common in Mexico. Certain edible weeds are encouraged through seeding in tilled fields and may be cultivated as in the cases of introduced mustard (*Brassica campestris*) and native peppergrass (*Lepidium virginicum*) in Chihuahua (Bye, 1979b). In various parts of Mexico, traditional farmers tolerate certain solanaceous plants that produce edible fruits (Davis & Bye, 1982; Williams, 1985). During the weeding process, they allow the spontaneous plants of sand-tomato (*Jaltomata procumbens*), husk-tomato (*Physalis chenopodifolia*) and coyote tomato (*Solanum mozinianum*) to remain with the crop. During and after the main harvest, the fruits are collected for consumption and sale in local markets.

Weeds are known to promote genetic diversification in domesticated relatives. Hybridization of weedy *Solanum* X *edinense* with the domesticated *S. tuberosum* in the central Mexico's Valley of Toluca is the cause of variability in the land races of cultivated potatoes (Ugent, 1968).

Domesticated Plants

Domesticated plants have been genetically altered to depend upon direct, conscious human intervention for their survival; and, in turn, human behavior has changed to accommodate the plant (Rindos, 1984). About 65 important world crops were domesticated in Mexico (Harlan, 1975). Mexico has been one of the three major world centers for agricultural origin along with the Middle East and China (Harlan, 1975). Hernández X. (this volume) documents the importance of Mexico as a center of domesticated plants. Of importance to Mexico and the world is the presence of land races and wild relatives upon which future plant breeding depends.

The biological diversity of domesticated plants is subject to four main processes (Wilkes, 1989a). The process of hybridization and genetic recombination increases diversity within species. The other three—genetic erosion, genetic vulnerability, and genetic wipeout—are linked with short term demands of growing human population and its rising expectations; they decrease biological diversity of crop plants and their wild relatives.

Cultivated Plants

A special note must be made about cultivated plants. All domesticated plants are cultivated but not all cultivated plants are domesticated. In many case, cultivated plants have not undergone drastic genetic modification but are propagated, planted and tended by people and may be undergoing evolutionary change. Many wild plants are brought into cultivation owing to their utility or attractiveness.

After the Spanish Conquest, Mexico's plants were highly popular throughout the world with visually attractive flowers such as those of *Dahlia, Cosmos, Zinnia,*and *Tagetes.* Based upon an analysis of the major compilation of horticultural plants, Hortus III (Bailey & Bailey, 1976), about 9% of cultivated species in temperate regions are native to Mexico. The most important families based upon the number of species are Cactaceae and Orchidaceae. Many of these plants in international commerce are derived directly from wild populations.

Mexico is well known for its tradition in the cultivation of plants. However, the role of introduced plants has become more important than that of the native plants. In an urban setting, most of the private garden plants are exotic or non-native in origin. In Mexico City 63.6% of the 750 garden plants are foreign in origin (Díaz-B. et al., 1987). On a national level, the SARH maintains 82 tree nurseries that supply material for national, state and local plantings. The propagation material is obtained each year by the nursery staff from natural and cultivated populations. Of the 115 species grown for reforestation, 60% are Mexican plants. Only 36% of the fruit trees (44 species) and ornamental plants (56 species) are native (Anonymous, 1985). Of the 55 families grown in federal Mexican nurseries, the most important family in terms of numerical production is the Pinaceae, with the pines (*Pinus*) dominating. By applying a diversity index to the list, the important families occur in this order: Fabaceae (61% native), Bignoniaceae (57% native), Pinaceae (75% native), Agavaceae, Boraginaceae, Chenopodiaceae, and Cupressaceae. Unfortunately the survival rate of transplants is only 20–60%, and specific data for areas reforested with these cultivated plants are not available. However, if we calculate that 20% of the 35 million saplings produced in 1987 are planted at one per square meter and survive, an additional 35,000 ha would have been reforested by 1990. In the southern Valley of Mexico, the area planted with eucalyptus (*Eucalyptus* spp.) from 1959 to 1977 increased by 467% to cover 136 ha (0.5% of the total area of the study region). These Australian trees are known to release substances that inhibit the growth of other plants. Therefore the specific composition of these new forests will probably be drastically altered in the future.

HUMAN ACTIVITIES AND HYBRIDIZATION

Hybridization combines two substantially different genetic systems resulting in (1) increased genetic variability in the intermediate offspring, and (2) gene products detected in the hybrids that are in neither parental population. Human activities not only transgress reproductive barriers of different plant species but also create intermediate habitats suitable for the survival of the hybrid offspring (Anderson, 1948). Often agricultural and deforestation activities weaken ecological, geographical, and temporal barriers between species and permit uni- or bidirectional gene flow. This situation is enhanced if the plants are outcrossers rather than inbreeders. On a geographic scale, hybrid suture zones are composed of hybrid zones of various pairs of interbreeding taxa, both plants and animals, and are caused by recent sympatry (Remmington, 1968). Six major suture-zones identified in the United States represent shifts in the flora and fauna as a result of climatic shifts as well as by the destruction of forested regions with the human settlement. Remmington (1968) indicated that suture-zones are to be expected in Mexico, but detailed documentation has not been conducted to date.

The few reported cases of hybridization in Mexican spontaneous plants are associated with human disturbances. The opening of forested areas is associated with hybrids between *Penstemon campanulatus* and *P. roseus* (Straw, 1963), *Iostephane heterophylla* and *I. madrensis* (Strother, 1983), *Odontotrichum multilobum* and *O. palmeri* (Pippen, 1968), and *Agastache aurantiaca* and *A. micrantha* (Sanders, 1987), to mention a few examples. In some cases, direct human intervention is required to maintain hybrid products. The medicinal "toronjil blanco" (*Agastache mexicana* subsp. *xolocotziana*) of central Mexico is probably a hybrid introgression between *A. mexicana* and *A. palmeri* and is vegetatively propagated in home gardens (Bye et al., 1987a). *Sedum amecamecanum*, a hybrid between *S. praealtum* and *Valladia batesii*, is cultivated in the mountains of the eastern Valley of Mexico (Clausen, 1975).

Hybridization between domesticated plants and weedy or wild relatives occurs in Mexico. The classic example has been the flow of maize (*Zea mays*) genes into weedy teosinte (*Zea mexicana*) in the Central Plateau, southern Chihuahua, and eastern Valley of Mexico (Wilkes, 1989b). Gene flow between these two wind-pollinated annuals is usually unidirectional (from maize to teosinte) and limited due to spatial and seasonal isolation of the two species, the lack of fitness of the hybrid (usually restricted to or near cultivated fields), and the different effects of human selection on the parental populations. Only in the Nobogame Valley of southwestern Chihuahua where the flowering periods of maize and teosinte are not seasonally separated is there bidirectional gene exchange (Wilkes, 1970). As a result there is evidence of teosinte introgression into cultivated maize. The diversity of contemporary maize crops and spontaneous teosinte populations in isolated areas of Mexico may reflect recent F_1 products as well as introgressions resulting from earlier sequential back-crosses. In central Mexico where cultivated "huauzontle" (*Chenopodium berlandieri* subsp. *nuttalliae*) and weedy goosefoot (*C. berlandieri* subsp. *berlandieri*) are sympatric, hybrids are usually found in fields (Wilson, 1990). At least five disturbed habitats in Oaxaca and Veracruz sustain free-growing hybrid populations of chayote (*Sechium edule* & *S. compositum*), which appear to be consequences of hybridization-differentiation cycles (Newstrom, 1989). Hybrid products of cultivated squash (*Cucurbita mixta*, now known as *C. argyrosperma*) and wild bitter gourd (*C. sororia*) along field margins have been documented in Sonora (Merrick & Nabhan, 1984) and observed in Guerrero. Hybrid trees of the edible guava (*Psidium guajava*) and the variable wild species (*P. guineense*) are known from the states of Chiapas, Colima, Guerrero, Nayarit, Oaxaca, Quintana Roo, Sinaloa, and Veracruz (L. Landrum, pers. comm.).

OTHER RESPONSES OF PLANTS TO HUMANS

Plant populations respond to human activities in a variety of ways. Populations of useful plants may survive over several generations in dwelling sites once occupied by people. Creation of new habitats promotes the establishment of new elements that migrate into the local flora. The formation of secondary vegetation after people remove the original green mantle can accelerate plant speciation.

Even after human activities have stopped, the altered habitat favors the persistence of populations derived from culturally important plants. At the archaeological site (abandoned in 1,300 AD) of Kohunlich, Quintana Roo, there is an unnatural concentration of the multipurpose "cohune" palm (*Orbignya cohune*). Maya ruins in Yucatan also present a conspicuous association of two useful trees (*Brosimum alicastrum* and *Protium copal*) which contrasts with surrounding vegetation (Lambert & Aranson, 1982). In the southwestern United States or northern "Mega-Mexico 3," Yarnell (1965) documented the 21 species whose contemporary distributions are restricted to archaeological sites abandoned over 800 years ago.

Changes in the composition of a local flora is another response to human alteration of the vegetation. Expansion of the agricultural activities creates new habitats for the establishment of plants that migrate into the area.

Rzedowski (1986) explained the presence of 12 calciphytes in the Valley of Mexico by their establishment on soil accumulated from erosion associated with the introduction of agriculture about 4,000 years ago. The plants were probably derived from populations growing on the dominant limestone soil to the north in the state of Hidalgo and entered through the Huehuetoca region.

The variability and diversification of tropical genera such as *Acalypha*, *Croton*, *Euphorbia*, *Miconia*, *Paspalum*, *Piper*, and *Psychotria* may be the consequences of human activities in the recent past (Gómez-Pompa, 1971). Tropical secondary vegetation created by humans provides greater niche diversity. The secondary vegetation behaves as a selective factor as well as an indirect ecological barrier and promotes speciation.

CONCLUSIONS

Mexico is a country favored with biological and cultural diversity. As a consequence of this heritage, it is one of the major world centers of agriculture—today's zenith of biological, cultural, and environmental modification. The impact made by humans on plants in a coevolutionary context is remarkable. Destructive actions also illustrate the negative effects people create in the plant world.

Culturally and botanically, Mexico is one of the richest in the New World. Despite the initial decline of the indigenous human population after the Conquest, it ranks first in the total number of native speakers, divided among 54 languages. The two major floristic kingdoms of the New World intersect in Mexico and generate ten basic vegetation types and a flora of about 30,000 species, the second largest in the Western Hemisphere.

The concordance of biological and cultural richness manifests a vibrant ethnobotanical diversity. However, the explanation for intricate synergism between plants and humans is not due to a simple overlay of richness centers. Further investigation is required to understand the causes of ethnobotanical diversity and its relationship to biological and cultural survival.

As in ecological studies, the richness component of ethnobotanical diversity is the most obvious. Even though about half of the ethnic groups have published ethnobotanical studies, Mexico still lacks a comprehensive inventory of culturally important plants. The potential for enlarging the register of useful plants is exemplified by a 34% increase in the number of medicinal plants recorded since the last major compilation during the mid-1970s. Folk classification is sensitive to plants that have major economic value or have been genetically modified. The recognition of important plant variation by indigenous peoples may be a practical guide to evaluating biological diversity.

Plants respond to humans through direct selection and environmental modification. Humans also change, especially in their behavior, when plants are domesticated. Alteration of plant populations as well as their habitats has produced marked changes in local flora, vegetation composition, and genetic structure of plants. In many cases, people have taken advantage of these variations and the human population has grown as a consequence. The new pressures, especially during the last 70 years, have dramatically modified Mexico's botanical diversity through agriculture and erosion. Specific actions on wild, weedy, and domesticated plants vary in their repercussions—sometimes favoring the plant, sometimes forcing it to extinction.

Human activities transgress reproductive and ecological barriers of plants by altering the natural spatial and temporal patterns. This influence is exhibited in the various types of hybrid products, some of which are beneficial to human economy. The persistence of useful plants long after the occupants have perished or the migration of plants with specific edaphic requirements into habitats created by people thousands of years ago attest to the long-term human impact upon our vegetal environment.

ACKNOWLEDGMENTS

I thank many colleagues and students who provided useful information and comments, especially: J. Alcorn, J. Arellano, B. Benz, V.L. Bohrer, C.H. Brown, P. Bretting, J. Caballero, A. Casas, L. Cervantes, P. Colunga, L. Corkidi Abud, T. Davis IV, E. Estrada, H.S. Gentry, E. Hernández Xolocotzi, E. Hunn, L. Landrum, W.J. Litzinger, C. Mapes, M.A. Martínez, G. Nabhan, L.E. Newstrom, J. Rzedowski, R. Ojeda Trejo, D. Rindos, N.J. Turner, M.C. Vázquez, J.L. Viveros, G. Wilkes, D.E. Williams, H. Wilson, R. Yarnell, and D. Zizumbo. Constructive criticism and encouragement were provided by T.P. Ramamoorthy, B. Simpson, M. Elliott, and E. Linares.

REFERENCES

Alcorn, J.B. 1984. Development policy, forests, and peasant farms: reflections on Huastec-managed forests' contributions to commercial production and resource conservation. Econ. Bot. 38:389–406.

Anderson, E. 1948. Hybridization of the habitat. Evolution 2:1–9.

Anonymous. 1983. Inventario de la areas erosionadas en el país. Manuscript, Dirección General de Conservación del Suelo y Agua, Secretaría de Agricultura y Recursos Hidráulicos, México, DF.

Anonymous. 1985. Relación de especies forestales que se propagan en los viveros de la SARH; 1985. Manuscript, Dirección General de Normatividad Forestal, Secretaría de Agricultura y Recursos Hidráulicos, México, DF.

Applezweig, N. 1977. *Dioscorea*—the pill crop. *In* D.S. Seigler (ed.), Crop Resources. New York: Academic Press. pp. 149–163.

Bailey, L.H. & E.Z. Bailey. 1976. Hortus Third. New York: Macmillan.

Baker, H.G. 1972. Human influences on plant evolution. Econ. Bot. 26:32–43.

Barreto, V.M., E. Linares & R. Bye. In prep. Comercialización de la cancerina (*Hippocratea excelsa* HBK.) en México. *In* L. Cervantes S. & R. Bye (eds.), Memorias de la Reunión Etnobotánica Ecológica Regional. Mexico City: UNAM, Inst. Biol.

Benítez B., G., A. Chacalo & I. Barois. 1987. Evaluación comparativa de la pérdida de la cubierta vegetal y cambios en el uso del suelo en el sur de la ciudad de México. *In* E.H. Rapoport & I.R. López-Moreno (eds.), Aportes a la Ecología Urbana de la Ciudad de México. Mexico City: Limusa. pp. 193–223.

Berlin, B., D.E. Breedlove & P.H. Raven. 1974. Principles of Tzeltal Plant Classification: An Introduction to the Botanical Ethnography of a Mayan-Speaking People of Highland Chiapas. New York: Academic Press.

Bohrer, V.L. 1975. The prehistoric and historic role of the cool-season grasses in the Southwest. Econ. Bot. 29:199–207.

Bohrer, V.L. 1978. Plants that have become locally extinct in the Southwest. New Mexico J. Sc. 18:10–19.

Brown, C.H. 1985. Mode of subsistence and folk biological taxonomy. Curr. Anthropol. 26:43–64.

Bye, R. 1979a. Hallucinogenic plants of the Tarahumara. J. Ethnopharmacol. 1(1):23–48.

—————. 1979b. Incipient domestication of mustards in northwest Mexico. Kiva 44:237–256.

—————. 1985. Botanical perspectives of ethnobotany of the Greater Southwest. Econ. Bot. 39:375–385.

————— & E. Linares. 1983. The role of plants found in the Mexican markets and their importance in ethnobotanical studies. J. Ethnobiol. 3:1–13.

—————, E. Linares, T.P. Ramamoorthy, F. García, O. Collera, G. Palomino and V. Corona. 1987a. *Agastache mexicana* subsp. *xolocotziana* (Lamiaceae), a new taxon from the Mexican medicinal plants. Phytologia 62(3):157–163.

—————, N. Meraz C. & C.C. Hernández Z. 1987b. Conservation and developmentof food and medicinal plants in the Sierra Tarahumara, Chihuahua, Mexico. *In* E.F. Aldon, C.E. Gonzales Vicente & W.H. Moir (eds.), Strategies for Classification and Management of Native Vegetation for Food Production in Arid Zones / Estrategias de Clasificación y Manejo de Vegetación Silvestre para la Producción de Alimentos en Zonas Aridas. General Technical Report RM-150. Fort Collins, CO: U.S. Department of Agriculture, Forest Service, Rocky Mountain Forest and Range Experiment Station. pp. 66–70, 246.

—————, E. Linares & C. Bonfil. 1989. Ethnobotany and markets. *In* W.E. Doolittle (ed.), Field Trip Guide (El Bajío), 1989 Conference of Latin American Geographers. Austin, TX: Department of Geography, Univ. of Texas, pp. 59–87.

—————, E. Estrada L. & E. Linares M. 1990. Recursos genéticos en plantas medicinales de México. *In* R. Ortega P., G. Palomino H., R. Castillo G., V.A. Gonzalez H. & M. Livera M. (eds.), Avances en el Estudio de los Recursos Fitogenéticos de México. Chapingo, México: Sociedad Mexicana de Fitogenética.

Caballero, J. 1987. Etnobotánica y desarrollo: la busqueda de nuevos recursos vegetales. *In* Memorias. IV Congreso Latinoamericano de Botánica. Simposio de Etnobotánica. Bogotá, Colombia: Instituto Colombiano para el Fomento de la Educación Superior. pp. 79–96.

—————, A. Rubluo, R. Bye, G. Palomino, M. Peña, H. Quero & L. Scheinvar. 1985. La Unidad de Investigación sobre Recursos Genéticos del Jardín Botánico del Instituto de Biología de la U.N.A.M. *In* Memoria de la 1a Reunión Nacional de Jardínes Botanicos. Mexico City: SEDUE/Asociación Mexicana de Jardínes Botánicos. pp. 44–53.

Callen, E.O. 1965. Food habits of some pre-Columbian Mexican Indians. Econ. Bot. 19:335–343.

———. 1967. Analysis of the Tehuacan coprolites. *In* D.S. Byers (ed.), The Prehistory of the Tehuacan Valley. 1(Environment and Subsistence):261–289. Austin, TX: Univ. of Texas Press.

———. 1973. Dietary patterns in Mexico between 6500 B.C. and 1580 A.D. *In* C.E. Smith, Jr. (ed.), Man and His Foods: Studies in the Ethnobotany of Nutrition— Contemporary, Primitive, and Prehistoric Non-European Diets. University, AL: Univ. of Alabama Press. pp. 29–49.

Cárdenas, J. & L. Coulston. 1967. Weeds of Mexico. Mimeo 67–3. Corvallis, OR: Oregon State Univ./U.S. Agency for International Development.

Cervantes S., L. & R. Bye. In prep. Plantas útiles de lal costa central de Jalisco. *In* L. Cervantes S. & R. Bye (eds.), Memorias de la Reunión Etnobotánica Ecológica Regional. Mexico City: UNAM., Inst. Biol.

Clausen, R.T. 1975. *Sedum* of North America North of the Mexican Plateau. Ithaca, NY: Cornell Univ. Press.

Colunga G-M., P., E. Hernández X. & A. Castillo M. 1986. Variación morfológica, manejo agricola tradicional y grado de domesticación de *Opuntia* spp. en El Bajío Guanajuatense. Agrociencia 65:7–49.

Davis, T. & R. Bye. 1982. Ethnobotany and progressive domestication of *Jaltomata* (Solanaceae) in Mexico and Central America. Econ. Bot. 36(2):225–241.

De la Cruz, M. & J. Badianus. 1964. Libellus de Medicinalibus Indorum Herbis. Manuscrito Azteca de 1552. Mexico City: Instituto Mexicano del Seguro Social.

De Wet, J.M.J. & J.R. Harlan. 1975. Weeds and domesticates: evolution in the man-made habitat. Econ. Bot. 29:99–107.

Del Amo, S. 1979. Plantas Medicinales del Estado de Veracruz. Xalapa, Veracruz: INIREB.

Díaz, J.L. 1976. Indice y Sinonimia de las Plantas Medicinales de México. Mexico City: Instituto Mexicano para el Estudio de las Plantas Medicinales.

Díaz-Betancourt, M., I. López-Moreno & E.H. Rapoport. 1987. Vegetación y ambiente urbano en la ciudad de México. Las plantas de los jardines privados. *In* E.H. Rapoport & I.R. López-Moreno (eds.), Aportes a la Ecología Urbana de la Ciudad de México. Mexico City: Limusa. pp. 13–72.

Durham, W.H. 1978. Towards a coevolutionary theory of human biology and behavior. *In* A.L. Caplan (ed.), The Sociobiology Debate. New York: Harper & Row. pp. 428–448.

Estrada L., E.I.J. 1989. El Códice Florentino: Su Información Etnobotánica. Montecillo, México: Colegio de Postgraduados.

Fernández del Castillo, F. 1961. Historia Bibliográfica del Instituto Médico Nacional de México (1888–1915) Antecesor del Instituto de Biología de la UNAM. Mexico City: Imprenta Universitaria.

García M., C. 1981. Naturaleza y Sociedad en Chalco-Amecameca. Toluca, México: Biblioteca Enciclopedica del Estado de México.

Gentry, H.S. 1982. Agaves of Continental North America. Tucson, AZ: Univ. of Arizona Press.

Gómez-Pompa, A. 1971. Posible papel de la vegetación secundaria en la evolución de la flora tropical. Biotropica 3:125–135.

——— & L.I. Nevling, Jr. 1988. Some reflections on floristic databases.Taxon 37:764–775.

Gonçalves de Lima, O. 1956. El Maguey y el Pulque en los Códices Mexicanos. Mexico City: Fondo de Cultura Económica.

Grant, V. 1985. The Evolutionary Process—A Critical Review of Evolutionary Theory. New York, NY: Columbia Univ. Press.

Harlan, J.R. 1975. Crops and Man. Madison, WS: American Society of Agronomy.
——— & J.M.J. de Wet. 1965. Some thoughts about weeds. Econ. Bot. 19:16–24.

Hawkes, J.G. 1983. The Diversity of Crop Plants. Cambridge, MA: Harvard Univ. Press.

Hernández, F. 1959. Historia Natural de las Plantas de Nueva España. Historia Natural de Nueva España. Vols. I & II. *In* F. Hernández, Obras Completas, Tomos II & III. Mexico City: UNAM.

Holm, L. 1969. Weed problems in developing countries. Weed Science 17:113–118.

Instituto Nacional de Estadística, Geografía e Informática. 1980. Atlas Nacional del Medio Físico. Mexico City: Secretaría de Programación y Presupuesto.

Instituto Nacional de Estadística, Geografía e Informática & Instituto Nacional de Antropología e Historia. 1986. Estadísticas Historicas de México, Tomos I & II. Mexico City: Secretaría de Programación y Presupuesto.

Knab, T. 1982. Usos tradicionales. Paper presented in symposium: Toloache, etno-farmacología de la solanaceas deliriogenas de México. VI Congreso Nacional de Farmacología, Durango, Durango, México.

Lambert, J.D.H. & J.T. Aranson. 1982. Ramon and Maya ruins: an ecological, not an economic, relation. Science 216:298–299.

Linares, E. & R. Bye. 1987. A study of four medicinal plant complexes of México and adjacent United States. J. Ethnopharmacol. 19:153–183.

Lobato, J.G. 1884. Estudio químico-industrial de various productos del maguey mexicano y análisis químico del aguamiel y el pulque. Mexico City: Secretaría de Fomento de México.

Magurran, A.E. 1988. Ecological Diversity and Its Measurement. Princeton, NJ: Princeton Univ. Press.

Marino A., A. 1966. The Pulque Agaves of Mexico. Ph.D. diss. Harvard Univ., Cambridge, MA.

Martínez, M. 1959. Plantas Utiles de la Flora Mexicana. Mexico City: Ediciones Botas.
———. 1969. Las Plantas Medicinales México. Mexico City: Ediciones Botas.
———. 1979. Catálogo de Nombres Vulgares y Científicos de Plantas Mexicanas. Mexico City: Fondo de Cultura Económica.

Martínez R., J. 1986. Diversidad Monolingüe de México en 1970. Mexico City: UNAM.

Mendieta, R.A. & S. del Amo R. 1981. Plantas Medicinales del Estado de Yucatán. Xalapa, Veracruz: Instituto Nacional de Investigaciones sobre Recursos Bióticos.

Merrick, L.C. & G.P. Nabhan. 1984. Natural hybridization of wild *Cucurbita sororia* group and domesticated *C. mixta* in southern Sonora, Mexico. Cucurbit Geneti. Coop. 6:74–75.

Moncrief, L.W. 1970. The cultural basis for our environmental crisis. Science 170:508–512.

Nabhan, G.P. 1985. Gathering the Desert. Tucson, AZ: Univ. of Arizona Press.
———. 1987. Nurse plant ecology of threatened desert plants. *In* T.S. Elias (ed.), Conservation and Management of Rare and Endangered Plants. Sacramento, CA: California Native Plant Society. pp. 377–383.

Newstrom, L.E. 1989. Reproductive biology and evolution of the cultivated chayote *Sechium edule*: Cucurbitaceae. *In* J.H. Bock & Y.B. Linhart (eds.), The Evolutionary Ecology of Plants. Boulder, CO: Westview Press. pp. 491–509.

Olivera, M., M.I. Ortiz & C. Valverde. 1982. La Población y las Lenguas Indigenas de México en 1970. Mexico City:UNAM.

Pielou. E.C. 1975. Ecological Diversity. New York: John Wiley & Sons.

Pippen, R.W. 1968. Mexican "Cacalioid" genera allied to *Senecio* (Compositae). Contr. U.S. Natl. Herb. 34:363–447.

Prescott-Allen, C. & R. Prescott-Allen. 1986. The First Resources. Wild Species in the North American Economy. New Haven, CT: Yale Univ. Press.

Rapoport, E.H., M.E. Díaz-Betancourt & I.R. López-Moreno. 1983. Aspectos de la Ecología Urbana en la Ciudad de México. Flora de las Calles y Baldíos. Mexico City: Limusa.

Reichenbacher, F.W. 1985. Conservation of Southwestern Agaves. Desert Plants 7:103–106, 88.

Remmington, C.L. 1968. Suture-zones of hybrid interactions between recently joined biota. Evol. Biol. 2:321–383.

Rindos, D. 1984. The Origins of Agriculture—An Evolutionary Process. New York: Academic Press.

Rose, J.N. 1899. Notes on useful plants of Mexico. Contr. U.S. Natl. Herb. 5:209–259.

Ruvalcaba M., J. 1983. El Maguey Manso: Historia y Presente de Epazoyucan, Hidalgo. Chapingo, México: Universidad Autónoma Chapingo.

Rzedowski, J. 1978. Vegetación de México. Mexico City: Limusa.

———. 1986. Las plantas calcicolas (incluyendo una gipsófita) del Valle de México y sus ligas con la erosión edáfica. Biotropica 18:12–15.

——— & M. Equihua. 1987. Atlas Cultural de México: Flora. Mexico City: SEP.

Sanders, R.W. 1987. Taxonomy of *Agastache* section *Brittonastrum* (Lamiaceae-Nepeteae). Syst. Bot. Monogr. 15.

Sahagún, B. 1963. Florentine Codex. General History of the Things of New Spain. Book 11—Earthly Things. Salt Lake City, UT: Univ. of Utah Press.

Schwanitz, F. 1967. The Origin of Cultivated Plants. Cambridge, MA: Harvard Univ. Press.

Smith, C.E., Jr. 1967. Plant Remains. *In* D.S. Byers (ed.), The Prehistory of the Tehuacan Valley. 1(Environment and Subsistence):220–255. Austin, TX: Univ. of Texas Press.

Sosa, V., J.S. Flores, V. Rico-Gray, R. Lira & J.J. Ortiz. 1985. Lista florística y sinónimia maya. Ethnoflora Yucatanense, Fascículo 1. Xalapa, Veracruz: INIREB.

Standley, P.C. 1920–1926. Trees and shrubs of Mexico. Contr. U.S. Natl. Herb. 23:1–1721.

Straw, R.M. 1963. The Penstemons of Mexico. III. Brittonia 15:49–64.

Strother, J.L. 1983. *Pionocarpus* becomes *Iostephane* (Compositae: Heliantheae): a synopsis. Madroño 30:34–38.

Toledo, V.M. 1987. La etnobotánica en Latinoamerica: vicisitudes, contextos, disafios. Memorias. IV Congreso Latinoamericano de Botánica. Simposio de Etnobotánica. Bogotá, Colombia: Instituto Colombiano para el Fomento de la Educación Superior. pp. 13–34

———. 1991. Bio-economic costs. *In* T. Downing (ed.), Cattle Ranching and Tropical Deforestation in Latin America. Boulder, CO: Westview Press. pp. 63–90

Turner, N.J. 1988. "The importance of a rose": evaluating the cultural significance of plants in Thompson and Lillooet Interior Salish. Amer. Anthropol. 90:272–290.

Ugent, D. 1968. The potato in Mexico: geography and primitive culture. Econ. Bot. 22:108–123.

Ulloa, M., T. Herrera & P. Lappe. 1987. Fermentaciones Tradicionales Indígenas de México. Mexico City: INI.

Valdés, J. & H. Flores. 1985. Comentarios a la Obra de Francisco Hernández; Historia de las Plantas de Nueva España. *In* F. Hernández, Obras Completas, Tomo VII. Mexico City: UNAM. pp. 7–222

Valdés, L.M. & M.T. Menéndez. 1987. Dinámica de la Población de Habla Indigena (1900–1980). Mexico City: INAH.

Vázquez R., M.C. 1986. El Uso de Plantas Silvestres y Semicultivadas en la Alimentación Tradicional en dos Comunidades Campesinas del Sur de Puebla. Thesis, UNAM.

Viveros S., J.L. & A. Casas F. 1985. Etnobotánica Mixteca: Alimentación y Subsistencia en la Montaña de Guerrero. Thesis, UNAM.

Werner, O. & J. Fenton. 1970. Method and theory in ethnoscience or ethnoepistemology. *In* R. Naroll & R. Cohen (eds.), A Handbook of Method in Cultural Anthropology. New York: Columbia Univ. Press. pp. 537–578.

Wilkes, H.G. 1970. Teosinte introgression in the maize of the Nobogame Valley. Bot. Mus. Leafl. 22:297–311.

Wilkes, G. 1989a. Germplasm preservation: objectives and needs. *In* L. Knutson & A.K. Stoner (eds.), Biotic Diversity and Germplasm Preservation, Global Imperatives. Boston, MA: Kluwer Academic Publishers. pp. 13–41.

———. 1989b. Maize: domestication, racial evolution, and spread. *In* D.R. Harris & G.C. Hillman (eds.), Foraging and Farming: The Evolution of Plant Exploitation. London: Unwin Hyman. pp. 440–455.

Williams, D.E. 1985. Tres Arvenses Solanáceas Comestibles y su Proceso de Domesticación en el Estado de Tlaxcala, México. M.S. thesis, Colegio de Postgraduados, Chapingo, México.

Wilson, H.D. 1990. Quinua and relatives (*Chenopodium* sect. *Chenopodium* subsect. *Cellulata*). Econ. Bot. 44(3 Suppl.):92–100.

Yarnell, R.A. 1965. Implications of distinctive flora on Pueblo ruins. Amer. Anthropol. 67:662–674.

York, J.C. & W.A. Dick-Peddie. 1969. Vegetation changes in southern New Mexico during the past hundred years. *In* W.G. McGinnies & B.J. Goldman (eds.), Arid Lands in Perspective. Tucson, AZ: Univ. of Arizona Press. pp. 157–166.

Zorrilla, L. 1988. El Maguey: "Arbol de las Maravillas." Mexico City: Museo Nacional de Culturas Populares.

25

Aspects of Plant Domestication in Mexico: A Personal View

EFRAIM HERNÁNDEZ XOLOCOTZI[1]

Aspects of plant domestication from the archaeological period to modern times in Mexico, which is one of the major centers of diversity of cultivated plants, and origins of agriculture are discussed. The process of domestication, the results of such processes, the persistence of domesticating factors, and the actual state of diversity of cultivated species are reviewed. The chapter summarizes research on some of these procedures and addresses the need for conservation of these natural genetic resources in the region.

During their early development, humans established a symbiotic and coevolutionary relationship with the surrounding vegetation, with humans as agents of dispersal and the vegetation providing the basic products for human subsistence (Rindos, 1984). The interaction was principally with gymnosperms and angiosperms, but the great use to which humans put the seeds of flowering plants during this coevolutive process made them "angiospermic organisms" for nutritional and other needs (Ames, 1939). It is possible that changes in floristic composition in and near habitats occupied by humans may have resulted from their preference of some species over others. A change in the coevolutive process occurred when humans developed physical and functional traits characteristic of human cultures, which led to human dominance of surrounding habitats (Childe, 1965).

It seems reasonable to speculate that early hominids, including australopithecines, probably appreciated biological phenomena and recognized organisms, particularly those useful to their survival. Although their needs may have included ceremonial and medicinal plants, it is possible that they emphasized species that met their nutritional needs and, later, species that provided firewood and fiber. The later descendants of the lineage of *Homo* (*H. sapiens*) most certainly began to alter terrestrial habitats about 100,000

[1]Deceased February 21, 1991.

years ago. During the last 50,000 years, radiating from an African center they settled all habitable parts of the world, including the polar regions and the Americas (Deevey, 1960). During this long period, humans provoked genetic changes in domesticated organisms by modifying their ecological background and increasing the availability of primary resources useful to humans; artificial selection by humans complemented natural selection (Hawkes, 1983; Schwanitz, 1967). Once this process was initiated, humans expanded the magnitude and depth of domestication through artificial selection, dictated by use, form of use, and the degree of acceptance of the product by humans. Selection and manipulation of the desirable organisms, for example, determined the course of agriculture in different human groups. Currently, research concerning domestication involves origin as well as loss and maintenance of diversity in cultivated plants.

During the cultural and agricultural development of humans, a certain dominance in the management of agronomic elements related with the production of favored organisms is noticeable. The long period of coexistence of humans and the favored organisms, on the other hand, has resulted in a broad and profound biological and ecological knowledge of these organisms. This knowledge was applied initially during a phase in which the principles of heredity were not known. Later, the scientific knowledge of reproductive mechanisms led to genetic breeding of organisms. Finally, a broader and deeper knowledge of genetics has enabled humans to manipulate the genetics of domesticated organisms as is seen at present times, culminating in the application of biotechnology and molecular biology. These steps have enriched the diversity of domesticated organisms and established the basis for conservation of these resources. Historical backgrounds of these processes for Mexico, whose biological diversity is of world significance (Appendix) is reviewed here from the archaeological period through the Prehispanic, Colonial, and present periods.

ARCHAEOLOGICAL BACKGROUND

The evidence of the origin and forms of development of agriculture rests with archaeological studies. In Mexico, such investigations have been notable in northwestern, northeastern, and central Mexico, but the most remarkable were carried out in Tehuacán, Puebla (MacNeish, 1961) and Mitla, Oaxaca (Flannery, 1986). MacNeish's aim was to locate, excavate, and analyze the remains found in sites where the beginnings of agriculture as well as its development in Mesoamerica have been recorded. Rich and diverse data directed him to the many caves in the limestone region at the northern foot of the Sierra Mazateca, where a semiarid climate prevails on the leeward side of the mountains. The excavations of several of the caves have led to the establishment of a chronology from around 7000 B.C. to A.D. 1500. The remains obtained included a rich collection of animal bones and a large number of botanical remains of numerous species, which were both nutritional as well as useful for making cordage. The excellent preservation of the archaeological remains allowed morphological and anatomical studies for their proper identifications, and the magnitude of excavation also enabled quantitative analysis. These studies included a number of collaborators. Among them were Mangelsdorf et al. (1967), who studied corn; Kaplan (1967), beans; Cutler (1967), squashes; Smith (1967), floristics and identification of wild useful plants; and Callen (1967), coprolites. A review of the list of plants presented by Smith (1967) reveals the floristic richness of this site. The results obtained suggest the beginnings of agriculture in different microniches in the region. Remains of several species of *Cucurbita*, *Amaranthus*, *Chenopodium*, and *Phaseolus* were found, genera that were later cultivated; the appearance of *Zea mays* occurs later in archaeological time and continued into the Spanish era. Of note are the great proliferation of races of corn (up to 15) and the persistence of the great part of these races until 1500 AD. The absence of wild teocintle in these remains has contributed to the polemics that persist on the origin of maize (Beadle, 1972; Iltis, 1987).

It is of interest that several problems of identification occurred during the study of the above-mentioned material. One example applies to the hundreds of discoid remains. The peeling process of "tuna," or prickly pear (*Opuntia* sp.), which persists in the arid zones today provides the key to the identification of these discs as the cut receptacles of the fruits. The other problem pertains to seeds of *Setaria* found in the remains. Initially they were not identified, but field observations revealed an abundance of *S. macrostachya* ("zacate tempranero"). The materials studied and their analysis left open several questions: (1) What was the ancestor of maize? (2) Where did this cereal originate? and more importantly (3) How did agriculture originate? Evidence from these sites are reviewed below.

Slopes of Puebla

Numerous niches associated with humidity are located on the slopes of the Sierra Mazateca above the valley floor. They are found above the caves studied by MacNeish, which are located in the openings of the arroyos and on the floor of the Tehuacan valley. Under these conditions, the evidence of modification of the environment to improve the desired vegetable species is great. For example, along the canyons and in more humid sites, fruit trees—*Persea* (avacado), *Spondias* (tropical plum), and *Cyrtocarpa* ("chupandilla")—have been favored. Annual species that prefer greater humidity (*Phaseolus* and *Cucurbita*) have been recorded. In semiarid zones, the presence of *Dioon* ("chamal") and *Setaria macrostachya* has been detected. In the openings of the arroyos (areas annually denuded of vegetation by torrential rains) annual species of *Chenopodium* and *Amaranthus*, which flourish in moist conditions, have been found (favoring production of seeds). The vegetation of the semiarid zone of the valley floor and the oaks and pines above the highest parts of the slopes were sources of other useful plants.

Oaxaca

The Guíla Naquitz caves are found at the edge of the semiarid valleys of Central Oaxaca, and in these sites there is an abundance of edible Agavaceae, Cactaceae, Euphorbiaceae, and Fabaceae. Oak and pine forests with their respective products are found on the upper slopes to the north.

The findings in Guíla Naquitz (near Mitla, Oaxaca) that date from the period 8900–6700 BC indicate that human populations were concentrated in the area of mesquite (*Prosopis*) and "huizache" (*Acacia*) in the lower zone with alluvial soil from the valley, where today mesquite (*Prosopis juliflora*) with edible fruits is abundant. Pinyon (*Pinus*) and acorns (*Quercus*) were probably obtained by way of excursions to the oak and pine sites of the highest zones. Fruits and pads of Cactaceae and leaves and stems of maguey were collected in the driest sites. During this period the diet was made up of a preponderance of carbohydrates and a reduced amount of proteins. The consumption of *Phaseolus coccineus* and a black-seeded bean suggests their cultivation but not domestication. Other agricultural precursors included *Cucurbita pepo and Lagenaria siceraria*. Later cultivated maize was introduced into the area.

These archaeological remains from Oaxaca, in conjunction with those from Puebla, provide grounds for the following considerations: (1) availability of time and leisure were not necessary to begin agricultural activity; (2) cultivation preceded the genetic change considered as an indicator of domestication; and (3) unpredictable climatic fluctuations accelerated the concentration of human groups in more favorable sites, as well as the beginnings of agriculture.

It must be noted that the development of several disciplines has broadened our knowledge and extended our ability to interpret these archaeological remains. It has permitted scientists to study biological remains excavated from archaeological sites. There were few serious attempts to recover the biological remains from sites in earlier times because of lack of techniques. Today they can be retrieved by diverse methods of recovery techniques such as flotation, which has enabled sampling of a larger range of biological material. Studies of contemporary plants in conjunction with archaeological remains often enrich the information available as well as their subsequent interpretation. For example, the cob and cupule characters provide critical data for the classification of contemporary maize in Mexico (Wellhausen et al., 1951) and interpretation of the evolution of maize through archaeological time. Additionally, the development of geochronology, using physical indicators (^{14}C), has allowed exact dating of biological remains.

One of the unresolved problems concerns the records of activities associated with agriculture and their evolutionary consequences on cultivated plants. It is desirable to include diverse data or methodologies in this study. For example, Flannery (1986) included ecological data from the acquisition sites of plant resources and their alteration to favor the development of the selected plants. Of the remains from Tehuacán, the seeds of various species of *Persea* (avocado) and *Mastichodendrum* ("tempesquite"), show morphological changes generally interpreted as evidence of domestication. The remains of beans during the long period of its recording do not show evidence of modification (Kaplan, 1967). Several other plant species could have been domesticated in these parts, as a rich local flora (Smith, 1967) was available.

DOMESTICATION DURING THE PRE-HISPANIC AND EARLY COLONIAL PERIOD

The following conditions favored the registry and conservation of scripts that recorded the regional conditions at the time of the Spanish arrival: (1) the recording of material in Mesoamerican codices (which survived the massive destruction of these documents by the Spanish); (2) the interest of the missionaries in understanding the material and metaphysical world of the recently conquered region (with a view to being able to complete their mission of proselytization); and (3) the need to inform the Spanish crown of the resources of New Spain.

The reference documents, which have been best conserved and studied, are the Códice Florentino by Fray Bernardino de Sahagún (1980), known by the title "Historia General de las Cosas de la Nueva España"; the work by the protomedico, Francisco Hernández (1959), titled "Historia Natural de la Nueva España," and others such as those by the soldier chronicler Bernal Díaz del Castillo (1976), Hernán Cortés (1970), and Fray Diego de Landa (1978).

These documents contain a wealth of information on the then extant organisms, their use, the processes of production, the implements used, and the concept that the Mesoamerican inhabitants had on use and form of use of the agricultural and medicinal products. This information allows us to reconstruct, specifically, the agricultural activity (Rojas & Sanders, 1985), and the wealth of domesticated plants (Torres, 1985) of which we might mention because of their great importance the following: maize (*Zea mays*), common bean (*Phaseolus vulgaris*), scarlet runner bean (*Phaseolus coccineus*), squashes (*Cucurbita* spp.), sweet potato (*Ipomoea batatus*), chiles (*Capsicum* spp.), cocoa (*Theobroma cacao*), tomato (*Lycopersicon esculentum*), peanut (*Arachis hypogea*), cassava (*Manihot esculenta*), yam bean (*Pachyrrhizus erosus*), vanilla (*Vanilla planifolia*), amaranth (*Amaranthus* spp.), "huauzontle" (*Chenopodium berlandieri* ssp. *nuttalliae*), and "chía" (*Salvia hispanica*).

Rojas (1983) reached the noteworthy conclusion that the agricultural implements suggest that the agricultural technology was simple and with little variation, although in itself it shows a diverse adaptation to many agroecological sites starting from the dry conditions to irrigated agriculture. The presence of these agricultural niches corresponds to a rich range of cultivated plants, which included annual nutritional cultivated plants, succulents (adapted to dry zones), medicinals, deciduous plants, and irrigated plants. Lack of species adapted to low temperatures led to practices of "almácigo" (plantations, nurseries, ceiling beds) and "arrope" (covered nurseries) especially in the "chinampas."

The study by Estrada L. et al. (1988), who analyzed the recorded useful plants in the Códice Florentino, listed the number of plants used in the first ten anthropocentric categories: medicinal (266), edible (229), ceremonial (81), esthetic (48), industrial (27), stimulants (20), firewood (14), construction material (14), foraging (14), and coloring (12).

The presence of several cultivated plants—*Capsicum* species (chile), *Lycopersicon esculentum* (tomato), *Arachis hypogea* (peanut), *Ananas comosus* (pineapple), and *Theobroma* species (cocoa)—whose center of diversity is outside Mesoamerica, suggests migrations of peoples and materials from the south to Mesoamerica, and from there to northeastern and northwestern Mexico. There is a marked flow of variations in maize, beans and squashes toward the northeast and northwest.

The cultivation of *Taxodium mucronatum* (Montezuma cypress, "ahuehuete"), the ceremonial tree, had by this time extended to the entire temperate zone of Mesoamerica.

DOMESTICATION DURING THE COLONIAL PERIOD

One of the important activities during the colonial period was the introduction of domesticated plants and animals to the new territories. Among plants, wheat (*Triticum aestivum*), rice (*Oryza sativa*), and other minor cereals; legumes including peas (*Pisum sativum*), garbanzo (*Cicer arietinum*), alfalfa (*Medicago sativa*), and fava (*Vicia faba*); sugar cane (*Saccharum officinarum*); banana (*Musa* spp.); coconut (*Cocos nucifera*); coffee (*Coffea arabica*); mango (*Mangifera indica*); deciduous fruit trees (e.g. Rosaceae); and crops tolerant of cold came to enrich the long list of domesticated plants in Mesoamerica. These new crops, e.g., wheat, sugar cane, and winter vegetables, filled ecological niches and made use of seasons not previously exploited.

The introduction of Old World domesticated animals had a profound impact on the natural pasturelands of the northern and central parts of the country. It was not limited to areas with grasses but extended to land with bushes of low and medium height. This use led to the establishment of great expanses for cattle farming (by only a few farmers) and a later interest in introducing species (pasture and forage plants) favored in other parts of the world.

There is little information on the cultivation of forest species; in contrast, there is information on the exploitation and use of these resources. A few are noteworthy, such as logwood (*Haematoxylum campechianum*), mahogany (*Swietenia macrophylla*), and tropical cedar (*Cedrela odorata*). All of these surely suffered great genetic and population changes during the period of their exploitation owing to fluctuating population changes and local extinctions.

CURRENT PERIOD

Germplasm of cultivated plants, which has now been introduced, includes such forest species as eucalyptus (*Eucalyptus* spp.), casuarina (*Casuarina* sp.), gmelina (*Gmelina*), Caribbean pine (*Pinus caribaea*), and teak (*Tectona grandis*).

The introduced useful plants include abaca (*Musa textilis*), soy bean (*Glycine max*), safflower (*Carthamus tinctorius*), cardamom (*Elettaria cardamomum*), sorghum (*Sorghum bicolor*), and rubber (*Hevea brasiliensis*). Several of these plants have come to occupy an important place in the list of cultivated plants in Mexico today.

The research that led to the "Green Revolution" enriched cultivars such as wheat, oats (*Avena sativa*), maize, soy bean, sorghum, cartamo, yucca, coffee, malanga (*Xanthosoma violaceum*), and banana (*Musa* spp.). Because of the introduction of improved varieties, part of the germplasm of older cultivars has been lost. The advances in the establishment of conservation banks have partially alleviated this problem.

AGROHABITAT

Human groups, through their everyday activities (e.g., clearing of fields, refuse, defecation, burnings) create disturbed niches that are invaded by pioneer plants that evolved beforehand in the natural disturbances of the environment. These pioneers, generally called weeds, have been the basis for the origin of agriculture according to Engelbrecht (1916) and Hawkes (1983), the latter of whom advocated the theory of human trash dumps. Although this theory is attractive, there is no archaeological evidence to support it. The archaeological remains from Oaxaca and Puebla and their interpretation suggest that agriculture arose as a result of the search for options in food production at more favorable sites in the face of unpredictable climate. The evidence points to intentionality of human behavior, which led to agricultural origins, a view not included in the model by Rindos (1984). The cited remains from Mexican sites indicate the beginnings of cultivation of plants prior to the appearance of agriculture as an organized system of production.

Agriculture as a series of practices favors the development and production of plants of interest. It requires empirical knowledge about the relationship between environment and plant response. The sequence of practices applied in cultivation consists of distribution of seeds in moist soil, elimination of competitive plants around the plant of interest, and eradication of predators.

In contrast, agriculture involves a series of 13 groups of practices (Cox & Atkins, 1979; Hernández X. et al., 1981) tending to secure and increase the desired production through (1) preparation of the soil (note that slash and burn is a form of soil preparation); (2) preparation and selection of the seed to be used (frequently it consists of a mixture to ensure production); (3) sowing at the optimal density and depth; (4) movement of the soil to give greater efficiency of use of the moisture; (5) elimination of competitive plants; (6) addition of fertilizing material; (7) elimination of predators; (8) harvest of production; (9) transport and storage of the product; (10) selection and storage of the seed; (11) ceremonies of prayer and thanks for a good harvest; (12) improvement of practices and tools; and (13) selection of the cultivars.

Traditional agriculture consists of the art and empirical knowledge of the application of these groups of practices to optimize production. Agriculture creates a special environment called the agrohabitat, which represents the preserve in which plants are domesticated. The domestication of cultivated plants has been characterized by genetic changes that permit improvement of the plant to the specific agrohabitat and a better matching of the product with human requirements. The result of this process is a great infraspecific variation of species selected for domestication (Hawkes, 1983; Rindos, 1984; Schwanitz, 1967).

One additional aspect of the agrohabitat is its invasion by populations of species correctly named weeds. These populations include species other than cultivated ones, such as populations of relatives. De Wet and Harlan (1975) studied the nature and role of these populations. In traditional agriculture, we find that they include species useful to humans principally for food, medicine and fodder, and sometimes they represent an important safeguard in the face of unpredictable climatological conditions (Zizumbo V. et al., 1988). From the agricultural point of view, these populations have been sources of secondary cultigens (Vavilov, 1951). On the other hand, the presence of weed populations within populations of congeneric cultivars marks them as genetic reservoirs that may be important in modern attempts to improve their relatives (Harlan, 1975; Hawkes, 1983); these populations and their persistence point to the possibility of conservation of germplasm in the sites of origin and of genetic dynamics of the primitive groups of the cultivated plants (Williams, 1985).

DOMESTICATION

Domestication of plants consists of their genetic modification for appropriate development within the agrohabitat created for its production. This process is related to the ecological environment, the existing flora, the forms of use of the produced material, human necessity, and the cultural adaptability of humans. Important aspects include the characteristics of the cultivated plants as a consequence of the ecological medium in which they develop. Attention has been called to the difference between domesticated species in the Middle East and Mesoamerica. Iltis (1987) has argued that the Mediterranean climate favors grasses (e.g., Hordeae) and pulses based on the plants' autoecological characteristics such as that of direct photosynthate translocation from the leaves to the grains during the wet growing season with solar energy increasing daily during the life cycle. In contrast, the domesticated plants in Mesoamerica developed during the rainy growing season when the solar radiation was declining during the cultivation cycle thus requiring intermediary storage for the early surplus carbohydrates until the later maturing grains could be filled. It has favored cultigens such as maize, beans, and squashes. In traditional agriculture, the farmers have a rich knowledge of the direct relationship between yield and the growth period of the plants. Their knowledge is reflected in the prior selection of precocious varieties and a medium period of growth and late maturing, representing choices to coordinate with the unpredictable periods of rain and to broaden the harvest period.

The flora of a region is a fundamental expression of the prevailing climate in the region. From this flora, humans have defined species that meet their needs. In Mitla, Oaxaca, the attention of the hunter-gatherers focused on Cactaceae, Agavaceae, the mesquite, and the "guaje" (*Leucaena*), as well on the acorns of the oaks (*Quercus*) and the seeds of the pinyons (*Pinus*). Later, attention was directed to Cucurbitaceae (*Lagenaria, Cucurbita,* and *Apodanthera*) and to beans (*Phaseolus*). It seems possible that maize was introduced from neighboring Tehuacán. Under these conditions, the agricultural system was based on sowing associated with maize, beans, and squashes, probably imitating the natural association of these plants in the flora (Miranda, 1966). From these conditions have emerged in Mesoamerica an aspect that confuses the results on domesticated species even now (Hernández X. et al., 1986); a number of plants in these areas that are in transition to their complete domestication. Examples are numerous Agavaceae, Cactaceae, Sapotaceae, some Rosaceae (*Crataegus*), Bignoniaceae (*Parmentiera* and *Crescentia*), Rutaceae (*Casimiroa*), and Fabaceae (*Crotalaria, Leucaena*). Humans do not consume solely for nutrition; they also exert considerable discretion differentiating the quality of food consumed. This quality is as much a product of nature as it is of the transformation wrought by humans. Thus we see that in the case of maize there are variants with different qualities (floury, popcorn, sweet, and crystalline), but at the same time these variants can be submitted to a large number of forms of transformation for their use, forms that affect the acceptance of the product consumed. The general forms of transformation of maize include "jilote," eaten tender with bracts and all; corn on the cob, boiled or broiled; dry grains, boiled, toasted, limed-ground, germinated, or fermented; and "enriado," boiled and dried or milled in flour form. These processes or similar ones are applied to various cultigens (beans, cassava, wheat, banana) influencing the vehicle of selection.

Cultural development of humans has been polarized by different concepts with relation to their cosmic vision and their relationship to nature. In *traditional agriculture*, for example, humans do not distinguish between the material and the metaphysical, nor do they perceive a need to dominate the material world. In contrast, *modern Western agriculture* is based on a cosmic vision with a clear distinction between the metaphysical and physical, and a desire to dominate nature. This difference of cosmic vision affects the

direction of cultural growth of humans. In both bases there is an effort to
better the capacity of environmental change, to acquire other edible cultivated plants of regions in the area, to break geographical barriers between the
populations of cultivars, and to diversify the use of plant products.

The fundamental difference between these two visions has consequences in the process of selection and in the search for new cultivated
plants through the process of genetic change. This point is illustrated by
traditional agriculture, which is limited in its ability to exert selection
pressure, as its pratitioners have lacked knowledge of the principles of
heredity. The sexual mechanisms of reproduction and the reappearance of
characteristics in the progenitors were known, but the mechanisms to
cause and exercise a selective pressure toward predetermined characteristics were not. The most that was accomplished has been the selection of
forms resulting from hybridization or mutations. In species of asexual
propagation the survival of variants resulting from chance sexual reproduction, which enriches variation, was permitted.

In occidental science, the process is characterized by an insistence on
quantitative observation of phenomena and the systematic recording of the
attempted experiences (called the scientific method). It has distinguished
the approach of European societies in their attempt to dominate and
impose their ethical and material values on nature. As a result, there has
been a constant search to understand the process of domestication and the
means to control the causal factors.

The development of two divergent systems of acquisition of knowledge
(the traditional and the scientific), gives a range of cultural capacities to
humans to intervene in the selective processes of the domestication of
plants and animals.

The traditional system is distinguished by symbiosis and organic coevolution, unconscious selection, and directed selection. In cultural development linked to occidental science, the following additional phases are
notable: selection based on empirical knowledge of hereditary mechanism;
selection based on phenotypical changes; and strict selection with elimination of undesirable offspring. The incipient knowledge of hereditary mechanisms provides an empirical basis for genetic breeding. A refinement of the
knowledge of hereditary mechanisms led to selection based ever increasingly on genotypes and breeding of uniform genotypes and search of hybrid
vigor. Finally, with the development of techniques that manipulate the
genetic structure, selection based on molecular genetics has resulted. During
each of these phases, amplitude and depth of genetic management results in
products of selection with greater degrees of differentiation.

At the turn of the twentieth century the knowledge of principles of
heredity as well as the careful empirical observations on the part of the
agriculturalists of the corn belt of the central United States served as
incentives to direct the course of plant breeding in that country. These
investigations led to the formation of hybrid double cross maize, based on
the careful formation of self-pollinating populations, its heterosis in yield,
and its recombination for the formation of the double cross. The results

inspired the application of these principle to other cultigens. During the following phase of research the genetic structures and methods of manipulation were expanded, leading to the search of sexually reproducing populations in cultigens that normally reproduce asexually, e.g., sugar cane, and the apocarpic formation of fruits (e.g., bananas).

Under these conditions, breeding consisted in the selection and formation of materials that draw closer a predetermined outcome and the subsequent loss of genetic diversity. By these standards, the traditional agriculturalists have not intervened in breeding, as they lack the predetermined goal toward which to direct selection.

The search for the material bases of the regulators of heredity and the development of instruments with ever increasing capacity resulted in the current knowledge of principles of heredity. These discoveries have opened up the fields of biotechnology and molecular genetics. In both cases, they have produced new organisms with new physiochemical capacities. In modern processes of selection of domesticated plants, the new systems of taxonomic differentiation based on both macro and micro characters have gained considerable importance. These modern taxonomic methods maintain relationships with the forces of biosystematics developed in 1940 (Anderson, 1949). Currently, diverse methodologies, which are more and more detailed in taxonomic differentiation, are used; these include numerical taxonomy, cytotaxonomy, electrophoresis, antibody reactions, phytochemical products, isoenzymes, and specific genetic markers (Goodman, 1972; McClintock et al., 1981; Longley & Kato, 1985; Yakoleff G. et al., 1982). The use of these techniques opens up possibilities for selection of genotypes with hidden variation that are totally new to plant breeders.

Ethnobotanical studies and exploration directed toward the search of potential materials to meet human needs have served as a basis for identification of plant products and species necessary to cure new illnesses generated by human activities, including overpopulation, overcrowding, nervous tensions from intensive activity, and environmental deterioration. Among these illnesses are arteriosclerosis, arterial pressure, mental deficiencies, nervous tension, cancer, and acquired immunodeficiency syndrome (AIDS). The result of these searches has been the discovery of new species susceptible to domestication to achieve desired ends. It is the case, for example, of Mexican yam (*Dioscorea composita*) for contraceptives, jimson weed (*Datura* sp.) for respiratory ailments, and periwinkle (*Catharanthus roseus*) for cancer. These searches have added to the larger pool of genetic diversity among domesticated plants.

Another motivation for domestication of plants has been the need for raw materials stemming from industrial activities. Examples of the new domesticated plants from this century include rubber (*Hevea brasiliensis*), sisal (*Agave sisalana*), marigolds (*Tagetes erecta*, and *T. patula*).

The traditional system has used family farms and gardens in the process of domestication. In Mexico, family farms show a general persistence throughout the territory. The diversity of plants included in these farms suggests the maintenance of potential variants for the greater cultivation and a strong interrelationship between the germplasm represented and the domesticating activity of families that maintain them. The formation of botanical gardens in Mesoamerica, which surely influenced domesticating activities, is noteworthy. In reality, the cultigens in the gardens and the associated weeds include species more susceptible to domestication.

In occidental science various institutions contribute to exploration, documentation, conservation, and exploration of genetic diversity of crop plants. Modern herbaria are a verified, identified source for useful plants (del Amo & Anaya, 1982). The gardens that continue ancient traditions can have an important function in the domestication of plants; an important example in Mexico is the Jardín Botánico del Instituto de Biología of UNAM in Mexico City. The expansion of agricultural activities has led to the establishment of a broad net of germplasm banks in the world. Currently, these banks in Mexico contain maize, beans, wheat, rice, yucca, potatoes, soy bean, and other garden vegetables. Generally, such banks are annexes, supporting the activities of experimental fields and converting them to centers of great diversity of cultigens. With greater support, these banks could be established as genetic banks of exceptional materials to use for plant breeding and biotechnology.

The study of floras and the collection of information on aspects regarding use of different species can be considered auxiliary to the studies of plant domestication. In the most direct form are ethnobotanical explorations that offer information and materials of cultivated plants. The need for such studies that complement each other cannot be overemphasized. These studies, however, naturally tend to concentrate information in a few hands, leading to troubling questions in many parts of the world. Will the accumulated information contribute to the improvement of the world at large, or will it be used for the social and economic domination of developing societies?

Mexico's biological diversity, one of the largest in the world, is in large measure due to the diverse ecological environments found in the country. Significantly, many ethnic groups exist in Mexico with their respective agricultural systems and management of natural resources. These two aspects have caught the attention of various researchers, who have carried out important studies on diversity, domestication, conservation, and loss of biological material in Mexico. The dynamics of the diversity of domesticated plant germplasm in the context of ecological variation and socioeconomic requirements can be illustrated by such studies as those on the increased diversity of maize in Chiapas and of beans in Aguascalientes, the intricacy of traditional agricultural systems, and the manipulation of prickly pear cacti in Guanajuato.

In relation to loss of genetic diversity in cultigens, especially maize and beans, there are two reports that compare the existing variation between two sampling periods within a 25-year interval. Ortega and Angel (1973) studied variation in maize and socioeconomic changes in Chiapas from 1946 to 1971. Their findings were as follows. (1) In the state of Chiapas, which is ecologically diverse, the need existed for early maturing variants suitable to ecological niches requiring early precocious producers, or additional harvest. Variants of different colors and textures were needed to satisfy the desire and dietary preferences of the people. Farmers strove to conserve the variants of maize they used from those of their neighbors. Religious ideas related to several specific characteristics, e.g., red grains or multiple ears, influenced their decisions. (2) There was an increase in races sown in 1971. Among them were Nal-tel, Zapalote chivo, Zapalote grande, Tepecintle, Olotone, Negro de Chimaltenango, Quicheño, Salpor, Juncana, Tehua, Olotillo, Tuxpeño, Vandeño, and Argentino. The Argentine race is a more recent introduction that was not available in 1946. (3) The improved maize (principally: H-503, H-520, and V-530 C) played a major part in maize cultivation. These races were sown in more than 22% of the tropical area, preferably on more fertile soils. The use of advanced generations, which was extensive,partially but not completely displaced some land races, principally Tuxpeño and Vandeño. In the areas where the improved maize dominated, the land races were still sown in areas with low productivity and early maturing variants. Cross-pollination in large populations of improved maize has resulted in significant genetic changes.

Andrade A. and Hernández X. (1988) recorded the following in a study of bean germplasm. (1) The variation of germplasm of beans (*Phaseolus vulgaris* and *P. coccineus*) in the state of Aguascalientes, collected in 1984, showed an inverse relationship to the economic resources of the producers. Thus the greatest variation was found in agricultural areas with the least economic resources, owing to the fact that the farmers sowed more heterogeneous variants to obtain higher yield. This situation had a certain relationship to the influence of development programs, which, in search of greater economic income, had favored the use of more homogeneous varieties among the clientele. (2) The morphological variation in beans of 1984 was higher than those of 1940. A combination of factors may be responsible for this change. Among them are genetic drift, natural selection, and selection by humans, as well as the effects of extraneous materials on the mutations and genetic combinations produced in geographically and reproductively isolated populations in that area. On the other hand, the greatest overlap of morphological variation of bean races of 1984 may have been due to genetic exchange between populations previously isolated but put in contact by way of migration processes and selection by man. By means of these last two processes new variants of bean have been introduced, considerably increasing the typical variants of Flor de Mayo, Canario, and Cacahuate. (3) The introduction of bean materials has contributed to the increase of variation at the level of cultivars, local names, and growth habit, the last most frequently seen with variants of the medium vine bean group. (4) The variants of bean in 1984 evidenced earlier maturation than those of 1940, without significant differences in yield and components of yield and quantitative morphological traits of the seed.

Zizumbo V. et al. (1988) studied the farming practices favored by traditional agriculture and its relationship to the conservation of germplasm. Their report recorded the practices carried out on traditional agriculture in the city of Yuríria, Guanajuato (on the slopes of the Cerro de Capulín) with a subhumid temperate climate (ACw' of García, 1973). It registered the use of (1) a high number of variants of cultigens (maize, squash, and bean); (2) inclusion and tolerance of numerous weeds of wild species of *Phaseolus* and teocintle (*Zea mays parviglumis*), among others; and (3) weeding practices that have been carried out usually after an increase in soil moisture. With these conditions the entire plant production is used for human or animal consumption.

Colunga G.-M. et al. (1986) explored the manipulative processes by farmers in the areas of steep slopes with rocky soils in a semiarid temperate climate in the zone of dry rocky agricultural fields ("huamil") of El Bajío region of Guanajuato. Here populations of prickly pear (*Opuntia* spp.) are found in diverse conditions of cultivation, apparently carefully tended by family farmers. These authors recognized in detail the morphological variation of the species of *Opuntia* prevalent in the region. The care given to different species is a function of the appreciation given to the material and the capacity of the terrain to favor its greater development. These observations are supported by multifactorial analyses, which in other cases helped define the taxa involved.

The above studies suggest that the practices related to traditional agriculture in Mexico and the socioeconomic environment in which this agriculture is practiced include processes that have been conserved in *in situ* populations of autochthonous cultigens and congeneric weeds.

CONCLUSIONS

Mexico's ecological and cultural background has promoted the formation of a center of rich natural floristic and human-influenced diversity. During the diverse historical stages of their development, cultigens from other parts of the world have been introduced. Changes that influenced the agricultural technology, e.g., traction animals and local economy, have repercussions in the diversity of domesticated plants and animals. Traditional agriculture provides a rich knowledge of the domestication process in conservation practices. The persistence of ethnic groups with prehistoric background and traditional technology has maintained the empiric processes of selection of domesticated plants and animals, and represents an important form of conservation *in situ* of the diversity of existing germplasm in the country.

ACKNOWLEDGMENTS

The constructive criticisms and positive observations by the reviewers are gratefully acknowledged.

APPENDIX

Representative plants domesticated in Mexico

Plant	Common name	Usage
GYMNOSPERMS		
PINACEAE		
Taxodium mucronatum Ten.	Ahuehuete	O
ANGIOSPERMS		
AGAVACEAE		
Agave fourcroydes Lem.	Henequen	T-I
A. atrovirens Karw. ex Salm	Maguey	EN-fj
A. mapisaga Trel.	Maguey	EN-fj
A. salmiana Otto ex Salm	Maguey pulquero, ixtle	EN-fj; T-I
A. sisalana Perrine	Sisal	T-I
A. tequilana Weber	Maguey	EN-dj; T-I
Yucca elephantipes Regel	Izote	F-fl; LF
AMARANTHACEAE		
Amaranthus cruentus L.	Alegría	F-s
A. hypochondriacus L.	Huauhtli	F-s
Amaranthus spp.	Quintonil	F-v
AMARYLLIDACEAE		
Bomarea edulis (Tuss.) Herb.	Coyolxóchitl	F-r
Polianthes tuberosa L.	Nardo	O
ANACARDIACEAE		
Anacardium occidentale L.	Maranon	F-f
Cyrtocarpa spp.	Chupandilla	F-f
Spondias mombin L.	Ciruela	F-f
S. purpurea L.	Jocote	F-f
ANNONACEAE		
Annona cherimola Mill.	Chirimoya	F-f
A. diversifolia Saff.	Ilama	F-f
A. glabra L.	Anona	F-f
A. muricata L.	Guanábana	F-f
A. reticulata L.	Anona	F-f
A. squamosa L.	Anona	F-f
ARECACEAE		
Chamaedorea tepejilote Liebm.	Tepejilote	F-infl
C. wendlandiana (Oerst.) Hemsl.	Pacaya	F-infl

Appendix (cont.)

Plant	Common name	Usage
ASTERACEAE		
Dahlia coccinea Cav.	Dalia	O
D. excelsa Benth.	Dalia	O
D. australis (Sherff) Sorensen var.		
liebmannii (Sherff) Sorensen	Dalia	LF; O
D. pinnata Cav.	Dalia	O
Helianthus annuus L.	Girasol	F-s
Montanoa spp.	Varablanca	O
Porophyllum tagetoides (HBK.) DC.	Papaloquelite	C-I
Tagetes erecta L.	Cempasúchil	O
T. patula L.	Cempasúchil	O
BIGNONIACEAE		
Crescentia cujete L.	Tecomate	U-f
Parmentiera edulis DC.	Caujilote	F-f
BIXACEAE		
Bixa orellana L.	Achiote	CL-s
BROMELIACEAE		
Ananas comosus (L.) Merr.	Piña	F-f
BURSERACEAE		
Protium copal (Schlecht. & Cham.)		
Engler	Copal	I-rs
CACTACEAE		
Hylocereus undatus (Haw.)		
Britt. & Rose	Pitahaya	F-f
Nopalea spp.	Nopalito	F-st
Nopalea cochenillifera (L.) Salm-Dyck	Nopal	H
Opuntia amyclaea Ten.	Nopal	F-st
O. atropes Rose	Nopal	F-st
O. crassa Haw.	Tuna	F-f
O. ficus-indica (L.) Mill.	Tuna	F-f
O. hyptiacantha Weber	Tuna	F-f
O. joconostle Weber	Joconostle	F-f
O. lasiacantha Pfeiffer	Joconostle	F-f
O. megacantha Salm-Dyck	Tuna	F-f
O. robusta Wendl. var. *larreyi* (Weber)		
Bravo	Tuna	F-f
O. streptacantha Lem.	Tuna	F-f
O. undulata Griffiths	Tuna	F-f
Opuntia spp.	Xoconoztle, nopal, tuna	F-f,st
Pachycereus marginatus (DC.)		
Burger & Buxb.	Organo	LF

Appendix (cont.)

Plant	Common name	Usage
CAPRIFOLIACEAE		
Sambucus mexicana Presl.	Sauco	F-f
CARICACEAE		
Carica papaya L.	Papaya	F-f
CHENOPODIACEAE		
Chenopodium ambrosioides L.	Epazote	C-l
C. berlandieri Moq. subsp. *nuttalliae*		
(Saff.) Wilson & Heiser	Huazontle	F-infr
CONVOLVULACEAE		
Ipomoea batatas (L.) Lam.	Camote	F-r
CUCURBITACEAE		
Cucurbita ficifolia Bouché	Chilacayote	F-f
C. mixta Pang.	Calabaza, pipiani	F-f,s
C. moschata (Duch.) Duch. ex Poir.	Calabaza	F-f
C. pepo L.	Calabaza	F-f
Lagenaria siceraria (Mol.) Standl.	Bule	U-f
Sechium edule (Jacq.) Swartz	Chayote, chinchayote	F-f,r
DIOSCOREACEAE		
Dioscorea spp.	Barbasco	M,F-r
EBENACEAE		
Diospyros digyna Jacq.	Zapote prieto	F-f
EUPHORBIACEAE		
Cnidoscolus chayamansa McVaugh	Chaya	F-l
Euphorbia pulcherrima		
Willd. ex Klotzsch	Nochebuena	O
Jatropha curcas L.	Piñoncillo	H; LF
Manihot esculenta Crantz	Guacamote	F-r
FABACEAE		
Arachis hypogaea L.	Cacahuate	F-s
Canavalia ensiformis (L.) DC.	Haba blanca	F-s
Crotalaria longirostrata Hook. & Arn.	Chipile	F-l
Erythrina americana Mill.	Colorín	LF
Gliricidia sepium (Jacq.)		
Kunth ex Walp.	Cacahuananache	LF
Indigofera suffruticosa Mill.	Añil	C-l,r
Leucaena spp.	Guaje	F-infl,l,s
Leucaena collinsii Britt. & Rose	Guaje	F-s
Pachyrhizus erosus (L.) Urb.	Jícama	F-r

Appendix (cont.)

Plant	Common name	Usage
Phaseolus acutifolius A. Gray	Tepari	F-s
P. coccineus L.	Ayocote	F-s
P. dumosus Macfadyen	Ibes	F-s
P. lunatus L.	Patashete	F-s
P. vulgaris L.	Frijol	F-s
IRIDACEAE		
Tigridia pavonia (L.f.) DC.	Oceloxóchitl	O
LAMIACEAE		
Hyptis suaveolens (L.) Poit.	Chía	F-s
Salvia hispanica L.	Chía	F-s
LAURACEAE		
Persea americana Mill.	Aguacate	F-f
P. schiedeana Nees	Chinine	F-f
MALPIGHIACEAE		
Byrsonima crassifolia (L.) HBK.	Nanche	F-f
MALVACEAE		
Gossypium hirsutum L.	Algodon	T-s
MARANTACEAE		
Maranta arundinacea L.	Sagú	F-r
MORACEAE		
Brosimum alicastrum Swartz	Ramón	F-f
MYRTACEAE		
Psidium guajava L.	Guayaba	F-f
P. sartorianum (Berg) Niedenzu	Guayabilla	F-f
ORCHIDACEAE		
Vanilla planifolia Andr.	Vainilla	C-f
PIPERACEAE		
Piper sanctum (Miquel) Schlecht.	Hoja santa	C-l
POACEAE		
Panicum sonorum Beal	Sauhui	F-s
Zea mays L.	Maíz	F-f
ROSACEAE		
Crataegus pubescens (HBK.) Steud.	Tejocote	F-f
Prunus serotina Ehrh. var. *capuli* (Cav.) McVaugh	Capulín	F-f

Appendix (cont.)

Plant	Common name	Usage
RUTACEAE		
Casimiroa edulis Llave & Lex.	Zapote blanco	F-f
C. sapota Oerst.	Matasano	F-f
C. viride Pitt.	Zapote blanco	F-f
SAPOTACEAE		
Manilkara zapota (L.) van Royen	Chicozapote	F-f
Mastichodendron spp.	Tempesquite	F-f
Pouteria campechiana (HBK.) Baehni	Zapote amarillo	F-f
P. hypoglauca (Standl.) Baehni	Zapote amarillo	F-f
P. sapota (Jacq.) H.E. Moore & Stearn	Zapote mamey	F-f
SOLANACEAE		
Capsicum annuum L.	Chile	C-f
C. frutescens L.	Chile	C-f
Lycopersicon esculentum Mill.	Jitomate	F-f
Physalis philadelphica Lam.	Tomate	F-f
STERCULIACEAE		
Theobroma angustifolium		
Moc. & Sessé ex DC.	Cacao	EN-s
T. bicolor H. & B.	Patashtle	EN-s
T. cacao L.	Cacao	EN-s

The information in this table is based on Colunga G.-M. et al. (1986); Harlan (1975); Hernández K. (1985); Miranda C. (1978).

The use of many of these plants by the peoples of the world underscores the extreme significance of Mexico's biological diversity. The Mexican common names of the domesticated species are given along with the usage of the plant.

C, condiment; CL, colorant; EN, inibrient beverage; F, food; H, host plant; I, incense; LF, living fence; M, medicinal; O, ornamental; T, textile; U, utensil; dj, distilled juice; f, fruit; fj, fermented juice; fl, flower; infl, inflorescence; infr, infructescence; l, leaf; r, root, rhizome; rs, resin; s, seed; st, stem; v, vegetable.

REFERENCES

Ames, O. 1939. Economic Annuals and Human Culture. Cambridge, MA: Botanical Museum, Harvard Univ.

Anderson, E. 1949. Introgressive Hybridization. New York: John Wiley and Sons.

Andrade Aguilar, J. A. & E. Hernández Xolocotzi. 1988. Comparación del plasma germinal de frijol (*Phaseolus vulgaris* L.) de Aguascalientes, México. 1940–1984. Agrociencia 71:257–273.

Beadle, G.W. 1972. The origin of Zea Mays. *In* C.A. Reed (ed.), Origins of Agriculture. The Hague: Mouton Press.

Callen, E.O. 1967. Analysis of Tehuacan coprolites. *In* D.S. Byers (ed.), The Prehistory of Tehuacan Valley. 1(Environment and Subsistence):261–289. Austin, TX: Univ. of Texas Press.

Colunga García-Marín, P., E. Hernández X. & A.C. Morales. 1986. Variación morfológica, manejo agricola tradicional y grado de domesticación de *Opuntia* spp. en el Bajío Guanajuatense. Agrociencia 65:7–50.

Cortés, H. 1970. Cartas de relación. Mexico City: Porrúa.

Cox, G.W. & M.D. Atkins. 1979. Agricultural Ecology. San Francisco, CA: W.H. Freeman.

Cutler, H.C. & T.W. Whitaker. 1967. Cucurbits from the Tehuacan caves. *In* D.S. Byers (ed.), The Prehistory of Tehuacan Valley. Vol. 1 (Environment and Subsistence). Austin, TX: Univ. of Texas Press.

Childe, G.G. 1965. Man Makes Himself. London: Fontana Library.

Deevey, Jr., E.S. 1960. The Human Population. Sc. Amer. 203(3):194–204.

De Landa, F.D. 1978. Relación de las cosas de Yucatán. Mexico City: Porrúa.

Del Amo R., S. & A.L. Anaya. 1982. Importancia de la sistematización de la información sobre plantas medicinales. Biotica 7(2):293–304.

De Wet, J.M.J. & J.R. Harlan. 1975. Weeds and domesticates: evolution in the man-made habitat. Econ. Bot. 29:99–107.

Díaz del Castillo, B. 1960. Historia de la conquista de la Nueva España. Vols. 1 & 2. Mexico City: Porrúa.

Engelbrecht, T.H. 1916. Uber die Entstehung einiger feldmassig angebauter Kultur Plansfen. Geogr. Z. 22:328–334.

Estrada Lugo, E., E. Hernández X., T.R. Rabiela, E.M. Engleman & M.A.C. Márquez. 1988. Códice Florentino: su información ethnobotánica. Agrociencia 71:272–286.

Flannery, K.V. 1986. Guíla Naquitz. Archaic foraging and early agriculture in Oaxaca, Mexico. New York: Academic Press.

García, E. 1973. Modificaciones al Sistema de Clasificación Climática de Köppen. Mexico City: UNAM, Inst. Geog.

Goodman, M.M. 1972. Distance analysis in biology. Syst. Zool. 21:174–186.

Harlan, J.R. 1975. Crops and Man. Madison, WS: American Society of Agronomy.

Hawkes, J.G. 1983. The Diversity of Crop Plants. Cambridge, MA: Harvard Univ. Press.

Hernández Xolocotzi, E. 1985. Biología agricola. Los Conocimientos Biologicos y su Aplicación a la Agricultura. Mexico City: Consejo Nacional para la Enseñanza de la Biología Continental.

————— & Collaborators. 1981. La technología de cultivo. Agroecosistemas 27(2):7–8; 28:5–7.

Hernández, F. 1959. Historia Natural de la Nueva España. Obras Completas Tomos II & III. Mexico City: UNAM.

Iltis, H.H. 1987. Maize evolution and agricultural origins. *In* T.R. Soderstrom, Hila S. Campbell & M.F. Barkworth (eds.), Grass Systematics and Evolution. Washington, DC: Smithsonian Institution Press.

Kaplan, L. 1967. Archaeological *Phaseolus* from Tehuacan. *In* D.S. Byers (ed.), The Prehistory of Tehuacan Valley. Vol. 1 (Environment and Subsistence). Austin, TX: Univ. of Texas Press.

Longley, A.E. & T.A. Kato. 1965. Chromosome morphology of certain races of maize in Latin America. Research Bullletin No. 1. El Batán, México: International Center for the Improvement of Maize and Wheat.

MacNeish, R.S. 1961. First Annual Report of the Tehuacan Archaeological-botanical project. Andover, MA: R.S. Peabody Foundation for Archaeology.

Mangelsdorf, P.C., R.S. MacNeish & W.C. Galinat. 1967. Prehistoric wild and cultivated maize. *In* D.S. Byers (ed.), The Prehistory of Tehuacan Valley. Vol. 1(Environment and Subsistence):178–200. Austin, TX: Univ. of Texas Press.

McClintock, B., T.A. Kato & A. Blumenschein. 1981. Chromosome constitution of races of maize. Chapingo, México: Colegio de Postgraduados.

Miranda C., S. 1966. Discusión sobre el origen y la evolución del maize. Memorias del Segundo Congreso Nacional de Fitogénetica. Chapingo, México: Colegio de Postgraduados.

———. 1978. Evolución de cultivares nativos de México. Ciencia y Desarrollo 21:130–131.

Ortega, P. & R. Angel, 1973. Variación en maize y cambios socioeconomicos en Chiapas, México. 1946–1971. M.S. thesis, Colegio de Postgraduados, Chapingo, México.

Rindos, D. 1984. The Origins of Agriculture: An Evolutionary Perspective. New York: Academic Press.

Rojas R., T. 1983. La Agricola Chinampera. Compilación historica. Chapingo, México: Univ. Autónomo Chapingo, Dirección de Difusión Cultural.

——— & W. Sanders (eds.). 1985. Historia de la agricultura. Epoca prehispanica. Siglo XVI. Vols. 1 & 2. Mexico City: INAH.

Sahagún, F.B. 1980. Códice Florentino. Edición Fascimil de Manuscrito 218–220 de la Colección Palatina de la Biblioteca Medicea Laurenziana. Vols. 1–3. Mexico City: Archivo General de Nación.

Smith Jr. C.E. 1967. Plant remains. *In* D.S. Byers (ed.), The Prehistory of Tehuacan Valley. Vol. 1(Environment and Subsistence):220–255. Austin, TX: Univ. of Texas Press.

Schwanitz, F. 1967. The Origin of Cultivated Plants. Cambridge, MA: Harvard Univ. Press.

Torres W., B. 1985. Las plantas útiles en el México antiguo según las fuentes del siglo XVI. *In* T. Rojas R. & W. Sanders (eds.), Historia de la agricultura. Epoca prehispanica. Siglo XVI. Vols. 1 & 2. Mexico City: INAH.

Vavilov, N.I. 1951. The origin, variation, immunity and breeding of cultivated plants. Chron. Bot. 13:1–366

Wellhausen, E.J., L.M. Roberts & E. Hernández X. In collaboration with P.C. Mangelsdorf. 1951. Razas de Maize en México. Su Origen, Caracteristicas y Distribución. Folleto técnica No. 5. Mexico City: Secretaria de Agricultura y Ganadaría.

Williams, D.E. 1985. Tres arvenses solanaceas comestibles y su proceso de domesticación en el estado de Tlaxcala, México. M.S. thesis, Colegio de Postgraduados, Chapingo, Mexico.

Yakoleff Greenhouse, V., E. Hernández X., C. Rodjind de Cuadra & C. Larralde. 1982. Electrophoretic and immunological characterization of pollen protein of *Zea mays* races. Econ. Bot. 36(1):113–123.

Zizumbo Villreal, D., E. Hernández X. & Heriberto Cuanalo de la Cerda. 1988. Estrategias agricolas tradicionales para el aprovechemiento del agua de lluvia durante el temporal: el caso de Yuríria, Guanajuato. Agrociencia 71:315–340.

VI
REVIEW OF HABITATS

26

The Biodiversity Scenario of Mexico: A Review of Terrestrial Habitats

VICTOR MANUEL TOLEDO
AND MA. DE JESÚS ORDÓÑEZ

It is estimated that Mexico's biodiversity comprises more than 12% of the biota of the world. They are located predominantly in the six principal terrestrial habitats (humid tropic, subhumid tropic, temperate humid, temperate subhumid, arid to semiarid and alpine zones) of the country. The status of these habitats and estimates of protected areas in each of these habitats are given, and estimations of the total areas of these habitats in Mexico and their principal land uses are provided. Provisional figures on species richness in these habitats are offered, and the nature of loss of biodiversity due to deterioration of these habitats is briefly discussed.

Biologically, Mexico is recognized as one of the most diverse countries in the world (Mittermeir, 1988; Toledo, 1988). This statement can be corroborated by a comparison of species numbers of the principal groups of organisms that have been documented and named on a worldwide level (Wilson, 1988) with the known or estimated number of existing species in Mexico (Table 26.1). According to the latest inventories the number of described species is around 1.4 million, of which nearly two-thirds correspond to species from temperate regions and the remaining one-third to species of intertropical regions (Raven & Johnson, 1986). Given that only two-thirds of the temperate species have been described and that 80% of the tropical species are still unknown to science, it can be estimated that a total of over 5 million organisms inhabit the Earth. This global estimate may be conservative, however, given the high number of species of invertebrates in the most recent inventories, such as that by Erwin (1988) on insects that inhabit the canopies of the tropical forests.

The data in Table 26.1 show Mexico as having more than 12% of the total extant biota of the world. This pattern is generally true for the better known groups of organisms, e.g., plants, fungi, terrestrial vertebrates, and butterflies (an exceptionally well-studied invertebrate group). The Mexican flowering plants account for 10–12% of the world total, whereas the crustaceans represent 23%. The Mexican biological inventory must wait for the identification and classification of between 400,000 and 500,000 taxa (of which most belong to the invertebrates, especially the insects and mollusks) to substantiate the above observation. This remarkable biodiversity, roughly one-tenth of the world's organisms, found in Mexico has led Mittermeir (1988) to place it among the seven most biologically diverse countries in the world—after Brazil and Colombia and ahead of Indonesia, Madagascar, Zaire, and Australia. The above appears to be corroborated when particular groups of organisms are analyzed. Mexico seems to be number one in terms of the number of reptile species (Flores-V., this volume) and possibly of mammals (compare, for example, the numbers in Eisenberg, 1981) and is exceptionally rich in flowering plants (Toledo, 1988), sea grasses (Lot, 1971), mosses (Delgadillo, this volume) and possibly crustaceans.

This unusual biodiversity, represented by a great variety of species, finds its expression concretely in the geographic and ecological distribution of the organisms, i.e., in their location in different spaces that make up the country's territory. The analysis of the status of each of the principal natural habitats (terrestrial and aquatic) of Mexico is thus an obligatory task. It would permit consideration of the degree of threat to the principal groups of organisms as well as the biological importance of each habitat. Keeping this point in view, the present chapter provides a conservation-motivated analysis of the biodiversity of Mexico from an ecogeographic perspective.

Table 26.1. Known and estimated number of species in Mexico for principal groups of organisms

Organism	Worldwide[1]	Mexico	%
PLANTS	266,000	29,000–34,000[2]	10.90–12.78
Anthophyta	235,000	25,000–30,000[2]	10.63–12.76
Orchidaceae	17,500	935[3]	5.34
Cactaceae	1,650[4]	900[5]	54.50
Leguminosae	16,400	1,706[6]	10.40
Cycadophyta	100	9[7]	9.00
Coniferophyta	550	80[8]	14.54
Pteridophyta	12,000	900–1,000[9]	8.33
Bryophyta	16,600	2,000[10]	12.04
ALGAE			
Marine	26,900	1,787[11]	6.64
Freshwater	?	ca. 400[11]	
FUNGI	46,983	4,000–5,000[12]	10.64
Myxomycetes	500	172[13]	34.40

Table 26.1. (cont.)

Organism	Worldwide[1]	Mexico	%
ANIMALS			
VERTEBRATES	45,202	4,361	9.60
Mammalia	4,500	439[14]	9.75
Aves	9,000	961[14]	10.67
Reptilia	5,965	693[14]	11.61
Amphibia	4,014	285[14]	7.10
Marine Fish	21,723	1,738[15]	7.37
Freshwater Fish	?	384[15]	?
INVERTEBRATES	(1,307,823)		
Porifera	5,000	?	
Cnidaria and Ctenophora	9,000	?	
Platyhelminthes	15,000	594[16]	3.96
Nematoda	12,000	217[16]	1.80
Annelida	12,000	?	
Mollusca	200,000[17]	10,000[17]	
Echinodermata	6,100	?	
Arthropoda			
Insects	750,000	?	
Apoidea	20,000	2,000[18]	10.00
Lepidoptera	220,000	25,000[19]	11.36
Papilionoidae	20,000	2,250[19]	11.25
Crustacea	38,723	8,915[20]	23.02
TOTAL	1.4 Million		

Sources: [1]Raven & Johnson (1986); Wilson (1988). [2]Toledo (1988); Rzedowski (1989). [3]Soto Arenas & Castillo Arguero (1989). [4]Mabberly (1987). [5]Rzedowski (this volume) [6]Sousa & Delgado (this volume). [7]Vovides et al. (1983). [8]Styles, (this volume). [9]Riba (this volume), [10]Delgadillo (this volume). [11]Martha Ortega (pers. comm.) [12]Gastón Guzman (pers. comm.) [13]Villareal & Pérez-Moreno (1988). [14]Nature Conservancy Data Bank. [15]Espinosa et al. (this volume; about 375 marine species are also found in continental waters of the country). [16]Lamothe (pers. comm.) [17]Cifuentes (1986). [18]Ayala et al. (this volume). [19]Llorente & Luis (this volume). [20]Villalobos et al. (1988).

PRINCIPAL NATURAL HABITATS OF MEXICO

Every species is represented in space three ways (Whittaker, 1975).

1. *Area*: its geographical distribution that is mappable
2. *Habitat*: refers to the type of environment in which the species is found and which tends to be defined by a type of biotic community (a species may occupy a whole range of habitats or more than one within its area of distribution)

3. *Niche*: the spatial-temporal position and the role the species plays within a particular habitat

Although in the strictest sense the criteria needed to define a habitat remain established by the specific nature of the species under study (it is not the same to speak of a tree or a vertebrate as it is of an insect or a moss), for the purposes of this essay the habitats are defined broadly. A distinction must also be made between *terrestrial* and *aquatic habitats*. The task is to divide Mexican territory (terrestrial and aquatic) into units that make sense from a biological point of view and that are valid for the wide spectrum of organisms comprising the biotic universe.

In the case of the terrestrial habitat, Mexican territory may be divided relatively easily based on the distribution of two elements: vegetation and climate. Several authors have divided the country into units based on these two criteria in order to study the culture (West, 1964), flora (Rzedowski, 1978), vertebrate fauna (Flores-V. & Gerez, 1989), "ecosystems" (Pérez-Gil et al., 1984), food production (Toledo et al., 1985) or to organizing the nation's territory (SEDUE, 1984). Even though many vegetation types can be distinguished (Miranda & Hernández-X. [1963] recognized 32 types; SPP [1981], 45; and the Comisión Técnica Consultiva para la Determinación de los Coeficientes de Agostadero, 71), the grouping by certain physiognomical and phytogeographical criteria and, above all, the correlation with the principal types of climate permits us to arrive at large environmental units in a form equivalent to the concept of *natural region*, *biome*, or (at another level) the *ecological zone* (Toledo et al., 1989). Six natural terrestrial habitats, defined by vegetation, climate, and biogeography (Table 26.2; Figs. 26.1 and 26.2) are recognized here. The methodology and the criteria used to define these six zones were described by Toledo (1991). This division of territory has been done following the climatological analysis most recently done by García (1989), and it coincides in general with those adopted by Rzedowski (this volume) and Flores-V. (this volume) in their respective analyses of the diversity of the flora and the herpetofauna of Mexico. To these six terrestrial habitats, here termed ecological zones, should be added at least two more that have special attributes defined differently: the *zone of continental waters* (including rivers, lakes, and swamps) and the *coastal zone*, which is an area of transition between the continental mass and the seas (see below). These two zones, however, possess particular components that should not be understood within the terrestrial portion of the country and thus should be considered separately.

The surface area of marine habitats exceeds the terrestrial area and can also be regionalized, although it has the peculiar attribute of not being separable from the coastal fringe of the continent for ecological reasons (Yañez-Arancibia, 1986). Hence the distinctions of natural habitats in the marine environments presuppose an overlapping with the regionalization of the terrestrial environments on the coastal fringes. Based on the propositions made by various authors (Benitez, 1988; Contreras, 1985;

Table 26.2. Principal characteristics of the six ecological zones in Mexico

Habitat	Estimated area[a] (in million ha)	Municipalities[b] >75%	Municipalities[b] <75%	Municipalities[b] Total No.	Dominant vegetation	Climate type[c]
Humid tropic	22	251	84	335	High and medium evergreen forests and savannahs	Am & Af
Subhumid tropic	40	578	247	825	Deciduous forests	Aw
Humid temperate	1	48	68	116	Mixed forests	A(C)m, C(A)m
Subhumid temperate	33	687	381	1,068	Pine, oak, and mixed forest	CW
Arid and semiarid	99	384	125	509	Scrub and grass land	Bs, Bw
Alpine	0.3	—	—	—	High, wide barren plains	E

[a]Toledo et al., 1989.
[b]Municipalities are roughly equivalent to counties in the United States.
[c]According to the Köeppen-García climatic system (see García, 1989).

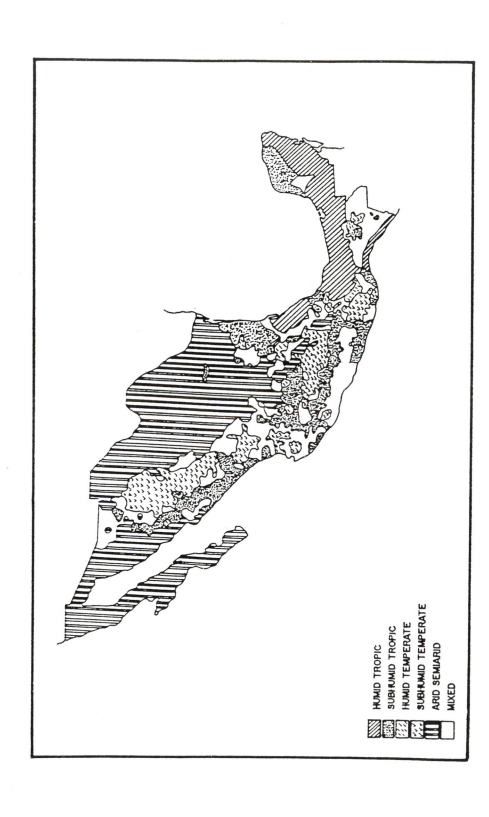

HUMID TROPIC

SUBHUMID TROPIC

HUMID TEMPERATE

SUBHUMID TEMPERATE

ARID SEMIARID

MIXED

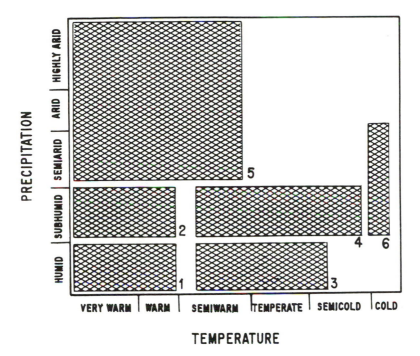

Figure 26.2. Terrestrial ecological zones of Mexico. (See text for details.)

Lankford, 1977; Ortuzar, 1978), six to eight principal ecologically defined zones can be distinguished in the marine section of Mexico. The most developed regionalization seems to be that of Benitez (1988), who proposed dividing the Mexican marine-estuarine environment into eight ecological zones distinguished based on criteria such as surface currents, tidal pattern, surface temperature and salinity, coastal geology, continental drainage, type of coastal lakes, characteristics of the continental platform, and climate of the coastal fringe. Each of these zones, according to Benitez, may be subdivided into three subzones: estuarine, neritic, and oceanic (greater and lesser). The zonification of the marine-estuarine environment may be analyzed with regard to distribution of the species of groups such as fish, crustaceans, mollusks, and echinoderms.

A complete analysis of the natural habitats of Mexico should realistically include the terrestrial as well as the marine habitats. This study, however, is restricted to the continental habitats. The analysis of the situation of the marine organisms of Mexico in their different habitats is nonetheless a task that should be done as soon as possible.

Figure 26.1. Terrestrial ecological zones of Mexico. The legend defines and illustrates the terrestrial ecological zones discussed in the chapter.

Table 26.3. Biological importance of the principal terrestrial habitats of Mexico

Inhabitants	Z1 (Humid tropic)	Z2 (Subhumid tropic)	Z3 (Humid temperate)	Z4 (Subhumid temperate)	Z5 (Arid & semiarid)	Z6 (Alpine)
Plants						
Anthophyta	R+	R++E++	R+E+	R+++E+++	R++E++	E
Orchidaceae	R	—	RE	—	—	—
Cactaceae	—	—	—	—	RE	—
Coniferophyta	—	—	—	RE	—	—
Pteridophyta	R	—	—	—	—	—
Bryophyta	R	—	R	R	—	E
Animals						
Terrestrial vertebrates	R+	R++	?	R+++	R++	—
Mammals	—	—	E	E	—	—
Birds	R	—	R	R	—	—
Reptiles	E+	E++	R-E+++	R-E+++	E+	—
Amphibians	E+	E++	R-E+++	R-E+++	E+	—
Lepidoptera (especially Papilionoidea)	R-E	—	R-E	—	—	—

R, richness in number of species; E, richness in number of endemisms.

STATUS OF THE TERRESTRIAL HABITATS

The terrestrial habitats of Mexico are of biological importance because they contain a great variety of species (on any of the three scales on which diversity can be measured) or they are occupied by groups of rare species or groups with a geographic distribution either restricted or circumscribed to Mexican territory (Table 26.3). An analysis of the status of each of the principal terrestrial habitats of the country therefore becomes an obligatory task. In the sections that follow, an evaluation of the degree of transformation suffered by the principal terrestrial habitats as a result of the human activities (agricultural, livestock, forest extraction, industrial) is provided. The consequences of these changes to the biodiversity of each of the habitats are discussed. The survey is based on the revision of the principal publications found in the literature and, in the case of the first five ecological zones, from two principal sources: (1) analysis of the statistical information contained in the VI Census of Agriculture, Livestock and Common Lands (1981) produced by Toledo et al. (1989) at the municipal level and by ecological zone; and (2) data from the digitalization of the Vegetation and Current Soil Use Map, Carta de Vegetación y Uso Actual del Suelo (scale 1:1,000,000) from the Atlas Nacional del Medio Fisico (SPP, 1980) which in turn was synthesized by the Programa Nacional de Desarrollo Ecológico de los Asentamientos Humanos (SAHOP, 1981) on a scale of 1:4,000,000. The sources in both provide information on the extension of the agricultural and livestock frontier and the "forest" area. Whereas the first derives its data from surveys directly done for the land, the second does it from a synthesis of information from aerial photographs and satellite images obtained on different dates over a decade during the 1970s. The first provides data on a more local level (municipalities) and the second on a more regional level. Despite the limitations, deficiencies, and inaccuracies, these materials constitute the *only* resources of available information that permit an evaluation at the *national level* on the degree of transformation of the principal terrestrial habitats in Mexico.

HUMID TROPIC

In Mexico the humid tropical zone, characterized by the highest thermal systems (average annual temperatures above 22°C), precipitation (2,000 mm or more per year), and an original covering of medium and high forest trees or savannahs, extends over nine southern and southeastern states. Estimated by authors to cover between 6% (Pennington & Sarukhán, 1968) and 12.8% (Leopold, 1950) of Mexico, this zone is comprised of 20.15 million hectares (ha) distributed in 251 municipalities where 75% or more of their territory corresponds to this ecological zone (Toledo et al., 1989). This area, obtained by detailed calculations from the vegetation and climatic maps at the municipal level, once adjusted, coincides with the estimate of 11% proposed by Rzedowski (1978).

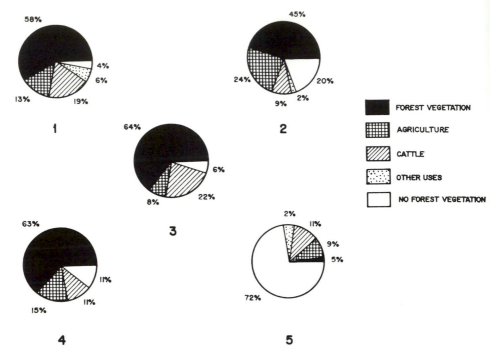

Figure 26.3. Land use in the five principal ecological zones of Mexico. *Source*: Toledo et al. (1989).

Table 26.4. Land use in the various zones

Zone and state	Total	Agriculture	Livestock	Forest	Other
Humid tropic					
Quintana Roo	4,496,100	143,840	122,063	3,648,804	290,738
Veracruz	4,678,935	1,175,471	1,384,683	1,860,224	57,177
Campeche	4,006,400	241,074	471,823	2,940,529	107,053
Tabasco	2,365,300	399,619	1,122,942	633,117	45,863
Oaxaca	2,017,702	283,606	151,805	1,257,081	106,567
Chiapas	1,807,800	279,880	471,100	923,644	41,126
Others	771,900	435,574	64,804	432,864	76,448
TOTAL	20,154,137	2,719,990	3,789,220	11,696,263	724,972
Subhumid tropic					
Yucatan	4,548,830	945,046	255,611	3,188,937	107,020
Sinaloa	3,965,180	797,876	77,737	2,539,050	445,211
Sonora	3,883,500	142,754	157,856	1,091,990	2,461,271
Michoacan	3,355,500	873,355	219,329	1,150,706	1,044,628
Jalisco	2,438,100	739,881	195,926	800,760	644,063
Tamaulipas	2,146,500	393,683	303,961	835,101	478,777
Guerrero	2,001,200	517,189	62,192	1,239,150	143,950
Oaxaca	1,919,590	306,776	282,578	846,962	423,607

Table 26.4. (cont.)

Zone and state	Total	Agriculture	Livestock	Forest	Other
Guanajuato	1,515,700	927,345	133,092	62,634	350,594
Veracruz	1,273,700	596,806	439,937	167,862	2,215
Puebla	1,012,700	332,596	336,870	284,295	55,934
Others	4,515,500	1,174,079	402,337	2,371,020	437,283
TOTAL	32,576,000	7,747,377	2,867,426	14,578,467	6,594,553
Subhumid temperate					
Chihuahua	7,885,400	742,819	997,136	5,525,317	583,732
Michoacan	4,158,230	416,719	65,682	3,436,172	212,817
Oaxaca	2,985,014	309,973	549,777	1,488,176	609,271
Durango	2,782,900	94,181	198,933	2,161,159	314,491
Jalisco	1,226,000	161,348	117,986	789,412	148,853
Puebla	1,217,000	492,026	239,506	398,872	77,296
Mexico	1,134,200	585,643	101,700	363,318	37,855
Others	5,389,200	1,196,179	639,214	2,660,471	841,012
TOTAL	26,777,944	3,998,888	2,909,934	16,822,897	2,825,327
Arid and semiarid					
Coahuila	14,606,900	514,634	733,538	329,980	12,737,598
Chihuahua	14,419,069	405,136	4,641,667	215,569	8,328,991
Sonora	13,815,507	890,204	1,253,884	265,549	9,734,981
Baja Cal. N.	6,761,730	482,237	143,314	58,595	5,647,165
Baja Cal. S.	6,689,800	80,581	2,324	36,295	6,323,748
Zacatecas	5,621,819	1,122,456	424,770	96,726	3,906,964
Durango	5,009,600	540,708	902,895	196,359	3,320,525
Nuevo Leon	4,912,880	348,901	284,660	239,850	4,022,075
Tamaulipas	3,898,400	892,184	440,133	84,347	2,146,361
S. L. Potosi	3,567,000	514,336	235,200	147,380	2,632,502
Others	4,721,730	1,997,938	601,604	333,735	1,653,802
TOTAL	84,024,435	7,789,315	9,663,989	2,004,385	60,454,712

Land measures are in hectares.
Source: Toledo et al. (1989).

The analysis by Toledo et al. (1989), based on the data from the Censo Agropecuario y Forestal of 1981, suggests that 13% and 19% of the humid tropical zone have been already transformed for agricultural (2.71 million ha) and livestock needs (3.78 million ha), respectively (Fig. 26.3). The conversion of forests to agricultural land has been done principally in the states of Veracruz, Tabasco, Oaxaca, and Chiapas, whereas the transformation to livestock pastures has taken place in Veracruz, Tabasco, and Chiapas (Table 26.4). These numbers leave a forest cover of 68% (13.66 million ha) including both primary and secondary forests. If the data derived from the census should be adjusted to 10% more (Toledo et al., 1989), the above figure coincides roughly with that obtained by the map

ENDEMISM   RICHNESS

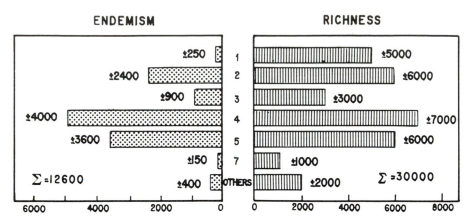

Figure 26.4. Floristic richness and endemism of the terrestrial habitats. 1. Humid tropic. 2. Subhumid tropic. 3. Humid temperate. 4. Subhumid temperate. 5. Arid/semiarid. 7. Aquatic vegetation. *Source*: Rzedowski (this volume).

representation of the Atlas Nacional del Medio Fisico (SPP, 1981) and whose area was calculated by way of digitalization by computer: 13.16 million ha. On the other hand, the numbers coincide in the agricultural-livestock transformation in Tabasco (humid tropical state with an area of 2.46 million ha), where the data of the census, the Atlas, and a state study (SARH, 1979) reported areas dedicated to agriculture and livestock at 1.52, 1.50, and 1.52 million ha respectively.

The above estimates suggest a deforested area equivalent to 40% of the total of the humid tropical zone by the end of the 1970s and beginning of the 1980s, and offer a panorama less unfavorable than that suspected. The large number of pronounced "estimates" without basis over the last few years give a skewed perception of the degree of destruction of the humid tropical zone of Mexico. Nonetheless, the following details should not be omitted. The figures obtained do not permit, however, an account of the portion of the area with forest cover that corresponds to the primary forests. Moreover during the last few years the expansion of the livestock frontier in the humid tropics has reached its highest levels (Toledo, 1987); in some states, the figures from certain studies on grazing area are far above those expected according to the data of the census, as in the case of Chiapas (Fernández Ortíz & García, 1983). It is not currently possible to provide exact figures regarding the area of this zone with vegetation that has not been affected. Estrada and Coates-Estrada (1988) have, however, estimated the areas of Mexico with well preserved areas of tall and medium forests and have given figures for the wet tropics. They did not, however, provide the original sources of their data, nor did they detail the methodology used for their analysis. They identified five areas as the last important redoubts of medium and tall forests in Mexico: Sierra Huasteca (with 100,000 ha), region of Los Tuxtlas in Veracruz (36,300 ha), region of

Uxpanapa, Veracruz (100,000 ha), Selva Lacandona in Chiapas (300,000 ha), and Reserva de Sian Ka'an in Quintana Roo (145,000 ha). These estimates would reduce the humid tropical zone to less than 5% of its original distribution, a figure that remains below the 10% estimated by Rzedowski (1978).

From the point of view of biodiversity, the humid tropical zone of Mexico is, as are its homologs in the rest of the world, the richest zone on a scale of micro or alfa diversity. One hectare alone of high forest in Bonampak, Chiapas contains 267 plant species (Meave del Castillo, 1983), whereas a site of about 1,000 ha supports more than 1,000 plant species, some 300 bird species, and 150 amphibian and reptile species (Toledo, 1991). On a broader scale, this ecological zone has a global richness lower than other zones (see below) and, although endemism is notable in groups such as amphibians, reptiles, and butterflies (Table 26.3), it is less significant in the case of vascular plants. In his most recent work, Rzedowski (this volume) shows that this zone in Mexico contains around 5,000 species of plants, of which about 250 are endemic (Fig. 26.4).

SUBHUMID TROPIC

Characterized by a hot climate with a dry period of five to nine months a year and by tropical deciduous forests or similar vegetation types, the subhumid tropical zone extends over 32.5 million ha (Table 26.4) in 578 municipalities (where 75% or more of its territory is found in this zone) and 21 states in Mexico (Toledo et al., 1989). In its distribution, this zone covers important portions of the Pacific coastal plain, the Yucatan Peninsula, central Veracruz, southern Tamaulipas, and western and southern Mexico. As with its homologs in Central America, this habitat has been extensively transformed (Fig. 26.3) for agricultural (24%), livestock (20%), and other (9%) uses so that only 45% of the forest cover had not been removed at the beginning of the 1980s (Toledo et al., 1989). The agricultural areas in this zone principally cover the states of Sinaloa, Guanajuato, Michoacan, Jalisco, and Yucatan, whereas the livestock areas cover Veracruz, Tamaulipas, and Oaxaca. In comparison with the humid tropical zone, the impact of transformation of this habitat comes more from agriculture (7.74 million ha) than from livestock (2.86 million ha).

From a biological point of view, the subhumid tropical zone, which in some areas has a floristic richness comparable to that in medium and tall forests, as in Chamela, Jalisco (Lott et al., 1987) on the scale of microdiversity, is of little importance in groups of hygrophilous plants such as orchids, ferns, and mosses, given the marked seasonal character with regard to the precipitation. On the other hand, this zone has greater indices of endemism, estimated at 40% by Rzedowski (this volume), than the preceding zone and presents a global floristic diversity assessed at 6,000 species (Fig. 26.4). The zone is also important with regard to species richness (Table

26.3) and amphibian and reptile endemisms, especially on the Pacific side (Flores-Villela, pers. comm.).

HUMID TEMPERATE

The cloud forests whose floristic composition includes both boreal and tropical elements (Miranda & Sharp, 1950) characterize the humid temperate zone in areas with subtropical climates. Located in the medium parts of the mountain chains (between 600 and 2,500 masl), it occupies restricted sites, especially on the slope of the Gulf of Mexico and extends from Tamaulipas to Chiapas. Despite the small area, this habitat merit the category of "ecological zone" because they constitute a phytoclimatic and biogeographic formation different from the mountains in Mexico. Distributed through 116 municipalities in eight states, this zone reaches an area of one million ha, principally distributed in the states of Oaxaca, Veracruz, and Chiapas, given that in the 48 municipalities with 75% or more of the territory in this zone, the total area is something more than 500,000 ha. According to the analysis of the census of 1981 (Toledo et al. 1989), this zone presents a forest cover of more than 60% with an agricultural area equivalent to 8% of the total land and a livestock area of 22% (Fig. 26.3).

Despite its relatively small area, the humid temperate zone is of great importance from the biological point of view. In the case of plants, it is a habitat high in species richness and endemisms, especially in some groups, e.g., orchids, ferns, and mosses. Taken globally, it presents a phanerogamic flora estimated at 3,000 species with an endemism of 30% (Rzedowski, this volume). Given the small area and the large number of species, it constitutes floristically the richest zone in Mexico by unit of area (Rzedowski, this volume). The zone is notable as well for its large number of endemisms among mammals, amphibians, reptiles, and butterflies—to such an extent that it is considered one of the principal centers of autochthonic species, resulting from its history and insular distribution.

SUBHUMID TEMPERATE

The subhumid temperate zone covers the greatest part of the mountainous areas of Mexico. Rich in pines, oaks, and mixed forests, the zone is distributed in areas with Cw climates (García, 1989) and covers an area of around 33 million ha in more than 1,000 municipalities of 20 states (Table 26.2). The area it covers in the 687 municipalities with 75% or more of its territory in this zone is 26.77 million ha, of which 15% constitute agricultural area, 11% livestock area, 11% nonforest vegetation, and 63% forests (Fig. 26.3). The states with the most extensive quantities of vegetation are, in order of importance, Chihuahua, Michoacan, Durango and Oaxaca (Table 26.6).

This insular "ecological zone" constitutes a habitat of enormous biological and biogeographical importance, as its main distribution is along the principal mountain chains in the country. Studies show that it is the most important biological zone. It is notable for its high species richness and endemism in flowering plants (Rzedowski, this volume), conifers (Styles, this volume), terrestrial vertebrates in general (Flores & Gerez, 1989), mammals, amphibians, and reptiles (Flores-Villela, this volume). Of special importance is the mammalian fauna that this zone holds along the Trans-Mexican Neovolcanic Belt, because it harbors one of the highest concentrations of species diversity and endemism presently known (Fa, 1989; Fa & Morales, this volume). Rzedowski (this volume) estimates 7,000 species in flowering plants (Fig. 26.4) of which 4,900 (or 70%) are endemic.

ARID AND SEMIARID ZONES

The arid and semiarid land comprises the most extensive ecological zone with an area almost equal to half of Mexican territory. Characterized by the scarce quantity of annual precipitation, this habitat includes in reality two large bioclimatic zones: (1) the arid zone, defined by an annual precipitation of 40 mm or less and 8–12 dry months; and (2) the semiarid zone with annual precipitation between 400 and 700 mm and 6–8 dry months, which correspond to the Bs and Bw climate types of the Koeppen-García system. This distinction permits a rough separation of the two principal types of vegetation that characterize the habitat: scrub and pasture land, although this "ecological zone" could be broken into numerous units (e.g., Miranda [1955] distinguished 43 based on the biotypes or biological forms of plant species).

Distributed in more than 500 municipalities in the principal northern and central states of Mexico and with a range near 100 million ha, this zone houses more than 20 types of desert or xerophytic scrub and a large variety of pasture land and halophytic vegetation. Of the 84 million ha corresponding to the 384 typically arid and semiarid municipalities (with more than 75% of its territory in this zone), an area of only a little less than 8 million ha was transformed into agricultural areas and more than 9 million ha was considered livestock area (Table 26.4, Fig. 26.3). These data nonetheless underestimate the true livestock area of the zone because the livestock areas of an extensive nature where cattle forage directly on the scrub species are not considered as such. It hampers estimation of the area of this habitat affected by livestock activities. To the contrary, revision of the livestock statistics indicates that toward 1980, the cattle herds of this "ecological zone" surpassed 8 million heads, occupying an estimated area of 57 million ha (Toledo, 1987), i.e., more than half of the total area of this habitat. The same may be said of the forest activities of extractive character that affect enormous areas of this habitat, as the exploitation of thousands of products of the desert plants such as the *candelilla* (*Euphorbia antisyphillitica*), *guayule* (*Parthenium argentatum*), *jojoba* (*Simmondsia chinensis*), *lechuguilla* (*Agave lechuguilla*), *mesquite* (*Prosopis juliflora*), and other species is not registered by this type of census. To this problem must be added the severe extraction and illegal traffic of numerous species of cacti from which broad northern arid areas suffer.

Rzedowski (this volume) estimates that the flora of the arid and semiarid zone of Mexico contains 6,000 species (Fig. 26.4), and it follows the temperate subhumid zone as the habitat with the greatest number of endemics (approximately 60% of the total flora). Explosive speciation has taken place in groups such as the cacti, composites and grasses. This habitat presents as well a relatively high diversity of vertebrate terrestrial species (Table 26.3) and is especially rich in amphibian and reptile endemism (Flores-Villela, this volume).

ALPINE ZONE

The areas found above the timber line at 4,000 masl in Mexico with a cold type of climate (E climate in the Koeppen system) are of notable biological and biogeographical importance. These areas are dominated by the so-called zacatonales or paramos located on the 12 highest mountains in the country. The climate of this zone is characterized by an average annual temperature of between 3° and 5°C, by the permanent presence of frost and snow, and by an annual precipitation of 600–800 mm (Rzedowski, 1978). From a floristic point of view, this zone shelters a flora of its own where 75% of the species are endemic to Mexico (Rzedowski, 1978; McDonald, this volume). Given its inaccessibility to human activity, these "ecological zones" are used by cattle, a practice that usually accompanies more or less regularly induced fires, started with the intention of stimulating the growth of tender stalks for the animals.

PROTECTED RESERVES IN MEXICO

These changes in Mexico's terrestrial habitats have affected the conservation status of the organisms that inhabit them in various ways. It is relevant in this context to review the status of Mexico's natural preserves. Although such preserves date back to the turn of the century in Mexico (Vargas, 1984), the establishment of the national network of protected areas through the creation of the Department of Urban Development and Ecology (Secretaria de Desarrollo Urbano y Ecología, SEDUE) in 1983 is recent. The governmental agency, the National System of Protected Natural Areas (Sistema Nacional de Areas Naturales Protegidas, SINAP) supervises and administers the reserves under the direction of SEDUE. As of November 1989, SINAP had under its administration 66 protected areas with an area of 5.73 million ha, which is 2.88% of the national territory. Of this land, 4.47 million

Table 26.5. Status of the protected reserves of Mexico

Kind of reserve	No.	Surface area (hectares)	Percent of national total
Protected areas of flora and fauna	10	486,365	0.24
National parks	43	688,953	0.34
Biospheric reserves	9	4,473,926	2.23
Special biospheric reserves	2	49,640	0.025
Natural monument	1	2,580	0.0001
Protected forest reserve and animal refuge	1	28,600	0.014
Total	66	5,730,064	2.88

Source: Data from SEDUE.

ha are in nine biospheric reserves, 658,953 ha in 43 national parks, and 486,365 ha in ten protected areas (Table 26.5).

Interesting patterns emerge when the theoretically protected surface area of Mexico is analyzed in light of the terrestrial habitats proposed in this report (Table 26.6). These patterns become clear when a map of each of the protected areas is superimposed on the ecological zones proposed by Toledo et al. (1989). The terrestrial habitat with more area under protection is the arid/semiarid zone, which has 15 protected areas with an area exceeding 3 million ha. It is a result of the recent creation of two biospheric reserves, Ojo de Liebre in Baja California and El Pinacate in Sonora. The humid tropic with six areas (1.246 million ha) and the temperate with 22 areas (325,000 ha) follow the arid/semiarid zone in surface area (Table 26.6). A total of 16 protected areas including Montes Azules (Chiapas 331,000 ha), Sierra de Manantlán (Jalisco 139,572 ha), and El Cielo (Tamaulipas 130,00 ha) include transitional or combinations of different ecological zones and hence may be considered multizonal in character (Table 26.6). It suggests that the data of the supposedly protected areas in each ecological zone should be adjusted, combining the corresponding areas from the multizonal areas. Various conclusions may be drawn from this analysis of the relative cover (percentages of the total surface area apparently protected in each of the zones). The humid tropic zone with about 6.5% of its total area and the humid temperate zone with 100,000 ha of its total area of about 1 million ha are two of the larger protected areas among the terrestrial habitats. The above contrasts sharply with the unhappy situation found in the subhumid tropic and temperate zones which are poorly represented in SINAP. Less than 0.5% of the approximately 32 million ha of the subhumid tropic is in the protected category, and this land is found in only three reserves in the northern Yucatan Peninsula (Celestún, Lagartos, and Dzibilchaltún). On the other hand, barely more than 1% of 27 million ha of the subhumid temperate zone,

Table 26.6. Distribution of areas by terrestrial ecological zones

Ecological zone	No. of areas	Surface area (hectares)	Percent of ecol. zone	Percent of national total
Humid tropic	6	1,246,470	6.1	0.62
Subhumid tropic	6	110,851	0.34	0.05
Humid temperate	1	30,000	5.3	0.015
Subhumid temperate	22	325,038	1.2	0.16
Arid/semiarid	15	3,172,332	3.7	1.58
Multizones	16	836,373	2.4	0.41

See text for further explanation.

whose diversity (species richness and endemisms) is high in vascular plants (Rzedowski, this volume), herpetofauna (Flores-Villela, this volume), mammals (Fa, 1989), bees (Ayala et al., this volume), and possibly butterflies, is theoretically protected. With the exception of La Michilía in Durango, which is a biosphere reserve, most of the supposedly protected areas of this zone are national parks (Nevado de Toluca, La Malinche, Iztaccihuatl-Popocatepetl, Cofre de Perote). Their small surface areas create logistical problems for conservation (Vargas, 1984).

CONCLUSIONS

The definition of the six principal terrestrial habitats, their biological characterization, the level of the ecological changes in each of them, and the status of their conservation in terms of protected areas are useful criteria for the analysis of the state and status of the biological diversity of Mexico.

Each of the six habitat zones of Mexico recognized in this chapter varies in terms of the area covered, degrees of species richness and endemism, and predominant land use patterns. The humid tropic zone located in southern and southeastern Mexico covers about 11% of the national territory. It is high in species richness but low in endemics among both vascular plants and animals. There appears to have been a drastic 40–90% alteration of this tropical forest since 1970 chiefly due to forest exploitation and livestock activities. The subhumid tropic zone covers about 17% of Mexico, principally along the Gulf and Pacific coastal plains and drainages. Species richness and endemism are notable for the flora and, especially, the herpetofauna. Over 55% of the dry deciduous forest cover has been removed in activities mostly associated with forest extraction and agriculture.

The humid temperate zone fringes the mountain slopes of the Gulf of Mexico and accounts for about 3% of the country. Because of the specialized microhabitats, several plant and animal groups have radiated extensively, resulting in high endemism and richness. Livestock and associated land uses have resulted in leaving altered 60% of the original cloud forest. The subhumid temperate zone covers the principal mountain chains throughout Mexico and occupies 14% of the total. The pine-oak forests in Mexico are remarkably rich in endemic taxa of both plants and animals, contrary to the general belief that they are less abundant than the tropical forests in Mexico. About 37% of this natural vegetation has been altered by agricultural practices.

The arid and semiarid zones of central and northern Mexico correspond to over 50% of the country. They are particularly high in plant and animal endemics. Livestock and plant extraction practices have altered more than half of this fragile ecological zone. The most limited ecological zone is the alpine region, located on the small mountain peaks in the eastern Sierra Madre and central Trans-Mexican Volcanic Belt. Biological endemism is particularly high in this area. Livestock activities have penetrated this highly restricted habitat.

Although any loss of habitat is to be regretted, the ecological alteration of humid tropic, subhumid tropic, and arid/semiarid zones continues to threaten the uniqueness of Mexican biodiversity. In particular, species richness in the humid tropic zone is diminishing while endemic taxa are disappearing in the subhumid tropic and arid/semiarid zones. The most common land use practice that alters Mexico's habitats is livestocking and associated activities.

A detailed and comprehensive analysis of percentages of organisms that are "protected" through reserves, parks, and their status (e.g., threatened, in danger of extinction) not only in these parks and terrestrial habitats but also in other habitats (e.g., wetlands, coastal zones, marine) is urgently needed. Such an analysis will contribute to strategy concerning research involving these organisms, their sensible exploitation, and conservation biology of Mexico's rich biological patrimony.

Acknowledgments

We are grateful to T.P. Ramamoorthy and Robert Bye for their help with the preparation of this paper. G. Guzmán, R. Lamothe, and M. Ortega provided unpublished data.

REFERENCES

Benítez, J. A. 1988. Los recursos bioticos marino-estuarinos de México: una tentativa de regionalización. Programa de Formación de Recursos Humanos, INIREB. Meconografiado.

Cifuentes, J. L. 1986. Los moluscos como alimento actual y futuro. en: Memorias de la II Reunión de Malacología y Conquiliología. Fac. de Ciencias. UNAM. Villahermosa, Tabasco. pp. 123–154.

Contreras, F. 1985. Las Lagunas Costeras Mexicanas. Mexico City: Centro de Ecodesarrollo/Secretaria de Pesca.

Eisenberg, J. F. 1981. The Mammalian Radiations: an Analysis of Trends in Evolution, Adaptation and Behavior. Chicago, IL: Univ. of Chicago Press.

Erwin, T. L. 1988. The tropical forest canopy: the heart of biotic diversity. *In* E. O. Wilson (ed.), Biodiversity: 123–129. Washington, DC.: National Academy Press.

Estrada, A. & R. Coates Estrada. 1988. Tropical rain forest conversion and perspectives in the conservation of wild primates (*Alouatta* and *Ateles*) in Mexico. Amer. J. Primat. 14:315–327.

Fa, J. E. 1989. Conservation-motivated analysis of mammalian biogeography in the Trans-Mexican Neovolcanic Belt. Nat. Geogr. Res. 5(3):296–316.

Fernández Ortíz, L. M. & M. Tarrio García. 1983. Ganadería y Estructura Agraria en Chiapas. Mexico City: UAMO.

Flores-Villea, O. & P. Gerez. 1989. Conservación en México: síntesis sobre vertebrados terrestres, vegetación y uso del suelo. INIREB/ Conservation International.

García, E. 1989. Diversidad climático vegetal en México. Simposio sobre Diversidad Biológica de México. Oaxtepec, Morelos.

Lankford, R. R., 1977. Coastal lagoons of Mexico: Their origin and classification. *In* M.L. Wiley (ed.), Estuarine Processes. New York, Academic Press. pp. 182–215.

Leopold, A. S. 1950. Vegetation zones of Mexico. Ecology 31:507–518.

Lot, A. 1971. Los pastos marinos de los arrecifes de Veracruz. An. Inst. Biol. UNAM 42. Ser. Bot. 1:1–44.

Lott, E., S. Bullock & J.A. Solis Magallanes. 1987. Floristic diversity and structure of a tropical deciduous forest of coastal Jalisco, Mexico. Biotropica 19:228–235.

Mabberly, 1987. The Plant Book. Cambridge: Cambridge Univ. Press.

Meave del Castillo, J. 1983. Estructura y composición de la selva alta perennifolia de Bonampak, Chiapas. Thesis, UNAM.

Miranda, F. 1955. Formas de vida vegetales y el problema de la delimitación de las zonas aridas de México. *In* IMERNAR, Mesas Redondas sobre las Zonas Aridas de México: pp. 85–109.

———— & A. J. Sharp. 1950. Characteristics of the vegetation in certain temperate regions of eastern Mexico. Ecology 31:313–333.

———— & E. Hernández X. 1963. Los tipos de Vegetación de México y su clasificación. Bol. Soc. Bot. Mex. 29:29–179.

Mittermeir, R.A. 1988. Primate diversity and the tropical forest. Case studies from Brazil and Madagascar and the importance of the megadiversity countries. *In* E.O. Wilson (ed.), Biodiversity. Washington, DC: National Academy Press. pp. 145–154.

Ortuzar, X. 1978. Conceptos fundamentales para el conocimiento adecuado de la zona costera: diagnosis ecológica. Ciencia Desarrollo 19:78–87.

Pennington, T. & J. Sarukhan. 1968. Los árboles tropicales de México. Rome: FAO.

Pérez-Gil, R., P. Robles-Gil & J. Robles-Gil. 1984. Ecosistemas de México. Mexico City: Banco BCH. Soc. Nal. de Crédito.

Raven, P.H. & G.B. Johnson. 1986. Biology. St. Louis: Times Mirror/Mosby College Publishing.

Rzedowski, J. 1978. Vegetación de México. Mexico City: Limusa.

SAHOP. 1981. Programa nacional del desarrollo ecológico de los asentamientos humanos. Carta Vegetación y uso do la secretaria de asentamientos humanos y Obras Públicas.

SEDUE. 1984. Regionalización ecológica del territorio. SEDUE. Subsecretaría Gral. de Ordenamiento Ecológico e Impacto Ambiental. Dirección de Area de Instrumentación.

Soto Arenas, M. A. & S. Castillo Arguero. 1989. Análisis de la Distribución de Orquideas en México. *In* Simposio sobre Diversidad Biológica de México. Oaxtepec, Morelos.

SPP. 1981. Atlas Nacional del Medio Físico. SPP. México.

Toledo, V.M. 1987. Vacas, cerdos, pollos y ecosistemas: ecología y ganadería in México. Ecol. Poli. Cult. 3:36–49.

———— 1988. La diversidad biológica de México. Ciencia Desarrollo. 81:17–30

———— 1991. Bio-economic costs of transforming forests to pastures in Latin America. *In* T. Downing (ed.), Cattle Ranching and Tropical Deforestation in Latin America. Westview Press. pp. 63–72.

———— 1992. Los habitats naturales: caracteristicas, destrucción y conservación de la flora y la vegetación de México. *In* Ceballos & D. Navarro (eds.), Fauna Mexicana en Peligro de Extinción. UNAM. In press.

————, J. Carabias, C. Mapes & C. Toledo. 1985. Ecología y Autosuficiencia Alimentaria. Mexico City: Siglo XXI Eds.

————, J. Carabias, C. Toledo & C. González Pacheco. 1989. La Producción Rural en México: Alternativas ecológicas. Colección medio ambiente No. 6 Mexico City: Fundación Universo XXI.

Vargas, F. 1984. Parques Nacional de Mexico, Reservas Equivalentes. Mexico City: UNAM, Inst. Invest. Econo.

Villalobos, J. L., J. C. Nates, & A. Cantú. 1988. Informe Técnico Colección de Carcinología del Instituto de Biología, UNAM. 1987–1988.

Villareal, L. & J. Pérez Moreno. 1988. Los Myxomycetes, parte de la diversidad ecológica en México. Simposio sobre Diversidad Biológica de México. Oaxtepec, Morelos.

Vovides, A., J. Rees & M. Vazquez Torres, 1983. Las Cicadaceas. Flora de Veracruz. No. 26. Xalapa, Veracruz: INIREB.

Wendt, T. 1987. Las selvas de Uxpanapa, Veracruz-Oaxaca, México: evidencia de refugios floristicos cenozoicos. An. Inst. Biol. UNAM Ser. Bot. 1:29–54.

West, R.C. 1964. The natural regions of Middle America. *In* R.C. West (ed.), Handbook of Middle American Indians, Vol. 1. Natural Environment and Early Cultures. Austin, TX: Univ. of Texas Press. pp. 363–383.

Whittaker, R.H. 1975. Communities and Ecosystems. New York: Macmillan.

Wilson, E.O. 1988. The current state of biological diversity. *In* E.O. Wilson (ed.), Biodiversity. Washington, DC: National Academy Press. pp. 2–18.

Yáñez Arancibia, A. 1986. Ecología de la Zona Costera. Análisis de siete topicos. Mexico City: AGT Editor.

Taxonomic Index

Numbers in italics refer to tables, figures, and appendixes. Names listed in appendixes and tables (e.g., checklist, list of genera, etc.) are not indexed below. Consult individual chapters for them.

e = extinct; f = fossil reference; MP = Morphotectonic Province

Subject Index

Numbers in italics refer to tables, figures, and appendixes.
f = fossil reference; MP = Morphotectonic Province; TI = Taxonomic Index